# Current Topics in Microbiology and Immunology

## Volume 391

More information about this series at http://www.springer.com/series/82

Christian Münz
Editor

# Epstein Barr Virus Volume 2

## One Herpes Virus: Many Diseases

Responsible Series Editor: Peter K. Vogt

*Editor*
Christian Münz
Institute of Experimental Immunology
University of Zürich
Zürich
Switzerland

ISSN 0070-217X ISSN 2196-9965 (electronic)
Current Topics in Microbiology and Immunology
ISBN 978-3-319-22833-4 ISBN 978-3-319-22834-1 (eBook)
DOI 10.1007/978-3-319-22834-1

Library of Congress Control Number: 2015948721

Springer Cham Heidelberg New York Dordrecht London


Printed on acid-free paper

Springer International Publishing AG Switzerland is part of Springer Science+Business Media
(www.springer.com)

# Contents

# Part I
# EBV Latency

# EBNA1

**Lori Frappier**

**Abstract** Epstein–Barr nuclear antigen 1 (EBNA1) plays multiple important roles in EBV latent infection and has also been shown to impact EBV lytic infection. EBNA1 is required for the stable persistence of the EBV genomes in latent infection and activates the expression of other EBV latency genes through interactions with specific DNA sequences in the viral episomes. EBNA1 also interacts with several cellular proteins to modulate the activities of multiple cellular pathways important for viral persistence and cell survival. These cellular effects are also implicated in oncogenesis, suggesting a direct role of EBNA1 in the development of EBV-associated tumors.

## Contents



L. Frappier (✉)
Department of Molecular Genetics, University of Toronto, 1 Kings College Circle, Toronto, ON M5S 1A8, Canada
e-mail: lori.frappier@utoronto.ca



C. Münz (ed.), *Epstein Barr Virus Volume 2*, Current Topics in Microbiology and Immunology 391, DOI 10.1007/978-3-319-22834-1_1

# 1 Introduction

Epstein–Barr nuclear antigen 1 (EBNA1) is expressed in all forms of EBV latency in proliferating cells and was the first reported EBV latency protein (Reedman and Klein 1973). EBNA1 has been extensively studied and shown to have multiple important roles in EBV infection. These include contributions to both the replication and mitotic segregation of EBV episomes that lead to stable persistence of EBV episomes in latent infection. EBNA1 also activates the transcription of other EBV latency genes important for cell immortalization. These functions require EBNA1 binding to specific DNA elements in the EBV latent origin of DNA replication (*oriP*). In recent years, it has become apparent that EBNA1 functions are not limited to its roles on EBV episomes but rather that EBNA1 also alters the cellular environment in multiple ways that contribute to cell survival and proliferation and viral persistence. EBNA1 lacks enzymatic activities but is able to affect many processes due to interactions with a variety of cellular proteins. This chapter reviews the multiple functions and mechanisms of action of EBNA1.

# 2 EBNA1 Functions at EBV Genomes

## *2.1 DNA Replication*

The origin of latent DNA replication, termed *oriP* (for plasmid origin), was identified by screening EBV DNA fragments for the ability to enable the replication and stable maintenance of plasmids in human cells that were latently infected with EBV (Yates et al. 1984). Subsequent studies showed that the only viral protein required for the replication of *oriP* plasmids was EBNA1 (Yates et al. 1985). Both EBV episomes and *oriP* plasmids were found to replicate once per cell cycle, mimicking cellular replication and providing a good model system for human DNA replication (Yates and Guan 1991; Sternas et al. 1990). Note that *oriP* is not the only origin of replication for EBV episomes, as replication forks have also been found to initiate from a poorly defined region outside of *oriP* that appears to be independent of EBNA1 (Little and Schildkraut 1995; Norio et al. 2000; Ott et al. 2011).

*OriP* contains two functional elements: the dyad symmetry (DS) element and the family of repeats (FR) (Reisman et al. 1985) (Fig. 1). The DS contains four EBNA1 recognition sites, two of which are located within a 65-bp DS sequence (Reisman et al. 1985; Rawlins et al. 1985). The DS element is the origin of

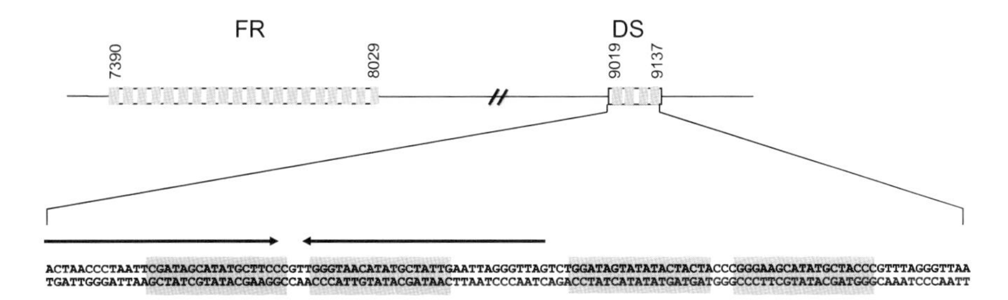


Fig. 1 Schematic representation of *oriP*. Organization of the *oriP* DS and FR elements showing genome nucleotide coordinates and EBNA1 binding sites (*gray boxes*). For the DS element, the positions of the four EBNA1 binding sites and 65-bp dyad symmetry sequence (*arrows*) are indicated

replication within *oriP* (Gahn and Schildkraut 1989) and has been shown to be both essential and sufficient for plasmid replication in the presence of EBNA1 (Wysokenski and Yates 1989; Harrison et al. 1994; Yates et al. 2000). Efficient replication from the DS element requires all four EBNA1 binding sites; however, a low level of DNA replication can be achieved with only two adjacent EBNA1 sites, provided that the 3-bp spacing between these sites is maintained (Koons et al. 2001; Harrison et al. 1994; Yates et al. 2000; Atanasiu et al. 2006; Bashaw and Yates 2001; Lindner et al. 2008). The FR element consists of 20 tandem copies of a 30-bp sequence, each of which contains an EBNA1 binding site (Rawlins et al. 1985; Reisman et al. 1985). The primary function of the FR element is in the mitotic segregation and transcriptional activation functions of EBNA1 as discussed below, although in some cell lines the FR also appears to be required for replication from the DS (Hodin et al. 2013). In addition, when bound by EBNA1, the FR can affect DNA replication by inhibiting the passage of replication forks, forming a major pause site (Gahn and Schildkraut 1989; Dhar and Schildkraut 1991; Norio and Schildkraut 2001, 2004; Ermakova et al. 1996).

The DNA replication activity of EBNA1 requires its DNA binding domain as well as additional sequences in the N-terminal half of EBNA1 (Fig. 2) (Yates and Camiolo 1988; Van Scoy et al. 2000; Kim et al. 1997; Kirchmaier and Sugden

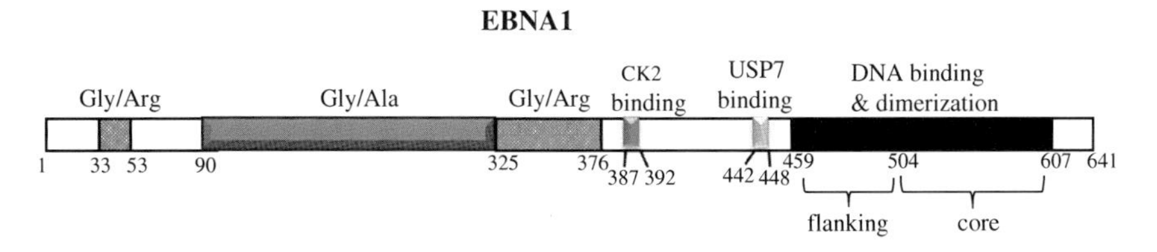


Fig. 2 Organization of the EBNA1 protein. Positions of some of the key elements of EBNA1 are indicating, including the two Gly–Arg-rich regions, variable Gly–Ala repeat, USP7 and CK2 binding sites, and DNA binding and dimerization domain (*black*), along with flanking and core subcomponents of the DNA binding and dimerization domain. Amino acid numbers are indicated *below*

1997; Mackey and Sugden 1999; Shire et al. 1999; Ceccarelli and Frappier 2000; Wu et al. 2002). Single localized deletions or mutations that disrupt EBNA1 DNA replication function have not been identified; rather, this function appears to involve redundant contributions of at least two EBNA1 regions (amino acids 8–67 and 325–376; Fig. 2). These N-terminal sequences can tether EBNA1 to cellular chromosomes (see Sect. 2.2), and the finding that they can be functionally replaced by a nucleosome interacting sequence suggests that chromosome tethering contributes to the replication function of EBNA1 (Hodin et al. 2013). In addition, deletion of EBNA1 amino acids 61–83 or 395–450, or point mutation of G81 or G425 within these regions, has been found to increase replication efficiency (Holowaty et al. 2003c; Wu et al. 2002; Deng et al. 2005). These point mutations disrupt EBNA1 binding to tankyrase, suggesting that tankyrase negatively regulates replication by EBNA1, possibly through the poly-ADP ribosylation of EBNA1 (Deng et al. 2005). In addition, the EBNA1 395–450 region is responsible for binding and recruiting to *oriP* the host ubiquitin-specific protease 7 (USP7), suggesting that USP7 may negatively regulate replication (Holowaty et al. 2003c). This contention was further supported by a study showing that the latent origin binding proteins of other gamma-herpesviruses also recruit USP7 to their origins and that disruption of this interaction increases DNA replication (Jager et al. 2012). Therefore, negative regulation of replication by USP7 appears to be a conserved feature of gamma-herpesvirus replication.

EBNA1 is bound to the *oriP* elements throughout the cell cycle, indicating that EBNA1 binding to the DS is not sufficient to activate DNA replication (Hsieh et al. 1993; Niller et al. 1995; Ritzi et al. 2003). While EBNA1 is the only EBV protein involved in latent-phase DNA replication, it lacks any enzymatic activities, including DNA helicase and origin melting activities present in the origin binding proteins of some viruses (Frappier and O'Donnell 1991b). Therefore, EBV depends heavily on host cellular proteins to replicate its episomes. Several studies have shown that the cellular origin recognition complex (ORC) and minichromosome maintenance (MCM) complex are associated with the DS element of *oriP*, implicating them in the initiation and licensing of EBV DNA replication (Schepers et al. 2001; Chaudhuri et al. 2001; Dhar et al. 2001). A functional role for ORC in *oriP* plasmid replication was shown by the failure of these plasmids to stably replicate in a cell line containing a hypomorphic *ORC2* mutation (Dhar et al. 2001). EBV replication was also found to be inhibited by geminin, a protein that inhibits rereplication from cellular origins by interacting with Cdt1 (Dhar et al. 2001). This suggests that Cdt1 loads the MCM complexes on EBV origins, as it does on cellular origins.

EBNA1 has been shown to be important for ORC recruitment to the DS (Schepers et al. 2001; Dhar et al. 2001; Julien et al. 2004). In addition, EBNA1 was found to interact with Cdc6 and this interaction increased ORC recruitment to the DS in vitro (Moriyama et al. 2012). Interestingly, ORC is not recruited by EBNA1 bound to the FR element, suggesting that the DS DNA sequence or arrangement of the EBNA1 binding sites are important for ORC recruitment (Schepers et al. 2001; Chaudhuri et al. 2001; Dhar et al. 2001; Moriyama et al.

2012). ORC recruitment by EBNA1 was initially reported to involve EBNA1N-terminal sequences including the Gly–Arg-rich regions, and in vitro studies suggested that this interaction was mediated by RNA molecules (Norseen et al. 2008; Moriyama et al. 2012). However, a second study found that, in the presence of Cdc6, EBNA1 could recruit ORC to the DS in an RNA-independent manner (Moriyama et al. 2012). In addition, it was recently reported that the EBNA1 DNA binding domain is sufficient for ORC recruitment to the DS (Hodin et al. 2013). EBNA1 may also facilitate the recruitment of telomere repeat binding factor 2 (TRF2) to site in the DS (Deng et al. 2002, 2003; Moriyama et al. 2012). TRF2 then appears to contribute to ORC recruitment to the DS in conjunction with EBNA1 (Julien et al. 2004; Atanasiu et al. 2006) and also affects the timing of replication in S phase through recruitment of additional proteins (Zhou et al. 2009, 2010).

EBNA1 has also been found to recruit template activating factor Iβ (TAF-Iβ also called SET) to both the DS and FR elements, through a direct interaction with the 325–376 Gly–Arg-rich region of EBNA1 (Holowaty et al. 2003b; Wang and Frappier 2009). TAF-Iβ appears to negatively regulate replication from *oriP* as TAF-Iβ depletion was found to increase *oriP* plasmid replication, while TAF-Iβ overexpression inhibited it (Wang and Frappier 2009). Since TAF-Iβ is a nucleosome-associated protein that can recruit either histone acetylases or deacetylases (Seo et al. 2001; Shikama et al. 2000), TAF-Iβ may negatively regulate replication from *oriP* by affecting the chromatin structure of the origin.

## 2.2 Mitotic Segregation

In latency, EBV episomes are present at low copy numbers that are stably maintained in proliferating cells. This stable maintenance requires a mechanism to ensure even partitioning of the episomes to the daughter cells during cell division. The mitotic segregation or partitioning of the EBV episomes requires two viral components, EBNA1 and the *oriP* FR element (Lupton and Levine 1985; Krysan et al. 1989; Lee et al. 1999). EBNA1 and the FR can also confer stability on a variety of constructs when combined with heterologous origin sequences (Krysan et al. 1989; Kapoor et al. 2001; Simpson et al. 1996). EBNA1 binding to its multiple recognition sites in the FR is crucial for its segregation function, as is the central Gly–Arg-rich region of EBNA1 (325–376) (Shire et al. 1999).

EBNA1 functions in segregation by tethering the EBV episomes to the cellular mitotic chromosomes. EBNA1, EBV episomes, and *oriP*-containing constructs have all been found to associate with mitotic chromosomes (Harris et al. 1985; Delecluse et al. 1993; Simpson et al. 1996; Grogan et al. 1983; Petti et al. 1990), and the association of *oriP* plasmids with mitotic chromosomes was shown to depend on the EBNA1-chromosome interaction (Kanda et al. 2001; Kapoor et al. 2005). In addition, EBNA1 mutants that are nuclear but defective in mitotic chromosome attachment fail to partition *oriP* plasmids (Hung et al. 2001; Shire et al. 1999; Wu et al. 2000). EBNA1 and EBV episomes are not localized to particular regions of mitotic chromosomes, but rather are widely distributed over the chromosomes, leading to the initial suggestion that EBNA1 and EBV episomes interact randomly with chromosomes (Harris et al. 1985). However, subsequent studies have indicated that initial pairing of EBV episomes on sister chromatids may ensure their equal distribution to the daughter cells and that this pairing may stem from the catenation of the newly replicated EBV plasmids (Delecluse et al. 1993; Kanda et al. 2007; Dheekollu et al. 2007; Nanbo et al. 2007). In addition, the FR element has been found to direct EBV genomes to chromatin regions with histone modifications typical of active chromatin (Deutsch et al. 2010).

Studies with EBNA1 deletion mutants showed that the central Gly–Arg-rich region of EBNA1 (amino acids 325–376) was critical for chromosome attachment and that N-terminal sequences (8–67) also contribute to this interaction (Wu et al. 2000, 2002; Shire et al. 1999, 2006; Marechal et al. 1999; Hung et al. 2001; Kanda et al. 2013). Interestingly, fusion proteins in which these EBNA1 regions have been replaced by other chromosome binding sequences are also able to support *oriP* plasmid maintenance (Hung et al. 2001; Sears et al. 2003). Both the central Gly–Arg repeat of EBNA1 and sequences spanning the smaller Gly–Arg-rich N-terminal sequence (amino acids 33–53) can cause proteins to associate with mitotic chromosomes when fused to them (Hung et al. 2001; Marechal et al. 1999; Sears et al. 2004). However, deletion of the N-terminal Gly–Arg sequence within EBNA1 does not affect EBNA1's ability to maintain *oriP* plasmids or to associate with mitotic chromosomes, indicating that it is the central Gly–Arg-rich region that is normally used by EBNA1 for chromosome interactions and segregation (Nayyar et al. 2009; Wu et al. 2002). This region contains a repeated GGRGRGGS sequence that is phosphorylated on the serines and methylated by PRMT1 or PRMT5 on the arginine residues (Laine and Frappier 1995; Shire et al. 2006).

The segregation of viral genomes by attachment to cellular chromosomes is not unique to EBV but is a strategy also used by Kaposi sarcoma associated herpesvirus (KSHV) and papillomavirus. In each case, the viral origin binding (LANA for KSHV and E2 for papillomavirus) tethers the viral plasmid to the cellular chromosome through interactions with one or more cellular proteins (You 2010; Krithivas et al. 2002; Barbera et al. 2006; Parish et al. 2006). For EBNA1, interactions with the cellular protein, EBP2, appear to be important for metaphase chromosome attachment and segregation function (Wu et al. 2000; Shire et al. 1999; Kapoor and Frappier 2003, 2005). EBP2 is largely nucleolar in interphase but redistributes to the chromosomes in mitosis (Wu et al. 2000). The EBNA1 325–376 region critical for chromosome attachment mediates EBP2 binding, and there is a close correspondence between the effect of EBNA1 mutations on EBP2 and metaphase chromosome interactions (Shire et al. 1999; Wu et al. 2000, 2002; Shire et al. 2006; Nayyar et al. 2009). In addition, EBP2 depletion in various cell lines, including the EBV-positive C666-1 nasopharyngeal carcinoma (NPC) cells, resulted in redistribution of EBNA1 from the metaphase chromosomes to the soluble cell fraction and a corresponding release of *oriP* plasmids from the chromosomes (Kapoor et al. 2005). EBP2 was also found to enable EBNA1 to segregate plasmids in budding yeast by facilitating EBNA1 attachment to the yeast mitotic chromosomes (Kapoor et al. 2001; Kapoor and Frappier 2003).

Detailed studies on the timing of chromosome association in human cells showed that EBNA1 associates with the chromosomes earlier in mitosis than EBP2 and that EBNA1 and EBP2 only associate on the chromosomes in metaphase to telophase (Nayyar et al. 2009). This suggests that EBNA1 initially contacts the chromosomes by an EBP2-independent mechanism and that subsequent interactions with EBP2 in mid-to-late mitosis might be important to maintain EBNA1 on chromosomes. The initial chromosome contact could involve direct DNA binding or interactions with chromosome-associated RNA molecules, since the 325–376 and N-terminal arginine-rich regions have been found to have some capacity to interact with DNA and RNA in vitro, and drugs that bind G-quadruplex RNA have been reported to decrease the mitotic chromosome association of EBNA1 (Sears et al. 2004; Norseen et al. 2008, 2009; Snudden et al. 1994). In addition, FRET analysis identified an interaction between EBNA1 and EBP2 in the nucleoplasm and nucleolus in interphase suggesting additional roles for this interaction, including the possibility that the EBNA1-EBP2 interaction in interphase is important for EBNA1-chromosome interactions in mitosis (Jourdan et al. 2012). This possibility is reminiscent of findings for bovine papillomavirus segregation, in which an interphase interaction between the viral E2 protein and host ChlR1 protein is required in order for E2 to associate with mitotic chromosomes and segregate papillomavirus genomes (Feeney et al. 2011; Parish et al. 2006).

## *2.3 EBV Transcriptional Activation*

Another function of EBNA1 at EBV episomes is in transcriptional activation. EBNA1 can act as a transcriptional activator when bound to the *oriP* FR element, enhancing the expression of reporter genes on FR-containing plasmids in a distance-independent manner (Lupton and Levine 1985; Reisman and Sugden 1986). The EBNA1-bound FR was also shown to activate expression from the viral Cp and LMP promoters, suggesting a role for EBNA1 in inducing the expression of the EBNA and LMP EBV latency genes in latent infection (Sugden and Warren 1989; Gahn and Sugden 1995). The EBNA1 residues required for transcriptional activation have been mapped to the 65–83 N-terminal sequence (Wu et al. 2002; Kennedy and Sugden 2003) as well as to the central Gly–Arg-rich region (residues 325–376) also required for segregation function (Yates and Camiolo 1988; Ceccarelli and Frappier 2000; Wang et al. 1997; Van Scoy et al. 2000). EBNA1 requires both of these regions to activate transcription, as deletion of either one abrogates the transcriptional activation function of EBNA1 (Wu et al. 2002; Ceccarelli and Frappier 2000). A Δ61–83 EBNA1 mutant was found to be fully active for replication and segregation functions, indicating that transcriptional activation is a distinct EBNA1 function (Wu et al. 2002). Similar conclusions were reached with a Δ65–89 EBNA1 mutant in the context of an infectious EBV, where EBNA1 Δ65–89 was shown to be defective in activating expression of the EBNA genes from the Cp promoter, but still supported stable plasmid replication (Altmann et al. 2006). EBV containing the Δ65–89 EBNA1 was also shown to be severely impaired in the ability to transform cells, indicating the importance of EBNA1-mediated transcriptional activation for EBV infection (Altmann et al. 2006).

Two cysteine residues within the N-terminal transactivation sequence (at positions 79 and 82) have been shown to be important for transactivation activity and to mediate an interaction with zinc (Aras et al. 2009). There is also evidence that the transcriptional activity of EBNA1 is zinc-dependent, suggesting that a zinc-dependent structure formed in the N-terminal transactivation region mediates the activity of this sequence (Aras et al. 2009). The 61–83 region also mediates an interaction with Brd4 (Lin et al. 2008), a cellular bromodomain protein that interacts with acetylated histones to regulate transcription (Wu and Chiang 2007). Within the EBV genome, Brd4 was shown to be preferentially localized to the EBNA1-bound FR enhancer element (Lin et al. 2008). Furthermore, Brd4 depletion inhibited EBNA1-mediated transcriptional activation, suggesting that EBNA1 uses Brd4 to activate transcription (Lin et al. 2008). Interestingly, an interaction between Brd4 and papillomavirus E2 proteins (the functional equivalent to EBNA1) has been shown to be important for transcriptional activation by E2 (Schweiger et al. 2006; McPhillips et al. 2006; Ilves et al. 2006), suggesting that EBNA1 and E2 may use common mechanisms to activate transcription. Whether or not the EBNA1–Brd4 interaction is zinc-dependent remains to be determined.

The EBNA1 325–376 region mediates interactions with several cellular proteins, some of which have been implicated in the transcriptional activity of EBNA1. For example, P32/TAP, which interacts with Arg-rich sequences, has been detected at *oriP* by chromatin immunoprecipitation, and its C-terminal region has some ability to activate a reporter gene when fused to the GAL4 DNA binding domain (Van Scoy et al. 2000; Wang et al. 1997). However, it is not clear whether P32/TAP is important for EBNA1-mediated transcriptional activation. The related nucleosome assembly proteins, NAP1, TAF-Iβ (also called SET), and nucleophosmin, also interact with the EBNA1 325–376 sequence and are known to affect transcription in multiple ways (Holowaty et al. 2003c; Wang and Frappier 2009; Park and Luger 2006; Malik-Soni and Frappier 2012, 2013). A role for NAP1, TAF-Iβ, and nucleophosmin in EBNA1-mediated transcriptional activation is supported by the finding that each protein is recruited to the FR element by EBNA1 and that EBNA1 transactivation activity is decreased upon depleting any of these proteins (Wang and Frappier 2009; Malik-Soni and Frappier 2013). Depletion of nucleophosmin had the biggest effect on EBNA1-mediated transcriptional activation, suggesting, either that the EBNA1-nucleophosmin interaction is the most important for transcription function, or that nucleophosmin is more limiting in the cell than NAP1 or TAF-Iβ (Malik-Soni and Frappier 2013). Note that another histone chaperone protein, nucleolin, was also recently reported to contribute to EBNA1 functions including transactivation, but this appears to be due to an effect on EBNA1 binding to *oriP* (Chen et al. 2014). As a whole, the data suggest that recruitment of both nucleosome assembly proteins and Brd4 are important for transcriptional activation by EBNA1, reflecting the requirement for the two transcriptional activation sequences.

It is expected that transcriptional activation by EBNA1 will involve changes to histone modifications and this may include ubiquitylation of histone H2B. This is suggested by the finding that EBNA1 binds to a complex of USP7 and GMP synthetase that functions to deubiquitylate H2B and recruits it to the FR (Sarkari et al. 2009). USP7 depletion results in increased levels of monoubiquitylated H2B at the FR and decreased transcriptional activation, suggesting that monoubiquitylation of H2B inhibits EBNA1-mediated transcriptional activation. In keeping with this result, an EBNA1 mutant defective in USP7 binding has decreased ability to activate transcription (Holowaty et al. 2003c).

### 2.4 Autoregulation

In addition to interactions with the *oriP* FR and DS elements, EBNA1 was found to bind a third region of the EBV genome near the Qp promoter that is used to express EBNA1 in the absence of other EBNAs (Jones et al. 1989; Sample et al. 1992; Nonkwelo et al. 1996). EBNA1 binding to two recognition sites located downstream of Qp was reported to repress EBNA1 expression from Qp (Sample et al. 1992). Since EBNA1 has lower affinity for these sites than either the DS or FR elements, EBNA1 would only bind the Qp sites when its levels are high enough to saturate the FR and DS elements, providing a feedback mechanism to shut off EBNA1 expression when EBNA1 levels are high (Jones et al. 1989; Ambinder et al. 1990). While EBNA1 was initially thought to inhibit expression from Qp by repressing transcription, a more recent study found that EBNA1 acts post- or co-transcriptionally to inhibit the processing of primary transcripts (Yoshioka et al. 2008).

## 3 EBNA1–DNA Interactions

### 3.1 Interactions with the EBV Genome

EBNA1 specifically recognizes an 18-bp palindromic sequence present in multiple copies in the *oriP* DS and FR elements as well as in the BamHI-Q fragment containing the Qp promoter (Rawlins et al. 1985; Jones et al. 1989; Ambinder et al. 1990, 1991; Frappier and O'Donnell 1991b; Shah et al. 1992). Sequence variation within the multiple copies of this palindrome results in different affinities of EBNA1 for the FR, DS, and BamHI-Q regions and for individual sites within these regions (Ambinder et al. 1990; Summers et al. 1996). EBNA1 has highest affinity for the FR and DS regions and remains bound to these sites throughout the cell cycle (Hsieh et al. 1993; Niller et al. 1995; Ritzi et al. 2003).

EBNA1 interacts with its recognition sites through its C-terminal domain (amino acids 459 and 607; Fig. 2), which also mediates the dimerization of EBNA1 (Ambinder et al. 1991; Chen et al. 1993; Summers et al. 1996; Shah et al. 1992). EBNA1 forms very stable homodimers both in solution and when bound to its recognition sites (Frappier and O'Donnell 1991b; Ambinder et al. 1991; Shah et al. 1992). The crystal structure of the DNA binding and dimerization domain was determined both in solution and bound to the EBNA1 consensus binding site (Bochkarev et al. 1995, 1996). The structure showed that dimerization was mediated by residues 504–604 (referred to as the core domain), which form an eight-stranded antiparallel β-barrel, comprised of four strands from each monomer and two α-helices per monomer (Fig. 3). This core domain is strikingly similar to the structure of the DNA binding domain of the E2 protein of papillomavirus, despite a complete lack of sequence homology (Edwards et al. 1998; Hegde et al. 1992). Residues 461–503 flank the core domain (flanking domain) and are comprised of an α-helix oriented perpendicular to the DNA and an extended chain that tunnels along the base of the minor groove of the DNA (Fig. 3). Both the helix and the extended chain make sequence-specific DNA contacts. In addition, a direct role of the core domain in DNA recognition was suggested by analogy to the E2 DNA binding domain and later confirmed by mutational analyses (Cruickshank et al. 2000). Combined, the structural and biochemical studies indicate that the core and flanking domains of EBNA1 work together to load EBNA1 on its recognition site, likely through a two-step DNA binding mechanism. In keeping with this model, thermodynamic and kinetic analyses of the EBNA1 DNA binding domain–DNA interaction revealed two DNA association and dissociation events (Oddo et al. 2006). In addition, the ability of EBNA1 to bind its recognition sites, both in vitro and in vivo, was found to be greatly stimulated by USP7 through its interaction

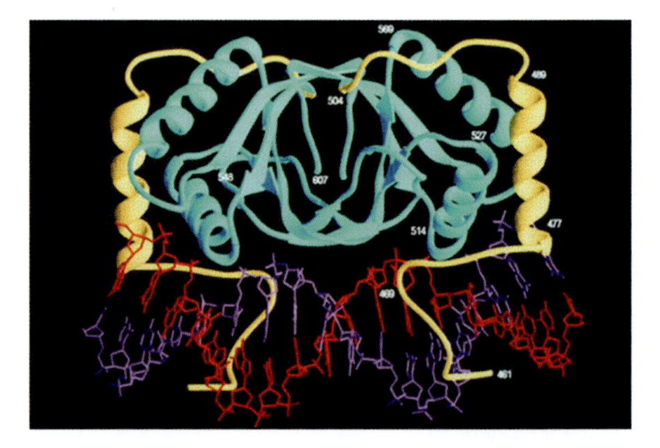

**Fig. 3** Crystal structure of the EBNA1 DNA binding and dimerization domain bound to DNA. The core and flanking components of the DNA binding and dimerization domain are shown in *green* and *yellow*, respectively. EBNA1 amino acid numbers are indicated. Reprinted with permission from Bochkarev et al. (1996)

with EBNA1 amino acids close to the flanking domain (442–448; Fig. 2) (Sarkari et al. 2009), suggesting that this USP7 interaction may facilitate the DNA loading of the flanking domain.

The interaction of the EBNA1 DNA binding and dimerization domain with a single recognition site causes the DNA to be smoothly bent and causes localized regions of helical overwinding and underwinding (Bochkarev et al. 1996). The overwinding is caused by the EBNA1 flanking domain residues that traverse along the minor groove (amino acids 463–468) (Bochkarev et al. 1998; Summers et al. 1997) and this results in the increased sensitivity of one T residue within the DS sites to permanganate oxidation (Frappier and O'Donnell 1992; Hearing et al. 1992; Hsieh et al. 1993; Summers et al. 1997). EBNA1 dimers assemble cooperatively on adjacent sites in the DS (Summers et al. 1996; Harrison et al. 1994), and this is predicted to induce additional changes in the DNA structure (such as unwinding), in order to accommodate the closely packed dimers (Bochkarev et al. 1996). The strict requirement for the 3-bp spacing that separates neighboring sites in the DS for origin function suggests that the proper interaction between the EBNA1 dimers bound to these sites is crucial for the initiation of DNA replication, possibly because of the DNA structural changes that it imparts (Bashaw and Yates 2001; Harrison et al. 1994). Interactions of EBNA1 dimers on the multiple sites within the DS and FR elements likely also contribute to the pronounced bending of these elements that have been observed and to the appearance of EBNA1 as a large single complex on each element (Frappier and O'Donnell 1991a; Goldsmith et al. 1993; Bashaw and Yates 2001).

EBNA1 complexes bound to the DS and FR elements of *oriP* can also interact with each other cause the looping out of the intervening DNA (when interaction occur within an *oriP* molecule) and the linking of multiple *oriP* molecules (when interactions occur between *oriP* molecules) (Frappier and O'Donnell 1991a; Goldsmith et al. 1993; Su et al. 1991; Middleton and Sugden 1992). The DNA looping and linking interactions stabilize EBNA1 binding to the DS and involve homotypic interactions mediated by two different regions of EBNA1; a stable interaction mediated by amino acids 327–377 and a less stable interaction mediated by residues 40–89 (Frappier et al. 1994; Laine and Frappier 1995; Mackey et al. 1995; Mackey and Sugden 1999; Avolio-Hunter and Frappier 1998). The looping/linking interactions of EBNA1 are not restricted to EBNA1 complexes formed on the DS or FR elements but also occur between single EBNA1 dimers bound to distant recognition sites (Goldsmith et al. 1993). The contribution of DNA looping and linking to EBNA1 functions remains unclear but the amino acids required for these interactions overlap with those required for EBNA1 replication, segregation, and transcriptional activation functions (Mackey and Sugden 1999; Wu et al. 2002; Shire et al. 1999).

In vivo EBV genomes are assembled into nucleosomes with a spacing similar to that in cellular chromatin (Shaw et al. 1979; Dyson and Farrell 1985). Since nucleosomes tend to inhibit sequence-specific DNA interactions, the ability of EBNA1 to bind its site in the DS in the context of a nucleosome was examined. Surprisingly, EBNA1 was able to access its recognition sites within the nucleosome and destabilized the nucleosome structure such that the histones could be displaced from the DNA (Avolio-Hunter et al. 2001). Efficient assembly of EBNA1 on the FR and DS elements was also observed on larger *oriP* templates containing physiologically spaced nucleosomes (Avolio-Hunter and Frappier 2003). The disruption of the DS-nucleosome by EBNA1 required all four recognition sites in the DS and was intrinsic to the DNA binding and dimerization domain of EBNA1 (Avolio-Hunter et al. 2001). The ability of EBNA1 to destabilize nucleosomes might be important for initiating DNA replication, a process known to be sensitive to nucleosome positioning. In addition, the ability of EBNA1 to access its sites within a nucleosome is likely to be important at times when chromatin is established prior to EBNA1 expression, for example, when latently infected resting cells (which do not express EBNA1) switch to proliferating forms of latency in which EBNA1 is expressed.

## 3.2 Interactions with Cellular DNA Sequences

The fact that EBNA1 can activate transcription, when bound to the EBV FR element, has prompted several studies to determine whether EBNA1 might also interact with specific sequences in cellular DNA to affect cellular gene expression. Chromatin IP (ChIP) experiments performed for EBNA1 from EBV-positive lymphoblastoid cell lines, followed by promoter array analysis, identified several EBNA1-associated DNA fragments, some of which were confirmed to be directly bound by EBNA1 in vitro (Dresang et al. 2009). While this approach identified a new EBNA1 recognition sequence (distinct from those in *oriP*), EBNA1 binding to this sequence did not activate reporter gene expression, so the significance of these EBNA1-cellular DNA interactions is not clear. ChIP combined with deep sequencing was also used to determine EBNA1 binding sites in B cells, identifying many EBNA1-associated sites, several of which were close to transcriptional start sites for cellular genes (Lu et al. 2010). The expression of some of these cellular genes was decreased upon EBNA1 depletion and induced by EBNA1 expression, suggesting that EBNA1 may affect their transcription. Like the previous study, these EBNA1 sites differed from those in *oriP*, but some were similar in sequence to those identified by Dresang et al. (2009). In addition, a cluster of high-affinity EBNA1 binding sites was identified on chromosome 11 between the divergent FAM55D and FAM55B genes, although the expression of these genes was not affected by EBNA1 (Lu et al. 2010). Canaan et al. (2009) conducted microarray experiments to compare cellular transcripts in B cells and 293 cells with and without EBNA1 and identified a small percentage of transcripts that were affected by EBNA1. In addition, EBNA1 was found to ChIP to most of these gene promoters, suggesting that it directly regulated them. However, whether or not EBNA1 bound directly to these promoters or was recruited through protein interactions was not determined.

The transcriptional activation function of EBNA1 on the EBV genome requires EBNA1 binding to multiple tandem recognition sites in the FR (Wysokenski and Yates 1989), and therefore, it seems unlikely that EBNA1 binding to any single recognition site would be sufficient to activate cellular transcription. To increase the probability of identifying functionally relevant EBNA1 interactions with cellular DNA, D'Herouel et al. (2010) used nearest neighbor position weight matrices to identify repeated EBNA1 binding sites in the human genome. The sites they identified had considerable overlap with those found by Dresang et al. (2009). Although the significance of the repeated EBNA1 sites that they identified remains to be determined, it is interesting that they include weak binding sites near the c-Jun and ATF promoters, which were previously shown to be activated by and associated with EBNA1 in NPC cells (O'Neil et al. 2008).

By comparing cell cycle-specific transcripts from EBV-negative B cells with and without EBNA1 expression, Lu et al. (2011) identified survivin (an inhibitor of apoptosis) as an EBNA1 target gene. EBNA1 increased the levels of survivin transcripts and protein and was shown to associate with the survivin promoter. Induction of survivin protein and transcripts required the EBNA1 residues 65–89 containing the N-terminal transcriptional activation sequence (Wu et al. 2002; Kennedy and Sugden 2003), suggesting that EBNA1 was activating the transcription of the survivin gene. However, since activation of the survivin promoter by EBNA1 involves the Sp1 binding sites, EBNA1 may be recruited to the promoter through the Sp1 host protein, as opposed to binding directly to the DNA. Similarly, Owen et al. (2010) found that EBNA1 increased the level of TFIIIC and ATF-2 transcripts and was associated with their promoter regions, consistent with a direct role in transcriptional activation. Finally, EBNA1 was recently reported to induce the expression of cellular let-7a microRNAs (miRNA) in nasopharyngeal and gastric carcinoma cells (Mansouri et al. 2014). EBNA1 increased the level of let-7a primary transcripts in a manner dependent on its N-terminal transactivation sequence, suggesting that EBNA1 directly induces their transcription. However, whether or not this involves a direct interaction of EBNA1 with DNA sequences regulating these primary transcripts remains to be determined.

Presumably any of the above direct interactions of EBNA1 with specific DNA sites would be mediated by the EBNA1 DNA binding domain. However, there have also been reports of less specific interactions of EBNA1 with DNA or chromatin through its Gly–Arg-rich regions, which resemble AT hooks (Sears et al. 2004; Coppotelli et al. 2013). In addition, EBNA1 has been reported to decondense heterochromatin through its Gly–Arg-rich sequences (Coppotelli et al. 2013). However, it is unclear whether this effect involves the association of the Gly–Arg sequences with DNA, chromatin-associated proteins (including the nucleosome assembly proteins known to bind to them), or another mechanism.

# 4 Cellular Effects of EBNA1

In addition to the roles of EBNA1 at the EBV genome, numerous reports suggest that EBNA1 directly contributes to cell proliferation and survival typical of latent EBV infection. The first implications came from the observations that EBNA1 is the only EBV protein expressed in all EBV-positive tumors and latency types in proliferating cells and is sometimes the only EBV protein expressed. EBNA1 was subsequently shown to be important for efficient B-cell immortalization by EBV (Hume et al. 2003; Altmann et al. 2006) and for the continued proliferation of some EBV-positive tumor cells (Kennedy et al. 2003; Hong et al. 2006; Yin and Flemington 2006). EBNA1 expression in various EBV-negative cancer cells has also been found to increase tumorigenicity (Sheu et al. 1996; Cheng et al. 2010; Kube et al. 1999; Kaul et al. 2007). In addition, EBNA1 expression in the B-cell compartment of a transgenic mouse has been reported to be sufficient to induce B-cell lymphomas (Wilson et al. 1996; Tsimbouri et al. 2002). However, these results were not reiterated in a second independent transgenic mouse study, suggesting that secondary events might contribute to the development of EBNA1-induced lymphomas (Kang et al. 2005, 2008). Nonetheless, the body of evidence indicates that EBNA1 contributes to oncogenesis, likely due to multiple effects on cellular proteins as discussed below and summarized in Fig. 4.

## *4.1 USP7 Interaction*

Proteomics methods identified several cellular proteins that are bound by EBNA1, including an interaction with the cellular ubiquitin-specific protease USP7 [also called HAUSP (Holowaty et al. 2003c; Malik-Soni and Frappier 2012)]. USP7

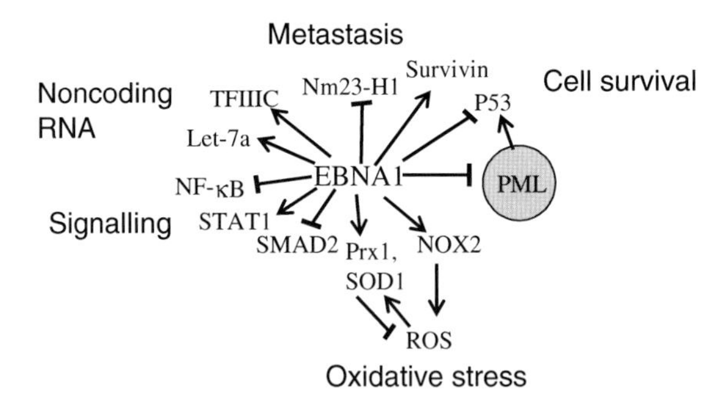


**Fig. 4** Summary of EBNA1 cellular effects. The cellular proteins whose functions or levels are affected by EBNA1 are shown, where *arrows* represent positive regulation and blunted *lines* represent negative regulation. Associated cellular processes are also indicated

was originally discovered as a binding partner of the ICP0 protein from herpes simplex virus type 1 and has since been shown to be targeted by proteins from several different herpesviruses (Everett et al. 1997; Salsman et al. 2012; Lee et al. 2011; Jager et al. 2012). USP7 has been reported to bind and regulate several cellular proteins including p53 and Mdm2 (an E3 ubiquitin ligase for p53), which USP7 stabilizes by removing the polyubiquitin chains that normally signal degradation (Li et al. 2002, 2004; Cummins et al. 2004; Nicholson and Suresh Kumar 2011; Frappier and Verrijzer 2011). EBNA1, p53, and Mdm2 compete for the same binding pocket in the N-terminal TRAF domain of USP7; however, EBNA1 was found to outcompete p53 or Mdm2 due to its higher affinity for USP7 (Holowaty et al. 2003a; Saridakis et al. 2005; Sheng et al. 2006). The EBNA1 region just N-terminal to the DNA binding domain was identified as the USP7 binding site, and a subsequent crystal structure of this EBNA1 peptide bound to the USP7 TRAF domain showed that EBNA1 amino acids 442–448 contact USP7 (Fig. 2) (Holowaty et al. 2003a; Saridakis et al. 2005; Sheng et al. 2006; Hu et al. 2006).

In theory, EBNA1 could destabilize either p53 or Mdm2 by blocking their interaction with USP7, resulting in opposite effects on p53 levels. In vivo EBNA1 has not been reported to lower Mdm2 levels, but has been confirmed to lower p53 levels at least in some cell backgrounds. For example, expression of EBNA1 but not a USP7-binding mutant of EBNA1 in U2OS cells was shown to reduce the accumulation of p53 in response to DNA damage and subsequent apoptosis (Saridakis et al. 2005). Similarly, EBNA1 expression in CNE2 NPC cells decreased the accumulation of p53 in response to DNA damage (Sivachandran et al. 2008), and the presence of EBNA1 or EBV in AGS or SCM1 gastric carcinoma cells decreased the steady-state levels of p53 (Sivachandran et al. 2012a; Cheng et al. 2010). This suggests that EBNA1 could promote cell survival by modulating p53 in EBV-infected epithelial cells.

## *4.2 Effects on PML Nuclear Bodies*

Promyelocytic leukemia (PML) nuclear bodies (also called ND10s) are nuclear foci for which PML tumor suppressor proteins form the structural basis. PML bodies are important for several cellular processes, including apoptosis, DNA repair, senescence, and p53 activation by acetylation (Salomoni et al. 2008; Bernardi and Pandolfi 2007; Takahashi et al. 2004; Guo et al. 2000; Wang et al. 1998; Pearson et al. 2000), and their loss has been associated with the development and/or progression of several tumors (Gurrieri et al. 2004; Salomoni et al. 2008). In addition, PML nuclear bodies suppress lytic viral infection as part of the innate antiviral response (Geoffroy and Chelbi-Alix 2011; Everett and Chelbi-Alix 2007; Reichelt et al. 2011). To counter this defense, many viruses encode proteins that disrupt PML nuclear bodies either by interfering with PML protein interactions need to form the bodies or by inducing the degradation of the PML proteins (Everett 2001).

EBNA1 was found to induce the loss of PML nuclear bodies in both NPC and gastric carcinoma cells, by inducing the degradation of the PML proteins (Sivachandran et al. 2008, 2012a). Consistent with known PML functions, EBNA1 expression in these cells was also found to decrease DNA repair efficiency, p53 acetylation, and apoptosis in response to DNA damaging agents (Sivachandran et al. 2008, 2012a). The results suggest that, as a result of EBNA1-induced PML loss, cells expressing EBNA1 are more likely to survive with DNA damage, which would be expected to contribute to the development of carcinomas. Importantly, these observations in cell lines appear to hold true in vivo, as a comparison of EBV-positive and EBV-negative gastric carcinoma tumor biopsies showed that PML levels were greatly reduced by the presence of EBV, presumably due to the action of EBNA1 (Sivachandran et al. 2012a).

The mechanism by which EBNA1 induces the degradation of PML proteins involves EBNA1 binding to both USP7 and the host casein kinase 2 (CK2) and recruitment of these proteins to the PML nuclear bodies (Sivachandran et al. 2008, 2010). EBNA1 was found to preferentially interact with PML isoform IV over the other five nuclear PML isoforms, and therefore, EBNA1 may localize to PML nuclear bodies through interactions with PML IV (Sivachandran et al. 2008, 2012b). EBNA1 mutants that fail to bind either USP7 or CK2 can still associate with PML bodies but do not induce their loss (Sivachandran et al. 2008, 2010). Similarly, wild-type EBNA1 does not affect PML nuclear bodies when USP7 or CK2 is depleted. In keeping with these observations, USP7 was subsequently shown to negatively regulate PML proteins (even in the absence of EBV or EBNA1), by a mechanism that is independent of its ubiquitin cleavage activity (Sarkari et al. 2011).

The interaction of EBNA1 with CK2 involves a direct interaction of EBNA1 amino acids 387–394 with the binding pocket in the β-regulatory subunit of CK2 and this interaction requires EBNA1 to be phosphorylated at S393 (Sivachandran et al. 2010; Cao et al. 2014). CK2 was previously identified as a negative regulator of PML and was shown to phosphorylate PML proteins at a particular serine residue that triggers polyubiquitylation and subsequent degradation (Scaglioni et al. 2006, 2008). Through its interaction with CK2, EBNA1 was shown to increase CK2-mediated phosphorylation of PML, which is expected to increase PML polyubiquitylation (Sivachandran et al. 2010). Since CK2 is involved in many cellular processes, it is possible that the interaction of EBNA1 with CK2 also affects additional pathways.

## *4.3 Modulation of Signaling Pathways*

EBNA1 has been reported to affect several signaling pathways. First, EBNA1 expression in three different carcinoma cell lines was found to increase the expression of STAT1 (Wood et al. 2007; Kim and Lee 2007). EBNA1 was subsequently shown to enhance STAT1 phosphorylation and nuclear localization in response to IFNγ (Wood et al. 2007). Second, EBNA1 expression was found to decrease the expression of TGF-β1-responsive genes suggesting that EBNA1 interferes with TGF-β signaling (Wood et al. 2007). This effect may be due to increased turnover of SMAD2 in the presence of EBNA1, resulting in decreased levels of SMAD complexes needed for TGF-β1-induced transcription (Wood et al. 2007; Flavell et al. 2008). Third, using NF-κB reporter plasmids in carcinoma cell lines, EBNA1 was found to inhibit NF-κB activity and DNA binding (Valentine et al. 2010). Additional experiments showed that the levels, nuclear localization, and phosphorylation of the p65 NF-κB subunit were all reduced in the presence of EBNA1 as was the phosphorylation of the p65 kinase, IKKα/β (Valentine et al. 2010). How EBNA1 elicits any of the above effects is presently unclear as no physical interaction has been detected between EBNA1 and STAT1, SMAD2, p65, or IKKα/β.

## 4.4 Induction of Oxidative Stress

EBV infection is associated with increased oxidative stress (Lassoued et al. 2008; Cerimele et al. 2005) and this may be at least partly due to EBNA1 expression. Stable or transient EBNA1 expression in B-cell lines was found to increase levels of reactive oxygen species (ROS), DNA damage foci and dysfunctional, uncapped telomeres, and these EBNA1 effects were decreased by ROS scavengers (Gruhne et al. 2009; Kamranvar and Masucci 2011). In addition, EBNA1 was found to increase the expression of the NOX2 NADPH oxidase which might account for the ROS induction (Gruhne et al. 2009). Similarly, a comparison of the nuclear proteome in NPC cells with and without EBNA expression showed that EBNA1 increased the levels of several oxidative stress response proteins including the antioxidants superoxide dismutase 1 and peroxiredoxin 1, known to be induced by ROS (Cao et al. 2011). Further studies confirmed that, in the presence of EBNA1, ROS levels were elevated and that NOX1 and NOX2 transcripts were increased (Cao et al. 2011). Therefore, EBNA1 appears to have multiple effects on the oxidative stress response, although the mechanisms of these effects are not yet known.

## 4.5 Effects on Noncoding RNA

EBNA1 expression has been reported to increase the transcript and protein levels of the RNA polymerase III transcription factor TFIIIC and, in keeping with this finding, increased the expression of several cellular pol III-transcribed genes (Owen et al. 2010). In the same study, EBNA1 was also found to induce the expression of ATF-2, a pol II transcription factor known to contribute to the expression of EBV EBER noncoding RNAs. The result prompted examination of the effect of EBNA1 on EBER levels and confirmed that EBNA1 induces EBER expression.

The effects of EBNA1 on cellular miRNAs have also been examined, using high-throughput sequencing to compare miRNA levels in two different

EBV-negative NPC cell lines with and without transient EBNA1 expression (Mansouri et al. 2014). A small percentage of miRNAs were found to be consistently affected by EBNA1 expression in the two cell lines; in particular, let-7 family miRNAs were increased by EBNA1. Further studies on let-7a miRNA in both EBV-positive and EBV-negative carcinoma cell lines confirmed that EBNA1 upregulates let-7a as well as its primary transcripts and that this required the EBNA1 61–83 transcriptional activation sequence (Mansouri et al. 2014). This induction of let-7a was shown to result in a corresponding decrease in the let-7a target, Dicer. The decreased Dicer levels did not result in global impairment of miRNA biogenesis, but rather the decreased Dicer and increased let-7a levels promoted EBV latency by inhibiting EBV reactivation. In addition, in another study EBNA1 overexpression was reported to decrease levels of miR-200a and miR-200b and this was suggested to contribute to induction of the epithelial–mesenchymal transition (EMT) (Wang et al. 2014).

### *4.6 Effects on Metastatic Potential*

A comparison of the nuclear proteomes of NPC cells with and without EBNA1 expression found that EBNA1 increased the nuclear levels of Nm23-H1, stathmin 1, and maspin, all of which have been found to be contribute to metastases (Cao et al. 2011). This effect on Nm23-H1 corroborated a previous study that showed that EBNA1 co-immunoprecipitated with Nm23-H1 from lymphoid cells and cause it to relocalize to the nucleus (Murakami et al. 2005). This interaction required EBNA1 amino acids 65–89, which are also important for transcriptional activation (Murakami et al. 2005). Nm23-H1 is a known suppressor of metastasis and cell migration, and EBNA1 was shown to counteract the ability of Nm23-H1 to suppress cell migration both in vitro and in a nude mouse model (Murakami et al. 2005; Kaul et al. 2007). The results suggest that EBNA1 contributes to the metastatic potential of EBV tumors. This contention is supported by a report by Sheu et al. (1996) that EBNA1 expression in HONE-1 NPC cells increased tumor metastases in nude mice. In addition, a recent study showed that EBNA1 expression in CNE NPC cell lines decreased expression of epithelial cell markers and increased expression of mesenchymal cell markers, consistent with induction of EMT (Wang et al. 2014). In keeping with this finding, EBNA1 expression increased the migration and colony formation of CNE1 cells, providing further support for a role of EBNA1 in metastasis.

## 5 Immune Evasion

While most of the EBV latency proteins elicit a strong immune response, cells that only express EBNA1 (referred to as latency I or EBNA1-only program) largely avoid immune detection (Babcock et al. 2000; Munz 2004; Khanna et al. 1995).

This is due to inefficient presentation of EBNA1 peptides on MHC class I molecules (Blake et al. 1997). Reduced EBNA1 presentation has been attributed to the central Gly–Ala repeat of EBNA1 which varies in size in different EBV isolates (~230 amino acid long in the B95-8 strain; Fig. 2). Removal of this repeat has been shown to restore EBNA1 presentation, while the addition of the Gly–Ala repeat to EBNA4 inhibits its recognition by cytotoxic T lymphocytes (Blake et al. 1997; Levitskaya et al. 1995).

Initially, it was thought that the Gly–Ala repeat inhibited EBNA1 presentation by interfering with its proteasomal processing. This was based on the studies in which insertion of the Gly–Ala repeat in other proteins inhibited their degradation (Levitskaya et al. 1997; Sharipo et al. 1998; Heessen et al. 2002; Dantuma et al. 2000; Zhang and Coffino 2004). However, the deletion of the Gly–Ala repeat from EBNA1 was not found to affect EBNA1 turnover, as EBNA1 is extremely stable with or without the Gly–Ala repeat (Daskalogianni et al. 2008). In addition, Tellam et al. (2004) found that EBNA1 turnover varied considerably in different cell backgrounds but that these rates did not correspond to the level of MHC I-restricted presentation of EBNA1 peptides. Moreover, EBNA1 presentation was shown to derive from newly synthesized protein, and the primary contribution of the Gly–Ala repeat on the presentation of EBNA1 peptides was found to be due to inhibition of its own translation (Tellam et al. 2004; Yin et al. 2003). This supported a model where the MHC I-restricted presentation of EBNA1 occurs through the generation of defective ribosomal products (DRiPs) which are reduced by the presence of the Gly–Ala repeat (Fahraeus 2005).

Further studies confirmed that MHC class I presentation and CTL recognition corresponded to the rate of EBNA1 translation and the levels of DRiPs and that all were inhibited by the Gly–Ala repeat (Tellam et al. 2007; Apcher et al. 2009, 2010; Tellam et al. 2008). The Gly–Ala repeat was also reported to interfere with the initiation of translation (Apcher et al. 2010). The importance of the translation elongation rate was further supported by the recent finding that the RNA encoding the Gly–Ala repeats forms clusters of G-quadruplexes that affect both translation and antigen presentation (Murat et al. 2014). Specifically, antisense oligonucleotides that destabilize these G-quadruplexes were shown to increase translation, ribosome association, and antigen presentation, while stabilization of G-quadruplexes by pyridostatin treatment inhibited translation. Since the EBNA1 protein has been reported to bind G-quadruplex RNA (Norseen et al. 2009), it is possible that EBNA1 may autoregulate its translation by binding to its own mRNA during translation, but this remains to be determined. It has also been reported that Hsp90 inhibitors decrease EBNA translation suggesting a role for Hsp90 in this process; however, the mechanism of this effect is unclear (Sun et al. 2010).

Unlike MHC class I presentation, EBNA1 elicits a strong CD4 response due to efficient MHC class II processing (Munz et al. 2000; Leen et al. 2001). The MHC class II-mediated presentation of EBNA1 results, at least in part, from the ability of endogenous EBNA1 to undergo lysosomal processing after entering the autophagy pathway (Paludan et al. 2005). In keeping with this conclusion, inhibition of autophagy decreased recognition of EBNA1 by $CD4^+$ T cells (Paludan et al. 2005). Subsequently, the range of EBNA1 CD4 epitopes generated was shown to be affected by the cellular localization of EBNA1, with nuclear localization limiting the displayed epitopes due to decreased accessibility to the macroautophagy pathway relative to cytoplasmic EBNA1 (Leung et al. 2010). How EBNA1-expressing cells are able to persist in the body despite recognition by $CD4^+$ T cells is unclear.

## 6 EBNA1 in Lytic EBV Infection

The above roles of EBNA1 all pertain to latent EBV infection. However, EBNA1 is also expressed in lytic infection from a lytic cycle-specific promoter (Fp), suggesting that it also contributes to productive infection (Brink et al. 2001; Lear et al. 1992; Schaefer et al. 1995). The contributions of EBNA1 to viral reactivation to the lytic cycle were examined in EBV-positive AGS gastric carcinoma cells (Sivachandran et al. 2012b). EBNA1 silencing was found to increase the frequency with which EBV spontaneously entered the lytic cycle, suggesting that EBNA1 can suppress reactivation. This may be due to ability of EBNA1 to induce let7 family miRNAs (including let-7a), which results in downregulation of Dicer, as both upregulation of let-7a and downregulation of Dicer have been shown to inhibit EBV reactivation (Iizasa et al. 2010; Mansouri et al. 2014).

EBNA1 has also been found to play a role in EBV lytic infection. When the lytic cycle is chemically induced, EBNA1 silencing decreases lytic gene expression and viral genome amplification indicating that EBNA1 can promote lytic infection. However, EBNA1 does not positively contribute to lytic infection in cells depleted in PML proteins, suggesting that the role of EBNA1 in lytic infection is in overcoming suppression by PML proteins. In keeping with this interpretation, PML proteins and nuclear bodies were found to suppress lytic infection by EBV (Sivachandran et al. 2012b; Sides et al. 2011).

## 7 Conclusion

In summary, EBNA1 makes multiple contributions to EBV infection due to its ability to interact with specific DNA sequences and multiple cellular proteins. Latency contributions include the replication and mitotic segregation of EBV episomes, contributions to viral transcription, and multiple effects on cellular proteins and pathways that promote cell survival and proliferation. In addition, EBNA1 impacts the reactivation of EBV and contributes to EBV lytic infection by overcoming suppression by PML nuclear bodies. This combination of diverse functions makes EBNA1 a key protein for EBV infection.

# References

Altmann M, Pich D, Ruiss R, Wang J, Sugden B, Hammerschmidt W (2006) Transcriptional activation by EBV nuclear antigen 1 is essential for the expression of EBV's transforming genes. Proc Natl Acad Sci USA 103(38):14188–14193

Ambinder RF, Shah WA, Rawlins DR, Hayward GS, Hayward SD (1990) Definition of the sequence requirements for binding of the EBNA-1 protein to its palindromic target sites in Epstein-Barr virus DNA. J Virol 64:2369–2379

Ambinder RF, Mullen M, Chang Y, Hayward GS, Hayward SD (1991) Functional domains of Epstein-Barr nuclear antigen EBNA-1. J Virol 65:1466–1478

Apcher S, Komarova A, Daskalogianni C, Yin Y, Malbert-Colas L, Fahraeus R (2009) mRNA translation regulation by the Gly-Ala repeat of Epstein-Barr virus nuclear antigen 1. J Virol 83(3):1289–1298

Apcher S, Daskalogianni C, Manoury B, Fahraeus R (2010) Epstein Barr virus-encoded EBNA1 interference with MHC class I antigen presentation reveals a close correlation between mRNA translation initiation and antigen presentation. PLoS Pathog 6(10):e1001151. doi:10.1371/journal.ppat.1001151

Aras S, Singh G, Johnston K, Foster T, Aiyar A (2009) Zinc coordination is required for and regulates transcription activation by Epstein-Barr nuclear antigen 1. PLoS Pathog 5(6):e1000469. doi:10.1371/journal.ppat.1000469

Atanasiu C, Deng Z, Wiedmer A, Norseen J, Lieberman PM (2006) ORC binding to TRF2 stimulates OriP replication. EMBO Rep 7(7):716–721

Avolio-Hunter TM, Frappier L (1998) Mechanistic studies on the DNA linking activity of the Epstein-Barr nuclear antigen 1. Nucl Acids Res 26:4462–4470

Avolio-Hunter TM, Frappier L (2003) EBNA1 efficiently assembles on chromatin containing the Epstein-Barr virus latent origin of replication. Virol 315:398–408

Avolio-Hunter TM, Lewis PN, Frappier L (2001) Epstein-Barr nuclear antigen 1 binds and destabilizes nucleosomes at the viral origin of latent DNA replication. Nucl Acids Res 29:3520–3528

Babcock GJ, Hochberg D, Thorley-Lawson DA (2000) The expression pattern of Epstein-Barr virus latent genes in vivo is dependent upon the differentiation stage of the infected B cell. Immunity 13:497–506

Barbera AJ, Chodaparambil JV, Kelley-Clarke B, Joukov V, Walter JC, Luger K, Kaye KM (2006) The nucleosomal surface as a docking station for Kaposi's sarcoma herpesvirus LANA. Science 311(5762):856–861

Bashaw JM, Yates JL (2001) Replication from oriP of Epstein-Barr virus requires exact spacing of two bound dimers of EBNA1 which bend DNA. J Virol 75:10603–10611

Bernardi R, Pandolfi PP (2007) Structure, dynamics and functions of promyelocytic leukaemia nuclear bodies. Nat Rev Mol Cell Biol 8:1006–1016

Blake N, Lee S, Redchenko I, Thomas W, Steven N, Leese A, Steigerwald-Mullen P, Kurilla MG, Frappier L, Rickinson A (1997) Human CD8+ T cell responses to EBV EBNA1: HLA class I presentation of the (Gly-ALA) containing protein requires exogenous processing. Immunity 7:791–802

Bochkarev A, Barwell J, Pfuetzner R, Furey W, Edwards A, Frappier L (1995) Crystal structure of the DNA binding domain of the Epstein-Barr virus origin binding protein EBNA1. Cell 83:39–46

Bochkarev A, Barwell J, Pfuetzner R, Bochkareva E, Frappier L, Edwards AM (1996) Crystal structure of the DNA-binding domain of the Epstein-Barr virus origin binding protein, EBNA1, bound to DNA. Cell 84:791–800

Bochkarev A, Bochkareva E, Frappier L, Edwards AM (1998) 2.2A structure of a permanganate-sensitive DNA site bound by the Epstein-Barr virus origin binding protein, EBNA1. J Mol Biol 284:1273–1278

Brink AA, Meijer CJ, Nicholls JM, Middeldorp JM, van den Brule AJ (2001) Activity of the EBNA1 promoter associated with lytic replication (Fp) in Epstein-Barr virus associated disorders. Mol Pathol 54(2):98–102

Canaan A, Haviv I, Urban AE, Schulz VP, Hartman S, Zhang Z, Palejev D, Deisseroth AB, Lacy J, Snyder M, Gerstein M, Weissman SM (2009) EBNA1 regulates cellular gene expression by binding cellular promoters. Proc Natl Acad Sci USA 106(52):22421–22426. doi:10.1073/pnas.0911676106 (0911676106 [pii])

Cao JY, Mansouri S, Frappier L (2011) Changes in the nasopharyngeal carcinoma nuclear proteome induced by the EBNA1 protein of Epstein-Barr virus reveal potential roles for EBNA1 in metastasis and oxidative stress responses. J Virol. doi:10.1128/JVI.05648-11 (JVI.05648-11 [pii])

Cao JY, Shire K, Landry C, Gish GD, Pawson T, Frappier L (2014) Identification of a novel protein interaction motif in the regulatory subunit of casein kinase 2. Mol Cell Biol 34(2): 246–258. doi:10.1128/MCB.00968-13MCB.00968-13 ([pii])

Ceccarelli DFJ, Frappier L (2000) Functional analyses of the EBNA1 origin DNA binding protein of Epstein-Barr virus. J Virol 74:4939–4948

Cerimele F, Battle T, Lynch R, Frank DA, Murad E, Cohen C, Macaron N, Sixbey J, Smith K, Watnick RS, Eliopoulos A, Shehata B, Arbiser JL (2005) Reactive oxygen signaling and MAPK activation distinguish Epstein-Barr Virus (EBV)-positive versus EBV-negative Burkitt's lymphoma. Proc Natl Acad Sci USA 102(1):175–179. doi:10.1073/pnas.0408381102 (0408381102 [pii])

Chaudhuri B, Xu H, Todorov I, Dutta A, Yates JL (2001) Human DNA replication initiation factors, ORC and MCM, associate with *oriP* of Epstein-Barr virus. Proc Natl Acad Sci USA 98:10085–10089

Chen M-R, Middeldorp JM, Hayward SD (1993) Separation of the complex DNA binding domain of EBNA-1 into DNA recognition and dimerization subdomains of novel structure. J Virol 67:4875–4885

Chen YL, Liu CD, Cheng CP, Zhao B, Hsu HJ, Shen CL, Chiu SJ, Kieff E, Peng CW (2014) Nucleolin is important for Epstein-Barr virus nuclear antigen 1-mediated episome binding, maintenance, and transcription. Proc Natl Acad Sci USA 111(1):243–248. doi:10.1073/pnas.1321800111 1321800111 ([pii])

Cheng TC, Hsieh SS, Hsu WL, Chen YF, Ho HH, Sheu LF (2010) Expression of Epstein-Barr nuclear antigen 1 in gastric carcinoma cells is associated with enhanced tumorigenicity and reduced cisplatin sensitivity. Int J Oncol 36(1):151–160

Coppotelli G, Mughal N, Callegari S, Sompallae R, Caja L, Luijsterburg MS, Dantuma NP, Moustakas A, Masucci MG (2013) The Epstein-Barr virus nuclear antigen-1 reprograms transcription by mimicry of high mobility group A proteins. Nucleic Acids Res 41(5):2950–2962. doi:10.1093/nar/gkt032 ([pii])

Cruickshank J, Davidson A, Edwards AM, Frappier L (2000) Two domains of the Epstein-Barr virus origin DNA binding protein, EBNA1, orchestrate sequence-specific DNA binding. J Biol Chem 275:22273–22277

Cummins JM, Rago C, Kohli M, Kinzler KW, Lengauer C, Vogelstein B (2004) Tumour suppression: disruption of HAUSP gene stabilizes p53. Nature 428:486–487

D'Herouel AF, Birgersdotter A, Werner M (2010) FR-like EBNA1 binding repeats in the human genome. Virology 405(2):524–529. doi:10.1016/j.virol.2010.06.040 (S0042-6822(10)00422-8 [pii])

Dantuma NP, Heessen S, Lindsten K, Jellne M, Masucci MG (2000) Inhibition of proteasomal degradation by the gly-Ala repeat of Epstein-Barr virus is influenced by the length of the repeat and the strength of the degradation signal. Proc Natl Acad Sci USA 97(15):8381–8385

Daskalogianni C, Apcher S, Candeias MM, Naski N, Calvo F, Fahraeus R (2008) Gly-Ala repeats induce position- and substrate-specific regulation of 26 S proteasome-dependent partial processing. J Biol Chem 283(44):30090–30100

Delecluse H-J, Bartnizke S, Hammerschmidt W, Bullerdiek J, Bornkamm GW (1993) Episomal and integrated copies of Epstein-Barr virus coexist in Burkitt's lymphoma cell lines. J Virol 67:1292–1299
Deng Z, Lezina L, Chen C-J, Shtivelband S, So W, Lieberman PM (2002) Telomeric proteins regulate episomal maintenance of Epstein-Barr virus origin of plasmid replication. Mol Cell 9:493–503
Deng Z, Atanasiu C, Burg JS, Broccoli D, Lieberman PM (2003) Telomere repeat binding factors TRF1, TRF2, and hRAP1 modulate replication of Epstein-Barr virus OriP. J Virol 77(22):11992–12001
Deng Z, Atanasiu C, Zhao K, Marmorstein R, Sbodio JI, Chi NW, Lieberman PM (2005) Inhibition of Epstein-Barr virus OriP function by tankyrase, a telomere-associated poly-ADP ribose polymerase that binds and modifies EBNA1. J Virol 79(8):4640–4650
Deutsch MJ, Ott E, Papior P, Schepers A (2010) The latent origin of replication of Epstein-Barr virus directs viral genomes to active regions of the nucleus. J Virol 84(5):2533–2546. doi:10.1128/JVI.01909-09 (JVI.01909-09 [pii])
Dhar V, Schildkraut CL (1991) Role of EBNA-1 in arresting replication forks at the Epstein-Barr virus *oriP* family of tandem repeats. Mol Cell Biol 11:6268–6278
Dhar SK, Yoshida K, Machida Y, Khaira P, Chaudhuri B, Wohlschlegel JA, Leffak M, Yates J, Dutta A (2001) Replication from oriP of Epstein-Barr virus requires human ORC and is inhibited by geminin. Cell 106:287–296
Dheekollu J, Deng Z, Wiedmer A, Weitzman MD, Lieberman PM (2007) A role for MRE11, NBS1, and recombination junctions in replication and stable maintenance of EBV episomes. PLoS ONE 2(12):e1257
Dresang LR, Vereide DT, Sugden B (2009) Identifying sites bound by Epstein-Barr virus nuclear antigen 1 (EBNA1) in the human genome: defining a position-weighted matrix to predict sites bound by EBNA1 in viral genomes. J Virol 83(7):2930–2940. doi:10.1128/JVI.01974-08 (JVI.01974-08 [pii])
Dyson PJ, Farrell PJ (1985) Chromatin structure of Epstein-Barr virus. J Gen Virol 66(Pt 9):1931–1940
Edwards AM, Bochkarev A, Frappier L (1998) Origin DNA-binding proteins. Curr Opin Struct Biol 8:49–53
Ermakova O, Frappier L, Schildkraut CL (1996) Role ot the EBNA-1 protein in pausing of replication forks in the Epstein-Barr virus genome. J Biol Chem 271:33009–33017
Everett RD (2001) DNA viruses and viral proteins that interact with PML nuclear bodies. Oncogene 20(49):7266–7273
Everett RD, Chelbi-Alix MK (2007) PML and PML nuclear bodies: implications in antiviral defence. Biochimie 89(6–7):819–830
Everett R, Meredith M, Orr A, Cross A, Kathoria M, Parkinson J (1997) A novel ubiquitin-specific protease is dynamically associated with the PML nuclear domain and binds to a herpesvirus regulatory protein. EMBO J 16:1519–1530
Fahraeus R (2005) Do peptides control their own birth and death? Nat Rev Mol Cell Biol 6(3):263–267
Feeney KM, Saade A, Okrasa K, Parish JL (2011) In vivo analysis of the cell cycle dependent association of the bovine papillomavirus E2 protein and ChlR1. Virology 414(1):1–9. doi:10.1016/j.virol.2011.03.015 (S0042-6822(11)00140-1 [pii])
Flavell JR, Baumforth KR, Wood VH, Davies GL, Wei W, Reynolds GM, Morgan S, Boyce A, Kelly GL, Young LS, Murray PG (2008) Down-regulation of the TGF-beta target gene, PTPRK, by the Epstein-Barr virus encoded EBNA1 contributes to the growth and survival of Hodgkin lymphoma cells. Blood 111(1):292–301
Frappier L, O'Donnell M (1991a) Epstein-Barr nuclear antigen 1 mediates a DNA loop within the latent replication origin of Epstein-Barr virus. Proc Natl Acad Sci USA 88:10875–10879
Frappier L, O'Donnell M (1991b) Overproduction, purification and characterization of EBNA1, the origin binding protein of Epstein-Barr virus. J Biol Chem 266:7819–7826

Frappier L, O'Donnell M (1992) EBNA1 distorts *oriP*, the Epstein-Barr virus latent replication origin. J Virol 66:1786–1790
Frappier L, Verrijzer CP (2011) Gene expression control by protein deubiquitinases. Curr Opin Genet Dev 21(2):207–213. doi:10.1016/j.gde.2011.02.005 (S0959-437X(11)00048-7 [pii])
Frappier L, Goldsmith K, Bendell L (1994) Stabilization of the EBNA1 protein on the Epstein-Barr virus latent origin of DNA replication by a DNA looping mechanism. J Biol Chem 269:1057–1062
Gahn TA, Schildkraut CL (1989) The Epstein-Barr virus origin of plasmid replication, *oriP*, contains both the initiation and termination sites of DNA replication. Cell 58:527–535
Gahn T, Sugden B (1995) An EBNA1 Dependent enhancer acts from a distance of 10 kilobase pairs to increase expression of the Epstien-Barr virus LMP gene. J Virol 69:2633–2636
Geoffroy MC, Chelbi-Alix MK (2011) Role of promyelocytic leukemia protein in host antiviral defense. J Interferon Cytokine Res 31(1):145–158. doi:10.1089/jir.2010.0111
Goldsmith K, Bendell L, Frappier L (1993) Identification of EBNA1 amino acid sequences required for the interaction of the functional elements of the Epstein-Barr virus latent origin of DNA replication. J Virol 67:3418–3426
Grogan EA, Summers WP, Dowling S, Shedd D, Gradoville L, Miller G (1983) Two Epstein-Barr viral nuclear neoantigens distinguished by gene transfer, serology and chromosome binding. Proc Natl Acad Sci USA 80:7650–7653
Gruhne B, Sompallae R, Marescotti D, Kamranvar SA, Gastaldello S, Masucci MG (2009) The Epstein-Barr virus nuclear antigen-1 promotes genomic instability via induction of reactive oxygen species. Proc Natl Acad Sci USA 106(7):2313–2318. doi:10.1073/pnas.0810619106 (0810619106 [pii])
Guo A, Salomoni P, Luo J, Shih A, Zhong S, Gu W, Pandolfi PP (2000) The function of PML in p53-dependent apoptosis. Nat Cell Biol 2(10):730–736
Gurrieri C, Capodieci P, Bernardi R, Scaglioni PP, Nafa K, Rush LJ, Verbel DA, Cordon-Cardo C, Pandolfi PP (2004) Loss of the tumor suppressor PML in human cancers of multiple histologic origins. J Natl Cancer Inst 96(4):269–279
Harris A, Young BD, Griffin BE (1985) Random association of Epstein-Barr virus genomes with host cell metaphase chromosomes in Burkitt's lymphoma-derived cell lines. J Virol 56:328–332
Harrison S, Fisenne K, Hearing J (1994) Sequence requirements of the Epstein-Barr virus latent origin of DNA replication. J Virol 68(3):1913–1925
Hearing J, Mulhaupt Y, Harper S (1992) Interaction of Epstein-Barr virus nuclear antigen 1 with the viral latent origin of replication. J Virol 66:694–705
Heessen S, Leonchiks A, Issaeva N, Sharipo A, Selivanova G, Masucci MG, Dantuma NP (2002) Functional p53 chimeras containing the Epstein-Barr virus Gly-Ala repeat are protected from Mdm2- and HPV-E6-induced proteolysis. Proc Natl Acad Sci USA 99(3):1532–1537
Hegde RS, Grossman SR, Laimins LA, Sigler PB (1992) Crystal structure at 1.7 Å of the bovine papillomavirus-1 E2 DNA-binding protein bound to its DNA target. Nature 359:505–512
Hodin TL, Najrana T, Yates JL (2013) Efficient replication of Epstein-Barr virus-derived plasmids requires tethering by EBNA1 to host chromosomes. J Virol 87(23):13020–13028. doi:10.1128/JVI.01606-13JVI.01606-13 ([pii])
Holowaty MN, Sheng Y, Nguyen T, Arrowsmith C, Frappier L (2003a) Protein interaction domains of the ubiqutin specific protease, USP7/HAUSP. J Biol Chem 278:47753–47761
Holowaty MN, Zeghouf M, Wu H, Tellam J, Athanasopoulos V, Greenblatt J, Frappier L (2003b) Protein profiling with Epstein-Barr nuclear antigen-1 reveals an interaction with the herpesvirus-associated ubiquitin-specific protease HAUSP/USP7. J Biol Chem 278(32):29987–29994. doi:10.1074/jbc.M303977200M303977200 ([pii])
Holowaty MN, Zeghouf M, Wu H, Tellam J, Athanasopoulos V, Greenblatt J, Frappier L (2003c) Protein profiling with Epstein-Barr nuclear antigen 1 reveals an interaction with the herpesvirus-associated ubiquitin-specific protease HAUSP/USP7. J Biol Chem 278:29987–29994
Hong M, Murai Y, Kutsuna T, Takahashi H, Nomoto K, Cheng CM, Ishizawa S, Zhao QL, Ogawa R, Harmon BV, Tsuneyama K, Takano Y (2006) Suppression of Epstein-Barr

nuclear antigen 1 (EBNA1) by RNA interference inhibits proliferation of EBV-positive Burkitt's lymphoma cells. J Cancer Res Clin Oncol 132(1):1–8
Hsieh D-J, Camiolo SM, Yates JL (1993) Constitutive binding of EBNA1 protein to the Epstein-Barr virus replication origin, oriP, with distortion of DNA structure during latent infection. EMBO J 12:4933–4944
Hu M, Gu L, Li M, Jeffrey PD, Gu W, Shi Y (2006) Structural basis of competitive recognition of p53 and MDM2 by HAUSP/USP7: implications for the regulation of the p53-MDM2 pathway. PLoS Biol 4(2):e27
Hume S, Reisbach G, Feederle R, Delecluse H-J, Bousset K, Hammerschmidt W, Schepers A (2003) The EBV nuclear antigen 1 (EBNA1) enhances B cell immortalization several thousandfold. Proc Natl Acad Sci 100:10989–10994
Hung SC, Kang M-S, Kieff E (2001) Maintenance of Epstein-Barr virus (EBV) oriP-based episomes requires EBV-encoded nuclear antigen-1 chromosome-binding domains, which can be replaced by high-mobility group-I or histone H1. Proc Natl Acad Sci USA 98:1865–1870
Iizasa H, Wulff BE, Alla NR, Maragkakis M, Megraw M, Hatzigeorgiou A, Iwakiri D, Takada K, Wiedmer A, Showe L, Lieberman P, Nishikura K (2010) Editing of Epstein-Barr virus-encoded BART6 microRNAs controls their dicer targeting and consequently affects viral latency. J Biol Chem 285(43):33358–33370. doi:10.1074/jbc.M110.138362M110.138362 ([pii])
Ilves I, Maemets K, Silla T, Janikson K, Ustav M (2006) Brd4 is involved in multiple processes of the bovine papillomavirus type 1 life cycle. J Virol 80(7):3660–3665
Jager W, Santag S, Weidner-Glunde M, Gellermann E, Kati S, Pietrek M, Viejo-Borbolla A, Schulz TF (2012) The ubiquitin-specific protease USP7 modulates the replication of Kaposi's sarcoma-associated herpesvirus latent episomal DNA. J Virol 86(12):6745–6757. doi:10.1128/JVI.06840-11 (JVI.06840-11 [pii])
Jones CH, Hayward SD, Rawlins DR (1989) Interaction of the lymphocyte-derived Epstein-Barr virus nuclear antigen EBNA-1 with its DNA-binding sites. J Virol 63:101–110
Jourdan N, Jobart-Malfait A, Dos Reis G, Quignon F, Piolot T, Klein C, Tramier M, Coppey-Moisan M, Marechal V (2012) Live-cell imaging reveals multiple interactions between Epstein-Barr virus nuclear antigen 1 and cellular chromatin during interphase and mitosis. J Virol 86(9):5314–5329. doi:10.1128/JVI.06303-11 (JVI.06303-11 [pii])
Julien MD, Polonskaya Z, Hearing J (2004) Protein and sequence requirements for the recruitment of the human origin recognition complex to the latent cycle origin of DNA replication of Epstein-Barr virus oriP. Virology 326(2):317–328
Kamranvar SA, Masucci MG (2011) The Epstein-Barr virus nuclear antigen-1 promotes telomere dysfunction via induction of oxidative stress. Leukemia 25(6):1017–1025. doi:10.1038/leu.2011.35leu201135 ([pii])
Kanda T, Otter M, Wahl GM (2001) Coupling of mitotic chromosome tethering and replication competence in Epstein-Barr virus-based plasmids. Mol Cell Biol 21:3576–3588
Kanda T, Kamiya M, Maruo S, Iwakiri D, Takada K (2007) Symmetrical localization of extrachromosomally replicating viral genomes on sister chromatids. J Cell Sci 120(Pt 9):1529–1539
Kanda T, Horikoshi N, Murata T, Kawashima D, Sugimoto A, Narita Y, Kurumizaka H, Tsurumi T (2013) Interaction between basic residues of Epstein-Barr virus EBNA1 protein and cellular chromatin mediates viral plasmid maintenance. J Biol Chem 288(33):24189–24199. doi:10.1074/jbc.M113.491167M113.491167 ([pii])
Kang MS, Lu H, Yasui T, Sharpe A, Warren H, Cahir-McFarland E, Bronson R, Hung SC, Kieff E (2005) Epstein-Barr virus nuclear antigen 1 does not induce lymphoma in transgenic FVB mice. Proc Natl Acad Sci USA 102(3):820–825
Kang MS, Soni V, Bronson R, Kieff E (2008) Epstein-Barr Virus Nuclear Antigen 1 does not cause lymphoma in C57BL/6J mice. J Virol 82:4180–4183
Kapoor P, Frappier L (2003) EBNA1 partitions Epstein-Barr virus plasmids in yeast by attaching to human EBNA1-binding protein 2 on mitotic chromosomes. J Virol 77:6946–6956
Kapoor P, Frappier L (2005) Methods for measuring the replication and segregation of Epstein-Barr virus-based plasmids. Methods Mol Biol 292:247–266

Kapoor P, Shire K, Frappier L (2001) Reconstitution of Epstein-Barr virus-based plasmid partitioning in budding yeast. EMBO J 20:222–230
Kapoor P, Lavoie BD, Frappier L (2005) EBP2 plays a key role in Epstein-Barr virus mitotic segregation and is regulated by aurora family kinases. Mol Cell Biol 25(12):4934–4945
Kaul R, Murakami M, Choudhuri T, Robertson ES (2007) Epstein-Barr virus latent nuclear antigens can induce metastasis in a nude mouse model. J Virol 81(19):10352–10361
Kennedy G, Sugden B (2003) EBNA-1, a bifunctional transcriptional activator. Mol Cell Biol 23(19):6901–6908
Kennedy G, Komano J, Sugden B (2003) Epstein-Barr virus provide a survival factor to Burkitt's lymphomas. Proc Natl Acad Sci 100:14269–14274
Khanna R, Burrows SR, Moss DJ (1995) Immune regulation in Epstein-Barr virus-associated diseases. Microbiol Rev 59(3):387–405
Kim HS, Lee MS (2007) STAT1 as a key modulator of cell death. Cell Signal 19(3):454–465. doi:10.1016/j.cellsig.2006.09.003 (S0898-6568(06)00269-5 [pii])
Kim AL, Maher M, Hayman JB, Ozer J, Zerby D, Yates JL, Lieberman PM (1997) An imperfect correlation between DNA replication activity of Epstein-Barr virus nuclear antigen 1 (EBNA1) and binding to the nuclear import receptor, Rch1/importin α. Virology 239:340–351
Kirchmaier AL, Sugden B (1997) Dominant-negative inhibitors of EBNA1 of Epstein-Barr virus. J Virol 71:1766–1775
Koons MD, Van Scoy S, Hearing J (2001) The replicator of the Epstein-Barr virus latent cycle origin of DNA replication, *oriP*, is composed of multiple functional elements. J Virol 75:10582–10592
Krithivas A, Fujimuro M, Weidner M, Young DB, Hayward SD (2002) Protein interactions targeting the latency-associated nuclear antigen of Kaposi's sarcoma-associated herpesvirus to cell chromosomes. J Virol 76:11596–11604
Krysan PJ, Haase SB, Calos MP (1989) Isolation of human sequences that replicate autonomously in human cells. Mol Cell Biol 9:1026–1033
Kube D, Vockerodt M, Weber O, Hell K, Wolf J, Haier B, Grasser FA, Muller-Lantzsch N, Kieff E, Diehl V, Tesch H (1999) Expression of Epstein-Barr virus nuclear antigen 1 is associated with enhanced expression of CD25 in the hodgkin cell line L428. J Virol 73:1630–1636
Laine A, Frappier L (1995) Identification of Epstein-Barr nuclear antigen 1 protein domains that direct interactions at a distance between DNA-bound proteins. J Biol Chem 270:30914–30918
Lassoued S, Ben Ameur R, Ayadi W, Gargouri B, Ben Mansour R, Attia H (2008) Epstein-Barr virus induces an oxidative stress during the early stages of infection in B lymphocytes, epithelial, and lymphoblastoid cell lines. Mol Cell Biochem 313(1–2):179–186. doi:10.1007/s11010-008-9755-z
Lear AL, Rowe M, Kurilla MG, Lee S, Henderson S, Kieff E, Rickinson AB (1992) The Epstein-Barr virus (EBV) nuclear antigen 1 BamHI F promoter is activated on entry of EBV-transformed B cells into the lytic cycle. J Virol 66(12):7461–7468
Lee MA, Diamond ME, Yates JL (1999) Genetic evidence that EBNA-1 is needed for efficient, stable latent infection by Epstein-Barr virus. J Virol 73(4):2974–2982
Lee HR, Choi WC, Lee S, Hwang J, Hwang E, Guchhait K, Haas J, Toth Z, Jeon YH, Oh TK, Kim MH, Jung JU (2011) Bilateral inhibition of HAUSP deubiquitinase by a viral interferon regulatory factor protein. Nat Struct Mol Biol 18(12):1336–1344. doi:10.1038/nsmb.2142nsmb.2142 ([pii])
Leen A, Meij P, Redchenko I, Middeldorp J, Bloemena E, Rickinson A, Blake N (2001) Differential immunogenicity of Epstein-Barr virus latent-cycle proteins for human CD4(+) T-helper 1 responses. J Virol 75(18):8649–8659
Leung CS, Haigh TA, Mackay LK, Rickinson AB, Taylor GS (2010) Nuclear location of an endogenously expressed antigen, EBNA1, restricts access to macroautophagy and the range of CD4 epitope display. Proc Natl Acad Sci USA 107(5):2165–2170. doi:10.1073/pnas.09094481070909448107 ([pii])

Levitskaya J, Coram M, Levitsky V, Imreh S, Steigerwald-Mullen P, Klein G, Kurilla MG, Masucci MG (1995) Inhibition of antigen processing by the internal repeat region of the Epstein-Barr virus nuclear antigen-1. Nature 375:685–688

Levitskaya J, Shapiro A, Leonchiks A, Ciechanover A, Masucci MG (1997) Inhibition of ubiquitin/proteasome-dependent protein degradation by the Gly-Ala repeat domain of the Epstein-Barr virus nuclear antigen 1. Proc Natl Acad Sci USA 94:12616–12621

Li M, Chen D, Shiloh A, Luo J, Nikolaev AY, Qin J, Gu W (2002) Deubiquitination of p53 by HAUSP is an important pathway for p53 stabilization. Nature 416(6881):648–653

Li M, Brooks CL, Kon N, Gu W (2004) A dynamic role of HAUSP in the p53-Mdm2 pathway. Mol Cell 13(6):879–886. doi:10.1016/S1097276504001571 ([pii])

Lin A, Wang S, Nguyen T, Shire K, Frappier L (2008) The EBNA1 protein of Epstein-Barr virus functionally interacts with Brd4. J Virol 82(24):12009–12019

Lindner SE, Zeller K, Schepers A, Sugden B (2008) The affinity of EBNA1 for its origin of DNA synthesis is a determinant of the origin's replicative efficiency. J Virol 82(12):5693–5702

Little RD, Schildkraut CL (1995) Initiation of latent DNA replicatoon in the Epstein-Barr virus genome can occur at sites other than the genetically defined origin. Mol Cell Biol 15:2893–2903

Lu F, Wikramasinghe P, Norseen J, Tsai K, Wang P, Showe L, Davuluri RV, Lieberman PM (2010) Genome-wide analysis of host-chromosome binding sites for Epstein-Barr Virus Nuclear Antigen 1 (EBNA1). Virol J 7:262. doi:10.1186/1743-422X-7-262 (1743-422X-7-262 [pii])

Lu J, Murakami M, Verma SC, Cai Q, Haldar S, Kaul R, Wasik MA, Middeldorp J, Robertson ES (2011) Epstein-Barr Virus nuclear antigen 1 (EBNA1) confers resistance to apoptosis in EBV-positive B-lymphoma cells through up-regulation of survivin. Virology 410(1):64–75. doi:10.1016/j.virol.2010.10.029 (S0042-6822(10)00676-8 [pii])

Lupton S, Levine AJ (1985) Mapping of genetic elements of Epstein-Barr virus that facilitate extrachromosomal persistence of Epstein-Barr virus-derived plasmids in human cells. Mol Cell Biol 5:2533–2542

Mackey D, Sugden B (1999) The linking regions of EBNA1 are essential for its support of replication and transcription. Mol Cell Biol 19:3349–3359

Mackey D, Middleton T, Sugden B (1995) Multiple regions within EBNA1 can link DNAs. J Virol 69:6199–6208

Malik-Soni N, Frappier L (2012) Proteomic profiling of EBNA1-Host Protein Interactions in latent and lytic Epstein-Barr virus infections. J Virol 86(12):6999–7002. doi:10.1128/JVI.00194-12 (JVI.00194-12 [pii])

Malik-Soni N, Frappier L (2013) Nucleophosmin contributes to the transcriptional activation function of the Epstein-Barr virus EBNA1 Protein. J Virol. doi:10.1128/JVI.02521-13 (JVI.02521-13 [pii])

Mansouri S, Pan Q, Blencowe BJ, Claycomb JM, Frappier L (2014) Epstein-Barr Virus EBNA1 protein regulates viral latency through effects on let-7 MicroRNA and dicer. J Virol 88(19):11166–11177. doi:10.1128/JVI.01785-14JVI.01785-14 ([pii])

Marechal V, Dehee A, Chikhi-Brachet R, Piolot T, Coppey-Moisan M, Nicolas J (1999) Mapping EBNA1 domains involved in binding to metaphse chromosomes. J Virol 73:4385–4392

McPhillips MG, Oliveira JG, Spindler JE, Mitra R, McBride AA (2006) Brd4 is required for E2-mediated transcriptional activation but not genome partitioning of all papillomaviruses. J Virol 80(19):9530–9543

Middleton T, Sugden B (1992) EBNA1 can link the enhancer element to the initiator element of the Epstein-Barr virus plasmid origin of DNA replication. J Virol 66:489–495

Moriyama K, Yoshizawa-Sugata N, Obuse C, Tsurimoto T, Masai H (2012) Epstein-Barr Nuclear Antigen 1 (EBNA1)-dependent recruitment of origin recognition complex (Orc) on oriP of Epstein-Barr virus with purified proteins: stimulation by Cdc6 through its direct interaction with EBNA1. J Biol Chem 287(28):23977–23994. doi:10.1074/jbc.M112.368456 (M112.368456 [pii])

Munz C (2004) Epstein-barr virus nuclear antigen 1: from immunologically invisible to a promising T cell target. J Exp Med 199(10):1301–1304
Munz C, Bickham KL, Subklewe M, Tsang ML, Chahroudi A, Kurilla MG, Zhang D, O'Donnell M, Steinman RM (2000) Human CD4(+) T lymphocytes consistently respond to the latent Epstein-Barr virus nuclear antigen EBNA1. J Exp Med 191(10):1649–1660
Murakami M, Lan K, Subramanian C, Robertson ES (2005) Epstein-Barr virus nuclear antigen 1 interacts with Nm23-H1 in lymphoblastoid cell lines and inhibits its ability to suppress cell migration. J Virol 79(3):1559–1568
Murat P, Zhong J, Lekieffre L, Cowieson NP, Clancy JL, Preiss T, Balasubramanian S, Khanna R, Tellam J (2014) G-quadruplexes regulate Epstein-Barr virus-encoded nuclear antigen 1 mRNA translation. Nat Chem Biol 10(5):358–364. doi:10.1038/nchembio.1479nchembio.1479 ([pii])
Nanbo A, Sugden A, Sugden B (2007) The coupling of synthesis and partitioning of EBV's plasmid replicon is revealed in live cells. EMBO J 26(19):4252–4262
Nayyar VK, Shire K, Frappier L (2009) Mitotic chromosome interactions of Epstein-Barr nuclear antigen 1 (EBNA1) and human EBNA1-binding protein 2 (EBP2). J Cell Sci 122(Pt 23):4341–4350
Nicholson B, Suresh Kumar KG (2011) The multifaceted roles of USP7: new therapeutic opportunities. Cell Biochem Biophys 60(1–2):61–68. doi:10.1007/s12013-011-9185-5
Niller HH, Glaser G, Knuchel R, Wolf H (1995) Nucleoprotein complexes and DNA 5′-ends at oriP of Epstein-Barr virus. J Biol Chem 270:12864–12868
Nonkwelo C, Skinner J, Bell A, Rickinson A, Sample J (1996) Transcription start site downstream of the Epstein-Barr virus (EBV) Fp promoter in early-passage Burkitt lymphoma cells define a fourth promoter for expression of the EBV EBNA1 protein. J Virol 70:623–627
Norio P, Schildkraut CL (2001) Visualization of DNA replication on individual Epstein-Barr virus episomes. Science 294:2361–2364
Norio P, Schildkraut CL (2004) Plasticity of DNA replication initiation in Epstein-Barr virus episomes. PLoS Biol 2(6):e152
Norio P, Schildkraut CL, Yates JL (2000) Initiation of DNA replication within *oriP* is dispensable for stable replication of the latent Epstein-Barr virus chromosome after infection of established cell lines. J Virol 74:8563–8574
Norseen J, Thomae A, Sridharan V, Aiyar A, Schepers A, Lieberman PM (2008) RNA-dependent recruitment of the origin recognition complex. Embo J 27(22):3024–3035. doi:10.1038/emboj.2008.221 (emboj2008221 [pii])
Norseen J, Johnson FB, Lieberman PM (2009) Role for G-quadruplex RNA binding by Epstein-Barr virus nuclear antigen 1 in DNA replication and metaphase chromosome attachment. J Virol 83(20):10336–10346. doi:10.1128/JVI.00747-09 (JVI.00747-09 [pii])
O'Neil JD, Owen TJ, Wood VH, Date KL, Valentine R, Chukwuma MB, Arrand JR, Dawson CW, Young LS (2008) Epstein-Barr virus-encoded EBNA1 modulates the AP-1 transcription factor pathway in nasopharyngeal carcinoma cells and enhances angiogenesis in vitro. J Gen Virol 89(Pt 11):2833–2842. doi:10.1099/vir.0.2008/003392-0 (89/11/2833 [pii])
Oddo C, Freire E, Frappier L, de Prat-Gay G (2006) Mechanism of DNA recognition at a viral replication origin. J Biol Chem 281(37):26893–26903
Ott E, Norio P, Ritzi M, Schildkraut C, Schepers A (2011) The dyad symmetry element of Epstein-Barr virus is a dominant but dispensable replication origin. PLoS ONE 6(5):e18609. doi:10.1371/journal.pone.0018609PONE-D-10-06467 ([pii])
Owen TJ, O'Neil JD, Dawson CW, Hu C, Chen X, Yao Y, Wood VH, Mitchell LE, White RJ, Young LS, Arrand JR (2010) Epstein-Barr virus-encoded EBNA1 enhances RNA polymerase III-dependent EBER expression through induction of EBER-associated cellular transcription factors. Mol Cancer 9:241. doi:10.1186/1476-4598-9-2411476-4598-9-241 ([pii])
Paludan C, Schmid D, Landthaler M, Vockerodt M, Kube D, Tuschl T, Munz C (2005) Endogenous MHC class II processing of a viral nuclear antigen after autophagy. Science 307(5709):593–596. doi:10.1126/science.1104904 (1104904 [pii])

Parish JL, Bean AM, Park RB, Androphy EJ (2006) ChlR1 is required for loading papillomavirus E2 onto mitotic chromosomes and viral genome maintenance. Mol Cell 24(6):867–876
Park YJ, Luger K (2006) Structure and function of nucleosome assembly proteins. Biochem Cell Biol 84(4):549–558
Pearson M, Carbone R, Sebastiani C, Cioce M, Fagioli M, Saito S, Higashimoto Y, Appella E, Minucci S, Pandolfi PP, Pelicci PG (2000) PML regulates p53 acetylation and premature senescence induced by oncogenic Ras. Nature 406(6792):207–210
Petti L, Sample C, Kieff E (1990) Subnuclear localization and phosphorylation or Epstein-Barr virus latent infection nuclear proteins. Virology 176:563–574
Rawlins DR, Milman G, Hayward SD, Hayward GS (1985) Sequence-specific DNA binding of the Epstein-Barr virus nuclear antigen (EBNA1) to clustered sites in the plasmid maintenance region. Cell 42:859–868
Reedman BM, Klein G (1973) Cellular localization of an Epstein-Barr virus (EBV)-associated complement-fixing antigen in producer and non-producer lymphoblastoid cell lines. Int J Cancer 11:499–520
Reichelt M, Wang L, Sommer M, Perrino J, Nour AM, Sen N, Baiker A, Zerboni L, Arvin AM (2011) Entrapment of viral capsids in nuclear PML cages is an intrinsic antiviral host defense against varicella-zoster virus. PLoS Pathog 7(2):e1001266. doi:10.1371/journal.ppat.1001266
Reisman D, Sugden B (1986) *trans* Activation of an Epstein-Barr viral transcripitonal enhancer by the Epstein-Barr viral nuclear antigen 1. Mol Cell Biol 6:3838–3846
Reisman D, Yates J, Sugden B (1985) A putative origin of replication of plasmids derived from Epstein-Barr virus is composed of two cis-acting components. Mol Cell Biol 5:1822–1832
Ritzi M, Tillack K, Gerhardt J, Ott E, Humme S, Kremmer E, Hammerschmidt W, Schepers A (2003) Complex protein-DNA dynamics at the latent origin of DNA replication of Epstein-Barr virus. J Cell Sci 116(Pt 19):3971–3984
Salomoni P, Ferguson BJ, Wyllie AH, Rich T (2008) New insights into the role of PML in tumour suppression. Cell Res 18(6):622–640
Salsman J, Jagannathan M, Paladino P, Chan PK, Dellaire G, Raught B, Frappier L (2012) Proteomic profiling of the human cytomegalovirus UL35 gene products reveals a role for UL35 in the DNA repair response. J Virol 86(2):806–820. doi:10.1128/JVI.05442-11JVI.05442-11 ([pii])
Sample J, Henson EBD, Sample C (1992) The Epstein-Barr virus nuclear protein 1 promoter active in type I latency is autoregulated. J Virol 66:4654–4661
Saridakis V, Sheng Y, Sarkari F, Holowaty MN, Shire K, Nguyen T, Zhang RG, Liao J, Lee W, Edwards AM, Arrowsmith CH, Frappier L (2005) Structure of the p53 binding domain of HAUSP/USP7 bound to Epstein-Barr nuclear antigen 1 implications for EBV-mediated immortalization. Mol Cell 18(1):25–36
Sarkari F, Sanchez-Alcaraz T, Wang S, Holowaty MN, Sheng Y, Frappier L (2009) EBNA1-mediated recruitment of a histone H2B deubiquitylating complex to the Epstein-Barr virus latent origin of DNA replication. PLoS Pathog 5(10):e1000624. doi:10.1371/journal.ppat.1000624
Sarkari F, Wang X, Nguyen T, Frappier L (2011) The herpesvirus associated ubiquitin specific protease, USP7, is a negative regulator of PML proteins and PML nuclear bodies. PLoS ONE 6(1):e16598. doi:10.1371/journal.pone.0016598
Scaglioni PP, Yung TM, Cai LF, Erdjument-Bromage H, Kaufman AJ, Singh B, Teruya-Feldstein J, Tempst P, Pandolfi PP (2006) A CK2-dependent mechanism for degradation of the PML tumor suppressor. Cell 126(2):269–283
Scaglioni PP, Yung TM, Choi SC, Baldini C, Konstantinidou G, Pandolfi PP (2008) CK2 mediates phosphorylation and ubiquitin-mediated degradation of the PML tumor suppressor. Mol Cell Biochem 316:149–154
Schaefer BC, Strominger JL, Speck SH (1995) The Epstein-Barr virus BamHI F promoter is an early lytic promoter: lack of correlation with EBNA 1 gene transcription in group 1 Burkitt's lymphoma cell lines. J Virol 69(8):5039–5047

Schepers A, Ritzi M, Bousset K, Kremmer E, Yates JL, Harwood J, Diffley JFX, Hammerschmidt W (2001) Human origin recognition complex binds to the region of the latent origin of DNA replication of Epstein-Barr virus. EMBO J 20:4588–4602

Schweiger MR, You J, Howley PM (2006) Bromodomain protein 4 mediates the papillomavirus E2 transcriptional activation function. J Virol 80(9):4276–4285

Sears J, Kolman J, Wahl GM, Aiyar A (2003) Metaphase chromosome tethering is necessary for the DNA synthesis and maintenance of oriP plasmids but is insufficient for transcription activation by Epstein-Barr nuclear antigen 1. J Virol 77(21):11767–11780

Sears J, Ujihara M, Wong S, Ott C, Middeldorp J, Aiyar A (2004) The amino terminus of Epstein-Barr Virus (EBV) nuclear antigen 1 contains AT hooks that facilitate the replication and partitioning of latent EBV genomes by tethering them to cellular chromosomes. J Virol 78(21):11487–11505

Seo S-B, McNamara P, Heo S, Turner A, Lane WS, Chakravarti D (2001) Regulation of histone acetylation and transcription by INHAT, a human cellular complex containing the Set oncoprotein. Cell 104:119–130

Shah WA, Ambinder RF, Hayward GS, Hayward SD (1992) Binding of EBNA-1 to DNA creates a protease-resistant domain that encompasses the DNA recognition and dimerization functions. J Virology 66:3355–3362

Sharipo A, Imreh M, Leonchiks A, Imreh S, Masucci MG (1998) A minimal glycine-alanine repeat prevents the interaction of ubiquitinated I kappaB alpha with the proteasome: a new mechanism for selective inhibition of proteolysis. Nat Med 4(8):939–944

Shaw J, Levinger L, Carter C (1979) Nucleosomal structure of Epstein-Barr virus DNA in transformed cell lines. J Virol 29:657–665

Sheng Y, Saridakis V, Sarkari F, Duan S, Wu T, Arrowsmith CH, Frappier L (2006) Molecular recognition of p53 and MDM2 by USP7/HAUSP. Nat Struct Mol Biol 13(3):285–291

Sheu LF, Chen A, Meng CL, Ho KC, Lee WH, Leu FJ, Chao CF (1996) Enhanced malignant progression of nasopharyngeal carcinoma cells mediated by the expression of Epstein-Barr nuclear antigen 1 in vivo. J Pathol 180(3):243–248

Shikama N, Chan HM, Krstic-Demonacos M, Smith L, Lee CW, Cairns W, La Thangue NB (2000) Functional interaction between nucleosome assembly proteins and p300/CREB-binding protein family coactivators. Mol Cell Biol 20(23):8933–8943

Shire K, Ceccarelli DFJ, Avolio-Hunter TM, Frappier L (1999) EBP2, a human protein that interacts with sequences of the Epstein-Barr nuclear antigen 1 important for plasmid maintenance. J Virol 73:2587–2595

Shire K, Kapoor P, Jiang K, Hing MN, Sivachandran N, Nguyen T, Frappier L (2006) Regulation of the EBNA1 Epstein-Barr virus protein by serine phosphorylation and arginine methylation. J Virol 80(11):5261–5272

Sides MD, Block GJ, Shan B, Esteves KC, Lin Z, Flemington EK, Lasky JA (2011) Arsenic mediated disruption of promyelocytic leukemia protein nuclear bodies induces ganciclovir susceptibility in Epstein-Barr positive epithelial cells. Virology 416(1–2):86–97. doi:10.1016/j.virol.2011.04.005S0042-6822(11)00173-5 ([pii])

Simpson K, McGuigan A, Huxley C (1996) Stable episomal maintenance of yeast artificial chromosomes in human cells. Mol Cell Biol 16:5117–5126

Sivachandran N, Sarkari F, Frappier L (2008) Epstein-Barr nuclear antigen 1 contributes to nasopharyngeal carcinoma through disruption of PML nuclear bodies. PLoS Pathog 4(10):e1000170. doi:10.1371/journal.ppat.1000170

Sivachandran N, Cao JY, Frappier L (2010) Epstein-Barr virus nuclear antigen 1 Hijacks the host kinase CK2 to disrupt PML nuclear bodies. J Virol 84(21):11113–11123

Sivachandran N, Dawson CW, Young LS, Liu FF, Middeldorp J, Frappier L (2012a) Contributions of the Epstein-Barr virus EBNA1 protein to gastric carcinoma. J Virol 86(1):60–68. doi:10.1128/JVI.05623-11 (JVI.05623-11[pii])

Sivachandran N, Wang X, Frappier L (2012b) Functions of the Epstein-Barr virus EBNA1 Protein in viral reactivation and lytic infection. J Virol 86(11):6146–6158. doi:10.1128/JVI.00013-12 (JVI.00013-12 [pii])

Snudden DK, Hearing J, Smith PR, Grasser FA, Griffin BE (1994) EBNA1, the major nuclear antigen of Epstein-Barr virus, resenbles 'RGG' RNA binding proteins. EMBO J 13:4840–4848
Sternas L, Middleton T, Sugden B (1990) The average number of molecules of Epstein-Barr nuclear antigen 1 per cell does not correlate with the average number of Epstein-Barr virus (EBV) DNA molecules per cell among different clones of EBV-immortalized cells. JVirol 64:2407–2410
Su W, Middleton T, Sugden B, Echols H (1991) DNA looping between the origin of replication of Epstein-Barr virus and its enhancer site: stabilization of an origin complex with Epstein-Barr nuclear antigen 1. Proc Natl Acad Sci USA 88:10870–10874
Sugden B, Warren N (1989) A promoter of Epstein-Barr virus that can function during latent infection can be transactivated by EBNA-1, a viral protein required for viral DNA replication during latent infection. J Virol 63(6):2644–2649
Summers H, Barwell JA, Pfuetzner RA, Edwards AM, Frappier L (1996) Cooperative assembly of EBNA1 on the Epstein-Barr virus latent origin of replication. J Virol 70:1228–1231
Summers H, Fleming A, Frappier L (1997) Requirements for EBNA1-induced permanganate sensitivity of the Epstein-Barr virus latent origin of DNA replication. J Biol Chem 272:26434–26440
Sun X, Barlow EA, Ma S, Hagemeier SR, Duellman SJ, Burgess RR, Tellam J, Khanna R, Kenney SC (2010) Hsp90 inhibitors block outgrowth of EBV-infected malignant cells in vitro and in vivo through an EBNA1-dependent mechanism. Proc Natl Acad Sci USA 107(7):3146–3151. doi:10.1073/pnas.0910717107 (0910717107 [pii])
Takahashi Y, Lallemand-Breitenbach V, Zhu J, de The H (2004) PML nuclear bodies and apoptosis. Oncogene 23(16):2819–2824
Tellam J, Connolly G, Green KJ, Miles JJ, Moss DJ, Burrows SR, Khanna R (2004) Endogenous presentation of CD8+ T cell epitopes from Epstein-Barr virus-encoded nuclear antigen 1. J Exp Med 199(10):1421–1431
Tellam J, Fogg MH, Rist M, Connolly G, Tscharke D, Webb N, Heslop L, Wang F, Khanna R (2007) Influence of translation efficiency of homologous viral proteins on the endogenous presentation of CD8+ T cell epitopes. J Exp Med 204(3):525–532
Tellam J, Smith C, Rist M, Webb N, Cooper L, Vuocolo T, Connolly G, Tscharke DC, Devoy MP, Khanna R (2008) Regulation of protein translation through mRNA structure influences MHC class I loading and T cell recognition. Proc Natl Acad Sci USA 105(27):9319–9324
Tsimbouri P, Drotar ME, Coy JL, Wilson JB (2002) bcl-xL and RAG genes are induced and the response to IL-2 enhanced in EmuEBNA-1 transgenic mouse lymphocytes. Oncogene 21(33):5182–5187
Valentine R, Dawson CW, Hu C, Shah KM, Owen TJ, Date KL, Maia SP, Shao J, Arrand JR, Young LS, O'Neil JD (2010) Epstein-Barr virus-encoded EBNA1 inhibits the canonical NF-kappaB pathway in carcinoma cells by inhibiting IKK phosphorylation. Mol Cancer 9:1. doi:10.1186/1476-4598-9-1 (1476-4598-9-1 [pii])
Van Scoy S, Watakabe I, Krainer AR, Hearing J (2000) Human p32: a coactivator for Epstein-Barr virus nuclear antigen-1-mediated transcriptional activation and possible role in viral latent cycle DNA replication. Virology 275:145–157
Wang S, Frappier L (2009) Nucleosome assembly proteins bind to Epstein-Barr virus nuclear antigen 1 and affect its functions in DNA replication and transcriptional activation. J Virol 83(22):11704–11714. doi:10.1128/JVI.00931-09 (JVI.00931-09 [pii])
Wang Y, Finan JE, Middeldorp JM, Hayward SD (1997) P32/TAP, a cellular protein that interacts with EBNA-1 of Epstein-Barr virus. Virology 236:18–29
Wang ZG, Ruggero D, Ronchetti S, Zhong S, Gaboli M, Rivi R, Pandolfi PP (1998) PML is essential for multiple apoptotic pathways. Nat Genet 20(3):266–272
Wang L, Tian WD, Xu X, Nie B, Lu J, Liu X, Zhang B, Dong Q, Sunwoo JB, Li G, Li XP (2014) Epstein-Barr virus nuclear antigen 1 (EBNA1) protein induction of epithelial-mesenchymal transition in nasopharyngeal carcinoma cells. Cancer 120(3):363–372. doi:10.1002/cncr.28418

Wilson JB, Bell JL, Levine AJ (1996) Expression of Epstein-Barr virus nuclar antigen-1 induces B cell neoplasia in transgenic mice. EMBO 15:3117–3126

Wood VH, O'Neil JD, Wei W, Stewart SE, Dawson CW, Young LS (2007) Epstein-Barr virus-encoded EBNA1 regulates cellular gene transcription and modulates the STAT1 and TGFbeta signaling pathways. Oncogene 26(28):4135–4147

Wu SY, Chiang CM (2007) The double bromodomain-containing chromatin adaptor Brd4 and transcriptional regulation. J Biol Chem 282(18):13141–13145. doi:10.1074/jbc.R700001200 (R700001200 [pii])

Wu H, Ceccarelli DFJ, Frappier L (2000) The DNA segregation mechanism of the Epstein-Barr virus EBNA1 protein. EMBO Rep 1:140–144

Wu H, Kapoor P, Frappier L (2002) Separation of the DNA replication, segregation, and transcriptional activation functions of Epstein-Barr nuclear antigen 1. J Virol 76(5):2480–2490

Wysokenski DA, Yates JL (1989) Multiple EBNA1-binding sites are required to form an EBNA1-dependent enhancer and to activate a minimal replicative origin within *oriP* of Epstein-Barr virus. J Virol 63:2657–2666

Yates JL, Camiolo SM (1988) Dissection of DNA replication and enhancer activation functions of Epstein-Barr virus nuclear antigen 1. Cancer Cells 6:197–205

Yates JL, Guan N (1991) Epstein-Barr virus-derived plasmids replicate only once per cell cycle and are not amplified after entry into cells. J Virol 65:483–488

Yates JL, Warren N, Reisman D, Sugden B (1984) A cis-acting element from the Epstein-Barr viral genome that permits stable replication of recombinant plasmids in latently infected cells. Proc Natl Acad Sci USA 81:3806–3810

Yates JL, Warren N, Sugden B (1985) Stable replication of plasmids derived from Epstein-Barr virus in various mammalian cells. Nature 313:812–815

Yates JL, Camiolo SM, Bashaw JM (2000) The minimal replicator of Epstein-Barr virus oriP. J Virol 74:4512–4522

Yin Q, Flemington EK (2006) siRNAs against the Epstein Barr virus latency replication factor, EBNA1, inhibit its function and growth of EBV-dependent tumor cells. Virology 346(2):385–393

Yin Y, Manoury B, Fahraeus R (2003) Self-inhibition of synthesis and antigen presentation by Epstein-Barr virus-encoded EBNA1. Science 301(5638):1371–1374

Yoshioka M, Crum MM, Sample JT (2008) Autorepression of Epstein-Barr virus nuclear antigen 1 expression by inhibition of pre-mRNA processing. J Virol 82(4):1679–1687

You J (2010) Papillomavirus interaction with cellular chromatin. Biochim Biophys Acta 1799(3–4):192–199. doi:10.1016/j.bbagrm.2009.09.009 (S1874-9399(09)00114-X [pii])

Zhang M, Coffino P (2004) Repeat sequence of Epstein-Barr virus-encoded nuclear antigen 1 protein interrupts proteasome substrate processing. J Biol Chem 279(10):8635–8641. doi:10.1074/jbc.M310449200M310449200 ([pii])

Zhou J, Snyder AR, Lieberman PM (2009) Epstein-Barr virus episome stability is coupled to a delay in replication timing. J Virol 83(5):2154–2162

Zhou J, Deng Z, Norseen J, Lieberman PM (2010) Regulation of Epstein-Barr virus origin of plasmid replication (OriP) by the S-phase checkpoint kinase Chk2. J Virol 84(10):4979–4987. doi:10.1128/JVI.01300-09 (JVI.01300-09 [pii])

# EBNA2 and Its Coactivator EBNA-LP

**Bettina Kempkes and Paul D. Ling**

**Abstract** While all herpesviruses can switch between lytic and latent life cycle, which are both driven by specific transcription programs, a unique feature of latent EBV infection is the expression of several distinct and well-defined viral latent transcription programs called latency I, II, and III. Growth transformation of B-cells by EBV in vitro is based on the concerted action of Epstein-Barr virus nuclear antigens (EBNAs) and latent membrane proteins(LMPs). EBV growth-transformed B-cells express a viral transcriptional program, termed latency III, which is characterized by the coexpression of EBNA2 and EBNA-LP with EBNA1, EBNA3A, -3B, and -3C as well as LMP1, LMP2A, and LMP2B. The focus of this review will be to discuss the current understanding of how two of these proteins, EBNA2 and EBNA-LP, contribute to EBV-mediated B-cell growth transformation.

## Contents



B. Kempkes (✉)
Department of Gene Vectors, Helmholtz Center Munich,
German Research Center for Environmental Health, Marchioninistr. 25,
81377 Munich, Germany
e-mail: kempkes@helmholtz-muenchen.de

P.D. Ling
Department of Molecular Virology and Microbiology,
Baylor College of Medicine, Houston, TX 77030, USA
e-mail: pling@bcm.edu



C. Münz (ed.), *Epstein Barr Virus Volume 2*, Current Topics in Microbiology and Immunology 391, DOI 10.1007/978-3-319-22834-1_2

## Abbreviation

| | |
|---|---|
| CBF1 | C-promoter binding factor |
| EBNA | Epstein-Barr virus nuclear antigen |
| EBNA-LP | Epstein-Barr virus leader protein |
| LMP | Latent membrane protein |
| LCL | Lymphoblastoid cell line |
| LCV | Lymphocryptovirus |
| TAD | Transactivation domain |
| PML-NB | Promyelocytic leukemia nuclear body |

## 1 EBNA2 and EBNA-LP Expression

In vivo, the latency III transcription program is expressed during a short-time window immediately after the infection of tonsillar B-cells of healthy individuals as well as in tonsillar EBV-infected B-cells in patients suffering from infectious mononucleosis (Kurth et al. 2003; Thorley-Lawson 2001). Expression of EBNA2 and EBNA-LP in EBV-related malignant diseases is confined to immunodeficient patients, who lack efficient T-cell immunosurveillance, a state that may be caused by immunosuppressive drug treatment after transplantation. Since the viral expression program in immunodeficient patients closely resembles the pattern seen in immortalized B-cells in vitro, EBNA2 and EBNA-LP are likely to drive the proliferation of these highly malignant cells. EBNA2 and EBNA-LP are not expressed in latently infected memory B-cells of healthy individuals or in EBV-associated malignancies of immunocompetent people as exemplified by patients suffering from Burkitt's lymphoma or Hodgkin's disease (for review see: Bornkamm and Hammerschmidt 2001; Macsween and Crawford 2003). Both EBNA2 and EBNA-LP are expressed from transcripts initiating from either of two promoters, Wp or Cp (Alfieri et al. 1991; Bodescot et al. 1987; Woisetschlaeger et al. 1990, 1991). While the EBNA2 protein is encoded by a single exon, transcription of EBNA-LP is more complex. For transcripts initiating from Wp, the first exon, W0, is joined to either the W1 exon, which does not contain an initiation codon for EBNA-LP, or via an alternate splice acceptor site to W1′ that does (Fig. 1) (Sample et al. 1986; Speck et al. 1986). W1 is then joined with the W2 exon and additional

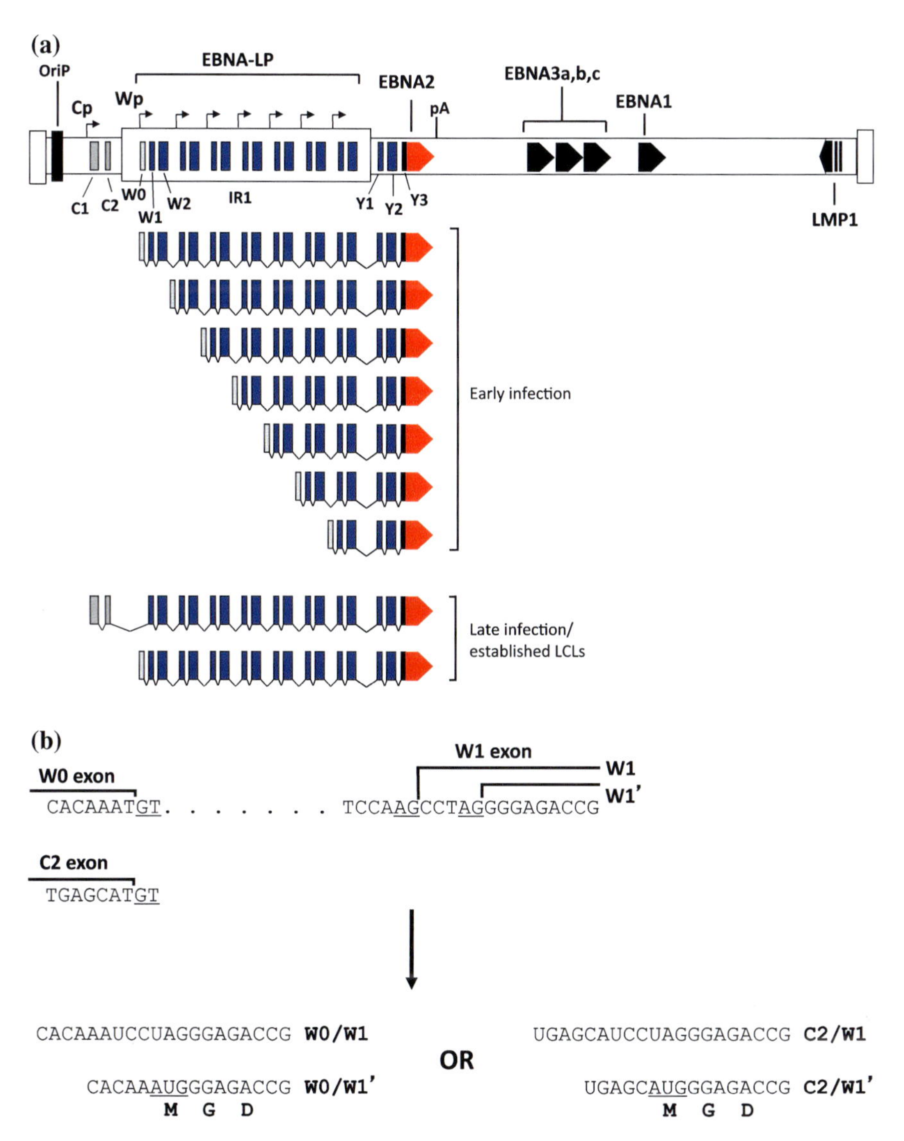


**Fig. 1** Exon organization of EBNA-LP and EBNA2 gene transcripts. **a** Transcription of the EBNA-LP gene initiates from either the W promoter (Wp) or C promoter (Cp). The different noncoding (C1, C2, W0) and coding exons for EBNA-LP (W1, W2, Y1, and Y2 in *blue*) and EBNA2 (Y3 in *red*) are indicated. During early stages of infection, transcription initiates from available Wp residing in each IR1 repeat, which results in the production of multiple EBNA-LP protein isoforms. During later stages of infection or in established LCLs, transcription from Cp is stimulated and there is a bias toward the OriP-proximal Wp. The level of Cp versus Wp-initiated transcription varies depending on several circumstances. The viral latent origin of replication (OriP), polyadenylation site (pA) for EBNA-LP/EBNA2 transcripts, and location of other EBNA genes and latent membrane proteins (LMP) are shown. **b** Alternative splicing generates an initiation codon for EBNA-LP. Splicing from W0 or C2 to a slightly shorter W1′ exon generates an AUG initiation codon for EBNA-LP, while splicing to the longer W1 exon does not. Splice donor and acceptor sites are *underlined*, and the resulting transcripts are shown below. The first three amino acid residues for EBNA-LP are shown below the transcripts that can translate EBNA-LP (adapted from Dr. S. Speck, with permission)

W1/W2 exons from each downstream repeat within major internal repeats (IR1). A complete transcript encoding EBNA-LP also joins unique exons Y1 and Y2 and the exon coding for EBNA2 (Bodescot et al. 1984; Sample et al. 1986; Speck et al. 1986). The final transcripts are bicistronic, encoding both EBNA-LP and EBNA2, or monocistronic for EBNA2 and containing a long 5′ untranslated leader sequence. Following initial infection of primary B-cells, transcription initiates first from Wp, but generally switches to Cp during later stages of infection or in established lymphoblastoid cell lines (LCLs), correlating with the detectable expression of EBNA2, which is also known to positively regulate Cp (see section on EBNA2 genetic and biochemical analysis). For transcripts initiating from Cp, the first two exons C1 and C2 can join either with W1 or W1′ to produce mRNAs encoding EBNA-LP or EBNA2 only, like those expressed from Wp (Rogers et al. 1990). EBNA-LP and EBNA2 are the first two latent proteins detectable following infection of primary B-cells (Alfieri et al. 1991; Allday et al. 1989).

The number of IR1 repeats determines the size of EBNA-LP, although during early stages of infection, multiple isoforms of EBNA-LP are detectable (Allan and Rowe 1989; Dillner et al. 1986; Finke et al. 1987). This phenomenon has not yet been explained, but one hypothesis is that because each IR1 repeat contains a Wp, there may be multiple sites for transcription initiation from different Wps. On the other hand, alternative splicing may occur across the IR1 repeats, generating transcripts encoding EBNA-LP proteins with varying repeat sequences. A methodical study of the number of IR1 repeats needed for optimal EBV-mediated B-cell immortalization has shown that at least two repeats are needed, but that 5 or more are associated with viruses that have optimal transforming activities (Tierney et al. 2011). This effect is apparently due to greater transcriptional activity from Wp rather than the size of EBNA-LP isoforms that are made. Sequencing of nonhuman primate lymphocryptoviruses (LCVs) confirms that these viruses encode the predicted EBNA2 and EBNA-LP proteins, and at least one study has shown that the complex transcriptional unit encoding EBNA-LP is well conserved (Peng et al. 2000a).

## 2 The EBNA2 Protein

EBNA2 forms granules (also called speckles) and localizes to the nucleoplasm, the chromatin fraction, and the nuclear matrix but excludes the nucleoli (Petti et al. 1990). Most studies on EBNA2 used the laboratory EBV strain B95-8 which encodes a 487 amino acid EBNA2 protein[1] (Baer et al. 1984; Skare et al. 1982) (Fig. 2). EBNA2, isolated from total cellular extracts, has an apparent molecular weight of approximately 84 kDa as judged by SDS-PAGE. Phosphorylation of

[1]In this review, we will refer to the primary structure of EBNA2 using the Swiss-Prot data entry of EBV laboratory strain 95-8:P12978.

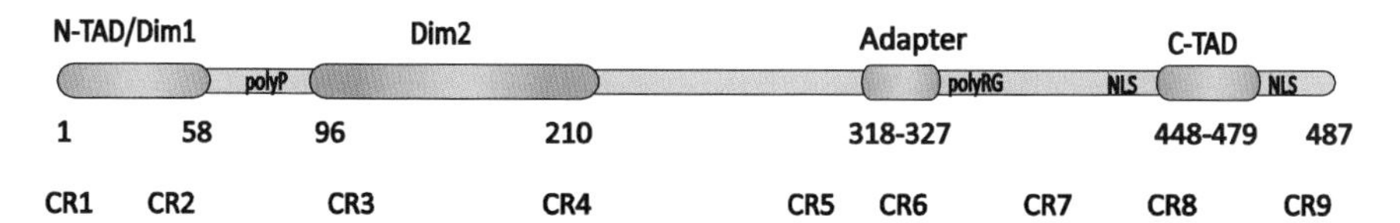


**Fig. 2** Schematic illustration showing the primary structure of EBNA2. Characteristic features of EBNA2 are a poly-proline (polyP) and a poly-arginine-glycine (RG) stretch and conserved regions (CR1–9), which have been defined by comparison of EBV strain types 1 and 2, baboon, and rhesus macaque lymphocryptoviruses. Regions of EBNA2, which mediate self-association, are labeled Dim1 and Dim2. CR5 and CR6 lie within the region, which mediates promoter targeting (adapter) by association with CBF1. Two transactivation domains (TAD) map to the amino (N-TAD) and carboxyl termini C-TAD, while the two nuclear localization signals (NLS) reside at the carboxyl terminus. Amino acid numbering refers to the B95-8 primary structure of EBNA2

EBNA2 may partially account for the discrepancy between the expected molecular weight and the electrophoretic mobility of the protein (Grasser et al. 1991). Notably, EBNA2 proteins from distinct nuclear compartments exhibit differential phosphorylation patterns, and these phosphorylation patterns vary during the cell cycle (Petti et al. 1990; Yue et al. 2004). A further modification of EBNA2 is the arginine methylation of the RG repeat (Barth et al. 2003), which is considered to be a modulator of EBNA2 activity. The RG repeat binds to poly G and histone H1 in vitro (Tong et al. 1994) and also serves as a substrate for arginine methyltransferase 5 (Liu et al. 2013). A poly-proline stretch (poly-P) separates two N-terminal dimerization domains, Dim1 (1–58) and Dim2 (96–210), which both mediate homotypic adhesion. In addition, Dim1 serves as an N-terminal transactivation domain (N-TAD) that may interact with EBNA-LP (Gordadze et al. 2004; Harada et al. 2001; Peng et al. 2004b). A second transactivation domain (C-TAD) resides in the C-terminus of EBNA2 (448–479) (Cohen and Kieff 1991). The core fragment of the transactivation domain (AA: 453–466) can be replaced by an acidic fragment of the herpesviral VP16 transactivation domain indicating that EBNA2 and VP16 share functional similarities (Cohen 1992; Cohen and Kieff 1991). In fact, both transactivation domains bind to TFIIB and TAF40, components of the transcription initiation complex, and TFIIH, a factor involved in promoter clearance. In addition, both bind to RPA70, the replication protein A (Tong et al. 1995a, b). In contrast to VP16, EBNA2 does not bind to TBP (Tong et al. 1995c). Both transactivation domains also recruit histone acetyltransferase activity by interacting with CBP, p300, and PCAF (Wang et al. 2000). The structure of C-TAD in complex with CBP/p300 or the TFB1/p62 subunit of the TFIIH complex has recently been solved by nuclear magnetic resonance (NMR) spectroscopy. C-TAD is an intrinsically unstructured region which folds into a 9-residue alpha helix upon complexation (Chabot et al. 2014). The structure of the entire EBNA2 protein has not been solved. The high proline content, the poly-P regions, and the RG repeats most likely prevent globular folding of the entire protein in the absence of cognate binding partners.

Mainly based on the sequence diversity of the EBNA2 alleles, EBV can be categorized into two individual strains called type 1 and 2 (or type A and B respectively). Type 1 and 2 EBV strains differ in their capacity to immortalize primary B-cells (Adldinger et al. 1985; Dambaugh et al. 1984) a feature that is predominantly determined by sequence variation in the C-terminus of EBNA2 (Tzellos et al. 2014; Tzellos and Farrell 2012). Lymphocryptoviruses have also been isolated from baboon and macaque. While the EBNA2 orthologs of baboon and macaque LCV show significant sequence similarity with EBNA2 protein encoded by the B95-8 strain (Cho et al. 1999; Peng et al. 2000a), sequence similarity with the positional EBNA2 homologue of marmoset LCV is below 20 % (reviewed in Wang 2013).

# 3 The EBNA-LP Protein

The EBNA-LP protein is composed of several 22 and 44 amino acid segments encoded by the W1 and W2 exons, which are joined with unique 11 and 34 amino acid segments at the carboxyl-terminal end that are encoded by the Y1 and Y2 exons. Once expressed, EBNA-LP localizes predominantly in the nucleus, which is facilitated by a bipartite nuclear localization signal (Peng et al. 2000b). Nuclear staining is diffuse early on after infection but then localizes in punctate structures known as promyelocytic leukemia nuclear bodies (PML NBs) (Bandobashi et al. 2001; Dillner et al. 1986; Ling et al. 2005; Nitsche et al. 1997; Szekely et al. 1996; Wang et al. 1987b). Transient expression of EBNA-LP from eukaryotic expression vectors tends to mimic the diffuse nuclear localization observed at early points following virus infection. Consistent with its association with PML NBs, biochemical fractionation studies on LCLs indicate that EBNA-LP is associated with the nuclear matrix (Petti et al. 1990; Yokoyama et al. 2001a). PML NBs are organized by the PML protein, which contains SUMO interaction motifs (SIMS) and is also posttranslationally modified by sumoylation. Many proteins associated with PML NBs also are SUMO-modified and have SIM domains, which presumably help mediate PML NB assembly or localization. Curiously, EBNA-LP is devoid of any obvious SIM motifs or lysine residues that are needed for SUMO conjugation, but as discussed later, localization of EBNA-LP to PML NBs may be mediated by a cellular factor. Other studies have also reported that a proportion of EBNA-LP can localize in the cytoplasm and this can be influenced by the number of W repeats (Garibal et al. 2007). Consistent with these studies, Ling et al. (2009) used heterokaryon assays to evaluate whether EBNA-LP shuttles between cytoplasmic and nuclear compartments. Only smaller isoforms with 2W repeats were found to shuttle in these assays, while larger isoforms did not, leading to speculations that the observed shuttling of the smaller isoforms was due to diffusion rather than through an active process. In contrast, EBNA-LP proteins with only a single W repeat (W1 and W2 exons) localized exclusively in the cytoplasm. Whether shuttling or cytoplasmic localization contributes significantly to EBNA-LP-mediated coactivation remains unknown.

EBNA-LP is a phosphoprotein (Petti et al. 1990). Phosphorylation appears to occur predominantly on serine residues, and while this can be detected throughout the cell cycle, it is hyperphosphorylated during G2/M and hypophosphorylated during G1/S (Kitay and Rowe 1996). There are three serine residues that are well conserved among human and nonhuman primate lymphocryptovirus (LCV) EBNA-LP homologs, and one of them is within a cyclin-dependent p34cdc2 site (Peng et al. 2000a, b). This serine is located within the 44 amino acid segment encoded by W2 and is critical for EBNA-LP-mediated coactivation (McCann et al. 2001; Peng et al. 2000b; Yokoyama et al. 2001b).

## 4 EBNA2-Associated Cellular Proteins, Which Mediate Chromatin Targeting

Like all transcription factors, EBNA2 carries a transactivation domain and a region that mediates DNA contact. Since EBNA2 cannot bind to DNA directly, it uses adaptor proteins to bind to cis-regulatory regions of its target genes and indirectly confers sequence-specific DNA contact.

So far, the best studied cellular DNA adaptor protein of EBNA2 is the DNA binding protein CBF1, which was first identified as a downstream effector molecule of EBNA2 in the context of viral promoter activation. CBF1 is a ubiquitously expressed protein and belongs to the group of CSL proteins (CBF1 for C-promoter binding protein, Su(H) in *Drosophila melanogaster*, Lag1 in *Caenorhabditis elegans*) also known as recombination binding protein-J (RBPJ, RBP, or RBPJκ). The minimal domain of EBNA2 that mediates CBF1 binding has been mapped to the EBNA2 fragment aa 318–327 (Ling and Hayward 1995). CBF1 is a sequence-specific DNA binding protein, which in the absence of EBNA2 recruits a corepressor complex to the promoter or enhancer of target genes. Constituents of this corepressor complex are SMRT/N-CoR, CIR, SKIP, Sin3A, SAP30, and HDAC1, which either directly or indirectly interfere with histone acetylation of target gene chromatin, thereby repressing transcription (reviewed in Lai 2002). Binding of EBNA2 relieves this repression by competition with corepressor binding as well as the recruitment of coactivators by virtue of its intrinsic transactivation domains (Hsieh and Hayward 1995).

Since CBF1 is also an important downstream element of the cellular Notch signal transduction pathway, the discovery of CBF1 in the context of the viral protein EBNA2 has provoked an intense search for potential parallels of Notch and EBNA2 signaling (Hayward et al. 2006). The crystal structure of the Notch/MAM/CBF1/DNA complex has been solved (Kovall 2007; Nam et al. 2006; Wilson and Kovall 2006). Although the structure of the EB viral proteins associated with CBF1 has not been published yet, there is a compelling biochemical and genetic evidence that EBNA2 and Notch contact the same hydrophobic pocket within the CSL protein, but in addition appear to bind to distinct amino acids in the vicinity of this hydrophobic pocket (Fuchs et al. 2001; Kovall and

Hendrickson 2004). Thus, EBNA2 or Notch binding to CBF-1 is mutually exclusive (Hsieh and Hayward 1995).

The second but less well characterized EBNA2 DNA adaptor is the PU.1 protein. Several laboratories have shown that PU.1 promoter binding is critical for activation of the viral LMP1 promoter (Johannsen et al. 1995; Laux et al. 1994a, b). However, complex formation of EBNA2 with endogenous PU.1 has only been reported once (Yue et al. 2004). The potential contact points of the interaction partners have not been mapped, and the interaction has been demonstrated in vitro using purified proteins.

## 5 The Genetic and Biochemical Analysis of Viral EBNA2-Responsive Promoter Elements Has Provided Major Insights into the Molecular Mechanisms of EBNA2 Action

Most of our knowledge on EBNA2 functions is based on the detailed genetic analysis of EBNA2-responsive cis-active elements within viral promoters, and the subsequent use of these insights was used to characterize the proteins that are involved biochemically.

In EBV-infected B-cells, the viral C promoter (Cp) as well as the promoters of the viral LMP1 (LMP1p), LMP2A (LMP2Ap), and LMP2B (LMP2Bp) genes is strongly activated by EBNA2 (Abbot et al. 1990; Fahraeus et al. 1990; Ghosh and Kieff 1990; Jin and Speck 1992; Sung et al. 1991; Wang et al. 1990). By deletion analysis of promoter reporter constructs or gel retardation assays, EBNA2-responsive elements (EBNA2-RE) have been identified. All these promoters carry at least one CBF1 binding site (Allday et al. 1993; Henkel et al. 1994; Laux et al. 1994b; Ling et al. 1994; Meitinger et al. 1994; Waltzer et al. 1994; Zimber-Strobl et al. 1991, 1993, 1994). The high-affinity CBF1 binding sites within the EBNA2-REs of Cp and LMP2Ap were used to identify CBF1 by four independent groups in 1994 (Grossman et al. 1994; Henkel et al. 1994; Waltzer et al. 1994; Zimber-Strobl et al. 1994).

All these CBF1 binding motifs are flanked by additional distinct transcription factor binding sites, which contribute to promoter activation but might not bind to EBNA2 directly.

Within the EBNA2-RE of Cp, a single CBF1 binding site is flanked by a binding site for the cyclic AMP-responsive AUF1/hnRNP D protein, also called CBF2. The binding sites for both factors are evolutionary conserved as shown by sequence comparison of EBV such as lymphocryptoviruses found in baboon and rhesus macaques (Fuentes-Panana and Ling 1998; Fuentes-Panana et al. 1999, 2000). Regions distal to the EBNA2-RE, both upstream and downstream, which bind additional cellular factors such as SP1, Egr-1, NF-Y, or the viral EBNA1/oriP complex modulate the basal activity of the C promoter in an EBNA2-independent fashion (Borestrom et al. 2003; Puglielli et al. 1996).

Analysis of LMP1p has revealed a complex pattern of transcription factor binding sites. PU.1/Spi-1 and CBF1 binding sites are both critical for EBNA2 transactivation (Johannsen et al. 1995; Laux et al. 1994a). In addition, ATF-2/c-Jun heterodimers enhance EBNA2 effects (Sjoblom et al. 1998). Further transcription factor binding sites such as an interferon-stimulated response element, a Sp1 binding site, and a yet undefined POU-Box protein contribute to LMP1p activity (Sjoblom et al. 1995a, b).

## 6 EBNA2 Binds to Cellular Promoter and Enhancer Regions and Can Promote the Formation of Chromatin Loops Within the Cellular Genome

In order to study the impact of EBNA2 on cellular target gene expression, genome-wide array-based screens or candidate approaches using either EBV-infected B-cells or EBNA2-expressing B-cell lines (Burgstahler et al. 1995; Calender et al. 1990; Johansen et al. 2003; Knutson 1990; Lucchesi et al. 2008; Maier et al. 2005; 2006; Mohan et al. 2006; Pegman et al. 2006; Sakai et al. 1998; Wang et al. 1987a; Zhao et al. 2006). These studies identified CD23, CD21, CCR7 (BLR2/EBI1), Hes-1, BATF, bfl-1, FcRH5, ABHD6, CCL3, CCL4, CDK5R1, DNASE1L3, MFN1, RAPGEF2, RHOH, SAMSN1, SLAMF1, and CXCR7 as EBNA2 target genes in EBV-negative B-cells. In EBV-infected B-cells, the proto-oncogene *MYC*, the p55α subunit of PIK3R1, CD21, CD23, AML-2, and FcRH5 were defined as a direct target gene, since their RNA can be induced by EBNA2 in the absence of de novo protein synthesis. In contrast, induction of cyclin D, cdk4 or tumor necrosis factor alpha (TNF-α), granulocyte colony-stimulating factor (G-CSF), and lymphotoxin (LT) requires additional cellular or viral functions (Kaiser et al. 1999; Mohan et al. 2006; Spender et al. 2001, 2002). Based on shRNA experiments, a small panel of selected EBNA2 target genes (CXCR7, Runx3 and p55α) has been identified that promote viability and proliferation of EBV-transformed B-cells (Lucchesi et al. 2008; Spender et al. 2005, 2006). For the majority of EBNA2 target genes, functional assays have not been performed. High-level expression of the EBNA2 target gene *MYC* in EBV-infected B-cells depleted for functional EBNA2 can promote cellular proliferation but leads to a switch of the viral and cellular transcription program from latency III to latency I (Pajic et al. 2000; Polack et al. 1996). Thus, it remains to be determined which of the EBNA2 target genes reflect the activated blast-like phenotype (latency III) but may not contribute to the success of immortalization process in vitro or the establishment of latency in vivo.

Since transcription factor binding sites in genomic regions can be mapped by chromatin immunoprecipitation combined with next-generation sequencing techniques, CBF1, EBNA2, and EBNA-LP binding to the cellular chromatin have been studied in B-cells (McClellan et al. 2012, 2013; Portal et al. 2013; Zhao et al. 2011). These studies mapped approximately 10 000 CBF1 binding site (Zhao et al.

2011) and 5000–20,000 EBNA2 binding sites (McClellan et al. 2013; Portal et al. 2013; Zhao et al. 2011). Obviously, the cellular background and alternative bioinformatics peak calling strategies can influence the results to a certain extent. 72 % of EBNA2 and CBF1 binding sites overlapped with each other confirming that CBF1 is the major DNA adaptor for EBNA2 (Zhao et al. 2011). All these studies benefitted from the comprehensive data sets provided by the ENCyclopedia Of DNA Elements (ENCODE) project on functional DNA elements obtained by the analysis of EBV-immortalized B-cells or primary B-cells. The comparative analysis of CBF1 and EBNA2 binding sites with regions annotated by the ENCODE project revealed that the majority of EBNA2/CBF1 binding sites were also enriched for B-cell transcription factors including ETS, RUNX, EBF, PU.1, and NκFB. Frequently, these cooccupied regions carried a characteristic enhancer chromatin signature that was also established in primary B-cells prior to infection indicating that EBNA2 is recruited to B-cell-specific open chromatin regions (Zhao et al. 2011).

A physical and functional link between an enhancer bound by EBNA2 to the promoter of the *MYC* target gene was recently established by chromatin conformation capture technologies. A chromatin loop links an EBNA2-bound enhancer more than 400 kb upstream of the *MYC* transcription start site in the presence of EBNA2 (Zhao et al. 2011). Novel technologies that integrate the analysis of the nuclear architecture with biochemical binding studies will be required for the correct assignment of cellular transcription initiation sites of target genes to remote EBNA2 binding enhancers.

Most recently binding sites of the coactivator of EBNA2, the EBNA-LP protein, in the chromatin of EBV-infected B-cells have been identified and mapped in a genome-wide ChIP-seq approach (Portal et al. 2013). These studies identified genomic binding sites shared by both viral factors, EBNA2 and EBNA-LP, but also identified sites bound by either factor, EBNA2 or EBNA-LP. According to ENCODE data sets, these binding sites colocalize to clusters of B-cell-specific transcription factor binding sites and exhibit chromatin signatures which characterize promoter and enhancer regions in the cellular genome of EBV-infected B-cells. In contrast to EBNA2, EBNA-LP binding sites preferentially occupied promoter rather than enhancer regions. Since EBNA-LP is not known to bind to DNA, the molecular mechanism by which EBNA-LP is targeted to DNA still needs to be explored.

## 7 Target Genes Which Are Down-Regulated in the Presence of EBNA2

EBNA2 not only induces but also actively down-regulates expression of target genes. Notably, EBNA2 interferes with the B-cell and germinal center phenotype by down-regulating IgM- or BCR-associated signal transduction moieties such as CD79A and CD79B, BCL6, TCL1A, and AID (Boccellato et al. 2007; Maier et al. 2005, 2006; Tobollik et al. 2006). Down regulation of IgM by EBNA2 appears to be at least partially independent of CBF1 signaling (Maier et al. 2005). In the context of Burkitt's lymphoma cell lines, which carry a chromosomal translocation that juxtaposes the IgM and the *MYC* gene locus on chromosome 8, repression of IgM coincides with *MYC* repression and a potent growth-inhibitory activity of EBNA2 (Jochner et al. 1996; Kempkes et al. 1996). This function of EBNA2 is mimicked by activated Notch (Strobl et al. 2000), which is somewhat surprising given that this EBNA2 function appeared to be partially CBF1-dependent.

Apparently, EBNA2- and *MYC*-driven proliferation programs are incompatible with each other (Pajic et al. 2001). In fact, EBNA2 downregulation is positively selected in Burkitt's lymphoma cells in vivo for at least two reasons: Firstly, it down-regulates a translocated *MYC* gene (Jochner et al. 1996), and secondly, it drives expression of LMP1 which promotes antigen presentation and T-helper-specific chemokines thus rendering the cells strongly immunogenic (Kelly et al. 2002).

## 8 EBNA-LP, the Coactivator of EBNA2

The most widely confirmed EBNA-LP function has been its ability to cooperate with EBNA2 and will be the focus in this review. However, EBNA-LP has been reported to mediate apoptosis and other cellular pathways through associations with a variety of cellular proteins. We refer the reader to another previous comprehensive review for details about these potential functions (Ling, P.D. EBNA-LP function. Epstein-Barr virus Latency 2010).

One of the first clues that EBNA-LP might have a role in gene regulation came from a study showing that expression of EBNA2 together with EBNA-LP in primary B-cells induced the expression of the cellular cyclin D2 gene (Sinclair et al. 1994). Subsequent studies by other investigators confirmed that EBNA-LP was a strong coactivator of EBNA2 (Harada and Kieff 1997; Nitsche et al. 1997; Peng et al. 2000a, b; Yokoyama et al. 2001a). These studies utilized two types of assays: (1) transient reporter gene assays with EBNA2-responsive reporter plasmids and (2) induction of endogenous EBNA2-responsive genes in Burkitt's lymphoma cell lines. Further validation of EBNA-LP coactivator function came from experiments showing that EBNA-LP from the rhesus LCV also coactivated EBNA2, demonstrating that this function was evolutionarily conserved (Peng et al. 2000a).

A major question is whether or not EBNA-LP is a global transcriptional coactivator. While independent studies from several groups have confirmed that EBNA-LP coactivates EBNA2-responsive genes LMP-1, LMP2B, and the Cp, other known EBNA2 target genes such as LMP2A, CD21, CD23, and Hes-1 appear not to be affected by EBNA-LP (Peng et al. 2005). There have been some reports that EBNA-LP can coactivate GAL4-EBNA2 fusion proteins or GAL4 acidic activation domain fusions in transient mammalian 2-hybrid systems

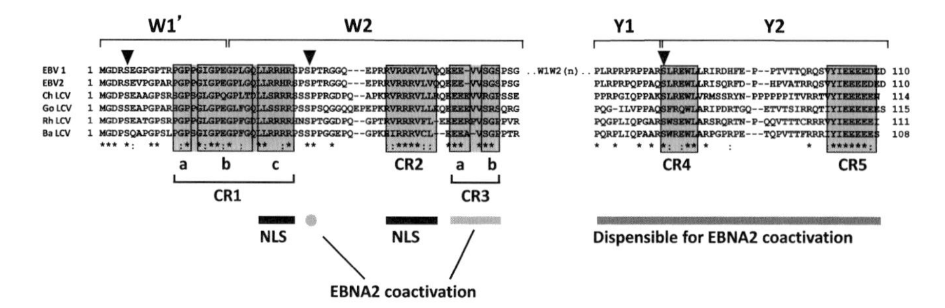


**Fig. 3** Sequence comparison of LCV EBNA-LP proteins. Conserved or similar amino acid residues are indicated by *asterisks* or *dotted lines*, respectively. The corresponding exons encoding EBNA-LP sequences and regions of conservation between the EBV and nonhuman primate LCVs are shown. Below, the *black* and *gray bars* indicate residues conferring nuclear localization (NLS), EBNA2 coactivation, and coactivation dispensable domains. For simplicity, amino acid numbering is for an EBNA-LP protein with only a single W1/W2 repeat. However, as indicated, the residues encoded by the W1/W2 repeats are present in multiple copies in wild-type EBNA-LP proteins

(Han et al. 2002; Harada and Kieff 1997), but subsequent studies have found that EBNA-LP acts as a repressor in some of these assays (Peng et al. 2004a). EBNA-LP was also unable to coactivate GAL4 fusions with other transcriptional activating proteins (Inman and Farrell 1995). A single report suggested that EBNA2 stimulation of endogenous Hes-1 in EBV-negative Burkitt lymphoma was enhanced twofold by EBNA-LP (Portal et al. 2011). Unlike a previous study, however, where the effects of EBNA2 and EBNA-LP were determined from multiple experiments in the same cells (Peng et al. 2005), the latter study was done by comparing cell lines constitutively expressing EBNA2 and EBNA-LP to cell lines expressing only one of these proteins and did not take into account clonal variation in these lines (Portal et al. 2011). Collectively, the available evidence suggests that EBNA-LP only coactivates EBNA2 on a subset of EBNA2-responsive genes.

The mechanism(s) by which EBNA-LP coactivates EBNA2 have not been fully elucidated. However, a methodical approach to understanding this process was to identify important functional domains within EBNA-LP that mediated this function. One helpful tool was the availability of known EBNA-LP protein sequences from human and nonhuman primate LCVs (McCann et al. 2001; Peng et al. 2000a, b). Comparison of these proteins revealed the presence of 5 conserved regions (CR1–5) (Fig. 3; McCann et al. 2001; Peng et al. 2000b). Deletion or alanine-scanning mutations introduced into these conserved regions showed that CR3 and a single serine residue within the W2 repeat were important for EBNA2 coactivation (McCann et al. 2001; Peng et al. 2000b). The mutational analyses also identified a nuclear localization signal (CR1 and CR2) (Peng et al. 2000b). The Y1- and Y2-encoded segments of EBNA-LP are not required for coactivation function, although it has been hypothesized that they might impose both negative and positive regulatory effects under certain conditions (Peng et al. 2007).

Further insight into the pathways utilized by EBNA-LP to mediate EBNA2 coactivation has come from the identification of associated cellular cofactors. At the current time, only a single cofactor has been identified that interacts with an EBNA-LP domain required for transcriptional coactivation. Ling et al. (2005) first identified a strong association between EBNA-LP CR3 and the cellular protein Sp100A, which is predominantly localized in PML NBs. EBNA-LP can displace Sp100A from PML NBs, and this correlates with its ability to coactivate EBNA2, while the noncoactivating EBNA-LP with a mutation in CR3 neither interacts with Sp100A nor displaces it from PML NBs. Furthermore, expression of an amino-terminal deletion mutant of Sp100A, which prevents it from entering PML NBs (i.e., it localizes as if EBNA-LPs were coexpressed), was sufficient to coactivate EBNA2 in the absence of EBNA-LP. Additionally, Sp100A is known to associate with the transcriptional repressor heterochromatin protein 1 (HP1). Deletion of the HP1 interaction domain ablates Sp100A coactivation function. The data are consistent with a model in which EBNA-LP-Sp100A interactions facilitate coactivation by a mechanism involving chromatin modification.

A second potential mechanism by which EBNA-LP might work to cooperate with EBNA2 is through direct interactions with EBNA2, which have been detected in vitro using small fragments of EBNA2 and EBNA-LP (Peng et al. 2004b). However, several investigators have been unable to detect EBNA2-EBNA-LP interactions in mammalian or yeast two-hybrid systems or using traditional coimmunoprecipitation assays (Kashuba et al. 2003; Peng et al. 2004a, 2005). In addition, it is unclear what EBNA-LP-associated cofactors or intrinsic property of EBNA-LP might be providing coactivation function in this context.

A third mechanism for EBNA-LP coactivation might involve displacement of NCoR-repressive complexes from enhancers (Portal et al. 2011). In addition, Chip-seq identified several thousand sites in which EBNA-LP was associated although interestingly, less than a third of these were also associated with EBNA2, consistent with the idea that EBNA-LP might not be a global EBNA2 coactivator (Portal et al. 2013). A limitation to these studies, as discussed previously, is that the DNA and transcription factor associations measured were done in an cellular environment where it is unclear whether EBNA-LP is functioning as an active coactivator or if this function is being masked because of pleotropic effects on transcription from other viral (e.g., LMP1) or cellular factors.

## 9 EBNA3A, B and C Proteins Can Counteract or Enhance EBNA2 Activity

The EBNA3 proteins, EBNA3A, 3B, and 3C, are all coexpressed with EBNA2 and EBNA-LP in latency III. The EBNA3 proteins score as transcriptional repressors when tethered to DNA by heterologous DNA adaptors like the GAL4 DNA binding domain (Bain et al. 1996; Cludts and Farrell 1998). All EBNA3 proteins bind to CBF1 and can interfere with EBNA2-mediated transactivation of the CBF1-dependent Cp, LMP2A, and LMP1 promoters in transient reporter assays (Le Roux et al. 1994). Further, it was reported that EBNA3C can cooperate with EBNA2 to activate PU.1-dependent transcription from LMP1p (Marshall and Sample 1995; Zhao and Sample 2000). However, expression of EBNA3C in EBV-positive Raji cells does not impair Cp-driven EBNA2 expression but significantly induces endogenous LMP1 levels indicating that the viral genome embedded in the context of the cellular chromatin reacts differently (Allday and Farrell 1994; Jimenez-Ramirez et al. 2006). The retrospect analysis of EBNA2 and EBNA3 target genes published by different laboratories showed a significant overlap indicating that EBNA2 and EBNA3 might indeed regulate similar target gene populations (Hertle et al. 2009; McClellan et al. 2013). In addition, EBNA2, EBNA3, and CBF1 binding sites in the cellular genome show a significant overlap (McClellan et al. 2013), and EBNA2 and EBNA3A can directly compete for CBF1 binding as shown for the CXCL9 and CXCL10 gene locus (Harth-Hertle et al. 2013).

An EBV–EBV interactome based on binary interactions identified by yeast two-hybrid high-throughput screening has been published (Calderwood et al. 2007). The study confirmed binding of EBNA2 to EBNA-LP and described EBNA2 binding of EBNA3A, BZLF1, the inducer of the lytic viral life cycle, and BDLF2, a tegument protein. To which extent the biological activity of EBNA2 is modulated by the viral context remains to be analyzed further using specific viral mutants during all stages of the growth transformation process and the lytic viral life cycle.

## 10 EBNA2-Associated Cellular Proteins, Which Highlight Additional Functions of EBNA2

Chromatin immunoprecipitation assays using EBNA2 and histones H3- and H4-specific antibodies proved that the LMP1p- and Cp-associated chromatin is differentially acetylated in the presence of EBNA2 (Alazard et al. 2003). A further histone acetylation-independent mechanism of Cp activation by EBNA2 is dependent on cdk9 activity, which phosphorylates Ser-5 of the C-terminal tail of polymerase II (Bark-Jones et al. 2006). In addition, EBNA2 forms a complex with a novel cellular coactivator, p100, which can bind to the general transcription factor TFIIE and thereby bridges STAT6/RNA polymerase II interactions (Tong et al. 1995b; Yang et al. 2002) (Table 1).

Apart from recruiting HAT activity and general transcription factors, phosphorylated EBNA2 also interacts with hSNF5/Ini, a component of the hSWI/SNF chromatin remodeling complex and potential tumor suppressor gene. EBNA2 recruits this protein to target promoters. This interaction is conferred by less conserved regions of EBNA2 and depends on the integrity of IPP285 and DQQ111 as well as phosphorylation of SS469 adjacent to the transactivation domain of EBNA2 (Kwiatkowski et al. 2004; Wu et al. 1996, 2000). The EBNA2 interaction

**Table 1** EBNA2-associated cellular proteins associated cellular proteins

| Protein | References |
|---|---|
| CBF1/RBP-J/ RBP-Jκ | Grossman et al. (1994), Henkel et al. (1994), Waltzer et al. (1994), Zimber-Strobl et al. (1994) |
| PU.1/Spi-1 | Yue et al. (2004) |
| TFIIB | Tong et al. (1995c) |
| TAF40 | Tong et al. (1995c) |
| CBP/p300 | Wang et al. (2000) |
| PCAF/GCN5 | Wang et al. (2000) |
| P100 | Tong et al. (1995b) |
| Nur77 | Lee et al. (2002) |
| SKIP | Zhou et al. (2000) |
| DP103 | Grundhoff et al. (1999) |
| SMN | Barth et al. (2003) |
| hSNF5/Ini1 | Wu et al. (1996) |
| BS69 | Ansieau and Leutz (2002) |
| p34cdc2 | Yue et al. (2004) |
| ATF-2/c-Jun | Sjoblom et al. (1998) |
| FOE | Kwiatkowski et al. (2004) |
| Nucleophosmin | Liu et al. (2012) |

with hSNF5/Ini1 could potentially serve a second function. It might interfere with the growth-suppressing activities of hSNF5/Ini1 in heterotrimeric complexes with GADD34 and PP-1 (Wu et al. 2002). A potential chromatin association of EBNA2 has been further suggested by the interaction of FOE (friend of EBNA), the human homologue of the *Drosophila* homologue of wap1, with EBNA2 (Kwiatkowski et al. 2004).

The carboxyl terminus, CR7 and CR8, can be further targeted by the Mynd domain protein and corepressor BS69, a cellular protein which was first described as one that binds to the adenovirus E1A protein (Hateboer et al. 1995). A potential function of EBNA2 in RNA processing has been suggested by the identification of the DEAD box protein DP103 (Gemin3/DDX20) which binds to (AA:121–213) (Grundhoff et al. 1999). The RG repeat region of EBNA2 is methylated at arginine residues and recruits the survival motor neuron (SMN) (Barth et al. 2003). SMN, a protein involved in RNA splicing, directly interacts with DP103 and can enhance LMP1 promoter activation by EBNA2 (Voss et al. 2001). The Ski-interacting protein (SKIP) is a multifunctional protein, which is a component of the spliceosome, a coactivator or corepressor of transcription and a pRB and E7 binding protein (reviewed in (Folk et al. 2004). SKIP binds to CBF1 and either facilitates binding of the SMRT, CIR, Sin3A, and HDAC2 corepressor complex or potentiates binding of EBNA2 to CBF1 by interacting with CR5 (Zhou et al. 2000). The chaperone nucleophosmin directly binds to EBNA2 and supports complex formation with CBF1 and promoter recruitment (Liu et al. 2012). Recently, a novel anti-apoptotic function of EBNA2, based on the finding that EBNA2 binds to Nur77, has been described (TR3, NGFI-B) (Lee et al. 2002). Nur77 is an orphan member of the nuclear hormone receptor superfamily and a bifunctional molecule. Nur77, a nuclear protein, either acts as a transcription factor or can be translocated from the nucleus into the cytoplasm and trigger cytochrome c release in response to apoptotic stimuli (Li et al. 2000; Philips et al. 1997). EBNA2 can protect cells from apoptotic cell death by retaining Nur77 in the nucleus upon apoptotic stimuli (Lee et al. 2002). Whether EBNA2 also modulates functions of Nur77 related to its role as a transcription factor has not been analyzed to date.

## 11 Final Conclusions and Open Questions

In summary, EBNA2 acts as a key determinant of the activated phenotype of EBV-infected B-cells. The systematic and in-depth analysis of EBNA2 viral target genes has provided important clues to the molecular mechanism by which EBNA2 exerts its function as a transcriptional activator and has shown that CBF1 is the central effector of EBNA2 function. Within the cellular genome, EBNA2 preferentially binds to enhancers of cellular target genes which also preferentially recruit multiple B-cell-specific transcription factors. Since promoter and enhancers can reside in distant parts of a chromosome, the assignment of functional pairs of promoters and enhancers that are activated by EBNA2 and EBNA-LP will require intensive further investigations that combine studies on transcription factor binding, chromatin state, and nuclear architecture. Activation of *MYC* by EBNA2 is the major rate-limiting step for initiation and maintenance of the proliferation of EBV-infected B-cell cultures. The potential contribution of further cellular EBNA2 target genes to the growth transformation process in vitro will need to be rigorously tested in large-scale RNAi-based screens. It is also to be expected that several EBNA2 target genes are critical for the establishment of viral latency in vivo. With regard to EBNA-LP, it appears that one of its principal functions is to coactivate a subset of EBNA2-regulated viral latency genes. Thus, EBNA-LP might be required to activate EBNA2 target genes, which exhibit a specific chromatin configuration in naïve B-cells. The exact mechanism remains to be elucidated, but the observed interactions with Sp100 and cellular repressors or repressor complexes suggest that it facilitates EBNA2 coactivation through the modulation of repressors or facultative heterochromatin. Through its interaction with Sp100, EBNA-LP shares features with other herpesvirus immediate early proteins that modulate PML NBs or PML NB-associated proteins. One hypothesis is that PML NBs exert a repressive effect on viral gene expression, referred to as an intrinsic antiviral defense mechanism, which is counteracted by viral immediate early or tegument proteins (Everett 2013; Everett and Chelbi-Alix 2007). An intriguing notion is that EBNA-LP provides similar function(s) to help jump start viral latency gene expression immediately following infection. Due to the complex nature of the IR1 repeats for both Wp and EBNA-LP functions, it has been technically challenging to generate EBNA-LP null EBV recombinants or recombinants that express

EBNA-LP coactivation mutants to interrogate its role in EBV-induced B-cell immortalization. Such reagents will be needed to confirm and extend the previous observations concerning cellular cofactor interactions under the physiological conditions of EBV infection in primary B-cells. Moreover, the emergence of CRISPR/Cas9 technology should enable investigators to generate targeted knockout of cellular genes to assess their importance for EBNA2 and EBNA-LP function in the near future. However, a serious limitation of all the results discussed in this review is the fact that there is no small animal model available, which allows assessing the specific contribution of a target gene to the pathogenesis of EBV-associated diseases.

**Acknowledgments** The research of BK is supported by the HELENA graduate school of the Helmholtz Center Munich and the Deutsche Krebshilfe (grant 109258). PDL is supported by NIH grant 5R01AI080681.

## References

Abbot SD, Rowe M, Cadwallader K, Ricksten A, Gordon J, Wang F, Rymo L, Rickinson AB (1990) Epstein-Barr virus nuclear antigen 2 induces expression of the virus-encoded latent membrane protein. J Virol 64:2126–2134

Adldinger HK, Delius H, Freese UK, Clarke J, Bornkamm GW (1985) A putative transforming gene of Jijoye virus differs from that of Epstein-Barr virus prototypes. Virology 141:221–234

Alazard N, Gruffat H, Hiriart E, Sergeant A, Manet E (2003) Differential hyperacetylation of histones H3 and H4 upon promoter-specific recruitment of EBNA2 in Epstein-Barr virus chromatin. J Virol 77:8166–8172

Alfieri C, Birkenbach M, Kieff E (1991) Early events in Epstein-Barr virus infection of human B lymphocytes. Virology 181:595–608

Allan GJ, Rowe DT (1989) Size and stability of the Epstein-Barr virus major internal repeat (IR-1) in Burkitt's lymphoma and lymphoblastoid cell lines. Virology 173:489–498

Allday MJ, Farrell PJ (1994) Epstein-Barr virus nuclear antigen EBNA3C/6 expression maintains the level of latent membrane protein 1 in G1-arrested cells. J Virol 68:3491–3498

Allday MJ, Crawford DH, Griffin BE (1989) Epstein-Barr virus latent gene expression during the initiation of B cell immortalization. J Gen Virol 70:1755–1764

Allday MJ, Crawford DH, Thomas JA (1993) Epstein-Barr virus (EBV) nuclear antigen 6 induces expression of the EBV latent membrane protein and an activated phenotype in Raji cells. J Gen Virol 74(Pt 3):361–369

Ansieau S, Leutz A (2002) The conserved Mynd domain of BS69 binds cellular and oncoviral proteins through a common PXLXP motif. J Biol Chem 277:4906–4910

Baer R, Bankier AT, Biggin MD, Deininger PL, Farrell PJ, Gibson TJ, Hatfull G, Hudson GS, Satchwell SC, Seguin C et al (1984) DNA sequence and expression of the B95-8 Epstein-Barr virus genome. Nature 310:207–211

Bain M, Watson RJ, Farrell PJ, Allday MJ (1996) Epstein-Barr virus nuclear antigen 3C is a powerful repressor of transcription when tethered to DNA. J Virol 70:2481–2489

Bandobashi K, Maeda A, Teramoto N, Nagy N, Szekely L, Taguchi H, Miyoshi I, Klein G, Klein E (2001) Intranuclear localization of the transcription coadaptor CBP/p300 and the transcription factor RBP-Jk in relation to EBNA-2 and -5 in B lymphocytes. Virology 288:275–282

Bark-Jones SJ, Webb HM, West MJ (2006) EBV EBNA 2 stimulates CDK9-dependent transcription and RNA polymerase II phosphorylation on serine 5. Oncogene 25:1775–1785

Barth S, Liss M, Voss MD, Dobner T, Fischer U, Meister G, Grasser FA (2003) Epstein-Barr virus nuclear antigen 2 binds via its methylated arginine-glycine repeat to the survival motor neuron protein. J Virol 77:5008–5013
Boccellato F, Anastasiadou E, Rosato P, Kempkes B, Frati L, Faggioni A, Trivedi P (2007) EBNA2 interferes with the germinal center phenotype by downregulating BCL6 and TCL1 in non-Hodgkin's lymphoma cells. J Virol 81:2274–2282
Bodescot M, Chambraud B, Farrell P, Perricaudet M (1984) Spliced RNA from the IR1-U2 region of Epstein-Barr virus: presence of an open reading frame for a repetitive polypeptide. EMBO J 3:1913–1917
Bodescot M, Perricaudet M, Farrell PJ (1987) A promoter for the highly spliced EBNA family of RNAs of Epstein-Barr virus. J Virol 61:3424–3430
Borestrom C, Zetterberg H, Liff K, Rymo L (2003) Functional interaction of nuclear factor y and sp1 is required for activation of the epstein-barr virus C promoter. J Virol 77:821–829
Bornkamm GW, Hammerschmidt W (2001) Molecular virology of Epstein-Barr virus. Philos Trans R Soc Lond B Biol Sci 356:437–459
Burgstahler R, Kempkes B, Steube K, Lipp M (1995) Expression of the chemokine receptor BLR2/EBI1 is specifically transactivated by Epstein-Barr virus nuclear antigen 2. Biochem Biophys Res Commun 215:737–743
Calderwood MA, Venkatesan K, Xing L, Chase MR, Vazquez A, Holthaus AM, Ewence AE, Li N, Hirozane-Kishikawa T, Hill DE et al (2007) Epstein-Barr virus and virus human protein interaction maps. Proc Natl Acad Sci USA 104:7606–7611
Calender A, Cordier M, Billaud M, Lenoir GM (1990) Modulation of cellular gene expression in B lymphoma cells following in vitro infection by Epstein-Barr virus (EBV). Int J Cancer 46:658–663
Chabot PR, Raiola L, Lussier-Price M, Morse T, Arseneault G, Archambault J, Omichinski JG (2014) Structural and functional characterization of a complex between the acidic transactivation domain of EBNA2 and the Tfb1/p62 subunit of TFIIH. PLoS Pathog 10:e1004042
Cho YG, Gordadze AV, Ling PD, Wang F (1999) Evolution of two types of rhesus lymphocryptovirus similar to type 1 and type 2 Epstein-Barr virus. J Virol 73:9206–9212
Cludts I, Farrell PJ (1998) Multiple functions within the Epstein-Barr virus EBNA-3A protein. J Virol 72:1862–1869
Cohen JI (1992) A region of herpes simplex virus VP16 can substitute for a transforming domain of Epstein-Barr virus nuclear protein 2. Proc Natl Acad Sci USA 89:8030–8034
Cohen JI, Kieff E (1991) An Epstein-Barr virus nuclear protein 2 domain essential for transformation is a direct transcriptional activator. J Virol 65:5880–5885
Dambaugh T, Hennessy K, Chamnankit L, Kieff E (1984) U2 region of Epstein-Barr virus DNA may encode Epstein-Barr nuclear antigen 2. Proc Natl Acad Sci USA 81:7632–7636
Dillner J, Kallin B, Alexander H, Ernberg I, Uno M, Ono Y, Klein G, Lerner RA (1986) An Epstein-Barr virus (EBV)-determined nuclear antigen (EBNA5) partly encoded by the transformation-associated Bam WYH region of EBV DNA: preferential expression in lymphoblastoid cell lines. Proc Natl Acad Sci USA 83:6641–6645
Everett RD (2013) The spatial organization of DNA virus genomes in the nucleus. PLoS Pathog 9:e1003386
Everett RD, Chelbi-Alix MK (2007) PML and PML nuclear bodies: implications in antiviral defence. Biochimie 89:819–830
Fahraeus R, Jansson A, Ricksten A, Sjoblom A, Rymo L (1990) Epstein-Barr virus-encoded nuclear antigen 2 activates the viral latent membrane protein promoter by modulating the activity of a negative regulatory element. Proc Natl Acad Sci USA 87:7390–7394
Finke J, Rowe M, Kallin B, Ernberg I, Rosen A, Dillner J, Klein G (1987) Monoclonal and polyclonal antibodies against Epstein-Barr virus nuclear antigen 5 (EBNA-5) detect multiple protein species in Burkitt's lymphoma and lymphoblastoid cell lines. J Virol 61:3870–3878
Folk P, Puta F, Skruzny M (2004) Transcriptional coregulator SNW/SKIP: the concealed tie of dissimilar pathways. Cell Mol Life Sci 61:629–640

Fuchs KP, Bommer G, Dumont E, Christoph B, Vidal M, Kremmer E, Kempkes B (2001) Mutational analysis of the J recombination signal sequence binding protein (RBP-J)/ Epstein-Barr virus nuclear antigen 2 (EBNA2) and RBP-J/notch interaction. Eur J Biochem 268:4639–4646

Fuentes-Panana EM, Ling PD (1998) Characterization of the CBF2 binding site within the Epstein-Barr virus latency C promoter and its role in modulating EBNA2-mediated transactivation. J Virol 72:693–700

Fuentes-Panana EM, Swaminathan S, Ling PD (1999) Transcriptional activation signals found in the Epstein-Barr virus (EBV) latency C promoter are conserved in the latency C promoter sequences from baboon and Rhesus monkey EBV-like lymphocryptoviruses (cercopithicine herpesviruses 12 and 15). J Virol 73:826–833

Fuentes-Panana EM, Peng R, Brewer G, Tan J, Ling PD (2000) Regulation of the Epstein-Barr virus C promoter by AUF1 and the cyclic AMP/protein kinase A signaling pathway. J Virol 74:8166–8175

Garibal J, Hollville E, Bell AI, Kelly GL, Renouf B, Kawaguchi Y, Rickinson AB, Wiels J (2007) Truncated form of the Epstein-Barr virus protein EBNA-LP protects against caspase-dependent apoptosis by inhibiting protein phosphatase 2A. J Virol 81:7598–7607

Ghosh D, Kieff E (1990) cis-acting regulatory elements near the Epstein-Barr virus latent-infection membrane protein transcriptional start site. J Virol 64:1855–1858

Gordadze AV, Onunwor CW, Peng R, Poston D, Kremmer E, Ling PD (2004) EBNA2 amino acids 3 to 30 are required for induction of LMP-1 and immortalization maintenance. J Virol 78:3919–3929

Grasser FA, Haiss P, Gottel S, Mueller-Lantzsch N (1991) Biochemical characterization of Epstein-Barr virus nuclear antigen 2A. J Virol 65:3779–3788

Grossman SR, Johannsen E, Tong X, Yalamanchili R, Kieff E (1994) The Epstein-Barr virus nuclear antigen 2 transactivator is directed to response elements by the J kappa recombination signal binding protein. Proc Natl Acad Sci USA 91:7568–7572

Grundhoff AT, Kremmer E, Tureci O, Glieden A, Gindorf C, Atz J, Mueller-Lantzsch N, Schubach WH, Grasser FA (1999) Characterization of DP103, a novel DEAD box protein that binds to the Epstein-Barr virus nuclear proteins EBNA2 and EBNA3C. J Biol Chem 274:19136–19144

Han I, Xue Y, Harada S, Orstavik S, Skalhegg B, Kieff E (2002) Protein kinase A associates with HA95 and affects transcriptional coactivation by Epstein-Barr virus nuclear proteins. Mol Cell Biol 22:2136–2146

Harada S, Kieff E (1997) Epstein-Barr virus nuclear protein LP stimulates EBNA-2 acidic domain-mediated transcriptional activation. J Virol 71:6611–6618

Harada S, Yalamanchili R, Kieff E (2001) Epstein-Barr virus nuclear protein 2 has at least two N-terminal domains that mediate self-association. J Virol 75:2482–2487

Harth-Hertle ML, Scholz BA, Erhard F, Glaser LV, Dolken L, Zimmer R, Kempkes B (2013) Inactivation of intergenic enhancers by EBNA3A initiates and maintains polycomb signatures across a chromatin domain encoding CXCL10 and CXCL9. PLoS Pathog 9:e1003638

Hateboer G, Gennissen A, Ramos YF, Kerkhoven RM, Sonntag-Buck V, Stunnenberg HG, Bernards R (1995) BS69, a novel adenovirus E1A-associated protein that inhibits E1A transactivation. EMBO J 14:3159–3169

Hayward SD, Liu J, Fujimuro M (2006) Notch and Wnt signaling: mimicry and manipulation by gamma herpesviruses. Sci STKE 2006:re4

Henkel T, Ling PD, Hayward SD, Peterson MG (1994) Mediation of Epstein-Barr virus EBNA2 transactivation by recombination signal-binding protein J kappa. Science 265:92–95

Hertle ML, Popp C, Petermann S, Maier S, Kremmer E, Lang R, Mages J, Kempkes B (2009) Differential gene expression patterns of EBV infected EBNA-3A positive and negative human B lymphocytes. PLoS Pathog 5:e1000506

Hsieh JJ, Hayward SD (1995) Masking of the CBF1/RBPJ kappa transcriptional repression domain by Epstein-Barr virus EBNA2. Science 268:560–563

Inman GJ, Farrell PJ (1995) Epstein-Barr virus EBNA-LP and transcription regulation properties of pRB, p107 and p53 in transfection assays. J Gen Virol 76(Pt 9):2141–2149
Jimenez-Ramirez C, Brooks AJ, Forshell LP, Yakimchuk K, Zhao B, Fulgham TZ, Sample CE (2006) Epstein-Barr virus EBNA-3C is targeted to and regulates expression from the bidirectional LMP-1/2B promoter. J Virol 80:11200–11208
Jin XW, Speck SH (1992) Identification of critical cis elements involved in mediating Epstein-Barr virus nuclear antigen 2-dependent activity of an enhancer located upstream of the viral BamHI C promoter. J Virol 66:2846–2852
Jochner N, Eick D, Zimber-Strobl U, Pawlita M, Bornkamm GW, Kempkes B (1996) Epstein-Barr virus nuclear antigen 2 is a transcriptional suppressor of the immunoglobulin mu gene: implications for the expression of the translocated c-myc gene in Burkitt's lymphoma cells. EMBO J 15:375–382
Johannsen E, Koh E, Mosialos G, Tong X, Kieff E, Grossman SR (1995) Epstein-Barr virus nuclear protein 2 transactivation of the latent membrane protein 1 promoter is mediated by J kappa and PU.1. J Virol 69:253–262
Johansen LM, Deppmann CD, Erickson KD, Coffin WF 3rd, Thornton TM, Humphrey SE, Martin JM, Taparowsky EJ (2003) EBNA2 and activated Notch induce expression of BATF. J Virol 77:6029–6040
Kaiser C, Laux G, Eick D, Jochner N, Bornkamm GW, Kempkes B (1999) The proto-oncogene c-myc is a direct target gene of Epstein-Barr virus nuclear antigen 2. J Virol 73:4481–4484
Kashuba E, Mattsson K, Pokrovskaja K, Kiss C, Protopopova M, Ehlin-Henriksson B, Klein G, Szekely L (2003) EBV-encoded EBNA-5 associates with P14ARF in extranucleolar inclusions and prolongs the survival of P14ARF-expressing cells. Int J Cancer 105:644–653
Kelly G, Bell A, Rickinson A (2002) Epstein-Barr virus-associated Burkitt lymphomagenesis selects for downregulation of the nuclear antigen EBNA2. Nat Med 8:1098–1104
Kempkes B, Zimber-Strobl U, Eissner G, Pawlita M, Falk M, Hammerschmidt W, Bornkamm GW (1996) Epstein-Barr virus nuclear antigen 2 (EBNA2)-oestrogen receptor fusion proteins complement the EBNA2-deficient Epstein-Barr virus strain P3HR1 in transformation of primary B cells but suppress growth of human B cell lymphoma lines. J Gen Virol 77:227–237
Kitay MK, Rowe DT (1996) Cell cycle stage-specific phosphorylation of the Epstein-Barr virus immortalization protein EBNA-LP. J Virol 70:7885–7893
Knutson JC (1990) The level of c-fgr RNA is increased by EBNA-2, an Epstein-Barr virus gene required for B-cell immortalization. J Virol 64:2530–2536
Kovall RA (2007) Structures of CSL, Notch and Mastermind proteins: piecing together an active transcription complex. Curr Opin Struct Biol 17:117–127
Kovall RA, Hendrickson WA (2004) Crystal structure of the nuclear effector of Notch signaling, CSL, bound to DNA. EMBO J 23:3441–3451
Kurth J, Hansmann ML, Rajewsky K, Kuppers R (2003) Epstein-Barr virus-infected B cells expanding in germinal centers of infectious mononucleosis patients do not participate in the germinal center reaction. Proc Natl Acad Sci USA 100:4730–4735
Kwiatkowski B, Chen SY, Schubach WH (2004) CKII site in Epstein-Barr virus nuclear protein 2 controls binding to hSNF5/Ini1 and is important for growth transformation. J Virol 78:6067–6072
Lai EC (2002) Keeping a good pathway down: transcriptional repression of Notch pathway target genes by CSL proteins. EMBO Rep 3:840–845
Laux G, Adam B, Strobl LJ, Moreau-Gachelin F (1994a) The Spi-1/PU.1 and Spi-B ets family transcription factors and the recombination signal binding protein RBP-J kappa interact with an Epstein-Barr virus nuclear antigen 2 responsive cis-element. EMBO J 13:5624–5632
Laux G, Dugrillon F, Eckert C, Adam B, Zimber-Strobl U, Bornkamm GW (1994b) Identification and characterization of an Epstein-Barr virus nuclear antigen 2-responsive cis element in the bidirectional promoter region of latent membrane protein and terminal protein 2 genes. J Virol 68:6947–6958

Le Roux A, Kerdiles B, Walls D, Dedieu JF, Perricaudet M (1994) The Epstein-Barr virus determined nuclear antigens EBNA-3A, -3B, and -3C repress EBNA-2-mediated transactivation of the viral terminal protein 1 gene promoter. Virology 205:596–602

Lee JM, Lee KH, Weidner M, Osborne BA, Hayward SD (2002) Epstein-Barr virus EBNA2 blocks Nur77- mediated apoptosis. Proc Natl Acad Sci USA 99:11878–11883 Epub 12002 Aug 11823

Li H, Kolluri SK, Gu J, Dawson MI, Cao X, Hobbs PD, Lin B, Chen G, Lu J, Lin F et al (2000) Cytochrome c release and apoptosis induced by mitochondrial targeting of nuclear orphan receptor TR3. Science 289:1159–1164

Ling PD, Hayward SD (1995) Contribution of conserved amino acids in mediating the interaction between EBNA2 and CBF1/RBPJk. J Virol 69:1944–1950

Ling PD, Hsieh JJ, Ruf IK, Rawlins DR, Hayward SD (1994) EBNA-2 upregulation of Epstein-Barr virus latency promoters and the cellular CD23 promoter utilizes a common targeting intermediate, CBF1. J Virol 68:5375–5383

Ling PD, Peng RS, Nakajima A, Yu JH, Tan J, Moses SM, Yang WH, Zhao B, Kieff E, Bloch KD et al (2005) Mediation of Epstein-Barr virus EBNA-LP transcriptional coactivation by Sp100. EMBO J 24:3565–3575

Ling PD, Tan J, Peng R (2009) Nuclear-cytoplasmic shuttling is not required for the Epstein-Barr virus EBNA-LP transcriptional coactivation function. J Virol 83:7109–7116

Liu CD, Chen YL, Min YL, Zhao B, Cheng CP, Kang MS, Chiu SJ, Kieff E, Peng CW (2012) The nuclear chaperone nucleophosmin escorts an Epstein-Barr Virus nuclear antigen to establish transcriptional cascades for latent infection in human B cells. PLoS Pathog 8:e1003084

Liu CD, Cheng CP, Fang JS, Chen LC, Zhao B, Kieff E, Peng CW (2013) Modulation of Epstein-Barr virus nuclear antigen 2-dependent transcription by protein arginine methyltransferase 5. Biochem Biophys Res Commun 430:1097–1102

Lucchesi W, Brady G, Dittrich-Breiholz O, Kracht M, Russ R, Farrell PJ (2008) Differential gene regulation by Epstein-Barr virus type 1 and type 2 EBNA2. J Virol 82:7456–7466

Macsween KF, Crawford DH (2003) Epstein-Barr virus-recent advances. Lancet Infect Dis 3:131–140

Maier S, Santak M, Mantik A, Grabusic K, Kremmer E, Hammerschmidt W, Kempkes B (2005) A somatic knockout of CBF1 in a human B-cell line reveals that induction of CD21 and CCR7 by EBNA-2 is strictly CBF1 dependent and that downregulation of immunoglobulin M is partially CBF1 independent. J Virol 79:8784–8792

Maier S, Staffler G, Hartmann A, Hock J, Henning K, Grabusic K, Mailhammer R, Hoffmann R, Wilmanns M, Lang R et al (2006) Cellular target genes of Epstein-Barr virus nuclear antigen 2. J Virol 80:9761–9771

Marshall D, Sample C (1995) Epstein-Barr virus nuclear antigen 3C is a transcriptional regulator. J Virol 69:3624–3630

McCann EM, Kelly GL, Rickinson AB, Bell AI (2001) Genetic analysis of the Epstein-Barr virus-coded leader protein EBNA-LP as a coactivator of EBNA2 function. J Gen Virol 82:3067–3079

McClellan MJ, Khasnis S, Wood CD, Palermo RD, Schlick SN, Kanhere AS, Jenner RG, West MJ (2012) Downregulation of integrin receptor-signaling genes by Epstein-Barr virus EBNA 3C via promoter-proximal and -distal binding elements. J Virol 86:5165–5178

McClellan MJ, Wood CD, Ojeniyi O, Cooper TJ, Kanhere A, Arvey A, Webb HM, Palermo RD, Harth-Hertle ML, Kempkes B et al (2013) Modulation of enhancer looping and differential gene targeting by epstein-barr virus transcription factors directs cellular reprogramming. PLoS Pathog 9:e1003636

Meitinger C, Strobl LJ, Marschall G, Bornkamm GW, Zimber-Strobl U (1994) Crucial sequences within the Epstein-Barr virus TP1 promoter for EBNA2-mediated transactivation and interaction of EBNA2 with its responsive element. J Virol 68:7497–7506

Mohan J, Dement-Brown J, Maier S, Ise T, Kempkes B, Tolnay M (2006) Epstein-Barr virus nuclear antigen 2 induces FcRH5 expression through CBF1. Blood 107:4433–4439

Nam Y, Sliz P, Song L, Aster JC, Blacklow SC (2006) Structural basis for cooperativity in recruitment of MAML coactivators to Notch transcription complexes. Cell 124:973–983

Nitsche F, Bell A, Rickinson A (1997) Epstein-Barr virus leader protein enhances EBNA-2-mediated transactivation of latent membrane protein 1 expression: a role for the W1W2 repeat domain. J Virol 71:6619–6628

Pajic A, Spitkovsky D, Christoph B, Kempkes B, Schuhmacher M, Staege MS, Brielmeier M, Ellwart J, Kohlhuber F, Bornkamm GW et al (2000) Cell cycle activation by c-myc in a Burkitt lymphoma model cell line. Int J Cancer 87:787–793

Pajic A, Staege MS, Dudziak D, Schuhmacher M, Spitkovsky D, Eissner G, Brielmeier M, Polack A, Bornkamm GW (2001) Antagonistic effects of c-myc and Epstein-Barr virus latent genes on the phenotype of human B cells. Int J Cancer 93:810–816

Pegman PM, Smith SM, D'Souza BN, Loughran ST, Maier S, Kempkes B, Cahill PA, Simmons MJ, Gelinas C, Walls D (2006) Epstein-Barr virus nuclear antigen 2 trans-activates the cellular antiapoptotic bfl-1 gene by a CBF1/RBPJ kappa-dependent pathway. J Virol 80:8133–8144

Peng R, Gordadze AV, Fuentes Panana EM, Wang F, Zong J, Hayward GS, Tan J, Ling PD (2000a) Sequence and functional analysis of EBNA-LP and EBNA2 proteins from nonhuman primate lymphocryptoviruses. J Virol 74:379–389

Peng R, Tan J, Ling PD (2000b) Conserved regions in the Epstein-Barr virus leader protein define distinct domains required for nuclear localization and transcriptional cooperation with EBNA2. J Virol 74:9953–9963

Peng CW, Xue Y, Zhao B, Johannsen E, Kieff E, Harada S (2004a) Direct interactions between Epstein-Barr virus leader protein LP and the EBNA2 acidic domain underlie coordinate transcriptional regulation. Proc Natl Acad Sci USA 101:1033–1038

Peng CW, Zhao B, Kieff E (2004b) Four EBNA2 domains are important for EBNALP coactivation. J Virol 78:11439–11442

Peng R, Moses SC, Tan J, Kremmer E, Ling PD (2005) The Epstein-Barr virus EBNA-LP protein preferentially coactivates EBNA2-mediated stimulation of latent membrane proteins expressed from the viral divergent promoter. J Virol 79:4492–4505

Peng CW, Zhao B, Chen HC, Chou ML, Lai CY, Lin SZ, Hsu HY, Kieff E (2007) Hsp72 up-regulates Epstein-Barr virus EBNALP coactivation with EBNA2. Blood 109:5447–5454

Petti L, Sample C, Kieff E (1990) Subnuclear localization and phosphorylation of Epstein-Barr virus latent infection nuclear proteins. Virology 176:563–574

Philips A, Lesage S, Gingras R, Maira MH, Gauthier Y, Hugo P, Drouin J (1997) Novel dimeric Nur77 signaling mechanism in endocrine and lymphoid cells. Mol Cell Biol 17:5946–5951

Polack A, Hortnagel K, Pajic A, Christoph B, Baier B, Falk M, Mautner J, Geltinger C, Bornkamm GW, Kempkes B (1996) c-myc activation renders proliferation of Epstein-Barr virus (EBV)-transformed cells independent of EBV nuclear antigen 2 and latent membrane protein 1. Proc Natl Acad Sci USA 93:10411–10416

Portal D, Zhao B, Calderwood MA, Sommermann T, Johannsen E, Kieff E (2011) EBV nuclear antigen EBNALP dismisses transcription repressors NCoR and RBPJ from enhancers and EBNA2 increases NCoR-deficient RBPJ DNA binding. Proc Natl Acad Sci USA 108:7808–7813

Portal D, Zhou H, Zhao B, Kharchenko PV, Lowry E, Wong L, Quackenbush J, Holloway D, Jiang S, Lu Y et al (2013) Epstein-Barr virus nuclear antigen leader protein localizes to promoters and enhancers with cell transcription factors and EBNA2. Proc Natl Acad Sci USA 110:18537–18542

Puglielli MT, Woisetschlaeger M, Speck SH (1996) oriP is essential for EBNA gene promoter activity in Epstein-Barr virus-immortalized lymphoblastoid cell lines. J Virol 70:5758–5768

Rogers RP, Woisetschlaeger M, Speck SH (1990) Alternative splicing dictates translational start in Epstein-Barr virus transcripts. EMBO J 9:2273–2277

Sakai T, Taniguchi Y, Tamura K, Minoguchi S, Fukuhara T, Strobl LJ, Zimber-Strobl U, Bornkamm GW, Honjo T (1998) Functional replacement of the intracellular region of the Notch1 receptor by Epstein-Barr virus nuclear antigen 2. J Virol 72:6034–6039

Sample J, Hummel M, Braun D, Birkenbach M, Kieff E (1986) Nucleotide sequences of mRNAs encoding Epstein-Barr virus nuclear proteins: a probable transcriptional initiation site. Proc Natl Acad Sci USA 83:5096–5100

Sinclair AJ, Palmero I, Peters G, Farrell PJ (1994) EBNA-2 and EBNA-LP cooperate to cause G0 to G1 transition during immortalization of resting human B lymphocytes by Epstein-Barr virus. EMBO J 13:3321–3328

Sjoblom A, Jansson A, Yang W, Lain S, Nilsson T, Rymo L (1995a) PU box-binding transcription factors and a POU domain protein cooperate in the Epstein-Barr virus (EBV) nuclear antigen 2-induced transactivation of the EBV latent membrane protein 1 promoter. J Gen Virol 76:2679–2692

Sjoblom A, Nerstedt A, Jansson A, Rymo L (1995b) Domains of the Epstein-Barr virus nuclear antigen 2 (EBNA2) involved in the transactivation of the latent membrane protein 1 and the EBNA Cp promoters. J Gen Virol 76:2669–2678

Sjoblom A, Yang W, Palmqvist L, Jansson A, Rymo L (1998) An ATF/CRE element mediates both EBNA2-dependent and EBNA2-independent activation of the Epstein-Barr virus LMP1 gene promoter. J Virol 72:1365–1376

Skare J, Edson C, Farley J, Strominger JL (1982) The B95-8 isolate of Epstein-Barr virus arose from an isolate with a standard genome. J Virol 44:1088–1091

Speck SH, Pfitzner A, Strominger JL (1986) An Epstein-Barr virus transcript from a latently infected, growth-transformed B-cell line encodes a highly repetitive polypeptide. Proc Natl Acad Sci USA 83:9298–9302

Spender LC, Cornish GH, Rowland B, Kempkes B, Farrell PJ (2001) Direct and indirect regulation of cytokine and cell cycle proteins by ebna-2 during Epstein-Barr virus infection. J Virol 75:3537–3546

Spender LC, Cornish GH, Sullivan A, Farrell PJ (2002) Expression of transcription factor AML-2 (RUNX3, CBF(alpha)-3) is induced by Epstein-Barr virus EBNA-2 and correlates with the B-cell activation phenotype. J Virol 76:4919–4927

Spender LC, Whiteman HJ, Karstegl CE, Farrell PJ (2005) Transcriptional cross-regulation of RUNX1 by RUNX3 in human B cells. Oncogene 24:1873–1881

Spender LC, Lucchesi W, Bodelon G, Bilancio A, Karstegl CE, Asano T, Dittrich-Breiholz O, Kracht M, Vanhaesebroeck B, Farrell PJ (2006) Cell target genes of Epstein-Barr virus transcription factor EBNA-2: induction of the p55alpha regulatory subunit of PI3-kinase and its role in survival of EREB2.5 cells. J Gen Virol 87:2859–2867

Strobl LJ, Hofelmayr H, Marschall G, Brielmeier M, Bornkamm GW, Zimber-Strobl U (2000) Activated Notch1 modulates gene expression in B cells similarly to Epstein-Barr viral nuclear antigen 2. J Virol 74:1727–1735

Sung NS, Kenney S, Gutsch D, Pagano JS (1991) EBNA-2 transactivates a lymphoid-specific enhancer in the BamHI C promoter of Epstein-Barr virus. J Virol 65:2164–2169

Szekely L, Pokrovskaja K, Jiang WQ, de The H, Ringertz N, Klein G (1996) The Epstein-Barr virus-encoded nuclear antigen EBNA-5 accumulates in PML-containing bodies. J Virol 70:2562–2568

Thorley-Lawson DA (2001) Epstein-Barr virus: exploiting the immune system. Nat Rev Immunol 1:75–82

Tierney RJ, Kao KY, Nagra JK, Rickinson AB (2011) Epstein-Barr virus BamHI W repeat number limits EBNA2/EBNA-LP coexpression in newly infected B cells and the efficiency of B-cell transformation: a rationale for the multiple W repeats in wild-type virus strains. J Virol 85:12362–12375

Tobollik S, Meyer L, Buettner M, Klemmer S, Kempkes B, Kremmer E, Niedobitek G, Jungnickel B (2006) Epstein-Barr virus nuclear antigen 2 inhibits AID expression during EBV-driven B-cell growth. Blood 108:3859–3864

Tong X, Yalamanchili R, Harada S, Kieff E (1994) The EBNA-2 arginine-glycine domain is critical but not essential for B-lymphocyte growth transformation; the rest of region 3 lacks essential interactive domains. J Virol 68:6188–6197
Tong X, Drapkin R, Reinberg D, Kieff E (1995a) The 62- and 80-kDa subunits of transcription factor IIH mediate the interaction with Epstein-Barr virus nuclear protein 2. Proc Natl Acad Sci USA 92:3259–3263
Tong X, Drapkin R, Yalamanchili R, Mosialos G, Kieff E (1995b) The Epstein-Barr virus nuclear protein 2 acidic domain forms a complex with a novel cellular coactivator that can interact with TFIIE. Mol Cell Biol 15:4735–4744
Tong X, Wang F, Thut CJ, Kieff E (1995c) The Epstein-Barr virus nuclear protein 2 acidic domain can interact with TFIIB, TAF40, and RPA70 but not with TATA-binding protein. J Virol 69:585–588
Tzellos S, Farrell PJ (2012) Epstein-Barr virus sequence variation-biology and disease. Pathogens 1:156–175
Tzellos S, Correia PB, Karstegl CE, Cancian L, Cano-Flanagan J, McClellan MJ, West MJ, Farrell PJ (2014) A single amino acid in EBNA-2 determines superior B lymphoblastoid cell line growth maintenance by Epstein-Barr virus type 1 EBNA-2. J Virol 88:8743–8753
Voss MD, Hille A, Barth S, Spurk A, Hennrich F, Holzer D, Mueller-Lantzsch N, Kremmer E, Grasser FA (2001) Functional cooperation of Epstein-Barr virus nuclear antigen 2 and the survival motor neuron protein in transactivation of the viral LMP1 promoter. J Virol 75:11781–11790
Waltzer L, Logeat F, Brou C, Israel A, Sergeant A, Manet E (1994) The human J kappa recombination signal sequence binding protein (RBP-J kappa) targets the Epstein-Barr virus EBNA2 protein to its DNA responsive elements. EMBO J 13:5633–5638
Wang F (2013). Nonhuman primate models for Epstein-Barr virus infection. Curr Opin Virol
Wang F, Gregory CD, Rowe M, Rickinson AB, Wang D, Birkenbach M, Kikutani H, Kishimoto T, Kieff E (1987a) Epstein-Barr virus nuclear antigen 2 specifically induces expression of the B-cell activation antigen CD23. Proc Natl Acad Sci USA 84:3452–3456
Wang F, Petti L, Braun D, Seung S, Kieff E (1987b) A bicistronic Epstein-Barr virus mRNA encodes two nuclear proteins in latently infected, growth-transformed lymphocytes. J Virol 61:945–954
Wang F, Tsang SF, Kurilla MG, Cohen JI, Kieff E (1990) Epstein-Barr virus nuclear antigen 2 transactivates latent membrane protein LMP1. J Virol 64:3407–3416
Wang L, Grossman SR, Kieff E (2000) Epstein-Barr virus nuclear protein 2 interacts with p300, CBP, and PCAF histone acetyltransferases in activation of the LMP1 promoter. Proc Natl Acad Sci USA 97:430–435
Wilson JJ, Kovall RA (2006) Crystal structure of the CSL-Notch-Mastermind ternary complex bound to DNA. Cell 124:985–996
Woisetschlaeger M, Yandava CN, Furmanski LA, Strominger JL, Speck SH (1990) Promoter switching in Epstein-Barr virus during the initial stages of infection of B lymphocytes. Proc Natl Acad Sci USA 87:1725–1729
Woisetschlaeger M, Jin XW, Yandava CN, Furmanski LA, Strominger JL, Speck SH (1991) Role for the Epstein-Barr virus nuclear antigen 2 in viral promoter switching during initial stages of infection. Proc Natl Acad Sci USA 88:3942–3946
Wu DY, Kalpana GV, Goff SP, Schubach WH (1996) Epstein-Barr virus nuclear protein 2 (EBNA2) binds to a component of the human SNF-SWI complex, hSNF5/Ini1. J Virol 70:6020–6028
Wu DY, Krumm A, Schubach WH (2000) Promoter-specific targeting of human SWI-SNF complex by Epstein-Barr virus nuclear protein 2. J Virol 74:8893–8903
Wu DY, Tkachuck DC, Roberson RS, Schubach WH (2002) The human SNF5/INI1 protein facilitates the function of the growth arrest and DNA damage-inducible protein (GADD34) and modulates GADD34-bound protein phosphatase-1 activity. J Biol Chem 277:27706–27715

Yang J, Aittomaki S, Pesu M, Carter K, Saarinen J, Kalkkinen N, Kieff E, Silvennoinen O (2002) Identification of p100 as a coactivator for STAT6 that bridges STAT6 with RNA polymerase II. EMBO J 21:4950–4958

Yokoyama A, Kawaguchi Y, Kitabayashi I, Ohki M, Hirai K (2001a) The conserved domain CR2 of Epstein-Barr virus nuclear antigen leader protein is responsible not only for nuclear matrix association but also for nuclear localization. Virology 279:401–413

Yokoyama A, Tanaka M, Matsuda G, Kato K, Kanamori M, Kawasaki H, Hirano H, Kitabayashi I, Ohki M, Hirai K et al (2001b) Identification of major phosphorylation sites of Epstein-Barr virus nuclear antigen leader protein (EBNA-LP): ability of EBNA-LP to induce latent membrane protein 1 cooperatively with EBNA-2 is regulated by phosphorylation. J Virol 75:5119–5128

Yue W, Davenport MG, Shackelford J, Pagano JS (2004) Mitosis-specific hyperphosphorylation of Epstein-Barr virus nuclear antigen 2 suppresses its function. J Virol 78:3542–3552

Zhao B, Sample CE (2000) Epstein-barr virus nuclear antigen 3C activates the latent membrane protein 1 promoter in the presence of Epstein-Barr virus nuclear antigen 2 through sequences encompassing an spi-1/Spi-B binding site. J Virol 74:5151–5160

Zhao B, Maruo S, Cooper A, Chase MR, Johannsen E, Kieff E, Cahir-McFarland E (2006) RNAs induced by Epstein-Barr virus nuclear antigen 2 in lymphoblastoid cell lines. Proc Natl Acad Sci USA 103:1900–1905

Zhao B, Zou J, Wang H, Johannsen E, Peng CW, Quackenbush J, Mar JC, Morton CC, Freedman ML, Blacklow SC et al (2011) Epstein-Barr virus exploits intrinsic B-lymphocyte transcription programs to achieve immortal cell growth. Proc Natl Acad Sci USA 108:14902–14907

Zhou S, Fujimuro M, Hsieh JJ, Chen L, Hayward SD (2000) A role for SKIP in EBNA2 activation of CBF1-repressed promoters. J Virol 74:1939–1947

Zimber-Strobl U, Suentzenich KO, Laux G, Eick D, Cordier M, Calender A, Billaud M, Lenoir GM, Bornkamm GW (1991) Epstein-Barr virus nuclear antigen 2 activates transcription of the terminal protein gene. J Virol 65:415–423

Zimber-Strobl U, Kremmer E, Grasser F, Marschall G, Laux G, Bornkamm GW (1993) The Epstein-Barr virus nuclear antigen 2 interacts with an EBNA2 responsive cis-element of the terminal protein 1 gene promoter. EMBO J 12:167–175

Zimber-Strobl U, Strobl LJ, Meitinger C, Hinrichs R, Sakai T, Furukawa T, Honjo T, Bornkamm GW (1994) Epstein-Barr virus nuclear antigen 2 exerts its transactivating function through interaction with recombination signal binding protein RBP-J kappa, the homologue of Drosophila Suppressor of Hairless. EMBO J 13:4973–4982

# The EBNA3 Family: Two Oncoproteins and a Tumour Suppressor that Are Central to the Biology of EBV in B Cells

**Martin J. Allday, Quentin Bazot and Robert E. White**

**Abstract** Epstein-Barr virus nuclear antigens EBNA3A, EBNA3B and EBNA3C are a family of three large latency-associated proteins expressed in B cells induced to proliferate by the virus. Together with the other nuclear antigens (EBNA-LP, EBNA2 and EBNA1), they are expressed from a polycistronic transcription unit that is probably unique to B cells. However, compared with the other EBNAs, hitherto the EBNA3 proteins were relatively neglected and their roles in EBV biology rather poorly understood. In recent years, powerful new technologies have been used to show that these proteins are central to the latency of EBV in B cells, playing major roles in reprogramming the expression of host genes affecting cell proliferation, survival, differentiation and immune surveillance. This indicates that the EBNA3s are critical in EBV persistence in the B cell system and in modulating B cell lymphomagenesis. EBNA3A and EBNA3C are necessary for the efficient proliferation of EBV-infected B cells because they target important tumour suppressor pathways—so operationally they are considered oncoproteins. In contrast, it is emerging that EBNA3B restrains the oncogenic capacity of EBV, so it can be considered a tumour suppressor—to our knowledge the first to be described in a tumour virus. Here, we provide a general overview of the EBNA3 genes and proteins. In particular, we describe recent research that has highlighted the complexity of their functional interactions with each other, with specific sites on the human genome and with the molecular machinery that controls transcription and epigenetic states of diverse host genes.

M.J. Allday (✉) · Q. Bazot · R.E. White
Molecular Virology, Division of Infectious Diseases, Department of Medicine,
Imperial College London, Norfolk Place, London W2 1PG, UK
e-mail: m.allday@imperial.ac.uk



C. Münz (ed.), *Epstein Barr Virus Volume 2*, Current Topics in Microbiology and Immunology 391, DOI 10.1007/978-3-319-22834-1_3

## Contents



## Abbreviations

| | |
|---|---|
| 4HT | 4-Hydroxytamoxifen |
| ABC | Activated B cell |
| BAC | Bacterial artificial chromosome |
| BARTs | BamH1-A rightward transcripts |
| BL | Burkitt's lymphoma |
| CaHV3 | Callitrichine herpesvirus 3 |
| CBF | C promoter binding factor |
| CCC | Chromatin conformation capture |
| CDK | Cyclin-dependent kinase |
| ChIP | Chromatin immunoprecipitation |
| ChIP-Seq | Chromatin immunoprecipitation coupled to high-throughput DNA sequencing |

| | |
|---|---|
| CtBP | C-terminal binding protein |
| CTL | Cytotoxic T lymphocyte |
| DDR | DNA damage response |
| DLBCL | Diffuse large B cell lymphoma |
| EBER | EBV-encoded RNA |
| ER | Oestrogen receptor |
| EBNA | Epstein-Barr virus nuclear antigen |
| EBV | Epstein-Barr virus |
| GST | Glutathione-S-transferase |
| HDAC | Histone deacetylase |
| HL | Hodgkin's lymphoma |
| IM | Infectious mononucleosis |
| IRES | Internal ribosome entry site |
| KO | Knockout |
| LCL | Lymphoblastoid cell line |
| LCV | Lymphocryptovirus |
| LZ | Leucine zipper |
| LMP | Latent membrane protein |
| MBC | Memory B cell |
| miRNA | MicroRNA |
| MIZ1 | Myc-interacting zinc-finger protein 1 |
| NLS | Nuclear localisation signal |
| NOD | Non-obese diabetic |
| Notch-IC | Intracellular Notch transcription factor |
| NSG | NOD-scid IL2 $\gamma$-null |
| OIS | Oncogene-induced senescence |
| ORF | Open reading frame |
| OSR | Oncogenic stress response |
| paHV1 | Papiine herpesvirus 1 |
| PcG | Polycomb group protein |
| Pol II | RNA polymerase II |
| PRC | Polycomb repressive complex |
| PTLD | Post-transplant lymphoproliferative disease |
| QPCR | Quantitative polymerase chain reaction |
| RB | Retinoblastoma protein |
| RBP | Recombining binding protein |
| REFs | Rat embryo fibroblasts |
| RhLCV | Rhesus lymphocryptovirus |
| SCID | Severe combined immunodeficiency |
| Su(H) | Suppressor of Hairless |
| TSS | Transcription start site |
| WT | Wild type |

# 1 The Biology of EBV in B Cells

EBV has co-evolved with its hominid and prehominid hosts for millions of years and is now a uniquely successful intracellular parasite that persists asymptomatically after childhood infection in >90 % of the human population. Despite this benign relationship with most humans, EBV is classified as a tumour virus because of its firmly established association with several cancers of B, NK/T and epithelial cell origin, its ability to induce B cell tumours in non-human primates and its capacity to growth transform primary human B cells in culture. This capacity of EBV to 'transform' resting B cells into continuously proliferating lymphoblastoid cell lines (LCLs) was discovered about 3 years after the virus was identified in samples of Burkitt's lymphoma (BL) 50 years ago (reviewed in Crawford 2001; Epstein 2001; Thorley-Lawson and Allday 2008; Young and Rickinson 2004). This remarkably efficient process has subsequently been used countless times to establish 'normal' human B cell lines for genetic, epigenetic and immunological studies (see, for example, ENCODE, the 1000 genomes project and the HapMap project). It has also been the focus of intensive research because the process produces limitless amounts of material for study and because of the general belief that the molecular mechanisms involved in the transformation of explanted B cells might explain aspects of EBV-associated cancer in humans. Today, after several decades of study, the general strategies utilised by EBV to drive mature B cells from quiescence into the cell cycle and maintain cell proliferation and survival (while preventing the activation of lytic virus replication) are largely understood—however, many of the biochemical details are poorly characterised.

When EBV induces the activation of resting B cells, it exploits the cellular transcription and translation machineries to express nine viral latency-associated proteins. There are six nuclear proteins (Epstein-Barr nuclear antigens (EBNAs) 1, 2, 3A, 3B, 3C and LP) and three membrane proteins (latent membrane proteins (LMPs) 1, 2A and 2B). In addition, two types of untranslated RNAs are expressed (EBV-encoded RNAs (EBERs) and the microRNAs (miRs) derived from the BamHI-A rightward transcripts (BARTs) and the BHRF1 transcript). The BART and BHRF1 RNAs can be processed into many miR species that probably target multiple host and viral mRNAs (comprehensively reviewed in the chapter authored by Skalsky and Cullen). This pattern of viral gene expression in B cells is known as latency III (sometimes referred to as the growth or proliferation programme) and is necessary to drive quiescent B cells into the cell cycle and sustain proliferation, while maintaining the EBV genome as extra-chromosomal 172 kb episomes (reviewed in Cai et al. 2006; Young and Rickinson 2004). The proliferating cells resemble, at least superficially, antigen-activated B cells (B blasts). As a consequence, it was a widely held view that when EBV transforms normal B cells into LCLs, the small number of viral latency-associated gene products exploits the physiological process of activation normally achieved by the interaction of a B cell with its specific antigen together with cognate T cell help. While in principle, this might be correct, it has gradually become apparent that LCLs are trapped at a B blast-like stage of differentiation, but they have a distinctive EBV-specific phenotype. In addition, the process of EBV-induced proliferation is 'sensed' by B cells as abnormal and so activates various innate responses that cells have evolved to prevent non-physiological DNA replication and proliferation and which have to be circumvented in order to establish latent infection. Several of the latency-associated viral factors, including EBNA3C—a major focus of this review—have evolved functions that overcome cellular checkpoints and pathways that trigger anti-proliferative or suicidal responses.

During lifelong persistence in healthy seropositive individuals, extra-chromosomal EBV episomes are largely found in non-dividing ($Ki67^{-ve}$), long-lived, class-switched ($IgD^{-ve}$), $CD27^{+ve}$ memory B cells (MBCs) in the peripheral circulation—and they do not appear to express the latency III proteins. However, in tonsils, naïve ($IgD^{+ve}$) B cells can be found that express the latency III programme and proliferate as LCL-like B blasts; so it seems that, in vivo as in vitro, EBV-infected naïve B cells undergo multiple divisions—at least transiently—as part of the normal viral life cycle. A compelling explanation of how some of the proliferating B blasts end up as memory cells was provided by Thorley-Lawson and colleagues and is reviewed elsewhere in this volume. Briefly, cells in the expanding B blast population are thought to enter a germinal centre in lymphoid tissue such as tonsil. In this microenvironment, these cells differentiate to become centroblasts, then centrocytes and finally resting memory B cells that enter the peripheral circulation (Babcock et al. 2000; Roughan and Thorley-Lawson 2009 and reviewed in Thorley-Lawson and Gross 2004; Thorley-Lawson et al. 2013). While the precise series of events that the EBV-positive B cells undergo to reach the memory compartment remains unknown, the consensus view is that it involves regulated shutdown of EBV latency-associated gene expression from the initial latency III state to latency II (in which EBNA2, EBNA-LP, and the EBNA3 family have been switched off by promoter switching from Cp to Qp—Fig. 1). In quiescent MBCs, Qp also becomes silenced and no EBV proteins can be detected—a state called latency 0. If MBCs periodically divide, then EBNA1 is switched back on to facilitate episome maintenance (this state is known as latency I). However, it is still unclear whether or not the differentiation of EBV-infected B blasts to memory B cells is absolutely dependent on the microenvironment of a germinal centre, although a role for T cells is indicated. T cell-derived cytokines provide signals that trigger the repression of the EBNA promoter, Cp, and reduce the expression of EBNA2 in LCLs and a mouse model (Kis et al. 2006, 2010; Nagy et al. 2012; Salamon et al. 2012)—however, it has yet to be proven that this occurs during persistence. In normal healthy individuals, the EBV-infected B blasts are targets for EBV-specific cytotoxic lymphocytes (CTLs) that can recognise and destroy these latently infected cells. Stable persistence depends, therefore, on the equilibrium established between the proliferation of B blasts—on the one hand—and immune elimination or differentiation to the resting memory compartment on the other (Babcock et al. 1999; Hawkins et al. 2013).

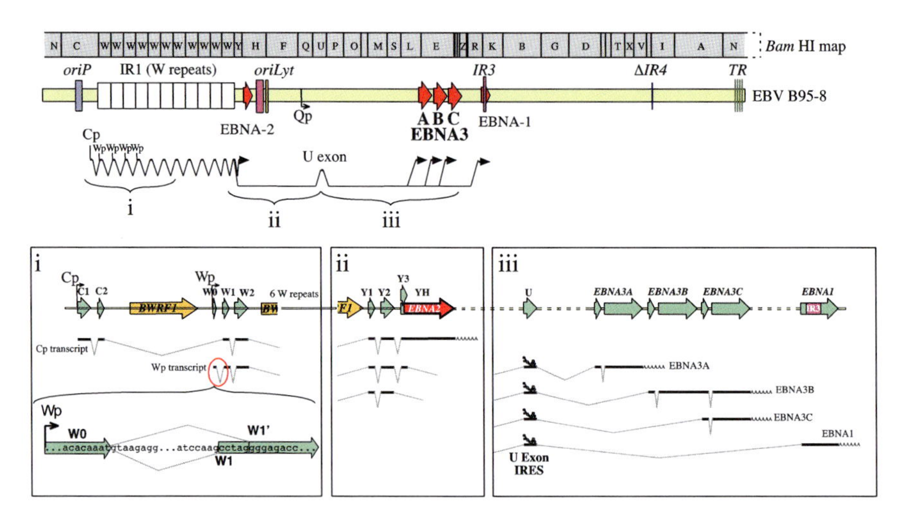

**Fig. 1** The B cell transcription unit. *Top* is an overview of the EBV genome (B95-8 strain), indicating major repeats and the nuclear antigens, with the Bam HI digest map (which defines the names of exons, promoters and most EBV ORFs) above. *Below* the map is a transcript summary, indicating the overall nature of the B cell transcription unit. Numbers (i)–(iii) indicate the major splicing 'decisions' that define the alternative transcripts that can be generated from the B cell transcription unit that are defined in detail in the boxes below. No coordination between these splicing events has been reported. *i* Early after infection, transcription begins from promoter Wp (exon W0), but promoter Cp is activated by EBNA2 (Exons C1C2). Exon W0 (and exon C2—not shown) can splice to either splice acceptor W1 or W1'. Splice acceptor W1' completes a start codon across the splice junction, allowing EBNA-LP translation from an array of exons W1W2 followed by Y1Y2. Splice acceptor Exon W1 does not generate an AUG codon, enabling translation from the next AUG codon that initiates EBNA2. *ii* Within exon YH (encoding EBNA2) is a splice donor (generating exon Y3) that can splice to the internal ribosome entry site within exon U. Exon Y2 can splice to either YH/Y3 or directly to exon U. Alternatively, the transcript is polyadenylated at the end of EBNA2 as shown. *iii* The IRES within the downstream U exon allows the translation of the EBNA3s and EBNA1 as shown. Polyadenylation has been observed after EBNA3A, EBNA3C and EBNA1. Curiously, EBNA3B transcripts also contain the EBNA3C open reading frame. Not only can the splice junction from exon U to the EBNA3s define the transcribed protein, but also failure to splice the internal introns within the EBNA3s has also been reported. Switching from the Cp/Wp to using Qp (not shown) results in the loss of expression of all EBNAs except the EBNA1 transcript. Theoretically, it would be possible to also generate EBNA3 transcripts from Qp, but this has not been reported

Primary infection with EBV during adolescence can sometimes result in the relatively benign illness infectious mononucleosis (IM), but disruption of components of the immune system by co-infections, genetic/epigenetic aberrations or iatrogenic causes can result in EBV-associated B cell neoplasia including post-transplant lymphoproliferative disease (PTLD), Burkitt's lymphoma (BL), Hodgkin's lymphoma (HL) and some diffuse large B cell lymphoma (DLBCL) (reviewed in Crawford 2001). Details of the pathogenesis of IM and these tumours are discussed elsewhere in the volume. Because, as far as we know, the EBNA3 family of proteins (along with EBNA2 and EBNA-LP) is only expressed in B cells from the unique B cell-specific transcription unit (Fig. 1 and below), here we have only described the biology of EBV latency in B cells. The biology and pathology of EBV in epithelial cells and NK/T cells are also described elsewhere in the volume.

## 2 The EBNA3 Family of Genes and Proteins

The EBV latency-associated genes EBNA3A, EBNA3B and EBNA3C (historically also called EBNA3, EBNA4 and EBNA6, respectively) are a family that probably arose during primate gamma-herpesvirus evolution by a series of gene duplication events since they have a similar gene structure, are arranged in tandem in the EBV genome and share partial sequence homology. EBNA3 transcripts are alternatively spliced from very long primary transcripts generally initiated at a single latency promoter and LCLs have only a few copies of these transcripts per cell, suggesting that their expression is tightly regulated. Although it is related to one another, there is nothing to suggest that the large nuclear proteins encoded by these genes have redundant functions; moreover, they possess no obvious similarities to any known cell or viral proteins—other than their localisation to the nucleus—that provide clues to their biological roles. Genetic studies using recombinant viruses originally indicated that EBNA3A and EBNA3C are essential for efficient in vitro transformation and immortalisation of B cells, whereas EBNA3B is dispensable. However, under the appropriate conditions with feeder cells in the culture, it has more recently been possible to establish EBNA3A-negative LCLs with relative ease (Hertle et al. 2009; Skalska et al. 2010). From an increasing number of studies that have made use of bacterial artificial chromosome (BAC)-derived EBV recombinants carrying deletions, fusions or mutations in the EBNA3 locus, the molecular details underlying these roles in B cell transformation are gradually emerging and will be reviewed herein.

### *2.1 Genomic Organisation, the B Cell Transcription Unit and Regulation of Expression*

The EBNA3 coding exons share a common structure, with a short 5′ exon, of approximately 360 nucleotides, and a longer 3′ exon of around 2.5 kb, separated by an intron of under 100 bp (Fig. 1). The coding regions of the longer exons are partially repetitive—with EBNA3B and EBNA3C containing variable length imperfect repeat regions (60 bp repeats for EBNA3B, and 45 bp repeats for EBNA3C). The C-terminal half of the EBNA3A gene comprises sequences that show evidence of historic duplication events followed by diversification, leaving regions with partial homology to each other (Baer et al. 1984; Hennessy et al. 1986). Expansions and contractions of these repetitive regions lead to EBV isolate-specific differences in the size, hence electrophoretic mobility, of each EBNA3 protein, allowing EBV clinical isolates to be distinguished by the apparent size of their EBNAs (EBNA-type—Yao et al. 1991).

All of the EBNAs are transcribed from extended transcripts originating at a common promoter. Initially after infection, the multiple copies of the B cell-specific promoter located in the BamH1W repeats (Wp) are used, but this soon switches to a single promoter just upstream of these repeats (Cp), which remains the dominant promoter during latency type III. Cp can be epigenetically silenced in the transition to latency states type II, type I and type 0 (reviewed in Tempera and Lieberman 2014 and Lieberman in this volume). Synthesis of the various EBNA proteins from this promoter is dependent in the first instance on a complex array of alternative splicing events. For clarity, these are labelled i, ii and iii in Fig. 1. Our current understanding is that the splicing event between the promoter exon (generated by either Cp or Wp) and the W1 exon (i in Fig. 1) can select either of two splice acceptors, 5 bases apart. The most downstream of these splice acceptors (W1′) creates the start codon of EBNA-LP across the splice site, facilitating the translation of EBNA-LP. The upstream splice acceptor (W1) does not generate an AUG, and as a result, the first start codon in the transcript is that of EBNA2. Regulation of this splice is thus believed to define the balance of EBNA-LP and EBNA2 translation in the cell.

A second alternative splicing event (ii in Fig. 1) links the upstream exons to the U exon. Both the Y2 and Y3 exons splice to the U exon with similar frequencies. In so doing, the splice to the U exon necessarily excises the EBNA2 ORF. Upstream of an open reading frame, the U exon is thought to function as an internal ribosome entry site (IRES) (Isaksson et al. 2003). Thus, its splicing to downstream exons (iii in Fig. 1) allows the translation of one of the EBNA3s or EBNA1 from a second cistron of the transcript (i.e. after EBNA-LP) with the EBNA1 splice site being favoured (Arvey et al. 2012). It has further been suggested that the internal intron of each EBNA3 may not splice with perfect efficiency in LCLs, which may constitute another mechanism to modulate EBNA3 protein levels (Kienzle et al. 1999). In addition to this already complex transcriptional profile, other splice variants have been observed at lower frequency (Arvey et al. 2012), but there is no indication that these are functionally relevant to the biology of the virus.

This cascade of alternative splicing results in the ORFs located more 3′ being transcribed less frequently than EBNA2 and EBNA-LP. This probably contributes to the low levels of EBNA3 mRNAs and proteins, such that their detection in infected B cells by immunofluorescence or immunocytochemistry has proven extremely difficult. However, the use of Western blotting has shown that a constant level of EBNA3 proteins is maintained across many LCLs, suggesting that there is tight control of EBNA3 protein levels in these cells. The mechanisms that balance the expression of the EBNAs are not fully understood, but EBNA3 homoeostasis may involve modulation of splicing, IRES function and protein turnover. Our experience of engineering EBNA3 recombinants has shown that this balance of expression is easily disrupted. Specifically, we find that introducing additional sequences onto the C-terminus of EBNA3B in the virus genome disrupts

EBNA3C splicing, while N-terminal fusions of EBNA3B can either increase or decrease EBNA3B protein levels (our unpublished results). Fortunately for genetic analysis of both EBNA3A and EBNA3C, C-terminal fusions are well tolerated and have produced no unpredictable or undesirable expression patterns in LCLs. As is described in the following sections, there appears to be considerable functional and perhaps physical interaction between the EBNA3 proteins and so maintaining balanced stoichiometry is probably vitally important for LCL fitness. However, mechanisms by which RNA splicing, translation and turnover of the EBNA3s mediate this careful regulation of protein levels remain largely uncharacterised.

## 2.2 Sequence Variation of the EBNA3 Proteins

There are two major clades of EBV—type 1 (aka type A) and type 2 (aka type B)—defined by the sequences of their EBNA2 and EBNA3 genes. Type 1 EBV is widespread worldwide, while type 2 EBV is largely confined to sub-Saharan Africa (Young et al. 1987). EBNA2 and EBNA3 represent the major sites of genome diversity of EBV, with approximately 80 % nucleotide identity between the prototype type 1 and type 2 EBNA3B and EBNA3C gene sequences, and 90 % for EBNA3A. It is not yet clear how precisely type 2 EBNA3s are functionally equivalent to their type 1 counterparts or—as with EBNA2—they have distinct capabilities in B cell transformation. The clearest differences between type 1 and 2 EBNA3s lie in the sizes of their repetitive regions, with distinct insertions/deletions permitting PCR-based identification of type 1 versus type 2 EBNA3s (Sample et al. 1990).

The diversity of EBNA3 sequences is further complicated by recombination between distinct virus strains. Examples of EBV from both Chinese and Korean samples have been identified with a type 1 EBNA2 and type 2 EBNA3s, or type 1 EBNA2 and EBNA3A with type 2 EBNA3B and EBNA3C (Kim et al. 2006; Midgley et al. 2000, 2003). An even more complex combination was found in the DNA isolated from the blood of an Austrian post-transplant lymphoproliferative disorder (PTLD) patient, identifying a crossover incorporating almost 1 kb of type 2 sequence encoding the last 200 amino acids of EBNA3A and the first exon of EBNA3B (Gorzer et al. 2006). We also observed a recombination between distinct type 1 genotypes in the EBNA3B of a Hodgkin's lymphoma sample, although in this case, it results in a small deletion within EBNA3B (White et al. 2012). It may be that this event is unique to this tumour, but it underlines the likelihood that interstrain recombination is a driver of EBV, and EBNA3, diversity.

Overall, it is clear that the EBNA3s are (with EBNA2) the most diverse of the EBV proteins and that, although it appears to be uncommon, linkage between EBNA3 and EBNA2 subtypes can be disrupted by recombination between diverse strains. Despite the significant amino acid sequence variation, the biological and immunological consequences of this diversity remain largely unexplored—for a more detailed consideration of EBV DNA and protein sequence variation, see the chapter authored by Paul Farrell.

## 2.3 The EBNA3 Counterparts in EBV-Like Primate Viruses and Their Evolution

The lymphocryptovirus (LCV) genus—to which Epstein-Barr virus belongs—appears to be restricted to primates. Evolutionary studies on LCV genes for the DNA-binding protein BALF2 and the fusion-mediating glycoprotein gB support the hypothesis that the viruses have evolved mainly through co-speciation with their hosts, with occasional horizontal transmission events, such as the apparent crossing of an ancestral LCV from macaques to hominoid apes (Orangutans; Gibbons) in Indonesia (Ehlers et al. 2003, 2010). Of the primate LCVs, only two—the Macacine herpesvirus 4 (better known as the rhesus lymphocryptovirus (RhLCV) as it was isolated from immunosuppressed rhesus macaques) and Callitrichine herpesvirus 3 (CaHV3, isolated from a common marmoset)—have been sequenced in full. *EBNA3A* and *EBNA3C* have also been cloned and sequenced from Papiine herpesvirus 1 (paHV1, host: Baboon). The *EBNA3* gene sequences of these herpesviruses are among the most divergent of all the LCV genes. The EBNA3s of paHV1 and RhLCV have only 30–50 % amino acid sequence conservation with their EBV homologues, and this is predominantly in the N-terminal half of the protein that includes the so-called homology domain (see Sect. 2.4 and Fig. 2) and—for EBNA3C—the terminal 100 amino acids (Jiang et al. 2000; Zhao et al. 2003). Initial analysis of the more distantly related CaHV3 genome—originating from a new world primate—failed to identify any *EBNA3* homologues. Only by determining the DNA sequence corresponding to the latency-associated transcripts of CaHV3 was the *EBNA3* paralogue identified, through the conserved transcript structure—an exon pair downstream of the EBNA-LP-like repeats. However, CaHV3 has only a single EBNA3 homologue, supporting the long-held supposition that the *EBNA3s* of EBV arose through gene duplication events in an ancestral virus with a single *EBNA3* gene (Rivailler et al. 2002). Functionally, the LCV EBNA3s share many properties of human EBV, such as the ability to compete with EBNA2 to suppress gene activation, and to bind RBP-JK/CBF1 (Jiang et al. 2000; Zhao et al. 2003; see Sects. 2.5 and 2.7.1). However, they are sufficiently distinct that the RhLCV EBNA3s could not support sustained transformation of human B cells when genetically substituted within the P3HR1 strain of EBV (Jiang et al. 2000).

## 2.4 Limited Homology but Shared Structural Features Between the EBNA3 Proteins

Immunofluorescence staining of cells ectopically expressing each of the EBNA3s revealed an exclusively nuclear distribution that at low expression levels can be speckled or punctate (particularly EBNA3C) and may spare the nucleolus (Allday et al. 1988; Hennessy et al. 1985; Krauer et al. 2004b; Petti et al. 1988, 1990;

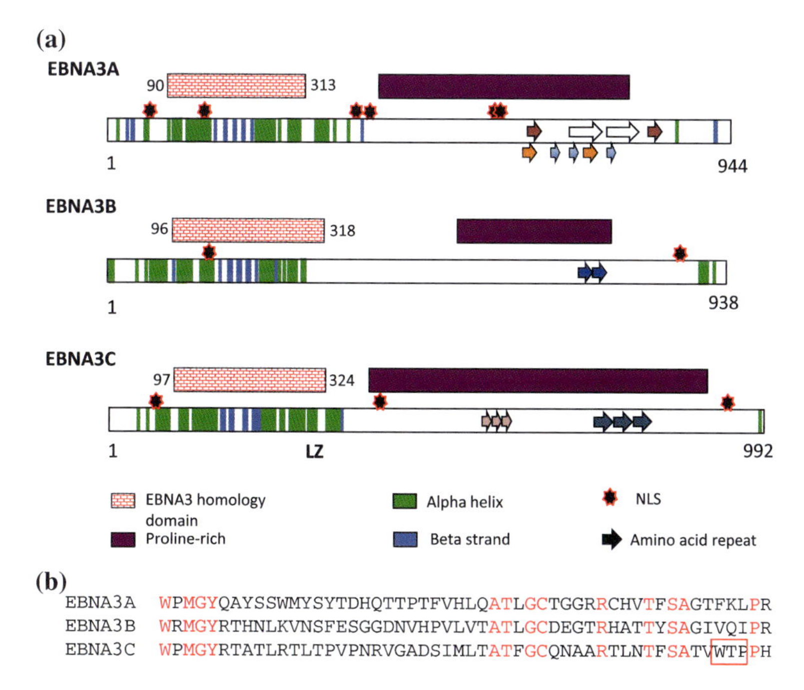


**Fig. 2** The EBNA3 family of proteins. **a** Schematic representation of the EBNA3 domains. The secondary structures of the EBNA3 proteins have been predicted using the **phyre²** protein fold recognition server. Domains and structural motifs are shown by the *filled green* and *blue rectangles* as indicated. The locations of nuclear localisation signals (NLS) that have been demonstrated in each protein are represented by *stars*. Similar *coloured arrows* represent similar amino acid repeats, and regions rich in *proline* are also indicated. The position of the putative leucine zipper (LZ) in EBNA3C is shown. These schematics are not drawn accurately to scale. **b** The central homology domain of the EBNA3s. Homologous amino acids are shown in *red*. The red rectangle highlights the WΦP motif of EBNA3C

Ricksten et al. 1988; Young et al. 2008 and our unpublished results). This is of no great surprise since it has been demonstrated that each of these large proteins (all are >900 aa) includes multiple nuclear localisation signals [NLS, (Fig. 2a)]: EBNA3A contains six NLSs (Buck et al. 2006; Le Roux et al. 1993), EBNA3B at least two (Burgess et al. 2006) and EBNA3C includes three (Allday et al. 1988; Krauer et al. 2004a). As well as having an exclusively nuclear distribution, the EBNA3s also associate tightly with chromatin and/or the nuclear matrix, even under high salt extraction conditions, but appear not to bind directly to DNA (Petti et al. 1990; Sample and Parker 1994 and our unpublished results). In LCLs, the proteins are remarkably stable, with a half-life of at least 24 h, and although it has been suggested that they are probably subject to ubiquitin-mediated proteosomal degradation in the production of peptides recognised by cytotoxic T cells (CTL) (Hislop et al. 2007), there is also evidence that the EBNA3s can directly bind to components of the 20S proteasome prior to degradation (Touitou et al. 2005).

Although the primary amino acid sequence of the EBNA3 proteins has diverged considerably during the course of their evolution from a common ancestor, the three proteins include an 'homology domain' near their N-terminus (Fig. 2). Within this region, EBNA3A and 3B share 28 % amino acid identity, EBNA3A and 3C share 23 %, and EBNA3B and 3C share 27 % identity (Le Roux et al. 1994). Comparative informatics analyses of the three proteins suggest that despite the very modest overall homology, there are similarities in their predicted secondary structures. Indeed, the predicted percentage similarity in secondary structure is 88 % between EBNA3A and EBNA3B, 72 % between EBNA3A and EBNAC and 72 % between EBNA3B and EBNA3C (Yenamandra et al. 2009). Predicted structural motifs are mainly alpha-helices, with few beta-strands, and are distributed primarily towards the N-terminus in the homology domain (Fig. 2a). Large regions of all three proteins have no obvious secondary structure. Other similarities between all three proteins include a proline-rich region (Fig. 2a) and repeat sequences as discussed in Sect. 2.2. A predicted structural feature that appears to be unique to EBNA3C, but for which no function has been established, is a leucine zipper (LZ (aa244-291) embedded within the homology domain (West 2006; West et al. 2004; Fig. 2a). Because of their large size and the prediction that much of each protein is disordered or flexible, structural analysis of the EBNA3s is going to be difficult. It is unlikely to be possible until functional domains can be precisely delineated, expressed and purified.

## *2.5 Relationship to EBNA2 and RBP-JK/CBF1—Initial Clues that the EBNA3s Are Regulators of Transcription*

When EBNAs 3A, 3B and 3C were identified as nuclear proteins and mapped to the BamHI-E region of the EBV genome (Allday et al. 1988; Hennessy et al. 1985, 1986; Petti and Kieff 1988; Petti et al. 1988; Ricksten et al. 1988), the EBNA2 protein was already characterised as an essential transactivator of viral and cellular genes during B cell transformation. It was soon established that although EBNA2 does not bind DNA directly, it can be targeted to genomic response elements by binding to cellular DNA sequence-specific transcription factors including RBP-JK [also called C promoter binding factor (CBF1)]; EBNA2 then recruits multiple co-activators of transcription to these binding sites (Zimber-Strobl and Strobl 2001). Subsequently, all three EBNA3 proteins were shown to bind to the same site on RBP-JK/CBF1 as EBNA2 (Robertson et al. 1995, 1996) and that EBNA3C can probably bind PU.1/SPI1 (Jimenez-Ramirez et al. 2006; Zhao and Sample 2000). When it was shown that, in transient reporter assays, each of the EBNA3s could inhibit EBNA2-mediated activation of viral promoters (LMP2A (aka TP-1) and Cp), it was proposed that all the EBNA3s might act in a regulatory loop as functional antagonists negatively regulating all genes activated by EBNA2 (Johannsen et al. 1996; Le Roux et al. 1994; Radkov et al. 1997; Robertson et al. 1995). For example, in the case of the EBV latency III promoter Cp, this would be a mechanism to prevent uncontrolled transactivation of the EBNA transcription unit and perhaps be involved in promoter switching during differentiation of B blasts to germinal centre B cells in vivo (Kis et al. 2010; Nagy et al. 2012; Thorley-Lawson and Allday 2008). Consistent with EBNA3s acting as repressive antagonists of EBNA2-mediated transactivation, it was discovered that when EBNA3A or EBNA3C was targeted directly to a reporter gene—by fusion to a GAL4 DNA-binding domain—both proteins exerted very robust repressor activity (Bain et al. 1996; Bourillot et al. 1998; Cludts and Farrell 1998). We have subsequently found that a Gal4-EBNA3B fusion has similar repressor activity (P. Young, PhD thesis, 2007, Imperial College London). Further support for the model came when it was shown that EBNA3A and EBNA3C interact with various repressive transcription co-factors including histone deacetylases (HDACs) HDAC-1 and HDAC-2, C-terminal binding protein (CtBP), Sin3A and NCoR (Hickabottom et al. 2002; Knight et al. 2003; Radkov et al. 1999; Touitou et al. 2001). However, the picture of EBNA3-mediated gene regulation became more complicated when it was shown in transient reporter assays that—in some circumstances—rather than exhibiting repressor activity, EBNA3C can activate transcription from viral (e.g. *LMP1* Jimenez-Ramirez et al. 2006; Zhao and Sample 2000) and cellular gene promoters (e.g. *COX-2* Kaul et al. 2006). This might be associated with its reported capacity to form complexes with transcriptional co-activators such as the histone acetyltransferase p300 (Cotter and Robertson 2000). Subsequent microarray studies have revealed the activation, as well as repression, of host genes by EBNA3A, EBNA3B and EBNA3C; so these proteins might all act as transactivators in some circumstances (see Sects. 3.2 and 4). However, in general, the molecular mechanisms of EBNA3-mediated transactivation are poorly understood.

Although there appear to be many sites across the human genome where EBNA2 and EBNA3s can apparently co-localise (see Sects. 3 and 4), it is still not known how widely EBNA2 and the EBNA3s have antagonistic roles in the regulation of host genes. Furthermore, no formal proof has been produced that the EBNA3s are involved in the negative regulation or silencing of the latency-associated EBNA promoter Cp, via RBP-JK/CBF1 response elements during viral infection and persistence. It is probable that the interaction of EBNA3s with RBP-JK/CBF1 plays an important role in the regulation of multiple genes, but until recently, no bona fide cellular targets that depend on this interaction had been identified (see Sects. 2.7.1 and 4.5). Even now, although genetic studies strongly indicate that the EBNA3A and EBNA3C binding to RBP-JK/CBF1 are important for B cell transformation (Lee et al. 2009; Maruo et al. 2005, 2009), the critical targets remain elusive and the molecular details of both protein:protein and protein:DNA interactions at regulatory loci are only now being explored (for more details, see Sects. 2.7.1 and 4).

## 2.6 Roles of EBNA3A and EBNA3C in Cell Cycle Regulation and as Potential Oncogenes

The initial indication that EBNA3C has activities associated with the cell cycle came from a study of its ectopic expression in Raji BL-derived cells that carry a virus with a deletion of *EBNA3C*. Complementation of the deleted gene by stable, constitutive expression from a transfected plasmid revealed that EBNA3C influences LMP1 expression only in G1 of the cell cycle (Allday and Farrell 1994). Although the basis of this phenotype was never established, it suggested that EBNA3C might have a G1-related activity. Shortly, thereafter EBNA3C was shown to ‘cooperate’ with oncogenic mutant Ras (Ha-Ras) in the immortalisation and transformation of primary rat embryo fibroblasts (REFs, Parker et al. 1996). In this type of assay, Ha-Ras alone induces exit from the cell cycle to a state called ‘premature senescence’ (Serrano et al. 1997) and other oncogenes that cooperate with Ha-Ras (e.g. adenovirus E1A; papillomavirus E7; SV40 TAg; cMyc) all subvert components of the G1 checkpoint modulated by the RB/p53 axis (reviewed in Lowe et al. 2004; Sherr 2012). This, therefore, indicated that EBNA3C might possess a similar anti-senescence activity in REFs, and perhaps also in B cells. The demonstration several years later that this activity depends on EBNA3C binding the transcriptional co-repressor CtBP (see Sect. 2.7.2) was consistent with the anti-senescence phenotype—at least in part—being linked to EBNA3C-mediated repression of transcription. The most direct and compelling evidence that EBNA3C modulates a G1 arrest checkpoint in B cells was to come from a study using an LCL established with a recombinant Akata EBV encoding a conditional EBNA3C (Maruo et al. 2006). Here, removing the inducer of EBNA3C activity from the cells resulted in the accumulation of p16$^{INK4a}$ mRNA and protein, accompanied by a substantial reduction in proliferation and considerable cell death. The molecular details and significance of these observations are discussed in detail in Sects. 4.2 and 5.

Additional evidence that EBNA3C can disrupt the regulation of the cell cycle came when it was shown that, in both rodent and human cells, constitutive overexpression of EBNA3C could induce aberrant cell division resulting in multi-nucleation, polyploidy and eventually cell death. EBNA3C also appeared to suppress the pro-metaphase arrest induced by drugs such as nocodazole that activate the mitotic spindle checkpoint (Parker et al. 2000). Together, these results suggested that unregulated expression of EBNA3C might disrupt cell cycle checkpoints occurring after DNA synthesis (S), perhaps in both G2 and mitosis. While these results raised interesting questions about the normal physiological role of EBNA3C in B cell proliferation, to date there have been no convincing genetic studies that have confirmed roles for EBNA3C in G2 or mitosis nor biochemical studies that have identified robust molecular interactions explaining these over-expression of phenotypes. A reported down-regulation of CHK2 by EBNA3C and interaction between the two proteins could have a role in facilitating the transition from G2 to mitosis (Choudhuri et al. 2007), but this has not been confirmed and it does not explain why, in LCLs, CHK2 is expressed at the same level as in mitogen-activated B cells and appears to function normally following DNA damage (O'Nions et al. 2006). Although it has been reported that EBNA3C might repress the transcription of the mitotic regulator BUBR1 in BJAB cells, this has not been extended to LCLs nor confirmed with EBV recombinants (Gruhne et al. 2009; Skalska et al. 2013 and our unpublished data). Several reports during the past decade indicate that EBNA3C can physically associate with a variety of other factors involved in the regulation of cell cycle progression and/or the G1/S checkpoint—most of these appear to be via amino acids located at the N-terminus of EBNA3C (aa100–200). The factors include the ubiquitin ligase $SCF^{SKP2}$, the tumour suppressor retinoblastoma (RB), the oncoprotein MYC, MDM2, p53, cyclin A, cyclin D1, E2F1, CHK2 and Aurora kinase B (Bajaj et al. 2008; Choudhuri et al. 2007; Jha et al. 2013; Kashuba et al. 2008; Knight and Robertson 2004; Knight et al. 2004, 2005a, b; Saha et al. 2009, 2011, 2012; Yi et al. 2009). It remains to be established, by reverse genetic studies of mutants that specifically and independently disrupt these binding sites, which of these interactions are functionally important in vitro during B cell transformation or in vivo during the establishment of persistence and/or in B cell lymphomagenesis.

EBNA3A was also shown to cooperate with Ha-Ras in the transformation and immortalisation of REFs, and again, there was a remarkably good correlation between EBNA3A binding to CtBP and rescue from Ha-Ras-induced senescence (Hickabottom et al. 2002). The earliest evidence that EBNA3A might affect cell cycle regulation in B cells came from two related studies in which EBNA3A was either over-expressed (by twofold–fivefold) in an LCL from an inducible plasmid (Cooper et al. 2003) or—as with EBNA3C—using a recombinant virus conditional for EBNA3A to establish an LCL (Maruo et al. 2003). The take-home message from these experiments was that a precisely controlled level of EBNA3A was critical for optimum LCL proliferation. An excess of EBNA3A appeared to produce the down-regulation of MYC and cyclin D2 leading to a prolonged G0/G1 cell cycle arrest, but when EBNA3A was inactivated in the conditional LCL, the outcome was a gradual decline in proliferation; however, the underlying molecular mechanisms were not identified (Cooper et al. 2003; Maruo et al. 2003). Also, consistent with EBNA3A having a role in cell cycle regulation, was the recent report of an interaction between EBNA3A and MYC-interacting zinc-finger protein-1 (MIZ1) being necessary for the down-regulation of the cyclin-dependent kinase inhibitor $p15^{INK4b}$ in LCLs. Because $p15^{INK4b}$ has the potential to inhibit cell cycle progression, this could contribute to the maintenance of EBV-infected B cell proliferation. However, currently—since the role of $p15^{INK4b}$ in human B cells is unknown—the biological significance in EBV biology is still subject to speculation (Bazot et al. 2014; also see Sect. 4.2 for further discussion).

## 2.7 Proteins that Can Interact with the EBNA3C

EBNA3C is by far the most studied member of the EBNA3 family and has been reported to interact with many cellular proteins. In most cases, this refers to the ability of a protein to co-immunoprecipitate with EBNA3C from EBV-infected B cells (or after co-transfection) using standard immunoprecipitation lysis buffer extraction and/or the ability to associate with EBNA3C in pull-down assays using polypeptides fused to glutathione-s-transferase (GST). To our knowledge, there is no case of a direct biochemical interaction being confirmed using proteins purified to homogeneity. Some proteins reported to associate with EBNA3C by one or both of the above criteria are listed in Table 1; however, due to constraints on the length of the review, only the interactions between EBNA3C and RBP-JK/CBF1 and CtBP will be considered in more detail. This is because (a) these two proteins associate with more than one EBNA3 (RBP-JK/CBF1 with all three; CtBP with EBNA3A and EBNA3C), (b) these are the only interactions for which essential EBNA3C amino acids have been precisely identified and—most importantly—(c) have binding mutants been analysed by reverse genetics or complementation to show a change in phenotype in EBV-infected B cells. It is perhaps no coincidence that these two proteins have also consistently been detected as 'interactors' in various EBNA3 yeast two-hybrid screens (e.g. Bazot et al. 2014; Lin et al. 2002; our unpublished results). It should be noted that EBNA3C can be co-immunoprecipitated with both EBNA3A and EBNA3B (Paschos et al. 2012; our unpublished results); however—although there is good evidence for functional crosstalk between these proteins (see Sects. 3.2 and 4)—the critical domains/residues have not yet been mapped; therefore, genetic analyses to test the significance of these potential biochemical associations have not yet been possible.

### 2.7.1 RBP-JK/CBF1

As indicated previously, EBNA2 and all three EBNA3 proteins can bind the conserved DNA-binding factor RBP-JK/CBF1, which is also an effector component of the Notch signalling pathway and is equivalent to *Drosophila melanogaster* Suppressor of Hairless, Su(H). In *Drosophila*, Notch pathways regulate cell fate determination, cell differentiation and developmental pattern formation. The pathways are highly conserved and are thought to play similar roles in human cells (Guruharsha et al. 2012; Zimber-Strobl and Strobl 2001). Once liberated from cellular membranes by proteolytic cleavage, the cleaved domains of Notch (Notch-IC, for intracellular) can bind to Su(H) to activate gene transcription. This process can be antagonised in *Drosophila* by competitive binding of Su(H). Hairless protein then itself recruits conserved co-repressors Groucho and dCtBP (Barolo et al. 2002; Morel et al. 2001). In EBV-transformed B cells, EBNA2 can mimic aspects of activated Notch signalling by binding—like Notch-IC—to RBP-JK/CBF1, recruiting co-activators and enhancing transcription of both viral and cellular genes that include RBP-JK/CBF1-binding sequences. In *Drosophila,* Notch signalling is antagonised by Hairless, so it has been proposed that EBNA3A, EBNA3B and EBNA3C could function in a similar manner to Hairless and repress activation of transcription mediated by EBNA2 bound to RBP-JK/CBF1 (Thorley-Lawson and Allday 2008; Zimber-Strobl and

**Table 1** Factors reported to interact with EBNA3C

| Interacting cellular proteins[a] | Proposed outcome | Genetic confirmation using recombinant EBV in B cells | References |
|---|---|---|---|
| RBP-JK (CBF1) | Directing EBNA3C to a subset of target genes | Yes | Robertson et al. (1996); Maruo et al. (2009); Lee et al. (2009); Calderwood et al. (2011); Harth-Hertle et al. (2013) |
| SPI1/PU1 | Directing EBNA3C to a subset of target genes | Indirect evidence | Zhao and Sample (2000); Jiang et al. (2014); McClellan et al. (2013) |
| CtBP | Repression of some target genes (e.g. $p16^{INK4a}$) | Yes | Touitou et al. (2001); Skalska et al. (2010) |
| HDACs 1/2 | Repression of target genes | Indirect evidence | Radkov et al. (1999); Knight et al. (2003); Paschos et al. (2009) |
| Sin3A | Repression of target genes | Indirect evidence | Knight et al. (2003); Jiang et al. (2014) |
| NCoR | Repression of target genes | No | Knight et al. (2003) |
| SUMO-1/3 | Sumoylation of EBNA3C? | No | Rosendorff et al. (2004); Touitou et al. (2005) |
| Cyclin A | Increased cdk activity | No | Knight et al. (2004); Knight and Robertson (2004) |
| Cyclin D1 | Cyclin D1 stabilised[b] | No | Saha et al. (2011) |
| $SCF^{SKP2}$ | Recruited to RB | No | Knight et al. (2005b) |
| RB | RB degraded and release of E2F1 | No | Knight et al. (2005a) |
| MYC | MYC stabilised | No | Bajaj et al. (2008) |
| p300 | Recruited in a transcription complex | No | Cotter and Robertson (2000) |
| Prothymosin-α | Recruited in a transcription complex | No | Cotter and Robertson (2000) |
| MRS18-2 | Releases E2F1 from RB | No | Kashuba et al. (2008) |
| NM23-H1 | Recruited to nucleus | No | Subramanian et al. (2001) |
| MDM2 | MDM2 stabilised | No | Saha et al. (2009) |
| ING4/5 | Blocks interaction of ING4/5 with p53 | No | Saha et al. (2011) |
| p53 | Inhibits p53-mediated transcription | No | Yi et al. (2009) |

(continued)

**Table 1** (continued)

| Interacting cellular proteins[a] | Proposed outcome | Genetic confirmation using recombinant EBV in B cells | References |
|---|---|---|---|
| CHK2 | Blocks CHK2 kinase activity, might restrict DDR | Indirect evidence | Choudhuri et al. (2007)<br>Nikitin et al. (2010) |
| GADD34 | Counteracts unfolded protein response | No | Garrido et al. (2009) |
| DP103 (Gemin3) | DP103 degraded | No | Cai et al. (2011) |
| Aurora Kinase B | Aurora B stabilised | No | Jha et al. (2013) |
| E2F1 | Inhibits E2F1 binding to DNA | No | Saha et al. (2012) |
| IRF4 | IRF4 stabilised, perhaps recruits EBNA3C to DNA | Indirect evidence | Banerjee et al. (2013); Jiang et al. (2014); McClellan et al. (2013) |
| H2AX | Relocalisation of H2AX, might restrict DDR | Indirect evidence | Jha et al. (2014), Nikitin et al. (2010) |
| EBNA3A | Collaboration with EBNA3C in gene regulation | Indirect evidence | Calderwood et al. (2007); Anderton et al. (2008); Skalska et al. (2010); White et al. (2010); Paschos et al. (2012); McClellan et al. (2013) |
| EBNA3B | Collaboration with or antagonism of EBNA3C in gene regulation | Indirect evidence | White et al. (2010); White et al. (2012); McClellan et al. (2013); our unpublished data |

[a]In some cases, this is probably direct, but for others, the interaction might involve a multi-protein complex—none of these interactions has been confirmed with purified proteins
[b]It should be noted that Cyclin D1 is not generally expressed in EBV-transformed B cells Palmero et al. (1993) and Pokrovskaja et al. (1996)

Strobl 2001). Furthermore, since EBNA3A and EBNA3C can recruit mammalian CtBP (discussed in the next section), EBV might mimic several components of the Notch effector pathway to precisely control the expression of host genes in an analogous manner to Notch in *Drosophila*. However, despite the functional similarities between Notch-IC/Hairless and EBNA2/3s, it has been shown that EBNA2 and Notch-IC are not interchangeable in regulating human genes (Kohlhof et al. 2009)—this is discussed further in the chapter authored by Kempkes and Ling.

Although there is not complete agreement on the precise details, there is an emerging consensus that the most important region of each EBNA3 in the interaction with RBP-JK/CBF1 is located in the middle of the homology domain (central homology domain) between amino acid 170-221 for EBNA3A, 176-227 for EBNA3B and 180-231 for EBNA3C (Fig. 2; Bourillot et al. 1998; Calderwood et al. 2011; Dalbies-Tran et al. 2001; Lee et al. 2009; Maruo et al. 2005, 2009; Robertson et al. 1996). EBNA3A and EBNA3C mutants deleted for those regions no longer interact with RBP-JK/CBF1 and fail to repress EBNA2-mediated transcriptional activation of reporter genes (Maruo et al. 2005, 2009). It should be noted, however, that mutations near, but not within, the central homology region might also affect the interaction between the EBNA3s and RBP-JK/CBF1 (Maruo et al. 2005; West et al. 2004). Sequence alignment of the EBNA3 central homology regions identified three amino acid clusters conserved between the EBNA3s (Fig. 2). A refined analysis of RBP-JK/CBF1 binding to EBNA3C performed by Calderwood and colleagues revealed that residues 211–233 of EBNA3C include an amino acid sequence (WTP) that resembles the WΦP motif of Notch-IC (WFP) and EBNA2 (WWP) that interacts with RBP-JK/CBF1 (Calderwood et al. 2011). They showed using EBNA3C mutated at ATFGC and/or WΦP that both regions are involved in the interaction with RBP-JK/CBF1 (Fig. 2).

An important observation that linked EBNA3:RBP-JK/CBF1 complexes to EBV biology was the demonstration that the interaction between RBP-JK/CBF1 and EBNA3A or EBNA3C is not only necessary for the regulation of EBNA2-mediated activation in reporter assays, but also essential for maintaining LCL proliferation. The ectopic expression of wild-type EBNA3A can maintain proliferation in an LCL conditional for EBNA3A when it is cultured in non-permissive conditions, whereas an EBNA3A mutant deleted for RBP-JK/CBF1 binding cannot (Maruo et al. 2005). In comparable complementation experiments using an EBNA3C-conditional LCL, the results were similar: no interaction with RBP-JK/CBF1 and no rescue of proliferation (Lee et al. 2009; Maruo et al. 2009). So it appears that EBNA3A and EBNA3C mutants that are deficient in their capacity to repress EBNA2-mediated activation of reporters are also unable to sustain LCL proliferation. However, it is not known whether these two functions are directly linked or precisely how EBNA3A and EBNA3C (and possibly EBNA3B) antagonise gene activation by EBNA2. Two mechanisms have been proposed, but neither confirmed: either the EBNA3s might destabilise the interaction of EBNA2:RBP-JK/CBF1 complexes with DNA (Robertson et al. 1995, 1996; Waltzer et al. 1996) or alternatively the EBNA3s could replace EBNA2 on DNA-bound RBP-JK/CBF1 and recruit co-repressors (see next section and Sect. 4). It is possible that both of these mechanisms could operate, but perhaps at different targets. These issues will only be resolved when the details of molecular complexes at the regulatory elements of fully validated target genes have been determined. Remarkably, it appears that only a few (about 16 %) of the many thousands of sites across the human genome that bind EBNA3s have been shown to coincide with reported RBP-JK/CBF1 sites. However, many coincide with sites where EBNA2 can bind (Jiang et al. 2014; McClellan et al. 2013; see Sect. 4 for more detailed discussion).

### 2.7.2 CtBP

EBNA3A and EBNA3C, but not EBNA3B, interact with the co-repressor CtBP (Hickabottom et al. 2002; Touitou et al. 2001), a protein initially discovered as a cellular factor interacting with the C-terminal region of the adenovirus E1A oncoprotein and subsequently identified as one of a highly conserved family of co-repressors of transcription (reviewed in Chinnadurai 2007). CtBP is now a generic term used to refer to two proteins, CtBP1 and CtBP2. These closely related regulators of transcription are encoded in vertebrates by separate genes (located on human chromosomes 4p16.3 and 10q23.13, respectively). While CtBP1 and CtBP2 share significant amino acid homology overall, CtBP2 includes a unique N-terminal domain that probably contributes to its nuclear localisation (Chinnadurai 2007; Zhao et al. 2014). It remains largely unexplored to what extent these proteins are functionally redundant, but there is evidence that at some developmental stages, they might play unique roles (Chinnadurai 2007). Both appear to act primarily as transcriptional co-repressors in the process of gene silencing and both interact with factors that have the conserved, prototypical CtBP-binding Pro-Leu-Asp-Leu-Ser ('PLDLS') motif. Proteomic analysis of mammalian CtBP1 and CtBP2 complexes has shown each can interact with a similar array of transcriptional co-repressors including HDAC-1 and HDAC-2, CoREST, G9a/GLP and LSD1; they have also been implicated in polycomb group (PcG) protein-mediated repression (Sewalt et al. 1999; Shi et al. 2003, 2004; Sundqvist et al. 1998; Zhao et al. 2014). Genetic evidence indicates that CtBP2 might also activate transcription in a context-dependent manner—this could account for some of its unique properties (Chinnadurai 2007). Since both CtBP1 and CtBP2 bind NAD+ and NADH, it has been proposed that their transcriptional regulatory capacity is modulated by the ratio of nuclear NAD+:NADH and that they play a role in monitoring the redox status of a cell (Zhang et al. 2006).

Touitou and colleagues identified a perfect PLDLS motif in the C-terminal region of EBNA3C and demonstrated that this motif (aa 728-732) was essential and sufficient for EBNA3C to interact with CtBP1 (Touitou et al. 2001). Subsequently, EBNA3A was shown to interact with CtBP1 through two non-consensus motifs, ALDLS (aa 857-861) and VLDLS (aas 891-895), also located at the carboxyl terminus. In vitro studies suggested that EBNA3A binds with a higher affinity than EBNA3C—perhaps because it includes the bipartite site—but this has not been confirmed (Hickabottom et al. 2002). Because EBNA3C and EBNA3A interact with CtBP1 via PLDLS (or a variant), it has been assumed that they can interact with both CtBP proteins, but—to our knowledge—this has not been formally tested. Although both CtBPs could be present simultaneously in human B cells, most available data suggest that CtBP2 is not expressed in EBV-infected B cells and this silencing is probably related to the expression of EBNA3A (Hertle et al. 2009; McClellan et al. 2012; White et al. 2010) (see Sect. 4.5 for a more detailed discussion).

The ability of EBNA3C to bind CtBP correlates only partially with the ability of EBNA3C to repress transcription when targeted to DNA in transient reporter assays, but correlates extremely well with EBNA3C's ability to behave as a cooperating nuclear oncoprotein when expressed in primary REFs with oncogenic Ha-ras (Touitou et al. 2001). Binding of CtBP to EBNA3A correlates well with both reporter repression and oncogenic cooperation (Hickabottom et al. 2002). Furthermore, since Marek's disease virus—a herpesvirus that induces T cell lymphoma in poultry—requires its nuclear oncoprotein MEQ to bind chicken CtBP for tumorigenesis (Brown et al. 2006), it seems likely that the interactions with EBNA3A and EBNA3C might be involved in human B cell transformation and lymphomagenesis. The importance of these EBNA3:CtBP interactions in B cell transformation became apparent when analysis of $p16^{INK4a}$ expression in LCLs established by using each of three different CtBP-binding mutant viruses ($3A^{CtBP}$; $3C^{CtBP}$ and a 3A/3C double CtBP-binding mutant called $E3^{CtBP}$) showed that CtBP is involved in the EBNA3A/3C-mediated epigenetic repression of the $p16^{INK4a}$-encoding gene locus (Skalska et al. 2010 and discussed in more detail in Sect. 4.2). Taken together, the data suggest that binding of EBNA3A and EBNA3C to CtBP augments transformation efficiency and LCL outgrowth at least in part by aiding the establishment or maintenance of repressive chromatin around the $p16^{INK4a}$ transcription start site (TSS). Both EBNA3A and EBNA3C can be readily immunoprecipitated from LCLs with CtBP (Hickabottom et al. 2002; Touitou et al. 2001); however, no EBNA3A/3C-CtBP complexes have been demonstrated on the $p16^{INK4a}$ promoter (Skalska et al. 2010). It is unclear whether this is because of technical limitations of the available reagents or whether the promoter is not actually a direct target of such complexes. Although we do not yet understand this requirement for CtBP binding, the data are consistent with reports that the C-terminus of EBNA3C—and specifically the PLDLS CtBP-binding site—is necessary to completely rescue proliferation in EBNA3C-conditional LCLs cultured without the appropriate activator (Lee et al. 2009; Maruo et al. 2009).

# 3 New Technologies and the Analysis of EBNA3 Function

## *3.1 BAC-Derived EBV Recombinants Facilitate Robust Genetic Analysis*

The earliest strategy for the genetic analysis of EBV functions entailed recombination between the non-transforming P3HR1 strain, which lacks EBNA2, and fragments of B95-8 virus. This system relied on selecting recombinants that had incorporated B95-8 EBNA2 that enabled the virus to transform B cells. Inclusion of second site targeting constructs induced a remarkably high proportion (around 10–15 %) of selected virus clones that also incorporated this second change, thereby allowing the modification of other EBV genes under selection of EBNA2 inclusion. Using this strategy, it was found that incorporating a type 1 region including the EBNA3s in place of the type 2 EBNA3s of P3HR1 had no significant effect on the virus's ability to transform B cells (Tomkinson and Kieff 1992b). Parallel attempts to generate EBNA3 knockouts—by incorporating stop codons into each EBNA3 independently within the type 1 EBNA3 cassette—found that EBNA3B knockouts could be generated with the same frequency as incorporation of the intact type 1 cassette (Tomkinson and Kieff 1992a). In contrast, recombinants containing either EBNA3A- or EBNA3C-knockout DNA grew out in <2 % of transformants, and then only where cells were co-infected with the non-transforming P3HR1 virus. In addition, these dually infected cells were prone to spontaneous reversion of the mutation through recombination with the complementing P3HR1 virus (Tomkinson et al. 1993). These data were interpreted to mean that EBNA3A and EBNA3C are essential for B cell transformation, whereas EBNA3B is dispensable.

The requirement for transformation- and replication-competent recombinant viruses, and the technical challenges of the first-generation recombination system, limited its use as a tool for the genetic analysis of EBV. However, cloning the genome of the B95-8 strain of EBV into an F factor-derived plasmid (Delecluse et al. 1998)—more commonly known as a bacterial artificial chromosome (BAC)—facilitated the generation and characterisation of EBV recombinants that lacked the ability to transform B cells (reviewed in detail in Feederle et al. 2010 and the chapter authored by H-J. Delecluse). Subsequently, the Akata strain of EBV was also cloned as a BAC, and both strains have now been used to analyse the functions of the EBNA3s. Two main modification strategies have been used to reveal functional contributions of the EBNA3s to the biology of EBV in B cells: first, introducing truncating mutations into the EBNA3s to create EBNA3 knockouts (KO) and second, creating proteins conditional for function by fusing EBNA3s to a modified oestrogen receptor (ER) in the viral genome. The resulting fusion protein is then functional only in the presence of oestrogen or its related analogue, 4-hydroxytamoxifen (4HT) that prevents the sequestration and degradation of the protein that would otherwise be mediated by the inactive ER (Maruo et al. 2006; Skalska et al. 2010). More recently, fusions of EBNA3s with epitope tags have proved to be very useful in locating sites to which individual EBNA3s are targeted on host chromatin using chromatin immunoprecipitation (ChIP) strategies (Jiang et al. 2014; Paschos et al. 2012; see below).

In the past, we and others have attempted to determine the roles of the EBNA3s in gene regulation and EBV biology by expressing single EBNA3 genes or cDNAs transiently, constitutively or from inducible vectors in B and non-B cells (for example Gruhne et al. 2009; Knight et al. 2005a; Parker et al. 2000; Yi et al. 2009; Young et al. 2008). However, while this approach provided some clues to EBNA3 functions, it was sometimes unhelpful and could potentially produce misleading data because the protein is generally over-expressed, often not in B cells and always in the absence of other EBV latency factors—that is out of its physiological context. Constructing recombinant viruses with mutations or deletions of the EBNA3 locus using EBV-BAC technology and using the resultant viruses to infect B cells has allowed the investigation of host phenotypes and transcriptomes associated with each EBNA3, in the context of latency-associated EBV gene expression in B cells. This strategy has already produced many surprising results and provided a wealth of information for future research.

### *3.2 Microarrays Reveal the Extent and Complexity of EBNA3-Mediated Regulation of Host Genes*

A second strategic advance in the study of EBNA3 function was the determination of genome-wide gene expression patterns using microarray technology in conjunction with the knockout, conditional or mutant EBNA3 viruses. To date, around 100 or more independent B cell lines have been analysed to produce unbiased views of how the host transcriptome is modified by EBNA3 expression (Chen et al. 2006; Hertle et al. 2009; Kelly et al. 2013; McClellan et al. 2012; Skalska et al. 2010; White et al. 2010; Zhao et al. 2011). Several of these data sets (derived from BL lines and LCLs in our and the Kempkes' laboratories) describing the expression of genes that are regulated by EBV in an EBNA3-dependent manner are available in a searchable format at the Website http://www.epstein-barrvirus.org.uk. Together, these analyses have revealed the extent and complexity of EBNA3-mediated host gene control. It is now well established that the regulation of well over a 1000 host genes by EBV requires EBNA3 expression, and a considerable number of these genes appear to be regulated by combinations of EBNA3s. Confirming this original observation, many genomic loci are bound by multiple EBNA3s, and also EBNA2, but surprisingly, relatively few of these locations are also bound by RBP-JK/CBF1 (see Sect. 3.3 for further discussion).

Using the EBV-BAC system and microarrays to analyse the regulation of host genes by EBNA3s has highlighted a number of underlying principles that should be considered when interpreting the gene expression analyses undertaken to explore the function of EBV proteins. First, using the same recombinant viruses in different cell backgrounds can give very different results. For instance, only a quarter of the genes identified as regulated by EBNA3B in LCLs were also EBNA3B regulated in the Burkitt's lymphoma cell line BL31 (White et al. 2010). There are many possible reasons for such differences. They may be due to the different differentiation states of the cell or changes associated with the mutations or other transformation events that produced the BL31 cell line. Since these analyses have been carried out on cell lines that have been established with mutant viruses, selection pressures during outgrowth can also distort the transcriptome. For instance, a loss/reduction of retinoblastoma protein (RB) expression appears to assist the outgrowth of EBNA3A-KO and CtBP-binding mutant LCLs (Hertle et al. 2009; Skalska et al. 2010). Also, differences in regulation of target genes due to host genetic background have also been observed. We, and others, have found that the regulation of CXCR4 by EBNA3B occurs only in a subset of donors (Chen et al. 2006; White et al. 2010). These issues are discussed in greater depth in the relevant reports, including Hertle et al. 2009; Kelly et al. 2013; McClellan et al. 2012; Skalska et al. 2010, 2013; White et al. 2010; and Zhao et al. 2011. Nevertheless, using microarray data as an indicator of potential EBNA3 targets—that are then rigorously validated—has proven to be remarkably fruitful, and existing data sets probably still have more to offer. Examples of some host genes that reveal different features of EBNA3-mediated gene regulation and provide novel insights into viral regulation of host transcription are considered in detail in Sect. 4 and are summarised in Table 2.

**Table 2** Host genes regulated by the EBNA3 family proteins

| Genes reported to be regulated by EBNA3s | EBNA3 binding to chromatin: ChIP | ChIPseq | Genetic confirmation: using recombinant EBV in B cells -KO | —conditional | Comments | References |
|---|---|---|---|---|---|---|
| *Alpha-V integrin*—repressed by EBNA3C | N/R | N/R | – | – | • Regulation involves NM23-H1, GATA1 and SPI1 | Choudhuri et al. (2006) |
| *COX-2*—activated by EBNA3C | N/R | N/R | – | – | • Regulation involves NM23-H1, GATA1 and SPI1 | Kaul et al. (2006) |
| *BIM (BCL2L11)*—repressed by EBNA3A and EBNA3C | + (proximal to TSS) | + | + | + | • Co-regulation by EBNA3A and EBNA3C<br>• Involves PcG complexes and H3K27me3 repressive chromatin mark<br>• Might involve inhibition of Pol II-$PO_4$ | Anderton et al. (2008); Paschos et al. (2009, 2012); White et al. (2010); McClellan et al. (2013); Jiang et al. (2014) |
| *TCL-1*—activated by EBNA3C plus EBNA3B | N/R | N/R | + | + | • Both EBNA3B and EBNA3C appear to be required for activation in LCLs<br>• Probably involves RBP-JK/CBF1 | Chen et al. (2006); Lee et al. (2009); White et al. (2010, 2012) |
| *p16$^{INK4a}$ (CDKN2A)*—repressed by EBNA3C (plus EBNA3A) | + (proximal to TSS) | + | + | + | • Co-regulation by EBNA3C and EBNA3A<br>• Involves polycomb protein complexes and H3K27me3 repressive chromatin mark<br>• Involves CtBP and might involve BATF/IRF4 or SPI1/IRF4 binding sites proximal to p14$^{ARF}$ TSS<br>• Whole p15$^{INK4b}$, p14$^{ARF}$, p16$^{INK4a}$ locus regulated coordinately? | Maruo et al. (2006); Hertle et al. (2009); Skalska et al. (2010, 2013); Zhao et al. (2011); Maruo et al. (2011); Jiang et al. (2014) |

(continued)

**Table 2** (continued)

| Genes reported to be regulated by EBNA3s | EBNA3 binding to chromatin: ChIP | ChIPseq | Genetic confirmation: using recombinant EBV in B cells -KO | —conditional | Comments | References |
|---|---|---|---|---|---|---|
| *p15*$^{INK4b}$ (*CDKN2B*)—repressed by EBNA3A | N/R | N/R | + | + | • Involves interaction between EBNA3A and the transcription factor MIZ1<br>• Probably co-regulated by EBNA3A and EBNA3C<br>• Involves PcG complexes and H3K27me3<br>• Whole p15$^{INK4b}$, p14$^{ARF}$, p16$^{INK4a}$ locus regulated coordinately? | Bazot et al. (2014); our unpublished results |
| *BUBR1*—repressed by EBNA3C | N/R | N/R | − | − | | Gruhne et al. (2009) |
| *p73*—repressed by EBNA3C | N/R | N/R | − | − | | Saha et al. (2012) |
| *ADAM28*—repressed by EBNA3A and EBNA3C | + (distal, intergenic) | + | + | + | • EBNA3 binding between *ADAM28* and *ADAMDEC1*—locus co-regulated by chromatin looping<br>• H3K27me3 involved | Hertle et al. (2009); McClellan et al. (2012, 2013) |
| *ADAMDEC1*—repressed by EBNA3A and EBNA3C | + (distal, intergenic) | + | + | + | • EBNA3 binding between *ADAM28* and *ADAMDEC1*—locus co-regulated by chromatin looping<br>• H3K27me3 involved | Hertle et al. (2009); McClellan et al. (2012, 2013); Skalska et al. (2013) |
| *IRF8*—repressed by EBNA3C | N/R | N/R | − | − | • EBNA3C reported to also bind IRF8 protein | Banerjee et al. (2013) |

(continued)

**Table 2** (continued)

| Genes reported to be regulated by EBNA3s | EBNA3 binding to chromatin: ChIP | ChIPseq | Genetic confirmation: using recombinant EBV in B cells -KO | —conditional | Comments | References |
|---|---|---|---|---|---|---|
| CtBP2—repressed by EBNA3A | + (intragenic enhancer) | + | + | – | • EBNA3A binding on intragenic enhancer and chromatin looping involved<br>• Co-regulation with EBNAs2, 3A and 3B that might bind the same site?<br>• Confirmed in EBNA3AKO LCL, but not EBNA3C-conditional LCL or EBNA3BKO LCL exon microarrays | Hertle et al. (2009); Skalska et al. (2013); McClellan et al. (2013) |
| *CXCL9/CXCL10*—repressed by EBNA3A | + (intergenic enhancer) | + | + | + | • *CXCL9* and *CXCL10* locus co-regulated<br>• EBNA3 binding on distal intergenic enhancer and RBP-JK/CBF1 involved<br>• Leads to H3K27me3 mark across locus<br>• Co-repression with EBNA3C<br>• EBNA2 activates, EBNA3B potentiates?<br>• Chromatin looping probably involved | Hertle et al. (2009); White et al. (2010, 2012); McClellan et al. (2012); Skalska et al. (2013); Harth-Hertle et al. (2013); our unpublished results |
| *Aurora kinase B*—repressed by EBNA3C | + (proximal to TSS?) | N/R | – | – | • EBNA3C reported to also bind Aurora kinase B protein | Jha et al. (2013) |
| *H2AX*—repressed by EBNA3C | N/R | N/R | – | – | • EBNA3C reported to also bind H2AX protein | Jha et al. (2014) |

N/R—Not reported

### 3.3 ChIP and ChIP-Seq Analyses and Mechanisms of Gene Regulation

The biological significance of interactions of nuclear proteins, such as the EBNA3s, with specific DNA fragments of defined sequence has been immensely enhanced by the advent of chromatin immunoprecipitation (ChIP) techniques. ChIP is a method by which proteins are selectively immunoprecipitated and the associated DNA sequences are determined—usually by methods involving quantitative PCR (QPCR) and/or high-throughput sequencing (ChIP-seq). This technique has been used to map the binding of transcription factors, co-factors and chromatin-modifying enzymes at particular loci and across the genome. It has also proven to be a very powerful tool for identifying the distribution of post-translationally modified histones and their spatial and functional relationship to individual genes.

Coupled with the use of EBV recombinants and transcriptome data, ChIP and ChIP-seq technologies have revolutionised the way we consider EBNA3 function. It has been possible to identify many thousands of specific genomic loci where the EBNA3s can be detected and, we assume, recruit cellular factors that exert effects on chromatin organisation. This has heralded some paradigm-shifting observations, including the EBNA3 manipulation of histone-modifying complexes to reprogram the epigenetic landscape and the modulation by EBNA3s of the three-dimensional architecture of chromatin during the regulation of gene expression (Harth-Hertle et al. 2013; Jiang et al. 2014; McClellan et al. 2012, 2013; Paschos et al. 2012; Skalska et al. 2013). Some of these studies will be considered in the next section.

## 4 Host Gene Regulation by the EBNA3s Involves Polycomb Proteins, Epigenetic Modifications and Chromatin Looping

Five host genes (or gene clusters) whose regulation by EBNA3s is particularly well characterised are described here in detail. Each reveals novel aspects of EBNA3 cooperation and/or mode of action in the regulation of transcription, and in some cases, how the identification of targets has provided remarkable insights into aspects of EBV biology and the virus's oncogenic potential.

### 4.1 BIM (BCL2L11)

The first evidence that EBNA3A and EBNA3C can cooperate to regulate specific host cell genes came using a panel of EBNA3-knockout recombinant

B95.8-derived EBVs to infect EBV-negative BL31 BL-derived cells. This revealed, among other things, that expression of both EBNA3A and EBNA3C is necessary to repress transcription of *BIM/BCL2L11* (Anderton et al. 2008).

BIM (Bcl2-interacting mediator) is a pro-apoptotic member of the BH3-only family of BCL2-like proteins and is encoded by the *BCL2L11* gene on human chromosome 2q13. BIM acts as a potent, direct initiator of apoptosis because it binds with high affinity to BCL2 and all the other pro-survival family members to inactivate them. BIM is particularly important in the immune system, acting as a major regulator of life-and-death decisions during lymphocyte development (Fischer et al. 2007; Strasser 2005). *Bim*-null mice accumulate excess lymphoid and myeloid cells, and loss of *Bim* accelerates B cell lymphomagenesis induced by an *Eμ-Myc* transgene in mice (Egle et al. 2004). The connection between deregulated *MYC* expression and the activity of BIM in human B cells was revealed by Hemann and colleagues (Hemann et al. 2005). They showed that MYC activates *BIM/BCL2L11* in EBV-negative BL, and this led to the proposal that MYC-induced apoptosis can be overridden by inactivation of any one of several MYC effectors—including p53, $p14^{ARF}$ or BIM (reviewed in Dang et al. 2005). MYC is induced and becomes constitutively expressed early after the infection of primary human B cells with EBV (Fig. 4 and Sect. 5); therefore, suppression of BIM expression by EBV is likely to be a contributory factor in B cell transformation and the development of any EBV-associated B cell lymphomas.

In cultured B cells, a reduction in BIM expression occurs very soon after infection with EBV (Anderton et al. 2008; Skalska et al. 2013). This did not involve detectable CpG methylation, but correlated with loss of histone acetylation and the deposition of the polycomb group (PcG) protein signature H3K27me3 on chromatin proximal to the transcription start site (TSS) for BIM (Paschos et al. 2009, 2012). Detailed chromatin immunoprecipitation (ChIP) analyses of the chromatin around the *BIM/BCL2L11* promoter revealed that latent EBV triggers the recruitment of polycomb repressive complex (PRC)2 core subunits and the trimethylation of histone H3 lysine 27 (H3K27me3) at this locus and represses transcription. In uninfected BL cells, PRC2 ancillary factors RbAp48 and JARID2 already associate with the chromatin proximal to the TSS, so the data suggest that at the *BIM/BCL2L11* locus, EBV infection is necessary to recruit core components SUZ12 and EZH2 (the histone methyl transferase) to establish functional PRC2. Assembly of PRC2 at the locus was absolutely dependent on both EBNA3A and EBNA3C being expressed, and, using a recombinant EBV expressing an epitope-tagged EBNA3C, ChIP showed that EBNA3C associates with chromatin near the TSS (Paschos et al. 2012). Subsequently, a broad EBNA3C-binding peak around the TSS was confirmed by ChIP-seq analyses (Jiang et al. 2014; McClellan et al. 2013; our unpublished results; Fig. 3a). It is therefore possible that EBNA3C (and/or EBNA3A) physically interacts with PRC2 and the transcription preinitiation complex, but this has not been formally demonstrated.

Since the activation mark H3K4me3 is largely unaltered at this locus irrespective of H3K27me3 or EBNA3 status, this indicates the establishment of a 'bivalent' chromatin domain. When the repressive histone modification H3K27me3 and

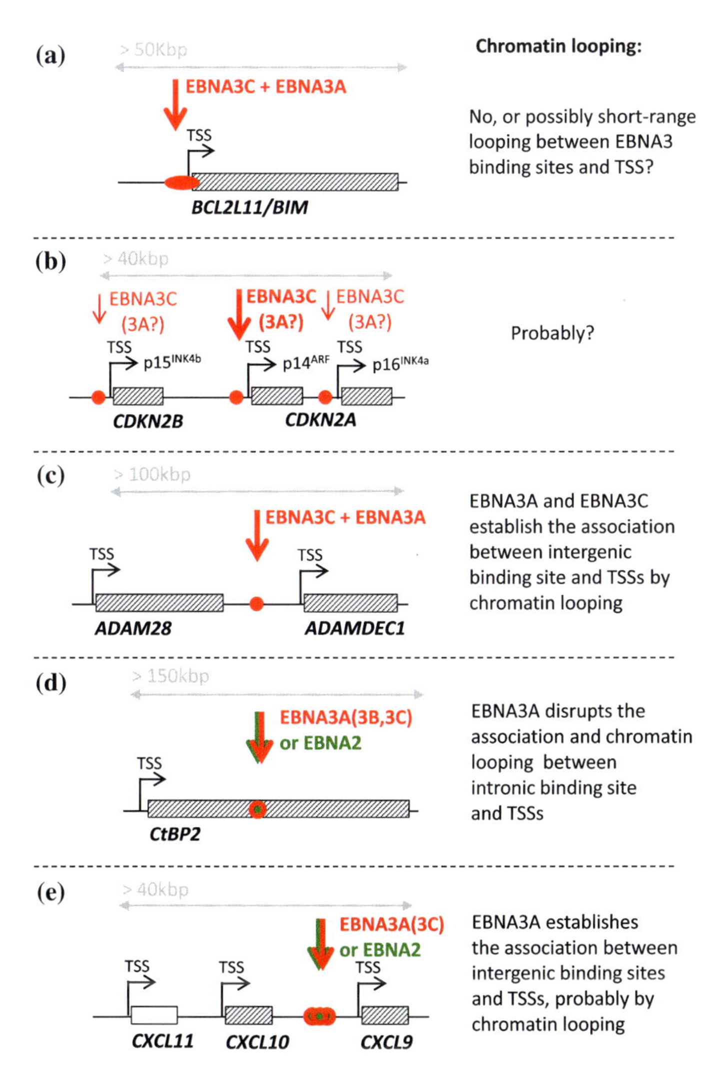


**Fig. 3** Examples of EBNA3 repressed gene loci. Schematic representations of several gene loci to which EBNA3 binding has been mapped by ChIP-seq and ChIP-QPCR (not drawn accurately to scale). In each case, the result of binding is repression of transcription—for details see text. **a** Repression of *BIM/BCL2L11* results from EBNA3A/EBNA3C binding proximal to the TSS. **b** Repression of $p15^{INK4b}$ (*CDKN2B*), $p14^{ARF}$ and $p16^{INK4a}$ (*CDKN2A*) probably involves a yet to be determined pattern of chromatin looping between the three TSSs, initiated by EBNA3C binding. **c** The *ADAM28/ADAMDEC1* locus is repressed by EBNA3A/EBNA3C binding to an intergenic site between the two genes and initiating chromatin looping to the TSSs. **d** At *CtBP2*, EBNA3A binding to an intragenic site (an intronic enhancer) is thought to displace bound trans-activator EBNA2 and disrupt chromatin looping between the enhancer and TSS, resulting in repression of transcription. **e** At the *CXCL9/10* locus, binding of EBNA3A to three intergenic enhancers is thought to displace the transactivator EBNA2 and initiate repression of transcription; this probably involves chromatin looping, but it has not been formally demonstrated

the permissive modification H3K4me3 occur at the same promoter, the gene is said to be in a bivalent state that can be either activated or silenced. This is associated with low-level transcription poised to increase or decrease (Bernstein et al. 2006; Stock et al. 2007). Consistent with the 'poised' nature of such domains, RNA polymerase II (Pol II) occupancy at the *BIM/BCL2L11* TSS was not diminished by EBV. However, analysis of phosphorylation of serine 5 on Pol II indicated that when EBNA3A and EBNA3C are both expressed, this phosphorylation step is inhibited and *BIM* transcripts are not initiated. It was not determined whether this involves the direct action of an EBV protein on the kinase CDK7 or is a consequence of the recruitment of PRC2 and/or PRC1 (the second PcG complex) to this particular locus. Conditional expression of EBNA3C revealed that this epigenetic repression of BIM expression was reversible, but took more than 30 days from when EBNA3C was inactivated to fully return to its active state, emphasising the stability of these chromatin modifications through rounds of cell division. Lentiviral delivery of shRNAs against polycomb complex subunits disrupted EBV repression of *BIM/BCL2L11,* thus confirming the requirement for the PcG system in maintaining the repression of BIM by EBV (Paschos et al. 2012). The mechanisms by which the PcG complexes are recruited to this locus are still unknown and the factor(s) responsible for the targeting of EBNA3C and/or EBNA3A to this specific site are also unidentified. The biological significance of *BIM/BCL2L11* repression is considered in more detail in Sect. 5.

## 4.2 *P16$^{INK4a}$ and the INK4a-ARF-INK4b Locus*

Maruo and colleagues, using a recombinant Akata EBV encoding a conditional EBNA3C fused to a modified oestrogen receptor, revealed that functional EBNA3C is essential to repress expression of the CDK inhibitor p16$^{INK4a}$—but not the CIP/KIP family inhibitors p21$^{WAF1}$ or p27$^{KIP1}$—in LCLs. Within the *INK4a-ARF-INK4b* locus at human chromosome 9p21, *CDKN2A* encodes two potent tumour suppressors, p16$^{INK4a}$ and p14$^{ARF}$ (p19$^{ARF}$ in mice); these proteins are crucial negative regulators of cell proliferation (Fig. 3b). Although exons 2 and 3 of *CDKN2A* are shared by *INK4a* and *ARF*, the proteins result from differential splicing but have no amino acid similarity because they are encoded in alternative reading frames (reviewed in Gil and Peters 2006; Sherr 2012). Adjacent to *CDKN2A* is the gene *CDKN2B* that encodes a protein closely related to p16$^{INK4a}$ called p15$^{INK4b}$. The cyclin-dependent kinase (CDK) inhibitor p16$^{INK4a}$ acts on the cyclin D-dependent kinases (CDK4 and CDK6) abrogating their binding to D-type cyclins and so inhibiting CDK4/6-mediated phosphorylation of the retinoblastoma protein (RB). By binding CDKs and blocking RB hyperphosphorylation, increased p16$^{INK4a}$ expression causes a G1 cell cycle arrest and senescence (Gil and Peters 2006; Sherr 2012). The CDK inhibitor p15$^{INK4b}$ has about 85 % amino acid similarity to p16$^{INK4a}$ and biochemically behaves in much the same way, but in mammalian hematopoietic cells—for reasons unknown—it has differentiation-associated functions that are distinct from those of p16$^{INK4a}$ and the two CDK inhibitors are not generally interchangeable (Humeniuk et al. 2013). In contrast to these CDK inhibitors, the p14 and p19 ARF proteins regulate p53 by inactivating MDM2, thus stabilising and activating of p53. This leads to cell cycle arrest by inducing the p53-dependent CDK regulator p21$^{WAF1}$ or apoptosis by inducing p53-dependent pro-apoptotic factors such as BAX and NOXA (Sherr 2012; Vousden and Prives 2009).

Inactivating the conditional EBNA3C resulted in an accumulation of both p16$^{INK4A}$ mRNA and protein, dephosphorylation of RB and concomitant cell cycle arrest (Maruo et al. 2006, 2011; Skalska et al. 2010, 2013). The *CDKN2A* locus encoding p16$^{INK4a}$ had been identified as a target of polycomb-mediated repression in proliferating fibroblasts and epithelial cells (Gil and Peters 2006), so it was not a surprise to discover that the EBNA3C-mediated repression of *CDKN2A* in cycling B cells was accompanied by the deposition of the repressive H3K27me3 epigenetic mark across the locus—primarily around the p16$^{INK4a}$ TSS (Maruo et al. 2011; Skalska et al. 2010, 2013). Analysis of LCLs established with EBNA3A knockout and conditional viruses revealed that EBNA3A also plays a role in creating the H3K27me3 modification and the repressed state of *CDKN2A* (Maruo et al. 2011; Skalska et al. 2010). In the absence of EBNA3A, cells consistently expressed higher levels of p16$^{INK4a}$ concomitant with lower levels of H3K27me3. These data were consistent with the earlier report of a microarray analysis that showed *CDKN2A* to be a target of EBNA3A-mediated repression (Hertle et al. 2009). Furthermore, establishing LCLs with recombinant EBV encoding CtBP-binding mutants of EBNA3C and/or EBNA3A showed that their interaction with this co-repressor was also necessary for the efficient deposition of H3K27me3 and repression of p16$^{INK4a}$ expression (Skalska et al. 2010). As with *BIM/BLC2L11*, B cell lines carrying EBV encoding the conditional EBNA3C modified oestrogen receptor fusion revealed that this epigenetic repression of *CDKN2A* was reversible by adding or removing 4HT from the medium (Skalska et al. 2010). The repression of *CDKN2A* by EBNA3C is likely to be direct because ChIP analysis of epitope-tagged EBNA3C expressed in LCLs identified binding peaks localised proximal to not only the TSS of p16$^{INK4A}$ and ARF, but also the *CDKN2B* gene encoding p15$^{INK4b}$ (Skalska et al. 2013). Subsequent ChIP-seq studies using tagged EBNA3C suggested that the major EBNA3C peak is adjacent to *ARF* (Jiang et al. 2014; our unpublished results; Fig. 3b). Taken together, all the various results indicate that EBV via EBNA3C (cooperating with EBNA3A) co-ordinately regulates the whole *INK4b-ARF-INK4a* locus by binding, repressing and directing the recruitment of PRC2 to sites near the three transcriptional start sites. Although this coordinated regulation might involve interactions between EBNA3C and IRF4/BATF-containing complexes (Jiang et al. 2014), the precise mechanism of targeting to the locus has not yet been identified. The role CtBP plays is also unknown. Consistent with all these data indicating that the whole 40 kb locus is co-regulated, we have found that p15$^{INK4b}$ mRNA is co-ordinately expressed with p16$^{INK4a}$ mRNA in EBNA3C-conditional and EBNA3A-conditional LCLs (our unpublished data) and it has recently been shown that EBNA3A can repress *CDKN2B* transcription via an interaction with MIZ1 and deposition of the H3K27me3 repressive histone modification around its TSS (Bazot et al. 2014).

Our current understanding of the *CDKN2A/CDKN2B* locus in EBV-infected B cells is far from complete. For instance, *CDKN2A/CDKN2B* regulation by the 'looping' of chromatin has been reported at the locus in various non-B cells (Kheradmand Kia et al. 2009), but this remains to be assessed after EBV infection or generally in B cells. Similarly, the roles of non-coding RNAs, such as ANRIL, that can regulate *CDKN2A* (Yap et al. 2010) require exploration in EBV-infected B cells. What determines the requirement for EBNA3A in addition to EBNA3C and what defines their relative contributions are recurring, unanswered questions. Regulation of the locus by EBNA3C in an RB-null LCL (Skalska et al. 2010) formally established that EBNA3C-mediated epigenetic control is quite independent of the degree of cell proliferation. The specific role of $p16^{INK4a}$ (rather than $p14^{ARF}$ or $p15^{INK4b}$) as a target for EBNA3C and as the major barrier to B cell transformation was further explored by making use of an 'experiment of nature' in the form of 'Leiden' B cells carrying a homozygous genomic deletion that specifically ablates production of functional $p16^{INK4a}$ (Brookes et al. 2002). A comparison of $p16^{INK4a}$-null LCLs with LCLs established from normal B cells showed that if $p16^{INK4a}$ is not functional, then active EBNA3C is unnecessary to sustain proliferation (Skalska et al. 2013). Consistent with these observations—and providing formal proof that $p16^{INK4a}$ is the main target of EBNA3C that enables the establishment of LCLs in vitro—it was possible to transform $p16^{INK4a}$-null primary B cells into stable LCLs with EBV lacking functional EBNA3C. Although EBV (via EBNA3A and EBNA3C) appears to epigenetically regulate expression of both $p15^{INK4a}$ and $p14^{ARF}$ together with $16^{INK4a}$, it is unclear precisely what role $p15^{INK4b}$ and $p14^{ARF}$ play in the inhibition of B cell transformation, in proliferating LCLs, or more generally in human B cell biology.

### 4.3 *ADAM28/ADAMDEC1* Locus

*ADAM28* and *ADAMDEC1* form part of a metalloproteinase gene cluster on chromosome 8p12 (McClellan et al. 2012, 2013; Fig. 3c). ADAM28 is a disintegrin/metalloproteinase expressed by lymphoid cells and in its cleaved, soluble form is thought to be a regulator of B cell adhesion and migration through endothelial cells. ADAMDEC1 (also known as Decysin-1) is also a metalloproteinase, but of unknown function. The available gene expression data derived from several microarray studies are largely consistent with both EBNA3A and EBNA3C playing a role in repressing these genes in EBV-infected B cells—probably by regulation of the whole locus (Hertle et al. 2009; McClellan et al. 2012; Skalska et al. 2013; White et al. 2010; http://www.epstein-barrvirus.org.uk). Furthermore, although in B cells infected with EBV, both EBNA3A and EBNA3C appear to be necessary for repression, when they are ectopically over-expressed, either can independently repress the expression of ADAM28 and ADAMDEC1 in EBV-negative B cells (McClellan et al. 2012). Since ADAM28 and ADAMDEC1 probably modulate cell:cell and cell:stroma interactions, their regulation is likely to be of greatest biological significance in vivo during viral persistence or possibly EBV-associated disease.

ChIP-seq analysis and confirmatory ChIP-QPCR, using both a latency III BL-derived cell line and LCLs, showed robust EBNA3C and EBNA3A (but not EBNA3B) binding peaks at the same site in an intergenic regulatory element located between the *ADAM28* and *ADAMDEC1* genes, approximately 70 kb downstream of the *ADAM28* promoter and about 16 kb upstream of the *ADAMDEC1* promoter (McClellan et al. 2012, 2013; our unpublished results; Fig. 3c). As with *BIM/BCL2L11* and *CDKN2A*, repression of this complete locus by EBNA3A and EBNA3C is associated with the distribution of H3K27me3 around the *ADAMDEC1* transcription start site and within *ADAM28*. Chromatin conformation capture (CCC—a method that reveals contact between two defined regions on a genome Naumova et al. 2012) was used to show that chromatin looping to bring the intergenic element that binds EBNA3s into close proximity with the *ADAM28* and *ADAMDEC1* promoters requires the presence of EBNA3C (McClellan et al. 2013). This coincides with the deposition of H3K27me3 that is presumably mediated by PcG repressor complexes and sustained repression of both genes. It is not known how or why EBNA3C and EBNA3A are both involved in EBV-infected cells, nor what cell factors are responsible for targeting the EBNA3s to the intergenic binding site, nor what facilitates enhancer 'looping' to specific promoters. Also, in common with the regulation of *BIM/BCL2L11* and *CDKN2A/CDKN2B,* we do not know what triggers recruitment of PcG complexes or the repression of transcription—or which comes first. Nevertheless, although the molecular details are still awaited, these important experiments have provided some of the first examples (along with genes described below) of host gene reprogramming via viral modulation of long-range interactions requiring chromatin looping between promoters and distal regulatory elements.

## 4.4 CtBP2

As described in Sect. 2.7.2, 'CtBP' consists of two proteins encoded by two separate genes—*CtBP1* and *CtBP2*—and therefore in theory, any vertebrate cell could express either protein or both proteins. Microarray expression studies of EBV-infected cells indicate that in human B cells expressing the complete latency III pattern of viral proteins (e.g. BL31-B95.8 and various LCLs), CtBP2 mRNA is transcribed at very low levels or the gene is silent. The exception to this was identified by Hertle and colleagues by analysing EBNA3A-deficient LCLs. Consistently, LCLs established with EBV recombinants carrying a deletion of the whole EBNA3A open reading frame expressed relatively high levels of CtBP2 mRNA (Hertle et al. 2009). In contrast, studies of LCLs deficient in EBNA3B or lacking functional EBNA3C did not produce this apparent derepression phenotype (Skalska et al. 2013; White et al. 2010; http://www.epstein-barrvirus.org.uk). Similar analyses of CtBP1 expression showed that in all EBV-infected cells investigated (irrespective of EBNA3 expression), the levels of CtBP1 transcription are relatively high and consistent with the level of protein detected by Western blotting.

Subsequently, McClellan and colleagues produced intriguing ChIP-seq and ChIP-QPCR data that partially explain this differential regulation of *CtBP1* and *CtBP2*. Specifically, they reported that while no EBNA3-enriched sites were seen at or anywhere near the *CtBP1* locus, EBNA3A, EBNA3B and EBNA3C could all be found at an intronic site (predicted by histone marks to be an enhancer) located in *CtBP2* (McClellan et al. 2013; Fig. 3d). Consistent with EBNA3A playing a central role in the repression of this locus, they were able to show by CCC that only in the absence of EBNA3A expression was there chromatin looping that juxtaposed the enhancer adjacent to the *CtBP2* promoter, thus facilitating transcription. That is, at this locus, EBNA3A prevents enhancer–promoter looping. Since not only EBNA3A, but also EBNA3B, EBNA3C and EBNA2 could be found binding to the intronic enhancer, a model was proposed in which the transactivator EBNA2 binding to the intragenic site would facilitate enhancer–promoter loop formation (although the role of EBNA2 was not experimentally assessed) and CtBP2 activation, only when an EBNA3 was absent. These observations provide another convincing example of EBNA3A (it was only demonstrated for EBNA3A) modulating the looping of chromatin between a distal site and a gene promoter—in this case inhibiting loop formation (McClellan et al. 2013). However, in the current model, it remains to be determined why EBNA3A acts negatively in the modulation of looping, while EBNA3B and 3C appear uninvolved despite their apparent binding to the locus. Because so little is known about the functional difference(s) between CtBP1 and CtBP2, the biological significance of this EBV-mediated pattern of expression cannot currently be assessed.

### 4.5 *CXCL9/10* Locus

*CXCL9*, *CXCL10* and *CXCL11* are a family of CXC chemokine genes tandemly located on chromosome 4q21.1 (Fig. 3e). The chemokines they encode are generally interferon-inducible and upon secretion attract leucocytes to sites of infection and inflammation and probably attract anti-tumour cytotoxic T cells to cancer tissue—which correlates with improved clinical outcome (Dufour et al. 2002; Mlecnik et al. 2010). Therefore, regulation of this locus is likely to be important in vivo, in EBV persistence and disease pathogenesis (see for example Sect. 6.3). The consensus of various microarray expression studies of this gene cluster has indicated that in LCLs carrying EBV that is wild type for EBNA3s, the locus is generally repressed, but in EBNA3A-deficient LCLs, *CXCL9/10* is strikingly derepressed; in EBNA3C-deficient conditional LCLs, it is slightly derepressed;

and in EBNA3B-deficient LCLs, it is more profoundly repressed. In all the LCLs, *CXCL11* transcripts are barely detectable and not regulated by EBNA3s (Hertle et al. 2009; Skalska et al. 2013; White et al. 2010; http://www.epstein-barrvirus.org.uk). These data are largely consistent with those derived from the tumour-derived B cell line, BJAB ectopically expressing individual EBNA3 genes—that is EBNA3A and EBNA3C both repress *CXCL10*, whereas EBNA3B induces modest activation (McClellan et al. 2012).

Studying in detail the regulation of CXCL9 and CXCL10 expression by EBNA3A, Kempkes and colleagues developed a model wherein EBNA3A binds three intergenic enhancers located between *CXCL9* and *CXCL10* that can be occupied by RBP-JK/CBF1 and the transactivator EBNA2 (Harth-Hertle et al. 2013; Fig. 3e). The binding of EBNA3A to RBP-JK/CBF1 displaces EBNA2 and by impairing enhancer activity rapidly inhibits recruitment of Pol II and transcription from both the *CXCL9* and *CXCL10* promoters. Studying the kinetics of repression and chromatin modifications indicated that here silencing by EBNA3A precedes the deposition of H3K27me3, which subsequently sustains the genes in an inactive configuration; this is the steady state in LCLs unless EBNA3A is absent in which case EBNA2 binds and substantially activates the locus (Harth-Hertle et al. 2013).

Summarising what is known about the action of the EBNA3s at this gene cluster, EBNA3A clearly plays a major role in silencing intergenic enhancers between *CXCL9* and *CXCL10*. EBNA3C can probably bind at the same locations and contribute to the repression (McClellan et al. 2012), although a binding peak has not yet been reported. As with most of the target genes we have considered here, it is not clear why EBNA3A and EBNA3C are not equivalent and interchangeable. In the absence of EBNA3A or EBNA3C, this enhancer can be bound by EBNA2, resulting in transactivation. EBNA3B appears to potentiate the activation of the locus, but whether to do this it binds at the same enhancer sites or acts by a novel mechanism is currently unknown. Since regulation could not be recapitulated in RBP-JK/CBF1-negative DG75 BL cells, the binding of EBNA3A (probably EBNA3C) and EBNA2 is likely to be mediated by RBP-JK/CBF1 binding to multiple response elements within the enhancers (Harth-Hertle et al. 2013). Although it has not been proven, the data strongly suggest that EBNA3A (and EBNA3C?) simultaneously inhibits chromatin looping between distal intergenic enhancers and the promoter of two genes. When the level of EBNA3A (or EBNA3C) is reduced, EBNA2 gains access to these regulatory elements and promotes enhancer–promoter looping and transcription—but this remains to be formally tested.

### 4.6 Overview of Gene Regulatory Mechanisms Mediated by the EBNA3s

The gene loci with specific EBNA binding sites described above are illustrated schematically in Fig. 3. They show how EBNA3-mediated repression can be achieved by a variety of topologically different mechanisms. Furthermore, putting these data on specific gene loci into perspective, analyses of genome-wide distributions have revealed >20,000 EBNA2 binding sites and (by using in ChIP-seq an antibody that recognises all three EBNA3 proteins) >7000 EBNA3 binding sites. Only a small proportion of all these binding sites are proximal to the TSS of genes, with the closest EBNA3 binding sites typically 10–50 kb from the nearest TSS—suggesting the modification of long-range interactions is common (Jiang et al. 2014; McClellan et al. 2013). Comparing EBNA2 and EBNA3 binding sites revealed that there was considerable overlap. At gene-proximal sites (<2 kb from TSS), about 60 % of EBNA3 binding peaks were overlapping with sites bound by EBNA2 and overall—including distal sites—about 80 % of genes closest to an EBNA3 binding peak were also closest genes to an EBNA2 binding peak. This is a strong indication that many genes targeted by EBNA3 proteins are co-regulated by EBNA2. It is interesting that (as indicated above in Sect. 2.7.1) probably only a small percentage of these overlapping sites correspond to RBP-JK/CBF1 response elements; however, many of the sites are enriched for multiple transcription factors, including PU1, SP1, EBF1, BATF, IRF4, PAX5 and p300. For detailed analyses of the available EBNA3/EBNA2 ChIP-seq data, the reader is referred to (Jiang et al. 2014; McClellan et al. 2012, 2013).

A comparison of EBNA3 binding sites with the genome-wide chromatin landscape (in LCL GM12878) available from the ENCODE database showed EBNA3s associated with active enhancers (defined as H3K4me1+ve; H3K27Ac+ve) or in many cases 'poised' enhancers (H3K4me1+ve; H3K27Ac-ve) consistent with EBNA3 functioning as both repressors and activators of transcription (McClellan et al. 2013). As we have seen above, specific repression mechanisms are starting to be understood; however, activation by EBNA3s remains largely unexplored. Adding to the global picture, Kempkes and colleagues performed a comprehensive data-mining analysis of genes regulated by EBNA3A (Harth-Hertle et al. 2013; Hertle et al. 2009). By comparing the data on EBNA3A-repressed genes with data from the ENCODE studies of genome-wide histone modifications, they made the following observations: (a) about 70 % of genes repressed by EBNA3A carried the PcG-associated signature H3K27me3; (b) 90 % of these genes were associated with PcG-mediated silencing in multiple cell types and contexts—that is, they may have an inherent capacity to undergo PcG-mediated silencing when repressed by a variety of mechanisms—and (c) about 20 % of genes repressed by EBNA3A were grouped in clusters of up to 4 genes and often H3K27me3 was enriched across the whole locus after repression by EBNA3A. It was concluded that EBNA3A-repressed genes were commonly associated with PcG modification of chromatin and that the target genes are often arranged in gene clusters (Harth-Hertle et al. 2013). As a result of these observations, the group performed the detailed study on the *CXCL9*, *CXCL10* and *CXCL11* gene cluster and its regulation by EBNA3A described above and in Fig. 3e.

These studies have highlighted the unexpected complexity of EBNA-mediated gene regulation, since *CXCL9 and CXCL10* appear to be genes co-ordinately regulated by the functional interaction of at least four EBNA proteins—EBNA2, EBNA3A, EBNA3B and EBNA3C. It is also possible that EBNA-LP might be involved, as it can in some circumstances cooperate with EBNA2 to activate transcription (Peng et al. 2005). The study of this locus has also raised the important question of whether polycomb-mediated chromatin modifications such as H3K27me3 are a cause of the gene repression induced by EBNA3s or a consequence. At the *CXCL9/10* locus, analysis of the kinetics and sequence of events made a strong case for the latter, with displacement of EBNA2 probably being the primary cause of repression (or strictly speaking deactivation). However, not all genes repressed by EBNA3 activity are transactivated by EBNA2. For example, at *BIM/BCL2L11,* there is no indication that EBNA2 is involved and—because this has all the characteristics of a poised 'bivalent' gene, where Pol II is always present—PcG complexes may play a much more direct role in repression by blocking transcription initiation (Bernstein et al. 2006; Stock et al. 2007). Consistent with this, at *BIM/BCL2L11,* the recruitment of core PRC2 factors is unambiguously dependent on the expression of both EBNA3A and EBNA3C. Nevertheless, because repression of *BIM/BCL2L11* also involves removal of histone H3 and H4 acetylation (Paschos et al. 2009, 2012), the association of EBNA3A and EBNA3C with HDACs is also likely to be important. So, although some of the details are emerging, the overall picture of EBNA3-mediated gene regulation gets even more complicated. It will be exciting to see what general principles emerge as we discover more about the interactions between EBNA3s and components of the multi-factor complexes that regulate epigenetic modifications of chromatin and programmes of transcription.

## 5 EBNA3C and EBNA3A as Modifiers of Oncogenic Stress Responses and the DDR

Viruses that establish a persistent latent infection commonly stimulate cellular DNA synthesis and sometimes cell division early after infection. However, cells of many metazoans have evolved responses that normally monitor unscheduled DNA synthesis and prevent cell proliferation when, for instance, cell proto-oncogenes are 'activated' by mutation and/or deregulated expression. These defence strategies that reduce the risk of neoplasia and cancer are collectively called oncogenic stress responses (OSR). Mechanisms include the activation of tumour suppressor genes that together trigger pathways leading to cell cycle arrest (e.g. oncogene-induced senescence, OIS) or complete elimination of cells (e.g. apoptosis). Because viruses that can induce cellular DNA synthesis and cell division are capable of triggering OSR/OIS, they often co-evolve countermeasures for inactivating or bypassing OSR/OIS (Bartek et al. 2007; Braig and Schmitt 2006; Nikitin and Luftig 2012). As cell proto-oncogenes generally control signalling pathways and/or gene networks that link proliferative signals to the cell cycle machinery, when they are deregulated, this can result in unscheduled and aberrant DNA synthesis (sometimes called 'replicative stress'). Consequently, oncogene activation can produce the stalling of DNA replication forks that results in DNA double-strand breaks. Such lesions can trigger, primarily via the ATM/CHK2-kinase signalling pathway, the stabilisation and activation of p53 and also the induction of p16$^{INK4a}$. Depending on the physiological and cellular context, this leads to DNA repair, cell death or senescence. This complex response is known as the DNA damage response (DDR). The links between DDR and OSR/OIS have been extensively reviewed (Acosta and Gil 2012; Bartek et al. 2007; Braig and Schmitt 2006; Halazonetis et al. 2008).

As detailed above, EBNA3C and EBNA3A appear to cooperate harnessing the PcG protein system for the sustained repression of two host tumour suppressor genes involved in OSR/OIS—*BCL2L11* encoding BIM and *CDKN2A* encoding p16$^{INK4a}$. An explanation for why EBV has evolved a mechanism for suppressing expression of the CDK inhibitor p16$^{INK4a}$ (and probably BIM) expression became apparent by comparing infections of normal B cells with EBNA3C-deficient EBV with infections by 'wild-type' (WT) EBV (Allday 2013; Skalska et al. 2013; Fig. 4). These experiments revealed that EBV infection led to a modest increase in p16$^{INK4a}$ transcription in the first few days after infection, when EBNA2 transactivates regulators of cell cycle progression (e.g. MYC, Cyclins D2 and E) to induce a period of hyperproliferation and concomitant activation of DDR pathways occurs (Allday et al. 1989; Nikitin et al. 2010; Shannon-Lowe et al. 2005; Sinclair et al. 1994; Spender et al. 1999; Thorley-Lawson and Strominger 1978). It is likely that entry into S phase, when uncoupled from normal signalling pathways, is interpreted as oncogenic or replicative stress and p16$^{INK4a}$ transcription is a consequence. When the infecting virus expressed functional EBNA3C (and EBNA3A), there was no further increase of p16$^{INK4a}$ transcripts from about day 7 onwards. In contrast, if functional EBNA3C was absent during infection, transcription from *CDKN2A* continued unrestrained and the level of mRNA progressively increased until most of the cells stopped proliferating and/or died. In a similar way, early after infection, BIM expression is down-regulated and within about 5 days reaches a steady state, unless EBNA3C is deleted or inactivated in the infecting EBV when the level of BIM mRNA also increases in parallel with that of p16$^{INK4a}$. This increase again continues until the cells arrest or die (Allday 2013; Skalska et al. 2013).

The EBNA3-mediated, sustained inhibition of $p16^{INK4a}$ and probably *BIM* transcription is therefore critical for EBV to bypass a host cell defence against oncogenic transformation triggered by EBNA2, acting through MYC (Nikitin et al. 2010). Thus, expression of EBNA3C (and EBNA3A) ensures expansion of the infected B cell population, LCL outgrowth and long-term latency. Strictly speaking, in this context, EBNA3C does not actually repress *CDKN2A* and *BIM/BCL2L11* transcription, but rather blocks their activation. We assume this involves the recruitment of PcG protein complexes to the loci, leading to H3K27me3 modifications on chromatin around the transcription start sites, as is seen in established LCLs; however, this has not yet been formally demonstrated in newly infected cells.

It was reported that EBNA3C might specifically reduce the DDR that is active in the first week after EBV infection of naïve B cells in vitro, when the cells are beginning to cycle very rapidly (Nikitin et al. 2010) and during the same period that p16$^{INK4a}$ is actively transcribed (Skalska et al. 2013). It may be that this increase in p16$^{INK4a}$ is a consequence of DNA damage caused by hyperproliferation, as has been variously reported in other types of cell (reviewed in Acosta and Gil 2012; Gil and Peters 2006). Nevertheless, an increase in p16$^{INK4a}$ should activate RB, which could then lead to the repression of E2F-regulated genes, while MYC is constitutively active. This might enable cells to enter S phase with suboptimal amounts of DNA precursors and/or replication enzymes and this would lead to stalled DNA replication that is 'read' as DNA damage, triggering phosphorylation of histone H2AX—a focal marker of DNA damage. When EBNA3C is functional, the increase in p16$^{INK4a}$ is soon attenuated and the DDR will apparently subside. So it could be that the lack of EBNA3C exacerbates the DDR because of the accumulating p16$^{INK4a}$. However, during primary infection, EBNA3C might also have a direct effect on the ATM/CHK2 signalling pathway or some component(s) of the DDR complex at the site of DNA strand breaks (Nikitin et al. 2010). Although EBNA3C can apparently associate with CHK2 and/or histone H2AX (Choudhuri et al. 2007; Jha et al. 2013), the possibility that these interactions are involved in the attenuation of the DDR early after infection requires further investigation. Only a systematic genetic analysis of EBNA3C and its actions early after infection will unravel the DDR and the p16$^{INK4a}$-mediated senescence response and establish what relative contributions each make to the inhibition of B cell transformation.

Through the action of EBNA3C (and EBNA3A) and interactions with the cellular PcG protein system, EBV appears to have evolved a very effective countermeasure to OSR/OIS and this is critical in the virus life cycle to establish a latent infection and therefore initiate long-term persistence in B cells. In vitro this mechanism manifestly overcomes a major early obstacle to B cell growth transformation, making EBV one of the most potent transforming/immortalising biological agents to have been identified. This strategy of utilising the PcG system to specifically regulate key tumour suppressor genes is—to our knowledge—unique among tumour viruses.

# 6 The EBNA3s In Vivo

## *6.1 EBNA3s in Asymptomatic Persistence*

If in order to establish a persistent infection in vivo, EBV initially commandeers resting (naïve) B cells and drives these to proliferate as activated B blasts, we assume that—by analogy to what is seen in culture—repression of p16$^{INK4a}$ is necessary to permit the transient proliferation of the infected population (Fig. 4). Furthermore, since in vitro EBNA3C and EBNA3A expression does not seem to be required for the early rounds of rapid cell division (up to about day 10), and these proteins have clearly evolved functions that extend the proliferative life of infected

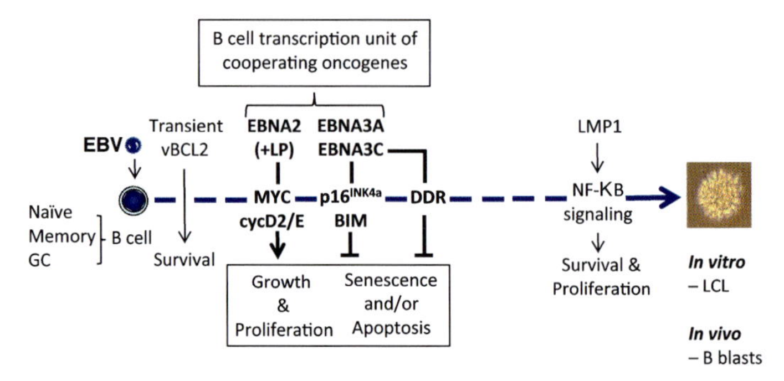


**Fig. 4** Events following infection of primary resting B cells by EBV that initiate transformation into continuously proliferating LCLs. Immediately after infection, viral anti-apoptotic factors normally associated with the EBV lytic cycle (vBCL2 factors: BHRF1 and BALF1) are transiently expressed to aid the survival of newly infected B cells (Altmann and Hammerschmidt 2005). During the first 2 days post-infection (pi) with EBV, cell genes associated with growth and cell cycle are transactivated and their products (e.g. MYC, cyclin D2, cyclin E) drive cells from G0 to G1, to become enlarged, activated and start proliferating. The whole process is driven by the EBV transactivator protein EBNA2, probably assisted by the co-factor EBNA-LP (Nikitin et al. 2010; Sinclair et al. 1994; Spender et al. 1999). Cells then undergo rounds of rapid cell division (doubling time 8–10 h) that in some cells results in damage to DNA that can activate the DNA damage response (DDR, Nikitin et al. 2010). If the full complement of nine EBV latency-associated proteins is expressed, the DDR becomes attenuated (in part by EBNA3C) and cells continue to proliferate to produce LCLs that have a population doubling time of about 24 h. Early after infection, expression of BIM is down-regulated, and although the level of $p16^{INK4a}$ expression increases slightly, this soon reaches a steady state. In both cases, EBNA3C and EBNA3A cooperate to restrain the transcription of these tumour suppressor genes (Paschos et al. 2012; Skalska et al. 2013 and described in the text). EBNA3C appears more important in this partnership, and for reasons we do not understand. Although transcripts can be detected earlier, Micah Luftig and colleagues have shown that LMP1 expression and concomitant NF-κB activity are not fully evident until 3 weeks after primary B cell infection, and then, they make further contributions to cell survival and proliferation (Price et al. 2012). If EBNA3C is deleted or functionally inactivated in the infecting EBV, levels of mRNAs corresponding to $p16^{INK4a}$ and BIM progressively increase until cells arrest and/or die (Skalska et al. 2013). In vitro, the sequence of events described here is probably similar whether the B cells infected are naïve, memory or germinal centre-derived (Babcock et al. 2000; Siemer et al. 2008), but this is not known for cells in vivo. With a B cell-specific transcription unit that encodes oncogenes that collaborate in vitro to transform normal B cells into continuously proliferating LCLs, why is EBV not an acutely transforming tumour virus in vivo? We would suggest that linking expression of this transcription unit to the state of B cell differentiation, having EBNA proteins that are immunodominant in CTL recognition and having one member of EBNA3 family that encourages immune surveillance and perhaps differentiation of infected cells, all contribute to a highly evolved state of equilibrium that is mutually beneficial to virus and host (see main text for a more detailed discussion)

cells, one is tempted to speculate that prolonged B blast proliferation in vivo must be an important component of the persistence programme. The survivors from this expanding population of activated B cells might then migrate into or initiate a germinal centre, where the cells differentiate to centroblasts and centrocytes and finally emerge as MBCs after the regulated shutdown of protein-coding EBV latency genes.

Since the repression of $p16^{INK4a}$ (and BIM) expression involves polycomb-mediated covalent histone modifications, it is possible the epigenetic memory (i.e. a heritable pattern of gene expression: Campos et al. 2014) of this programme will be passed to progeny cells and carried through into the MBC population, even if the initiators, EBNA3A and EBNA3C, are no longer expressed. It is well established that polycomb-mediated gene repression is also a common precursor to promoter DNA methylation in cancer (reviewed in Cedar and Bergman 2009), so it is reasonable to hypothesise that the B cells reprogrammed in vivo by EBV will be particularly prone to aberrant DNA methylation at the *CDKN2A* and *BIM/BCL2L11* loci during tumorigenesis. This is consistent with reports describing promoter methylation of these genes in EBV-positive B cell lymphomas and derived cell lines (Klangby et al. 1998; Nagy et al. 2005; Paschos et al. 2009; Richter-Larrea et al. 2010). The role of EBNA3B in persistence and B cell lymphomagenesis is discussed in Sect. 6.3.

### 6.2 Immune Responses to EBNA3 Proteins

EBV infection in humans is tightly regulated by the immune system. This has been extensively reviewed elsewhere (Hislop et al. 2007) and also in several chapters of this volume. The role that the EBNA3 family proteins play in T cell-mediated immunity is very well established and is generally consistent with the notion that immune regulation of EBV-infected B blasts is critical in maintaining the virus–host balance in persistence. Indeed, the EBNA3s are the primary targets of $CD8^{+ve}$ T cells (cytotoxic T lymphocytes or CTL) in the peripheral circulation of healthy EBV carriers (Hislop et al. 2007) and have been shown to be, among all the latency-associated viral proteins, the most immunogenic. CTL responses against cells latently infected with EBV are primarily targeted against EBNA3A/3B/3C-derived epitopes that are recognised by $CD8^{+ve}$ CTL in association with specific HLA molecules (Khanna et al. 1992, 1997; Murray et al. 1992; Steven et al. 1996), but the EBNA3 proteins are also a source of epitopes for $CD4^{+ve}$ T cells, with EBNA3C-derived epitopes being more abundant than EBNA3A and 3B. A comprehensive summary of the $CD8^{+ve}$ and $CD4^{+ve}$ T cell epitopes derived from the EBNA3 proteins is shown in (Hislop et al. 2007).

The strong CTL-mediated response against the EBNA3 proteins has allowed Moss and colleagues to develop a $CD8^{+ve}$ T cell epitope-based EBV vaccine strategy—to protect against infectious mononucleosis—using EBNA3A-derived epitopes (Elliott et al. 2008). Furthermore, patients with PTLD have also been successfully treated with infusions of EBV-specific CTLs primarily directed against the EBNA3s (reviewed in Heslop et al. 2010; Nikiforow and Lacasce 2014 and the chapter in this volume authored by Gottschalk and Rooney). This focusing of CTL-mediated immune surveillance on the EBNA3s ensures that potentially oncogenic, EBV-driven, proliferating B blasts are removed before they cause morbidity or possibly mortality. It has been proposed that evolutionary pressures on the CTL epitopes in these proteins (and therefore DNA sequence variation) may be towards their conservation rather than their inactivation, thus helping maintain a stable relationship between virus and host that favours both (Khanna et al. 1997). Given the diversity of HLA molecules in the human population, this may be one explanation of the considerable sequence variation seen in the EBNA3 genes. Conservation of some CTL epitopes across EBV subtypes and diversification of others may be driven by MHC variation in host populations to either enhance presentation or avoid it.

## 6.3 EBNA3B: Manipulation of Immune Surveillance and Role as Tumour Suppressor

Exon microarray analysis of the genes differentially regulated between WT (B95.8) and EBNA3B-KO LCLs identified about 200 EBNA3B-regulated genes, including many expressed at the cell surface or as secreted chemokines—for example CD28, CD305 (LAIR-1), CXCR4, CXCL10, IL10 and IL19 (White et al. 2010, 2012; http://www.epstein-barrvirus.org.uk). This led to the speculation that although these changes did not have gross effects on EBV-infected B cells in vitro, they might alter heterotypic cell:cell interactions and possibly migration or homing properties of cells in vivo. The observation that EBNA3B has not been counter-selected over millions of years of virus–host co-evolution suggested an important role for this gene in vivo during long-term EBV persistence. Furthermore, in our hands, EBNA3B-negative LCLs grow particularly robustly in vitro, and after long-term culture of LCLs, there appears to be a propensity for deletions in EBNA3B to arise. This applies to at least two widely used cord blood-derived LCLs—IB4 (Chen et al. 2005) and X50-7 (Allday and Farrell 1994)—and has been observed to arise spontaneously in adult B cell-derived LCLs established using B95.8 EBV by us and others (unpublished observations).

In order to test the hypothesis that deletion of EBNA3B would significantly alter the behaviour of EBV-infected B cells in vivo, mice engineered to be susceptible to EBV infection were used. For this purpose irradiated, newborn NOD-scid *IL2*$\gamma^{null}$ (NSG) immunodeficient mice were engrafted with fetal liver-derived CD34 + haematopoietic progenitor cells (reviewed in the chapter by Christian Münz). After approximately 3 months, these mice (hu-NSG mice) were validated for successful engraftment of human immune cells and infected with similar doses of WT, EBNA3B-KO or revertant EBV-BAC-derived viruses (White et al. 2012). After a month, mice were killed and found to exhibit splenomegaly, with large tumour masses in over half of the EBNA3B-KO-infected mice, but not the WT or revertant virus-infected animals. Routine histology and immunocytochemistry confirmed that the EBNA3B-KO-infected mice had developed monomorphic, highly proliferative tumours (>90 % Ki67+ve)—described as DLBCL-like activated B cell (ABC) sub-type—that destroyed the architecture of the spleen. In contrast, spleens from the WT and revertant groups were not so enlarged and had markedly less proliferative (30–60 % Ki-67+ve), polymorphous infiltrates with significant numbers of immunoblasts and plasma cells—features similar to human polymorphic PTLD and to the plasmacytoid lesions described previously in both hu-NSG mice (Strowig et al. 2009) and in SCID mice implanted with LCLs (Rochford et al. 1993). Furthermore, it was particularly striking that in the spleens of WT- and revertant-infected animals, there were substantial T cell infiltrates, whereas the EBNA3B-KO tumours exhibited very few infiltrating T cells. The derepression of CXCL10 by EBNA3B indicated by microarray was confirmed in ELISA to facilitate CXCL10 induction by interferon. Restoration of CXCL10 levels in culture media partially restored a migratory deficit for T cells towards EBNA3B-KO LCLs (White et al. 2012; see also Sect. 4.6).

Overall, these remarkable results indicated that failure to express EBNA3B: (a) produces aggressive B cell tumours resembling DLBCL, (b) may somehow reduce immune cell trafficking and T cell surveillance of the tumour cells, perhaps partially through loss of CXCL10 secretion, but likely also involving additional migratory signals and (c) could perhaps be inhibiting B cell/plasmacytoid differentiation. It is significant that there were at least two cases of transplant patients with PTLD that fail to express EBNA3B because of a genomic deletion in the resident EBV, one of which was an aggressive fatal PTLD (Gottschalk et al. 2001), much like those seen in the animal model. Furthermore, screening of a further eleven EBV+ve ABC-DLBCLs revealed another tumour carrying a truncated EBNA3B gene. Additional tumours carried point changes that were unique to that tumour sample (across 100 sequences)—including small internal deletions in one of 10 BL and one of 11 HLs. Without knowing either the original sequence of the virus, or having a much clearer picture of EBNA3B sequence diversity, we cannot tell whether these also represent mutations, or merely polymorphisms (White et al. 2012). It is therefore currently impossible to judge how prevalent EBNA3B mutations are in lymphoma development.

Nevertheless, these data identified EBNA3B as a bone fide tumour suppressor that could have evolved to encourage T cell interactions via chemokine secretion. T cell contact is an essential component of the germinal centre reaction and B cell differentiation, so EBNA3B may also be paving the way to allow B cell differentiation and the switch to latency II. This is supported by the observation that EBNA3B-regulated genes include several that are repressed upon entry to the germinal centre (White et al. 2010). Equally, enhanced T cell surveillance clearly helps to limit the morbidity or mortality from lymphoproliferative disease caused by the proliferating B blast-like cells induced by the latency III growth programme. Coordinating the expression of EBNA3B with the oncoproteins EBNA2, EBNA3A and EBNA3C from a single transcription unit would represent an evolutionary adaptation to minimise oncogenic risk to the host (Fig. 4).

## 6.4 EBNA3s in B Cell Lymphomagenesis

As described in various chapters of this volume, and briefly above, different EBV-associated tumours exhibit different viral latency gene expression profiles, according to the type—and differentiation state—of the cell of origin of the tumour. Since the viral gene expression profile of diverse tumours in clinical samples is typically defined by the immunohistochemical detection of EBNA2 and LMP1 (and in situ hybridisation of EBERs), the expression of the EBNA3s is assumed to correlate with EBNA2 expression. By this logic, to the best of our knowledge, it is only the immunoblastic lymphomas (seen in the immunosuppressed) and a subset of DLBCLs—that express the full latency III B cell transcription unit (Fig. 1)—in which the EBNA3s are probably contributing to pathogenesis. The iatrogenic (i.e. immune-suppressive drug-induced) immunoblastic lymphomas can often be treated by removal of immunosuppression, allowing the patient's immune response to destroy the malignant cells. Where this fails, infusions of autologous or partially HLA-matched CTLs raised against EBV antigens are also able to establish an immune response that destroys the malignant cells (Heslop et al. 2010; Hislop et al. 2007). Typically, these CTLs are predominantly responsive to the EBNA3s, due to the apparent immunodominance of EBNA3-derived epitopes. The observation that latency III cells are scarce in immunocompetent individuals (Babcock et al. 1998, 1999), combined with the success of immunotherapy for latency III tumours, suggests that the EBNA3s make excellent candidates for immunotherapeutic targets, particularly once the diversity of EBNA3 CTL epitopes across viral strains and host HLA types has been fully defined (discussed in chapters by Rajiv Khanna and Gottschalk and Rooney in this volume).

The role of EBNA3s in lymphomagenesis is complicated by the apparently opposing behaviours of EBNA3A/C and EBNA3B. Where EBNA3A and EBNA3C seem to play crucial oncogenic roles in suppressing senescence and apoptosis, and perhaps overriding cell cycle checkpoints (discussed above), EBNA3B appears to have an opposing, tumour-suppressing role, as indicated by its mutation in some immunoblastic lymphomas. Mechanistically, this may be mediated by enhancing the expression of factors that promote T cell interaction, enhancing their surveillance by the immune system (see Sect. 6.3 and White et al. 2010) or some other as-yet unidentified function. Thus, inactivating mutation of EBNA3B appears to contribute to some lymphomas (White et al. 2012), whereas (at least for immunoblastic lymphomas) EBNA3C and EBNA3A seem likely to be essential.

The role of the EBNA3s in other malignancies remains much less clearly defined. The widely held supposition that the EBNA3s are absent from the latency I and latency II tumours—largely based on the assumption that EBNA3s are not present where EBNA2 is also absent—is supported by only a few early studies looking at virus gene expression in tumours or tissues. For instance, EBNA3s were not detected by immunoblotting in 24 nasopharyngeal cancers that had been passaged in nude mice (Young et al. 1988). An analysis of the latency II profile in the normal germinal centre failed to detect EBNA3 transcripts (Babcock et al. 2000), similarly the whole IgD$^{-ve}$ fraction of tonsils was negative for EBNA3 transcripts (Joseph et al. 2000). Attempts to detect EBNA3s in HL are hard to find in the literature, with the original description of the latency II status of HL not addressing the EBNA3s (Deacon et al. 1993). Certainly, no comprehensive survey exists to conclusively exclude the expression in HL of any of the EBNA3s.

Nevertheless, topologically, there is no reason that those latency states that use Qp should not express the EBNA3s, although Qp-driven EBNA3 has not been reported. Indeed, related transcripts (using the lytic Fp immediately upstream of Qp) have been found in latency I BL cell lines induced to enter the lytic cycle (Touitou et al. 2003). Additionally, a subset of BLs that carry a deletion of *EBNA2* express the EBNA3s from the Wp promoter (Kelly et al. 2002). Cell lines freshly established from these 'Wp-restricted' BLs exhibit a generally BL-like gene signature, but clearly represent a distinct subset, exhibiting a less germinal centre-type expression profile, that was more plasmacytoid (Kelly et al. 2013). Central players in this profile (reduced BCL6 and raised BLIMP1 and IRF4) are all regulated by EBNA3s in a BL background (White et al. 2010; http://www.epstein-barrvirus.org.uk). Notably, however, the Wp-restricted BLs are considerably more resistant to apoptosis in vitro than latency I BLs, through the effects of the BCL2 homologue BHRF1, combined with the EBNA3s and other factors acting on pro-apoptotic proteins including BIM and NOXA (Anderton et al. 2008; Kelly et al. 2009; Vereide and Sugden 2011; Watanabe et al. 2010; Yee et al. 2011). This increased robustness of the Wp-restricted BL may have implications for treatment strategies and success, but this has not yet been tested. It also remains formally possible that a Wp-restricted viral gene expression profile (or something similar) may also be found in other cancers. The expression of EBNA3s is only rarely assessed in cancers, and their low expression level (both of transcript and protein) and intimate association with chromatin make their detection potentially more challenging than for other antigens.

A few recent observations lend some credence to the possibility that EBNA3s may contribute to the onset of malignancies that do not exhibit typical latency III gene transcription. Analysis of the viral gene expression of a number of DLBCLs of the elderly identified several showing reverse transcription PCR and/or immunohistochemical evidence of tumours expressing EBNA3A but not EBNA2 (Nguyen-Van et al. 2011). Banked CTLs, whose specificity was predominantly against the EBNA3s, were able to successfully treat transplant-associated EBV lymphomas including one BL and five out of six HLs that were negative for EBNA2. Unfortunately, the authors were unable to assess EBNA3 expression in these lymphomas (McAulay et al. 2009). Furthermore, while potentially oncogenic EBNA3B truncations were identified in DLBCLs (that can adopt latency III), small internal deletions were also seen in EBNA3B genes isolated from an HL and a BL, which are not conventionally thought to express EBNA3B (White et al. 2012).

As indicated previously, the epigenetic nature of gene regulation mediated by the EBNA3s may continue after they are (presumably) no longer expressed. For instance, *BIM/BCL2L11*, repressed by EBNA3A and EBNA3C, is usually methylated in latency I EBV-positive BL, but often mutated or deleted in EBV-negative BL (Paschos et al. 2009; Richter-Larrea et al. 2010). Similar phenomena have been observed for $p16^{INK4a}$ (Klangby et al. 1998; Nagy et al. 2005). In the absence of validated reagents for their detection, the expression of EBNA3s in lymphomas remains inconclusive, but even the absence of the proteins themselves does not preclude essential roles, through epigenetic imprinting, in the development of a variety of EBV-associated tumours.

## 7 Summary and Outlook

Eukaryotic gene regulation is remarkably complicated, involving very large multiprotein complexes with enzymatic and structural functions, covalent modifications to histones and DNA, changes in the 3D organisation of chromatin and even the distribution of genes within nuclear space (Bickmore 2013; Bickmore and van Steensel 2013). Added to this enormous complexity is the problem of transferring expression patterns to progeny cells after DNA replication and mitosis—that is, creating some kind of epigenetic memory of gene expression programmes (Campos et al. 2014). Our current understanding of how the EBNA3 proteins fit into this broad picture of gene regulation has barely scratched the surface—but the results so far are interesting and thought provoking. Already, it appears that these viral factors manipulate multiple facets of chromatin organisation and function, but it is obvious that many issues remain unresolved. For example:

1. What is the explanation for the variety of EBNA3-regulated host genes? Is it merely a reflection of the regulatory elements that EBNA3-targeting factors bind? Or is it a more coordinated manipulation of a physiological B cell transcription programme(s) determined by differentiation-specific combinations and permutations from a group of host transcription factors (that probably includes among others RBP-JK/CBF1, PU1, SPI1, RUNX3, ATF2, BATF and IRF4)?
2. What are the features of EBNA3A, EBNA3B and EBNA3C that are responsible for their unique activities? This is particularly important for EBNA3A and EBNA3C that often cooperate, but do not appear to be interchangeable.
3. How do EBNA3s interact with the PcG protein system—is it by direct protein:protein contacts or by less direct means involving chromatin modifications or may be non-coding RNAs?
4. Are all the EBNA3 binding sites identified across the genome functionally significant (i.e. does localisation of a particular EBNA3 as revealed by ChIP-seq always result in a change in gene expression that is physiologically relevant)?
5. Where multiple EBNAs apparently bind to the same genomic locus, what determines which is functionally dominant or are they always mutually exclusive on a single site? What are the molecular interactions responsible for specific patterns of chromatin looping?
6. Do the EBNA3s bind to and regulate chromatinised EBV episomes? Although there are models suggesting this, so far there is little direct evidence.
7. Do the EBNA3s have functions distinct from their ability to regulate gene expression? Of particular importance is their suggested role(s) in modulation of the DDR.

8. Do the EBNA3s function in the same way during B cell infection in vivo as they do in vitro? For EBNA3B, there are limited data to suggest this is the case, but for EBNA3A and EBNA3C so far nothing is known. Further experiments with mice reconstituted with components of the human immune system will hopefully go some way to answer this question.

Striving for complete biochemical descriptions of EBNA3 function, coupled to precise genetic dissection of each protein will undoubtedly answer many of these questions and provide unique insights into EBV biology and disease pathogenesis, also into fundamental mechanisms of gene regulation. However, there is an important caveat that should not be overlooked—most of the studies described here have utilised the B95.8 strain of EBV or genes derived from it. This should be considered a 'laboratory-adapted' strain, chosen for its effectiveness in the ex vivo transformation of B cells, so a long-term goal must be to determine whether the EBNA network of interactions is conserved in virus strains that could be considered more representative of a 'wild-type' EBV. One should keep in mind the selection pressures that have created and 'moulded' these proteins: EBV did not evolve over millions of years to transform B cells into LCLs, nor to create B cell lymphomas. We should aim to determine what aspects of the fine virus–host equilibrium the EBNA3 proteins have evolved to maintain.

## References


Acosta JC, Gil J (2012) Senescence: a new weapon for cancer therapy. Trends Cell Biol 22:211–219

Allday MJ (2013) EBV finds a polycomb-mediated, epigenetic solution to the problem of oncogenic stress responses triggered by infection. Front Genet 4:212

Allday MJ, Crawford DH, Griffin BE (1988) Prediction and demonstration of a novel Epstein-Barr virus nuclear antigen. Nucleic Acids Res 16:4353–4367

Allday MJ, Crawford DH, Griffin BE (1989) Epstein-Barr virus latent gene expression during the initiation of B cell immortalization. J Gen Virol 70(Pt 7):1755–1764

Allday MJ, Farrell PJ (1994) Epstein-Barr virus nuclear antigen EBNA3C/6 expression maintains the level of latent membrane protein 1 in G1-arrested cells. J Virol 68:3491–3498

Altmann M, Hammerschmidt W (2005) Epstein-Barr virus provides a new paradigm: a requirement for the immediate inhibition of apoptosis. PLoS Biol 3:e404

Anderton E, Yee J, Smith P, Crook T, White RE, Allday MJ (2008) Two Epstein-Barr virus (EBV) oncoproteins cooperate to repress expression of the proapoptotic tumour-suppressor Bim: clues to the pathogenesis of Burkitt's lymphoma. Oncogene 27:421–433

Arvey A, Tempera I, Tsai K, Chen HS, Tikhmyanova N, Klichinsky M, Leslie C, Lieberman PM (2012) An atlas of the Epstein-Barr virus transcriptome and epigenome reveals host-virus regulatory interactions. Cell Host Microbe 12:233–245

Babcock GJ, Decker LL, Freeman RB, Thorley-Lawson DA (1999) Epstein-Barr virus-infected resting memory B cells, not proliferating lymphoblasts, accumulate in the peripheral blood of immunosuppressed patients. J Exp Med 190:567–576

Babcock GJ, Decker LL, Volk M, Thorley-Lawson DA (1998) EBV persistence in memory B cells in vivo. Immunity 9:395–404

Babcock GJ, Hochberg D, Thorley-Lawson AD (2000) The expression pattern of Epstein-Barr virus latent genes in vivo is dependent upon the differentiation stage of the infected B cell. Immunity 13:497–506

Baer R, Bankier AT, Biggin MD, Deininger PL, Farrell PJ, Gibson TJ, Hatfull G, Hudson GS, Satchwell SC, Seguin C et al (1984) DNA sequence and expression of the B95-8 Epstein-Barr virus genome. Nature 310:207–211

Bain M, Watson RJ, Farrell PJ, Allday MJ (1996) Epstein-Barr virus nuclear antigen 3C is a powerful repressor of transcription when tethered to DNA. J Virol 70:2481–2489

Bajaj BG, Murakami M, Cai Q, Verma SC, Lan K, Robertson ES (2008) Epstein-Barr virus nuclear antigen 3C interacts with and enhances the stability of the c-Myc oncoprotein. J Virol 82:4082–4090

Banerjee S, Lu J, Cai Q, Saha A, Jha HC, Dzeng RK, Robertson ES (2013) The EBV latent antigen 3C inhibits apoptosis through targeted regulation of interferon regulatory factors 4 and 8. PLoS Pathog 9:e1003314

Barolo S, Stone T, Bang AG, Posakony JW (2002) Default repression and Notch signaling: hairless acts as an adaptor to recruit the corepressors Groucho and dCtBP to suppressor of hairless. Genes Dev 16:1964–1976

Bartek J, Bartkova J, Lukas J (2007) DNA damage signalling guards against activated oncogenes and tumour progression. Oncogene 26:7773–7779

Bazot Q, Deschamps T, Tafforeau L, Siouda M, Leblanc P, Harth-Hertle ML, Rabourdin-Combe C, Lotteau V, Kempkes B, Tommasino M, Gruffat H, Manet E (2014) Epstein-Barr virus nuclear antigen 3A protein regulates CDKN2B transcription via interaction with MIZ-1. Nucleic Acids Res 42:9700–9716

Bernstein BE, Mikkelsen TS, Xie X, Kamal M, Huebert DJ, Cuff J, Fry B, Meissner A, Wernig M, Plath K, Jaenisch R, Wagschal A, Feil R, Schreiber SL, Lander ES (2006) A bivalent chromatin structure marks key developmental genes in embryonic stem cells. Cell 125:315–326

Bickmore WA (2013) The spatial organization of the human genome. Annu Rev Genomics Hum Genet 14:67–84

Bickmore WA, van Steensel B (2013) Genome architecture: domain organization of interphase chromosomes. Cell 152:1270–1284

Bourillot PY, Waltzer L, Sergeant A, Manet E (1998) Transcriptional repression by the Epstein-Barr virus EBNA3A protein tethered to DNA does not require RBP-Jkappa. J Gen Virol 79(Pt 2):363–370

Braig M, Schmitt CA (2006) Oncogene-induced senescence: putting the brakes on tumor development. Cancer Res 66:2881–2884

Brookes S, Rowe J, Ruas M, Llanos S, Clark PA, Lomax M, James MC, Vatcheva R, Bates S, Vousden KH, Parry D, Gruis N, Smit N, Bergman W, Peters G (2002) INK4a-deficient human diploid fibroblasts are resistant to RAS-induced senescence. EMBO J 21:2936–2945

Brown AC, Baigent SJ, Smith LP, Chattoo JP, Petherbridge LJ, Hawes P, Allday MJ, Nair V (2006) Interaction of MEQ protein and C-terminal-binding protein is critical for induction of lymphomas by Marek's disease virus. Proc Natl Acad Sci USA 103:1687–1692

Buck M, Burgess A, Stirzaker R, Krauer K, Sculley T (2006) Epstein-Barr virus nuclear antigen 3A contains six nuclear-localization signals. J Gen Virol 87:2879–2884

Burgess A, Buck M, Krauer K, Sculley T (2006) Nuclear localization of the Epstein-Barr virus EBNA3B protein. J Gen Virol 87:789–793

Cai Q, Guo Y, Xiao B, Banerjee S, Saha A, Lu J, Glisovic T, Robertson ES (2011) Epstein-Barr virus nuclear antigen 3C stabilizes Gemin3 to block p53-mediated apoptosis. PLoS Pathog 7:e1002418

Cai X, Schafer A, Lu S, Bilello JP, Desrosiers RC, Edwards R, Raab-Traub N, Cullen BR (2006) Epstein-Barr virus microRNAs are evolutionarily conserved and differentially expressed. PLoS Pathog 2:e23

Calderwood MA, Lee S, Holthaus AM, Blacklow SC, Kieff E, Johannsen E (2011) Epstein-Barr virus nuclear protein 3C binds to the N-terminal (NTD) and beta trefoil domains (BTD) of RBP/CSL; only the NTD interaction is essential for lymphoblastoid cell growth. Virology 414:19–25

Calderwood MA, Venkatesan K, Xing L, Chase MR, Vazquez A, Holthaus AM, Ewence AE, Li N, Hirozane-Kishikawa T, Hill DE, Vidal M, Kieff, Johannsen E (2007) Epstein-Barr virus and virus human protein interaction maps. Proc Natl Acad Sci USA 104:7606–7611
Campos EI, Stafford JM, Reinberg D (2014) Epigenetic inheritance: histone bookmarks across generations. Trends Cell Biol 24:664–674
Cedar H, Bergman Y (2009) Linking DNA methylation and histone modification: patterns and paradigms. Nat Rev Genet 10:295–304
Chen A, Divisconte M, Jiang X, Quink C, Wang F (2005) Epstein-Barr virus with the latent infection nuclear antigen 3B completely deleted is still competent for B-cell growth transformation in vitro. J Virol 79:4506–4509
Chen A, Zhao B, Kieff E, Aster JC, Wang F (2006) EBNA-3B- and EBNA-3C-regulated cellular genes in Epstein-Barr virus-immortalized lymphoblastoid cell lines. J Virol 80:10139–10150
Chinnadurai G (2007) Transcriptional regulation by C-terminal binding proteins. Int J Biochem Cell Biol 39:1593–1607
Choudhuri T, Verma SC, Lan K, Murakami M, Robertson ES (2007) The ATM/ATR signaling effector Chk2 is targeted by Epstein-Barr virus nuclear antigen 3C to release the G2/M cell cycle block. J Virol 81:6718–6730
Choudhuri T, Verma SC, Lan K, Robertson ES (2006) Expression of alpha V integrin is modulated by Epstein-Barr virus nuclear antigen 3C and the metastasis suppressor Nm23-H1 through interaction with the GATA-1 and Sp1 transcription factors. Virology 351:58–72
Cludts I, Farrell PJ (1998) Multiple functions within the Epstein-Barr virus EBNA-3A protein. J Virol 72:1862–1869
Cooper A, Johannsen E, Maruo S, Cahir-McFarland E, Illanes D, Davidson D, Kieff E (2003) EBNA3A association with RBP-Jkappa down-regulates c-myc and Epstein-Barr virus-transformed lymphoblast growth. J Virol 77:999–1010
Cotter MA 2nd, Robertson ES (2000) Modulation of histone acetyltransferase activity through interaction of epstein-barr nuclear antigen 3C with prothymosin alpha. Mol Cell Biol 20:5722–5735
Crawford DH (2001) Biology and disease associations of Epstein-Barr virus. Philos Trans R Soc Lond B Biol Sci 356:461–473
Dalbies-Tran R, Stigger-Rosser E, Dotson T, Sample CE (2001) Amino acids of Epstein-Barr virus nuclear antigen 3A essential for repression of Jkappa-mediated transcription and their evolutionary conservation. J Virol 75:90–99
Dang CV, O'Donnell KA, Juopperi T (2005) The great MYC escape in tumorigenesis. Cancer Cell 8:177–178
Deacon EM, Pallesen G, Niedobitek G, Crocker J, Brooks L, Rickinson AB, Young LS (1993) Epstein-Barr virus and Hodgkin's disease: transcriptional analysis of virus latency in the malignant cells. J Exp Med 177:339–349
Delecluse HJ, Hilsendegen T, Pich D, Zeidler R, Hammerschmidt W (1998) Propagation and recovery of intact, infectious Epstein-Barr virus from prokaryotic to human cells. Proc Natl Acad Sci USA 95:8245–8250
Dufour JH, Dziejman M, Liu MT, Leung JH, Lane TE, Luster AD (2002) IFN-gamma-inducible protein 10 (IP-10; CXCL10)-deficient mice reveal a role for IP-10 in effector T cell generation and trafficking. J Immunol 168:3195–3204
Egle A, Harris AW, Bouillet P, Cory S (2004) Bim is a suppressor of Myc-induced mouse B cell leukemia. Proc Natl Acad Sci USA 101:6164–6169
Ehlers B, Ochs A, Leendertz F, Goltz M, Boesch C, Matz-Rensing K (2003) Novel simian homologues of Epstein-Barr virus. J Virol 77:10695–10699
Ehlers B, Spiess K, Leendertz F, Peeters M, Boesch C, Gatherer D, McGeoch DJ (2010) Lymphocryptovirus phylogeny and the origins of Epstein-Barr virus. J Gen Virol 91:630–642
Elliott SL, Suhrbier A, Miles JJ, Lawrence G, Pye SJ, Le TT, Rosenstengel A, Nguyen T, Allworth A, Burrows SR, Cox J, Pye D, Moss DJ, Bharadwaj M (2008) Phase I trial of a CD8+T-cell peptide epitope-based vaccine for infectious mononucleosis. J Virol 82:1448–1457
Epstein MA (2001) Historical background. Philos Trans R Soc Lond B Biol Sci 356:413–420

Feederle R, Bartlett EJ, Delecluse HJ (2010) Epstein-Barr virus genetics: talking about the BAC generation. Herpesviridae 1:6

Fischer SF, Bouillet P, O'Donnell K, Light A, Tarlinton DM, Strasser A (2007) Proapoptotic BH3-only protein Bim is essential for developmentally programmed death of germinal center-derived memory B cells and antibody-forming cells. Blood 110:3978–3984

Garrido JL, Maruo S, Takada K, Rosendorff A (2009) EBNA3C interacts with Gadd34 and counteracts the unfolded protein response. Virol J 6:231

Gil J, Peters G (2006) Regulation of the INK4b-ARF-INK4a tumour suppressor locus: all for one or one for all. Nat Rev Mol Cell Biol 7:667–677

Gorzer I, Niesters HG, Cornelissen JJ, Puchhammer-Stockl E (2006) Characterization of Epstein-Barr virus Type I variants based on linked polymorphism among EBNA3A, -3B, and -3C genes. Virus Res 118:105–114

Gottschalk S, Ng CY, Perez M, Smith CA, Sample C, Brenner MK, Heslop HE, Rooney CM (2001) An Epstein-Barr virus deletion mutant associated with fatal lymphoproliferative disease unresponsive to therapy with virus-specific CTLs. Blood 97:835–843

Gruhne B, Sompallae R, Masucci MG (2009) Three Epstein-Barr virus latency proteins independently promote genomic instability by inducing DNA damage, inhibiting DNA repair and inactivating cell cycle checkpoints. Oncogene 28:3997–4008

Guruharsha KG, Kankel MW, Artavanis-Tsakonas S (2012) The Notch signalling system: recent insights into the complexity of a conserved pathway. Nat Rev Genet 13:654–666

Halazonetis TD, Gorgoulis VG, Bartek J (2008) An oncogene-induced DNA damage model for cancer development. Science 319:1352–1355

Harth-Hertle ML, Scholz BA, Erhard F, Glaser LV, Dolken L, Zimmer R, Kempkes B (2013) Inactivation of intergenic enhancers by EBNA3A initiates and maintains polycomb signatures across a chromatin domain encoding CXCL10 and CXCL9. PLoS Pathog 9:e1003638

Hawkins JB, Delgado-Eckert E, Thorley-Lawson DA, Shapiro M (2013) The cycle of EBV infection explains persistence, the sizes of the infected cell populations and which come under CTL regulation. PLoS Pathog 9:e1003685

Hemann MT, Bric A, Teruya-Feldstein J, Herbst A, Nilsson JA, Cordon-Cardo C, Cleveland JL, Tansey WP, Lowe SW (2005) Evasion of the p53 tumour surveillance network by tumour-derived MYC mutants. Nature 436:807–811

Hennessy K, Fennewald S, Kieff E (1985) A third viral nuclear protein in lymphoblasts immortalized by Epstein-Barr virus. Proc Natl Acad Sci USA 82:5944–5948

Hennessy K, Wang F, Bushman EW, Kieff E (1986) Definitive identification of a member of the Epstein-Barr virus nuclear protein 3 family. Proc Natl Acad Sci USA 83:5693–5697

Hertle ML, Popp C, Petermann S, Maier S, Kremmer E, Lang R, Mages J, Kempkes B (2009) Differential gene expression patterns of EBV infected EBNA-3A positive and negative human B lymphocytes. PLoS Pathog 5:e1000506

Heslop HE, Slobod KS, Pule MA, Hale GA, Rousseau A, Smith CA, Bollard CM, Liu H, Wu MF, Rochester RJ, Amrolia PJ, Hurwitz JL, Brenner MK, Rooney CM (2010) Long-term outcome of EBV-specific T-cell infusions to prevent or treat EBV-related lymphoproliferative disease in transplant recipients. Blood 115:925–935

Hickabottom M, Parker GA, Freemont P, Crook T, Allday MJ (2002) Two nonconsensus sites in the Epstein-Barr virus oncoprotein EBNA3A cooperate to bind the co-repressor carboxyl-terminal-binding protein (CtBP). J Biol Chem 277:47197–47204

Hislop AD, Taylor GS, Sauce D, Rickinson AB (2007) Cellular responses to viral infection in humans: lessons from Epstein-Barr virus. Annu Rev Immunol 25:587–617

Humeniuk R, Rosu-Myles M, Fares J, Koller R, Bies J, Wolff L (2013) The role of tumor suppressor p15Ink4b in the regulation of hematopoietic progenitor cell fate. Blood Cancer J 3:e99

Isaksson A, Berggren M, Ricksten A (2003) Epstein-Barr virus U leader exon contains an internal ribosome entry site. Oncogene 22:572–581

Jha HC, Aj MP, Saha A, Banerjee S, Lu J, Robertson ES (2014) Epstein-Barr virus essential antigen EBNA3C attenuates H2AX expression. J Virol 88:3776–3788
Jha HC, Lu J, Saha A, Cai Q, Banerjee S, Prasad MA, Robertson ES (2013) EBNA3C-mediated regulation of aurora kinase B contributes to Epstein-Barr virus-induced B-cell proliferation through modulation of the activities of the retinoblastoma protein and apoptotic caspases. J Virol 87:12121–12138
Jiang H, Cho YG, Wang F (2000) Structural, functional, and genetic comparisons of Epstein-Barr virus nuclear antigen 3A, 3B, and 3C homologues encoded by the rhesus lymphocryptovirus. J Virol 74:5921–5932
Jiang S, Willox B, Zhou H, Holthaus AM, Wang A, Shi TT, Maruo S, Kharchenko PV, Johannsen EC, Kieff E, Zhao B (2014) Epstein-Barr virus nuclear antigen 3C binds to BATF/IRF4 or SPI1/IRF4 composite sites and recruits Sin3A to repress CDKN2A. Proc Natl Acad Sci USA 111:421–426
Jimenez-Ramirez C, Brooks AJ, Forshell LP, Yakimchuk K, Zhao B, Fulgham TZ, Sample CE (2006) Epstein-Barr virus EBNA-3C is targeted to and regulates expression from the bidirectional LMP-1/2B promoter. J Virol 80:11200–11208
Johannsen E, Miller CL, Grossman SR, Kieff E (1996) EBNA-2 and EBNA-3C extensively and mutually exclusively associate with RBPJkappa in Epstein-Barr virus-transformed B lymphocytes. J Virol 70:4179–4183
Joseph AM, Babcock GJ, Thorley-Lawson DA (2000) Cells expressing the Epstein-Barr virus growth program are present in and restricted to the naive B-cell subset of healthy tonsils. J Virol 74:9964–9971
Kashuba E, Yurchenko M, Yenamandra SP, Snopok B, Isaguliants M, Szekely L, Klein G (2008) EBV-encoded EBNA-6 binds and targets MRS18-2 to the nucleus, resulting in the disruption of pRb-E2F1 complexes. Proc Natl Acad Sci USA 105:5489–5494
Kaul R, Verma SC, Murakami M, Lan K, Choudhuri T, Robertson ES (2006) Epstein-Barr virus protein can upregulate cyclo-oxygenase-2 expression through association with the suppressor of metastasis Nm23-H1. J Virol 80:1321–1331
Kelly G, Bell A, Rickinson A (2002) Epstein-Barr virus-associated Burkitt lymphomagenesis selects for downregulation of the nuclear antigen EBNA2. Nat Med 8:1098–1104
Kelly GL, Long HM, Stylianou J, Thomas WA, Leese A, Bell AI, Bornkamm GW, Mautner J, Rickinson AB, Rowe M (2009) An Epstein-Barr virus anti-apoptotic protein constitutively expressed in transformed cells and implicated in burkitt lymphomagenesis: the Wp/BHRF1 link. PLoS Pathog 5:e1000341
Kelly GL, Stylianou J, Rasaiyaah J, Wei W, Thomas W, Croom-Carter D, Kohler C, Spang R, Woodman C, Kellam P, Rickinson AB, Bell AI (2013) Different patterns of Epstein-Barr virus latency in endemic Burkitt lymphoma (BL) lead to distinct variants within the BL-associated gene expression signature. J Virol 87:2882–2894
Khanna R, Burrows SR, Kurilla MG, Jacob CA, Misko IS, Sculley TB, Kieff E, Moss DJ (1992) Localization of Epstein-Barr virus cytotoxic T cell epitopes using recombinant vaccinia: implications for vaccine development. J Exp Med 176:169–176
Khanna R, Slade RW, Poulsen L, Moss DJ, Burrows SR, Nicholls J, Burrows JM (1997) Evolutionary dynamics of genetic variation in Epstein-Barr virus isolates of diverse geographical origins: evidence for immune pressure-independent genetic drift. J Virol 71:8340–8346
Kheradmand Kia S, Solaimani Kartalaei P, Farahbakhshian E, Pourfarzad F, von Lindern M, Verrijzer CP (2009) EZH2-dependent chromatin looping controls INK4a and INK4b, but not ARF, during human progenitor cell differentiation and cellular senescence. Epigenetics Chromatin 2:16
Kienzle N, Young DB, Liaskou D, Buck M, Greco S, Sculley TB (1999) Intron retention may regulate expression of Epstein-Barr virus nuclear antigen 3 family genes. J Virol 73:1195–1204
Kim SM, Kang SH, Lee WK (2006) Identification of two types of naturally-occurring intertypic recombinants of Epstein-Barr virus. Mol Cells 21:302–307

Kis LL, Takahara M, Nagy N, Klein G, Klein E (2006) Cytokine mediated induction of the major Epstein-Barr virus (EBV)-encoded transforming protein, LMP-1. Immunol Lett 104:83–88
Kis LL, Salamon D, Persson EK, Nagy N, Scheeren FA, Spits H, Klein G, Klein E (2010) IL-21 imposes a type II EBV gene expression on type III and type I B cells by the repression of C- and activation of LMP-1-promoter. Proc Natl Acad Sci USA 107:872–877
Klangby U, Okan I, Magnusson KP, Wendland M, Lind P, Wiman KG (1998) p16/INK4a and p15/INK4b gene methylation and absence of p16/INK4a mRNA and protein expression in Burkitt's lymphoma. Blood 91:1680–1687
Knight JS, Lan K, Subramanian C, Robertson ES (2003) Epstein-Barr virus nuclear antigen 3C recruits histone deacetylase activity and associates with the corepressors mSin3A and NCoR in human B-cell lines. J Virol 77:4261–4272
Knight JS, Robertson ES (2004) Epstein-Barr virus nuclear antigen 3C regulates cyclin A/p27 complexes and enhances cyclin A-dependent kinase activity. J Virol 78:1981–1991
Knight JS, Sharma N, Kalman DE, Robertson ES (2004) A cyclin-binding motif within the amino-terminal homology domain of EBNA3C binds cyclin A and modulates cyclin A-dependent kinase activity in Epstein-Barr virus-infected cells. J Virol 78:12857–12867
Knight JS, Sharma N, Robertson ES (2005a) Epstein-Barr virus latent antigen 3C can mediate the degradation of the retinoblastoma protein through an SCF cellular ubiquitin ligase. Proc Natl Acad Sci USA 102:18562–18566
Knight JS, Sharma N, Robertson ES (2005b) SCFSkp2 complex targeted by Epstein-Barr virus essential nuclear antigen. Mol Cell Biol 25:1749–1763
Kohlhof H, Hampel F, Hoffmann R, Burtscher H, Weidle UH, Holzel M, Eick D, Zimber-Strobl U, Strobl LJ (2009) Notch1, Notch2, and Epstein-Barr virus-encoded nuclear antigen 2 signaling differentially affects proliferation and survival of Epstein-Barr virus-infected B cells. Blood 113:5506–5515
Krauer K, Buck M, Flanagan J, Belzer D, Sculley T (2004a) Identification of the nuclear localization signals within the Epstein-Barr virus EBNA-6 protein. J Gen Virol 85:165–172
Krauer KG, Buck M, Belzer DK, Flanagan J, Chojnowski GM, Sculley TB (2004b) The Epstein-Barr virus nuclear antigen-6 protein co-localizes with EBNA-3 and survival of motor neurons protein. Virology 318:280–294
Le Roux A, Berebbi M, Moukaddem M, Perricaudet M, Joab I (1993) Identification of a short amino acid sequence essential for efficient nuclear targeting of the Epstein-Barr virus nuclear antigen 3A. J Virol 67:1716–1720
Le Roux A, Kerdiles B, Walls D, Dedieu JF, Perricaudet M (1994) The Epstein-Barr virus determined nuclear antigens EBNA-3A, -3B, and -3C repress EBNA-2-mediated transactivation of the viral terminal protein 1 gene promoter. Virology 205:596–602
Lee S, Sakakibara S, Maruo S, Zhao B, Calderwood MA, Holthaus AM, Lai CY, Takada K, Kieff E, Johannsen E (2009) Epstein-Barr virus nuclear protein 3C domains necessary for lymphoblastoid cell growth: interaction with RBP-Jkappa regulates TCL1. J Virol 83:12368–12377
Lin J, Johannsen E, Robertson E, Kieff E (2002) Epstein-Barr virus nuclear antigen 3C putative repression domain mediates coactivation of the LMP1 promoter with EBNA-2. J Virol 76:232–242
Lowe SW, Cepero E, Evan G (2004) Intrinsic tumour suppression. Nature 432:307–315
Maruo S, Johannsen E, Illanes D, Cooper A, Kieff E (2003) Epstein-Barr virus nuclear protein EBNA3A is critical for maintaining lymphoblastoid cell line growth. J Virol 77:10437–10447
Maruo S, Johannsen E, Illanes D, Cooper A, Zhao B, Kieff E (2005) Epstein-Barr virus nuclear protein 3A domains essential for growth of lymphoblasts: transcriptional regulation through RBP-Jkappa/CBF1 is critical. J Virol 79:10171–10179
Maruo S, Wu Y, Ishikawa S, Kanda T, Iwakiri D, Takada K (2006) Epstein-Barr virus nuclear protein EBNA3C is required for cell cycle progression and growth maintenance of lymphoblastoid cells. Proc Natl Acad Sci USA 103:19500–19505

Maruo S, Wu Y, Ito T, Kanda T, Kieff ED, Takada K (2009) Epstein-Barr virus nuclear protein EBNA3C residues critical for maintaining lymphoblastoid cell growth. Proc Natl Acad Sci USA 106:4419–4424

Maruo S, Zhao B, Johannsen E, Kieff E, Zou J, Takada K (2011) Epstein-Barr virus nuclear antigens 3C and 3A maintain lymphoblastoid cell growth by repressing p16INK4A and p14ARF expression. Proc Natl Acad Sci USA 108:1919–1924

McAulay KA, Haque T, Urquhart G, Bellamy C, Guiretti D, Crawford DH (2009) Epitope specificity and clonality of EBV-specific CTLs used to treat posttransplant lymphoproliferative disease. J Immunol 182:3892–3901

McClellan MJ, Khasnis S, Wood CD, Palermo RD, Schlick SN, Kanhere AS, Jenner RG, West MJ (2012) Downregulation of integrin receptor-signaling genes by Epstein-Barr virus EBNA 3C via promoter-proximal and -distal binding elements. J Virol 86:5165–5178

McClellan MJ, Wood CD, Ojeniyi O, Cooper TJ, Kanhere A, Arvey A, Webb HM, Palermo RD, Harth-Hertle ML, Kempkes B, Jenner RG, West MJ (2013) Modulation of enhancer looping and differential gene targeting by Epstein-Barr virus transcription factors directs cellular reprogramming. PLoS Pathog 9:e1003636

Midgley RS, Bell AI, McGeoch DJ, Rickinson AB (2003) Latent gene sequencing reveals familial relationships among Chinese Epstein-Barr virus strains and evidence for positive selection of A11 epitope changes. J Virol 77:11517–11530

Midgley RS, Blake NW, Yao QY, Croom-Carter D, Cheung ST, Leung SF, Chan AT, Johnson PJ, Huang D, Rickinson AB, Lee SP (2000) Novel intertypic recombinants of epstein-barr virus in the chinese population. J Virol 74:1544–1548

Mlecnik B, Tosolini M, Charoentong P, Kirilovsky A, Bindea G, Berger A, Camus M, Gillard M, Bruneval P, Fridman WH, Pages F, Trajanoski Z, Galon J (2010) Biomolecular network reconstruction identifies T-cell homing factors associated with survival in colorectal cancer. Gastroenterology 138:1429–1440

Morel V, Lecourtois M, Massiani O, Maier D, Preiss A, Schweisguth F (2001) Transcriptional repression by suppressor of hairless involves the binding of a hairless-dCtBP complex in Drosophila. Curr Biol 11:789–792

Murray RJ, Kurilla MG, Brooks JM, Thomas WA, Rowe M, Kieff E, Rickinson AB (1992) Identification of target antigens for the human cytotoxic T cell response to Epstein-Barr virus (EBV): implications for the immune control of EBV-positive malignancies. J Exp Med 176:157–168

Nagy E, Veress G, Szarka K, Csoma E, Beck Z (2005) Frequent methylation of p16INK4A/p14ARF promoters in tumorigenesis of Epstein-Barr virus transformed lymphoblastoid cell lines. Anticancer Res 25:2153–2160

Nagy N, Adori M, Rasul A, Heuts F, Salamon D, Ujvari D, Madapura HS, Leveau B, Klein G, Klein E (2012) Soluble factors produced by activated CD4+T cells modulate EBV latency. Proc Natl Acad Sci USA 109:1512–1517

Naumova N, Smith EM, Zhan Y, Dekker J (2012) Analysis of long-range chromatin interactions using chromosome conformation capture. Methods 58:192–203

Nguyen-Van D, Keane C, Han E, Jones K, Nourse JP, Vari F, Ross N, Crooks P, Ramuz O, Green M, Griffith L, Trappe R, Grigg A, Mollee P, Gandhi MK (2011) Epstein-Barr virus-positive diffuse large B-cell lymphoma of the elderly expresses EBNA3A with conserved CD8 T-cell epitopes. Am J Blood Res 1:146–159

Nikiforow S, Lacasce AS (2014) Targeting Epstein-Barr virus-associated lymphomas. J Clin Oncol 32:830–832

Nikitin PA, Luftig MA (2012) The DNA damage response in viral-induced cellular transformation. Br J Cancer 106:429–435

Nikitin PA, Yan CM, Forte E, Bocedi A, Tourigny JP, White RE, Allday MJ, Patel A, Dave SS, Kim W, Hu K, Guo J, Tainter D, Rusyn E, Luftig MA (2010) An ATM/Chk2-mediated DNA damage-responsive signaling pathway suppresses Epstein-Barr virus transformation of primary human B cells. Cell Host Microbe 8:510–522

O'Nions J, Turner A, Craig R, Allday MJ (2006) Epstein-Barr virus selectively deregulates DNA damage responses in normal B cells but has no detectable effect on regulation of the tumor suppressor p53. J Virol 80:12408–12413

Palmero I, Holder A, Sinclair AJ, Dickson C, Peters G (1993) Cyclins D1 and D2 are differentially expressed in human B-lymphoid cell lines. Oncogene 8:1049–1054

Parker GA, Crook T, Bain M, Sara EA, Farrell PJ, Allday MJ (1996) Epstein-Barr virus nuclear antigen (EBNA)3C is an immortalizing oncoprotein with similar properties to adenovirus E1A and papillomavirus E7. Oncogene 13:2541–2549

Parker GA, Touitou R, Allday MJ (2000) Epstein-Barr virus EBNA3C can disrupt multiple cell cycle checkpoints and induce nuclear division divorced from cytokinesis. Oncogene 19:700–709

Paschos K, Parker GA, Watanatanasup E, White RE, Allday MJ (2012) BIM promoter directly targeted by EBNA3C in polycomb-mediated repression by EBV. Nucleic Acids Res 40:7233–7246

Paschos K, Smith P, Anderton E, Middeldorp JM, White RE, Allday MJ (2009) Epstein-barr virus latency in B cells leads to epigenetic repression and CpG methylation of the tumour suppressor gene Bim. PLoS Pathog 5:e1000492

Peng R, Moses SC, Tan J, Kremmer E, Ling PD (2005) The Epstein-Barr virus EBNA-LP protein preferentially coactivates EBNA2-mediated stimulation of latent membrane proteins expressed from the viral divergent promoter. J Virol 79:4492–4505

Petti L, Kieff E (1988) A sixth Epstein-Barr virus nuclear protein (EBNA3B) is expressed in latently infected growth-transformed lymphocytes. J Virol 62:2173–2178

Petti L, Sample C, Kieff E (1990) Subnuclear localization and phosphorylation of Epstein-Barr virus latent infection nuclear proteins. Virology 176:563–574

Petti L, Sample J, Wang F, Kieff E (1988) A fifth Epstein-Barr virus nuclear protein (EBNA3C) is expressed in latently infected growth-transformed lymphocytes. J Virol 62:1330–1338

Pokrovskaja K, Ehlin-Henriksson B, Bartkova J, Bartek J, Scuderi R, Szekely L, Wiman KG, Klein G (1996) Phenotype-related differences in the expression of D-type cyclins in human B cell-derived lines. Cell Growth Differ 7:1723–1732

Price AM, Tourigny JP, Forte E, Salinas RE, Dave SS, Luftig MA (2012) Analysis of Epstein-Barr virus-regulated host gene expression changes through primary B-cell outgrowth reveals delayed kinetics of latent membrane protein 1-mediated NF-kappaB activation. J Virol 86:11096–11106

Radkov SA, Bain M, Farrell PJ, West M, Rowe M, Allday MJ (1997) Epstein-Barr virus EBNA3C represses Cp, the major promoter for EBNA expression, but has no effect on the promoter of the cell gene CD21. J Virol 71:8552–8562

Radkov SA, Touitou R, Brehm A, Rowe M, West M, Kouzarides T, Allday MJ (1999) Epstein-Barr virus nuclear antigen 3C interacts with histone deacetylase to repress transcription. J Virol 73:5688–5697

Richter-Larrea JA, Robles EF, Fresquet V, Beltran E, Rullan AJ, Agirre X, Calasanz MJ, Panizo C, Richter JA, Hernandez JM, Roman-Gomez J, Prosper F, Martinez-Climent JA (2010) Reversion of epigenetically mediated BIM silencing overcomes chemoresistance in Burkitt lymphoma. Blood 116:2531–2542

Ricksten A, Kallin B, Alexander H, Dillner J, Fahraeus R, Klein G, Lerner R, Rymo L (1988) BamHI E region of the Epstein-Barr virus genome encodes three transformation-associated nuclear proteins. Proc Natl Acad Sci USA 85:995–999

Rivailler P, Cho YG, Wang F (2002) Complete genomic sequence of an Epstein-Barr virus-related herpesvirus naturally infecting a new world primate: a defining point in the evolution of oncogenic lymphocryptoviruses. J Virol 76:12055–12068

Robertson ES, Grossman S, Johannsen E, Miller C, Lin J, Tomkinson B, Kieff E (1995) Epstein-Barr virus nuclear protein 3C modulates transcription through interaction with the sequence-specific DNA-binding protein J kappa. J Virol 69:3108–3116

Robertson ES, Lin J, Kieff E (1996) The amino-terminal domains of Epstein-Barr virus nuclear proteins 3A, 3B, and 3C interact with RBPJ(kappa). J Virol 70:3068–3074

Rochford R, Hobbs MV, Garnier JL, Cooper NR, Cannon MJ (1993) Plasmacytoid differentiation of Epstein-Barr virus-transformed B cells in vivo is associated with reduced expression of viral latent genes. Proc Natl Acad Sci USA 90:352–356

Rosendorff A, Illanes D, David G, Lin J, Kieff E, Johannsen E (2004) EBNA3C coactivation with EBNA2 requires a SUMO homology domain. J Virol 78:367–377

Roughan JE, Thorley-Lawson DA (2009) The intersection of Epstein-Barr virus with the germinal center. J Virol 83:3968–3976

Saha A, Bamidele A, Murakami M, Robertson ES (2011) EBNA3C attenuates the function of p53 through interaction with inhibitor of growth family proteins 4 and 5. J Virol 85:2079–2088

Saha A, Lu J, Morizur L, Upadhyay SK, Aj MP, Robertson ES (2012) E2F1 mediated apoptosis induced by the DNA damage response is blocked by EBV nuclear antigen 3C in lymphoblastoid cells. PLoS Pathog 8:e1002573

Saha A, Murakami M, Kumar P, Bajaj B, Sims K, Robertson ES (2009) Epstein-Barr virus nuclear antigen 3C augments Mdm2-mediated p53 ubiquitination and degradation by deubiquitinating Mdm2. J Virol 83:4652–4669

Salamon D, Adori M, Ujvari D, Wu L, Kis LL, Madapura HS, Nagy N, Klein G, Klein E (2012) Latency type-dependent modulation of Epstein-Barr virus-encoded latent membrane protein 1 expression by type I interferons in B cells. J Virol 86:4701–4707

Sample C, Parker B (1994) Biochemical characterization of Epstein-Barr virus nuclear antigen 3A and 3C proteins. Virology 205:534–539

Sample J, Young L, Martin B, Chatman T, Kieff E, Rickinson A (1990) Epstein-Barr virus types 1 and 2 differ in their EBNA-3A, EBNA-3B, and EBNA-3C genes. J Virol 64:4084–4092

Serrano M, Lin AW, McCurrach ME, Beach D, Lowe SW (1997) Oncogenic ras provokes premature cell senescence associated with accumulation of p53 and p16INK4a. Cell 88:593–602

Sewalt RG, Gunster MJ, van der Vlag J, Satijn DP, Otte AP (1999) C-Terminal binding protein is a transcriptional repressor that interacts with a specific class of vertebrate Polycomb proteins. Mol Cell Biol 19:777–787

Shannon-Lowe C, Baldwin G, Feederle R, Bell A, Rickinson A, Delecluse HJ (2005) Epstein-Barr virus-induced B-cell transformation: quantitating events from virus binding to cell outgrowth. J Gen Virol 86:3009–3019

Sherr CJ (2012) Ink4-arf locus in cancer and aging. Wiley Interdiscip Rev Dev Biol 1:731–741

Shi Y, Lan F, Matson C, Mulligan P, Whetstine JR, Cole PA, Casero RA, Shi Y (2004) Histone demethylation mediated by the nuclear amine oxidase homolog LSD1. Cell 119:941–953

Shi Y, Sawada J, Sui G, el Affar B, Whetstine JR, Lan F, Ogawa H, Luke MP, Nakatani Y, Shi Y (2003) Coordinated histone modifications mediated by a CtBP co-repressor complex. Nature 422:735–738

Siemer D, Kurth J, Lang S, Lehnerdt G, Stanelle J, Kuppers R (2008) EBV transformation overrides gene expression patterns of B cell differentiation stages. Mol Immunol 45:3133–3141

Sinclair AJ, Palmero I, Peters G, Farrell PJ (1994) EBNA-2 and EBNA-LP cooperate to cause G0 to G1 transition during immortalization of resting human B lymphocytes by Epstein-Barr virus. EMBO J 13:3321–3328

Skalska L, White RE, Franz M, Ruhmann M, Allday MJ (2010) Epigenetic repression of p16(INK4A) by latent Epstein-Barr virus requires the interaction of EBNA3A and EBNA3C with CtBP. PLoS Pathog 6:e1000951

Skalska L, White RE, Parker GA, Sinclair AJ, Paschos K, Allday MJ (2013) Induction of p16(INK4a) is the major barrier to proliferation when Epstein-Barr virus (EBV) transforms primary B cells into lymphoblastoid cell lines. PLoS Pathog 9:e1003187

Spender LC, Cannell EJ, Hollyoake M, Wensing B, Gawn JM, Brimmell M, Packham G, Farrell PJ (1999) Control of cell cycle entry and apoptosis in B lymphocytes infected by Epstein-Barr virus. J Virol 73:4678–4688

Steven NM, Leese AM, Annels NE, Lee SP, Rickinson AB (1996) Epitope focusing in the primary cytotoxic T cell response to Epstein-Barr virus and its relationship to T cell memory. J Exp Med 184:1801–1813

Stock JK, Giadrossi S, Casanova M, Brookes E, Vidal M, Koseki H, Brockdorff N, Fisher AG, Pombo A (2007) Ring1-mediated ubiquitination of H2A restrains poised RNA polymerase II at bivalent genes in mouse ES cells. Nat Cell Biol 9:1428–1435

Strasser A (2005) The role of BH3-only proteins in the immune system. Nat Rev Immunol 5:189–200

Strowig T, Gurer C, Ploss A, Liu YF, Arrey F, Sashihara J, Koo G, Rice CM, Young JW, Chadburn A, Cohen JI, Munz C (2009) Priming of protective T cell responses against virus-induced tumors in mice with human immune system components. J Exp Med 206:1423–1434

Subramanian C, Cotter MA 2nd, Robertson ES (2001) Epstein-Barr virus nuclear protein EBNA-3C interacts with the human metastatic suppressor Nm23-H1: a molecular link to cancer metastasis. Nat Med 7:350–355

Sundqvist A, Sollerbrant K, Svensson C (1998) The carboxy-terminal region of adenovirus E1A activates transcription through targeting of a C-terminal binding protein-histone deacetylase complex. FEBS Lett 429:183–188

Tempera I, Lieberman PM (2014) Epigenetic regulation of EBV persistence and oncogenesis. Semin Cancer Biol 26:22–29

Thorley-Lawson DA, Allday MJ (2008) The curious case of the tumour virus: 50 years of Burkitt's lymphoma. Nat Rev Microbiol 6:913–924

Thorley-Lawson DA, Gross A (2004) Persistence of the Epstein-Barr virus and the origins of associated lymphomas. N Engl J Med 350:1328–1337

Thorley-Lawson DA, Hawkins JB, Tracy SI, Shapiro M (2013) The pathogenesis of Epstein-Barr virus persistent infection. *Curr Opin Virol*

Thorley-Lawson DA, Strominger JL (1978) Reversible inhibition by phosphonoacetic acid of human B lymphocyte transformation by Epstein-Barr virus. Virology 86:423–431

Tomkinson B, Kieff E (1992a) Second-site homologous recombination in Epstein-Barr virus: insertion of type 1 EBNA 3 genes in place of type 2 has no effect on in vitro infection. J Virol 66:780–789

Tomkinson B, Kieff E (1992b) Use of second-site homologous recombination to demonstrate that Epstein-Barr virus nuclear protein 3B is not important for lymphocyte infection or growth transformation in vitro. J Virol 66:2893–2903

Tomkinson B, Robertson E, Kieff E (1993) Epstein-Barr virus nuclear proteins EBNA-3A and EBNA-3C are essential for B-lymphocyte growth transformation. J Virol 67:2014–2025

Touitou R, Arbach H, Cochet C, Feuillard J, Martin A, Raphael M, Joab I (2003) Heterogeneous Epstein-Barr virus latent gene expression in AIDS-associated lymphomas and in type I Burkitt's lymphoma cell lines. J Gen Virol 84:949–957

Touitou R, Hickabottom M, Parker G, Crook T, Allday MJ (2001) Physical and functional interactions between the corepressor CtBP and the Epstein-Barr virus nuclear antigen EBNA3C. J Virol 75:7749–7755

Touitou R, O'Nions J, Heaney J, Allday MJ (2005) Epstein-Barr virus EBNA3 proteins bind to the C8/alpha7 subunit of the 20S proteasome and are degraded by 20S proteasomes in vitro, but are very stable in latently infected B cells. J Gen Virol 86:1269–1277

Vereide DT, Sugden B (2011) Lymphomas differ in their dependence on Epstein-Barr virus. Blood 117:1977–1985

Vousden KH, Prives C (2009) Blinded by the light: the growing complexity of p53. Cell 137:413–431

Waltzer L, Perricaudet M, Sergeant A, Manet E (1996) Epstein-Barr virus EBNA3A and EBNA3C proteins both repress RBP-J kappa-EBNA2-activated transcription by inhibiting the binding of RBP-J kappa to DNA. J Virol 70:5909–5915

Watanabe A, Maruo S, Ito T, Ito M, Katsumura KR, Takada K (2010) Epstein-Barr virus-encoded Bcl-2 homologue functions as a survival factor in Wp-restricted Burkitt lymphoma cell line P3HR-1. J Virol 84:2893–2901

West MJ (2006) Structure and function of the Epstein-Barr virus transcription factor, EBNA 3C. Curr Protein Pept Sci 7:123–136

West MJ, Webb HM, Sinclair AJ, Woolfson DN (2004) Biophysical and mutational analysis of the putative bZIP domain of Epstein-Barr virus EBNA 3C. J Virol 78:9431–9445

White RE, Groves IJ, Turro E, Yee J, Kremmer E, Allday MJ (2010) Extensive co-operation between the Epstein-Barr virus EBNA3 proteins in the manipulation of host gene expression and epigenetic chromatin modification. PLoS ONE 5:e13979

White RE, Ramer PC, Naresh KN, Meixlsperger S, Pinaud L, Rooney C, Savoldo B, Coutinho R, Bodor C, Gribben J, Ibrahim HA, Bower M, Nourse JP, Gandhi MK, Middeldorp J, Cader FZ, Murray P, Munz C, Allday MJ (2012) EBNA3B-deficient EBV promotes B cell lymphomagenesis in humanized mice and is found in human tumors. J Clin Invest 122:1487–1502

Yao QY, Rowe M, Martin B, Young LS, Rickinson AB (1991) The Epstein-Barr virus carrier state: dominance of a single growth-transforming isolate in the blood and in the oropharynx of healthy virus carriers. J Gen Virol 72(Pt 7):1579–1590

Yap KL, Li S, Munoz-Cabello AM, Raguz S, Zeng L, Mujtaba S, Gil J, Walsh MJ, Zhou MM (2010) Molecular interplay of the noncoding RNA ANRIL and methylated histone H3 lysine 27 by polycomb CBX7 in transcriptional silencing of INK4a. Mol Cell 38:662–674

Yee J, White RE, Anderton E, Allday MJ (2011) Latent Epstein-Barr virus can inhibit apoptosis in B cells by blocking the induction of NOXA expression. PLoS ONE 6:e28506

Yenamandra SP, Sompallae R, Klein G, Kashuba E (2009) Comparative analysis of the Epstein-Barr virus encoded nuclear proteins of EBNA-3 family. Comput Biol Med 39:1036–1042

Yi F, Saha A, Murakami M, Kumar P, Knight JS, Cai Q, Choudhuri T, Robertson ES (2009) Epstein-Barr virus nuclear antigen 3C targets p53 and modulates its transcriptional and apoptotic activities. Virology 388:236–247

Young LS, Dawson CW, Clark D, Rupani H, Busson P, Tursz T, Johnson A, Rickinson AB (1988) Epstein-Barr virus gene expression in nasopharyngeal carcinoma. J Gen Virol 69(Pt 5):1051–1065

Young LS, Rickinson AB (2004) Epstein-Barr virus: 40 years on. Nat Rev Cancer 4:757–768

Young LS, Yao QY, Rooney CM, Sculley TB, Moss DJ, Rupani H, Laux G, Bornkamm GW, Rickinson AB (1987) New type B isolates of Epstein-Barr virus from Burkitt's lymphoma and from normal individuals in endemic areas. J Gen Virol 68(Pt 11):2853–2862

Young P, Anderton E, Paschos K, White R, Allday MJ (2008) Epstein-Barr virus nuclear antigen (EBNA) 3A induces the expression of and interacts with a subset of chaperones and co-chaperones. J Gen Virol 89:866–877

Zhang Q, Wang SY, Nottke AC, Rocheleau JV, Piston DW, Goodman RH (2006) Redox sensor CtBP mediates hypoxia-induced tumor cell migration. Proc Natl Acad Sci USA 103:9029–9033

Zhao B, Dalbies-Tran R, Jiang H, Ruf IK, Sample JT, Wang F, Sample CE (2003) Transcriptional regulatory properties of Epstein-Barr virus nuclear antigen 3C are conserved in simian lymphocryptoviruses. J Virol 77:5639–5648

Zhao B, Mar JC, Maruo S, Lee S, Gewurz BE, Johannsen E, Holton K, Rubio R, Takada K, Quackenbush J, Kieff E (2011) Epstein-Barr virus nuclear antigen 3C regulated genes in lymphoblastoid cell lines. Proc Natl Acad Sci USA 108:337–342

Zhao B, Sample CE (2000) Epstein-barr virus nuclear antigen 3C activates the latent membrane protein 1 promoter in the presence of Epstein-Barr virus nuclear antigen 2 through sequences encompassing an spi-1/Spi-B binding site. J Virol 74:5151–5160

Zhao LJ, Subramanian T, Vijayalingam S, Chinnadurai G (2014) CtBP2 proteome: role of CtBP in E2F7-mediated repression and cell proliferation. Genes Cancer 5:31–40

Zimber-Strobl U, Strobl LJ (2001) EBNA2 and Notch signalling in Epstein-Barr virus mediated immortalization of B lymphocytes. Semin Cancer Biol 11:423–434

# The Latent Membrane Protein 1 (LMP1)

**Arnd Kieser and Kai R. Sterz**

**Abstract** Almost exactly twenty years after the discovery of Epstein-Barr virus (EBV), the latent membrane protein 1 (LMP1) entered the EBV stage, and soon thereafter, it was recognized as the primary transforming gene product of the virus. LMP1 is expressed in most EBV-associated lymphoproliferative diseases and malignancies, and it critically contributes to pathogenesis and disease phenotypes. Thirty years of LMP1 research revealed its high potential as a deregulator of cellular signal transduction pathways leading to target cell proliferation and the simultaneous subversion of cell death programs. However, LMP1 has multiple roles beyond cell transformation and immortalization, ranging from cytokine and chemokine induction, immune modulation, the global alteration of gene and microRNA expression patterns to the regulation of tumor angiogenesis, cell–cell contact, cell migration, and invasive growth of tumor cells. By acting like a constitutively active receptor, LMP1 recruits cellular signaling molecules associated with tumor necrosis factor receptors such as tumor necrosis factor receptor-associated factor (TRAF) proteins and TRADD to mimic signals of the costimulatory CD40 receptor in the EBV-infected B lymphocyte. LMP1 activates NF-κB, mitogen-activated protein kinase (MAPK), phosphatidylinositol 3-kinase (PI3-K), IRF7, and STAT pathways. Here, we review LMP1's molecular and biological functions, highlighting the interface between LMP1 and the cellular signal transduction network as an important factor of virus–host interaction and a potential therapeutic target.

A. Kieser (✉) · K.R. Sterz
Helmholtz Center Munich—German Research Center for Environmental Health, Munich, Germany
e-mail: a.kieser@helmholtz-muenchen.de

A. Kieser
German Center for Infection Research (DZIF), Partner Site Munich, Munich, Germany



C. Münz (ed.), *Epstein Barr Virus Volume 2*, Current Topics in Microbiology and Immunology 391, DOI 10.1007/978-3-319-22834-1_4

## Contents



## Abbreviations

AP1 Activator protein 1
CTAR C-terminal activating region
cIAP Cellular inhibitor of apoptosis
EBV Epstein-Barr virus
ERK Extracellular signal-regulated kinase
GCM Germinal center model
Id Inhibitor of DNA binding
IFNγ Interferon-γ
IL Interleukin
IRF7 Interferon regulatory factor 7
IκB Inhibitor of NF-κB
IKK IκB kinase
IRAK1 Interleukin 1 receptor-associated kinase 1
JAK3 Janus kinase 3
JNK c-Jun N-terminal kinase
LMP1 Latent membrane protein 1
LCL Lymphoblastoid cell line
MAPK Mitogen-activated protein kinase
MEF Mouse embryonic fibroblast
NF-κB Nuclear factor of kappa light polypeptide gene enhancer in B cells
NIK NF-κB-inducing kinase
NPC Nasopharyngeal carcinoma
UPR Unfolded protein response

| | |
|---|---|
| PI3-K | Phosphatidylinositol 3-kinase |
| PKCδ | Protein kinase C δ |
| PTLD | Post-transplant lymphoproliferative disease |
| Rb | Retinoblastoma protein |
| RIP1 | Receptor-interacting protein 1 |
| RING | Really interesting new gene, protein domain with E3 ubiquitin ligase activity |
| STAT | Signal transducers and activators of transcription |
| SUMO | Small ubiquitin-like modifier |
| TAK1 | Transforming growth factor β-activated kinase 1 |
| TAB | TAK1-binding protein |
| TES | Transformation effector site |
| TNF-R | Tumor necrosis factor receptor |
| TNIK | TRAF2- and Nck-interacting kinase |
| TM | Transmembrane domain |
| TRAF | Tumor necrosis factor receptor-associated factor |

# 1 LMP1 Expression

The first hint to latent membrane protein 1 (LMP1)'s existence arose in the year 1984 when sequencing of an mRNA transcribed in primary lymphocytes, which had been latently infected with B95.8 Epstein-Barr virus (EBV), suggested the existence of a so far unknown viral transmembrane protein. The protein translated from this mRNA was predicted to consist of a short cytoplasmic N-terminus of 24 amino acids, six transmembrane spanning domains of 162 amino acids, and a cytoplasmic C-terminus of 200 amino acids (Fennewald et al. 1984). The first available LMP1 antibodies allowed the determination of LMP1's apparent molecular weight to approximately 63 kDa and confirmed its localization within the cell membrane (Hennessy et al. 1984; Mann et al. 1985; Modrow and Wolf 1986). Later, it was recognized that the largest fraction of LMP1 actually locates to intracellular membranes where it is biologically active as well (Lam and Sugden 2003). LMP1 is a phosphoprotein, although the functional relevance of LMP1 phosphorylation at serine and threonine residues remains unclear (Baichwal and Sugden 1987; Mann and Thorley-Lawson 1987; Moorthy and Thorley-Lawson 1993b). LMP1 expression is detectable as early as four days after infection of human B lymphocytes with EBV, the viral LMP1 promoter being induced by the transcription factor Epstein-Barr virus nuclear antigen 2 (EBNA2) (Allday et al. 1989; Fahraeus et al. 1990).

Once expressed, several positive and negative autoregulatory loops balance LMP1 protein levels. In cells with low amounts of LMP1, the unfolded protein response (UPR) pathway, which is triggered at the transmembrane domains of LMP1, induces activation transcription factor 4 (ATF4). ATF4, in turn, promotes transcription of the LMP1 promoter (Lee and Sugden 2008b). Furthermore, cellular signal transduction pathways activated at the carboxy-terminal signaling domain of LMP1 such as interferon regulatory factor 7 (IRF7), NF-κB, and p38 mitogen-activated protein kinase (MAPK) contribute to the positive autoregulation of LMP1 expression (Ning et al. 2003, 2005; Demetriades and Mosialos 2009; Johansson et al. 2009, 2010). At high physiological levels, LMP1 causes cytostasis and autophagy, which limits cellular LMP1 amounts (Hammerschmidt et al. 1989; Lee and Sugden 2008a). Overexpression of LMP1 is deleterious for the cell. The LMP1 transmembrane domains alone have the potential to trigger apoptosis via the UPR pathway, which is suppressed at physiological LMP1 levels by signal transduction events initiated at LMP1's signaling domain (Pratt et al. 2012). LMP1 overexpression also induces high cellular levels of the death receptor Fas, leading to Fas autoactivity and apoptosis (Le Clorennec et al. 2008). LMP1 remains expressed throughout viral latency type III, which is found in EBV-transformed B cell lines, so-called lymphoblastoid cell lines (LCLs), post-transplant lymphoproliferative disease (PTLD), X-linked proliferative disease, and infectious mononucleosis. EBV-associated Hodgkin's disease, peripheral T cell lymphoma, and undifferentiated nasopharyngeal carcinoma (NPC), which display viral type II latency, are characterized by the presence of LMP1 protein as well (Niedobitek 1999; Cohen 2000; Yoshizaki et al. 2013). During type II latency, where EBNA2 is absent, cytokine-induced activity of signal transducers and activators of transcription (STAT) is responsible for LMP1 expression (Chen et al. 2001, 2003; Kis et al. 2006, 2010, 2011). Also, transcription factors of the C/EBP family are involved in maintaining LMP1 expression in the absence of EBNA2 (Noda et al. 2011). Apart from pleiotropic effects on its target cell, which will be discussed in the forthcoming paragraphs, LMP1 contributes to the stabilization of viral latency by suppressing activation of the EBV-derived transcription factor BZLF1, which controls EBV's entry into the lytic cycle, and by inducing IRF7, which represses the Qp promoter of EBV in type III latency (Zhang and Pagano 2000; Adler et al. 2002; Prince et al. 2003).

# 2 Biological Functions

## 2.1 *Transformation of Epithelial Cells and Fibroblasts*

The first breakthrough toward LMP1's biological functions was the discovery of its transforming potential in rodent fibroblasts, identifying LMP1 as a genuine herpesviral oncogene. Mouse and rat fibroblasts transfected with LMP1 show phenotypic changes typical for transformed cells such as anchorage- and serum-independent cell growth. Moreover, LMP1-transformed rat fibroblasts turned out to be tumorigenic in nude mice (Wang et al. 1985; Baichwal and Sugden 1988). Also, terminal differentiation of human epithelial cells is inhibited by LMP1, a mechanism by which LMP1 might contribute to multistep pathogenesis of EBV-associated undifferentiated NPC (Dawson et al. 1990). In nasopharyngeal epithelial cells and rodent fibroblasts, LMP1 upregulates basic helix-loop-helix (bHLH) transcription factors of the inhibitor of DNA-binding (Id) family such as Id-1, which mediates epithelial cell cycle progression by downregulation of the retinoblastoma protein (Rb)/p16$^{INK4a}$ pathway (Everly et al. 2004; Li et al. 2004). The tumor necrosis factor receptor-associated factor (TRAF)-binding domain located within the signaling domain is required for rat fibroblast transformation and upregulates cell cycle markers linked to G1/S-phase transition of the cell cycle (Mainou et al. 2005, 2007). Notably, fibroblast transformation by LMP1 is dependent on the PI3-K/AKT and c-Jun N-terminal kinase (JNK) pathways, but independent of NF-κB (Mainou et al. 2005; Kutz et al. 2008). Cell immortalization is fostered by LMP1 through telomerase activation via the p16$^{INK4a}$/Rb, PI3-K/AKT, and JNK pathways (Ding et al. 2007; Yang et al. 2014). Moreover, LMP1 aids NPC cells to evade apoptosis, for instance induced by TNF-related apoptosis-inducing ligand (TRAIL), a factor with potential therapeutic relevance for cancer (Li et al. 2011).

## *2.2 B cell Transformation*

EBV primarily infects and transforms human B lymphocytes (Young and Rickinson 2004). Virions lacking functional LMP1 loose the capability of efficiently transforming B cells, underscoring LMP1's critical role as the primary oncogene of EBV (Kaye et al. 1993; Dirmeier et al. 2003). LMP1 is also oncogenic in the B cell compartment of transgenic mice. Expression of LMP1 from the immunoglobulin heavy chain promoter/enhancer results in the sporadic development of B cell lymphomas of follicular center cell phenotype (Kulwichit et al. 1998). The conditional activation of LMP1 from the pro/pre-B cell stage in mice causes a rapid and fatal lymphoproliferation with high efficiency, if T lymphocytes are depleted at the same time. This result highlights the transforming potential of LMP1 in vivo as well as its role in immune surveillance of EBV-positive B cells (Zhang et al. 2012).

Activation of naive B cells depends on the stimulation of their B cell receptor by antigen and the costimulatory receptor CD40 by T helper cells. Several in vitro and in vivo studies have demonstrated that LMP1 mimics signals that are physiologically induced by CD40, a member of the TNF-R family (Zimber-Strobl et al. 1996; Hatzivassiliou et al. 1998; Kilger et al. 1998; Busch and Bishop 1999; Uchida et al. 1999; Stunz et al. 2004; Rastelli et al. 2008; Zhang et al. 2012). Accordingly, primary human B cells infected with LMP1-deficient EBV are only able to induce lymphomas in immunodeficient mice if they are supplied with CD40-mediated T cell help (Ma et al. 2015). Together with LMP2A, which generates B cell receptor-like signals, LMP1 drives B cell proliferation and survival, rendering the infected B cell independent of B cell receptor and CD40 stimulation (Mancao and Hammerschmidt 2007). In particular, activation of the NF-κB pathway by LMP1 contributes essential survival signals for the transformation of B cells in type III latency (Cahir-McFarland et al. 1999; Feuillard et al. 2000; Cahir-McFarland et al. 2004). LMP1 even possesses general antiapoptotic potential as it protects EBV-negative Burkitt's lymphoma cells from programmed cell death by upregulation of antiapoptotic genes such as *bcl-2*, *mcl-1*, and *bfl*-1, or downregulation of proapoptotic genes such as *bax* (Henderson et al. 1991; D'Souza et al. 2004; Grimm et al. 2005). Moreover, LMP1 disrupts expression or function of tumor suppressors such as DOK1 in EBV-infected primary B cells (Li et al. 2012; Siouda et al. 2014).

Simultaneous promotion of proliferation, cell cycle progression, and cell growth of B cells can be attributed to LMP1's potential of inducing the expression of c-Myc and the epidermal growth factor receptor (EGF-R) and to upregulate cell cycle-regulating kinases such as Cdk2 and Cdc2 (Miller et al. 1995; Dirmeier et al. 2005; Shair et al. 2007; Kutz et al. 2008). Splenic B cells of LMP1 transgenic mice display increased levels of Cdk2 and phosphorylated Rb protein, a key substrate of Cdk2 and master regulator of G1- to S-phase transition of the cell cycle (Shair et al. 2007). Cdc2, which is essential for G2- to M-phase transition, is upregulated in LCLs by LMP1 via the JNK signaling pathway (Kutz et al. 2008). Notably, chemical inactivation of JNK by a small molecule inhibitor caused a proliferation defect in LCLs in vitro and strongly retarded tumor growth of EBV-transformed human B cells in a tumor xenograft model in mice (Kutz et al. 2008). Recent data point to an additional role of IRF7 in B cell transformation by EBV, because the knockdown of this transcription factor in LCLs causes a defect in cell growth (Xu et al. 2015).

### 2.3 Global Effects on Chromatin and Gene Expression

Accumulating evidence suggests that LMP1 also impacts the chromatin of its target cell. A first hint toward this direction came from studies, revealing that LMP1 upregulates expression of the DNA methyltransferase DNMT1 via the JNK pathway. LMP1-induced DNMT1 forms a transcriptional repression complex together with histone acetylase at the E-cadherin promoter (Tsai et al. 2006). LMP1 also induces recruitment of a DNMT1-containing repression complex to the promoter of the tumor suppressor DOK1, resulting in histone H3 trimethylation at lysine 27 (H3K27me3) and gene silencing (Siouda et al. 2014). Moreover, LMP1 stimulates phosphorylation of histone H3 at serine 10, which is an important factor in carcinogenesis of NPC (Li et al. 2013). It is tempting to speculate that additional regulatory links between LMP1 and the chromatin will emerge with more studies to be conducted.

Considering the diversity of biological processes and signal transduction pathways regulated by LMP1, it was not surprising to learn that LMP1 has an extensive impact on the gene expression pattern of its target cell. Several studies reported global analyses of LMP1-regulated gene expression in different cell types ranging from NPC cells to primary B cells infected with EBV. LMP1-regulated genes include for instance transcription factors, cytokines, chemokines, growth factors, receptors and signaling mediators, genes involved in apoptosis and survival regulation, metabolism, and structural proteins as well as genes involved in cell motility and immune modulation (Kwok Fung Lo et al. 2001; Cahir-McFarland et al. 2004; Dirmeier et al. 2005; Morris et al. 2008; Vockerodt et al. 2008; Faumont et al. 2009; Gewurz et al. 2011; Shair and Raab-Traub 2012; Xiao et al. 2014). In primary human germinal center B cells, LMP1 induces a transcriptional pattern characteristic for Hodgkin/Reed–Sternberg tumor cells including the downregulation of B cell receptor components (Vockerodt et al. 2008). LMP1 affects cell metabolism by deregulation of glycolytic genes such as hexokinase 2, which results in increased glycolysis (Xiao et al. 2014). Moreover, it has become evident recently that not only genes translated into proteins but also genes encoding small non-coding RNA species such as microRNAs or vault RNA are regulated by LMP1 (Motsch et al. 2007; Amort et al. 2015).

## 2.4 Functions Beyond Cell Transformation

Apart from affecting cell survival and proliferation pathways, LMP1 influences many other processes involved, for instance, in immune modulation or the dissemination of tumor cells in the body. Due to space limitations, only some examples can be given here. The first transmembrane domain of LMP1 harbors the immunosuppressive peptide LALLFWL, which strongly inhibits T cell proliferation and NK T cell cytotoxicity, likely upon LMP1 release from EBV-positive cells via exosomes (Dukers et al. 2000; Middeldorp and Pegtel 2008).

Mice expressing LMP1 as a transgene in the skin develop epidermal hyperplasia associated with inflammatory processes, which are mediated by LMP1-induced cytokine and chemokine expression (Wilson et al. 1990; Curran et al. 2001; Hannigan et al. 2011). Like many other tumors, also NPC is accompanied by heavy chronic inflammation that adds to the severity of the disease. Several studies have shown that upregulation of proinflammatory cytokines such as interleukin (IL)-1$\alpha$/$\beta$, IL-6, IL-8, chemokine (C-X-C) motif ligand 1 (CXCL-1), or granulocyte-macrophage colony-stimulating factor (GM-CSF) is related to EBV infection and LMP1 expression in undifferentiated NPC (Eliopoulos et al. 1999b; Huang et al. 1999; Li et al. 2007; Lai et al. 2010; Hannigan et al. 2011). Moreover, LMP1 activity is linked to tumor angiogenesis by upregulating angiogenic factors such as vascular endothelial growth factor (VEGF), fibroblast growth factor-2 (FGF-2), and IL-8 (Yoshizaki et al. 2001; Murono et al. 2001; Wakisaka et al. 2002).

LMP1 promotes cell motility, migration, and tumor metastasis by inducing anchorage-independent and invasive growth of human epithelial and nasopharyngeal cells (Tsao et al. 2002; Yoshizaki 2002; Chew et al. 2010). Cell motility and migration are fostered through activation of the MAPK extracellular-regulated kinase (ERK) and PI3-K/AKT pathways and of the Rho-GTPase Cdc42 (Dawson et al. 2008; Shair et al. 2008; Liu et al. 2012). LMP1 furthermore interacts with the SUMO-conjugating enzyme Ubc9, which mediates LMP1-induced SUMOylation of cellular target proteins. Disruption of the LMP1::Ubc9 interaction affected cell migration (Bentz et al. 2011). In NPC tissues, LMP1 expression is directly correlated with ezrin phosphorylation, a linker between the cell membrane and the actin cytoskeleton that mediates cell migration. LMP1 increases ezrin phosphorylation and its activation through a protein kinase C (PKC)-dependent pathway, which is required for LMP1-induced cell motility and invasion of nasopharyngeal cells (Endo et al. 2009). Also, upregulation of alpha(v) integrins by LMP1 contributes to cell migration (Huang et al. 2000). Recent data further showed that LMP1 induces the TNFα-induced protein 2 (TNFAIP2), whose expression correlates with metastasis and poor survival of NPC patients, via the NF-κB pathway. TNFAIP2 associates with actin and promotes the formation of membrane protrusions (Chen et al. 2014).

Tumor cells must be able to degrade and invade the extracellular matrix to migrate through surrounding tissue and to penetrate lymphatic or blood vessels, a prerequisite for tumor metastasis. LMP1 upregulates a panel of factors that are associated with cell invasiveness such as matrix metalloproteinases (Lu et al. 2003; Yoshizaki et al. 1998; Kondo et al. 2005). Furthermore, LMP1 increases invasive properties of tumor cells by downregulation of E-cadherin, a $Ca^{2+}$-dependent adhesion molecule at the cell surface responsible for cell–cell contact (Tsai et al. 2002). LMP1 also regulates invasive migration of lymphocytes by upregulating the tumor marker Fascin, a stabilizer of filamentous actin in filopodia of migrating cells, via the NF-κB pathway (Mohr et al. 2014).

## *2.5 The Role of LMP1 for EBV Infection*

The germinal center model (GCM) of EBV infection explains EBV biology as well as the pathogenesis of lymphoma (Thorley-Lawson et al. 2013). According to this model, EBV exploits normal B cell biology to establish a lifelong latency in the body. After infection of naive B cells, EBV pushes the cells into proliferating lymphoblasts expressing LMP1, LMP2, and EBNA2, which then transit the germinal center as centroblasts and centrocytes to finally become long-lived resting memory B cells (Babcock et al. 2000). From the centroblast stage, EBV-infected cells express a restricted pattern of latent genes comprising LMP1, LMP2, and EBNA1, but not EBNA2 (Babcock et al. 2000). It has long been assumed that LMP1 (together with LMP2) allows the EBV-infected cells to pass the germinal center in the absence of antigen, T cell help and CD40 signaling. However, EBV-positive memory B cells apparently underwent antigen selection as their EBV-negative counterparts, suggesting that the impact of LMP1 (and LMP2) on the cells during their passage through the germinal center may be moderate (Souza et al. 2005). LMP1 likely supports the survival of EBV-infected cells in this competitive environment and favors their differentiation into memory B cells rather than plasma cells (Thorley-Lawson et al. 2013). So far, this model is largely based upon expression analyses of EBV genes in B cell subsets of infected individuals. Due to the lack of appropriate in vivo models for EBV infection, it has been nearly impossible to test this model experimentally. This may now change due to the availability of infection models in mice carrying a reconstituted human immune system (Chatterjee et al. 2014). Recent experiments in highly immunocompetent hNSG(thy) mice engrafted with human fetal CD34-positive cells and human thymus showed that EBV lacking LMP1 can still establish long-term viral latency in vivo, but is unable to induce lymphomas (Ma et al. 2015). However, long-term latency in this model may rather resemble latency III of LCLs than the latency program found in resting memory B cells of EBV-positive individuals, because the latently infected B cells in the spleens of these animals still expressed EBNA2. This finding is in line with in vitro data demonstrating that LMP1-deficient EBV supports the establishment of LCLs at a very low frequency if the cells are supplied with a favorable environment such as a fibroblast feeder layer (Dirmeier et al. 2003).

# 3 Structure–Function Relationship

## *3.1 Amino-terminus and Transmembrane Domain*

The LMP1 molecule consists of a short cytoplasmic amino-terminus (amino acids 1–24), six transmembrane domains (amino acids 25–186), and a carboxy-terminal signaling domain (amino acids 187–386), which is located in the cytoplasm (Fig. 1). There are no crystal or NMR structures of the LMP1 protein available, which is

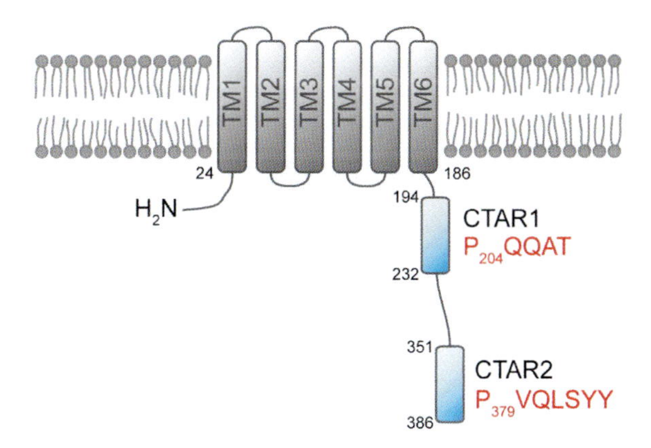


**Fig. 1** Structure of the LMP1 molecule. Six transmembrane domains connect a short amino-terminal domain with a carboxy-terminal signaling domain which is located in the cytoplasm of the infected cell. The signaling domain contains the two effector sites CTAR1 and CTAR2, also called TES1 and TES2, respectively. CTAR1 harbors the TRAF interaction motif PxQxT and CTAR2 the motif PYQLSYY that is critical for JNK and canonical NF-κB signaling

partly due to its unfavorable chemical properties as a hydrophobic transmembrane protein with a signaling domain that seems to be highly unstructured, at least in the absence of a binding partner. LMP1 functions like a constitutively active receptor that is independent of ligand binding. The transmembrane domain of LMP1 has the intrinsic property to form homo-oligomers in the membrane, which replaces cross-linking by a ligand and leads to the initiation of signal transduction events at the signaling domain of the molecule (Gires et al. 1997). Intermolecular interactions between transmembrane domains (TM) 3–6 and an FWLY motif in TM1 mediate oligomerization and are thus required for LMP1 signaling (Yasui et al. 2004; Soni et al. 2006). LMP1 is distributed between lipid rafts and non-raft regions of the membrane (Ardila-Osorio et al. 1999; Higuchi et al. 2001). Also, lipid raft localization of LMP1 is mediated by the FWLY motif (Coffin et al. 2003; Yasui et al. 2004). In addition, a leucine heptad motif located within TM1 contributes to the efficient homing of LMP1 to lipid rafts (Lee and Sugden 2007). Lipid rafts are membrane subdomains enriched in sphingolipids and cholesterol that are actively involved in signal transduction processes. LMP1 specifically recruits cellular signaling molecules such as TRAF2 and 3, the TNF receptor-associated death domain protein (TRADD), or PI3-K into lipid rafts, suggesting a role for these membrane domains in LMP1 signaling (Ardila-Osorio et al. 1999; Brown et al. 2001; Schneider et al. 2008; Meckes et al. 2013).

The amino-terminus is responsible for correct insertion and orientation of LMP1 within the membrane, mediates association of LMP1 with the cytoskeleton, and is involved in the regulation of LMP1 degradation and turnover (Wang et al. 1988b; Martin and Sugden 1991; Izumi et al. 1994; Aviel et al. 2000; Coffin et al. 2001). Ubiquitination and proteasome-dependent degradation of LMP1 is further dependent on the TRAF-binding site within the carboxy-terminus of LMP1 (Rothenberger et al. 2003; Hau et al. 2011). Mutation of the amino-terminus showed that this domain does not contribute critical signals for B cell transformation (Izumi et al. 1994; Dirmeier et al. 2003).

## 3.2 Carboxy-terminal Signaling Domain

Extensive mutational analysis revealed the central importance of the carboxy-terminal domain and the two functional subdomains it harbors, for cell transformation and the initiation of signal transduction. The so-called transformation effector sites (TES) 1 (amino acids 187–231) and 2 (amino acids 351–386) have been defined by their essential functions in initial B cell transformation or long-term outgrowth of EBV-infected B cells, respectively (Kaye et al. 1995, 1999; Dirmeier et al. 2003). Transformation of rat fibroblasts depends on these effector sites as well (Moorthy and Thorley-Lawson 1993a; Mainou et al. 2005). Soon, it has been recognized that TES1 and 2 coincide with two regions named C-terminal activating regions (CTAR) 1 and 2, respectively, that are responsible for interaction of LMP1 with cellular signaling molecules of the TRAF family and TRADD, and the

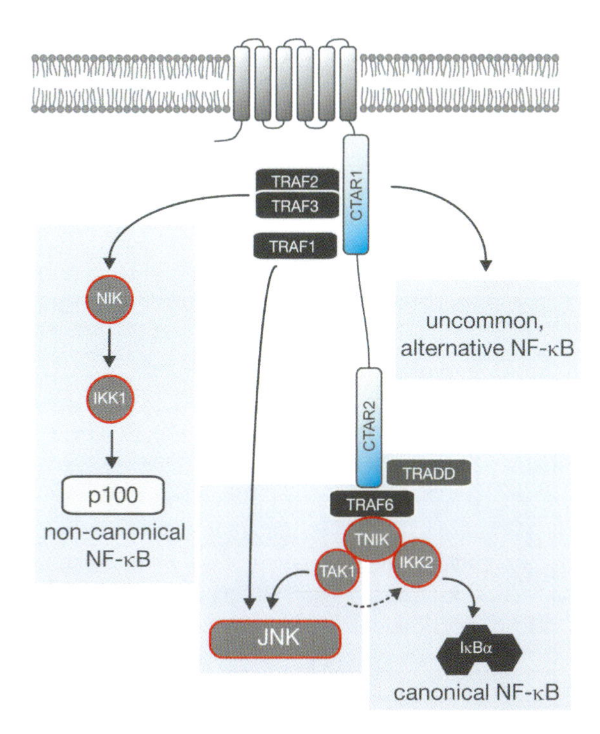


**Fig. 2** NF-κB and JNK signaling by LMP1

activation of NF-κB and JNK/AP1 signaling (Fig. 2) (Huen et al. 1995; Mitchell and Sugden 1995; Mosialos et al. 1995; Devergne et al. 1996; Brodeur et al. 1997; Izumi et al. 1997; Izumi and Kieff 1997; Kieser et al. 1997; Eliopoulos and Young 1998).

The TRAF interaction motif $P_{204}xQxT$ is critical for CTAR1 domain function and was shown to directly bind TRAF1, 2, and 3, whereas the molecular basis of TRAF5 interaction with CTAR1 remains unclear (Devergne et al. 1996; Brodeur et al. 1997; Sandberg et al. 1997; Miller et al. 1998). The crystal structure of TRAF2 in complex with the CTAR1-derived peptide $P_{204}QQATDD$ revealed a conserved binding mode of TRAF2 to LMP1 and its cellular counterpart, CD40 (Ye et al. 1999).

CTAR2 activity depends on the presence of a minor TRAF-binding motif $P_{379}VQLSYY$ in order to induce signaling, although direct binding of TRAF proteins to this site has never been demonstrated (Floettmann and Rowe 1997; Kieser et al. 1999; Schneider et al. 2008). Instead, it was the death domain protein TRADD that was first described to interact directly with CTAR2, the motif $Y_{384}Y$ of CTAR2 being essential for TRADD recruitment (Izumi and Kieff 1997). This interaction, however, is unique and differs from cellular TRADD-interacting receptors, as it (i) does not require the death domain of TRADD and (ii) dictates an LMP1-specific, transferable, and non-apoptotic type of TRADD signaling (Kieser et al. 1999; Kieser 2008; Schneider et al. 2008). LMP1 is thus another impressive example of a viral protein that reprograms and thereby exploits cellular proteins and functions for its own purposes, in this case the subversion of cellular pathways regulating cell survival. In contrast to CTAR1, CTAR2 relies on TRAF6 as the essential signaling mediator (Schultheiss et al. 2001; Luftig et al. 2003; Wan et al. 2004; Schneider et al. 2008). However, despite its central role in LMP1 signaling, it has been unclear by which mechanism LMP1 is recruiting TRAF6. Our own unpublished data now suggest that TRAF6 in fact binds to the $P_{379}$VQLSYY motif of CTAR2 directly. Other factors such as BS69 or TRADD might then further stabilize the complex at CTAR2 (Wan et al. 2006; Schneider et al. 2008).

The region between CTAR1 and CTAR2 is dispensable for cell transformation (Izumi et al. 1999). The functional relevance of CTAR3 (amino acids 275–330) comprising two box 1 and one box 2 Janus kinase 3 (JAK3) interaction and activation motifs, respectively, for JAK/STAT activation by LMP1 remains controversial (Gires et al. 1999; Higuchi et al. 2002). However, LMP1 can induce STAT3 via autocrine loops involving cytokines such as IL-6 (Chen et al. 2003). STAT1 activation has been linked to LMP1-dependent IFNγ secretion in PTLD-derived B cells (Vaysberg et al. 2009). Recent studies demonstrated that CTAR3 in fact exists and that it regulates Ubc9-mediated SUMO modification of cellular proteins including the transcription factor IRF7 (Bentz et al. 2011, 2012).

Most experimental data on molecular biology and the biological functions of LMP1 have been generated using the "prototype" LMP1 derived from the B95.8 strain of EBV (Hudson et al. 1985). In addition to B95.8-LMP1, a large number of LMP1 sequence variants have been isolated predominantly from NPC samples or healthy carriers of different geographical regions of the world (Hu et al. 1991; Miller et al. 1994; Sandvej et al. 1997; Hatton et al. 2014; Lorenzetti et al. 2012; Renzette et al. 2014). Due to space limitations, a detailed description of such LMP1 variants cannot be given here. Some of these variants, for instance the CAO-LMP1 variant derived from a Chinese NPC, exhibit altered signaling properties as compared to B95.8-LMP1 (Hu et al. 1991; Blake et al. 2001; Fielding et al. 2001; Stevenson et al. 2005; Mainou and Raab-Traub 2006).

# 4 Signal Transduction by LMP1

## *4.1 NF-κB Pathway*

The first cellular signaling pathway identified as an LMP1 target was NF-κB (Fig. 2) (Laherty et al. 1992). The transcription factor NF-κB is a key regulator of lymphocyte development, immunity, inflammation, and cancer. The NF-κB family comprises five proteins, p65 (RelA), c-Rel, RelB, p50 (and its precursor p105), and p52 (and its precursor p100), which are kept inactive in the cytoplasm either by complex formation with inhibitory proteins of the inhibitor of NF-κB (IκB) family or as precursors (Vallabhapurapu and Karin 2009). Two major NF-κB pathways have been described, the canonical and the non-canonical pathway (Fig. 2). The canonical pathway involves phosphorylation of IκB proteins by IκB kinase 2

(IKK2) and the subsequent degradation of IκB, which liberates NF-κB dimers usually including p65. The non-canonical pathway depends on IKK1 and leads to the processing of p100 to p52 and the subsequent translocation of p52::RelB dimers to the nucleus (Vallabhapurapu and Karin 2009). Early studies revealed that both CTAR1 and CTAR2 contribute to the activation of NF-κB (Huen et al. 1995; Mitchell and Sugden 1995). However, CTAR1 induces multiple forms of NF-κB dimers, whereas CTAR2 activates complexes containing p65 (Paine et al. 1995). Later, it became evident that CTAR1 primarily activates the non-canonical NF-κB pathway including p100 to p52 processing and p50::p52 and p52::p65 dimers as well as atypical p50::p50::Bcl3 complexes, while CTAR2 is responsible for canonical NF-κB activation (Atkinson et al. 2003; Eliopoulos et al. 2003a; Saito et al. 2003; Thornburg et al. 2003; Luftig et al. 2004; Thornburg and Raab-Traub 2007). The canonical NF-κB pathway is essential for most CTAR2-dependent gene regulation (Gewurz et al. 2011). However, the recent global ChIP-seq analysis of the NF-κB-binding landscape in LCLs revealed a pattern of NF-κB complexes, which does not fully reflect the paradigm of canonical and non-canonical NF-κB pathways (Zhao et al. 2014). This result may indicate that our current view of LMP1-induced NF-κB activity still requires further refinement.

In the uninduced state, TRAF2 and TRAF3 are key inhibitors of the non-canonical NF-κB pathway leading to permanent degradation of NF-κB-inducing kinase (NIK). Upon activation, TRAF3 is cleared from the cytosol in a TRAF2 and cIAP1/2-dependent mechanism, which leads to NIK stabilization and NIK-mediated activation of IKK1 (Vallabhapurapu and Karin 2009). Overexpression of either TRAF was shown to reduce CTAR1-dependent NF-κB activation as well as processing of p100, and dominant-negative TRAF2 inhibited CTAR1 signaling, indicating that both TRAF2 and TRAF3 are involved in the regulation of non-canonical NF-κB by LMP1 (Devergne et al. 1996; Kaye et al. 1996; Song and Kang 2010). Accordingly, the siRNA-mediated knockdown of TRAF2 caused a marked reduction of total NF-κB activity and the concomitant induction of apoptosis in LCLs (Guasparri et al. 2008). However, and in contrast to CD40 signaling, TRAF3 is eliminated by LMP1 through a mechanism independent of the proteasome (Brown et al. 2001). Also, a non-essential role for TRAF6 in CTAR1 signaling has been suggested but still requires further validation (Schultheiss et al. 2001; Arcipowski et al. 2011).

Although dominant-negative TRAF2 had mild negative effects on CTAR2-induced NF-κB as well (Kaye et al. 1996), it is nowadays widely accepted that CTAR2 signaling does not depend on TRAF2 or its close relative TRAF5. Ligation of a CD40-LMP1 chimera in TRAF2-deficient mouse B cells still resulted in the degradation of IκBα (Xie and Bishop 2004). The double knockout of TRAF2 and TRAF5 in mouse embryonic fibroblasts did not impair CTAR2-induced IKK2 activation or translocation of canonical p65 NF-κB into the nucleus, excluding the possibility that TRAF2 deficiency was rescued by TRAF5 (Luftig et al. 2003; Wu et al. 2006). Instead of TRAF2, activation of canonical NF-κB by LMP1 requires TRAF6 and its function as an E3 ubiquitin ligase (Schultheiss et al. 2001). CTAR2-dependent NF-κB activation is blocked upon expression of dominant-negative TRAF6 lacking its RING domain, and the deficiency of TRAF6 significantly reduces p65 translocation and NF-κB-dependent gene transcription induced by CTAR2 (Schultheiss et al. 2001; Luftig et al. 2003; Schneider et al. 2008; Boehm et al. 2010). TRADD was one of the first factors found to be involved in CTAR2-induced NF-κB signaling, albeit its interaction with LMP1 does not require the TRADD death domain (Izumi and Kieff 1997; Kieser et al. 1999; Schneider et al. 2008). Accordingly, a TRADD deletion mutant lacking the TRADD death domain, which is required for downstream signaling, acted as a dominant-negative allele in NF-κB signaling by LMP1 (Kieser et al. 1999). The genetic knockout of TRADD in human B cells delivered the definitive proof for an important role of TRADD in canonical NF-κB signaling by LMP1, because LMP1 was unable to activate IKK2 in the absence of TRADD in these cells (Schneider et al. 2008). However, the precise molecular function of TRADD in CTAR2 signaling is still not understood. TRADD might aid to stabilize the holocomplex at CTAR2, probably in a cell type-dependent manner depending on the abundance of other factors involved in CTAR2 signaling. In human B cells, TRADD seems to be required for the recruitment of IKK2 into the LMP1 complex (Schneider et al. 2008).

Further downstream in the cascade, TRAF6 recruits the germinal center kinase family member TRAF2- and Nck-interacting kinase (TNIK) to LMP1, which mediates canonical NF-κB and JNK signaling by the viral oncoprotein (Fig. 2) (Shkoda et al. 2012). TNIK is involved in the assembly of the TRAF6/TAB/TAK1/IKK2 complex and the bifurcation of the NF-κB and JNK pathways. The amino-terminal kinase domain of TNIK is required for NF-κB but not JNK signaling, whereas its carboxy-terminal germinal center kinase homology domain mediates JNK but not NF-κB activation (Shkoda et al. 2012). Interestingly, CTAR2 alone is able to induce an association between TNIK and TRAF2 as well, an interaction whose relevance for CTAR2 function is unclear (Shkoda et al. 2012). The TAK1 protein, but not its kinase activity, seems to be important for canonical NF-κB induction by LMP1, because the knockdown of TAK1, but not the chemical inhibition of its kinase activity, interfered with IKK2 activation by CTAR2 (Uemura et al. 2006; Wu et al. 2006). The IKK complex consisting of IKK1, IKK2, and the regulatory subunit NEMO (IKKγ) is the central mediator of the NF-κB pathway (Vallabhapurapu and Karin 2009). This is also true for LMP1 signaling, although the mode of NEMO utilization seems to differ from cellular receptors such as TNF-R1 or CD40 (Boehm et al. 2010). NEMO lacking its Zn-finger domain or part of the coiled coil region is capable of mediating NF-κB activation upon LMP1 expression, but not upon CD40 or TNF-R1 ligation in Jurkat T cells. However, the complete lack of NEMO blocks LMP1-induced canonical NF-κB in different cell types, underscoring its critical role in LMP1 signaling (Boehm et al. 2010). Likewise, inhibition or lack of IKK2 abolishes LMP1-induced NF-κB activity, confirming the essential role of this IKK isoform in LMP1 signaling (Luftig et al. 2003; Boehm et al. 2010). However, recent knockdown experiments suggested that IKK1 and IKK2 have partially redundant roles in this pathway (Gewurz et al. 2012).

Interleukin 1 receptor-associated kinase 1 (IRAK1), a coplayer of TRAF6 in Toll-like/IL-1 receptor signaling, is also involved in canonical NF-κB activation by LMP1, albeit its kinase activity seems to be dispensable for this pathway (Luftig et al. 2003; Song et al. 2006). IRAK1 deficiency does not interfere with IKK2 activation or p65 translocation induced by LMP1 but rather with p65 phosphorylation at serine 536, suggesting that IRAK1 plays a different role in LMP1 signaling compared to cellular receptors (Song et al. 2006). IRAK1 function in this pathway requires its association with $Ca^{2+}$/Calmodulin-dependent kinase II (CaMKII), which phosphorylates p65 (Kim et al. 2014).

## *4.2 MAPK Pathways—JNK, ERK, and P38 MAPK*

It was in the year 1997 when JNK/AP1 was identified as the second major pathway activated by LMP1 in epithelial cells and LCLs (Kieser et al. 1997). AP1 is a dimeric transcription factor composed of members of the Jun and Fos proto-oncoprotein families. AP1 was found induced by LMP1 through the JNK signaling cascade, involving JNK1-mediated phosphorylation and activation of c-Jun (Kieser et al. 1997). JNK belongs, together with ERK and p38 MAPK, to the MAPK family of kinases. JNK1 activation critically relies on CTAR2 and its $P_{379}$VQLSYY motif (Fig. 2) (Kieser et al. 1997, 1999; Eliopoulos and Young 1998). It has long been unclear which signaling mediators at CTAR2 are involved in JNK activation. The use of dominant-negative TRAF2 alleles yielded conflicting results regarding a potential role of TRAF2 in this pathway (Eliopoulos et al. 1999a; Kieser et al. 1999). Several years later, experiments in TRAF2-deficient fibroblasts and B cells finally excluded a critical function of TRAF2 in JNK activation by LMP1, because the pathway was fully functional in cells lacking TRAF2 (Wan et al. 2004; Xie et al. 2004). JNK signaling is also intact in mouse embryonic fibroblasts devoid of both TRAF2 and TRAF5 (own unpublished data). A potential function of TRADD in JNK signaling by LMP1 had been subject of discussion for quite some time. Although overexpression of TRADD enhanced JNK signaling by LMP1, dominant-negative TRADD did not block the pathway in HEK293 cells (Eliopoulos et al. 1999a; Kieser et al. 1999). The issue was clarified when neither the knockdown of TRADD in HeLa cells nor the genetic knockout of TRADD in human B cells impaired JNK signaling by LMP1, demonstrating that TRADD is not critical for this pathway (Wan et al. 2004; Schneider et al. 2008). As with canonical NF-κB signaling, it turned out that TRAF6 is in fact the essential mediator of CTAR2-induced JNK activity (Wan et al. 2004). The pathway further involves TAB1, TAK1, and TNIK, the latter orchestrating the bifurcation of canonical NF-κB and JNK (Wan et al. 2004; Uemura et al. 2006; Shkoda et al. 2012).

It seems as if LMP1 is able to eventually make use of different panels of TRAF molecules to establish its signaling network, likely depending on the investigated cell type. Usually, JNK activation by LMP1 is exclusively triggered at CTAR2.

In some cell lines, however, CTAR1 contributes to JNK activity as well. JNK1 activation in Rat-1 cells was equally mapped to CTAR1 and CTAR2 (Kutz et al. 2008). Likewise, CTAR1 alone can induce JNK in BJAB cells, which is due to the very strong expression of TRAF1 in these cells. Accordingly, JNK1 activation by CTAR1 was possible after overexpression of TRAF1 in HEK293 cells (Eliopoulos et al. 2003b). Also, TRAF3 can eventually have a role in JNK signaling, as was demonstrated in TRAF3-deficient mouse B lymphoblastic leukemia cells (Xie et al. 2004). Likely, LMP1 constitutes a versatile viral oncoprotein that is variable in establishing its signaling network dependent on the cellular context it comes across.

LMP1 induces the ERK pathway (Fig. 3) (Liu et al. 2003). ERK activity in epithelial cells is mainly initiated at CTAR1 via the canonical Raf-MEK-ERK pathway independent of Ras (Dawson et al. 2008). The mechanism of ERK activation at CTAR1 is still unresolved. An involvement of TRAF2 and TRAF3 in ERK activation has been proposed, because expression of dominant-negative TRAF2 and TRAF3 reduced ERK phosphorylation induced by LMP1 (Mainou et al. 2007). At the same time, mutation of $P_{204}xQxT$ to $A_{204}xAxA$ abolished ERK activation (Mainou et al. 2007). However, mutation of this motif to $A_{204}xAxT$, which is incapable of binding TRAF2 (Devergne et al. 1998), was not sufficient to reduce pathway activation, indicating that the mechanism of ERK activation might involve other molecules besides of TRAFs. Another component of the pathway is the isoform δ of protein kinase C (PKCδ). Inhibition of this kinase by the PKCδ inhibitor Rottlerin abolished ERK activation after LMP1-CTAR1 expression in C33A cells

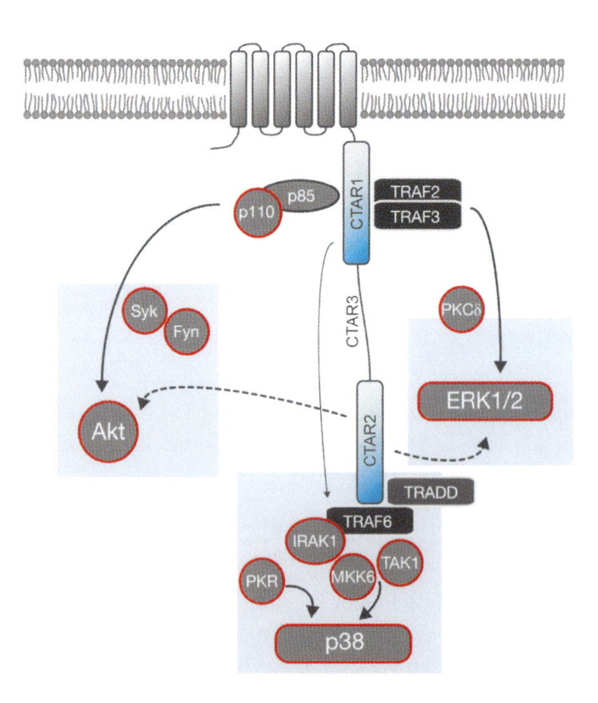


**Fig. 3** PI3-K/AKT, p38 MAPK, and ERK signaling by LMP1

(Kung et al. 2011). In HEK293 cells, also CTAR2 can activate ERK (Gewurz et al. 2011). Particularly, the latter result suggests that also the ERK pathway may be regulated by different mechanisms in different cell types.

The p38 MAPK pathway is induced by LMP1 as well, mediating upregulation of IL-6, IL-8, and IL-10 (Fig. 3) (Eliopoulos et al. 1999b; Vockerodt et al. 2001). Both CTAR1 and CTAR2 are involved in the pathway, although the contribution of CTAR2 to p38 MAPK activation is more prevalent (Schultheiss et al. 2001). TRAF6 is the key component of this pathway as was demonstrated by defective LMP1 signaling to p38 MAPK in TRAF6-deficient MEFs (Schultheiss et al. 2001). Also, the lack of IRAK1 causes a defect in p38 MAPK signaling by LMP1 (Song et al. 2006). Other kinases involved in p38 MAPK signaling by LMP1 include MKK6, TAK1, and PKR (Schultheiss et al. 2001; Wan et al. 2004; Lin et al. 2010).

## *4.3 PI3-K/AKT Pathway*

The PI3-K/AKT pathway is induced by a wide range of cellular receptors and is an important mitogenic stimulus in many cell types. Activation of the PI3-K/AKT pathway was first linked to LMP1 in the year 2003, when Dawson and colleagues immunoprecipitated the p85α regulatory subunit of PI3-K together with LMP1 from HeLa cells (Fig. 3) (Dawson et al. 2003). Although it is still unclear whether this interaction is direct, expression of LMP1-CTAR1 caused (i) an enrichment of the PI3-K substrate phosphatidylinositol-3,4,5-trisphosphate (PIP3) in the plasma membrane and (ii) the phosphorylation and activation of AKT kinase, also known as protein kinase B, at serine 473 (Dawson et al. 2003). Dominant-negative p85 or the PI3-K inhibitor LY294002 blocked LMP1-induced stress fiber formation and actin remodeling as well as transformation of rat fibroblasts (Dawson et al. 2003; Mainou et al. 2005). It is accepted that CTAR1 and the TRAF-binding site $P_{204}$xQxT are involved in activation of the PI3-K/AKT pathway (Dawson et al. 2003; Mainou et al. 2007; Lambert and Martinez 2007). However, CTAR2 alone was as efficient as CTAR1 in inducing AKT phosphorylation in C666-1 cells, indicating the existence of a CTAR2-dependent PI3-K/AKT pathway (Shair et al. 2008). The precise mechanism of PI3-K/AKT activation by LMP1 has not been fully elucidated but may involve the Src family tyrosine kinases Fyn and Syk (Hatton et al. 2012).

## *4.4 IRF7 Pathway*

The transcription factor IRF7 is a critical regulator of type I interferon and thus adaptive and innate immune responses as well as of EBV latency (Zhang and Pagano 2001). LMP1 induces IRF7 expression and facilitates its ubiquitination, phosphorylation, and nuclear translocation (Zhang and Pagano 2000; Ning et al. 2003; Huye et al. 2007). CTAR2 is involved in recruiting IRF7 to LMP1, which is a function of LMP1 that is independent of the TRADD and TRAF6 interaction site $Y_{384}Y$ (Song et al. 2008). CTAR2 then mediates IRF7 ubiquitination in a receptor-interacting protein 1 (RIP1)- and TRAF6-dependent mechanism (Huye et al. 2007; Ning et al. 2008; Song et al. 2008). TRAF6 is involved in IRF7 activation by acting as E3 ubiquitin ligase, which mediates linkage of ubiquitin chains at the carboxy-terminal lysines 444, 446, and 452 of IRF7 (Ning et al. 2008). Two major mechanisms have been identified, aiding the negative regulation of IRF7 activity. First, LMP1 induces expression of A20, which acts as a deubiquitinase for IRF7 and thus negatively regulates its activity as a transcription factor (Ning and Pagano 2010). Furthermore, SUMOylation of IRF7 at lysine 452 was shown to decrease DNA-binding and transcriptional activity of IRF7. Interestingly, this mechanism is dependent on CTAR3 and its interaction with the SUMO-conjugating enzyme Ubc9 (Bentz et al. 2011, 2012).

## 5 Future Perspectives

Despite thirty years of research and more than 2000 papers on the molecular and biological functions of LMP1, many important questions still remain unanswered. The molecular signaling mechanisms at the effector sites CTAR1 and CTAR2 have not been fully elucidated. Particularly, the composition and precise architecture of the signaling complex at CTAR2 including the intermolecular interactions between the components of this complex is still unresolved. The situation is getting more complex by the fact that LMP1 molecules reside within large clusters and several LMP1 molecules might be involved in forming functional CTAR1 or CTAR2 complexes of higher order. This could mean that not all components of the CTAR2 complex must necessarily interact with the same LMP1 molecule to form an active signaling complex. Also, molecular interactions between CTAR1 and CTAR2 have been suggested, although both domains can function independently from each other (Floettmann et al. 1998). The elucidation of an LMP1 crystal structure, possibly in complex with LMP1-binding partner(s), will certainly be very important to answer these questions.

Other urgent questions concern the functions of LMP1 during EBV infection and the establishment of latency in vivo. Usually and due to the lack of appropriate in vivo infection models for EBV, the analysis of LMP1's functions has largely been restricted to its role in cell transformation in cell culture. It is likely that some LMP1-induced pathways have so far unrecognized functions in the establishment of latent infection in vivo and in immune modulation. The role of LMP1 during early infection of B cells is also unclear. LMP1 expression starts four days post-infection with EBV. Surprisingly, LCL expression levels are reached only three weeks after infection, and LMP1-induced NF-κB seems to be dispensable during that time (Price et al. 2012).

LMP1 is present in exosomes that are secreted from EBV-positive B cell and NPC lines (Dukers et al. 2000; Meckes et al. 2010). The functions of secreted LMP1 in immune modulation, cell–cell communication, tumor microenvironment, and potential other processes will also be interesting topics of future research.

Another enigma are the biological and molecular functions of a truncated variant of LMP1, the so-called lytic LMP1, which is highly expressed during the lytic cycle from the ED-L1A promoter located within the first intron of the LMP1 gene (Hudson et al. 1985; Modrow and Wolf 1986; Baichwal and Sugden 1987; Erickson and Martin 1997). Due to the lack of the first four transmembrane domains, lytic LMP1 is incapable of oligomerization and, thus, likely unable to signal (Mitchell and Sugden 1995; Gires et al. 1997). Lytic LMP1 rather inhibits the activity of full-length LMP1 (Erickson and Martin 2000). In contrast to full-length LMP1, the lytic variant has no transforming capacity in fibroblasts and is not essential for B cell transformation by EBV (Wang et al. 1988a; Dirmeier et al. 2003).

Due to its proven critical importance for cell transformation and pathogenesis, LMP1 constitutes an excellent therapeutic target for intervention strategies against EBV-associated malignancies. By inhibiting LCL tumor growth with a small molecule JNK inhibitor, we have already delivered proof of principle that targeting of LMP1's signaling network is a feasible strategy (Kutz et al. 2008). However, to be more specific for LMP1, future approaches shall rather target the LMP1 molecule itself or exploit the unique interface between LMP1 and its critical cellular signaling molecules such as CTAR1::TRAF2 or CTAR2::TRAF6 as targets for small molecule inhibitors. For instance, DNAzymes targeting LMP1 expression have antitumor effects and sensitize NPC cells for radiotherapy (Cao et al. 2014). Our own laboratory is involved in the screening for inhibitors of CTAR1::TRAF2 interaction, which shall block this interaction and uncouple LMP1 from critical parts of its transforming signaling network.

## References

Adler B, Schaadt E, Kempkes B, Zimber-Strobl U, Baier B, Bornkamm GW (2002) Control of Epstein-Barr virus reactivation by activated CD40 and viral latent membrane protein 1. Proc Natl Acad Sci USA 99(1):437–442

Allday MJ, Crawford DH, Griffin BE (1989) Epstein-Barr virus latent gene expression during the initiation of B cell immortalization. J Gen Virol 70(Pt 7):1755–1764

Amort M, Nachbauer B, Tuzlak S, Kieser A, Schepers A, Villunger A, Polacek A (2015) Expression of the vault RNA protects cells from undergoing apoptosis. Nat Com 6:7030

Arcipowski KM, Stunz LL, Graham JP, Kraus ZJ, Vanden Bush TJ, Bishop GA (2011) Molecular mechanisms of TNFR-associated factor 6 (TRAF6) utilization by the oncogenic viral mimic of CD40, latent membrane protein 1 (LMP1). J Biol Chem 286(12):9948–9955

Ardila-Osorio H, Clausse B, Mishal Z, Wiels J, Tursz T, Busson P (1999) Evidence of LMP1-TRAF3 interactions in glycosphingolipid-rich complexes of lymphoblastoid and nasopharyngeal carcinoma cells. Int J Cancer 81(4):645–649

Atkinson PG, Coope HJ, Rowe M, Ley SC (2003) Latent membrane protein 1 of Epstein-Barr virus stimulates processing of NF-kappa B2 p100 to p52. J Biol Chem 278(51):51134–51142

Aviel S, Winberg G, Massucci M, Ciechanover A (2000) Degradation of the Epstein-Barr virus latent membrane protein 1 (LMP1) by the ubiquitin-proteasome pathway. Targeting via ubiquitination of the N-terminal residue. J Biol Chem 275(31):23491–23499

Babcock GJ, Hochberg D, Thorley-Lawson AD (2000) The expression pattern of Epstein-Barr virus latent genes in vivo is dependent upon the differentiation stage of the infected B cell. Immunity 13(4):497–506

Baichwal VR, Sugden B (1987) Posttranslational processing of an Epstein-Barr virus-encoded membrane protein expressed in cells transformed by Epstein-Barr virus. J Virol 61(3):866–875

Baichwal VR, Sugden B (1988) Transformation of Balb 3T3 cells by the BNLF-1 gene of Epstein-Barr virus. Oncogene 2(5):461–467

Bentz GL, Whitehurst CB, Pagano JS (2011) Epstein-Barr virus latent membrane protein 1 (LMP1) C-terminal-activating region 3 contributes to LMP1-mediated cellular migration via its interaction with Ubc9. J Virol 85(19):10144–10153

Bentz GL, Shackelford J, Pagano JS (2012) Epstein-Barr virus latent membrane protein 1 regulates the function of interferon regulatory factor 7 by inducing its sumoylation. J Virol 86(22):12251–12261

Blake SM, Eliopoulos AG, Dawson CW, Young LS (2001) The transmembrane domains of the EBV-encoded latent membrane protein 1 (LMP1) variant CAO regulate enhanced signalling activity. Virology 282(2):278–287

Boehm D, Gewurz BE, Kieff E, Cahir-McFarland E (2010) Epstein-Barr latent membrane protein 1 transformation site 2 activates NF-kappaB in the absence of NF-kappaB essential modifier residues 133–224 or 373–419. Proc Natl Acad Sci USA 107(42):18103–18108

Brodeur SR, Cheng G, Baltimore D, Thorley-Lawson DA (1997) Localization of the major NF-kappaB-activating site and the sole TRAF3 binding site of LMP-1 defines two distinct signaling motifs. J Biol Chem 272(32):19777–19784

Brown KD, Hostager BS, Bishop GA (2001) Differential signaling and tumor necrosis factor receptor-associated factor (TRAF) degradation mediated by CD40 and the Epstein-Barr virus oncoprotein latent membrane protein 1 (LMP1). J Exp Med 193(8):943–954

Busch LK, Bishop GA (1999) The EBV transforming protein, latent membrane protein 1, mimics and cooperates with CD40 signaling in B lymphocytes. J Immunol 162(5):2555–2561

Cahir-McFarland ED, Izumi KM, Mosialos G (1999) Epstein-barr virus transformation: involvement of latent membrane protein 1-mediated activation of NF-kappaB. Oncogene 18(49):6959–6964

Cahir-McFarland ED, Carter K, Rosenwald A, Giltnane JM, Henrickson SE, Staudt LM, Kieff E (2004) Role of NF-kappa B in cell survival and transcription of latent membrane protein 1-expressing or Epstein-Barr virus latency III-infected cells. J Virol 78(8):4108–4119

Cao Y, Yang L, Jiang W, Wang X, Liao W, Tan G, Liao Y, Qiu Y, Feng D, Tang F, Hou BL, Zhang L, Fu J, He F, Liu X, Jiang W, Yang T, Sun LQ (2014) Therapeutic evaluation of Epstein-Barr virus-encoded latent membrane protein-1 targeted DNAzyme for treating of nasopharyngeal carcinomas. Mol Ther 22(2):371–377

Chatterjee B, Leung CS, Munz C (2014) Animal models of Epstein Barr virus infection. J Immunol Methods 410:80–87

Chen H, Lee JM, Zong Y, Borowitz M, Ng MH, Ambinder RF, Hayward SD (2001) Linkage between STAT regulation and Epstein-Barr virus gene expression in tumors. J Virol 75(6):2929–2937

Chen H, Hutt-Fletcher L, Cao L, Hayward SD (2003) A positive autoregulatory loop of LMP1 expression and STAT activation in epithelial cells latently infected with Epstein-Barr virus. J Virol 77(7):4139–4148

Chen CC, Liu HP, Chao M, Liang Y, Tsang NM, Huang HY, Wu CC, Chang YS (2014) NF-kappaB-mediated transcriptional upregulation of TNFAIP2 by the Epstein-Barr virus oncoprotein, LMP1, promotes cell motility in nasopharyngeal carcinoma. Oncogene 33(28):3648–3659

Chew MM, Gan SY, Khoo AS, Tan EL (2010) Interleukins, laminin and Epstein-Barr virus latent membrane protein 1 (EBV LMP1) promote metastatic phenotype in nasopharyngeal carcinoma. BMC Cancer 10:574

Coffin WF 3rd, Erickson KD, Hoedt-Miller M, Martin JM (2001) The cytoplasmic amino-terminus of the Latent Membrane Protein-1 of Epstein-Barr virus: relationship between transmembrane orientation and effector functions of the carboxy-terminus and transmembrane domain. Oncogene 20(38):5313–5330

Coffin WF 3rd, Geiger TR, Martin JM (2003) Transmembrane domains 1 and 2 of the latent membrane protein 1 of Epstein-Barr virus contain a lipid raft targeting signal and play a critical role in cytostasis. J Virol 77(6):3749–3758

Cohen JI (2000) Epstein-Barr virus infection. N Engl J Med 343(7):481–492

Curran JA, Laverty FS, Campbell D, Macdiarmid J, Wilson JB (2001) Epstein-Barr virus encoded latent membrane protein-1 induces epithelial cell proliferation and sensitizes transgenic mice to chemical carcinogenesis. Cancer Res 61(18):6730–6738

Dawson CW, Rickinson AB, Young LS (1990) Epstein-Barr virus latent membrane protein inhibits human epithelial cell differentiation. Nature 344:777–780

Dawson CW, Tramountanis G, Eliopoulos AG, Young LS (2003) Epstein-Barr virus latent membrane protein 1 (LMP1) activates the phosphatidylinositol 3-kinase/Akt pathway to promote cell survival and induce actin filament remodeling. J Biol Chem 278(6):3694–3704

Dawson CW, Laverick L, Morris MA, Tramoutanis G, Young LS (2008) Epstein-Barr virus-encoded LMP1 regulates epithelial cell motility and invasion via the ERK-MAPK pathway. J Virol 82(7):3654–3664

Demetriades C, Mosialos G (2009) The LMP1 promoter can be transactivated directly by NF-kappaB. J Virol 83(10):5269–5277

Devergne O, Hatzivassiliou E, Izumi KM, Kaye KM, Kleijnen MF, Kieff E, Mosialos G (1996) Association of TRAF1, TRAF2, and TRAF3 with an Epstein-Barr virus LMP1 domain important for B-lymphocyte transformation: role in NF-kappaB activation. Mol Cell Biol 16(12):7098–7108

Devergne O, Cahir-McFarland ED, Mosialos G, Izumi KM, Ware CF, Kieff E (1998) Role of the TRAF binding site and NF-kappaB activation in Epstein-Barr virus latent membrane protein 1-induced cell gene expression. J Virol 72(10):7900–7908

Ding L, Li L, Yang J, Zhou S, Li W, Tang M, Shi Y, Yi W, Cao Y (2007) Latent membrane protein 1 encoded by Epstein-Barr virus induces telomerase activity via p16INK4A/Rb/E2F1 and JNK signaling pathways. J Med Virol 79(8):1153–1163

Dirmeier U, Neuhierl B, Kilger E, Reisbach G, Sandberg ML, Hammerschmidt W (2003) Latent membrane protein 1 is critical for efficient growth transformation of human B cells by Epstein-Barr virus. Cancer Res 63(11):2982–2989

Dirmeier U, Hoffmann R, Kilger E, Schultheiss U, Briseno C, Gires O, Kieser A, Eick D, Sugden B, Hammerschmidt W (2005) Latent membrane protein 1 of Epstein-Barr virus coordinately regulates proliferation with control of apoptosis. Oncogene 24(10):1711–1717

D'Souza BN, Edelstein LC, Pegman PM, Smith SM, Loughran ST, Clarke A, Mehl A, Rowe M, Gelinas C, Walls D (2004) Nuclear factor kappa B-dependent activation of the antiapoptotic bfl-1 gene by the Epstein-Barr virus latent membrane protein 1 and activated CD40 receptor. J Virol 78(4):1800–1816

Dukers DF, Meij P, Vervoort MB, Vos W, Scheper RJ, Meijer CJ, Bloemena E, Middeldorp JM (2000) Direct immunosuppressive effects of EBV-encoded latent membrane protein 1. J Immunol 165(2):663–670

Eliopoulos AG, Young LS (1998) Activation of the cJun N-terminal kinase (JNK) pathway by the Epstein-Barr virus-encoded latent membrane protein 1 (LMP1). Oncogene 16(13):1731–1742

Eliopoulos AG, Blake SM, Floettmann JE, Rowe M, Young LS (1999a) Epstein-Barr virus-encoded latent membrane protein 1 activates the JNK pathway through its extreme C terminus via a mechanism involving TRADD and TRAF2. J Virol 73(2):1023–1035

Eliopoulos AG, Gallagher NJ, Blake SM, Dawson CW, Young LS (1999b) Activation of the p38 mitogen-activated protein kinase pathway by Epstein-Barr virus-encoded latent membrane protein 1 coregulates interleukin-6 and interleukin-8 production. J Biol Chem 274(23):16085–16096

Eliopoulos AG, Caamano JH, Flavell J, Reynolds GM, Murray PG, Poyet JL, Young LS (2003a) Epstein-Barr virus-encoded latent infection membrane protein 1 regulates the processing of p100 NF-kappaB2 to p52 via an IKKgamma/NEMO-independent signalling pathway. Oncogene 22(48):7557–7569

Eliopoulos AG, Waites ER, Blake SM, Davies C, Murray P, Young LS (2003b) TRAF1 is a critical regulator of JNK signaling by the TRAF-binding domain of the Epstein-Barr virus-encoded latent infection membrane protein 1 but not CD40. J Virol 77(2):1316–1328

Endo K, Kondo S, Shackleford J, Horikawa T, Kitagawa N, Yoshizaki T, Furukawa M, Zen Y, Pagano JS (2009) Phosphorylated ezrin is associated with EBV latent membrane protein 1 in nasopharyngeal carcinoma and induces cell migration. Oncogene 28(14):1725–1735

Erickson KD, Martin JM (1997) Early detection of the lytic LMP-1 protein in EBV-infected B-cells suggests its presence in the virion. Virology 234(1):1–13

Erickson KD, Martin JM (2000) The late lytic LMP-1 protein of Epstein-Barr virus can negatively regulate LMP-1 signaling. J Virol 74(2):1057–1060

Everly DN Jr, Mainou BA, Raab-Traub N (2004) Induction of Id1 and Id3 by latent membrane protein 1 of Epstein-Barr virus and regulation of p27/Kip and cyclin-dependent kinase 2 in rodent fibroblast transformation. J Virol 78(24):13470–13478

Fahraeus R, Jansson A, Ricksten A, Sjoblom A, Rymo L (1990) Epstein-Barr virus-encoded nuclear antigen 2 activates the viral latent membrane protein promoter by modulating the activity of a negative regulatory element. Proc Natl Acad Sci USA 87(19):7390–7394

Faumont N, Durand-Panteix S, Schlee M, Gromminger S, Schuhmacher M, Holzel M, Laux G, Mailhammer R, Rosenwald A, Staudt LM, Bornkamm GW, Feuillard J (2009) c-Myc and Rel/NF-kappaB are the two master transcriptional systems activated in the latency III program of Epstein-Barr virus-immortalized B cells. J Virol 83(10):5014–5027

Fennewald S, van Santen V, Kieff E (1984) Nucleotide sequence of an mRNA transcribed in latent growth-transforming virus infection indicates that it may encode a membrane protein. J Virol 51(2):411–419

Feuillard J, Schuhmacher M, Kohanna S, Asso-Bonnet M, Ledeur F, Joubert-Caron R, Bissieres P, Polack A, Bornkamm GW, Raphael M (2000) Inducible loss of NF-kappaB activity is associated with apoptosis and Bcl-2 down-regulation in Epstein-Barr virus-transformed B lymphocytes. Blood 95(6):2068–2075

Fielding CA, Sandvej K, Mehl A, Brennan P, Jones M, Rowe M (2001) Epstein-Barr virus LMP-1 natural sequence variants differ in their potential to activate cellular signaling pathways. J Virol 75(19):9129–9141

Floettmann JE, Rowe M (1997) Epstein-Barr virus latent membrane protein-1 (LMP1) C-terminus activation region 2 (CTAR2) maps to the far C-terminus and requires oligomerisation for NF-kappaB activation. Oncogene 15(15):1851–1858

Floettmann JE, Eliopoulos AG, Jones M, Young LS, Rowe M (1998) Epstein-Barr virus latent membrane protein-1 (LMP1) signalling is distinct from CD40 and involves physical cooperation of its two C-terminus functional regions. Oncogene 17(18):2383–2392

Gewurz BE, Mar JC, Padi M, Zhao B, Shinners NP, Takasaki K, Bedoya E, Zou JY, Cahir-McFarland E, Quackenbush J, Kieff E (2011) Canonical NF-kappaB activation is essential for Epstein-Barr virus latent membrane protein 1 TES2/CTAR2 gene regulation. J Virol 85(13):6764–6773

Gewurz BE, Towfic F, Mar JC, Shinners NP, Takasaki K, Zhao B, Cahir-McFarland ED, Quackenbush J, Xavier RJ, Kieff E (2012) Genome-wide siRNA screen for mediators of NF-kappaB activation. Proc Natl Acad Sci USA 109(7):2467–2472

Gires O, Zimber-Strobl U, Gonnella R, Ueffing M, Marschall G, Zeidler R, Pich D, Hammerschmidt W (1997) Latent membrane protein 1 of Epstein-Barr virus mimics a constitutively active receptor molecule. EMBO J 16(20):6131–6140

Gires O, Kohlhuber F, Kilger E, Baumann M, Kieser A, Kaiser C, Zeidler R, Scheffer B, Ueffing M, Hammerschmidt W (1999) Latent membrane protein 1 of Epstein-Barr virus interacts with JAK3 and activates STAT proteins. EMBO J 18(11):3064–3073
Grimm T, Schneider S, Naschberger E, Huber J, Guenzi E, Kieser A, Reitmeir P, Schulz TF, Morris CA, Sturzl M (2005) EBV latent membrane protein-1 protects B cells from apoptosis by inhibition of BAX. Blood 105(8):3263–3269
Guasparri I, Bubman D, Cesarman E (2008) EBV LMP2A affects LMP1-mediated NF-κB signaling and survival of lymphoma cells by regulating TRAF2 expression. Blood 111(7):3813–3820
Hammerschmidt W, Sugden B, Baichwal VR (1989) The transforming domain alone of the latent membrane protein of Epstein-Barr virus is toxic to cells when expressed at high levels. J Virol 63(6):2469–2475
Hannigan A, Qureshi AM, Nixon C, Tsimbouri PM, Jones S, Philbey AW, Wilson JB (2011) Lymphocyte deficiency limits Epstein-Barr virus latent membrane protein 1 induced chronic inflammation and carcinogenic pathology in vivo. Mol Cancer 10(1):11
Hatton O, Lambert SL, Krams SM, Martinez OM (2012) Src kinase and Syk activation initiate PI3K signaling by a chimeric latent membrane protein 1 in Epstein-Barr virus (EBV)+ B cell lymphomas. PLoS ONE 7(8):e42610
Hatton OL, Harris-Arnold A, Schaffert S, Krams SM, Martinez OM (2014) The interplay between Epstein-Barr virus and B lymphocytes: implications for infection, immunity, and disease. Immunol Res 58(2–3):268–276
Hatzivassiliou E, Miller WE, Raab-Traub N, Kieff E, Mosialos G (1998) A fusion of the EBV latent membrane protein-1 (LMP1) transmembrane domains to the CD40 cytoplasmic domain is similar to LMP1 in constitutive activation of epidermal growth factor receptor expression, nuclear factor-kappa B, and stress-activated protein kinase. J Immunol 160(3):1116–1121
Hau PM, Tsang CM, Yip YL, Huen MS, Tsao SW (2011) Id1 interacts and stabilizes the Epstein-Barr virus latent membrane protein 1 (LMP1) in nasopharyngeal epithelial cells. PLoS ONE 6(6):e21176
Henderson S, Rowe M, Gregory C, Croom-Carter D, Wang F, Longnecker R, Kieff E, Rickinson A (1991) Induction of bcl-2 expression by Epstein-Barr virus latent membrane protein 1 protects infected B cells from programmed cell death. Cell 65(7):1107–1115
Hennessy K, Fennewald S, Hummel M, Cole T, Kieff E (1984) A membrane protein encoded by Epstein-Barr virus in latent growth-transforming infection. Proc Natl Acad Sci USA 81(22):7207–7211
Higuchi M, Izumi KM, Kieff E (2001) Epstein-Barr virus latent-infection membrane proteins are palmitoylated and raft-associated: protein 1 binds to the cytoskeleton through TNF receptor cytoplasmic factors. Proc Natl Acad Sci USA 98(8):4675–4680
Higuchi M, Kieff E, Izumi KM (2002) The Epstein-Barr virus latent membrane protein 1 putative Janus kinase 3 (JAK3) binding domain does not mediate JAK3 association or activation in B-lymphoma or lymphoblastoid cell lines. J Virol 76(1):455–459
Hu LF, Zabarovsky ER, Chen F, Cao SL, Ernberg I, Klein G, Winberg G (1991) Isolation and sequencing of the Epstein-Barr virus BNLF-1 gene (LMP1) from a Chinese nasopharyngeal carcinoma. J Gen Virol 72(Pt 10):2399–2409
Huang YT, Sheen TS, Chen CL, Lu J, Chang Y, Chen JY, Tsai CH (1999) Profile of cytokine expression in nasopharyngeal carcinomas: a distinct expression of interleukin 1 in tumor and CD4+ T cells. Cancer Res 59(7):1599–1605
Huang S, Stupack D, Liu A, Cheresh D, Nemerow GR (2000) Cell growth and matrix invasion of EBV-immortalized human B lymphocytes is regulated by expression of alpha(v) integrins. Oncogene 19(15):1915–1923
Hudson GS, Farrell PJ, Barrell BG (1985) Two related but differentially expressed potential membrane proteins encoded by the EcoRI Dhet region of Epstein-Barr virus B95-8. J Virol 53(2):528–535

Huen DS, Henderson SA, Croom-Carter D, Rowe M (1995) The Epstein-Barr virus latent membrane protein-1 (LMP1) mediates activation of NF-kappa B and cell surface phenotype via two effector regions in its carboxy-terminal cytoplasmic domain. Oncogene 10(3):549–560

Huye LE, Ning S, Kelliher M, Pagano JS (2007) Interferon regulatory factor 7 is activated by a viral oncoprotein through RIP-dependent ubiquitination. Mol Cell Biol 27(8):2910–2918

Izumi KM, Kieff ED (1997) The Epstein-Barr virus oncogene product latent membrane protein 1 engages the tumor necrosis factor receptor-associated death domain protein to mediate B lymphocyte growth transformation and activate NF-kappaB. Proc Natl Acad Sci USA 94(23):12592–12597

Izumi KM, Kaye KM, Kieff ED (1994) Epstein-Barr virus recombinant molecular genetic analysis of the LMP1 amino-terminal cytoplasmic domain reveals a probable structural role, with no component essential for primary B-lymphocyte growth transformation. J Virol 68(7):4369–4376

Izumi KM, Kaye KM, Kieff ED (1997) The Epstein-Barr virus LMP1 amino acid sequence that engages tumor necrosis factor receptor associated factors is critical for primary B lymphocyte growth transformation. Proc Natl Acad Sci USA 94(4):1447–1452

Izumi KM, Cahir-McFarland ED, Riley EA, Rizzo D, Chen Y, Kieff E (1999) The residues between the two transformation effector sites of Epstein-Barr virus latent membrane protein 1 are not critical for B-lymphocyte growth transformation. J Virol 73(12):9908–9916

Johansson P, Jansson A, Ruetschi U, Rymo L (2009) Nuclear factor-kappaB binds to the Epstein-Barr virus LMP1 promoter and upregulates its expression. J Virol 83(3):1393–1401

Johansson P, Jansson A, Ruetschi U, Rymo L (2010) The p38 signaling pathway upregulates expression of the Epstein-Barr virus LMP1 oncogene. J Virol 84(6):2787–2797

Kaye KM, Izumi KM, Kieff E (1993) Epstein-Barr virus latent membrane protein 1 is essential for B-lymphocyte growth transformation. Proc Natl Acad Sci USA 90(19):9150–9154

Kaye KM, Izumi KM, Mosialos G, Kieff E (1995) The Epstein-Barr virus LMP1 cytoplasmic carboxy terminus is essential for B-lymphocyte transformation; fibroblast cocultivation complements a critical function within the terminal 155 residues. J Virol 69(2):675–683

Kaye KM, Devergne O, Harada JN, Izumi KM, Yalamanchili R, Kieff E, Mosialos G (1996) Tumor necrosis factor receptor associated factor 2 is a mediator of NF-kappa B activation by latent infection membrane protein 1, the Epstein-Barr virus transforming protein. Proc Natl Acad Sci USA 93(20):11085–11090

Kaye KM, Izumi KM, Li H, Johannsen E, Davidson D, Longnecker R, Kieff E (1999) An Epstein-Barr virus that expresses only the first 231 LMP1 amino acids efficiently initiates primary B-lymphocyte growth transformation. J Virol 73(12):10525–10530

Kieser A (2008) Pursuing different 'TRADDes': TRADD signaling induced by TNF-receptor 1 and the Epstein-Barr virus oncoprotein LMP1. Biol Chem 389(10):1261–1271

Kieser A, Kilger E, Gires O, Ueffing M, Kolch W, Hammerschmidt W (1997) Epstein-Barr virus latent membrane protein-1 triggers AP-1 activity via the c-Jun N-terminal kinase cascade. EMBO J 16(21):6478–6485

Kieser A, Kaiser C, Hammerschmidt W (1999) LMP1 signal transduction differs substantially from TNF receptor 1 signaling in the molecular functions of TRADD and TRAF2. EMBO J 18(9):2511–2521

Kilger E, Kieser A, Baumann M, Hammerschmidt W (1998) Epstein-Barr virus-mediated B-cell proliferation is dependent upon latent membrane protein 1, which simulates an activated CD40 receptor. EMBO J 17(6):1700–1709

Kim JE, Kim SY, Lim SY, Kieff E, Song YJ (2014) Role of Ca2+/Calmodulin-dependent Kinase II-IRAK1 interaction in LMP1-induced NF-kappaB activation. Mol Cell Biol 34(3):325–334

Kis LL, Takahara M, Nagy N, Klein G, Klein E (2006) IL-10 can induce the expression of EBV-encoded latent membrane protein-1 (LMP-1) in the absence of EBNA-2 in B lymphocytes and in Burkitt lymphoma- and NK lymphoma-derived cell lines. Blood 107(7):2928–2935

Kis LL, Salamon D, Persson EK, Nagy N, Scheeren FA, Spits H, Klein G, Klein E (2010) IL-21 imposes a type II EBV gene expression on type III and type I B cells by the repression of C- and activation of LMP-1-promoter. Proc Natl Acad Sci USA 107(2):872–877

Kis LL, Gerasimcik N, Salamon D, Persson EK, Nagy N, Klein G, Severinson E, Klein E (2011) STAT6 signaling pathway activated by the cytokines IL-4 and IL-13 induces expression of the Epstein-Barr virus-encoded protein LMP-1 in absence of EBNA-2: implications for the type II EBV latent gene expression in Hodgkin lymphoma. Blood 117(1):165–174

Kondo S, Wakisaka N, Schell MJ, Horikawa T, Sheen TS, Sato H, Furukawa M, Pagano JS, Yoshizaki T (2005) Epstein-Barr virus latent membrane protein 1 induces the matrix metalloproteinase-1 promoter via an Ets binding site formed by a single nucleotide polymorphism: Enhanced susceptibility to nasopharyngeal carcinoma. Int J Cancer 115(3):368–376

Kulwichit W, Edwards RH, Davenport EM, Baskar JF, Godfrey V, Raab-Traub N (1998) Expression of the Epstein-Barr virus latent membrane protein 1 induces B cell lymphoma in transgenic mice. Proc Natl Acad Sci USA 95(20):11963–11968

Kung CP, Meckes DG Jr, Raab-Traub N (2011) Epstein-Barr virus LMP1 activates EGFR, STAT3, and ERK through effects on PKCdelta. J Virol 85(9):4399–4408

Kutz H, Reisbach G, Schultheiss U, Kieser A (2008) The c-Jun N-terminal kinase pathway is critical for cell transformation by the latent membrane protein 1 of Epstein-Barr virus. Virology 371(2):246–256

Laherty CD, Hu HM, Opipari AW, Wang F, Dixit VM (1992) The Epstein-Barr virus LMP1 gene product induces A20 zinc finger protein expression by activating nuclear factor kappa B. J Biol Chem 267(34):24157–24160

Lai HC, Hsiao JR, Chen CW, Wu SY, Lee CH, Su IJ, Takada K, Chang Y (2010) Endogenous latent membrane protein 1 in Epstein-Barr virus-infected nasopharyngeal carcinoma cells attracts T lymphocytes through upregulation of multiple chemokines. Virology 405(2):464–473

Lam N, Sugden B (2003) LMP1, a viral relative of the TNF receptor family, signals principally from intracellular compartments. EMBO J 22(12):3027–3038

Lambert SL, Martinez OM (2007) Latent membrane protein 1 of EBV activates phosphatidylinositol 3-kinase to induce production of IL-10. J Immunol 179(12):8225–8234

Le Clorennec C, Ouk TS, Youlyouz-Marfak I, Panteix S, Martin CC, Rastelli J, Adriaenssens E, Zimber-Strobl U, Coll J, Feuillard J, Jayat-Vignoles C (2008) Molecular basis of cytotoxicity of Epstein-Barr virus (EBV) latent membrane protein 1 (LMP1) in EBV latency III B cells: LMP1 induces type II ligand-independent autoactivation of CD95/Fas with caspase 8-mediated apoptosis. J Virol 82(13):6721–6733

Lee J, Sugden B (2007) A membrane leucine heptad contributes to trafficking, signaling, and transformation by latent membrane protein 1. J Virol 81(17):9121–9130

Lee DY, Sugden B (2008a) The latent membrane protein 1 oncogene modifies B-cell physiology by regulating autophagy. Oncogene 27(20):2833–2842

Lee DY, Sugden B (2008b) The LMP1 oncogene of EBV activates PERK and the unfolded protein response to drive its own synthesis. Blood 111(4):2280–2289

Li HM, Zhuang ZH, Wang Q, Pang JC, Wang XH, Wong HL, Feng HC, Jin DY, Ling MT, Wong YC, Eliopoulos AG, Young LS, Huang DP, Tsao SW (2004) Epstein-Barr virus latent membrane protein 1 (LMP1) upregulates Id1 expression in nasopharyngeal epithelial cells. Oncogene 23(25):4488–4494

Li J, Zhang XS, Xie D, Deng HX, Gao YF, Chen QY, Huang WL, Masucci MG, Zeng YX (2007) Expression of immune-related molecules in primary EBV-positive Chinese nasopharyngeal carcinoma: associated with latent membrane protein 1 (LMP1) expression. Cancer Biol Ther 6(12):1997–2004

Li SS, Yang S, Wang S, Yang XM, Tang QL, Wang SH (2011) Latent membrane protein 1 mediates the resistance of nasopharyngeal carcinoma cells to TRAIL-induced apoptosis by activation of the PI3K/Akt signaling pathway. Oncol Rep 26(6):1573–1579

Li L, Li W, Xiao L, Xu J, Chen X, Tang M, Dong Z, Tao Q, Cao Y (2012) Viral oncoprotein LMP1 disrupts p53-induced cell cycle arrest and apoptosis through modulating K63-linked ubiquitination of p53. Cell Cycle 11(12):2327–2336

Li B, Huang G, Zhang X, Li R, Wang J, Dong Z, He Z (2013) Increased phosphorylation of histone H3 at serine 10 is involved in Epstein-Barr virus latent membrane protein-1-induced carcinogenesis of nasopharyngeal carcinoma. BMC Cancer 13:124

Lin SS, Lee DC, Law AH, Fang JW, Chua DT, Lau AS (2010) A role for protein kinase PKR in the mediation of Epstein-Barr virus latent membrane protein-1-induced IL-6 and IL-10 expression. Cytokine 50(2):210–219

Liu LT, Peng JP, Chang HC, Hung WC (2003) RECK is a target of Epstein-Barr virus latent membrane protein 1. Oncogene 22(51):8263–8270

Liu HP, Chen CC, Wu CC, Huang YC, Liu SC, Liang Y, Chang KP, Chang YS (2012) Epstein-Barr virus-encoded LMP1 interacts with FGD4 to activate Cdc42 and thereby promote migration of nasopharyngeal carcinoma cells. PLoS Pathog 8(5):e1002690

Lo AKF, Liu Y, Wang X, Wong YC, Lee CKF, Huang DP, Tsao SW (2001) Identification of downstream target genes of latent membrane protein 1 in nasopharyngeal carcinoma cells by suppression subtractive hybridization. Biochim Biophys Acta 1520(2):131–140

Lorenzetti MA, Gantuz M, Altcheh J, De Matteo E, Chabay PA, Preciado MV (2012) Distinctive Epstein-Barr virus variants associated with benign and malignant pediatric pathologies: LMP1 sequence characterization and linkage with other viral gene polymorphisms. J Clin Microbiol 50(3):609–618

Lu J, Chua HH, Chen SY, Chen JY, Tsai CH (2003) Regulation of matrix metalloproteinase-1 by Epstein-Barr virus proteins. Cancer Res 63(1):256–262

Luftig M, Prinarakis E, Yasui T, Tsichritzis T, Cahir-McFarland E, Inoue J, Nakano H, Mak TW, Yeh WC, Li X, Akira S, Suzuki N, Suzuki S, Mosialos G, Kieff E (2003) Epstein-Barr virus latent membrane protein 1 activation of NF-kappaB through IRAK1 and TRAF6. Proc Natl Acad Sci USA 100(26):15595–15600

Luftig M, Yasui T, Soni V, Kang MS, Jacobson N, Cahir-McFarland E, Seed B, Kieff E (2004) Epstein-Barr virus latent infection membrane protein 1 TRAF-binding site induces NIK/IKK alpha-dependent noncanonical NF-kappaB activation. Proc Natl Acad Sci USA 101(1):141–146

Ma SD, Xu X, Plowshay J, Ranheim EA, Burlingham WJ, Jensen JL, Asimakopoulos F, Tang W, Gulley ML, Cesarman E, Gumperz JE, Kenney SC (2015) LMP1-deficient Epstein-Barr virus mutant requires T cells for lymphomagenesis. J Clin Invest 125(1):304–315

Mainou BA, Raab-Traub N (2006) LMP1 strain variants: biological and molecular properties. J Virol 80(13):6458–6468

Mainou BA, Everly DN, Raab-Traub N (2005) Epstein-Barr virus latent membrane protein 1 CTAR1 mediates rodent and human fibroblast transformation through activation of PI3K. Oncogene 24(46):6917–6924

Mainou BA, Everly DN Jr, Raab-Traub N (2007) Unique signaling properties of CTAR1 in LMP1-mediated transformation. J Virol 81(18):9680–9692

Mancao C, Hammerschmidt W (2007) Epstein-Barr virus latent membrane protein 2A is a B-cell receptor mimic and essential for B-cell survival. Blood 110(10):3715–3721

Mann KP, Thorley-Lawson D (1987) Posttranslational processing of the Epstein-Barr virus-encoded p63/LMP protein. J Virol 61(7):2100–2108

Mann KP, Staunton D, Thorley-Lawson DA (1985) Epstein-Barr virus-encoded protein found in plasma membranes of transformed cells. J Virol 55(3):710–720

Martin J, Sugden B (1991) Transformation by the oncogenic latent membrane protein correlates with its rapid turnover, membrane localization, and cytoskeletal association. J Virol 65(6):3246–3258

Meckes DG Jr, Shair KH, Marquitz AR, Kung CP, Edwards RH, Raab-Traub N (2010) Human tumor virus utilizes exosomes for intercellular communication. Proc Natl Acad Sci USA 107(47):20370–20375

Meckes DG Jr, Menaker NF, Raab-Traub N (2013) Epstein-Barr virus LMP1 modulates lipid raft microdomains and the vimentin cytoskeleton for signal transduction and transformation. J Virol 87(3):1301–1311

Middeldorp JM, Pegtel DM (2008) Multiple roles of LMP1 in Epstein-Barr virus induced immune escape. Semin Cancer Biol 18(6):388–396

Miller WE, Edwards RH, Walling DM, Raab-Traub N (1994) Sequence variation in the Epstein-Barr virus latent membrane protein 1. J Gen Virol 75:2729–2740

Miller WE, Earp HS, Raab-Traub N (1995) The Epstein-Barr virus latent membrane protein 1 induces expression of the epidermal growth factor receptor. J Virol 69(7):4390–4398

Miller WE, Cheshire JL, Raab-Traub N (1998) Interaction of tumor necrosis factor receptor-associated factor signaling proteins with the latent membrane protein 1 PXQXT motif is essential for induction of epidermal growth factor receptor expression. Mol Cell Biol 18(5):2835–2844

Mitchell T, Sugden B (1995) Stimulation of NF-kappa B-mediated transcription by mutant derivatives of the latent membrane protein of Epstein-Barr virus. J Virol 69(5):2968–2976

Modrow S, Wolf H (1986) Characterization of two related Epstein-Barr virus-encoded membrane proteins that are differentially expressed in Burkitt lymphoma and in vitro-transformed cell lines. Proc Natl Acad Sci USA 83(15):5703–5707

Mohr CF, Kalmer M, Gross C, Mann MC, Sterz KR, Kieser A, Fleckenstein B, Kress AK (2014) The tumor marker Fascin is induced by the Epstein-Barr virus-encoded oncoprotein LMP1 via NF-kappaB in lymphocytes and contributes to their invasive migration. Cell Commun Signal 12(1):46

Moorthy RK, Thorley-Lawson DA (1993a) All three domains of the Epstein-Barr virus-encoded latent membrane protein LMP-1 are required for transformation of rat-1 fibroblasts. J Virol 67(3):1638–1646

Moorthy RK, Thorley-Lawson DA (1993b) Biochemical, genetic, and functional analyses of the phosphorylation sites on the Epstein-Barr virus-encoded oncogenic latent membrane protein LMP-1. J Virol 67(5):2637–2645

Morris MA, Dawson CW, Wei W, O'Neil JD, Stewart SE, Jia J, Bell AI, Young LS, Arrand JR (2008) Epstein-Barr virus-encoded LMP1 induces a hyperproliferative and inflammatory gene expression programme in cultured keratinocytes. J Gen Virol 89(Pt 11):2806–2820

Mosialos G, Birkenbach M, Yalamanchili R, VanArsdale T, Ware C, Kieff E (1995) The Epstein-Barr virus transforming protein LMP1 engages signaling proteins for the tumor necrosis factor receptor family. Cell 80(3):389–399

Motsch N, Pfuhl T, Mrazek J, Barth S, Grasser FA (2007) Epstein-Barr Virus-encoded latent membrane protein 1 (LMP1) induces the expression of the cellular microRNA miR-146a. RNA Biol 4(3):131–137

Murono S, Inoue H, Tanabe T, Joab I, Yoshizaki T, Furukawa M, Pagano JS (2001) Induction of cyclooxygenase-2 by Epstein-Barr virus latent membrane protein 1 is involved in vascular endothelial growth factor production in nasopharyngeal carcinoma cells. Proc Natl Acad Sci USA 98(12):6905–6910

Niedobitek G (1999) The Epstein-Barr virus: a group 1 carcinogen? Virchows Arch 435(2):79–86

Ning S, Pagano JS (2010) The A20 deubiquitinase activity negatively regulates LMP1 activation of IRF7. J Virol 84(12):6130–6138

Ning S, Hahn AM, Huye LE, Pagano JS (2003) Interferon regulatory factor 7 regulates expression of Epstein-Barr virus latent membrane protein 1: a regulatory circuit. J Virol 77(17):9359–9368

Ning S, Huye LE, Pagano JS (2005) Interferon regulatory factor 5 represses expression of the Epstein-Barr virus oncoprotein LMP1: braking of the IRF7/LMP1 regulatory circuit. J Virol 79(18):11671–11676

Ning S, Campos AD, Darnay BG, Bentz GL, Pagano JS (2008) TRAF6 and the three C-terminal lysine sites on IRF7 are required for its ubiquitination-mediated activation by the tumor necrosis factor receptor family member latent membrane protein 1. Mol Cell Biol 28(20):6536–6546

Noda C, Murata T, Kanda T, Yoshiyama H, Sugimoto A, Kawashima D, Saito S, Isomura H, Tsurumi T (2011) Identification and characterization of CCAAT enhancer-binding protein (C/EBP) as a transcriptional activator for Epstein-Barr virus oncogene latent membrane protein 1. J Biol Chem 286(49):42524–42533

Paine E, Scheinman RI, Baldwin AS Jr, Raab-Traub N (1995) Expression of LMP1 in epithelial cells leads to the activation of a select subset of NF-kappa B/Rel family proteins. J Virol 69(7):4572–4576

Pratt ZL, Zhang J, Sugden B (2012) The latent membrane protein 1 (LMP1) oncogene of Epstein-Barr virus can simultaneously induce and inhibit apoptosis in B cells. J Virol 86(8):4380–4393

Price AM, Tourigny JP, Forte E, Salinas RE, Dave SS, Luftig MA (2012) Analysis of Epstein-Barr virus-regulated host gene expression changes through primary B-cell outgrowth reveals delayed kinetics of latent membrane protein 1-mediated NF-kappaB activation. J Virol 86(20):11096–11106

Prince S, Keating S, Fielding C, Brennan P, Floettmann E, Rowe M (2003) Latent membrane protein 1 inhibits Epstein-Barr virus lytic cycle induction and progress via different mechanisms. J Virol 77(8):5000–5007

Rastelli J, Homig-Holzel C, Seagal J, Muller W, Hermann AC, Rajewsky K, Zimber-Strobl U (2008) LMP1 signaling can replace CD40 signaling in B cells in vivo and has unique features of inducing class-switch recombination to IgG1. Blood 111(3):1448–1455

Renzette N, Somasundaran M, Brewster F, Coderre J, Weiss ER, McManus M, Greenough T, Tabak B, Garber M, Kowalik TF, Luzuriaga K (2014) Epstein-barr virus latent membrane protein 1 genetic variability in peripheral blood B cells and oropharyngeal fluids. J Virol 88(7):3744–3755

Rothenberger S, Burns K, Rousseaux M, Tschopp J, Bron C (2003) Ubiquitination of the Epstein-Barr virus-encoded latent membrane protein 1 depends on the integrity of the TRAF binding site. Oncogene 22(36):5614–5618

Saito N, Courtois G, Chiba A, Yamamoto N, Nitta T, Hironaka N, Rowe M, Yamaoka S (2003) Two carboxyl-terminal activation regions of Epstein-Barr virus latent membrane protein 1 activate NF-kappaB through distinct signaling pathways in fibroblast cell lines. J Biol Chem 278(47):46565–46575

Sandberg M, Hammerschmidt W, Sugden B (1997) Characterization of LMP-1's association with TRAF1, TRAF2, and TRAF3. J Virol 71(6):4649–4656

Sandvej K, Gratama JW, Munch M, Zhou XG, Bolhuis RL, Andresen BS, Gregersen N, Hamilton-Dutoit S (1997) Sequence analysis of the Epstein-Barr virus (EBV) latent membrane protein-1 gene and promoter region: identification of four variants among wild-type EBV isolates. Blood 90(1):323–330

Schneider F, Neugebauer J, Griese J, Liefold N, Kutz H, Briseno C, Kieser A (2008) The viral oncoprotein LMP1 exploits TRADD for signaling by masking its apoptotic activity. PLoS Biol 6(1):e8

Schultheiss U, Puschner S, Kremmer E, Mak TW, Engelmann H, Hammerschmidt W, Kieser A (2001) TRAF6 is a critical mediator of signal transduction by the viral oncogene latent membrane protein 1. EMBO J 20(20):5678–5691

Shair KH, Raab-Traub N (2012) Transcriptome changes induced by Epstein-Barr virus LMP1 and LMP2A in transgenic lymphocytes and lymphoma. mBio 3(5). doi:10.1128/mBio.00288-12

Shair KH, Bendt KM, Edwards RH, Bedford EC, Nielsen JN, Raab-Traub N (2007) EBV latent membrane protein 1 activates Akt, NFkappaB, and Stat3 in B cell lymphomas. PLoS Pathog 3(11):e166

Shair KH, Schnegg CI, Raab-Traub N (2008) EBV latent membrane protein 1 effects on plakoglobin, cell growth, and migration. Cancer Res 68(17):6997–7005

Shkoda A, Town JA, Griese J, Romio M, Sarioglu H, Knofel T, Giehler F, Kieser A (2012) The germinal center kinase TNIK is required for canonical NF-kappaB and JNK signaling in B-cells by the EBV oncoprotein LMP1 and the CD40 receptor. PLoS Biol 10(8):e1001376

Siouda M, Frecha C, Accardi R, Yue J, Cuenin C, Gruffat H, Manet E, Herceg Z, Sylla BS, Tommasino M (2014) Epstein-Barr virus down-regulates tumor suppressor DOK1 expression. PLoS Pathog 10(5):e1004125
Song YJ, Kang MS (2010) Roles of TRAF2 and TRAF3 in Epstein-Barr virus latent membrane protein 1-induced alternative NF-kappaB activation. Virus Genes 41(2):174–180
Song YJ, Jen KY, Soni V, Kieff E, Cahir-McFarland E (2006) IL-1 receptor-associated kinase 1 is critical for latent membrane protein 1-induced p65/RelA serine 536 phosphorylation and NF-kappaB activation. Proc Natl Acad Sci USA 103(8):2689–2694
Song YJ, Izumi KM, Shinners NP, Gewurz BE, Kieff E (2008) IRF7 activation by Epstein-Barr virus latent membrane protein 1 requires localization at activation sites and TRAF6, but not TRAF2 or TRAF3. Proc Natl Acad Sci USA 105(47):18448–18453
Soni V, Yasui T, Cahir-McFarland E, Kieff E (2006) LMP1 transmembrane domain 1 and 2 (TM1-2) FWLY mediates intermolecular interactions with TM3-6 to activate NF-kappaB. J Virol 80(21):10787–10793
Souza TA, Stollar BD, Sullivan JL, Luzuriaga K, Thorley-Lawson DA (2005) Peripheral B cells latently infected with Epstein-Barr virus display molecular hallmarks of classical antigen-selected memory B cells. Proc Natl Acad Sci USA 102(50):18093–18098
Stevenson D, Charalambous C, Wilson JB (2005) Epstein-Barr virus latent membrane protein 1 (CAO) up-regulates VEGF and TGF alpha concomitant with hyperlasia, with subsequent up-regulation of p16 and MMP9. Cancer Res 65(19):8826–8835
Stunz LL, Busch LK, Munroe ME, Sigmund CD, Tygrett LT, Waldschmidt TJ, Bishop GA (2004) Expression of the cytoplasmic tail of LMP1 in mice induces hyperactivation of B lymphocytes and disordered lymphoid architecture. Immunity 21(2):255–266
Thorley-Lawson DA, Hawkins JB, Tracy SI, Shapiro M (2013) The pathogenesis of Epstein-Barr virus persistent infection. Curr Opin Virol 3(3):227–232
Thornburg NJ, Raab-Traub N (2007) Induction of epidermal growth factor receptor expression by Epstein-Barr virus latent membrane protein 1 C-terminal-activating region 1 is mediated by NF-kappaB p50 homodimer/Bcl-3 complexes. J Virol 81(23):12954–12961
Thornburg NJ, Pathmanathan R, Raab-Traub N (2003) Activation of nuclear factor-kappaB p50 homodimer/Bcl-3 complexes in nasopharyngeal carcinoma. Cancer Res 63(23):8293–8301
Tsai CN, Tsai CL, Tse KP, Chang HY, Chang YS (2002) The Epstein-Barr virus oncogene product, latent membrane protein 1, induces the downregulation of E-cadherin gene expression via activation of DNA methyltransferases. Proc Natl Acad Sci USA 99(15):10084–10089
Tsai CL, Li HP, Lu YJ, Hsueh C, Liang Y, Chen CL, Tsao SW, Tse KP, Yu JS, Chang YS (2006) Activation of DNA methyltransferase 1 by EBV LMP1 Involves c-Jun NH(2)-terminal kinase signaling. Cancer Res 66(24):11668–11676
Tsao SW, Tramoutanis G, Dawson CW, Lo AK, Huang DP (2002) The significance of LMP1 expression in nasopharyngeal carcinoma. Semin Cancer Biol 12(6):473–487
Uchida J, Yasui T, Takaoka-Shichijo Y, Muraoka M, Kulwichit W, Raab-Traub N, Kikutani H (1999) Mimicry of CD40 signals by Epstein-Barr virus LMP1 in B lymphocyte responses. Science 286(5438):300–303
Uemura N, Kajino T, Sanjo H, Sato S, Akira S, Matsumoto K, Ninomiya-Tsuji J (2006) TAK1 is a component of the Epstein-Barr virus LMP1 complex and is essential for activation of JNK but not of NF-kappaB. J Biol Chem 281(12):7863–7872
Vallabhapurapu S, Karin M (2009) Regulation and function of NF-kappaB transcription factors in the immune system. Annu Rev Immunol 27:693–733
Vaysberg M, Lambert SL, Krams SM, Martinez OM (2009) Activation of the JAK/STAT pathway in Epstein Barr virus+-associated posttransplant lymphoproliferative disease: role of interferon-gamma. Am J Transplant 9(10):2292–2302
Vockerodt M, Haier B, Buttgereit P, Tesch H, Kube D (2001) The Epstein-Barr virus latent membrane protein 1 induces interleukin-10 in Burkitt's lymphoma cells but not in Hodgkin's cells involving the p38/SAPK2 pathway. Virology 280(2):183–198
Vockerodt M, Morgan SL, Kuo M, Wei W, Chukwuma MB, Arrand JR, Kube D, Gordon J, Young LS, Woodman CB, Murray PG (2008) The Epstein-Barr virus oncoprotein, latent

membrane protein-1, reprograms germinal centre B cells towards a Hodgkin's Reed-Sternberg-like phenotype. J Pathol 216(1):83–92
Wakisaka N, Murono S, Yoshizaki T, Furukawa M, Pagano JS (2002) Epstein-barr virus latent membrane protein 1 induces and causes release of fibroblast growth factor-2. Cancer Res 62(21):6337–6344
Wan J, Sun L, Mendoza JW, Chui YL, Huang DP, Chen ZJ, Suzuki N, Suzuki S, Yeh WC, Akira S, Matsumoto K, Liu ZG, Wu Z (2004) Elucidation of the c-Jun N-terminal kinase pathway mediated by Estein-Barr virus-encoded latent membrane protein 1. Mol Cell Biol 24(1):192–199
Wan J, Zhang W, Wu L, Bai T, Zhang M, Lo KW, Chui YL, Cui Y, Tao Q, Yamamoto M, Akira S, Wu Z (2006) BS69, a specific adaptor in the latent membrane protein 1-mediated c-Jun N-terminal kinase pathway. Mol Cell Biol 26(2):448–456
Wang D, Liebowitz D, Kieff E (1985) An EBV membrane protein expressed in immortalized lymphocytes transforms established rodent cells. Cell 43(3 Pt 2):831–840
Wang D, Liebowitz D, Kieff E (1988a) The truncated form of the Epstein-Barr virus latent-infection membrane protein expressed in virus replication does not transform rodent fibroblasts. J Virol 62(7):2337–2346
Wang D, Liebowitz D, Wang F, Gregory C, Rickinson A, Larson R, Springer T, Kieff E (1988b) Epstein-Barr virus latent infection membrane protein alters the human B-lymphocyte phenotype: deletion of the amino terminus abolishes activity. J Virol 62(11):4173–4184
Wilson JB, Weinberg W, Johnson R, Yuspa S, Levine AJ (1990) Expression of the BNLF-1 oncogene of Epstein-Barr virus in the skin of transgenic mice induces hyperplasia and aberrant expression of keratin 6. Cell 61:1315–1327
Wu L, Nakano H, Wu Z (2006) The C-terminal activating region 2 of the Epstein-Barr virus-encoded latent membrane protein 1 activates NF-kappaB through TRAF6 and TAK1. J Biol Chem 281(4):2162–2169
Xiao L, Hu ZY, Dong X, Tan Z, Li W, Tang M, Chen L, Yang L, Tao Y, Jiang Y, Li J, Yi B, Li B, Fan S, You S, Deng X, Hu F, Feng L, Bode AM, Dong Z, Sun LQ, Cao Y (2014) Targeting Epstein-Barr virus oncoprotein LMP1-mediated glycolysis sensitizes nasopharyngeal carcinoma to radiation therapy. Oncogene 33(37):4568–4578
Xie P, Bishop GA (2004) Roles of TNF receptor-associated factor 3 in signaling to B lymphocytes by carboxyl-terminal activating regions 1 and 2 of the EBV-encoded oncoprotein latent membrane protein 1. J Immunol 173(9):5546–5555
Xie P, Hostager BS, Bishop GA (2004) Requirement for TRAF3 in signaling by LMP1 but not CD40 in B lymphocytes. J Exp Med 199(5):661–671
Xu D, Zhang Y, Zhao L, Cao M, Lingel A, Zhang L (2015) Interferon regulatory factor 7 is involved in the growth of Epstein-Barr virus-transformed human B lymphocytes. Virus Res 195:112–118
Yang L, Xu Z, Liu L, Luo X, Lu J, Sun L, Cao Y (2014) Targeting EBV-LMP1 DNAzyme enhances radiosensitivity of nasopharyngeal carcinoma cells by inhibiting telomerase activity. Cancer Biol Ther 15(1):61–68
Yasui T, Luftig M, Soni V, Kieff E (2004) Latent infection membrane protein transmembrane FWLY is critical for intermolecular interaction, raft localization, and signaling. Proc Natl Acad Sci USA 101(1):278–283
Ye H, Park YC, Kreishman M, Kieff E, Wu H (1999) The structural basis for the recognition of diverse receptor sequences by TRAF2. Mol Cell 4(3):321–330
Yoshizaki T (2002) Promotion of metastasis in nasopharyngeal carcinoma by Epstein-Barr virus latent membrane protein-1. Histol Histopathol 17(3):845–850
Yoshizaki T, Sato H, Furukawa M, Pagano JS (1998) The expression of matrix metalloproteinase 9 is enhanced by Epstein-Barr virus latent membrane protein 1. Proc Natl Acad Sci USA 95(7):3621–3626
Yoshizaki T, Horikawa T, Qing-Chun R, Wakisaka N, Takeshita H, Sheen TS, Lee SY, Sato H, Furukawa M (2001) Induction of interleukin-8 by Epstein-Barr virus latent membrane

protein-1 and its correlation to angiogenesis in nasopharyngeal carcinoma. Clin Cancer Res 7(7):1946–1951

Yoshizaki T, Kondo S, Wakisaka N, Murono S, Endo K, Sugimoto H, Nakanishi S, Tsuji A, Ito M (2013) Pathogenic role of Epstein-Barr virus latent membrane protein-1 in the development of nasopharyngeal carcinoma. Cancer Lett 337:1–7

Young LS, Rickinson AB (2004) Epstein-Barr virus: 40 years on. Nat Rev Cancer 4(10):757–768

Zhang L, Pagano JS (2000) Interferon regulatory factor 7 is induced by Epstein-Barr virus latent membrane protein 1. J Virol 74(3):1061–1068

Zhang L, Pagano JS (2001) Interferon regulatory factor 7: a key cellular mediator of LMP-1 in EBV latency and transformation. Semin Cancer Biol 11(6):445–453

Zhang B, Kracker S, Yasuda T, Casola S, Vanneman M, Homig-Holzel C, Wang Z, Derudder E, Li S, Chakraborty T, Cotter SE, Koyama S, Currie T, Freeman GJ, Kutok JL, Rodig SJ, Dranoff G, Rajewsky K (2012) Immune surveillance and therapy of lymphomas driven by Epstein-Barr virus protein LMP1 in a mouse model. Cell 148(4):739–751

Zhao B, Barrera LA, Ersing I, Willox B, Schmidt SC, Greenfeld H, Zhou H, Mollo SB, Shi TT, Takasaki K, Jiang S, Cahir-McFarland E, Kellis M, Bulyk ML, Kieff E, Gewurz BE (2014) The NF-kappaB genomic landscape in lymphoblastoid B cells. Cell reports 8(5):1595–1606

Zimber-Strobl U, Kempkes B, Marschall G, Zeidler R, Van Kooten C, Banchereau J, Bornkamm GW, Hammerschmidt W (1996) Epstein-Barr virus latent membrane protein (LMP1) is not sufficient to maintain proliferation of B cells but both it and activated CD40 can prolong their survival. EMBO J 15(24):7070–7078

# Latent Membrane Protein 2 (LMP2)

**Osman Cen and Richard Longnecker**

**Abstract** LMP2A is an EBV-encoded protein with three domains: (a) an N-terminal cytoplasmic domain, which has PY motifs that bind to WW domain-containing E3 ubiquitin ligases and an ITAM that binds to SH2 domain-containing proteins, (b) a transmembrane domain with 12 transmembrane segments that localizes LMP2A in cellular membranes, and (c) a 27-amino acid C-terminal domain which mediates homodimerization and heterodimerization of LMP2 protein isoforms. The most prominent two isoforms of the protein are LMP2A and LMP2B. The LMP2B isoform lacks the 19-amino acid N-terminal domain found in LMP2A, which modulates cellular signaling resulting in a baseline activation of B cells and degradation of cellular kinases leading to the downregulation of normal B cell signaling pathways. These two seemingly contradictory processes allow EBV to establish and maintain latency. LMP2 is expressed in many EBV-associated malignancies. While its antigenic properties may be useful in developing LMP2-specific immunity, the LMP2A N-terminal motifs also provide a basis to target LMP2A-modulated cellular kinases for the development of treatment strategies.

## Contents



O. Cen · R. Longnecker (✉)
Department of Microbiology and Immunology, Feinberg School of Medicine,
Northwestern University, Chicago, IL, USA
e-mail: r-longnecker@northwestern.edu



C. Münz (ed.), *Epstein Barr Virus Volume 2*, Current Topics in Microbiology
and Immunology 391, DOI 10.1007/978-3-319-22834-1_5

## Abbreviations

| | |
|---|---|
| Abl, c-Abl | Abelson murine leukemia viral oncogene homolog 1, a proto-oncogene, a prominent non-receptor tyrosine kinase within the B cell receptor signaling pathway, whose deregulation is generally associated with myelogenous leukemia |
| AIP4, ITCH | Atrophin-1-interacting protein-4, also known as ITCH or WWP2, is a member of E3 ubiquitin ligases |
| Akata cells | An EBV-positive Burkitt lymphoma cell line |
| Akt | Also known as protein kinase B, it is a serine/threonine kinase downstream of PI3 K and involves in cellular metabolism such as survival, activation, proliferation, and migration |
| Bcl-2 | B cell lymphoma 2, a defining member of Bcl-2 family of proteins that regulate cell survival or cell death. Bcl-2 is an anti-apoptotic and a proto-oncogene |
| Bcl-xL | B cell lymphoma-extra large, another anti-apoptotic pro-survival member of Bcl-2 family of proteins |
| BCR | B cell receptor, provides survival and activation cellular signaling upon antigen binding |
| BJAB cell line | An EBV-negative Burkitt lymphoma cell line |
| BL | Burkitt lymphoma |
| BLNK | B cell linker protein, an adaptor protein that bridges the BCR signaling to downstream signaling pathways |
| BTK | Bruton's tyrosine kinase, an important proximal BCR signaling kinase |
| BZLF1 | Transcriptional activator protein of EBV, an immediate early gene that regulates the switch from latent to lytic cycle of the virus |
| Cbl | A member of E3 ubiquitin protein ligases |
| CD19 | B cell-specific cell surface antigen that cluster with the BCR and help decrease threshold for antigen induced cell signaling in B cells |

| | |
|---|---|
| Crk | A signaling adaptor molecule that associates with various signaling molecules through its SH2 and SH3 domains leading to the modulation of involved signaling pathways |
| Csk | A tyrosine kinase that inhibits activation of Src family tyrosine kinases by phosphorylating their conserved C-terminal regulatory tyrosine residue |
| E3 ubiquitin ligases | A class of ubiquitin ligases that transfer the ubiquitin from E2 to target proteins leading to their degradation proteasome |
| EBNA-2 | One of the EBV-encoded nuclear viral transactivator that involves regulation of viral and host gene expression |
| EBV | Epstein-Barr virus |
| EGR1 | A zinc-finger domain-containing transcription factor whose activity is increased by LMP2A |
| Erk1/2 | Extracellular signal-regulated kinases are members of serine/threonine kinases in the Raf/Ras/MEK/Erk signaling pathway that are activated by mitogen-induced cell survival and cell activation signaling |
| Fyn | A member of Src family tyrosine kinases |
| Gp350/220 | The EBV surface glycoprotein, the viral receptor that mediates B cell infection through binding to cellular CR2 receptor CD21. It is a part of fusion machinery that also involves viral gp42 and host MHC II molecules |
| Grb2 | Growth factor receptor-bound 2 is an adaptor molecule that binds to membrane proximal signaling molecules through its SH2 and SH3 domains to orchestrate cellular signaling pathways |
| Hck | A member of Src family tyrosine kinases prominent in hematopoietic cells |
| HIV | Human Immunodeficiency virus |
| HRS cells | Malignant Hodgkin/Reed–Sternberg cells, germinal center B cell-derived giant cells that are characteristics of Hodgkin's lymphoma. They are also known as lacunar histiocytes, have multi- or bi-lobed nuclei, and lack functional BCR |
| ITAM | Immunoreceptor tyrosine-based activation motif with the consensus sequence of YxxL/I-$X_{6\text{-}8}$-YxxL/I |
| ITCH, Itchy, AIP4 | A member of family of HECT domain E3 ubiquitin ligases that ubiquitinate associated proteins leading to their proteasome-mediated degradation |
| ITSN1 | Intersectin-1 protein or SH3-interacting protein coordinates endocytic membrane traffic with actin assembly machinery |
| Kit, c-Kit | A proto-oncogene, also known as stem cell growth factor receptor |
| Lck | A T cell-specific member of Src family tyrosine kinases |
| LCL | Lymphoblastoid cell lines obtained by EBV infection of human peripheral blood mononuclear cells |

| | |
|---|---|
| LMP1 | A latent membrane protein 1 of EBV, it mimics TNF-α/CD40 signaling |
| LMP2 | EBV gene and its encoded latent membrane protein 2, it mimics BCR signaling |
| LPD | Lymphoproliferative diseases |
| Lyn | A member of Src family tyrosine kinases, it is more specific in B cells, granulocytes, and mast cells |
| MAPK | (Mitogen-activated protein kinases) are serine/threonine kinases in the Raf/Ras/MEK/Erk signaling pathway activated by mitogenic signals |
| MHC | Major histocompatibility complex molecules that present digested and processed antigens to T cells |
| miR-BART22 | An EBV-encoded microRNA, it targets LMP2A |
| mTOR | Mammalian target of rapamycin, a serine/threonine kinase, is a central regulator of cellular metabolism |
| Myc | A proto-oncogene is a master regulatory transcription factor, which regulates the expression of many other signaling molecules involved in cell cycle, apoptosis, survival, and transformation |
| Nck-1 | A signaling adaptor molecule that modulates cell signaling pathways through interaction with other signaling molecules with its SH2 and SH3 domains |
| Nedd4 | A member of E3 ubiquitin ligase family, they contain HECT and WW domain and associate with various signaling molecules leading to their proteasome-mediated degradation through ubiquitination |
| NFkB | A well-regulated transcription factor complex that was first discovered as the transcription regulator of antibody light chain gene in B cells. Its disregulation has been associated with cancers |
| NKT | Natural killer T cells, a very small unusual subset of T cells that recognizes lipid antigens presented by CD1 molecules |
| Notch | A family of transmembrane proteins whose interactions on the adjacent cells induce Notch signaling network that is important in cell differentiation and development |
| NPC | Nasopharyngeal carcinoma, with high rate of EBV association |
| p27 | Cyclin-dependent kinase (Cdk) inhibitor, also know as Kip1, controls the cell cycle progression at G1 phase |
| p53 | A member of tumor suppressor proteins, it has been deactivated in various tumors |
| Pax-5 | A member of paired box family of transcription factors important in regulation of early B cell and neural development |

| | |
|---|---|
| PI3K | Phosphoinositide 3-kinase, a member of lipid kinase family that regulates cell growth, proliferation, differentiation, and survival of the cells, and its deregulated forms have been associated with various cancers |
| PKC | A large family of serine/threonine kinases that regulate various cellular signaling pathways |
| PLC-$\gamma$ | A member of Phospholipase C kinases that regulate lipid molecules |
| PTLD | Post-transplant lymphoproliferative disorders which is characterized by abnormally proliferating B cells due to immunosuppression or EBV infection |
| PU.1 | An ETS domain transcription factor that regulates gene expression during myeloid and B cell development |
| PY motif | Proline-rich tyrosine-containing motif with the consensus sequence of PPxY that binds to WW domains found in various proteins |
| Raf | A proto-oncogene in the Raf/Ras/MAPK pathway |
| Rag1 | Recombination-activating gene is one of the two genes that catalyze the VDJ recombination of antigen receptors in B and T cells |
| RBP-J | (also known as CBF1) Recombination signal binding protein for immunoglobulin kappa J region is human homolog of the Drosophila suppressor of hairless gene |
| Ras | A superfamily of proteins called GTPases important in regulating activation/inactivation of various kinases by exchanging GTP and GDP molecules |
| SH2 domain | Src homology 2 domain, protein–protein interaction domain of signaling proteins that binds to phosphorylated tyrosine (Y) residues within ITAMs |
| SH3 domain | Src homology 3 domain, protein–protein interaction domain of signaling proteins that binds to PxxP motifs found in various signaling molecules |
| SH3 motif | Motifs with the general PxxP motif that binds to SH3 domains of various proteins |
| Shb | SH2 domain-containing adaptor protein B, it binds to various kinases and signaling molecules |
| Shc-1 | SHC transforming protein is a signaling adaptor protein with three different isoforms involving in apoptosis and survival |
| Src | The defining member of Src family tyrosine kinases |
| siRNAs | Small inhibitory RNA molecules that specifically initiates degradation of target mRNAs |
| Syk | A tyrosine kinase important in the B cell signaling B counterpart of ZAP70 in T cells |

| | |
|---|---|
| TCR | T cell receptor generates antigen-induced T cell signaling upon binding to antigens presented by MHC molecules |
| TGF-b | A member of transforming growth factor family that controls proliferation and differentiation of cells |
| TNF-a | Tumor necrosis factor alpha is cell secreted signaling molecule that involves in systemic inflammation |
| TR | Terminal repeat sequences at the end of the linear genome of some viruses, including EBV |
| Unc119 | A signaling adaptor molecule specific in the retinal photoreceptors and hematopoietic cells, it involves in activation of Src kinases upon receptor stimulation |
| Vav | A family of adaptor proteins involved in cell signaling |
| Wnt | A conserved family of lipid-modified transmembrane signaling glycoproteins involved in the initiation of developmental signal transduction pathways by binding to its transmembrane ligands, Frizzled family receptors |
| WW domain | A modular protein domain with two conserved tryptophan amino acids spaced 20-22 amino acid aparts that binds to proline-rich PPxY motif |
| WWP2 | WW2 domain-containing protein 2, also known as AIP4 or ITCH, is a member of E3 ubiquitin ligases |
| ZAP70 | CD3 Zeta-chain-associated protein kinase 70, it is Syk counterpart of T cells that is essential in T cell signaling |

# 1 Introduction

The LMP2 gene of EBV is unique in that it spans both ends of linear genome and is expressed only when the EBV genome circularizes as an episome during latent infection. It expresses two different predominant transcripts from two separate promoters. The LMP2A isoform is the longest transcript, which encodes a protein of 497 amino acids (Laux et al. 1988, 1989; Rowe et al. 1990; Sample et al. 1989). The N-terminal cytoplasmic domain of LMP2A has two proline-rich PY motifs and several phosphorylated tyrosine residues including a consensus immunoreceptor tyrosine-based activation motif (ITAM) that binds to cellular signaling molecules and alters the cellular signal transduction pathways. Tyrosine residues within the ITAM contained within the N-terminal domain of LMP2A interact with BCR signaling components and mimic BCR signaling (Engels et al. 2012; Fotheringham and Raab-Traub 2013). The LMP2A PY motifs interact with E3 ubiquitin ligases such as Nedd4 and initiate ubiquitination-dependent proteasomal degradation of targeted cellular proteins. By doing so, LMP2A helps establish and maintain latency. LMP2B, which lacks the LMP2A N-terminal cytoplasmic domain, regulates the distribution and function of LMP2A also helping to maintain the latency of the virus. LMP2 is expressed in various EBV-associated malignancies and likely contributes to their development by inducing cell survival, proliferation, and activation. The LMP2 proteins also induce an immune response, which might help the immune system to keep the EBV infection under control. Since LMP2A is expressed in malignancies and induces cell proliferation, it is a viable target for the development of therapeutics for EBV-associated malignancies.

# 2 LMP2 Gene Structure

## *2.1 Gene Structure*

The latent membrane protein 2 (LMP2) gene of EBV spans the circularized form of the conventional EBV genome (B95-8/Raji strain), covering the terminal repeats (TR) at both ends in the linear form. It starts from the nucleotide 166,040, crosses the circularized ends at the terminal repeats, and ends at the nucleotide 5856 of the 172 Kb (171,823 bp) EBV genome of the B95-8 strain (BioProject PRJNA14413; NCBI accession no: NC_007605.1; GI: 82503188) (http://www.ncbi.nlm.nih.gov/nuccore/82503188, retrieved on 08/12/14). It comprises 9 exons with the first exon at the end (nucleotides 166040 through 166458), while the exons 2–9 are at the beginning of the linear EBV genome (nucleotides 58-5856) (Tables 1 and 2; Fig. 1) (Laux et al. 1988, 1989; Rowe et al. 1990; Sample et al. 1989).

**Table 1** The location of LMP2A's exons on EBV genome B. The LMP2A isoforms

| LMP2 gene | Genomic location of exons | | Length (nt) | Comments |
|---|---|---|---|---|
| Exon | First nucleotide | Last nucleotide | | |
| Ex-1 (2A) | 166,040* | 166,458 | 419 | LMP2A |
| Ex-1 (2B) | 169,294 | 169,448 | 155 | LMP2B |
| Ex-1 (2-TR) | 171,619 | | | LMP2-TR1 |
| | 171,791 | | | LMP2-TR2 |
| Ex-2 | 58 | 272 | 215 | |
| Ex-3 | 360 | 458 | 99 | |
| Ex-4 | 540 | 788 | 249 | |
| Ex-5 | 871 | 951 | 81 | |
| Ex-6 | 1026 | 1196 | 171 | |
| Ex-7 | 1280 | 1495 | 216 | |
| Ex-8 | 1574 | 1682 | 109 | |
| Ex-9 | 5408 | 5856 | 449 | |
| Total: Ex 1–9 | | | | |

**Table 2** The LMP2A isoforms

| LMP2 isoform | Exons | Protein length | Comments |
|---|---|---|---|
| LMP2A | Ex1A-Ex9 | 497 aa | Full length |
| LMP2B | Ex1B-Ex9 | 378 aa | Lacks the initial translation initiation codon |
| LMP2-TR | ExTR1-Ex9<br>ExTR2-Ex9 | ? | Lack the initial translation initiation codon<br>Induces cytotoxic CD8 T cell response |
| Alternative splice isoforms | Ex1, Ex3-Ex9<br>Ex1, Ex6-Ex9<br>Ex1, Ex7-Ex9 | ? | Include frameshift mutation due to alternative splicing |

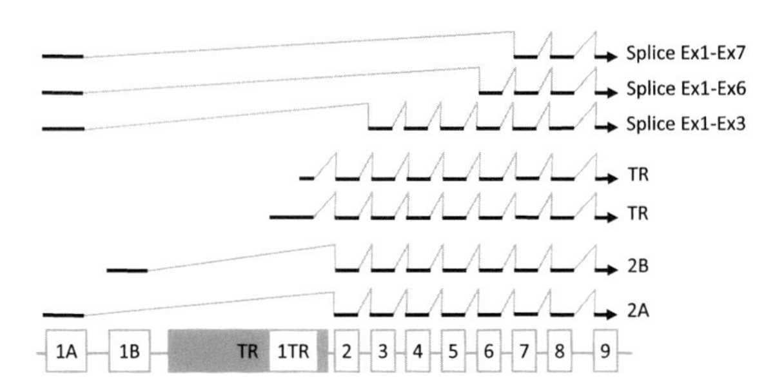


**Fig. 1** The LMP2 gene and its transcripts. The LMP2 gene spans the TR of the circularized EBV genome. It has nine exons the first of which is located at the end of linear genome and has three alternatives: Exon 1A, 1B, or 1TR, each of which starts from a separate promoter. LMP2A isoform starts from Exon 1A and translates the longest protein isoform containing all nine exons. The alternatively spliced isoforms starts with exon1 and lack the intervening exons. The nucleotide location of each exon is given in Tables 1 and 2

## *2.2 Expressed Transcripts*

Several different transcripts have been identified for the LMP2 gene (Fig. 1). The two most prominent and well-studied transcripts encode for the LMP2A and LMP2B isoforms (Laux et al. 1988, 1989; Rowe et al. 1990; Sample et al. 1989). These two transcripts are controlled by two different promoters separated by 3 kb (Sample et al. 1989). Both promoters have an EBNA-2-responsive element and RBP-Jk and PU.1 binding sites. The LMP2B promoter is also shared with the LMP1 promoter but on the complementary strand, which transcribes LMP1 in the opposite direction (Laux et al. 1988, 1989; Rowe et al. 1990; Sample et al. 1989). The transcripts from each promoter have 9 exons differing only in their first 5' exon (Fig. 1) (Laux et al. 1988, 1989; Rowe et al. 1990; Sample et al. 1989).

The LMP2A open reading frame (ORF) is 1494 nucleotides in length and encodes for a protein of 497 amino acids (GI: 82503189; YP_401631; Uniprot no: P13285-1). LMP2B ORF (GI: 82503190; YP_401632; Uniprot no: P13285-2) encodes for a protein of 378 amino acids. The first 5' exon of the LMP2B transcript lacks the initial translation initiation site leading to the absence of the 119-amino acid N-terminal domain present in LMP2A (Laux et al. 1988, 1989; Rowe et al. 1990; Sample et al. 1989).

Recently, additional novel LMP2 transcripts have been identified (Tables 1 and 2). These transcripts, like LMP2B, lack the first exon found in LMP2A. Two such transcripts were detected in the EBV-positive NK and T cell tumors through functional analysis (Fox et al. 2010). In this study, the LMP2-specific effector CD8-positive T cells recognized and killed the NK and T cell tumor cell lines despite the absence of any LMP2A or LMP2B transcripts or protein using conventional PCR primers or LMP2A antibody. These transcripts start at nucleotides 171619 and 171791 within the TR and may be transcribed from yet another unidentified promoter. In addition, three other alternatively spliced transcripts have also been identified using the RNA-Seq (Concha et al. 2012). These isoforms are spliced from exon 1 of LMP2A to exon 3, exon 6, or exon 7, respectively, and lack the intervening exons (Concha et al. 2012). The junction after splicing of these isoforms may generate a frameshift in the LMP2 ORF, resulting in truncation of the predicted protein products (Concha et al. 2012). However, the protein products of these transcripts and their importance have not been studied. In addition, in the same study, incompletely spliced polyadenylated transcripts were detected in the cytoplasm of Akata cells. The mechanism and physiological importance of these transcripts have not been studied. Since these isoforms were abundantly detected after induction of lytic phase through B cell receptor stimulation, they might be specific to lytic replication.

The generally used consensus PCR primers detect only LMP2A and LMP2B isoforms, and therefore, these newly identified isoforms have escaped detection until only recently. The available antibodies recognize epitopes within the N-terminal cytoplasmic domain present only in the LMP2A isoform. Therefore, the generally used PCR primers and LMP2 antibodies do not provide an accurate detection assay for all LMP2 transcripts and protein isoforms. Therefore, more elaborate analyses of these different transcripts require application of various techniques including sequencing, RT-PCR, and functional assays to determine their importance in EBV pathogenesis.

## 3 Regulation and Distribution of LMP2 Proteins in the Cell

### *3.1 Regulation of LMP2A Expression*

The expression of LMP2A is regulated through viral and host cell factors. The LMP2 promoter has an EBNA-2-responsive element, which leads to LMP2 expression when EBNA-2 is expressed, such as in type-2 and type-3 latency (Laux et al. 1994b; Zimber-Strobl et al. 1991, 1993). This site also has binding sites for RBP-Jk and PU.1 transcription factors (Laux et al. 1994a). LMP2 expression can also be activated by host cell proteins such as the Notch signaling pathway. The Notch signaling pathway can be activated by LMP2A-induced signaling, independently of EBNA-2. Activation of Notch pathway leads to increased expression of LMP2A (Anderson and Longnecker 2008b). This may provide an autoregulatory forward activation loop for LMP2A expression, in which activation of Notch signaling pathway leads to expression of LMP2A, which in turn induces activation of Notch signaling.

Another viral regulator of LMP2A expression is the EBV-encoded miR-BART22 microRNA. The miR-BART22 downregulates the expression of LMP2A as it is highly expressed in NPC (Lung et al. 2009). In these same samples, even though the LMP2A transcript is detected, the LMP2A protein level is decreased with the increased expression of miR-BART22 indicating a posttranscriptional regulation of LMP2A protein expression by the viral miRNA.

The expression of LMP2 gene is also induced by the activation of lytic phase of the viral life cycle. B cell receptor activation, which induces lytic viral replication in infected cells, increases LMP2 gene expression a few hundred folds in latently infected cells (Concha et al. 2012). Such induction also leads to increased exons 1–6 and 1–7 alternatively spliced transcripts and causes nuclear retention of LMP2 transcripts (Concha et al. 2012). The importance of this alternative splicing and nuclear retention in the EBV pathogenesis requires further investigation.

## *3.2 Distribution of LMP2A Protein in the Cell*

The LMP2 proteins localize to cellular membranes by virtue of the multiple alpha-helical transmembrane segments and by palmitoylation. LMP2 proteins are also palmitoylated on multiple cysteine residues within the C-terminal end, but this palmitoylation is not required for its association with the cellular membranes (Katzman and Longnecker 2004; Matskova et al. 2001). However, palmitoylation of LMP2A has been suggested to be important for the localization and transport of LMP2A among membrane structures such as vesicles (Matskova et al. 2001). In the membrane, LMP2A is associated with lipid rafts (Higuchi et al. 2001; Katzman and Longnecker 2004; Matskova et al. 2001). The association of LMP2A with lipids may provide a platform for LMP2A to associate with signaling proteins such as the Lyn and Syk kinases and other membrane proximal signaling molecules enriched in lipid rafts.

While LMP2A localizes to the cellular plasma membrane, LMP2B localizes to the perinuclear area of the infected cells and can sequester LMP2A (Lynch et al. 2002). This indicates that LMP2A and LMP2B might form heterodimers through which the modification of cellular signaling pathway by LMP2A might be inhibited by LMP2B (Rovedo and Longnecker 2007) which lacks the N-terminal signaling domain and therefore signaling motifs located within. It will be interesting to know whether the protein isoforms encoded by the LMP2-TR and alternatively spliced transcripts regulate the localization of the LMP2A and whether this alters the role of LMP2A in the modification of cellular signaling pathways.

# 4 Protein Structure, Domain, and Motifs

The LMP2A is a 497-amino acid membrane protein with the cytoplasmic N- and C-terminal ends. Its exact structure is not known, as its crystal structure has not yet been solved. It has three distinct functional domains: The cytoplasmic N-terminal, the transmembrane, and the cytoplasmic C-terminal domains (Fig. 2).

## *4.1 N-Terminal Cytoplasmic Domain*

The N-terminal cytoplasmic domain has 119 amino acids and is absent in the LMP2B (Laux et al. 1988; Rowe et al. 1990; Sample et al. 1989). This domain contains functional motifs such as multiple phosphorylated tyrosine residues, an ITAM, and multiple proline-rich motifs (PY motifs) (Beaufils et al. 1993). Through these motifs, LMP2A associates with and modifies the host cellular signaling molecules.

### 4.1.1 ITAM

Within the N-terminal end of LMP2A (Fig. 3), there are several tyrosine residues that are potential binding and phosphorylation targets of tyrosine kinases. Two of these tyrosine residues (Y74 and Y85) constitute the LMP2A ITAM located at aa 74-88 (Beaufils et al. 1993; Engels et al. 2012; Fotheringham and Raab-Traub 2013). ITAMs are found in the cytoplasmic tails of the antigen receptors of the immune cells. It consists of a pair of ITAM signatures, each with the YxxL/I sequence separated by 6–8 spacer amino acids (YxxL/I-$X_{6-8}$-YxxL/I) (Harwood and Batista 2010; Reth 1989). Generally, both of the conserved tyrosine residues within ITAM are phosphorylated by the membrane-juxtaposed tyrosine kinases such as Src family kinases (Dal Porto et al. 2004; Harwood and Batista 2010; Reth 1989). The SH2 domain-containing downstream signaling molecules are then recruited to the phosphorylated tyrosine residues within the ITAM leading to the

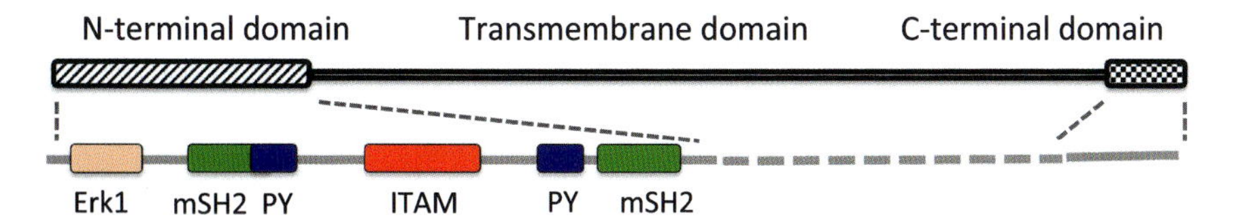


**Fig. 2** The domains and the motifs of LMP2A. LMP2A protein has three domains: a 119-amino acid N-terminal cytoplasmic domain, a transmembrane domain with 12 transmembrane segments, and a 27-amino acid C-terminal cytoplasmic domain. The N-terminal domain has several signaling motifs: ITAM, PY, SH2-binding, and Erk-binding motifs. LMP2A binds to the cellular signaling molecules through these motifs

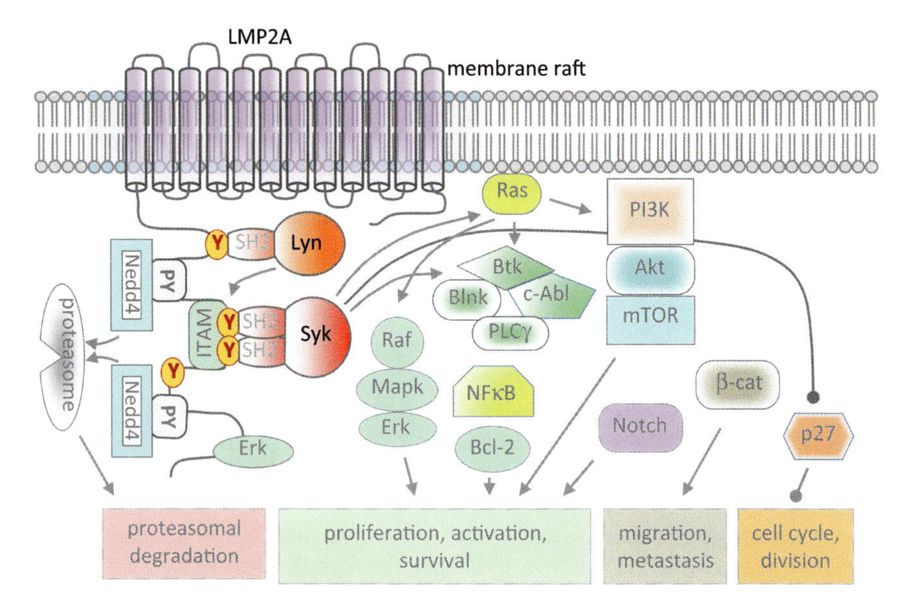


**Fig. 3** LMP2A modifies cellular signaling pathways by recruiting their components to it specific motifs found in its N-terminal domain. Lyn and Syk kinases are recruited to the phosphorylated tyrosine residues where they are activated and then initiate activation of various pathways leading to cell proliferation, division, activation, survival, migration, and metastasis. It also recruits E3 ubiquitin ligases such as Nedd4 to the PY motifs and initiates proteasomal degradation pathway to downregulate the levels of some of the recruited signaling molecules to fine-tune the cellular milieu to establish and maintain the viral latency

phosphorylation and activation of recruited molecules. This provides a signaling cascade that fulfills the response of immune cells to the signal-generating antigens.

The LMP2A ITAM (aa 74-88) functions like ITAMs in the B and T cell receptors (Dal Porto et al. 2004; Reth 1989). It recruits the SH2 domain-containing signaling proteins including tyrosine kinases Lyn, Fyn, Syk, and Csk (Beaufils et al. 1993; Burkhardt et al. 1992; Longnecker et al. 1991; Scholle et al. 1999). Lyn kinase can bind to LMP2A at the DQSL sequence (aa 81-84) within the LMP2A ITAMs and on the Y112 within the membrane proximal ITAM-like SH2-binding motif (aa 105-119) (Fruehling and Longnecker 1997; Fruehling et al. 1998; Merchant et al. 2000). This initial recruitment of Lyn might then lead to the phosphorylation of LMP2A by Lyn on multiple tyrosine residues, including LMP2A ITAM, which forms a docking site for SH2 domain-containing kinases and signaling adaptor molecules (Beaufils et al. 1993; Burkhardt et al. 1992; Longnecker et al. 1991; Scholle et al. 1999). One such molecule is Syk kinase, which is recruited to phosphorylated Y74 and Y85 of the LMP2A. Upon binding to LMP2A, Syk then phosphorylates the other proteins in the proximity of LMP2A (Fotheringham et al. 2012; Hatton et al. 2013; Lu et al. 2006). This association and ensuing phosphorylation events lead to the activation of cells, calcium mobilization, and cytokine production (Fruehling et al. 1996, 1998; Rovedo and

Longnecker 2007). LMP2A also modifies T cell signaling in a similar way it modifies B cell signaling. When expressed in T cells, LMP2A also associates with the T cell-specific Src kinase Lck on phosphorylated Y112, in a very similar way it associates with Lyn in B cells, and initiates recruitment of Syk/ZAP70 to the phosphorylated ITAM (Ingham et al. 2005b).

LMP2A has two other ITAM-like motifs that recruit SH2 domain-containing signaling proteins (Fig. 2). The tyrosine residues within these motifs are phosphorylated by Lyn and Syk kinases after being recruited to LMP2A complexes (Beaufils et al. 1993). These ITAM-like motifs may also recruit, in addition to Src kinases, other kinases such as Btk (Fig. 3) and lead to their phosphorylation and activation (Merchant and Longnecker 2001). Another possible target of the activated Src family tyrosine kinases is c-Abl, a seminal kinase in the B cell signaling pathway (Meyn et al. 2006). Therefore, LMP2A mimics B cell receptor signaling by recruiting immediate BCR signaling components Lyn and Syk kinases through which it initiates activation of downstream signaling molecules such as Btk and c-Abl.

#### 4.1.2 Proline-Rich Motifs

The N-terminal domain harbors 5 proline-rich regions, two of which conform to WW domain-binding proline-rich PY motifs (Fig. 2 and 3). The PY motifs have a consensus sequence of PPxY and strongly interact with WW domain-containing proteins, such as E3 ubiquitin ligases Nedd4, Nedd4-2/KIAA0439, AIP4/Itchy, Cbl, and WWP2 (Ikeda et al. 2000, 2002, 2003; Ingham et al. 2005a; Longnecker et al. 2000; Winberg et al. 2000). The recruited ubiquitin ligases can ubiquitinate proteins such as LMP2A and Lyn, initiating proteasomal degradation processes (Ikeda et al. 2000, 2003; Winberg et al. 2000).

The other 3 proline-rich regions within the N-terminal domain of LMP2A have a PxxP sequence and conform to the minimal SH3 domain-binding motif. The SH3 domains mediate protein–protein interaction by binding to the SH3-binding motifs with canonical PxxP consensus sequence (Nguyen et al. 1998). This interaction is phosphorylation independent and may lead to autophosphorylation of Src kinases leading to their activation, the initial event in the activation cascade of signaling pathways (Moarefi et al. 1997). Lyn, through its SH3 domain, might be initially recruited to one of these PxxP motifs resulting in Lyn activation (Cen et al. 2003). However, experimental data with GST-fused SH3 domains of Src, Lyn, Hck, Abl, or Grb2 did not firmly confirm a high-affinity association with the N-terminal domain of LMP2A (Longnecker et al. 2000). This indicates that the PxxP motifs within the N-terminal end of LMP2A may be of weak affinity or not interact with the SH3 domain-containing kinases or adaptor molecules. However, a weak or transient interaction of SH3 ligands with the SH3 domains of Src kinases is enough to induce their phosphorylation and activation (Moarefi et al. 1997; Nguyen et al. 1998; Sicheri et al. 1997). A detailed analysis of general PxxP motifs reveals different binding motifs for the class I and class II SH3 domains with the RKxxPxxP and PxxPxR sequences, respectively. In addition to those in the N-terminal domain, the C-terminal domain of LMP2A also has a predicted class II SH3-binding sequence (aa 489-494, RPPTPYR). This motif may recruit Src kinases including Lyn and Hck as shown for the interaction of Hck with HIV Nef protein (Alvarado et al. 2014; Moarefi et al. 1997) and of Lyn with Unc119 (Cen et al. 2003). However, no experimental data have shown the importance of this C-terminal SH3 motif of LMP2A in the activation of Src kinases.

### 4.2 Transmembrane Domain

The LMP2A transmembrane domain consists of twelve alpha-helical transmembrane segments embedded in the membrane. This domain is essential for the membrane localization of LMP2 proteins (Laux et al. 1988; Rowe et al. 1990; Sample et al. 1989). The transmembrane segments are located between amino acids 120–470 with short alternating extracellular and cytoplasmic segments. Although these 12 transmembrane segments resemble those found in the membrane transport molecules, no such function has been described for LMP2A. Truncation of LMP2A after the third transmembrane domain results in a protein that is diffusely located in the membrane (Lynch et al. 2002; Tomaszewski-Flick and Rowe 2007). Truncation after the sixth transmembrane domain results in LMP2A that has some characteristic staining as seen with wild-type LMP2A but also has some diffuse staining similar to the larger LMP2A truncation. The smaller truncation is still able to block BCR signaling (Lynch et al. 2002; Tomaszewski-Flick and Rowe 2007).

### 4.3 C-Terminal Cytoplasmic Domain

The C-terminal cytoplasmic domain has 27 amino acids, is palmitoylated, and mediates the self-association of LMP2 proteins (Laux et al. 1988; Rechsteiner et al. 2008a; Rovedo and Longnecker 2007; Rowe et al. 1990; Sample et al. 1989; Tomaszewski-Flick and Rowe 2007). The aggregation may occur as homodimers (LMP2A/LMP2A and LMP2B/LMP2B) or heterodimers of LMP2A and LMP2B (LMP2A/LMP2B). The aggregation of LMP2A with LMP2B may inhibit cellular signaling by the N-terminal domain of LMP2A. The expression of LMP2A with LMP2B can restore the calcium mobilization prevented by LMP2A and reverse the reduced Lyn level when only LMP2A is expressed (Matskova et al. 2001; Rovedo and Longnecker 2007). LMP2B was shown to negatively regulate LMP2A activity when these genes were expressed in the EBV-negative Burkitt lymphoma (BL) cell line BJAB. The inhibitory effect of LMP2A on signaling may require LMP2A aggregation, which may be modulated by the presence of LMP2A by limiting the formation of LMP2A homodimers. LMP2B colocalizes with LMP2A in the membrane, where the C termini of both variants can interact and regulate the activity of each other (Matskova et al. 2001; Rovedo and Longnecker 2007; Tomaszewski-Flick and Rowe 2007).

## 5 Modification of Cellular Signaling Pathways

Expression of LMP2A in the cell alters developmental, survival, and apoptotic pathways through modification of cellular signaling cascades. LMP2A modifies cellular signaling through recruiting components of signaling pathways to the LMP2A ITAM and PY motifs within the LMP2A N-terminal domain (Fig. 3). This modification culminates in two differing outcomes:

1. Dominant-positive effect: By acting as a BCR mimic, LMP2A provides a constitutive activation of a BCR-like signaling through the recruited Lyn and Syk kinases.
2. Dominant-negative effect: It deprives the BCR signaling complex of its components, which it can accomplish in two different ways:

   (a) Appropriation of the B cell signaling components, such as Lyn and Syk, from the BCR complex.
   (b) Proteasomal degradation of the associated BCR signaling molecules through the ubiquitination and degradation by recruited E3 ubiquitin ligases.

These two differing functional outcomes of LMP2A interference are reflective of how LMP2A behaves when expressed in a relevant cell type. The LMP2A recruitment of cellular kinases involved in the BCR signaling pathway can both provide an activating signal (the LMP2A complexes themselves) (Burkhardt et al. 1992; Caldwell et al. 1998; Lu et al. 2006; Merchant et al. 2000; Swanson-Mungerson et al. 2006) and inhibiting signal (Dykstra et al. 2001; Engels et al. 2001; Schaadt et al. 2005). When Lyn and Syk kinases are recruited to LMP2A, the kinases are phosphorylated and activated resulting in a signal that mimics a BCR signal (Fruehling et al. 1998; Ikeda et al. 2001; Merchant et al. 2000; Swart et al. 2000). While this leads to the activation of a BCR-like downstream signaling pathways, it is not physiologically needed for the cell since it is not a response to the environmental stimuli such as antigens and as such LMP2A is acting as a constitutively active BCR.

By recruiting proteins that are normally activated by the BCR signaling pathway, LMP2A deprives the BCR from its signaling components needed for functional signaling. This deprivation leads to the absence or downregulation of antigen-induced cellular signaling (Fruehling et al. 1996) and the development of BCR-negative B cells in the periphery (Caldwell et al. 2000; Winberg et al. 2000). This is consistent with the notion that the LMP2A is a negative regulator of BCR signaling (Ikeda and Longnecker 2009). Similarly, when ectopically expressed in T cells, LMP2A recruits and downregulates the activity of Lck in T cells (Ingham et al. 2005b). In this sense, LMP2A functions as a "sponge" attracting signaling proteins resulting in the lack or inadequate signaling events that would otherwise be initiated through antigen binding. In this respect, LMP2A acts like a dominant-negative BCR by scavenging the main BCR signaling components, such as Lyn and Syk kinases. This blocking effect of the LMP2A might provide the virus an opportunity to isolate the cells from its milieu and to secure viral persistence since cellular activation is usually associated with a shift from EBV latent infection to lytic infection.

The recruitment of E3 ubiquitin ligases via interaction with LMP2A PY motifs similarly modifies cellular signaling. The recruitment of ubiquitin ligases to LMP2A results in the ubiquitination of signaling components resulting in their degradation (Ikeda et al. 2003; Ikeda and Longnecker 2009). This degradation may be important in modulating LMP2A-mediated signaling as well as modifying normal signal transduction. From this perspective, recruitment of kinases, adaptor molecules, and ubiquitin ligases provides complementary checkpoints for continuation of the viral latency: to prevent the induction of lytic phase to evade activation of the immune system and to provide the cell with a baseline survival signal that will enable the virus to continue its existence within the surviving cell.

The functional outcome of the LMP2A is accomplished through the recruited signal-activating kinases and protease-activating ubiquitin ligases. Another level of regulation is accomplished through C-terminal Src kinase (Csk), which is also recruited to LMP2A (Scholle et al. 1999). Csk is a negative regulator of Src family tyrosine kinase (Nada et al. 1993; Okada 2012; Young et al. 2001). The association of Csk with LMP2A results in Csk phosphorylation in the epithelial cells (Scholle et al. 1999). Normally, Csk phosphorylates the C-terminal regulatory tyrosine residue of the Src family tyrosine kinases, which inhibits their activation. This trio interaction of LMP2A-Src-Csk provides another layer of regulation of LMP2A in modifying cellular signaling pathways.

Another important cellular signaling molecule activated through LMP2A expression is Ras likely by the colocalization of Ras with LMP2A in membrane microdomains (Portis and Longnecker 2004b). Ras is prenylated and palmitoylated, which localizes it to the membrane (Booden et al. 1999; Cuiffo and Ren 2010; Porfiri et al. 1994). Increased Ras activity, either through mutations or by posttranslational modification, is observed in many human cancers. Its association with the membrane microdomains that also contain LMP2A in EBV-infected cells provides an environment similar to the antigen receptor activation (Booden et al. 1999; Cadwallader et al. 1994; Cuiffo and Ren 2010; Nomura et al. 2004). The presence of kinases recruited to the LMP2A-containing microdomains may then activate Ras.

LMP2A signaling complexes culminate in activation of downstream signaling pathways such as Ras/Raf/MAP, PI3 K/Akt/mTOR, and NFkB pathways (Allen et al. 2005; Caldwell et al. 1998; Portis and Longnecker 2004b; Swart et al. 2000). Furthermore, the activity of PI3 K/Akt/mTOR, PLC$\gamma$2, Abl, Crk, and Nck is increased with the LMP2A expression (Cheng et al. 2011; Engels et al. 2001; Shishido et al. 2001; Stewart et al. 2004). In transgenic mice expressing LMP2A in B cells, the increased activation of Ras, PI3 K, and Akt correlates with the increased expression of Bcl-xL (Portis and Longnecker 2004b). Pharmacological inhibition of these kinases increased apoptosis in these cells, indicating that the induction of cell survival by LMP2A is mediated through increased Ras, PI3 K/Atk, and Bcl-xL activity (Pan et al. 2008; Portis and Longnecker 2004b).

The full phosphorylation and activation of PI3 K/Akt by LMP2A is mediated at least in part by the scaffolding protein Shb, which is recruited to the phosphorylated LMP2A (Matskova et al. 2007). The endocytic adaptor protein ITSN1 is also recruited to the LMP2A complex probably through Shb. Shb binds to phosphorylated tyrosine residues of LMP2A and binds to SH3 domain of ITSN1 making at least a triple complex (Dergai et al. 2013). Syk kinase is recruited to the complex, which then phosphorylates both Sbh and ITSN1. Shb requires, in addition to Syk, Lyn for full phosphorylation (Dergai et al. 2013).

The LMP2A N-terminal domain is predicted to have motifs targeted for the phosphorylation by serine/threonine kinases (Anderson and Longnecker 2008a; Panousis and Rowe 1997). LMP2A is phosphorylated on S15 and S102 by Erk1 (Panousis and Rowe 1997). In addition, LMP2A upregulates UDP glucose dehydrogenase activation through PI3 K/Akt and Erk pathways in EBV-positive nasopharyngeal carcinoma (NPC) (Pan et al. 2008). LMP2A expression also leads to phosphorylation, activation, and stability of c-Jun transcription factor possibly through Erk activation (Chen et al. 2002). The activation of Erk by LMP2A also leads to increased invasiveness of these cells (Chen et al. 2002). Another downstream target gene regulated by LMP2A is EGR1, which is activated by LMP2A- and BCR-induced viral replication (Vockerodt et al. 2013). Suppression of EGR1 prevents BCR-induced LMP2A-driven viral replication in malignant Hodgkin/Reed–Sternberg (HRS) cells (Vockerodt et al. 2013). In Myc-induced lymphoma, EGR1 expression is upregulated by LMP2A, but it is not required for maintenance of the tumor growth or survival (Bieging et al. 2011).

In gene expression analysis, LMP2A has shown to alter a vast array of host genes in non-tumor cells. The B cell-specific genes such as E2A, EBF, and Pax-5 genes as well as c-Kit, Notch, Wnt, NFkB, TNF-a, TGF-b, apoptosis, and integrin pathways have been significantly altered (Anderson and Longnecker 2008b, 2009; Portis and Longnecker 2003, 2004a; Shair and Raab-Traub 2012; Stewart et al. 2004; Strobl et al. 2000). Interestingly, in a tumor model dependent on the expression of Myc and LMP2A, there is not a dramatic difference in cellular gene expression which may indicate that the contribution of LMP2A in tumorigenesis may be transient and may not be needed for tumor maintenance (Bieging et al. 2009, 2011).

## 6 Functions of LMP2 Proteins

The modulation of cellular signaling pathways exerts various functional outcomes such as increased cell survival, activation, proliferation, migration, and transformation. For example, the recruitment of Syk kinase is essential in LMP2A-induced epithelial cell transformation (Fotheringham et al. 2012; Lu et al. 2006). LMP2A ITAM mutants or inhibition of Syk abrogated the transforming feature of the LMP2A in these cells (Fukuda and Kawaguchi 2014). In non-transformed epithelial cells, expression of LMP2A increases the stem-like cell population with high transformation ability (Nakaya et al. 2013). The activation of Syk kinase through LMP2A expression induces epithelial cell migration, which is inhibited with Syk-specific inhibitors or Syk-specific siRNA (Lu et al. 2006). Expression of LMP2A and LMP2B in epithelial cells also promotes cell spreading through activation of integrin. LMP2A enhances the global tyrosine phosphorylation in response to adhesion in epithelial cells (Allen et al. 2005).

In lymphoblastoid cell lines, LMP2A interferes with the EBV activation of lytic replication by blocking the activation of protein tyrosine kinases (Miller et al. 1994b, 1995). This may allow persistence of EBV in latently infected B cells (Portis et al. 2002). Indeed, LMP2A is expressed in EBV-infected tonsillar memory B cells (Babcock et al. 2000; Babcock and Thorley-Lawson 2000). LMP2B as well as LMP2A expression promotes spread and motility of epithelial cells (Allen et al. 2005). These observations suggest a central role for LMP2A to maintain EBV latency within the infected cell and that LMP2B, despite lacking the N-terminal domain, still can modify functional cellular signaling pathways.

LMP2A expression in BL cells prevents BCR-induced lytic viral replication by downregulating the expression of immediate early lytic viral gene BZLF1 and late lytic viral gene gp350/220. In EBV-positive BL Akata cells, silencing LMP2A increased the lytic replication of the virus (Rechsteiner et al. 2007). However, the coexpression of LMP2A and LMP2B abrogated this effect of LMP2A and resulting in increased lytic viral replication (Rechsteiner et al. 2008a). This indicates that the LMP2B counteracts the role of LMP2A in establishing and maintaining viral latency (Rechsteiner et al. 2008b). This notion is further supported by the recent data showing sequestration of LMP2 transcripts in the nucleus after induction of lytic EBV replication (Concha et al. 2012). Similarly, the expression of LMP2B along with the LMP2A in BJAB cells restored the LMP2A-inhibited calcium mobilization and phosphorylation of Lyn kinase (Rovedo and Longnecker 2007).

The activation of the cellular kinases is therefore a mechanism by which EBV LMP2A plays a role in tumor development. LMP2A transgenic mice, in which LMP2A expression is directed to B cells, do not have increased tumor incidence, but when crossed with Myc transgenic mice, splenomegaly at early age and an acceleration of Myc-induced lymphoma development are observed (Bultema et al. 2009; Caldwell et al. 1998). LMP2A expression in B cells leads to the bypass of the B cell developmental checkpoints and the escape of immature B cells into periphery (Caldwell et al. 1998, 2000). With these seemingly opposing effects, providing constitutive baseline activation for survival but prevention of viral activation, LMP2A prevents activation of anti-EBV immunity and provides an environment for viral persistence.

Constitutive ectopic expression of LMP2A in BL lymphoblasts blocks the signal transduction events initiated through cross-linking of surface receptors such as BCR, CD19, or class II MHC (Miller et al. 1993, 1994a, b). This blockade is similar to the desensitization that occurs after BCR signal transduction (Buhl and Cambier 1997; Cambier et al. 1993). Mutant LMP2A unable to associate with Lyn and Syk does not block signal transduction (Fruehling et al. 1996; Fruehling and

Longnecker 1997; Fruehling et al. 1998). This inhibition/desensitization may be due to three or combination of three different mechanisms:

(a) Degradation of signaling proteins through ubiquitination.
(b) Receptor endocytosis after ligand–receptor interaction.
(c) Relocalization of signaling molecules after initiation of signaling cascades and rearrangement of these molecules within the lipid rafts (Longnecker et al. 2013).

LMP2A may deplete the cell of essential signaling kinases such as Lyn and Syk by their degradation through ubiquitination, induction of desensitization, or depletion of kinases in the lipid rafts. All of these three mechanisms will lead to the absence or decreased amount of Lyn and Syk recruited to the BCR (Dykstra et al. 2001). In addition, the baseline phosphorylation of B cell signaling molecules, including Lyn, Syk, and the p85 subunit of PI3 K, is higher in lymphoblastoid cell lines (LCLs) transformed by recombinant EBV expressing wild-type LMP2A compared to the LCLs transformed with the recombinant EBV lacking LMP2A (Miller et al. 1994a, 1995). Furthermore, cross-linking of B cell receptor on LCLs transformed with the wild-type EBV abrogated phosphorylation of Src family and Syk kinases phosphorylation, while cross-linking B cell receptor on LCLs transfected with EBV with null-mutant LMP2A induced phosphorylation of Lyn, Fyn, Syk, p85, Vav, PLC-$\gamma$, Shc, and Grb2 (Miller et al. 1995). Consistent with the desensitization model, the N-terminal domain of LMP2A fused to CD8 results in a transiently increased intracellular $Ca^{2+}$ (Beaufils et al. 1993). Cross-linking of BCR on primary B lymphocytes transformed by LMP2 null-mutant recombinants results in normal BCR-mediated signal transduction and activation of EBV replication (Miller et al. 1994b). Furthermore, activating protein kinase C (PKC) and raising intracellular free calcium with phorbol ester and calcium ionophore treatment can break the LMP2A block on the lytic replication (Miller et al. 1994b; Schaadt et al. 2005).

Another role for LMP2A is to provide survival and differentiation signals to B cells. LMP2A expression in transgenic mice under the control of the IgH promoter and enhancer (Eμ) leads to the presence of BCR-negative B cells in the periphery indicating the bypass of important B cell developmental checkpoints (Anderson and Longnecker 2008a; Caldwell et al. 1998, 2000; Rovedo and Longnecker 2008). Thus, LMP2A can mediate sufficient constitutive forward signaling to alter normal B cell survival. This forward activating signaling through LMP2A is also observed in RAG 1(–/–) mice, and it depends on the activation of Syk, Btk, and BLNK through ITAM in the N-terminal end of LMP2A (Engels et al. 2001; Merchant et al. 2000; Merchant and Longnecker 2001). Thus, LMP2A expression could have a role in latency type 2-infected lymphocytes in enhancing B cell survival through activation of Ras, PI3K, and Akt (Fukuda and Longnecker 2005; Portis and Longnecker 2004b). LMP2A can block LMP1 expression and indirectly inhibit the expression of IL-6 through inhibiting the JAK–STAT pathway (Stewart et al. 2004), indicating LMP2A's ability to alter cytokine signaling by interfering with the JAK–STAT pathway.

A recent finding indicates that LMP2A can also alter cell cycle by interfering with the function of p27. The level of p27 is decreased in Myc-induced lymphoma cells expressing LMP2A (Fish et al. 2014). The precise mechanism LMP2A utilizes to decrease p27 is not known, but it is possible that the p27 is degraded through the activation of proteasomal pathway through recruited ubiquitin ligases.

The levels of LMP2A protein can also alter BCR signaling resulting in different developmental outcomes for B cells. Transgenic mice with high-level LMP2A expression have B-1 cells in bone marrow, spleen, and periphery, whereas low-level LMP2A expression results in spontaneous germinal centers (Casola et al. 2004; Ikeda et al. 2004). Also, the level of LMP2A expression is partially regulated by N-terminal monoubiquitination (Ikeda et al. 2002). In murine models of BL, LMP2A has been shown to dramatically accelerate disease, and this is in part due to bypassing the requirement of mutation of the p53 pathway (Bieging et al. 2009; Bultema et al. 2009). Remarkably, tumors derived from LMP2A/Myc mice share a similar pattern of gene expression with Myc tumors derived from transgenic mice despite the absence of p53 pathway mutations in the LMP2A/Myc mice (Bieging et al. 2009).

LMP2A's constitutive forward signaling can also be important in the pathogenesis of cancers such as Hodgkin's lymphoma and NPC. LMP2A expression during B cell development in mice results in decreased expression transcription factors important for B cell development, such as E2A, EBF, and Pax-5 (Portis et al. 2003; Portis and Longnecker 2003, 2004a, b). In addition, LMP2A, by activating Notch signaling, may alter B cell identity (Anderson and Longnecker 2009). LMP2A expression in epithelial cell lines results in hyperproliferation in raft cultures, alterations in differentiation, and increased cloning efficiency in soft agar and promotes epithelial cell spreading and migration (Kong et al. 2010; Lu et al. 2006; Scholle et al. 1999, 2000, 2001). AKT activation increases beta-catenin and cell growth through the LMP2A ITAM and PY motifs (Morrison et al. 2003; Morrison and Raab-Traub 2005; Pegtel et al. 2004, 2005). LMP2A has also been shown to activate the human endogenous retrovirus HERV-K18 in infected B cells, which encodes a superantigen that strongly stimulates T cells (Hsiao et al. 2006, 2009).

It is worth to notice that LMP2A has an important role in the development of malignant BCR-negative B cells such as HRS cells in Hodgkin's lymphoma by providing a survival signal and preventing the default apoptotic pathway in cells with faulty somatic hypermutation that would otherwise undergo apoptosis. The LMP2A-provided survival and proliferation will rescue the "crippled" germinal center B cells (Mancao et al. 2005) and may lead to the accumulation of secondary mutations that will provide a ground for the development of malignancy.

## 7 Expression in Malignancies

LMP2 isoforms are expressed in various types of EBV infection. In particular, type-2 and type-3 latent infections show a high-expression level of LMP2A (Longnecker et al. 2013; Ocheni and Oyekunle 2010). LMP2A is expressed in

EBV-immortalized LCLs and is also expressed in Hodgkin's and non-Hodgkin's lymphomas as well as in various lymphoproliferative malignancies (Longnecker et al. 2013). LMP2A is expressed in BL samples (Bell et al. 2006; Tao et al. 1998; Xue et al. 2002). However, the expression of LMP2 may be underappreciated as exemplified by recent discovery of novel LMP2 transcripts, such as LMP2-TR and alternatively spliced isoforms (Concha et al. 2012; Fox et al. 2010). The standard detection probes (primers and antibodies) are directed to detect only LMP2A or LMP2B but not other possible isoforms. LMP2A is also expressed in a high percent of lymphoproliferative disorders (LPDs) (Qu et al. 2000; Yoshioka et al. 2001). Furthermore, LMP2A is expressed in non-lymphoid EBV-associated carcinoma such as in NPCs and gastric carcinoma (Han et al. 2012; Luo et al. 2005; Tanaka et al. 1999). The LMP2A gene in these cancers has various polymorphisms, but the ITAM and PY motifs are well conserved indicating their functional importance in the biology of EBV infections (Han et al. 2012). Another type of malignancies that have been shown to express LMP2 transcripts is the EBV-positive T/NKT tumors, even though the T cells and NKT cells are not traditional host cells for EBV infection (Fox et al. 2010).

# 8 Targeting for Therapeutic Purposes

With its role in the modification of cellular signaling pathways, LMP2A may provide an attractive therapeutic target for the treatment of EBV-associated cancers. Three different approaches may be applied to target LMP2A for therapeutic purposes:

1. Harnessing the LMP2-specific immune response
2. Targeting the expression of LMP2
3. Targeting the cellular signaling pathways activated by LMP2A

LMP2A has antigenic properties, which may be harnessed for induction of specific immune response. It may be used to load dendritic cells for induction of LMP2A-specific CD8 T cell-dependent immunity. LMP2-TR but not LMP2A or LMP2B has been shown to be expressed in EBV-positive NK and NKT cell lymphoma. Interestingly, there was a strong LMP2-specific CD8 T response to these tumors in the absence of apparent LMP2A or LMP2B expression, indicating that LMP2 has immune eliciting antigenic properties in parts other than its N-terminal domain (Fox et al. 2010). Even though not predicted to be highly antigenic, the transmembrane segments of LMP2A have been found to be recognized by CD8 T cells (Hislop et al. 2007). These indicate that LMP2A expression can be used to develop strategies aiming to boost specific cytotoxic CD8 T cell response.

Targeting the expression of LMP2 may provide another alternative approach for therapeutic strategies. For example, it can be targeted for knockdown through miRNAs. However, this approach may or may not be effective enough as knocking out the LMP2A completely may not be feasible through miRNAs.

Targeting LMP2A-modified cellular signaling molecules has been shown to be effective in prevention of LMP2A-expressing murine lymphoma development in preclinical in vivo models (Cen and Longnecker 2011; Dargart et al. 2012; Moody et al. 2005). For example, dasatinib, a specific inhibitor of BCR-Abl, c-Kit, and Lyn kinases, and rapamycin, an mTOR inhibitor, are very effective in inhibiting LMP2A-associated lymphoma development and splenomegaly (Cen and Longnecker 2011; Dargart et al. 2012; Moody et al. 2005). Similarly, the specific Syk kinase inhibitors fostamatinib and TAK-659 have shown superior therapeutic effect in PTLD cells and LMP2A positive lymphoma tumors (Hatton et al. 2013)

## 9 Summary, Conclusion, and Outlook

EBV is a ubiquitous human pathogen that has been well adapted to the human host. This successful adaption requires viral strategies that neither kill the host cells nor induce a robust immune response that may clear the virus. One of the genes that help the virus to accomplish this viral strategy is LMP2A, a virally encoded membrane protein. On the one hand, LMP2A induces a cellular signaling pathway similar to that of B cells, driving B cells into proliferation and activation state providing a milieu for viral replication. On the other hand, it activates ubiquitination-dependent proteasomal degradation of cellular proteins. These two counterbalancing activities help the virus stay latent without inducing an effective immune response. This is the success of the virus to stay in the host without being cleared by the immune system. Therefore, developing strategies to either target cellular kinases/proteins modulated by LMP2A (Table 3) or boost immune system against LMP2 proteins will be viable approaches for the development of therapeutic strategies for EBV-positive malignancies. Yet, antigenic properties of LMP2 may also be exploited for the development of LMP2-specific vaccines.

**Table 3** Cellular proteins/kinases modulated by LMP2A

| Group | Kinase/protein |
|---|---|
| Tyrosine kinases | Lyn, Fyn, Lck, Syk, Btk, Csk |
| Lipid kinases | PI3K, PLCγ |
| S/T kinases | Akt, mTOR, Erk1/2 |
| Adaptors | BLNK, Shb |
| Transcription factors | NFkB, Notch |
| GTPase | Ras |
| Ubiquitin ligases | Nedd4, Cbl |
| Cell cycle regulators | P27 |
| Apoptosis | Bcl-2, Bcl-xL |

## References


Allen MD, Young LS, Dawson CW (2005) The Epstein-Barr virus-encoded LMP2A and LMP2B proteins promote epithelial cell spreading and motility. J Virol 79(3):1789–1802. doi:10.1128/jvi.79.3.1789-1802.2005

Alvarado JJ, Tarafdar S, Yeh JI, Smithgall TE (2014) Interaction with the SH3-SH2 region of the Src-family kinase Hck structures the HIV-1 Nef dimer for kinase activation and effector recruitment. J Biol Chem. doi:10.1074/jbc.M114.600031

Anderson LJ, Longnecker R (2008a) EBV LMP2A provides a surrogate pre-B cell receptor signal through constitutive activation of the ERK/MAPK pathway. J Gen Virol 89(7):1563–1568. doi:10.1099/vir.0.2008/001461-0 [pii] 89/7/1563

Anderson LJ, Longnecker R (2008b) An auto-regulatory loop for EBV LMP2A involves activation of notch. Virology 371(2):257–266. doi:10.1016/j.virol.2007.10.009 [pii] S0042-6822(07)00666-6

Anderson LJ, Longnecker R (2009) Epstein-Barr virus latent membrane protein 2A exploits Notch1 to alter B-cell identity in vivo. Blood 113(1):108–116. doi:10.1182/blood-2008-06-160937 [pii] blood-2008-06-160937

Babcock GJ, Thorley-Lawson DA (2000) Tonsillar memory B cells, latently infected with Epstein-Barr virus, express the restricted pattern of latent genes previously found only in Epstein-Barr virus-associated tumors. Proc Natl Acad Sci USA 97(22):12250–12255. doi:10.1073/pnas.200366597

Babcock GJ, Hochberg D, Thorley-Lawson AD (2000) The expression pattern of Epstein-Barr virus latent genes in vivo is dependent upon the differentiation stage of the infected B cell. Immunity 13(4):497–506

Beaufils P, Choquet D, Mamoun RZ, Malissen B (1993) The (YXXL/I)2 signalling motif found in the cytoplasmic segments of the bovine leukaemia virus envelope protein and Epstein-Barr virus latent membrane protein 2A can elicit early and late lymphocyte activation events. EMBO J 12(13):5105–5112

Bell AI, Groves K, Kelly GL, Croom-Carter D, Hui E, Chan AT, Rickinson AB (2006) Analysis of Epstein-Barr virus latent gene expression in endemic Burkitt's lymphoma and nasopharyngeal carcinoma tumour cells by using quantitative real-time PCR assays. J Gen Virol 87(10):2885–2890. doi:10.1099/vir.0.81906-0 [pii] 87/10/2885

Bieging KT, Amick AC, Longnecker R (2009) Epstein-Barr virus LMP2A bypasses p53 inactivation in a MYC model of lymphomagenesis. Proc Natl Acad Sci USA 106(42):17945–17950. doi:10.1073/pnas.0907994106 [pii] 0907994106

Bieging KT, Fish K, Bondada S, Longnecker R (2011) A shared gene expression signature in mouse models of EBV-associated and non-EBV-associated Burkitt lymphoma. Blood 118(26):6849–6859. doi:10.1182/blood-2011-02-338434 [pii] blood-2011-02-338434

Booden MA, Baker TL, Solski PA, Der CJ, Punke SG, Buss JE (1999) A non-farnesylated Ha-Ras protein can be palmitoylated and trigger potent differentiation and transformation. J Biol Chem 274(3):1423–1431

Buhl AM, Cambier JC (1997) Co-receptor and accessory regulation of B-cell antigen receptor signal transduction. Immunol Rev 160:127–138

Bultema R, Longnecker R, Swanson-Mungerson M (2009) Epstein-Barr virus LMP2A accelerates MYC-induced lymphomagenesis. Oncogene 28(11):1471–1476. doi:10.1038/onc.2008.492 [pii] onc2008492

Burkhardt AL, Bolen JB, Kieff E, Longnecker R (1992) An Epstein-Barr virus transformation-associated membrane protein interacts with src family tyrosine kinases. J Virol 66(8):5161–5167

Cadwallader KA, Paterson H, Macdonald SG, Hancock JF (1994) N-terminally myristoylated Ras proteins require palmitoylation or a polybasic domain for plasma membrane localization. Mol Cell Biol 14(7):4722–4730

Caldwell RG, Wilson JB, Anderson SJ, Longnecker R (1998) Epstein-Barr virus LMP2A drives B cell development and survival in the absence of normal B cell receptor signals. Immunity 9(3):405–411 [pii] S1074-7613(00)80623-8

Caldwell RG, Brown RC, Longnecker R (2000) Epstein-Barr virus LMP2A-induced B-cell survival in two unique classes of EmuLMP2A transgenic mice. J Virol 74(3):1101–1113

Cambier JC, Bedzyk W, Campbell K, Chien N, Friedrich J, Harwood A, Jensen W, Pleiman C, Clark MR (1993) The B-cell antigen receptor: structure and function of primary, secondary, tertiary and quaternary components. Immunol Rev 132:85–106

Casola S, Otipoby KL, Alimzhanov M, Humme S, Uyttersprot N, Kutok JL, Carroll MC, Rajewsky K (2004) B cell receptor signal strength determines B cell fate. Nat Immunol 5(3):317–327. doi:10.1038/ni1036

Cen O, Longnecker R (2011) Rapamycin reverses splenomegaly and inhibits tumor development in a transgenic model of Epstein-Barr virus-related Burkitt's lymphoma. Mol Cancer Ther 10(4):679–686. doi:10.1158/1535-7163.MCT-10-0833 [pii] 1535-7163.MCT-10-0833

Cen O, Gorska MM, Stafford SJ, Sur S, Alam R (2003) Identification of UNC119 as a novel activator of SRC-type tyrosine kinases. J Biol Chem 278(10):8837–8845. doi:10.1074/jbc.M208261200 [pii] M208261200

Chen SY, Lu J, Shih YC, Tsai CH (2002) Epstein-Barr virus latent membrane protein 2A regulates c-Jun protein through extracellular signal-regulated kinase. J Virol 76(18):9556–9561

Cheng S, Coffey G, Zhang XH, Shaknovich R, Song Z, Lu P, Pandey A, Melnick AM, Sinha U, Wang YL (2011) SYK inhibition and response prediction in diffuse large B-cell lymphoma. Blood 118(24):6342–6352. doi:10.1182/blood-2011-02-333773 [pii] blood-2011-02-333773

Concha M, Wang X, Cao S, Baddoo M, Fewell C, Lin Z, Hulme W, Hedges D, McBride J, Flemington EK (2012) Identification of new viral genes and transcript isoforms during Epstein-Barr virus reactivation using RNA-Seq. J Virol 86(3):1458–1467. doi:10.1128/jvi.06537-11

Cuiffo B, Ren R (2010) Palmitoylation of oncogenic NRAS is essential for leukemogenesis. Blood 115(17):3598–3605. doi:10.1182/blood-2009-03-213876

Dal Porto JM, Gauld SB, Merrell KT, Mills D, Pugh-Bernard AE, Cambier J (2004) B cell antigen receptor signaling 101. Mol Immunol 41(6–7):599–613. doi:10.1016/j.molimm.2004.04.008

Dargart JL, Fish K, Gordon LI, Longnecker R, Cen O (2012) Dasatinib therapy results in decreased B cell proliferation, splenomegaly, and tumor growth in a murine model of lymphoma expressing Myc and Epstein-Barr virus LMP2A. Antiviral Res. doi:10.1016/j.antiviral.2012.05.003

Dergai O, Dergai M, Skrypkina I, Matskova L, Tsyba L, Gudkova D, Rynditch A (2013) The LMP2A protein of Epstein-Barr virus regulates phosphorylation of ITSN1 and Shb adaptors by tyrosine kinases. Cell Signal 25(1):33–40. doi:10.1016/j.cellsig.2012.09.011

Dykstra ML, Longnecker R, Pierce SK (2001) Epstein-Barr virus coopts lipid rafts to block the signaling and antigen transport functions of the BCR. Immunity 14(1):57–67 [pii] S1074-7613(01)00089-9

Engels N, Merchant M, Pappu R, Chan AC, Longnecker R, Wienands J (2001) Epstein-Barr virus latent membrane protein 2A (LMP2A) employs the SLP-65 signaling module. J Exp Med 194(3):255–264

Engels N, Yigit G, Emmerich CH, Czesnik D, Schild D, Wienands J (2012) Epstein-Barr virus LMP2A signaling in statu nascendi mimics a B cell antigen receptor-like activation signal. Cell commun signal CCS 10:9. doi:10.1186/1478-811x-10-9

Fish K, Chen J, Longnecker R (2014) Epstein-Barr virus latent membrane protein 2A enhances MYC-driven cell cycle progression in a mouse model of B lymphoma. Blood 123(4):530–540. doi:10.1182/blood-2013-07-517649

Fotheringham JA, Raab-Traub N (2013) Epstein-Barr virus latent membrane protein 2 effects on epithelial acinus development reveal distinct requirements for the PY and YEEA motifs. J Virol 87(24):13803–13815. doi:10.1128/jvi.02203-13

Fotheringham JA, Coalson NE, Raab-Traub N (2012) Epstein-Barr virus latent membrane protein-2A induces ITAM/Syk- and Akt-dependent epithelial migration through alphav-integrin membrane translocation. J Virol 86(19):10308–10320. doi:10.1128/jvi.00853-12
Fox CP, Haigh TA, Taylor GS, Long HM, Lee SP, Shannon-Lowe C, O'Connor S, Bollard CM, Iqbal J, Chan WC, Rickinson AB, Bell AI, Rowe M (2010) A novel latent membrane 2 transcript expressed in Epstein-Barr virus-positive NK- and T-cell lymphoproliferative disease encodes a target for cellular immunotherapy. Blood 116(19):3695–3704. doi:10.1182/blood-2010-06-292268
Fruehling S, Longnecker R (1997) The immunoreceptor tyrosine-based activation motif of Epstein-Barr virus LMP2A is essential for blocking BCR-mediated signal transduction. Virology 235(2):241–251. doi:10.1006/viro.1997.8690 [pii] S0042-6822(97)98690-6
Fruehling S, Lee SK, Herrold R, Frech B, Laux G, Kremmer E, Grasser FA, Longnecker R (1996) Identification of latent membrane protein 2A (LMP2A) domains essential for the LMP2A dominant-negative effect on B-lymphocyte surface immunoglobulin signal transduction. J Virol 70(9):6216–6226
Fruehling S, Swart R, Dolwick KM, Kremmer E, Longnecker R (1998) Tyrosine 112 of latent membrane protein 2A is essential for protein tyrosine kinase loading and regulation of Epstein-Barr virus latency. J Virol 72(10):7796–7806
Fukuda M, Kawaguchi Y (2014) Role of the immunoreceptor tyrosine-based activation motif of latent membrane protein 2A (LMP2A) in Epstein-Barr virus LMP2A-induced cell transformation. J Virol 88(9):5189–5194. doi:10.1128/jvi.03714-13
Fukuda M, Longnecker R (2005) Epstein-Barr virus (EBV) latent membrane protein 2A regulates B-cell receptor-induced apoptosis and EBV reactivation through tyrosine phosphorylation. J Virol 79(13):8655–8660. doi:10.1128/JVI.79.13.8655-8660.2005 [pii] 79/13/8655
Han J, Chen JN, Zhang ZG, Li HG, Ding YG, Du H, Shao CK (2012) Sequence variations of latent membrane protein 2A in Epstein-Barr virus-associated gastric carcinomas from Guangzhou, southern China. PLoS One 7(3):e34276. doi:10.1371/journal.pone.0034276
Harwood NE, Batista FD (2010) Early events in B cell activation. Annu Rev Immunol 28:185–210. doi:10.1146/annurev-immunol-030409-101216
Hatton O, Lambert SL, Phillips LK, Vaysberg M, Natkunam Y, Esquivel CO, Krams SM, Martinez OM (2013) Syk-induced phosphatidylinositol-3-kinase activation in Epstein-Barr virus posttransplant lymphoproliferative disorder. Am J Transplant 13(4):883–890. doi:10.1111/ajt.12137
Higuchi M, Izumi KM, Kieff E (2001) Epstein-Barr virus latent-infection membrane proteins are palmitoylated and raft-associated: protein 1 binds to the cytoskeleton through TNF receptor cytoplasmic factors. Proc Natl Acad Sci USA 98(8):4675–4680. doi:10.1073/pnas.081075298
Hislop AD, Taylor GS, Sauce D, Rickinson AB (2007) Cellular responses to viral infection in humans: lessons from Epstein-Barr virus. Annu Rev Immunol 25:587–617. doi:10.1146/annurev.immunol.25.022106.141553
Hsiao FC, Lin M, Tai A, Chen G, Huber BT (2006) Cutting edge: Epstein-Barr virus transactivates the HERV-K18 superantigen by docking to the human complement receptor 2 (CD21) on primary B cells. J Immunol 177(4):2056–2060
Hsiao FC, Tai AK, Deglon A, Sutkowski N, Longnecker R, Huber BT (2009) EBV LMP-2A employs a novel mechanism to transactivate the HERV-K18 superantigen through its ITAM. Virology 385(1):261–266. doi:10.1016/j.virol.2008.11.025 [pii] S0042-6822(08)00766-6
Ikeda M, Longnecker R (2009) The c-Cbl proto-oncoprotein downregulates EBV LMP2A signaling. Virology 385(1):183–191. doi:10.1016/j.virol.2008.11.018 [pii] S0042-6822(08)00758-7
Ikeda M, Ikeda A, Longan LC, Longnecker R (2000) The Epstein-Barr virus latent membrane protein 2A PY motif recruits WW domain-containing ubiquitin-protein ligases. Virology 268(1):178–191. doi:10.1006/viro.1999.0166 [pii] S0042-6822(99)90166-6

Ikeda M, Ikeda A, Longnecker R (2001) PY motifs of Epstein-Barr virus LMP2A regulate protein stability and phosphorylation of LMP2A-associated proteins. J Virol 75(12):5711–5718. doi:10.1128/JVI.75.12.5711-5718.2001

Ikeda M, Ikeda A, Longnecker R (2002) Lysine-independent ubiquitination of Epstein-Barr virus LMP2A. Virology 300(1):153–159 [pii] S004268220291562X

Ikeda A, Caldwell RG, Longnecker R, Ikeda M (2003) Itchy, a Nedd4 ubiquitin ligase, downregulates latent membrane protein 2A activity in B-cell signaling. J Virol 77(9):5529–5534

Ikeda A, Merchant M, Lev L, Longnecker R, Ikeda M (2004) Latent membrane protein 2A, a viral B cell receptor homologue, induces CD5 + B-1 cell development. J Immunol 172(9):5329–5337

Ingham RJ, Colwill K, Howard C, Dettwiler S, Lim CS, Yu J, Hersi K, Raaijmakers J, Gish G, Mbamalu G, Taylor L, Yeung B, Vassilovski G, Amin M, Chen F, Matskova L, Winberg G, Ernberg I, Linding R, O'Donnell P, Starostine A, Keller W, Metalnikov P, Stark C, Pawson T (2005a) WW domains provide a platform for the assembly of multiprotein networks. Mol Cell Biol 25(16):7092–7106. doi:10.1128/MCB.25.16.7092-7106.2005 [pii] 25/16/7092

Ingham RJ, Raaijmakers J, Lim CS, Mbamalu G, Gish G, Chen F, Matskova L, Ernberg I, Winberg G, Pawson T (2005b) The Epstein-Barr virus protein, latent membrane protein 2A, co-opts tyrosine kinases used by the T cell receptor. J Biol Chem 280(40):34133–34142. doi:10.1074/jbc.M507831200 [pii] M507831200

Katzman RB, Longnecker R (2004) LMP2A does not require palmitoylation to localize to buoyant complexes or for function. J Virol 78(20):10878–10887. doi:10.1128/JVI.78.20.10878-10887.2004 [pii] 78/20/10878

Kong QL, Hu LJ, Cao JY, Huang YJ, Xu LH, Liang Y, Xiong D, Guan S, Guo BH, Mai HQ, Chen QY, Zhang X, Li MZ, Shao JY, Qian CN, Xia YF, Song LB, Zeng YX, Zeng MS (2010) Epstein-Barr virus-encoded LMP2A induces an epithelial-mesenchymal transition and increases the number of side population stem-like cancer cells in nasopharyngeal carcinoma. PLoS Pathog 6(6):e1000940. doi:10.1371/journal.ppat.1000940

Laux G, Perricaudet M, Farrell PJ (1988) A spliced Epstein-Barr virus gene expressed in immortalized lymphocytes is created by circularization of the linear viral genome. EMBO J 7(3):769–774

Laux G, Economou A, Farrell PJ (1989) The terminal protein gene 2 of Epstein-Barr virus is transcribed from a bidirectional latent promoter region. J Gen Virol 70(Pt 11):3079–3084

Laux G, Adam B, Strobl LJ, Moreau-Gachelin F (1994a) The Spi-1/PU.1 and Spi-B ets family transcription factors and the recombination signal binding protein RBP-J kappa interact with an Epstein-Barr virus nuclear antigen 2 responsive cis-element. EMBO J 13(23):5624–5632

Laux G, Dugrillon F, Eckert C, Adam B, Zimber-Strobl U, Bornkamm GW (1994b) Identification and characterization of an Epstein-Barr virus nuclear antigen 2-responsive cis element in the bidirectional promoter region of latent membrane protein and terminal protein 2 genes. J Virol 68(11):6947–6958

Longnecker R, Druker B, Roberts TM, Kieff E (1991) An Epstein-Barr virus protein associated with cell growth transformation interacts with a tyrosine kinase. J-Virol 65(7):3681–3692

Longnecker R, Merchant M, Brown ME, Fruehling S, Bickford JO, Ikeda M, Harty RN (2000) WW- and SH3-domain interactions with Epstein-Barr virus LMP2A. Exp Cell Res 257(2):332–340. doi:10.1006/excr.2000.4900 [pii] S0014-4827(00)94900-0

Longnecker R, Kieff E, Cohen JI (2013) Epstein-Barr virus replication. In: Knipe DM, Howley PT (eds) Fields virology, vol II, 6th edn. Lippincott Williams & Wilkins, Philadelphia

Lu J, Lin WH, Chen SY, Longnecker R, Tsai SC, Chen CL, Tsai CH (2006) Syk tyrosine kinase mediates Epstein-Barr virus latent membrane protein 2A-induced cell migration in epithelial cells. J Biol Chem 281(13):8806–8814. doi:10.1074/jbc.M507305200 [pii] M507305200

Lung RW, Tong JH, Sung YM, Leung PS, Ng DC, Chau SL, Chan AW, Ng EK, Lo KW, To KF (2009) Modulation of LMP2A expression by a newly identified Epstein-Barr virus-encoded microRNA miR-BART22. Neoplasia (New York, NY) 11(11):1174–1184

Luo B, Wang Y, Wang XF, Liang H, Yan LP, Huang BH, Zhao P (2005) Expression of Epstein-Barr virus genes in EBV-associated gastric carcinomas. World J Gastroenterol WJG 11(5):629–633

Lynch DT, Zimmerman JS, Rowe DT (2002) Epstein-Barr virus latent membrane protein 2B (LMP2B) co-localizes with LMP2A in perinuclear regions in transiently transfected cells. J Gen Virol 83(Pt 5):1025–1035

Mancao C, Altmann M, Jungnickel B, Hammerschmidt W (2005) Rescue of "crippled" germinal center B cells from apoptosis by Epstein-Barr virus. Blood 106(13):4339–4344. doi:10.1182/blood-2005-06-2341

Matskova L, Ernberg I, Pawson T, Winberg G (2001) C-terminal domain of the Epstein-Barr virus LMP2A membrane protein contains a clustering signal. J Virol 75(22):10941–10949. doi:10.1128/JVI.75.22.10941-10949.2001

Matskova LV, Helmstetter C, Ingham RJ, Gish G, Lindholm CK, Ernberg I, Pawson T, Winberg G (2007) The Shb signalling scaffold binds to and regulates constitutive signals from the Epstein-Barr virus LMP2A membrane protein. Oncogene 26(34):4908–4917. doi:10.1038/sj.onc.1210298 [pii] 1210298

Merchant M, Longnecker R (2001) LMP2A survival and developmental signals are transmitted through Btk-dependent and Btk-independent pathways. Virology 291(1):46–54. doi:10.1006/viro.2001.1187 [pii] S0042-6822(01)91187-0

Merchant M, Caldwell RG, Longnecker R (2000) The LMP2A ITAM is essential for providing B cells with development and survival signals in vivo. J Virol 74(19):9115–9124

Meyn MA 3rd, Wilson MB, Abdi FA, Fahey N, Schiavone AP, Wu J, Hochrein JM, Engen JR, Smithgall TE (2006) Src family kinases phosphorylate the Bcr-Abl SH3-SH2 region and modulate Bcr-Abl transforming activity. J Biol Chem 281(41):30907–30916. doi:10.1074/jbc.M605902200

Miller CL, Longnecker R, Kieff E (1993) Epstein-Barr virus latent membrane protein 2A blocks calcium mobilization in B lymphocytes. J Virol 67:3087–3094

Miller CL, Lee JH, Kieff E, Burkhardt AL, Bolen JB, Longnecker R (1994a) Epstein-Barr virus protein LMP2A regulates reactivation from latency by negatively regulating tyrosine kinases involved in sIg-mediated signal transduction. Infect Agents Dis 3(2–3):128–136

Miller CL, Lee JH, Kieff E, Longnecker R (1994b) An integral membrane protein (LMP2) blocks reactivation of Epstein-Barr virus from latency following surface immunoglobulin crosslinking. Proc Natl Acad Sci USA 91(2):772–776

Miller CL, Burkhardt AL, Lee JH, Stealey B, Longnecker R, Bolen JB, Kieff E (1995) Integral membrane protein 2 of Epstein-Barr virus regulates reactivation from latency through dominant negative effects on protein-tyrosine kinases. Immunity 2(2):155–166 [pii] S1074-7613(95)80040-9

Moarefi I, LaFevre-Bernt M, Sicheri F, Huse M, Lee CH, Kuriyan J, Miller WT (1997) Activation of the Src-family tyrosine kinase Hck by SH3 domain displacement. Nature 385(6617):650–653. doi:10.1038/385650a0

Moody CA, Scott RS, Amirghahari N, Nathan CO, Young LS, Dawson CW, Sixbey JW (2005) Modulation of the cell growth regulator mTOR by Epstein-Barr virus-encoded LMP2A. J Virol 79(9):5499–5506. doi:10.1128/JVI.79.9.5499-5506.2005 [pii] 79/9/5499

Morrison JA, Raab-Traub N (2005) Roles of the ITAM and PY motifs of Epstein-Barr virus latent membrane protein 2A in the inhibition of epithelial cell differentiation and activation of {beta}-catenin signaling. J Virol 79(4):2375–2382. doi:10.1128/JVI.79.4.2375-2382.2005 [pii] 79/4/2375

Morrison JA, Klingelhutz AJ, Raab-Traub N (2003) Epstein-Barr virus latent membrane protein 2A activates beta-catenin signaling in epithelial cells. J Virol 77(22):12276–12284

Nada S, Yagi T, Takeda H, Tokunaga T, Nakagawa H, Ikawa Y, Okada M, Aizawa S (1993) Constitutive activation of Src family kinases in mouse embryos that lack Csk. Cell 73(6):1125–1135

Nakaya T, Kikuchi Y, Kunita A, Ishikawa S, Matsusaka K, Hino R, Aburatani H, Fukayama M (2013) Enrichment of stem-like cell population comprises transformation ability of Epstein-Barr virus latent membrane protein 2A for non-transformed cells. Virus Res 174(1–2):108–115. doi:10.1016/j.virusres.2013.03.009

Nguyen JT, Turck CW, Cohen FE, Zuckermann RN, Lim WA (1998) Exploiting the basis of proline recognition by SH3 and WW domains: design of N-substituted inhibitors. Science 282(5396):2088–2092

Nomura K, Kanemura H, Satoh T, Kataoka T (2004) Identification of a novel domain of Ras and Rap1 that directs their differential subcellular localizations. J Biol Chem 279(21):22664–22673. doi:10.1074/jbc.M314169200

Ocheni SOD, Oyekunle AA (2010) EBV-associated malignancies. Open Infect Dis J 4(1):101–122

Okada M (2012) Regulation of the SRC family kinases by Csk. Int J Biol Sci 8(10):1385–1397. doi:10.7150/ijbs.5141

Pan YR, Vatsyayan J, Chang YS, Chang HY (2008) Epstein-Barr virus latent membrane protein 2A upregulates UDP-glucose dehydrogenase gene expression via ERK and PI3 K/Akt pathway. Cell Microbiol 10(12):2447–2460. doi:10.1111/j.1462-5822.2008.01221.x [pii] CMI1221

Panousis CG, Rowe DT (1997) Epstein-Barr virus latent membrane protein 2 associates with and is a substrate for mitogen-activated protein kinase. J Virol 71(6):4752–4760

Pegtel DM, Middeldorp J, Thorley-Lawson DA (2004) Epstein-Barr virus infection in ex vivo tonsil epithelial cell cultures of asymptomatic carriers. J Virol 78(22):12613–12624. doi:10.1128/JVI.78.22.12613-12624.2004 [pii] 78/22/12613

Pegtel DM, Subramanian A, Sheen TS, Tsai CH, Golub TR, Thorley-Lawson DA (2005) Epstein-Barr-virus-encoded LMP2A induces primary epithelial cell migration and invasion: possible role in nasopharyngeal carcinoma metastasis. J Virol 79(24):15430–15442. doi:10.1128/jvi.79.24.15430-15442.2005

Porfiri E, Evans T, Chardin P, Hancock JF (1994) Prenylation of Ras proteins is required for efficient hSOS1-promoted guanine nucleotide exchange. J Biol Chem 269(36):22672–22677

Portis T, Longnecker R (2003) Epstein-Barr virus LMP2A interferes with global transcription factor regulation when expressed during B-lymphocyte development. J Virol 77(1):105–114

Portis T, Longnecker R (2004a) Epstein-Barr virus (EBV) LMP2A alters normal transcriptional regulation following B-cell receptor activation. Virology 318(2):524–533. doi:10.1016/j.virol.2003.09.017 [pii] S0042682203007049

Portis T, Longnecker R (2004b) Epstein-Barr virus (EBV) LMP2A mediates B-lymphocyte survival through constitutive activation of the Ras/PI3 K/Akt pathway. Oncogene 23(53):8619–8628. doi:10.1038/sj.onc.1207905 [pii] 1207905

Portis T, Cooper L, Dennis P, Longnecker R (2002) The LMP2A signalosome–a therapeutic target for Epstein-Barr virus latency and associated disease. Front Biosci 7:d414–d426

Portis T, Dyck P, Longnecker R (2003) Epstein-Barr Virus (EBV) LMP2A induces alterations in gene transcription similar to those observed in Reed-Sternberg cells of Hodgkin lymphoma. Blood 102(12):4166–4178. doi:10.1182/blood-2003-04-1018 [pii] 2003-04-1018

Qu L, Green M, Webber S, Reyes J, Ellis D, Rowe D (2000) Epstein-Barr virus gene expression in the peripheral blood of transplant recipients with persistent circulating virus loads. J Infect Dis 182(4):1013–1021. doi:10.1086/315828

Rechsteiner MP, Berger C, Weber M, Sigrist JA, Nadal D, Bernasconi M (2007) Silencing of latent membrane protein 2B reduces susceptibility to activation of lytic Epstein-Barr virus in Burkitt's lymphoma Akata cells. J Gen Virol 88(Pt 5):1454–1459. doi:10.1099/vir.0.82790-0

Rechsteiner MP, Berger C, Zauner L, Sigrist JA, Weber M, Longnecker R, Bernasconi M, Nadal D (2008a) Latent membrane protein 2B regulates susceptibility to induction of lytic Epstein-Barr virus infection. J Virol 82(4):1739–1747. doi:10.1128/JVI.01723-07 [pii] JVI.01723-07

Rechsteiner MP, Bernasconi M, Berger C, Nadal D (2008b) Role of latent membrane protein 2 isoforms in Epstein-Barr virus latency. Trends Microbiol 16(11):520–527. doi:10.1016/j.tim.2008.08.007

Reth M (1989) Antigen receptor tail clue. Nature 338(6214):383–384. doi:10.1038/338383b0

Rovedo M, Longnecker R (2007) Epstein-barr virus latent membrane protein 2B (LMP2B) modulates LMP2A activity. J Virol 81(1):84–94. doi:10.1128/JVI.01302-06 [pii] JVI.01302-06

Rovedo M, Longnecker R (2008) Epstein-Barr virus latent membrane protein 2A preferentially signals through the Src family kinase Lyn. J Virol 82(17):8520–8528. doi:10.1128/JVI.00843-08 [pii] JVI.00843-08

Rowe DT, Hall L, Joab I, Laux G (1990) Identification of the Epstein-Barr virus terminal protein gene products in latently infected lymphocytes. J Virol 64(6):2866–2875

Sample J, Liebowitz D, Kieff E (1989) Two related Epstein-Barr virus membrane proteins are encoded by separate genes. J Virol 63(2):933–937

Schaadt E, Baier B, Mautner J, Bornkamm GW, Adler B (2005) Epstein-Barr virus latent membrane protein 2A mimics B-cell receptor-dependent virus reactivation. J Gen Virol 86(Pt 3):551–559. doi:10.1099/vir.0.80440-0

Scholle F, Longnecker R, Raab-Traub N (1999) Epithelial cell adhesion to extracellular matrix proteins induces tyrosine phosphorylation of the Epstein-Barr virus latent membrane protein 2: a role for C-terminal Src kinase. J Virol 73(6):4767–4775

Scholle F, Bendt KM, Raab-Traub N (2000) Epstein-Barr virus LMP2A transforms epithelial cells, inhibits cell differentiation, and activates Akt. J Virol 74(22):10681–10689

Scholle F, Longnecker R, Raab-Traub N (2001) Analysis of the phosphorylation status of Epstein-Barr virus LMP2A in epithelial cells. Virology 291(2):208–214. doi:10.1006/viro.2001.1197 [pii] S0042-6822(01)91197-3

Shair KH, Raab-Traub N (2012) Transcriptome changes induced by Epstein-Barr virus LMP1 and LMP2A in transgenic lymphocytes and lymphoma. mBio 3 (5):e00288–12. doi:10.1128/mBio.00288-12

Shishido T, Akagi T, Chalmers A, Maeda M, Terada T, Georgescu MM, Hanafusa H (2001) Crk family adaptor proteins trans-activate c-Abl kinase. Genes to Cells 6(5):431–440 [pii] gtc431

Sicheri F, Moarefi I, Kuriyan J (1997) Crystal structure of the Src family tyrosine kinase Hck. Nature 385(6617):602–609. doi:10.1038/385602a0

Stewart S, Dawson CW, Takada K, Curnow J, Moody CA, Sixbey JW, Young LS (2004) Epstein-Barr virus-encoded LMP2A regulates viral and cellular gene expression by modulation of the NF-kappaB transcription factor pathway. Proc Natl Acad Sci USA 101(44):15730–15735. doi:10.1073/pnas.0402135101

Strobl LJ, Hofelmayr H, Marschall G, Brielmeier M, Bornkamm GW, Zimber-Strobl U (2000) Activated Notch1 modulates gene expression in B cells similarly to Epstein-Barr viral nuclear antigen 2. J Virol 74(4):1727–1735

Swanson-Mungerson M, Bultema R, Longnecker R (2006) Epstein-Barr virus LMP2A enhances B-cell responses in vivo and in vitro. J Virol 80(14):6764–6770. doi:10.1128/JVI.00433-06 [pii] 80/14/6764

Swart R, Ruf IK, Sample J, Longnecker R (2000) Latent membrane protein 2A-mediated effects on the phosphatidylinositol 3-Kinase/Akt pathway. J Virol 74(22):10838–10845

Tanaka M, Kawaguchi Y, Yokofujita J, Takagi M, Eishi Y, Hirai K (1999) Sequence variations of Epstein-Barr virus LMP2A gene in gastric carcinoma in Japan. Virus Genes 19(2):103–111

Tao Q, Robertson KD, Manns A, Hildesheim A, Ambinder RF (1998) Epstein-Barr virus (EBV) in endemic Burkitt's lymphoma: molecular analysis of primary tumor tissue. Blood 91(4):1373–1381

Tomaszewski-Flick MJ, Rowe DT (2007) Minimal protein domain requirements for the intracellular localization and self-aggregation of Epstein-Barr virus latent membrane protein 2. Virus Genes 35(2):225–234. doi:10.1007/s11262-007-0118-8

Vockerodt M, Wei W, Nagy E, Prouzova Z, Schrader A, Kube D, Rowe M, Woodman CB, Murray PG (2013) Suppression of the LMP2A target gene, EGR-1, protects Hodgkin's lymphoma cells from entry to the EBV lytic cycle. J Pathol 230(4):399–409. doi:10.1002/path.4198

Winberg G, Matskova L, Chen F, Plant P, Rotin D, Gish G, Ingham R, Ernberg I, Pawson T (2000) Latent membrane protein 2A of Epstein-Barr virus binds WW domain E3 protein-ubiquitin ligases that ubiquitinate B-cell tyrosine kinases. Mol Cell Biol 20(22):8526–8535

Xue SA, Labrecque LG, Lu QL, Ong SK, Lampert IA, Kazembe P, Molyneux E, Broadhead RL, Borgstein E, Griffin BE (2002) Promiscuous expression of Epstein-Barr virus genes in Burkitt's lymphoma from the central African country Malawi. Int J Cancer 99(5):635–643. doi:10.1002/ijc.10372

Yoshioka M, Ishiguro N, Ishiko H, Ma X, Kikuta H, Kobayashi K (2001) Heterogeneous, restricted patterns of Epstein-Barr virus (EBV) latent gene expression in patients with chronic active EBV infection. J Gen Virol 82(Pt 10):2385–2392

Young MA, Gonfloni S, Superti-Furga G, Roux B, Kuriyan J (2001) Dynamic coupling between the SH2 and SH3 domains of c-Src and Hck underlies their inactivation by C-terminal tyrosine phosphorylation. Cell 105(1):115–126

Zimber-Strobl U, Suentzenich KO, Laux G, Eick D, Cordier M, Calender A, Billaud M, Lenoir GM, Bornkamm GW (1991) Epstein-Barr virus nuclear antigen 2 activates transcription of the terminal protein gene. J Virol 65(1):415–423

Zimber-Strobl U, Kremmer E, Grasser F, Marschall G, Laux G, Bornkamm GW (1993) The Epstein-Barr virus nuclear antigen 2 interacts with an EBNA2 responsive cis-element of the terminal protein 1 gene promoter. EMBO J 12(1):167–175

# EBV Noncoding RNAs

**Rebecca L. Skalsky and Bryan R. Cullen**

**Abstract** EBV expresses a number of viral noncoding RNAs (ncRNAs) during latent infection, many of which have known regulatory functions and can post-transcriptionally regulate viral and/or cellular gene expression. With recent advances in RNA sequencing technologies, the list of identified EBV ncRNAs continues to grow. EBV-encoded RNAs (EBERs), the BamHI-A rightward transcripts (BARTs), a small nucleolar RNA (snoRNA), and viral microRNAs (miRNAs) are all expressed during EBV infection in a variety of cell types and tumors. Recently, additional novel EBV ncRNAs have been identified. Viral miRNAs, in particular, have been under extensive investigation since their initial identification over ten years ago. High-throughput studies to capture miRNA targets have revealed a number of miRNA-regulated viral and cellular transcripts that tie into important biological networks. Functions for many EBV ncRNAs are still unknown; however, roles for many EBV miRNAs in latency and in tumorigenesis have begun to emerge. Ongoing mechanistic studies to elucidate the functions of EBV ncRNAs should unravel additional roles for ncRNAs in the viral life cycle. In this chapter, we will discuss our current knowledge of the types of ncRNAs expressed by EBV, their potential roles in viral latency, and their potential involvement in viral pathogenesis.

R.L. Skalsky (✉)
Vaccine and Gene Therapy Institute, Oregon Health and Science University, Beaverton, OR, USA
e-mail: skalsky@ohsu.edu

B.R. Cullen
Department of Molecular Genetics and Microbiology, Duke University Medical Center, Durham, NC, USA
e-mail: bryan.cullen@duke.edu



C. Münz (ed.), *Epstein Barr Virus Volume 2*, Current Topics in Microbiology and Immunology 391, DOI 10.1007/978-3-319-22834-1_6

## Contents



## Abbreviations

| | |
|---|---|
| EBV | Epstein-Barr virus |
| LCV | Lymphocryptovirus |
| miRNA | microRNA |
| snoRNA | Small nucleolar RNA |
| sisRNA | Stable intronic-sequence RNA |
| ncRNA | Noncoding RNA |
| EBERs | EBV-encoded RNAs |
| BARTs | BamHI-A rightward transcripts |
| BHRF1 | BamHI right forward 1 |
| UTR | Untranslated region |
| HL | Hodgkin's lymphoma |
| BL | Burkitt's lymphoma |
| DLBCL | Diffuse large B cell lymphoma |
| PTLD | Post-transplant lymphoproliferative disease |
| NPC | Nasopharyngeal carcinoma |
| LCL | Lymphoblastoid cell line |
| LMP | Latent membrane protein |
| EBNA | EBV nuclear antigen |
| RNA | Ribonucleic acid |
| RISC | RNA-induced silencing complex |
| PAR-CLIP | Photoactivatable ribonucleoside-enhanced cross-linking and immunoprecipitation |
| HITS-CLIP | High-throughput sequencing of RNA isolated by cross-linking and immunoprecipitation |

# 1 Introduction

Latent EBV infection is causally linked to a variety of lymphoid and epithelial malignancies in vivo including Burkitt's lymphoma (BL), Hodgkin's lymphoma(HL) and non-Hodgkin's lymphoma (NHL), rare NK and T cell lymphomas, nasopharyngeal carcinoma (NPC), and a subset of gastric carcinomas (GC). Over 80 protein-coding open reading frames (ORFs) have been identified within the EBV genome as well as ~30 different ncRNAs (Fig. 1). The EBV transcriptome is complex and consists of many alternatively spliced transcripts which yield various gene products. During latency, only a subset of viral genes is expressed. At least three distinct latency programs have been described for EBV, which are characterized by different patterns of coding and noncoding viral gene expression. During latency III, which occurs in vivo predominantly in the context of immunosuppression, such as in AIDS-associated NHL and in post-transplant lymphoproliferative disease (PTLD), as well as in B cells infected in vitro, nine latent viral genes and all EBV ncRNAs described to date are expressed. Interestingly, the majority of EBV ncRNAs are also expressed in the other latency stages, while EBV protein expression is more limited, suggesting that ongoing EBV ncRNA expression contributes to the maintenance of viral latency. In latency II, commonly observed in EBV+ NPCs, in EBV+ HL, and in EBV+ primary effusion lymphoma (PEL) cells that are co-infected with another herpesvirus, Kaposi's sarcoma-associated herpesvirus (KSHV), the episomal maintenance protein EBV nuclear antigen 1 (EBNA-1), three latent membrane proteins (LMP1, LMP2A, and LMP2B), and abundant levels of Epstein-Barr virus-encoded RNAs (EBERs), BamHI-A rightward transcripts (BARTs), and BART microRNAs (miRNAs) are expressed. In the most restricted stage of latency, latency I, which is the characteristic of BL, only the EBNA-1 protein is expressed, while low levels of BART miRNAs are also detectable.

Exact functions for many EBV ncRNAs remain unknown; however, many studies suggest that a number of EBV ncRNAs—particularly viral miRNAs—may contribute to the establishment and persistence of viral latency, EBV-driven B cell immortalization in vitro, and potentially, the development of cancer in vivo. Here, we discuss the types of ncRNAs expressed by EBV and their known functions.

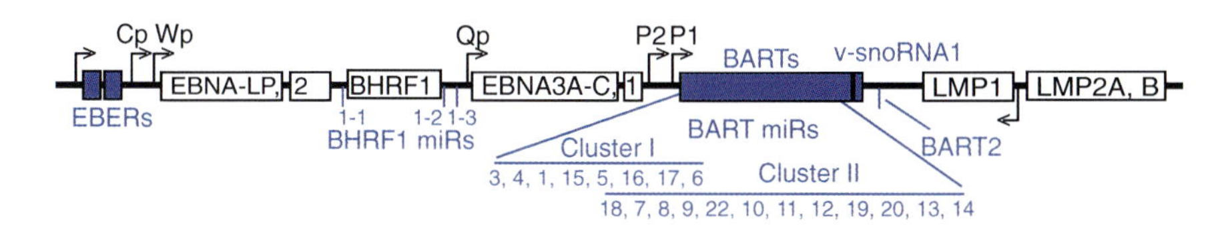


**Fig. 1** Genomic origin of latent EBV transcripts, including EBV noncoding RNAs (*shaded*). The noncoding RNAs include the EBERs, miRNAs, BART transcripts, and an EBV snoRNA. The 25 EBV precursor miRNAs are clustered in the BHRF1 and BART regions of the genome

## 2 Epstein-Barr Virus-Encoded RNAs (EBERs)

The EBERs (EBER1 and EBER2), separated by ~160 nt in the EBV genome, are expressed individually as non-polyadenylated RNA polymerase III transcripts that remain stable within the nucleus of EBV-infected cells. Both EBER1 and EBER2 contain intragenic A and B box transcriptional control elements that are characteristic of many RNA pol-III transcripts; upstream Sp1 and ATF binding sites as well as TATA-like sequences are necessary for efficient transcription (Howe and Shu 1989). The EBER transcripts are highly abundant and expressed between one to five million copies per cell (Lerner et al. 1981), making them the most abundant viral RNA species present in presumably all EBV-infected cell types. Due to their high expression levels, the EBERs are often utilized as in situ biomarkers for EBV infection in clinical samples. Their presence in virtually all EBV-associated tumors makes the EBERs potential therapeutic targets for EBV-associated cancers.

Exact function(s) of the EBERs remain both controversial and obscure. Despite their prevalence in all EBV-infected cells, tumors, and other clinical samples, their biological roles, particularly in vivo, are poorly understood. Since the initial identification of EBERs nearly thirty-five years ago, studies from a number of groups have provided conflicting results as to their activities in B cells and epithelial cells during EBV infection. Recombinant Akata-derived viruses in which the EBERs are mutationally inactivated exhibit an ~100-fold decrease in their ability to induce LCL outgrowth compared to wild-type virus, a phenotype which may or may not be specific to EBER2 (Yajima et al. 2005; Wu et al. 2007; Gregorovic et al. 2011). The role of the EBERs in lymphomagenesis has also been tested in vivo. EBER1-expressing transgenic mice develop lymphoid hyperplasia, some of which progress to B cell lymphomas (Repellin et al. 2010). Contradicting these results, deletion of the EBERs in the B95-8 background had no measurable effect on human B cell transformation efficiency or LCL growth rates in vitro (Swaminathan et al. 1991). The inconsistencies in reported EBER-associated phenotypes are possibly due to the differences in EBV strains utilized; EBER-related phenotypic effects have been observed for Akata-derived viruses, which contain additional viral transcripts and miRNAs that are absent from the EBV B95-8 (utilized by Gregorovic et al. 2011) and P3HR1 strains (utilized by Swaminathan et al. 1991). Regardless, all of these results have suggested that EBERs are not essential for, but likely contribute to, B cell transformation and the establishment and maintenance of latency.

The EBERs were also shown to be dispensable for lytic replication (Swaminathan et al. 1991), suggesting that they exert their functions primarily during latency. Furthermore, deletion of EBER1 or EBER2 individually in EBV B95-8 correlates with specific gene expression changes in LCLs. Among EBER dependent, differentially expressed genes were genes with functional roles in membrane signaling, the regulation of apoptosis, and interferon responses (Gregorovic et al. 2011). Consistent with these data, the EBERs can protect EBV-infected BL cells from interferon alpha-induced apoptosis (Ruf et al. 2005; Nanbo et al. 2002).

Further insight into EBER function may come from the unique secondary structures adopted by these two RNAs, which can facilitate interactions with host proteins. EBER1 (167 nt) and EBER2 (172 nt) adopt well-defined, evolutionarily conserved RNA secondary structures consisting of multiple stem-loop domains and have an uncapped 5′ tri-phosphate and a 3′ polyuridylate region that are characteristics of RNA pol-III transcripts (Fig. 2a). The 3′ stretch of U nucleotides is thought to facilitate interactions with cellular factors. A number of cellular proteins are known to interact with the EBERs to form ribonucleoprotein complexes. Both EBERs have been shown to bind the auto-antigen La (Lerner et al. 1981), a nuclear phosphoprotein that is known to bind many RNA pol-III transcripts, and the retinoic acid-inducible gene I (RIG-I) protein (Samanta et al. 2006), a detector of double-stranded RNAs and activator of type I interferon signaling. EBER2 may provide additional structured RNA elements for binding to other as yet undefined host factors. Indeed, new studies demonstrate interactions between EBER2, the EBV terminal repeats, and the B cell transcription factor PAX5 which can mediate LMP expression (Lee et al. 2015).

EBER1 can bind human ribosomal protein L22, which results in the relocalization of L22 from the cytoplasm to the nucleoplasm (Fok et al. 2006b; Toczyski et al. 1994). EBER1 also binds host hnRNPs (A1, A2/B1, and AUF1/D) (Lee et al. 2012). AUF1 (AU-rich element (ARE) binding factor 1) interactions with ARE-containing mRNAs usually result in enhanced mRNA destabilization and decay (reviewed in White et al. 2013). High levels of EBER1 can interfere with AUF1 binding to ARE-rich mRNAs, and ~15 % of the transcripts that are down-regulated

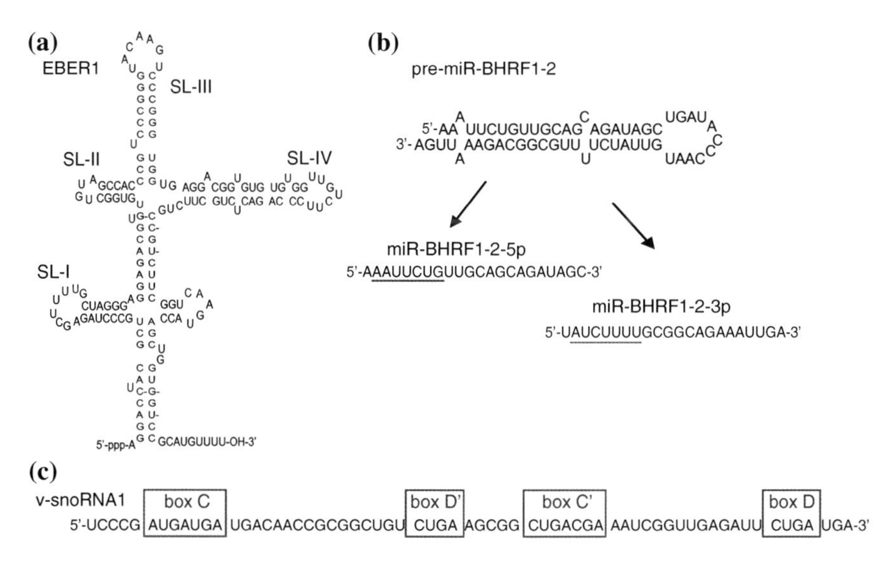


**Fig. 2** Structural features of select EBV ncRNAs. **a** EBER1 structure. SL indicates the four major "stem-loops." **b** Predicted folding structure of the precursor miRNA for miR-BHRF1-2; *highlighted* are the mature miRNAs derived from the 5p and 3p arms of the pre-miRNA with their underlined seed sequences that mediate initial interactions between RISC and target mRNAs. **C** v-snoRNA1 sequence with outlined canonical C/D boxes

upon deletion of EBER1 contain AREs, suggesting that EBER1 may contribute to the stability of these mRNAs (Lee et al. 2012; Gregorovic et al. 2011).

The EBERs are similar in size and structure to two well-characterized adenoviral small ncRNAs, termed VAI and VAII, which are essential for adenovirus replication and have been shown to inhibit PKR-mediated shutdown of translation. Intriguingly, the EBERs can functionally substitute for VAI/II and partly rescue replication of adenoviruses lacking VAI/II (Bhat and Thimmappaya 1983). EBERs have been reported to interact with the normally cytoplasmic PKR protein in vitro (Clarke et al. 1991); however, given their nuclear localization and the lack of an effect on PKR phosphorylation in EBV-infected BL cells (Ruf et al. 2005), such interactions seem unlikely to occur in vivo.

Both VAI and VAII are processed by Dicer into functional miRNAs (Furuse et al. 2013; Andersson et al. 2005; Aparicio et al. 2006; Lu and Cullen 2004), which has raised the question of whether the EBERs might also undergo miRNA processing. Small RNAs mapping to the EBERs have been reported in a number of small RNA sequencing experiments (Lung et al. 2009; Skalsky et al. 2012, 2014; Riley et al. 2012). Some of these EBER-derived RNA fragments are associated with RISC (Riley et al. 2012; Skalsky et al. 2012, 2014); however, current evidence argues against these EBER-derived RNA fragments as *bona fide* miRNAs. EBER fragments lack the precise 5′ end and specific length that is characteristic of miRNAs (Skalsky et al. 2014). Both EBERs are confined to the nucleus (Fok et al. 2006a) and therefore lack access to miRNA biogenesis machinery, namely Dicer, in the cytoplasm. Lastly, in vitro experiments have revealed that EBER1 is resistant to Dicer cleavage (Sano et al. 2006). A more likely explanation for the observed RISC-associated EBER fragments is that these small RNAs arise from EBER RNA breakdown products.

## 3 BamHI-A Rightward Transcripts (BARTs)

The BARTs represent another abundant, stable viral RNA species present in all infected cell types. The BARTs were originally identified in the C15 NPC xenograft tumor that is serially propagated in nude mice (Gilligan et al. 1990; Hitt et al. 1989). The transcripts were readily detectable by Northern blot in a number of NPC cell lines and patient biopsies (Gilligan et al. 1991) and found to arise from regions antisense to several lytic genes, including BALF5 (Karran et al. 1992). Additional studies have revealed that the BARTs are detectable during lytic infection and latent infection, in the peripheral blood of EBV-infected individuals, in B and T cell lymphomas, in B cell lines infected in vitro, and in epithelial carcinomas, especially NPCs (Edwards et al. 2008; Al-Mozaini et al. 2009; Chen et al. 2005). Interestingly, the B95-8 laboratory strain of EBV bears a deletion that removes almost the entire BART region in EBV yet B95-8 fully retains the ability to immortalize primary B cells in culture, arguing that the BARTs are dispensable for B cell transformation (Robertson et al. 1994). While the level of BART transcripts can vary dramatically between cell types, BARTs are consistently highly abundant in latency II NPCs (Marquitz and Raab-Traub 2012) and thus are thought to play a contributing role in NPC pathogenesis.

The BARTs are a complex family of alternatively spliced, polyadenylated RNAs that remain stable in the nucleus following processing of longer primary transcript(s). The entire BART locus is ~20 kbp and includes seven exons; all BARTs contain exon VII and therefore share the same 3′ end (reviewed in Marquitz and Raab-Traub 2012). Two TATA-less promoters, P1 and P2 (Fig. 1), located ~400 nt upstream of BART exon 1 are responsible for BART mRNA transcription (Chen et al. 2005). These promoters can be regulated by a number of transcription factors. P1 is upregulated by Jun family members, which can bind to a consensus AP-1 site directly upstream of P1, and is suppressed by IRF-5 and IRF-7, which bind an IRF site following their induction by type I interferon (Chen et al. 2005). P2 contains putative binding sites for c-Myc and C/EBP family members, and it has been suggested that high levels of C/EBP proteins in NPCs may contribute to the high levels of BARTs detectable in these tumors (Chen et al. 2005). Epigenetic mechanisms, including methylation of the P2 region, further regulate BART transcription, and levels of the BART miRNAs that arise from the first four BART introns (see Sect. 6) have been shown to correlate with the level of promoter methylation. BL cell lines express comparatively low levels of BART miRNAs and exhibit high BART promoter methylation, while LCLs express higher levels of BART miRNAs and lower promoter methylation (Kim do et al. 2011). Consistent with these observations, treatment of EBV-infected B cells with a DNA methyltransferase inhibitor leads to the induction of BART miRNA expression (Kim do et al. 2011). The BART promoter region in NPC tumors and cell lines is reportedly hypomethylated (Al-Mozaini et al. 2009; de Jesus et al. 2003), which may further explain the abundance of BARTs in NPCs.

The protein-coding potential of the BARTs remains controversial. Several ORFs have been suggested for the BARTs including BARF0, RK-BARF0, RPMS1, and A73 (Gilligan et al. 1990; Smith et al. 2000). Exon VII contains a small, putative ORF predicted to encode a 174 amino acid protein called BARF0 (BamH1 A rightward frame 0). An alternatively spliced transcript encompassing exon V and exon VII was predicted to encode a larger, 279 aa protein termed RK-BARF0 (Kienzle et al. 1999). BARF0, generated via in vitro translation, could be detected by Western blot using serum from NPC patients, which initially suggested that NPC patients might produce antibodies to BARF0 (Gilligan et al. 1991). Contradicting these results, BL cells expressing a BARF0 recombinant protein failed to elicit a cytotoxic T cell (CTL) response when using CTLs from EBV-seropositive patients (Kienzle et al. 1998). Additional experiments have been inconsistent in providing firm evidence for the existence of BARF0 and RK-BARF0 protein products. Antibodies raised against BARF0 peptides can detect in vitro generated BARF0 and RK-BARF0; however, these antibodies also cross-react with cellular HLA-DR in EBV-negative cells (reviewed in Marquitz and Raab-Traub 2012). Recombinant proteins RPMS1 and A73, and 103 aa and 126 aa, respectively, can be artificially expressed in *Escherichia coli* from spliced BART exons (Smith et al. 2000). Binding assays and yeast two-hybrid screens with these recombinant proteins indicate interactions with cellular proteins such as RBP-Jk/CBF1 for RPMS1 and the calcium-regulator RACK1 for A73 (Smith et al. 2000; Zhang et al. 2001); such interactions have not been confirmed in the context of infection.

Despite the ability to experimentally generate recombinant BART-origin proteins in vitro as well as observe phenotypes associated with their expression in tissue culture (reviewed in Marquitz and Raab-Traub 2012), no BART protein products have been detected in naturally infected cells in vivo to date (Smith et al. 2000; Al-Mozaini et al. 2009) and it remains unclear whether any of the predicted BART ORFs are indeed translated. Given the predominant nuclear localization of the BARTs and lack of clear evidence for BART protein products, it is possible that the BARTs represent noncoding regulatory RNAs that function similar to cellular long ncRNAs (lncRNAs) to regulate viral and/or cellular gene expression. A major role for the intronic regions of these transcripts may also be to produce the viral BART miRNAs (Sect. 6). While the function of BART transcripts therefore remains undefined, they are indisputably abundant in EBV epithelial tumors and therefore thought to be a contributing factor in NPC pathogenesis.

# 4 Viral snoRNA1

Small nucleolar RNAs (snoRNAs) are ~60–200-nt stable, noncoding RNAs that localize to the nucleolus, a sub-nuclear compartment, and form snoRNA:protein complexes (snoRNPs) that guide the chemical modifications of other RNAs. EBV encodes a single, ~65-nt canonical box C/D snoRNA (Fig. 2c), termed v-snoRNA1, that is located within the BART region, ~100 bp downstream of miR-BART2 (Fig. 1), and is detectable by Northern blot in latently infected B cell lines (Hutzinger et al. 2009). V-snoRNA1 binds canonical core ribonucleoproteins including fibrillarin, Nop65, and Nop58 and is thus thought to assemble with these proteins into a functional snoRNP to guide RNA modifications (Hutzinger et al. 2009). A 24-nt viral RNA with miRNA-like activity has been proposed to be processed from v-snoRNA (Hutzinger et al. 2009) and is detectable by Northern blot at varying levels in EBV-infected B cells and epithelial cells (Lung et al. 2013). Contrary to these studies, RISC immunoprecipitation/deep sequencing experiments in LCLs and EBV+ BLs and deep sequencing experiments in NPCs have failed to capture a v-snoRNA-derived RNA species with miRNA-like features (Riley et al. 2012; Skalsky et al. 2012, 2014; Chen et al. 2010). Nucleotide differences in the v-snoRNA1 region have been noted in different EBV strains (Lung et al. 2013) which may account for differences in these studies. Additionally, the 24-nt v-snoRNA-derived RNA may be present only in epithelial cells during lytic infection (Lung et al. 2013).

## 5 EBV-sisRNA-1 and Other Viral ncRNAs with Potential Regulatory Activities

RNA sequencing analysis of nuclear RNAs from latently infected B cells recently uncovered several novel EBV ncRNAs including a stable intronic-sequence RNA (ebv-sisRNA-1) that arises from the W repeat region in the genome and is predicted to form a conserved loop structure with two small hairpins (Moss and Steitz 2013). Spliced introns are normally degraded in the nucleus; however, the unusually stable 81-nt ebv-sisRNA-1 is abundantly detectable in latently infected cells by Northern blot at levels comparable to EBER2. Functions for the newly described ebv-sisRNA-1 have yet to be determined. Studies on cellular sisRNAs in *Xenopus* oocytes have revealed that many sisRNAs are stable for at least 48 h post-transcription (Gardner et al. 2012; Talhouarne and Gall 2014). Interestingly, injection of *Xenopus* oocytes with SV40 polyomavirus DNA yields a similar, unusually stable nuclear, non-capped, non-polyadenylated, and intronic viral RNA (Michaeli et al. 1988), raising the possibility that other DNA tumor viruses might encode sisRNAs. In fact, lariat-derived, stable intronic RNAs have been described for other herpesviruses, such as the ~2-kb latency-associated transcript (LAT) expressed by herpes simplex virus (Bloom 2004), a conserved ~5-kb intron expressed by human cytomegalovirus (hCMV) (Kulesza and Shenk 2004), and a 7.2-kb ortholog, RNA7.2, expressed by mouse CMV, which facilitates persistent viral replication in vivo (Kulesza and Shenk 2006).

Recent RNA sequencing analysis of poly(A)-selected or rRNA-depleted RNA from Mutu I BL cells and lytically reactivated Akata BL cells revealed bidirectional transcription in many regions of the EBV genome (Concha et al. 2012; O'Grady et al. 2014). Hundreds of novel viral transcripts and stable introns arising from complex, alternative splicing events were detectable during lytic infection. Most of these RNAs lack predicted protein-coding potential. While many of these RNAs may result from RNA degradation, some of the transcripts may represent authentic viral ncRNAs and play a role in the viral transcriptional program or epigenetic regulation similar to what has been reported for cellular long ncRNAs. Additional studies are needed to characterize and define roles for these RNAs during EBV infection.

## 6 EBV microRNAs

One of the most recently identified and now widely studied forms of EBV ncRNA are the viral miRNAs. miRNAs are an important class of small, ~22-nt regulatory ncRNAs that post-transcriptionally regulate gene expression by guiding the RNA-induced silencing complex (RISC) to partially complementary sequences on target mRNAs. Depending on the degree of complementarity between the miRNA and the target mRNA sequence, miRNA-loaded RISC binding can induce the immediate degradation of a target mRNA or result in translational inhibition often followed by mRNA destabilization (Ambros 2004). miRNAs are expressed by all metazoans and more recently have been identified in many viruses—in particular, the herpesviruses (reviewed in Skalsky and Cullen 2010). miRNAs require only limited sequence complementarity in order to interact with a target mRNA. Predominantly, complementarity to nt 2–7 or 8 of the mature miRNA, termed the miRNA "seed" sequence, is required for target interactions (Bartel 2009). As such, individual miRNAs are able to regulate upward of 200 different transcripts (Friedman et al. 2009) and collectively regulate >30 % of protein-coding transcripts (Carthew and Sontheimer 2009), and thus have a wide impact on gene expression. Cellular miRNAs are implicated in a number of critical cell signaling networks and biological processes including homeostasis, hematopoiesis, and the development of immunological responses. Furthermore, deregulated miRNA expression has been causatively linked to a number of cancers and disease states (Adams et al. 2014).

EBV was the first virus shown to encode viral miRNAs (Pfeffer et al. 2004). Five viral miRNAs were originally identified during cloning of small RNAs from EBV B95-8-infected Burkitt's lymphoma (BL) cells (Pfeffer et al. 2004). The EBV B95-8 strain has a 12-kb deletion within the BART region and thus lacks many EBV miRNAs. Additional sequencing efforts to examine miRNAs in EBV/KSHV-infected BC-1 cells (Cai et al. 2006; Gottwein et al. 2011), EBV B95-8 and wild-type EBV LCLs (Skalsky et al. 2012, 2014), and NPC samples infected with wild-type EBV strains, supplemented with bioinformatics analysis (Zhu et al. 2009; Chen et al. 2010; Grundhoff et al. 2006; Edwards et al. 2008), have revealed a total of 25 EBV precursor miRNAs (pre-miRNAs) from which ~40 mature miRNAs are processed. Three BHRF1 pre-miRNAs are encoded adjacent to the BHRF1 ORF, which encodes a viral Bcl2 homolog. The remaining pre-miRNAs are in introns located in the BART region; these consist of two large clusters together encompassing 21 BART miRNAs, as well as the more isolated pre-miRNA for miR-BART2, which lies antisense to the EBV BALF5 gene that encodes the viral DNA polymerase (Fig. 1).

Notably, the EBV miRNAs are highly conserved in other lymphocryptoviruses (LCVs) of the gamma-herpesvirus family (Riley et al. 2010; Walz et al. 2009; Cai et al. 2006; Skalsky et al. 2014). The closely related rhesus LCV, which is separated from EBV by ~13 million years of evolution, encodes 34 pre-miRNAs, twenty one of which share extensive sequence identity with EBV miRNAs and are located in homologous regions of the viral genome (Cai et al. 2006; Walz et al. 2009). Three additional LCVs that infect Old World non-human primates, *Herpesvirus pan, H. papio,* and *Pan paniscus LCV1*, encode homologs of the BHRF1 and BART cluster I miRNAs (Skalsky et al. 2014; Aswad and Katzourakis 2014; R.L. Skalsky, unpublished). Additional LCV sequences are currently lacking; however, other LCVs are also predicted to encode BART cluster II miRNA homologs based on the analysis of LCV LMP1 3′UTRs, which contain evolutionarily conserved binding sites for multiple BART miRNA homologs (Skalsky et al. 2012, 2014; Riley et al. 2012; Lo et al. 2007). With the exception of several miRNA seed-sequence mimics (see below and Fig. 4), none of the EBV miRNAs exhibit homology otherwise to known cellular miRNAs.

## *6.1 miRNA Biogenesis*

EBV miRNAs are dependent entirely on the cellular miRNA processing machinery for their biogenesis. To date, no viral factors are known to be involved in EBV miRNA processing. Comparable to their cellular counterparts, EBV miRNAs arise from long, nuclear RNA polymerase II primary miRNA (pri-miRNA) transcripts, which form stem-loop structures that are cleaved by the microprocessor complex, a heterodimer consisting of the RNAse III-like enzyme Drosha and its cofactor DGCR8 (Fig. 3). This cleavage results in a ~60-nt pre-miRNA, which contains an imperfect ~22 bp RNA stem with a 2 nt 3′ overhang and a terminal loop of at least 10 nt (Figs. 2b and 3). Pre-miRNA export from the nucleus into the cytoplasm is mediated by Exportin-5. Subsequent cleavage of the pre-miRNA by cytoplasmic Dicer yields a ~22-bp duplex intermediate, with 2 nt 3′ overhangs, consisting of the mature miRNA and the miRNA passenger strand. One strand of the duplex is incorporated into RISC, which minimally consists of an Argonaute (Ago) family protein, such as Ago2, and a mature miRNA (miRNA processing is reviewed in Ambros 2004; Bartel 2004; Skalsky and Cullen 2010). Ago proteins have two RNA-binding domains: a PIWI domain that binds the miRNA 5′ end and a PAZ domain that binds the 3′ end of the miRNA (Yang and Yuan 2009). Mature, functional miRNAs can be derived from either the 5′ or 3′ arm of a pre-miRNA and are denoted 5p or 3p based on their origin (Ambros et al. 2003) (Fig. 2b). miRNA-loaded RISC subsequently binds to sites on target mRNAs, preferentially in

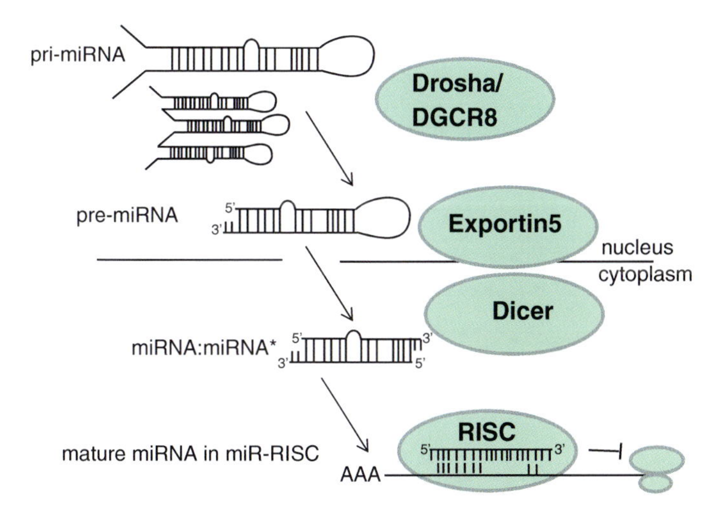


**Fig. 3** Canonical miRNA biogenesis. EBV miRNAs utilize the cellular miRNA biogenesis machinery and arise from long primary miRNA (pri-miRNA) transcripts in the nucleus that are cleaved into precursor miRNAs (pre-miRNAs) and exported into the cytoplasm. Subsequent cleavage by Dicer yields a miRNA duplex, one strand of which is incorporated into the RNA-induced silencing complex (RISC) to guide RISC to sites in 3′UTRs of target mRNAs. miR-RISC binding results in translational silencing of the target mRNA

3′UTRs, and attenuates mRNA translation and/or stability. Ago2 has endonuclease activity and is able to cleave target transcripts directly, depending on the level of miRNA complementarity (Pillai et al. 2007).

## 6.2 Expression of EBV miRNAs

High-throughput sequencing and PCR-based miRNA arrays have significantly advanced our understanding of the types of EBV miRNAs expressed in virally infected tissues and tumors. The BHRF1 and BART miRNA clusters are transcribed by different promoters: the major latency promoters, Cp and Wp, for the BHRF1 miRNAs and the BART P1 and P2 promoters for the BART miRNAs (Fig. 1), and thus, the miRNAs are differentially expressed depending on the particular viral latency program within the infected cell type.

Latency III occurs in LCLs in vitro and in many EBV+ B cell tumors, including a subset of NHL and EBV+ AIDS diffuse large B cell lymphomas. High levels of BHRF1 miRNAs are detectable by qRT-PCR and by next-generation sequencing in all latency III EBV+ B cells where the Cp and Wp promoters are active (Amoroso et al. 2011; Xia et al. 2008; Skalsky et al. 2012; Pratt et al. 2009). Additionally, BHRF1 miRNAs are detectable in Wp-restricted BL cell lines, which may contribute to the observed resistance of Wp-restricted BL to cell death stimuli compared to latency I BL cells (Amoroso et al. 2011). Unlike in latency II DLBCLs, the BHRF1 miRNAs constitute a substantial portion of the miRNA population in latency III AIDS-DLBCLs, suggesting they may contribute to ongoing pathogenesis in a subset of lymphomas (Imig et al. 2011; Barth et al. 2011; R.L. Skalsky et al., unpublished observations).

Northern blot assays indicate that the BHRF1 miRNAs are most likely derived from introns following splicing of latent Cp and/or Wp transcripts that also generate LTIII BHRF1 mRNAs during latent infection (Xing and Kieff 2007, 2011). During lytic replication, an alternative promoter (BHRF1p) drives BHRF1 mRNA expression. Kinetic studies show that miR-BHRF1-2 and miR-BHRF1-3 levels, in particular, increase during lytic reactivation and correlate with expression of the lytic BHRF1 transcript, suggesting that these two BHRF1 miRNAs can additionally be generated from Drosha cleavage of the lytic BHRF1 transcript (Xing and Kieff 2007, 2011; Amoroso et al. 2011). miR-BHRF1-1, on the other hand, overlaps the BHRF1 promoter region, and thus, expression of miR-BHRF1-1 appears to be entirely dependent on Cp/Wp activity (Amoroso et al. 2011). Coordination between Drosha processing and the splicing machinery that processes LTIII BHRF1 RNAs is further required for miR-BHRF1-1 expression (Xing and Kieff 2011).

BART miRNAs are detectable during all forms of latency; however, their expression levels vary dramatically depending on the infected cell type (Qiu et al. 2011). As the BART miRNAs derive from introns of the longer BART ncRNAs (Edwards et al. 2008), they are thought to share the BART ncRNA expression pattern, being found at especially high levels in latency II that occurs in transformed epithelial cells, including NPCs and GCs, as well as in EBV+/KSHV+ PELs (Cai et al. 2006; Gottwein et al. 2011; Chen et al. 2010; Lung et al. 2009; Zhu et al. 2009; Marquitz et al. 2014). EBV-associated B cell tumors in vivo undergoing latency I and II programs express the BART but not the BHRF1 miRNAs, and low levels of BART miRNAs are detectable in latency I BL cell lines in vitro (Vereide et al. 2014; Pratt et al. 2009; Amoroso et al. 2011; Qiu et al. 2011). Intriguingly, even though BART miRNAs are expressed as clusters, the relative abundance of individual BART miRNAs within each cluster can vary dramatically (reviewed in Marquitz and Raab-Traub 2012). Some BART miRNAs are reported to be present at thousands of copies per cell, while others are present at less than one hundred copies per cell. There are inconsistencies in the specific abundance of a given BART miRNA in different studies, which is likely due to the different EBV strains and cell types examined as well as methods utilized to detect miRNA expression (Edwards et al. 2008; Marquitz et al. 2014; Skalsky et al. 2012, 2014; Amoroso et al. 2011; Cai et al. 2006; Zhu et al. 2009; Chen et al. 2010; Pratt et al. 2009; Gottwein et al. 2011; Qiu et al. 2011). Nevertheless, BART miRNAs are consistently abundant in EBV-infected epithelial cells, and similar to BHRF1 miRNAs, the BART miRNAs can be upregulated in response to lytic reactivation (Cai et al. 2006; Amoroso et al. 2011).

### *6.3 Sequence Polymorphisms in EBV miRNAs*

RNA secondary structure plays an important role in miRNA expression. Nucleotide changes that disrupt stem pairing in the pre-miRNA can abrogate Drosha or Dicer cleavage and thereby miRNA expression (Gottwein et al. 2006). Many EBV pre-miRNAs are able to tolerate nucleotide changes within their terminal loops, and cell line-dependent sequence differences for BHRF1 miRNAs, in particular, have been noted (Amoroso et al. 2011); however, these sequence changes do not affect the overall pre-miRNA stem-loop structure that acts as a substrate for Dicer; thus, expression of the mature miRNA is unaffected. miR-BHRF1-3, for example, exhibits sequence variations in the seed sequence (notably a C to U change that affects seed pairing) in several BL cell lines (Ava, Kem, Glor, Sal, and Sav) that are compensated by changes in the opposite passenger strand (Amoroso et al. 2011). Interestingly, miR-BHRF1-3 is not as well conserved at the miRNA sequence level as the other lymphocryptovirus BHRF1 miRNAs (Skalsky et al. 2014), although the flanking regions surrounding all three BHRF1 pre-miRNAs are highly conserved, suggesting that additional sequence requirements may contribute to viral miRNA processing. Nucleotide polymorphisms have also been reported in the regions flanking miR-BART21 and miR-BART22, which may affect the expression levels of these miRNAs in certain cell types (Lung et al. 2009).

Post-transcriptional editing events can also have an effect on miRNA expression. The primary transcripts for four viral miRNAs, miR-BHRF1-1, miR-BART6, miR-BART8, and miR-BART16, were reported to undergo editing at specific nucleotide sites, none of which are located in the miRNA seed regions (Iizasa et al. 2010). Pri-miR-BART6, in particular, undergoes A-to-I editing—most likely by the ADAR1 enzyme—at specific adenosine residues that affect Drosha cleavage and miRNA processing when accompanied by uridine nucleotide deletions in the pre-miRNA terminal loop (Iizasa et al. 2010). Such uridine deletions are observed in the primary BART6 transcripts for Daudi and C666-1 viral strains compared to other wild-type EBV strains and may regulate the ability of miR-BART6 to target host mRNAs encoding Dicer (Iizasa et al. 2010). Notably, the miR-BART6 editing events are also detectable in the Akata and MutuI BL cell lines (Lin et al. 2013b). A second EBV miRNA, miR-BART3-5p, has recently been reported to undergo A-to-I editing in epithelial carcinoma cells, which may alter its ability to target the DICE1 mRNA (Lei et al. 2013); however, deep sequencing experiments have not detected significant levels of edited miR-BART3-5p in EBV-infected AGS epithelial carcinoma cells, NPCs, or LCLs (Skalsky et al. 2012; Chen et al. 2010; Marquitz et al. 2014). Further studies are needed to understand how these editing events, which appear to occur at quite low levels, might contribute to miRNA phenotypes in vivo.

## *6.4 Viral miRNAs as Biomarkers and in Exosomes*

Multiple studies have shown that miRNAs have tremendous potential as biomarkers for disease states and for monitoring responses to therapies. qRT-PCR analysis of the level of miR-BART2-5p, miR-BART6-5p, and miR-BART17-5p in serum from NPC patients and healthy control individuals demonstrated that EBV miRNA expression correlates with NPC status, strongly indicating that the presence of circulating viral miRNAs could be used as a non-invasive diagnostic or prognostic biomarker for EBV-associated tumors (Wong et al. 2012). miRNAs are highly stable in serum and plasma samples, perhaps due in part to their association with membrane-bound exosomes, which can protect miRNAs from nuclease degradation.

Exosomes are small, extracellular membrane vesicles that contain mRNAs, miRNAs, and proteins, arise through endosome and vesicular trafficking pathways, and are secreted by many different cell types (Valadi et al. 2007). Pegtel et al. first demonstrated that EBV miRNAs were present in CD63+ exosomes purified from the supernatant of EBV-infected cell cultures (Pegtel et al. 2010); the secreted, exosome-associated viral miRNAs were reported to be internalized by recipient cells and present at physiological levels high enough to inhibit a 3′UTR reporter in these cells. Although studies are still in their early stages, exosome-mediated delivery of miRNAs and other virus products, including the EBERs, has been proposed as a method for intercellular communication during infection, and exosomal transfer of viral miRNAs, in particular, could dampen signals in neighboring cells related to immune activation (Pegtel et al. 2010; Gourzones et al. 2010; Meckes et al. 2010; Valadi et al. 2007). However, given that an expression level of at least 0.1 % of the total miRNA pool in a given cell is needed to exert a detectable effect on target mRNA expression (Mullokandov et al. 2012), exosome-delivered viral miRNAs may only be present at limiting levels in recipient cells. Notably, contact-dependent intercellular transfer of EBV miRNAs from latency III Raji BL cells to CD3+ T cells has also been observed, which may introduce a greater amount of viral miRNAs into recipient cells and potentially result in the downregulation of EBV miRNA-targeted transcripts (Rechavi et al. 2009). Thus, it seems that EBV has hijacked or exploited multiple intercellular communication systems in order to alter external signals from neighboring, non-infected cells.

## *6.5 Functions for EBV miRNAs*

Functions for EBV miRNAs have begun to slowly emerge, and it is now apparent that many viral miRNAs may contribute to and/or promote viral latency by targeting viral and cellular factors involved in host cell growth, survival, and signaling pathways as well as cellular factors involved in anti-viral immune responses. Some of the viral miRNA targets involved in these processes are now known (see next sections).

A number of studies point to a role for EBV miRNAs during the initial stages of infection. Unlike KSHV miRNA mutant viruses (Lei et al. 2010), EBV mutants lacking miRNAs do not spontaneously reactivate (Seto et al. 2010; Feederle et al. 2011a, b), arguing that EBV miRNAs, at least in B cells, exert some of their regulation during events leading to latent infection. BHRF1 and BART miRNAs can be detected by two days post-infection of primary B cells in vitro, and EBV miRNA levels continue to increase during the first week of infection as B cells progress to LCLs (Pratt et al. 2009; Amoroso et al. 2011). miR-BHRF1-1 and miR-BHRF1-2, in particular, peak by 3 dpi (Amoroso et al. 2011). Loss-of-function studies show that the EBV BHRF1 miRNAs contribute to LCL outgrowth, B cell immortalization, and cell cycle progression in vitro (Feederle et al. 2011a, b; Seto et al. 2010; Wahl et al. 2013). In line with a role for BHRF1 miRNAs early after infection, a 20-fold to 30-fold reduction in LCL outgrowth was observed for EBV miR-BHRF1 mutants (Feederle et al. 2011a, b; Seto et al. 2010). Established LCLs lacking all three BHRF1 miRNAs continue to exhibit a reduced growth rate when compared to “wild-type” LCLs and exhibit a significant reduction in their ability to enter S-phase (Feederle et al. 2011b). After four weeks in culture, LCLs lacking BHRF1 miRNAs also exhibit an increase in Cp/Wp promoter activity and higher levels of EBNA-LP (Feederle et al. 2011a, b). Since BHRF1 miRNAs do not target EBNA-LP mRNAs, the higher EBNA-LP levels must be an indirect consequence of miR-BHRF1 inactivation. Humanized mouse model studies using CD34+ human fetal liver cell transplants to reconstitute the immune system in

NSG mice show that the BHRF1 miRNAs play a role in acute, systemic infection in vivo; mutational inactivation of the BHRF1 miRNAs results in significant delays in viremia (Wahl et al. 2013), further supporting an important role for the BHRF1 miRNAs early after infection.

Recently, phenotypic studies with miRNA mutant recombinant viruses have implicated EBV miRNAs in cell growth and survival, which is relevant for cancer. BART miRNAs downregulate pro-apoptotic and tumor suppressor gene products, enhance the growth transforming properties of EBV-infected epithelial cells, and can enhance metastatic potential in epithelial carcinomas (Marquitz et al. 2011, 2012; Hsu et al. 2014; Choy et al. 2008; Skalsky et al. 2012; Kang et al. 2015; Cai et al. 2015). Expression of BART miRNAs in early-stage LCLs or BL cell lines protects cells from apoptosis (Vereide et al. 2014; Seto et al. 2010). Additionally, individual BART miRNAs can influence NF-kB signaling through the regulation of LMP1 and cellular transcripts that control IkBa stability (Skalsky et al. 2012, 2014; Lo et al. 2007). Thus, EBV-encoded miRNAs can influence multiple signaling pathways in infected cells.

EBV laboratory strains lacking the BART region (i.e., B95-8) and BHRF1 miRNA knockout viruses remain able to effectively immortalize B cell in vitro, demonstrating that EBV miRNAs are not essential for B cell transformation. Furthermore, absence of the BHRF1 miRNAs and the majority of BART miRNAs had little effect on the oncogenic potential of EBV in humanized immunodeficient mice (Walsh et al. 2010). In immunocompetent human patients in vivo, however, EBV miRNAs may promote viral latency or oncogenesis by modulating cellular mRNAs, especially in cell types and tumors where the viral miRNAs are expressed at high levels, and play a key role in the persistence of latently infected cells by attenuating host immune responses (see Sect. 6.8.1).

Studies on related oncogenic herpesviruses have linked viral miRNAs to cancer formation in vivo. KSHV, a human g-herpesvirus linked to KS and PEL, encodes 12 viral pre-miRNAs, eight of which enhance tumor incidence in nude mouse models (presumably, the viral miRNAs are targeting conserved cellular genes within this context) (Moody et al. 2013). MDV-1, a chicken herpesvirus that causes T cell lymphomas, encodes a viral mimic of miR-155 (Yao et al. 2008). Deletion of the viral miR-155 mimic within the context of the viral genome fully abrogates tumor formation in chickens (Zhao et al. 2011), demonstrating that viral miRNAs can exert robust phenotypes within the context of their natural hosts.

## 6.6 Identifying Viral miRNA Targets

miRNA target identification continues to be a major hurdle for the field. Methods to first bioinformatically predict miRNA targets based on seed pairing (Lewis et al. 2005), combined with assays to examine transcriptional or translational changes in response to miRNA gain or loss of function, have been successful in determining a handful of viral miRNA targets (Skalsky et al. 2007b; Gottwein et al. 2007; Xia et al. 2008; Lo et al. 2007, 2012; Marquitz et al. 2011; Choy et al. 2008). Transcriptome-wide studies, such as RISC immunoprecipitation followed by microarray-based transcriptome profiling (RIP-Chip), have significantly increased the list of potential viral miRNA targets (Dolken et al. 2010). While techniques such as RIP-Chip have been instrumental in identifying the mRNAs that are RISC-associated during viral infection, these techniques are still unable to distinguish mRNAs targeted by viral miRNAs from those mRNAs that are targeted by cellular miRNAs.

More recently, high-throughput approaches to experimentally isolate and sequence RNAs that are cross-linked to RISC have been highly successful in capturing hundreds of direct viral miRNA targets. Two such methods, PAR-CLIP (photoactivatable ribonucleoside-enhanced cross-linking and immunoprecipitation) and HITS-CLIP (high-throughput sequencing of RNA isolated by cross-linking and immunoprecipitation), have been applied to EBV-infected B cells including LCLs (Skalsky et al. 2012, 2014), latency II BC-1 PEL cells co-infected with EBV and KSHV (Gottwein et al. 2011), and the latency III BL cell line Jijoye (Riley et al. 2012). The HITS-CLIP method utilizes a cross-linking wavelength of UV 245 nm and has an advantage in that it can be applied to both cell lines in vitro as well as tissue samples obtained in vivo (Chi et al. 2009; Haecker et al. 2012; Riley et al. 2012). The PAR-CLIP method relies on growing cells in the presence of a photoactivatable nucleoside analog, such as 4-thiouridine (4SU), that is incorporated into nascent RNAs and allows for efficient cross-linking at a UV wavelength of 365 nm (Hafner et al. 2010; Skalsky et al. 2012; Gottwein et al. 2011). For both methods, following UV cross-linking, RISC-associated RNAs are immunopurified using antibodies to an Ago protein and the complexes are digested with an RNase to leave only RISC-protected RNAs, representing miRNA-interaction sites. These RNAs are subsequently ligated to adapters and PCR-amplified prior to high-throughput sequencing. Computational algorithms are then applied to reconstruct miRNA-interaction sites from the CLIP data.

Key to PAR-CLIP is the addition of the 4SU; not only does this enhance cross-linking, but it also marks the cross-linked site, giving the method an advantage in defining a miRNA-interaction site. During PCR amplification of the sequencing library, 4SU pairs with a "G" instead of an "A" which causes a T-to-C conversion in the sequencing read at the cross-linked site (Hafner et al. 2010). Thus, PAR-CLIP, combined with the PARalyzer algorithm designed specifically to extract RNA-binding protein sites from PAR-CLIP sequencing datasets (Corcoran et al. 2011), has a high success rate in defining miRNA binding sites. In fact, over 80 % of miRNA targets captured via PAR-CLIP represent *bona fide* targets and can be experimentally confirmed (Skalsky et al. 2012; Gottwein et al. 2011).

While both PAR-CLIP and HITS-CLIP techniques are technically and computationally challenging, they have clearly yielded novel insights into the types of miRNA regulation occurring during EBV infection. Consistent with studies examining cellular miRNA function (Lewis et al. 2005; Bartel 2009; Hafner et al. 2010), the majority of EBV miRNA binding sites occur in 3′UTRs, and to a lesser extent, in coding regions, and occur predominantly in cellular mRNAs (Skalsky et al. 2012, 2014; Riley et al. 2012; Gottwein et al. 2011). HITS-CLIP experiments in Jijoye cells indicate that approximately half of the EBV miRNA targets are co-targeted by members of the miR-17/92 cluster (Riley et al. 2012), which is abundantly expressed in BL cells and has a well-established role in BL pathogenesis. Notably, over 75 % of viral targets captured by PAR-CLIP in LCLs are also targeted by B cell miRNAs, and the majority of these co-targeted mRNAs are evolutionarily conserved (Skalsky et al. 2012, 2014), suggesting that viral miRNAs are tying into existing miRNA regulatory networks, such as those involved in B cell activation.

PAR-CLIP experiments with miRNA mutant viruses have further revealed unique features of viral miRNA targeting. For example, analysis of high-confidence cellular target sites for BHRF1 miRNAs revealed that miR-BHRF1-1 utilizes canonical seed-based targeting (pairing with nt 1–8 of the mature miRNA), while other viral miRNAs may tolerate bulge pairing in the seed regions and bind non-canonical sites in addition to seed-match sites (Skalsky et al. 2012; Majoros et al. 2013). While additional experiments are required to confirm these observations, these studies demonstrate that hundreds of both canonical and non-canonical sites on cellular transcripts are bound by EBV miRNAs during infection.

## 6.7 EBV Transcripts Targeted by miRNAs

EBV miRNAs have described roles during the viral life cycle and can target the 3′UTRs of multiple viral protein-coding mRNAs (Table 1). miR-BART2-5p is perfectly complementary to the BALF5 3′UTR and inhibits expression of the viral DNA polymerase by inducing cleavage of the BALF5 mRNA during lytic reactivation (Pfeffer et al. 2004; Barth et al. 2008). BZLF1 and BRLF1, two lytic gene products that share 3′UTR sequences, were recently reported to be targeted by miR-BART20-5p (Jung et al. 2014). PAR-CLIP experiments captured sites for BART miRNAs in the 3′UTRs of BNRF1 and BALF2 (Skalsky et al. 2014).

**Table 1** EBV transcripts targeted by miRNAs

| EBV target | miRNA(s) | References |
|---|---|---|
| LMP1 | Multiple BART miRNAs, miR-17/106/20/93 family | Lo et al. (2007); Skalsky et al. (2012); Riley et al. (2012); Skalsky et al. (2014) |
| BHRF1 | miR-BART10-3p, miR-17/106/20/93 family | Skalsky et al. (2012); Riley et al. (2012); Skalsky et al. (2014) |
| BALF5 | miR-BART2-5p | Pfeffer et al. (2004); Barth et al. (2008) |
| BZLF1/BRLF1 | miR-BART20-5p | Jung et al. (2014) |
| LMP2A | miR-BART22 | Lung et al. (2009) |
| EBNA2 | ? | Skalsky et al. (2012); Riley et al. (2012) |
| BNRF1 | miR-BART5-5p, miR-BART17-3p | Skalsky et al. (2014) |
| BALF2 | miR-BART1-5p | Skalsky et al. (2014) |

Analysis of miRNA targets for the closely related rLCV revealed that homologs of BZLF1, BALF2, and other lytic gene products are targeted by multiple LCV miRNAs (Skalsky et al. 2014), and notably, many of the viral miRNA binding sites are conserved. A number of other herpesvirus miRNAs have been shown to target and inhibit expression of immediate early genes involved in lytic viral replication (reviewed in Skalsky and Cullen 2010). It has been proposed that viral miRNA regulation of lytic mRNAs is a mechanism for stabilizing, maintaining, and/or establishing latency.

Latency-associated viral gene products are also regulated by viral miRNAs. The BHRF1 3′UTR is targeted by miR-BART10-3p (Skalsky et al. 2014; Riley et al. 2012). Multiple BART miRNAs target the 3′UTRs of LMP1 and LMP2A, which are two highly immunogenic viral proteins that contribute to the proliferation and transformation of EBV-infected cells by activating specific cell signaling pathways (see other Chapters). LMP2A protein levels were reported to be reduced in response to miR-BART22 expression in epithelial cells (Lung et al. 2009). Likewise, BART cluster I miRNAs can downregulate LMP1 protein levels following ectopic expression (Lo et al. 2007; Skalsky et al. 2014). Stringent analysis of the LMP1 3′UTR recently revealed that miR-BART3 and miR-BART5, in particular, directly target the LMP1 3′UTR (Skalsky et al. 2014). Intriguingly, these miRNAs are conserved in rLCV, and their homologs target the rLCV LMP1 3′UTR in rLCV-infected B cells (Skalsky et al. 2014). PAR-CLIP analysis of rLCV LCLs further revealed that additional viral miRNAs, including a miR-BART20 homolog, interact with the LMP1 3′UTR (Skalsky et al. 2014). The consequences of these interactions have partly been examined; modulation of LMP1 by EBV miRNAs reduces epithelial cell sensitivity to apoptotic stimuli (Lo et al. 2007) and modulates LMP1-mediated NF-kB activation (Skalsky et al. 2014; Lo et al. 2007).

### 6.7.1 Viral Transcripts Are Targeted by Cellular miRNAs at Conserved Sites

LMP1 has pleiotropic activities to activate multiple cell signaling pathways, such as NF-kB (Young and Rickinson 2004), that promote B cell activation and LMP1 can also induce apoptosis when expressed at high levels (Pratt et al. 2012). In fact, many EBV-induced cellular gene expression changes can be attributed to LMP1 (Cahir-McFarland et al. 2004; Luftig et al. 2003; Soni et al. 2007). Given these activities, is it therefore not surprising that the LMP1 mRNA is subject to intensive regulation by not only viral BART miRNAs, but also conserved cellular miRNAs, namely members of the miR-17 seed family, which includes miR-17, miR-20a/b, miR-106a/b, and miR-93 (Skalsky 2012, 2014; Riley 2012). Inhibition of endogenous miR-17/20/106 activity in LCLs leads to an increase in LMP1 protein levels, demonstrating canonical miRNA-mediated regulation of LMP1 by miR-17 family members (Skalsky et al. 2012). The miR-17/20/106/93 miRNAs arise from three Myc-regulated miRNA clusters (miR-17/92, miR-106b/25, miR-106a/363) and are upregulated in response to transient Myc induction, attributed to

EBNA2 expression, shortly following de novo B cell infection (Nikitin et al. 2010; Price et al. 2012). Studies examining LMP1 mRNA and protein levels following de novo infection show a delay in LMP1 protein expression as well as a delay in expression of downstream NF-kB target genes despite the LMP1 transcript being present (Price et al. 2012). This delay may be explained in part due to the activities of miR-17, which may play an important role in transitioning infected B cells between an early Myc-dependent growth program and a later NF-kB-dependent growth program (Faumont et al. 2009; Price et al. 2012).

Regulation of LMP1 by Myc-regulated miRNAs is likely important to lymphocryptovirus biology since the miR-17/20/106/93 binding site is evolutionarily conserved in the LMP1 3′UTRs of several other LCVs (Skalsky et al. 2014). Intriguingly, another conserved binding site for the miR-17 seed family is present in the 3′UTR of BHRF1, a viral Bcl2 homolog with a role in inhibiting apoptosis (Skalsky et al. 2012, 2014; Riley et al. 2012). BHRF1 is also reported to be regulated by cellular miR-142 (Riley et al. 2012), a miRNA highly expressed in B cells, although this target site is not conserved in other LCVs (Skalsky et al. 2014). The evolutionary conservation of the miR-17 family target sites in LMP1 and BHRF1 indicates that the activities of these two viral gene products are intricately linked to the cellular regulatory pathways controlled by this miRNA family. PAR-CLIP and HITS-CLIP studies of miR-17 family targets as well as studies with transgenic miR-17/92-expressing mice demonstrate that these miRNAs negatively regulate canonical NF-kB activation and inhibit pro-apoptotic genes (Skalsky et al. 2012, 2014; Jin et al. 2013; Riley et al. 2012). Thus, in EBV-infected cells, LMP1, BHRF1, and miR-17 family members have both antagonistic and synergistic roles in relation to one another, which may be resolved through miR-17 control of the viral gene products.

## 6.8 Cellular Targets of Viral miRNAs

### 6.8.1 miRNA Targets Involved in Immune Evasion

EBV has evolved multiple strategies to escape recognition by host immune defenses, thereby permitting the virus to latently persist in cells throughout the life of the host. Recently, viral miRNAs have been shown to contribute to EBV-mediated immune evasion strategies by targeting a number of cellular factors involved in immune responses (Table 2). One of the first reported EBV miRNA cellular targets, CXCL11/I-TAC, is an interferon-inducible T cell-attracting chemokine that selectively binds to a T cell chemokine receptor, CXCR3, expressed on NK and Th1 cells. Three non-canonical binding sites in the CXCL11 3′UTR were bioinformatically predicted for miR-BHRF1-3 (Pfeffer et al. 2004), and inhibition of miR-BHRF1-3 in EBV-infected BL cells enhanced CXCL11 mRNA levels (Xia et al. 2008), indicating that miR-BHRF1-3, either directly or perhaps indirectly, regulates CXCL11 expression. TBX21/T-bet, a direct

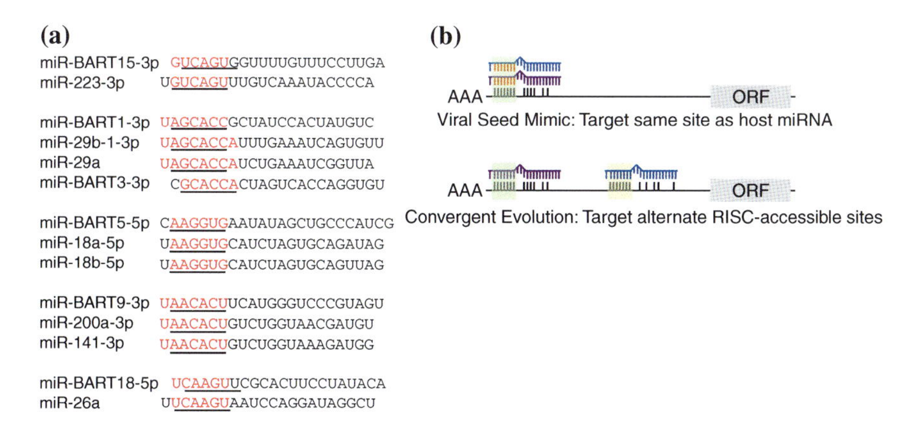


**Fig. 4** **a** EBV miRNAs exhibit full as well as offset seed-sequence homology to human miRNAs. The seed (nt 2–7) of each mature miRNA is underlined. **b** EBV can usurp existing miRNA regulatory networks by (i) encoding mimics of cellular miRNAs (see **a**), (ii) perturbing cellular miRNA expression patterns, and (iii) targeting RISC-accessible sites on cellular RNAs involved in conserved biological pathways (convergent evolution)

transcriptional activator of IFN-gamma and regulator of IL-2 and Th2 cytokine production, was reported as a target of miR-BART20-5p in invasive EBV+ nasal NK/T cell lymphomas; inhibition of T-bet expression by miR-BART20-5p may contribute to tumor development or allow for EBV replication by inhibiting cytokine production (Lin et al. 2013a). In a similar manner, NLRP3 inflammasome activation and pro-inflammatory cytokine production (IL-1beta) were shown to be inhibited in the presence of miR-BART15-3p (Haneklaus et al. 2012). miR-BART15-3p can target the NLRP3 3′UTR directly, and interestingly, this binding occurs at the miR-223 binding site due to part sequence homology between these two miRNAs (Fig. 4a). MICB, a stress-induced NK cell ligand that is recognized by the NKG2D receptor on NK cells and CD8+ T cells, was reported to be a target of miR-BART2-5p (Nachmani et al. 2009); however, PAR-CLIP and HITS-CLIP studies indicate binding sites for other BART miRNAs in the MICB 3′UTR (Skalsky et al. 2012; Riley et al. 2012). The MICB 3′UTR also contains binding sites for KSHV and HCMV miRNAs; viral miRNA expression leads to a decrease in cell surface expression of MICB and subsequently a reduced cytolytic response following NKG2D activation (Nachmani et al. 2009).

PAR-CLIP experiments in LCLs have also revealed several EBV miRNA targets related to immune evasion, including SP100, ZNF451, LY75/CD205, PDE7A, and CLEC2D. Both SP100 and ZNF451 are involved in promyelocytic leukemia-nuclear (PML) body formation that occurs in the nucleus during antiviral innate immune responses. A number of herpesviruses use multiple strategies to target PML body formation (Tavalai and Stamminger 2009). LY75 (lymphocyte antigen 75) is a transmembrane receptor that is involved in antigen transport from the cell surface to late endosomes containing MHC class I and II receptors (Gurer et al. 2008). PDE7A is involved in cytokine production and the proliferation of NK cells (Goto et al. 2009). Finally, CLEC2D is another NK ligand expressed on B cell surfaces following B cell receptor signaling or activation of toll-like receptor signaling. CD161, present on NK cells and T cells, recognizes CLEC2D, resulting in the production of IFN-gamma (Germain et al. 2011). EBV miRNA targeting of these transcripts has been confirmed by luciferase reporter assays; however, functional studies are required to understand the biological significance of these interactions—in particular, how inhibition of these transcripts by viral miRNAs might attenuate host immunological responses to EBV infection.

### 6.8.2 miRNA Targets Involved in Apoptosis

Many of the identified and confirmed targets for BART miRNAs are pro-apoptotic (Table 2) which supports the observed role of BART miRNAs in protection from apoptotic stimuli (Vereide et al. 2014; Marquitz et al. 2011; Kang et al. 2015). The mRNA of the BH3-only protein, Bim (BCL2L11), was confirmed as a target for multiple BART miRNAs (Marquitz et al. 2011), and subsequent PAR-CLIP experiments in LCLs revealed binding sites for miR-BART-4 and miR-BART15 in the BCL2L11 3′UTR (Skalsky et al. 2012). Bim inhibits the anti-apoptotic function of Bcl-2; consequently, inhibition of Bim expression by BART miRNAs should confer enhanced cell survival (Marquitz et al. 2011). PUMA (BBC3), another BH-3 only protein that facilitates release of cytochrome C from the mitochondria in response to apoptotic stimuli, has been reported as a target of miR-BART5-5p (Choy et al. 2008); however, recent experiments examining miR-BART5 functions in epithelial carcinoma cells have failed to confirm these findings (Kang et al. 2015), and BBC3 transcripts are not enriched in RISC in BART-expressing EBV-negative cells (Vereide et al. 2014).

PAR-CLIP screens identified CASP3 as a potential target of multiple BART miRNAs (Gottwein et al. 2011; Skalsky et al. 2012). Caspase 3, a member of the cysteine-aspartic acid protease family, has an extensively documented role in apoptosis and functions as the executioner caspase. The CASP3 mRNA is decreased in response to BART miRNA expression in epithelial cells (Marquitz et al. 2012). Recent studies using RISC-IP experiments and luciferase reporter assays indicate that CASP3 is a target of miR-BART16 and miR-BART1-3p in BL cells (Vereide et al. 2014) although this could not be confirmed in epithelial cells (Kang et al. 2015). PAR-CLIP and HITS-CLIP studies have identified multiple other cellular factors involved in apoptotic pathways, such as APAF1 (a component of the apoptosome), BCLAF1 (pro-apoptotic), and CAPRIN2 [also related to Wnt signaling (see below)], as well as CASZ1, DICE1, OCT1, CREBBP, SH2B3, PAK2 and TP53INP1 that are targets of EBV miRNAs, further supporting an anti-apoptotic role for EBV miRNAs (Skalsky et al. 2012; Riley et al. 2012; Kang et al. 2015).

### 6.8.3 miRNA Targets Involved in Multiple Signal Transduction Pathways

As noted above, BART miRNAs can alter NF-kB signaling pathways by directly inhibiting LMP1 expression. Ectopic expression of miR-BART3 or miR-BART1 in the absence of LMP1 disrupts NF-kB activation and stabilizes IkBa (Skalsky et al. 2014), indicating that these miRNAs also target cellular gene products involved in these pathways. Targets potentially involved in IkBa stability include CAND1, an exchange factor for the Skip/Cullin/F-box ubiquitin ligase complex that regulates IkBa, and FBXW9, an evolutionarily conserved F-box protein (Skalsky et al. 2014). Additional targets related to NF-kB signaling include PELI1, a confirmed target of miR-BART2-5p, and E3 ligases cIAP1/XIAP and cIAP2/BIRC3, the deubiquitinase CYLD, A20/TNFAIP3, IKKa/CHUK, and NFKBIZ, all of which are targeted by multiple EBV miRNAs in LCLs (Skalsky et al. 2012, 2014). Notably, these targets are both activators and repressors of NF-kB signaling. An attractive hypothesis is that EBV miRNAs direct and maintain the level of NF-kB activation within a threshold suitable for viral persistence (Skalsky et al. 2014). PELI1 is an E3 ubiquitin ligase that is activated following IL-1beta signaling through IL-1R or through MyD88 (reviewed in Moynagh 2009). PELI1 activity leads to NF-kB activation and the induction of pro-inflammatory cytokines. Thus, knockdown of PELI1 by miR-BART2-5p may also impair innate immune responses (Table 2).

Similar to what is observed for EBV miRNA targets involved in NF-kB signaling, both inhibitors and enhancers of Wnt signaling are regulated by EBV miRNAs. CAPRIN2, targeted by miR-BART13-3p (Riley et al. 2012), promotes activation of canonical Wnt signaling by stabilizing beta-catenin (Ding et al. 2008). siRNA knockdown of CAPRIN2 decreased Wnt3a-induced LEF1/TCF promoter activity and expression of Wnt target genes (Ding et al. 2008). PAR-CLIP studies in LCLs and microarray experiments in BART-expressing epithelial cell lines identified DAZAP2 as a target of miR-BART3 (Skalsky et al. 2012; Gottwein et al. 2011; Marquitz et al. 2012). DAZAP2 can interact with Tcf/Lef family members, including TCF-4, which transcriptionally activates Wnt-responsive genes. Knockdown of Dazap2 expression reduced Tcf-mediated transcription and responsiveness to Wnt stimulation (Lukas et al. 2009), similar to what has been reported for CAPRIN2 knockdown.

Inhibitors of Wnt signaling are regulated by several BART miRNAs and may play an important role in the proliferation of EBV-infected epithelial cells. Three Wnt antagonists were identified as EBV miRNA targets following profiling of EBV+ NPC tumors (Wong et al. 2012). WIF1 (targeted by miR-BART19-3p), APC (targeted by miR-BART7, miR-BART19-3p, and miR-BART17-5p), and NLK (targeted by miR-BART19-3p, miR-BART14, and miR-BART18-5p) protein and transcript levels were reduced following transient expression of BART miRNA mimics in EBV-negative epithelial cell lines (Wong et al. 2012). Presumably, downregulation of these Wnt antagonists by BART miRNAs would activate Wnt signaling. NLK, activated via MAPK signaling, can block Tcf/Lef

**Table 2** Cellular transcripts targeted by EBV miRNAs

| Cellular target | Function | miRNA(s) | References |
|---|---|---|---|
| BACH1 | Oxidative stress | miR-BHRF1-2 | Skalsky et al. (2012) |
| KDM4B | Histone demethylase | miR-BHRF1-2 | Skalsky et al. (2012) |
| OTUD1 | NF-kB signaling | miR-BART2-5p | Skalsky et al. (2012) |
| PELI1 | NF-kB signaling | miR-BART2-5p | Skalsky et al. (2012); Kang et al. (2015) |
| PDE7A | Cytokine production | Two BART miRNAs | Skalsky et al. (2012) |
| CLEC2D | Immune responses | Two BART miRNAs | Skalsky et al. (2012) |
| LY75 | Immune responses | miR-BART1-5p | Skalsky et al. (2012) |
| SP100 | PML bodies | miR-BART1-5p | Skalsky et al. (2012) |
| ZNF451 | PML bodies | miR-BHRF1-2 | Skalsky et al. (2012) |
| CLIP1 | Immune responses | miR-BART1-5p | Skalsky et al. (2012) |
| GUF1 | GTPase | miR-BHRF1-1 | Skalsky et al. (2012) |
| SCRN1 | Exocytosis | miR-BHRF1-1 | Skalsky et al. (2012) |
| NAT12 | acetyltransferase | miR-BHRF1-1 | Skalsky et al. (2012) |
| DAZAP2 | Wnt signaling | miR-BART3-3p | Skalsky et al. (2012); Kang et al. (2015) |
| CAPRIN2 | Wnt signaling | miR-BART13-3p | Riley et al. (2012) |
| CAND1 | NF-kB signaling | miR-BART3 | Skalsky et al. (2014) |
| FBXW9 | Adaptor protein | miR-BART3 | Skalsky et al. (2014) |
| DICE1 | Tumor suppressor | miR-BART3-5p | Lei et al. (2013); Kang et al. (2015) |
| MAP3K2 | MAPK signaling | miR-BART18-5p | Qiu and Thorley-Lawson (2014) |
| CXCL11 | Immune responses | miR-BHRF1-3 | Xia et al. (2008) |
| DICER | miRNA biogenesis | miR-BART6-5p | Iizasa et al. (2010); Kang et al. (2015) |
| BBC3/PUMA | Apoptosis | miR-BART5? | Choy et al. (2008) |
| BCL2L11/Bim | Apoptosis | Multiple BART miRs | Marquitz et al. (2011) |
| TBX21/T-bet | Immune responses | miR-BART20-5p | Lin et al. (2013a, b) |
| IPO7 | Transport | Two BART miRNAs | Dolken et al. (2010); Skalsky et al. (2012); Riley et al. (2012); Vereide et al. (2014); Kang et al. (2015) |
| TOMM22 | Transport | miR-BART16 | Dolken et al. (2010) |
| MICB | Immune responses | miR-BART2-5p | Nachmani et al. (2009) |
| CASP3 | Apoptosis | Two BART miRNAs | Vereide et al. (2014) |
| WIF1 | Wnt signaling | miR-BART19-3p | Wong et al. (2012) |
| APC | Wnt signaling | Multiple BART miRs | Wong et al. (2012) |
| NLK | Wnt signaling | Multiple BART miRs | Wong et al. (2012) |
| YWHAZ | Protein modifier | miR-BART14 | Grosswendt et al. (2014) |
| NLRP3 | Inflammasome | miR-BART15-3p | Haneklaus et al. (2012) |
| BRUCE | Apoptosis | miR-BART15-3p | Choi et al. (2013) |
| CDH1 | Cell migration | miR-BART9 | Hsu et al. (2014) |
| PTEN | Tumor suppressor | Multiple BART miRs | Cai et al. (2015) |

transcriptional activation, APC is a direct inhibitor of beta-catenin, and WIF1 can block the induction of the Wnt pathway (Wong et al. 2012). Interestingly, WIF1 is also a positive regulator of miR-200 family members (Ramachandran et al. 2014), which inhibit ZEB1 and ZEB2 expression (Ellis-Connell et al. 2010); both cellular proteins control the switch between latency and lytic reactivation by repressing the BZLF1 promoter. Thus, BART miRNAs may also indirectly regulate lytic viral gene expression through WIF1 targeting.

More recent studies show that targeting of MAP3K2 by miR-BART18-5p plays a role in repressing lytic viral reactivation (Qiu and Thorley-Lawson 2014). MAP3K2 is a central player in multiple signal transduction pathways. Activation of the BZLF1 promoter and initiation of the lytic cascade, at least following B cell receptor cross-linking, occur through several different pathways, including PI3K, Ras, Rac1, and phospholipase C (PLC). Both Rac1 and PLC pathways converge on MAP3K2 to activate p38 and JNK transcriptional regulators. Overexpression of miR-BART18-5p in EBV-infected BL cells inhibited viral reactivation and conversely expression of MAP3K2 in B95-8 LCLs upregulated lytic gene expression (Qiu and Thorley-Lawson 2014), thereby demonstrating a role for miR-BART18-5p in promoting latency by targeting a key signaling molecule.

Several mRNAs are now known to be targeted by multiple, different EBV miRNAs, indicating that EBV miRNAs can synergistically downregulate the expression of specific gene products. In addition to targeting by miR-BART18-5p, the 3′UTR of MAP3K2, for example, is targeted by miR-BART2-5p, BART4, BART5, and BART19-3p (Gottwein et al. 2011; Skalsky et al. 2012). Wnt-inhibitory transcripts as well as pro-apoptotic transcripts, in particular, appear to be co-targeted by multiple EBV miRNAs. Furthermore, multiple components with both activating and repressing potential within a given signal transduction pathway can be collectively targeted by multiple viral miRNAs (Skalsky et al. 2012; Riley et al. 2012; Gottwein et al. 2011). These observations indicate that viral miRNAs cooperatively facilitate a complex, yet finely tuned and directed outcome for signaling events and the transcriptional environment in an infected cell. Future functional studies must therefore consider the combinatorial effects of EBV miRNAs on the full spectrum of targets related to a given pathway.

## 6.9 EBV Exploits Intrinsic Cellular miRNA Regulatory Networks

Target identification studies have shown that multiple biological processes and cell signaling pathways can be regulated by viral miRNAs, most of which are also intrinsically regulated by cellular miRNAs. EBV is able to tie into these intrinsic miRNA regulatory networks via several means, including (i) by mimicking cellular miRNA sequences (Fig. 4a), (ii) by occupying RISC-accessible sites that are not otherwise occupied by cellular miRNAs (Fig. 4b), and (iii) by altering cellular miRNA expression patterns.

Seed mimicking by a viral miRNA, which presumably enables the viral miRNA to compete for identical seed-match sites on cellular transcripts, was first noted for KSHV miR-K11 and the cellular oncomiR, miR-155 (Skalsky et al. 2007b; Gottwein et al. 2007). miR-155 is upregulated in many B cell lymphomas and critically required for B cell activation and the formation of germinal center reactions (Thai et al. 2007; Rodriguez et al. 2007). The first ten nucleotides of KSHV miR-K11 and miR-155 are identical, and as a result, these miRNAs share a highly overlapping mRNA target repertoire (Skalsky et al. 2007b; Gottwein et al. 2007). To demonstrate that miR-K11 can indeed function as a mimic of miR-155 in vivo, NOD/SCID IL2Rgamma-null mice were injected with CD34+ human cord blood progenitor cells transduced with miR-K11 or miR-155 expression vectors (Boss et al. 2011). Expression of either miR-155 or KSHV miR-K11 led to downregulation of C/EBPbeta, increased production of IL-6, and enhanced B cell proliferation in the spleen, demonstrating that seed mimicking by a viral miRNA can translate into significant consequences in vivo (Boss et al. 2011). Notably, MDV-1, a virus linked to T cell lymphomas in chickens, also encodes a mimic of miR-155, miR-M4, which is involved in tumorigenesis (Zhao et al. 2011) and targets of miR-M4 overlap with targets of miR-155 (Parnas et al. 2014). EBV does not encode a miR-155 seed mimic, but instead strongly induces miR-155 expression directly, which is critically required for the survival and proliferation of EBV-infected cells (Linnstaedt et al. 2010).

Several EBV miRNAs mimic the seed sequences of cellular miRNAs with described tumor suppressor and/or oncomiR activities (Skalsky et al. 2007a; Skalsky and Cullen 2010) (Fig. 4a) and are thus thought to bind cognate sites, although the extent to which these mimics might compete with the cellular miRNAs for miRNA binding sites is not yet known. The list includes miR-BART9-3p and the miR-200a family, which regulates cell migration and invasion (Bracken et al. 2014; Burk et al. 2008), miR-BART1-3p and the miR-29 family, which is linked to B cell tumors (Pekarsky and Croce 2010), miR-BART5-5p and miR-18-5p, which is a member of the miR-17/92 oncomiR cluster, and lastly, two BART miRNAs which exhibit offset seed homology to cellular miRNAs (miR-BART15-3p and miR-223-3p; miR-BART18-5p and miR-26a-5p) (Haneklaus et al. 2012; Qiu and Thorley-Lawson 2014). MiR-223 is a hematopoietic miRNA with roles in myeloid lineage development and can also inhibit the inflammasome-induced production of IL-1beta. Both miR-223 and miR-BART15-3p reportedly target the NLRP3 3′UTR at the same site, which consequently attenuates inflammasome-mediated activation of IL-1beta production and subsequently reduces inflammation (Haneklaus et al. 2012). miR-BART18-5p reportedly targets a characterized miR-26a binding site in the 3′UTR of MAP3K2 (Qiu and Thorley-Lawson 2014), although this site does not appear to be bound by miR-26a in EBV-infected B cells (Skalsky et al. 2012; Gottwein et al. 2011). miR-BART5 binds a site within the LMP1 3′UTR and can inhibit LMP1 expression (Skalsky et al. 2014; Riley et al. 2012). Experiments with miR-18-5p suggest that only miR-BART5 and not miR-18-5p can bind this site in the LMP1 3′UTR, despite having the same seed sequence (Riley et al. 2012). The lack of LMP1 3′UTR targeting by the miR-18-5p mimic may be attributed to requirements for additional base pairing between the miRNA and the mRNA target outside of the seed region (Riley et al. 2012). Interestingly, a RISC-associated site overlapping this region was identified in LCLs that completely lack miR-BART5, suggesting that miR-18-5p might bind this site in the absence of miR-BART5 (Skalsky et al. 2012).

Common targets for miR-BART9-3p/miR-200a and miR-BART1-3p/miR-29 have not yet been investigated. The miR-200 family is considered to have tumor suppressor activity since miR-200a maintains the epithelial phenotype by suppressing epithelial to mesenchymal transition (EMT) and can regulate a spectrum of targets related to actin cytoskeleton dynamics, focal adhesion, Rho GTPase signaling, and metalloprotease activity (Burk et al. 2008; Bracken et al. 2014). In EBV-infected epithelial cells and carcinomas, miR-200 family members are downregulated (Lin et al. 2010; Marquitz et al. 2014; Shinozaki et al. 2010) and have been shown to play a role in the EBV latent/lytic switch via the targeting of ZEB1 and ZEB2; induction of miR-200 contributes to lytic replication (Ellis-Connell et al. 2010). Intriguingly, and opposite to what has been documented for miR-200a, miR-BART9 expression can induce a mesenchymal-like phenotype and has been linked to increased metastasis and invasiveness by NPC cells in vitro (Hsu et al. 2014). Thus, while miR-BART9 and miR-200a exhibit homology in the seed regions, miR-BART9 might not be able to bind cognate miR-200a target sites due to sequence differences outside of the miRNA seed region.

miRNA targetome studies indicate that EBV miRNAs predominantly regulate cellular, not viral, gene products (Skalsky et al. 2012, 2014; Gottwein et al. 2011; Riley et al. 2012; Kang et al. 2015). With some exceptions, the sites that are occupied by viral miRNAs on cellular mRNAs are largely not occupied by cellular miRNAs, leading to the hypothesis that viral miRNAs have evolved in part to bind alternate, RISC-accessible sites on cellular gene products that have effector functions pertinent to the viral life cycle. What is further intriguing is that many of these RISC-accessible sites are evolutionarily conserved (Majoros et al. 2013), despite not being occupied by a cellular miRNA—at least, in PAR-CLIP or HITS-CLIP studies. Thus, while a viral miRNA may not bind to the same site(s) as a cellular miRNA, a viral miRNA or combination of viral miRNAs could interact with a set of transcripts specifically regulated by and related to the function of a given cellular miRNA (Fig. 4b). Examples of this include the co-targeting, at distinct sites, of a set of cellular transcripts by EBV miRNAs and members of the miR-17/92 cluster (Riley et al. 2012).

In addition to encoding viral miRNAs, EBV infection dramatically perturbs the cellular miRNA environment and EBV proteins can induce the expression of specific cellular miRNAs that can play pertinent roles in the EBV life cycle (Cameron et al. 2008a, b; Kieff 2007). Most EBV-induced cellular gene expression changes following de novo infection can be attributed to EBNA2 and the CD40 mimic, LMP1. Both EBNA2 and LMP1 initiate dichotomous cell proliferation programs, facilitated through c-Myc and NF-kB activation (Young and Rickinson 2004; Faumont et al. 2009), which also activates expression of known oncomiRs such as the miR-17/92 cluster and miR-155 (Price et al. 2012; Forte and Luftig 2011; Nikitin et al. 2010; Kieff 2007). Recent studies show that transient c-Myc induction is an inherent and critical property of normal B cells to initiate germinal center formation (Dominguez-Sola et al. 2012; Calado et al. 2012), while miR-155 expression is required for completion of GC reactions (Thai et al. 2007; Rodriguez et al. 2007; Xiao and Rajewsky 2009). miR-155, in particular—a miRNA which is linked to hematopoietic malignancies in vivo—is highly induced following EBV infection, in part by LMP1 activation of NF-kB, and is critically required for the growth of LCLs in vitro (Cameron et al. 2008a, b; Linnstaedt et al. 2010). Consequently, EBV-induction of miR-155 can allow EBV to tie into the existing miR-155-regulated pathways. c-Myc can also antagonize NF-kB signaling (Faumont et al. 2009), which may be facilitated in part through Myc-regulated miR-17/92 targeting of NF-kB components (i.e., A20, CYLD, TRAF3, RNF11) (Skalsky et al. 2012; Jin et al. 2013). EBV thus hijacks these miRNA signals to elicit B cell proliferation, which may inadvertently contribute to lymphomagenesis in the absence of appropriate regulatory signals.

# 7 Summary and Outlook

EBV employs multiple strategies to enable long-term persistence within a host, including expressing viral ncRNAs with described roles in a variety of biological processes. Many EBV ncRNAs are highly evolutionarily conserved among LCVs at both the sequence and structural level, indicating that they have important roles in the virus life cycle and likely favorably shape the host environment to promote viral fitness and facilitate viral persistence.

Since their initial identification over ten years ago, EBV miRNAs and their contributions to viral pathogenesis and oncogenesis have been the subject of intense investigation. Several key signaling pathways, such as the NF-kB and Wnt pathways, and processes, such as apoptosis and immune activation, are now known to be regulated by EBV miRNAs. The next steps are to understand how miRNA-mediated regulation of targets involved in these pathways can contribute to viral infection and to determine the most critical miRNA targets that may be amenable to therapeutic intervention in EBV-associated diseases.

A number of open questions remain regarding the functions for many EBV ncRNAs. For example, do viral ncRNAs act in concert with viral proteins to exert a biological effect? Do EBV miRNAs synergistically target a specific pathway, as has been observed for other herpesvirus miRNAs? What, if any, are the important targets of viral miRNAs during lytic replication? And finally, how do the other viral ncRNAs contribute to the establishment of latent infection?

Future studies should provide mechanistic insight into how viral miRNAs and other ncRNAs might contribute to EBV-mediated oncogenic processes, help elucidate when EBV ncRNAs exert their functions during the natural viral life cycle, and may lead to the rational design of novel, EBV ncRNA-targeted therapies for EBV-associated diseases.

## References


Adams BD, Kasinski AL, Slack FJ (2014) Aberrant regulation and function of MicroRNAs in cancer. Curr Biol 24:R762–R776

Al-Mozaini M, Bodelon G, Karstegl CE, Jin B, Al-Ahdal M, Farrell PJ (2009) Epstein-Barr virus BART gene expression. J Gen Virol 90:307–316

Ambros V (2004) The functions of animal microRNAs. Nature 431:350–355

Ambros V, Bartel B, Bartel DP, Burge CB, Carrington JC, Chen X, Dreyfuss G, Eddy SR, Griffiths-Jones S, Marshall M, Matzke M, Ruvkun G, Tuschl T (2003) A uniform system for microRNA annotation. RNA 9:277–279

Amoroso R, Fitzsimmons L, Thomas WA, Kelly GL, Rowe M, Bell AI (2011) Quantitative studies of Epstein-Barr virus-encoded microRNAs provide novel insights into their regulation. J Virol 85:996–1010

Andersson MG, Haasnoot PC, Xu N, Berenjian S, Berkhout B, Akusjarvi G (2005) Suppression of RNA interference by adenovirus virus-associated RNA. J Virol 79:9556–9565

Aparicio O, Razquin N, Zaratiegui M, Narvaiza I, Fortes P (2006) Adenovirus virus-associated RNA is processed to functional interfering RNAs involved in virus production. J Virol 80:1376–1384

Aswad A, Katzourakis A (2014) The first endogenous herpesvirus, identified in the tarsier genome, and novel sequences from primate rhadinoviruses and lymphocryptoviruses. PLoS Genet 10:e1004332

Bartel DP (2004) MicroRNAs: genomics, biogenesis, mechanism, and function. Cell 116:281–297

Bartel DP (2009) MicroRNAs: target recognition and regulatory functions. Cell 136:215–233

Barth S, Pfuhl T, Mamiani A, Ehses C, Roemer K, Kremmer E, Jaker C, Hock J, Meister G, Grasser FA (2008) Epstein-Barr virus-encoded microRNA miR-BART2 down-regulates the viral DNA polymerase BALF5. Nucleic Acids Res 36:666–675

Barth S, Meister G, Grasser FA (2011) EBV-encoded miRNAs. Biochim Biophys Acta 1809:631–640

Bhat RA, Thimmappaya B (1983) Two small RNAs encoded by Epstein-Barr virus can functionally substitute for the virus-associated RNAs in the lytic growth of adenovirus 5. Proc Natl Acad Sci USA 80:4789–4793

Bloom DC (2004) HSV LAT and neuronal survival. Int Rev Immunol 23:187–198

Boss IW, Nadeau PE, Abbott JR, Yang Y, Mergia A, Renne R (2011) A Kaposi's sarcoma-associated herpesvirus-encoded ortholog of microRNA miR-155 induces human splenic B-cell expansion in NOD/LtSz-scid IL2Rgammanull mice. J Virol 85:9877–9886

Bracken CP, Li X, Wright JA, Lawrence D, Pillman KA, Salmanidis M, Anderson MA, Dredge BK, Gregory PA, Tsykin A, Neilsen C, Thomson DW, Bert AG, Leerberg JM, Yap AS, Jensen KB, Khew-Goodall Y, Goodall GJ (2014) Genome-wide identification of miR-200 targets reveals a regulatory network controlling cell invasion. EMBO J 33:2040–2056

Burk U, Schubert J, Wellner U, Schmalhofer O, Vincan E, Spaderna S, Brabletz T (2008) A reciprocal repression between ZEB1 and members of the miR-200 family promotes EMT and invasion in cancer cells. EMBO Rep 9:582–589

Cahir-Mcfarland ED, Carter K, Rosenwald A, Giltnane JM, Henrickson SE, Staudt LM, Kieff E (2004) Role of NF-kappa B in cell survival and transcription of latent membrane protein 1-expressing or Epstein-Barr virus latency III-infected cells. J Virol 78:4108–4119

Cai X, Schafer A, Lu S, Bilello JP, Desrosiers RC, Edwards R, Raab-Traub N, Cullen BR (2006) Epstein-Barr virus microRNAs are evolutionarily conserved and differentially expressed. PLoS Pathog 2:e23

Cai L, Ye Y, Jiang Q, Chen Y, Lyu X, Li J, Wang S, Liu T, Cai H, Yao K, Li JL, Li X (2015) Epstein-Barr virus-encoded microRNA BART1 induces tumour metastasis by regulating PTEN-dependent pathways in nasopharyngeal carcinoma. Nat Commun 6:7353

Calado DP, Sasaki Y, Godinho SA, Pellerin A, Kochert K, Sleckman BP, de Alboran IM, Janz M, Rodig S, Rajewsky K (2012) The cell-cycle regulator c-Myc is essential for the formation and maintenance of germinal centers. Nat Immunol 13:1092–1100

Cameron JE, Fewell C, Yin Q, McBride J, Wang X, Lin Z, Flemington EK (2008a) Epstein-Barr virus growth/latency III program alters cellular microRNA expression. Virology 382:257–266

Cameron JE, Yin Q, Fewell C, Lacey M, McBride J, Wang X, Lin Z, Schaefer BC, Flemington EK (2008b) Epstein-Barr virus latent membrane protein 1 induces cellular MicroRNA miR-146a, a modulator of lymphocyte signaling pathways. J Virol 82:1946–1958

Carthew RW, Sontheimer EJ (2009) Origins and mechanisms of miRNAs and siRNAs. Cell 136:642–655

Chen H, Huang J, Wu FY, Liao G, Hutt-Fletcher L, Hayward SD (2005) Regulation of expression of the Epstein-Barr virus BamHI-A rightward transcripts. J Virol 79:1724–1733

Chen SJ, Chen GH, Chen YH, Liu CY, Chang KP, Chang YS, Chen HC (2010) Characterization of Epstein-Barr virus miRNAome in nasopharyngeal carcinoma by deep sequencing. PLoS One 5:e12745

Chi SW, Zang JB, Mele A, Darnell RB (2009) Argonaute HITS-CLIP decodes microRNA-mRNA interaction maps. Nature 460:479–486

Choy EY, Siu KL, Kok KH, Lung RW, Tsang CM, To KF, Kwong DL, Tsao SW, Jin DY (2008) An Epstein-Barr virus-encoded microRNA targets PUMA to promote host cell survival. J Exp Med 205:2551–2560

Clarke PA, Schwemmle M, Schickinger J, Hilse K, Clemens MJ (1991) Binding of Epstein-Barr virus small RNA EBER-1 to the double-stranded RNA-activated protein kinase DAI. Nucleic Acids Res 19:243–248

Concha M, Wang X, Cao S, Baddoo M, Fewell C, Lin Z, Hulme W, Hedges D, McBride J, Flemington EK (2012) Identification of new viral genes and transcript isoforms during Epstein-Barr virus reactivation using RNA-Seq. J Virol 86:1458–1467

Corcoran DL, Georgiev S, Mukherjee N, Gottwein E, Skalsky RL, Keene JD, Ohler U (2011) PARalyzer: definition of RNA binding sites from PAR-CLIP short-read sequence data. Genome Biol 12:R79

de Jesus O, Smith PR, Spender LC, Elgueta Karstegl C, Niller HH, Huang D, Farrell PJ (2003) Updated Epstein-Barr virus (EBV) DNA sequence and analysis of a promoter for the BART (CST, BARF0) RNAs of EBV. J Gen Virol 84:1443–1450

Ding Y, Xi Y, Chen T, Wang JY, Tao DL, Wu ZL, Li YP, Li C, Zeng R, Li L (2008) Caprin-2 enhances canonical Wnt signaling through regulating LRP5/6 phosphorylation. J Cell Biol 182:865–872

Dolken L, Malterer G, Erhard F, Kothe S, Friedel CC, Suffert G, Marcinowski L, Motsch N, Barth S, Beitzinger M, Lieber D, Bailer SM, Hoffmann R, Ruzsics Z, Kremmer E, Pfeffer S, Zimmer R, Koszinowski UH, Grasser F, Meister G, Haas J (2010) Systematic analysis of viral and cellular microRNA targets in cells latently infected with human gamma-herpesviruses by RISC immunoprecipitation assay. Cell Host Microbe 7:324–334

Dominguez-Sola D, Victora GD, Ying CY, Phan RT, Saito M, Nussenzweig MC, Dalla-Favera R (2012) The proto-oncogene MYC is required for selection in the germinal center and cyclic reentry. Nat Immunol 13:1083–1091

Edwards RH, Marquitz AR, Raab-Traub N (2008) Epstein-Barr virus BART microRNAs are produced from a large intron prior to splicing. J Virol 82:9094–9106

Ellis-Connell AL, Iempridee T, Xu I, Mertz JE (2010) Cellular microRNAs 200b and 429 regulate the Epstein-Barr virus switch between latency and lytic replication. J Virol 84:10329–10343

Faumont N, Durand-Panteix S, Schlee M, Gromminger S, Schuhmacher M, Holzel M, Laux G, Mailhammer R, Rosenwald A, Staudt LM, Bornkamm GW, Feuillard J (2009) c-Myc and Rel/NF-kappaB are the two master transcriptional systems activated in the latency III program of Epstein-Barr virus-immortalized B cells. J Virol 83:5014–5027

Feederle R, Haar J, Bernhardt K, Linnstaedt SD, Bannert H, Lips H, Cullen BR, Delecluse HJ (2011a) The members of a viral miRNA cluster co-operate to transform B lymphocytes. J Virol 85:9801–9810

Feederle R, Linnstaedt SD, Bannert H, Lips H, Bencun M, Cullen BR, Delecluse HJ (2011b) A viral microRNA cluster strongly potentiates the transforming properties of a human herpesvirus. PLoS Pathog 7:e1001294
Fok V, Friend K, Steitz JA (2006a) Epstein-Barr virus noncoding RNAs are confined to the nucleus, whereas their partner, the human La protein, undergoes nucleocytoplasmic shuttling. J Cell Biol 173:319–325
Fok V, Mitton-Fry RM, Grech A, Steitz JA (2006b) Multiple domains of EBER 1, an Epstein-Barr virus noncoding RNA, recruit human ribosomal protein L22. RNA 12:872–882
Forte E, Luftig MA (2011) The role of microRNAs in Epstein-Barr virus latency and lytic reactivation. Microbes Infect 13:1156–1167
Friedman RC, Farh KK, Burge CB, Bartel DP (2009) Most mammalian mRNAs are conserved targets of microRNAs. Genome Res 19:92–105
Furuse Y, Ornelles DA, Cullen BR (2013) Persistently adenovirus-infected lymphoid cells express microRNAs derived from the viral VAI and especially VAII RNA. Virology 447:140–145
Gardner EJ, Nizami ZF, Talbot CC Jr, Gall JG (2012) Stable intronic sequence RNA (sisRNA), a new class of noncoding RNA from the oocyte nucleus of Xenopus tropicalis. Genes Dev 26:2550–2559
Germain C, Meier A, Jensen T, Knapnougel P, Poupon G, Lazzari A, Neisig A, Hakansson K, Dong T, Wagtmann N, Galsgaard ED, Spee P, Braud VM (2011) Induction of lectin-like transcript 1 (LLT1) protein cell surface expression by pathogens and interferon-gamma contributes to modulate immune responses. J Biol Chem 286:37964–37975
Gilligan K, Sato H, Rajadurai P, Busson P, Young L, Rickinson A, Tursz T, Raab-Traub N (1990) Novel transcription from the Epstein-Barr virus terminal EcoRI fragment, DIJhet, in a nasopharyngeal carcinoma. J Virol 64:4948–4956
Gilligan KJ, Rajadurai P, Lin JC, Busson P, Abdel-Hamid M, Prasad U, Tursz T, Raab-Traub N (1991) Expression of the Epstein-Barr virus BamHI A fragment in nasopharyngeal carcinoma: evidence for a viral protein expressed in vivo. J Virol 65:6252–6259
Goto M, Murakawa M, Kadoshima-Yamaoka K, Tanaka Y, Inoue H, Murafuji H, Hayashi Y, Miura K, Nakatsuka T, Nagahira K, Chamoto K, Fukuda Y, Nishimura T (2009) Phosphodiesterase 7A inhibitor ASB16165 suppresses proliferation and cytokine production of NKT cells. Cell Immunol 258:147–151
Gottwein E, Cai X, Cullen BR (2006) A novel assay for viral microRNA function identifies a single nucleotide polymorphism that affects Drosha processing. J Virol 80:5321–5326
Gottwein E, Mukherjee N, Sachse C, Frenzel C, Majoros WH, Chi JT, Braich R, Manoharan M, Soutschek J, Ohler U, Cullen BR (2007) A viral microRNA functions as an orthologue of cellular miR-155. Nature 450:1096–1099
Gottwein E, Corcoran DL, Mukherjee N, Skalsky RL, Hafner M, Nusbaum JD, Shamulailatpam P, Love CL, Dave SS, Tuschl T, Ohler U, Cullen BR (2011) Viral microRNA targetome of KSHV-infected primary effusion lymphoma cell lines. Cell Host Microbe 10:515–526
Gourzones C, Gelin A, Bombik I, Klibi J, Verillaud B, Guigay J, Lang P, Temam S, Schneider V, Amiel C, Baconnais S, Jimenez AS, Busson P (2010) Extra-cellular release and blood diffusion of BART viral micro-RNAs produced by EBV-infected nasopharyngeal carcinoma cells. Virol J 7:271
Gregorovic G, Bosshard R, Karstegl CE, White RE, Pattle S, Chiang AK, Dittrich-Breiholz O, Kracht M, Russ R, Farrell PJ (2011) Cellular gene expression that correlates with EBER expression in Epstein-Barr Virus-infected lymphoblastoid cell lines. J Virol 85:3535–3545
Grosswendt S, Filipchyk A, Manzano M, Klironomos F, Schilling M, Herzog M, Gottwein E, Rajewsky N (2014) Unambiguous identification of miRNA:target site interactions by different types of ligation reactions. Mol Cell 54:1042–1054
Grundhoff A, Sullivan CS, Ganem D (2006) A combined computational and microarray-based approach identifies novel microRNAs encoded by human gamma-herpesviruses. RNA 12:733–750

Gurer C, Strowig T, Brilot F, Pack M, Trumpfheller C, Arrey F, Park CG, Steinman RM, Munz C (2008) Targeting the nuclear antigen 1 of Epstein-Barr virus to the human endocytic receptor DEC-205 stimulates protective T-cell responses. Blood 112:1231–1239

Haecker I, Gay LA, Yang Y, Hu J, Morse AM, McIntyre LM, Renne R (2012) Ago HITS-CLIP expands understanding of Kaposi's Sarcoma-associated herpesvirus miRNA function in primary effusion lymphomas. PLoS Pathog 8:e1002884

Hafner M, Landthaler M, Burger L, Khorshid M, Hausser J, Berninger P, Rothballer A, Ascano M Jr, Jungkamp AC, Munschauer M, Ulrich A, Wardle GS, Dewell S, Zavolan M, Tuschl T (2010) Transcriptome-wide identification of RNA-binding protein and microRNA target sites by PAR-CLIP. Cell 141:129–141

Haneklaus M, Gerlic M, Kurowska-Stolarska M, Rainey AA, Pich D, McInnes IB, Hammerschmidt W, O'Neill LA, Masters SL (2012) Cutting edge: miR-223 and EBV miR-BART15 regulate the NLRP3 inflammasome and IL-1beta production. J Immunol 189:3795–3799

Hitt MM, Allday MJ, Hara T, Karran L, Jones MD, Busson P, Tursz T, Ernberg I, Griffin BE (1989) EBV gene expression in an NPC-related tumour. EMBO J 8:2639–2651

Howe JG, Shu MD (1989) Epstein-Barr virus small RNA (EBER) genes: unique transcription units that combine RNA polymerase II and III promoter elements. Cell 57:825–834

Hsu CY, Yi YH, Chang KP, Chang YS, Chen SJ, Chen HC (2014) The Epstein-Barr virus-encoded microRNA MiR-BART9 promotes tumor metastasis by targeting E-cadherin in nasopharyngeal carcinoma. PLoS Pathog 10:e1003974

Hutzinger R, Feederle R, Mrazek J, Schiefermeier N, Balwierz PJ, Zavolan M, Polacek N, Delecluse HJ, Huttenhofer A (2009) Expression and processing of a small nucleolar RNA from the Epstein-Barr virus genome. PLoS Pathog 5:e1000547

Iizasa H, Wulff BE, Alla NR, Maragkakis M, Megraw M, Hatzigeorgiou A, Iwakiri D, Takada K, Wiedmer A, Showe L, Lieberman P, Nishikura K (2010) Editing of EBV-encoded BART6 microRNAs controls their dicer targeting and consequently affects viral latency. J Biol Chem 285:33358–33370

Imig J, Motsch N, Zhu JY, Barth S, Okoniewski M, Reineke T, Tinguely M, Faggioni A, Trivedi P, Meister G, Renner C, Grasser FA (2011) microRNA profiling in Epstein-Barr virus-associated B-cell lymphoma. Nucleic Acids Res 39:1880–1893

Jin HY, Oda H, Lai M, Skalsky RL, Bethel K, Shepherd J, Kang SG, Liu WH, Sabouri-Ghomi M, Cullen BR, Rajewsky K, Xiao C (2013) MicroRNA-17 ~ 92 plays a causative role in lymphomagenesis by coordinating multiple oncogenic pathways. EMBO J 32:2377–2391

Jung YJ, Choi H, Kim H, Lee SK (2014) MicroRNA miR-BART20-5p stabilizes Epstein-Barr virus latency by directly targeting BZLF1 and BRLF1. J Virol 88:9027–9037

Kang D, Skalsky RL, Cullen BR (2015) EBV BART MicroRNAs target multiple pro-apoptotic cellular genes to promote epithelial cell survival. PLos Path 11:e1004979

Karran L, Gao Y, Smith PR, Griffin BE (1992) Expression of a family of complementary-strand transcripts in Epstein-Barr virus-infected cells. Proc Natl Acad Sci USA 89:8058–8062

Kieff E (2007) Epstein-Barr virus and its replication. In: Knipe DMaH PM (ed) Fields virology, 5 edn

Kienzle N, Sculley TB, Poulsen L, Buck M, Cross S, Raab-Traub N, Khanna R (1998) Identification of a cytotoxic T-lymphocyte response to the novel BARF0 protein of Epstein-Barr virus: a critical role for antigen expression. J Virol 72:6614–6620

Kienzle N, Buck M, Greco S, Krauer K, Sculley TB (1999) Epstein-Barr virus-encoded RK-BARF0 protein expression. J Virol 73:8902–8906

Kim Do N, Song YJ, Lee SK (2011) The role of promoter methylation in Epstein-Barr virus (EBV) microRNA expression in EBV-infected B cell lines. Exp Mol Med 43:401–410

Kulesza CA, Shenk T (2004) Human cytomegalovirus 5-kilobase immediate-early RNA is a stable intron. J Virol 78:13182–13189

Kulesza CA, Shenk T (2006) Murine cytomegalovirus encodes a stable intron that facilitates persistent replication in the mouse. Proc Natl Acad Sci USA 103:18302–18307

Lee N, Pimienta G, Steitz JA (2012) AUF1/hnRNP D is a novel protein partner of the EBER1 noncoding RNA of Epstein-Barr virus. RNA 18:2073–2082

Lee N, Moss WN, Yario TA, Steitz JA (2015) EBV noncoding RNA binds nascent RNA to drive host PAX5 to viral DNA. Cell 160:607–618

Lei X, Bai Z, Ye F, Xie J, Kim CG, Huang Y, Gao SJ (2010) Regulation of NF-kappaB inhibitor IkappaBalpha and viral replication by a KSHV microRNA. Nat Cell Biol 12:193–199

Lei T, Yuen KS, Xu R, Tsao SW, Chen H, Li M, Kok KH, Jin DY (2013) Targeting of DICE1 tumor suppressor by Epstein-Barr virus-encoded miR-BART3* microRNA in nasopharyngeal carcinoma. Int J Cancer 133:79–87

Lerner MR, Andrews NC, Miller G, Steitz JA (1981) Two small RNAs encoded by Epstein-Barr virus and complexed with protein are precipitated by antibodies from patients with systemic lupus erythematosus. Proc Natl Acad Sci USA 78:805–809

Lewis BP, Burge CB, Bartel DP (2005) Conserved seed pairing, often flanked by adenosines, indicates that thousands of human genes are microRNA targets. Cell 120:15–20

Lin Z, Wang X, Fewell C, Cameron J, Yin Q, Flemington EK (2010) Differential expression of the miR-200 family microRNAs in epithelial and B cells and regulation of Epstein-Barr virus reactivation by the miR-200 family member miR-429. J Virol 84:7892–7897

Lin TC, Liu TY, Hsu SM, Lin CW (2013a) Epstein-Barr virus-encoded miR-BART20-5p inhibits T-bet translation with secondary suppression of p53 in invasive nasal NK/T-cell lymphoma. Am J Pathol 182:1865–1875

Lin Z, Wang X, Strong MJ, Concha M, Baddoo M, Xu G, Baribault C, Fewell C, Hulme W, Hedges D, Taylor CM, Flemington EK (2013b) Whole-genome sequencing of the Akata and Mutu Epstein-Barr virus strains. J Virol 87:1172–1182

Linnstaedt SD, Gottwein E, Skalsky RL, Luftig MA, Cullen BR (2010) Virally induced cellular miR-155 plays a key role in B-cell immortalization by EBV. J Virol 84:11670–11678

Lo AK, To KF, Lo KW, Lung RW, Hui JW, Liao G, Hayward SD (2007) Modulation of LMP1 protein expression by EBV-encoded microRNAs. Proc Natl Acad Sci USA 104:16164–16169

Lo AK, Dawson CW, Jin DY, Lo KW (2012) The pathological roles of BART miRNAs in nasopharyngeal carcinoma. J Pathol 227:392–403

Lu S, Cullen BR (2004) Adenovirus VA1 noncoding RNA can inhibit small interfering RNA and MicroRNA biogenesis. J Virol 78:12868–12876

Luftig M, Prinarakis E, Yasui T, Tsichritzis T, Cahir-Mcfarland E, Inoue J, Nakano H, Mak TW, Yeh WC, Li X, Akira S, Suzuki N, Suzuki S, Mosialos G, Kieff E (2003) Epstein-Barr virus latent membrane protein 1 activation of NF-kappaB through IRAK1 and TRAF6. Proc Natl Acad Sci USA 100:15595–15600

Lukas J, Mazna P, Valenta T, Doubravska L, Pospichalova V, Vojtechova M, Fafilek B, Ivanek R, Plachy J, Novak J, Korinek V (2009) Dazap2 modulates transcription driven by the Wnt effector TCF-4. Nucleic Acids Res 37:3007–3020

Lung RW, Tong JH, Sung YM, Leung PS, Ng DC, Chau SL, Chan AW, Ng EK, Lo KW, To KF (2009) Modulation of LMP2A expression by a newly identified Epstein-Barr virus-encoded microRNA miR-BART22. Neoplasia 11:1174–1184

Lung RW, Tong JH, To KF (2013) Emerging roles of small Epstein-Barr virus derived non-coding RNAs in epithelial malignancy. Int J Mol Sci 14:17378–17409

Majoros WH, Lekprasert P, Mukherjee N, Skalsky RL, Corcoran DL, Cullen BR, Ohler U (2013) MicroRNA target site identification by integrating sequence and binding information. Nat Methods 10:630–633

Marquitz AR, Raab-Traub N (2012) The role of miRNAs and EBV BARTs in NPC. Semin Cancer Biol 22:166–172

Marquitz AR, Mathur A, Nam CS, Raab-Traub N (2011) The Epstein-Barr Virus BART microRNAs target the pro-apoptotic protein Bim. Virology 412:392–400

Marquitz AR, Mathur A, Shair KH, Raab-Traub N (2012) Infection of Epstein-Barr virus in a gastric carcinoma cell line induces anchorage independence and global changes in gene expression. Proc Natl Acad Sci USA

Marquitz AR, Mathur A, Chugh PE, Dittmer DP, Raab-Traub N (2014) Expression profile of microRNAs in Epstein-Barr virus-infected AGS gastric carcinoma cells. J Virol 88:1389–1393

Meckes DG Jr, Shair KH, Marquitz AR, Kung CP, Edwards RH, Raab-Traub N (2010) Human tumor virus utilizes exosomes for intercellular communication. Proc Natl Acad Sci USA 107:20370–20375

Michaeli T, Pan ZQ, Prives C (1988) An excised SV40 intron accumulates and is stable in Xenopus laevis oocytes. Genes Dev 2:1012–1020

Moody R, Zhu Y, Huang Y, Cui X, Jones T, Bedolla R, Lei X, Bai Z, Gao SJ (2013) KSHV microRNAs mediate cellular transformation and tumorigenesis by redundantly targeting cell growth and survival pathways. PLoS Pathog 9:e1003857

Moss WN, Steitz JA (2013) Genome-wide analyses of Epstein-Barr virus reveal conserved RNA structures and a novel stable intronic sequence RNA. BMC Genom 14:543

Moynagh PN (2009) The Pellino family: IRAK E3 ligases with emerging roles in innate immune signalling. Trends Immunol 30:33–42

Mullokandov G, Baccarini A, Ruzo A, Jayaprakash AD, Tung N, Israelow B, Evans MJ, Sachidanandam R, Brown BD (2012) High-throughput assessment of microRNA activity and function using microRNA sensor and decoy libraries. Nat Methods 9:840–846

Nachmani D, Stern-Ginossar N, Sarid R, Mandelboim O (2009) Diverse herpesvirus microRNAs target the stress-induced immune ligand MICB to escape recognition by natural killer cells. Cell Host Microbe 5:376–385

Nanbo A, Inoue K, Adachi-Takasawa K, Takada K (2002) Epstein-Barr virus RNA confers resistance to interferon-alpha-induced apoptosis in Burkitt's lymphoma. EMBO J 21:954–965

Nikitin PA, Yan CM, Forte E, Bocedi A, Tourigny JP, White RE, Allday MJ, Patel A, Dave SS, Kim W, Hu K, Guo J, Tainter D, Rusyn E, Luftig MA (2010) An ATM/Chk2-mediated DNA damage-responsive signaling pathway suppresses Epstein-Barr virus transformation of primary human B cells. Cell Host Microbe 8:510–522

O'Grady T, Cao S, Strong MJ, Concha M, Wang X, Splinter Bondurant S, Adams M, Baddoo M, Srivastav SK, Lin Z, Fewell C, Yin Q, Flemington EK (2014) Global bidirectional transcription of the Epstein-Barr virus genome during reactivation. J Virol 88:1604–1616

Parnas O, Corcoran DL, Cullen BR (2014) Analysis of the mRNA targetome of microRNAs expressed by Marek's disease virus. MBio 5:e01060-13

Pegtel DM, Cosmopoulos K, Thorley-Lawson DA, van Eijndhoven MA, Hopmans ES, Lindenberg JL, de Gruijl TD, Wurdinger T, Middeldorp JM (2010) Functional delivery of viral miRNAs via exosomes. Proc Natl Acad Sci USA 107:6328–6333

Pekarsky Y, Croce CM (2010) Is miR-29 an oncogene or tumor suppressor in CLL? Oncotarget 1:224–227

Pfeffer S, Zavolan M, Grasser FA, Chien M, Russo JJ, Ju J, John B, Enright AJ, Marks D, Sander C, Tuschl T (2004) Identification of virus-encoded microRNAs. Science 304:734–736

Pillai RS, Bhattacharyya SN, Filipowicz W (2007) Repression of protein synthesis by miRNAs: how many mechanisms? Trends Cell Biol 17:118–126

Pratt ZL, Kuzembayeva M, Sengupta S, Sugden B (2009) The microRNAs of Epstein-Barr virus are expressed at dramatically differing levels among cell lines. Virology 386:387–397

Pratt ZL, Zhang J, Sugden B (2012) The latent membrane protein 1 (LMP1) oncogene of Epstein-Barr virus can simultaneously induce and inhibit apoptosis in B cells. J Virol 86:4380–4393

Price AM, Tourigny JP, Forte E, Salinas RE, Dave SS, Luftig MA (2012) Analysis of Epstein-Barr virus-regulated host gene expression changes through primary B-cell outgrowth reveals delayed kinetics of latent membrane protein 1-mediated NF-kappaB activation. J Virol 86:11096–11106

Qiu J, Thorley-Lawson DA (2014) EBV microRNA BART 18-5p targets MAP3K2 to facilitate persistence in vivo by inhibiting viral replication in B cells. Proc Natl Acad Sci USA 111:11157–11162

Qiu J, Cosmopoulos K, Pegtel M, Hopmans E, Murray P, Middeldorp J, Shapiro M, Thorley-Lawson DA (2011) A novel persistence associated EBV miRNA expression profile is disrupted in neoplasia. PLoS Pathog 7:e1002193

Ramachandran I, Ganapathy V, Gillies E, Fonseca I, Sureban SM, Houchen CW, Reis A, Queimado L (2014) Wnt inhibitory factor 1 suppresses cancer stemness and induces cellular senescence. Cell Death Dis 5:e1246

Rechavi O, Erlich Y, Amram H, Flomenblit L, Karginov FV, Goldstein I, Hannon GJ, Kloog Y (2009) Cell contact-dependent acquisition of cellular and viral nonautonomously encoded small RNAs. Genes Dev 23:1971–1979

Repellin CE, Tsimbouri PM, Philbey AW, Wilson JB (2010) Lymphoid hyperplasia and lymphoma in transgenic mice expressing the small non-coding RNA, EBER1 of Epstein-Barr virus. PLoS ONE 5:e9092

Riley KJ, Rabinowitz GS, Steitz JA (2010) Comprehensive analysis of Rhesus lymphocryptovirus microRNA expression. J Virol 84:5148–5157

Riley KJ, Rabinowitz GS, Yario TA, Luna JM, Darnell RB, Steitz JA (2012) EBV and human microRNAs co-target oncogenic and apoptotic viral and human genes during latency. EMBO J 31:2207–2221

Robertson ES, Tomkinson B, Kieff E (1994) An Epstein-Barr virus with a 58-kilobase-pair deletion that includes BARF0 transforms B lymphocytes in vitro. J Virol 68:1449–1458

Rodriguez A, Vigorito E, Clare S, Warren MV, Couttet P, Soond DR, van Dongen S, Grocock RJ, Das PP, Miska EA, Vetrie D, Okkenhaug K, Enright AJ, Dougan G, Turner M, Bradley A (2007) Requirement of bic/microRNA-155 for normal immune function. Science 316:608–611

Ruf IK, Lackey KA, Warudkar S, Sample JT (2005) Protection from interferon-induced apoptosis by Epstein-Barr virus small RNAs is not mediated by inhibition of PKR. J Virol 79:14562–14569

Samanta M, Iwakiri D, Kanda T, Imaizumi T, Takada K (2006) EB virus-encoded RNAs are recognized by RIG-I and activate signaling to induce type I IFN. EMBO J 25:4207–4214

Sano M, Kato Y, Taira K (2006) Sequence-specific interference by small RNAs derived from adenovirus VAI RNA. FEBS Lett 580:1553–1564

Seto E, Moosmann A, Gromminger S, Walz N, Grundhoff A, Hammerschmidt W (2010) Micro RNAs of Epstein-Barr virus promote cell cycle progression and prevent apoptosis of primary human B cells. PLoS Pathog 6

Shinozaki A, Sakatani T, Ushiku T, Hino R, Isogai M, Ishikawa S, Uozaki H, Takada K, Fukayama M (2010) Downregulation of microRNA-200 in EBV-associated gastric carcinoma. Cancer Res 70:4719–4727

Skalsky RL, Cullen BR (2010) Viruses, microRNAs, and host interactions. Annu Rev Microbiol 64:123–141

Skalsky RL, Hu J, Renne R (2007a) Analysis of viral cis elements conferring Kaposi's sarcoma-associated herpesvirus episome partitioning and maintenance. J Virol 81:9825–9837

Skalsky RL, Samols MA, Plaisance KB, Boss IW, Riva A, Lopez MC, Baker HV, Renne R (2007b) Kaposi's sarcoma-associated herpesvirus encodes an ortholog of miR-155. J Virol 81:12836–12845

Skalsky RL, Corcoran DL, Gottwein E, Frank CL, Kang D, Hafner M, Nusbaum JD, Feederle R, Delecluse HJ, Luftig MA, Tuschl T, Ohler U, Cullen BR (2012) The viral and cellular microRNA targetome in lymphoblastoid cell lines. PLoS Pathog 8:e1002484

Skalsky RL, Kang D, Linnstaedt SD, Cullen BR (2014) Evolutionary conservation of primate lymphocryptovirus microRNA targets. J Virol 88:1617–1635

Smith PR, de Jesus O, Turner D, Hollyoake M, Karstegl CE, Griffin BE, Karran L, Wang Y, Hayward SD, Farrell PJ (2000) Structure and coding content of CST (BART) family RNAs of Epstein-Barr virus. J Virol 74:3082–3092

Soni V, Cahir-Mcfarland E, Kieff E (2007) LMP1 TRAFficking activates growth and survival pathways. Adv Exp Med Biol 597:173–187

Swaminathan S, Tomkinson B, Kieff E (1991) Recombinant Epstein-Barr virus with small RNA (EBER) genes deleted transforms lymphocytes and replicates in vitro. Proc Natl Acad Sci USA 88:1546–1550

Talhouarne GJ, Gall JG (2014) Lariat intronic RNAs in the cytoplasm of Xenopus tropicalis oocytes. RNA 20:1476–1487

Tavalai N, Stamminger T (2009) Interplay between herpesvirus infection and host defense by PML nuclear bodies. Viruses 1:1240–1264

Thai TH, Calado DP, Casola S, Ansel KM, Xiao C, Xue Y, Murphy A, Frendewey D, Valenzuela D, Kutok JL, Schmidt-Supprian M, Rajewsky N, Yancopoulos G, Rao A, Rajewsky K (2007) Regulation of the germinal center response by microRNA-155. Science 316:604–608

Toczyski DP, Matera AG, Ward DC, Steitz JA (1994) The Epstein-Barr virus (EBV) small RNA EBER1 binds and relocalizes ribosomal protein L22 in EBV-infected human B lymphocytes. Proc Natl Acad Sci USA 91:3463–3467

Valadi H, Ekstrom K, Bossios A, Sjostrand M, Lee JJ, Lotvall JO (2007) Exosome-mediated transfer of mRNAs and microRNAs is a novel mechanism of genetic exchange between cells. Nat Cell Biol 9:654–659

Vereide DT, Seto E, Chiu YF, Hayes M, Tagawa T, Grundhoff A, Hammerschmidt W, Sugden B (2014) Epstein-Barr virus maintains lymphomas via its miRNAs. Oncogene 33:1258–1264

Wahl A, Linnstaedt SD, Esoda C, Krisko JF, Martinez-Torres F, Delecluse HJ, Cullen BR, Garcia JV (2013) A cluster of virus-encoded microRNAs accelerates acute systemic Epstein-Barr virus infection but does not significantly enhance virus-induced oncogenesis in vivo. J Virol 87:5437–5446

Walsh KB, Marsolais D, Welch MJ, Rosen H, Oldstone MB (2010) Treatment with a sphingosine analog does not alter the outcome of a persistent virus infection. Virology 397:260–269

Walz N, Christalla T, Tessmer U, Grundhoff A (2009) A global analysis of evolutionary conservation among known and predicted gammaherpesvirus microRNAs. J Virol 84:716–728

White EJ, Brewer G, Wilson GM (2013) Post-transcriptional control of gene expression by AUF1: mechanisms, physiological targets, and regulation. Biochim Biophys Acta 1829:680–688

Wong AM, Kong KL, Tsang JW, Kwong DL, Guan XY (2012) Profiling of Epstein-Barr virus-encoded microRNAs in nasopharyngeal carcinoma reveals potential biomarkers and oncomirs. Cancer 118:698–710

Wu Y, Maruo S, Yajima M, Kanda T, Takada K (2007) Epstein-Barr virus (EBV)-encoded RNA 2 (EBER2) but not EBER1 plays a critical role in EBV-induced B-cell growth transformation. J Virol 81:11236–11245

Xia T, O'Hara A, Araujo I, Barreto J, Carvalho E, Sapucaia JB, Ramos JC, Luz E, Pedroso C, Manrique M, Toomey NL, Brites C, Dittmer DP, Harrington WJ Jr (2008) EBV microRNAs in primary lymphomas and targeting of CXCL-11 by ebv-mir-BHRF1-3. Cancer Res 68:1436–1442

Xiao C, Rajewsky K (2009) MicroRNA control in the immune system: basic principles. Cell 136:26–36

Xing L, Kieff E (2007) Epstein-Barr virus BHRF1 micro- and stable RNAs during latency III and after induction of replication. J Virol 81:9967–9975

Xing L, Kieff E (2011) cis-Acting effects on RNA processing and Drosha cleavage prevent Epstein-Barr virus latency III BHRF1 expression. J Virol 85:8929–8939

Yajima M, Kanda T, Takada K (2005) Critical role of Epstein-Barr Virus (EBV)-encoded RNA in efficient EBV-induced B-lymphocyte growth transformation. J Virol 79:4298–4307

Yang J, Yuan YA (2009) A structural perspective of the protein-RNA interactions involved in virus-induced RNA silencing and its suppression. Biochim Biophys Acta 1789:642–652
Yao Y, Zhao Y, Xu H, Smith LP, Lawrie CH, Watson M, Nair V (2008) MicroRNA profile of Marek's disease virus-transformed T-cell line MSB-1: predominance of virus-encoded microRNAs. J Virol 82:4007–4015
Young LS, Rickinson AB (2004) Epstein-Barr virus: 40 years on. Nat Rev Cancer 4:757–768
Zhang J, Chen H, Weinmaster G, Hayward SD (2001) Epstein-Barr virus BamHi-a rightward transcript-encoded RPMS protein interacts with the CBF1-associated corepressor CIR to negatively regulate the activity of EBNA2 and NotchIC. J Virol 75:2946–2956
Zhao Y, Xu H, Yao Y, Smith LP, Kgosana L, Green J, Petherbridge L, Baigent SJ, Nair V (2011) Critical role of the virus-encoded microRNA-155 ortholog in the induction of Marek's disease lymphomas. PLoS Pathog 7:e1001305
Zhu JY, Pfuhl T, Motsch N, Barth S, Nicholls J, Grasser F, Meister G (2009) Identification of novel Epstein-Barr virus microRNA genes from nasopharyngeal carcinomas. J Virol 83:3333–3341

# Part II
# Lytic EBV Infection

# Viral Entry

**Liudmila S. Chesnokova, Ru Jiang and Lindsey M. Hutt-Fletcher**

**Abstract** Epstein-Barr virus primarily, though not exclusively, infects B cells and epithelial cells. Many of the virus and cell proteins that are involved in entry into these two cell types in vitro have been identified, and their roles in attachment and fusion are being explored. This chapter discusses what is known about entry at the cellular level in vitro and describes what little is known about the process in vivo. It highlights some of the questions that still need to be addressed and considers some models that need further testing.

## Contents



L.S. Chesnokova · R. Jiang · L.M. Hutt-Fletcher (✉)
Department of Microbiology and Immunology, Center for Molecular and Tumor Virology, Feist-Weiller Cancer Center, Louisiana State University Health Sciences Center, 1501 Kings Highway, Shreveport, LA 71130, USA
e-mail: lhuttf@lsuhsc.edu

Present R. Jiang
Department of Clinical Teaching and Training, Tianjin University of Traditional Chinese Medicine, 312 West Anshan Road, 300193 Nankai District, Tianjin, China



C. Münz (ed.), *Epstein Barr Virus Volume 2*, Current Topics in Microbiology and Immunology 391, DOI 10.1007/978-3-319-22834-1_7

## Abbreviations

| | |
|---|---|
| EBV | Epstein-Barr virus |
| NK | Natural killer |
| CR2 | Complement receptor type 2 |
| CR1 | Complement receptor type 1 |
| SCR | Short consensus repeat |
| CTLD | C-type lectin domain |

## 1 Introduction

Epstein-Barr virus (EBV) is shed in the saliva of persistently infected healthy carriers and is generally described as being orally transmitted. It ultimately establishes latency in the memory B cell compartment, but is also thought to replicate productively in epithelial cells, can, at least rarely, infect T cells and natural killer (NK) cells, and is found in the muscle cells of leiomyosarcomas. A full comprehension of virus entry, which encompasses access to vulnerable tissues at the organismal level as well as delivery of virus DNA to the nucleus at the cellular level, is nowhere near a reality. However, progress has been made, particularly in deciphering early events in entry into the two major target cells of the virus, B cells and epithelial cells. This chapter discusses what we know about these early events and starts to put them in the context of the broader picture of infection, transmission, and spread.

## 2 Early Events in B Cell Entry

### *2.1 Proteins Involved in Attachment*

It was recognized early on that EBV is a B lymphotropic virus, and this, together with the fact that human B cells were relatively easily obtained and intensively studied, focused attention on identification of the virus and cell proteins important for B cell entry. Investigations were also simplified and facilitated by the efficient infection of B cells that could be achieved with cell-free virus.

Virus was first shown to attach to B cells as a result of a high-affinity interaction (Moore et al. 1989) between the abundant virion glycoprotein gp350 and complement receptor type 2, CR2 or CD21 (Fingeroth et al. 1984; Nemerow et al. 1987; Tanner et al. 1987). Glycoprotein gp350 is a single-pass, heavily glycosylated, type I membrane protein of 907 amino acids, and the CD21 binding site has been mapped to a glycan-free surface in the membrane-distal amino-terminal domain of the protein (Martin et al. 1991; Szakonyi et al. 2006). It includes a peptide sequence with similarities to the natural ligand of CD21, the C3dg fragment of complement (Nemerow et al. 1989). The ectodomain of CD21 consists of tandem repeats of modules of 60–75 amino acids known as short consensus repeats (SCRs), and the binding site for gp350 is contained in SCR1 and SCR2 (Martin et al. 1991). These SCRs are those furthest from the cell membrane, and, somewhat problematically, attachment initially positions the virus approximately 50 nm from the cell surface (Nemerow and Cooper 1984). However, the EBV virion also contains a splice variant of gp350, gp220, missing residues 500–757, but retaining the CD21 binding site (Beisel et al. 1985). Whether there is a switch from the use of gp350 to gp220 is not known, but perhaps this, together with the segmental flexibility of CD21 provided by the tandemly repeated SCRs, allows the virus to move closer to the cell following its initial attachment.

On primary B cells CD21 can exist independently, but is additionally found in a trimeric complex with CD19 and CD81 and in a complex with another complement receptor, complement receptor type 1, CR1 or CD35 (Tuveson et al. 1991). CD35 is also a ligand for gp350 that is capable of initiating infection (Ogembo et al. 2013). It possibly binds to the same glycan-free surface as has been shown to interact with CD21, though, since CD35 is lost when B cells are transformed with EBV, its properties as a receptor have only been discovered recently and have not been extensively studied. The trimeric CD21/CD19/CD81 complex is particularly interesting for two reasons. First, CD19 functions as a signal transducer. Cross-linking of CD21 by gp350, as virus attaches, can then activate NF-κB (Sinclair and Farrell 1995; Sugano et al. 1997) and protein kinase C pathways and induce interleukin-6 production (D'Addario et al. 2001; Tanner et al. 1996). Although signaling may have no relevance to entry per se, it may profoundly influence downstream events. Second, CD81 and CD19 can both associate with HLA class II (Reem et al. 2004), a cell protein required for the next step in entry as discussed below.

## 2.2 *Proteins Involved in Internalization*

Once attached to the B cell surface via CD21 virus is endocytosed into a low pH compartment from which it must ultimately escape by fusing its envelope with the vesicle membrane (Miller and Hutt-Fletcher 1992). Fusion requires four virus glycoproteins, gH, gL, gp42 and gB, and one cell protein, HLA class II (reviewed in (Hutt-Fletcher 2007). Glycoprotein gH, which has an apparent molecular weight of 85 kilodaltons and in the older literature is referred to as gp85, is, like gp350, a single-pass type I membrane protein. Glycoprotein gL (formerly gp25) is a peripheral membrane protein from which the signal peptide is cleaved and it dimerizes with gH. The crystal structure of gH and gL (Matsuura et al. 2010) reveals a four-domain cylindrical complex in which the membrane-distal globular domain I is comprised of the amino terminus of gH and the entire cleaved gL. Coexpression of gH and gL, from hereon referred to as a single entity, gHgL, is thus required for the correct folding of the structure.

Herpesvirus glycoprotein nomenclature has now settled on naming of molecules conserved throughout the entire herpesvirus family by letters of the alphabet. Proteins not found in all subfamilies are referred to by either their apparent molecular weights or their gene names. The beta and gammaherpesviruses, but not apparently the alphaherpesviruses, have proteins that associate with gHgL. The first described was EBV glycoprotein gp42 (Li et al. 1995). Glycoprotein gp42 is a type II membrane protein which exists in virions in a cleaved and an uncleaved form, although the cleaved form is functionally optimal (Sorem et al. 2009). It associates non-covalently with gHgL, probably with domain II and the domain I/domain II interface (Chen et al. 2012; Sathiyamoorthy et al. 2014), through residues 36–82 in its amino-terminal region (Kirschner et al. 2007). It binds to the β1 domain of HLA class II, at the side of the peptide binding groove, via a C-type lectin domain (CTLD) at the carboxyl terminus (Mullen et al. 2002), and the interaction with HLA class II is required for the initiation of fusion (Haan et al. 2000; Li et al. 1997). A hydrophobic pocket within the CTLD domain also makes contact with gHgL at the junction of domains II and III (Sathiyamoorthy et al. 2014).

Like gH and gp350, glycoprotein gB (referred to in the early literature as gp110 or gp125, the glycosylated form carrying complex sugars) is a single-pass type I transmembrane protein. It exists as a trimer, and the crystal structure of gB closely resembles not only that of its HSV counterpart, but also the class III fusion proteins of vesicular stomatitis virus and baculovirus (Backovic et al. 2009). The gB homologs of all herpesviruses are now generally considered to be the final executors of fusion, as discussed below.

# 3 Early Events in Epithelial Cell Entry

## *3.1 Proteins Involved in Attachment*

Attachment of EBV to an epithelial cell is a more complicated issue than B cell attachment as a variety of virus glycoproteins, and cell proteins have been implicated in the process. Some epithelial cells in culture express at least low levels of CD21 (Fingeroth et al. 1999), and both these and cells engineered to express CD21 can be infected at high levels (Borza et al. 2004; Li et al. 1992; Valencia and Hutt-Fletcher 2012). Determination of whether epithelial cells in vivo express CD21 has, however, been confounded by the fact that only a subset of antibodies that recognize CD21 on B cells is reactive with all epithelial cells, reports from different groups using them have not always been consistent (Levine et al. 1990; Niedobitek et al. 1989; Talacko et al. 1991; Thomas and Crawford 1989), and one commonly used antibody cross-reacts with an unrelated epithelial protein (Young et al. 1989). Analysis of expression of transcripts in epithelial cells isolated by laser capture microdissection found CD21 mRNA in tonsil and adenoid epithelium, but not elsewhere, although tissues that when normal failed to express CD21 became positive as they became dysplastic and expression levels correlated with an increase in the grade of dysplasia (Jiang et al. 2008, 2012). This may have implications for the development of nasopharyngeal carcinoma. Individuals who develop these tumors have increases in antibodies to lytic cycle proteins several years prior to diagnosis (Zeng 1985; Zeng et al. 1985), and the role of EBV has been postulated to be that of a tumor promoter rather than a tumor initiator (Lo et al. 2012).

In the absence of CD21, EBV can use gHgL, which includes a KGD motif, to bind to any one of a subset of αv integrins, αvβ5, αvβ6, or αvβ8 (Chesnokova and Hutt-Fletcher 2011; Chesnokova et al. 2009). However, infection by virus attached via gHgL and an integrin is not at all efficient, for reasons that are not yet clear. One possibility is that the use of gHgL for attachment compromises its ability to function in fusion (Borza et al. 2004). A second is that CD21 does more than simply tether virus to the cell surface. Cross-linking of CD21 on an epithelial cell by virus has been reported to result in an interaction between its cytoplasmic tail and the formin homolog overexpressed in spleen (FHOS/FHOD1) (Gill et al. 2004). Formins are scaffolding proteins that nucleate actin and link signal transduction to actin reorganization, which is required for intracellular transport of EBV to the nucleus. However, both a truncated form of CD21 lacking its cytoplasmic domain (Valencia and Hutt-Fletcher 2012) and a construct in which both the cytoplasmic domain and the transmembrane domain were replaced with those of gH (unpublished) support infection as efficiently as the full-length protein. Whether there are interactions between the ectodomain of CD21 and other epithelial surface molecules needs further exploration. There is precedent for this, for while neither CD35 nor CD19 is found on epithelial cells, and their interactions with CD21 on a B cell can occur through the ectodomains of each (Fearon and Carter 1995).

A second interaction between EBV and an integrin has also been implicated in attachment, this time between an RGD motif in the BMRF2 gene product, a multi-span membrane glycoprotein, and αv, α3, α5, and β1 integrins (Tugizov et al. 2003; Xiao et al. 2008). Its major impact seems to be in infection of polarized cells, although how much the role of pBMRF2 reflects an essential involvement in attachment and how much it reflects effects on cell-to-cell spread (Xiao et al. 2009) are not entirely clear. pBMRF2 forms a dimeric complex with the BDLF2 gene product, a type II membrane protein (Gore and Hutt-Fletcher 2008). Homologs of these two proteins are found in other gammaherpesviruses, and, in the murine gammaherpesvirus, MHV68, they have been implicated in a membrane remodeling that is important to virus spread (Gill et al. 2008).

In immune individuals, virus coated with immunoglobulin A can attach to the basolateral surfaces of epithelial cells via the polymeric IgA receptor (Sixbey and Yao 1992), although in polarized cells this leads to transcytosis rather than infection (Gan et al. 1997). Virus bound to CD21 on the surface of a B cell can also be transferred to a CD21-negative epithelial cell (Shannon-Lowe and Rowe 2011; Shannon-Lowe et al. 2006). Following EBV attachment, CD21 co-caps with and activates adhesion molecules which allow the B cell to form a virologic synapse with an epithelial cell. Finally, there is a very low affinity, but saturable interaction, possibly involving hydrogen and ionic bonds, between the BDLF3 gene product gp150 and an unknown molecule(s) on the surface of an epithelial cell (unpublished). However, virus lacking gp150, which is a highly glycosylated mucin-like protein, is very slightly more infectious for an epithelial cell than is wild-type virus (Borza and Hutt-Fletcher 1998) so this particular interaction would not appear to be productive.

## 3.2 Virus Proteins Involved in Internalization

Definitive information about internalization of virus has also been difficult to obtain because of the plasticity of epithelial cells in culture, because of the different behaviors of unpolarized, polarized, and stratified epithelial cells, and because, since it is often not feasible to obtain primary cultures, cancer cell lines have frequently been used. Fusion with primary foreskin epithelial cells occurs at neutral pH and unlike B cell fusion does not require endocytosis (Miller and Hutt-Fletcher 1992). Entry into the SVKCR2 cell line, SV40-transformed keratinocytes engineered to express CD21 (Li et al. 1992), the AGS gastric cancer cell line, and hTERT-immortalized normal oral keratinocytes does not require actin remodeling, which is also consistent with fusion at the cell surface (Valencia and Hutt-Fletcher 2012). However, whether or not entry into polarized or stratified epithelial cells follows the same route has not yet been reported.

It is known, however, that fusion with an epithelial cell requires a complement of virus and cell proteins different from those used for B cell fusion. This has allowed EBV to evolve an elegant strategy for switching tropism and cycling between its two major target cells. Glycoproteins gB and gHgL, sometimes referred to as the "core fusion machinery" of herpesviruses, are necessary for virus fusion with an epithelial cell, as they are for fusion with a B cell, but gp42 is not. Fusion is triggered not by HLA class II, which is not constitutively expressed on an epithelial cell, but by an interaction between gHgL and one of the three αv integrins to which it binds (Chesnokova and Hutt-Fletcher 2011; Hutt-Fletcher and Chesnokova 2010). The interaction between gHgL and an integrin is blocked if gp42 is present, and access to both cell types is only possible because EBV carries both three-part gHgLgp42 complexes and two-part gHgL complexes in the virion envelope (Wang et al. 1998; Chen et al. 2012). Virus lacking gp42 can only infect a B cell if a soluble form of gp42, which can reform trimeric complexes, is added in trans, and the same soluble gp42, if added in saturating amounts to wild-type virus, can block infection of an epithelial cell (Wang et al. 1998). Virus produced in B cells carries a reduced number of the trimeric complexes because some bind to HLA class II in the endoplasmic reticulum and travel with HLA class II to the peptide-loading compartment, which is rich in proteases. This does not happen in HLA class II-negative epithelial cells, which produce virus that is enriched for trimeric complexes and which is as much as a hundred-fold more infectious for B cells than virus produced by B cells themselves. In turn, B cell virus is about five-fold more infectious for an epithelial cell than virus replicated in an epithelial cell. HLA class II and gp42 thus provide the mechanism for alternating replication in the two cell types (Borza and Hutt-Fletcher 2002).

## 4 Mechanisms of Fusion

Although still incompletely understood, the individual roles of the virus glycoproteins involved in fusion are becoming clearer and models of fusion are being developed and tested. Probably, the major breakthrough came when the crystal structures of first herpes simplex virus gB and then EBV gB were solved (Backovic et al. 2009; Heldwein et al. 2006). The striking resemblance of both proteins to the post-fusion conformations of other class III fusion proteins strongly supported the nascent hypothesis that gB was central and proximal to the event. Class III fusion proteins exist as rod-shaped trimers, and each monomer of gB consists of five domains (Backovic and Jardetzky 2009). Domain I, modeled as closest to the virus membrane, is described as a fusion module and contains two fusion loops which are thought to insert into the cell membrane as fusion progresses. Mutations in the putative fusion loops of EBV gB abrogate fusion (Backovic et al. 2007a), and the same hydrophobic residues enable a truncated form of gB to form rosettes that are also typical of class I and class II fusion proteins in their post-fusion form (Backovic et al. 2007b). Domains II and IV contain β-sheets, and domain IV of EBV gB, which is found at the top of the trimeric spike, is thought to be flexible. Domain III has a long α-helix which in the trimer is part of a central coiled-coil. Domain V is an extended segment that inserts between the other two units in the trimer. The class III fusion protein of vesicular stomatitis virus, which is triggered by exposure to low pH, has been crystallized in both its post-fusion and its pre-fusion conformation and significant, but reversible, refolding, such as is seen in class I fusion proteins, occurs as one transitions to the other (Roche et al. 2008). Exposure of EBV to a triggering integrin leads to a change in its proteolytic digestion pattern, suggesting that a, perhaps analogous, conformational shift occurs during EBV fusion (Chesnokova et al. 2014).

The role of gHgL with or without gp42 is now generally thought to be as a regulator rather than an effector of fusion per se. The structure of gHgL resembles that of no known fusion protein, a gB truncation mutant can mediate some epithelial cell fusion in the absence of gHgL (McShane and Longnecker 2004), and heat can act as a partial surrogate for a gHgL interaction with an integrin, triggering the same change in the proteolytic digestion pattern of gB (Chesnokova et al. 2014). Clearly, however, under normal circumstances, an interaction between gHgLgp42 and HLA class II or an interaction between gHgL and an αv integrin is essential.

The crystal structure of gp42 has been solved in both the presence and absence of HLA class II. Mutational analysis of gp42 had highlighted the functional importance of a hydrophobic pocket within the CTLD (Silva et al. 2004), and comparison of the liganded and unliganded structures of gp42 indicated a small change in the hydrophobic pocket in the liganded form (Kirschner et al. 2009). It was originally suggested that this change, triggered by HLA class II binding, might affect a second interaction between gp42 and either gB or gHgL. The second interaction is now known to be with gHgL, and the introduction of an N-liked glycosylation site in gHgL at this interface reduces membrane fusion (Sathiyamoorthy et al. 2014). A model is developing in which binding of gp42 to HLA class II causes the hydrophobic pocket in the protein to widen, and this change is transmitted via gHgL to gB. There is evidence from work done with herpes simplex virus and human cytomegalovirus that gHgL interacts with gB under conditions where fusion is possible (Atanasiu et al. 2007; Avitabile et al. 2007; Cairns et al. 2011; Vanarsdall et al. 2008), although whether this interaction occurs before or as fusion is triggered is equivocal. Whether or not EBV gB and gHgL interact has not yet been assessed.

A conformational change in gHgL between the domain I/domain II interface was also suggested when mutations affecting fusion were mapped to the crystal structure (Matsuura et al. 2010). There is a single unpaired cysteine residue in the groove between the two domains which allowed coupling of thiol-reactive, environmentally sensitive, fluorescent probes to a soluble truncated form of gHgL. Addition of a soluble αv integrin, which on its own can trigger virus fusion with an epithelial cell, produced a conformational change that could then be detected by fluorescence spectroscopy (Chesnokova and Hutt-Fletcher 2011). Subsequently, additional mutations were made in gHgL which introduced a novel disulfide bond linking domain I and domain II, presumably constraining such a conformational change (Chen et al. 2013). The mutated protein could still mediate fusion with a B cell, but lost the ability to fuse with an epithelial cell. This parallels other mutations made in domain IV of gHgL which differentially affect fusion with a B cell and an epithelial cell (Wu et al. 2005; Wu and Hutt-Fletcher 2007). Though all mutations have been functionally tested only in the context of cell-based fusion assays, three monoclonal antibodies to gHgL, one which binds to domain IV, one which binds close to the domain I/domain II interface, and one which has not yet been mapped, all neutralize epithelial infection very efficiently, but have little to no effect on B cell infection (Chesnokova and Hutt-Fletcher 2011; Wu et al. 2005) suggesting that the observations are probably relevant to virus entry as well.

The further implication of the observations would seem to be that, as they interact with their cellular partners, the architecture of the two fusion complexes, gB and gHgL or gB and gHgLgp42, the latter of which has been partially described (Sathiyamoorthy et al. 2014), is different. What this means for how gHgL interfaces with gB is unclear, although it seems likely that the final events in fusion are similar for both cell types. Fusion can be achieved if gHgL and gB are expressed in trans as well as in cis, and a virus lacking gHgL can enter an epithelial cell expressing gHgL. Infection at low levels can also be achieved in a B cell expressing gHgL without gp42, if a soluble integrin is added, providing some support for the concept that, once activated for fusion, gB proceeds similarly in both cell types (Chesnokova et al. 2014).

## 5 Transit to the Nucleus

The occurrence of fusion at different sites in a B cell and an epithelial cell, at least in the epithelial cells studied so far, has potential implications for transport of virus into the nucleus. A virus entering by endocytosis avoids the problem of crossing the actin cortex and is in a somewhat protected environment until it fuses out of the vesicle. There is precedent in other virus systems for vesicular transport well into the cell, although where precisely EBV fuses, is not known. Microtubules are required for delivery of virus DNA to the nucleus in both B cells and epithelial cells, the actin cytoskeleton also appears to be needed in epithelial, but not B cells, and it is clear that the process is much more efficient in a B cell than in an epithelial cell (Shannon-Lowe et al. 2009; Valencia and Hutt-Fletcher 2012). Approximately half of the virus delivered into an epithelial cell is degraded within hours. What is perhaps surprising is that although only a small percentage of virus that binds to a cell actually makes it to the nucleus, that small amount that does infect either a B cell or an epithelial cell expressing CR2 is capable of efficient gene expression (Borza et al. 2004; Shannon-Lowe et al. 2005).

## 6 Entry in Vivo

Translating what we know about entry at the cellular level into the reality of what happens in vivo is difficult. Beside the general assumption that virus transmitted in saliva first infects cells somewhere in the oropharynx, there is little or no concrete information on the subject. Both cell-free virus and cell-associated virus, or at least cell-associated virus DNA, are found in saliva (Haque and Crawford 1997), but if the virus is cell-associated, which cell type it is associated with is unclear. There is certainly evidence for virus-producing desquamating epithelial cells in saliva (Lemon et al. 1977; Sixbey et al. 1984), and cell-free virus in saliva of carriers has the characteristics of virus shed from an HLA class II-negative cell, implying that it is being shed from an epithelial cell. It is higher in gp42 than virus made in a B cell from the same individual and binds via gHgL to αv integrins very poorly (Jiang et al. 2006). Modeling studies have also suggested that the levels of virus in saliva can only be accounted for by amplification of virus in epithelial cells (Hadinoto et al. 2009). However, whether or not virus-producing B cells are also present in saliva, or uninfected B cells to which virus is bound, both of which might allow for more efficient infection of mucosal epithelium (Shannon-Lowe et al. 2006; Tugizov et al. 2003), is not known.

Beyond this, there are several questions that can be raised. Does virus replicate its way through an epithelial barrier to reach B cells (Temple et al. 2014), is it transcytosed across epithelium (Tugizov et al. 2013), is it picked up by macrophages or dendritic cells (Tugizov et al. 2007), and does it gain access as a result of breaks in the epithelial barrier, as is assumed for human papillomaviruses?

Given that the only symptomatic primary infection associated with EBV is infectious mononucleosis, which has an estimated incubation period of several weeks, it is likely that we will have to rely on future studies with non-human primate lymphocryptoviruses to provide answers to these questions. Examination of the path of virus reactivated in persistent carriers may, however, be more amenable to study.

Work done on oral hairy leukoplakia in the 1990s reinforced the assumption that, while B cells are the reservoir of latent virus, epithelial cells are normally the site of lytic infection and virus is latent only in malignant or pre-malignant epithelial tissue. Although a model akin to that of human papillomaviruses, where virus establishes latency in basal epithelial cells and replication is differentiation-linked, had been proposed (Allday and Crawford 1988; Sixbey 1989), it was reported that while productive replication was indeed probably differentiation-linked (Young et al. 1991), there was no detectable virus in basal epithelium (Niedobitek et al. 1991). Reexamination by real-time reverse transcriptase PCR of RNA in cells isolated by laser capture microdissection from sections of oral hairy leukoplakia, however, not only confirmed the presence of lytic transcripts in middle and upper layers of epithelium, but also found EBER expression in the absence of lytic transcripts in basal cells (unpublished). Similarly, basal epithelial cells in normal tonsils, identified by faint EBER in situ hybridization staining, expressed EBERs in the absence of lytic transcripts, although, in the sections examined to date, no lytic transcripts have been found in cells in the layers above. This does, however, suggest that perhaps the "papilloma virus" model should be revisited. If virus is shed by a terminally differentiating infected B cell in, for example, the tonsil, then on its path out into saliva, it would plausibly first encounter the basal surface of a basal epithelial cell. B cell-mediated transfer infection of polarized cells occurs only through the basal surface (Shannon-Lowe and Rowe 2011), and the integrins, needed for fusion, are primarily in the basolateral membrane. The virus that is shed from a B cell is also low in gp42 and hence epithelial-tropic. Differentiation of latently infected basal epithelial cells might then sporadically lead to production of virus at the epithelial surface. Virus shed from the epithelial cell is of course highly lymphotropic, and release at the epithelial surface is perhaps more likely to result in shedding than in reinfection at the apical surface. Such a model is most consistent with the idea that access to a new B cell occurs where the epithelial barrier is damaged.

## 7 Conclusions

Progress has clearly been made in understanding virus entry. We know many, though probably not all, of the proteins initially involved at the cellular level, and we are beginning to understand how some of them function. However, we know little about intracellular transport and delivery of DNA to the nucleus, which marks the completion of the successful entry process. Also, most efforts have focused on just two cell types, when it is clear that EBV can infect more than B lymphocytes and epithelial cells. We notice this principally when things go wrong, when, for example, leiomyosarcomas or NK/T cell lymphomas develop, but does it happen in uneventful infections? If not, what changes make these cells accessible? We need to be continually mindful of the fact that almost all we know comes from the study of virus cell interactions out of context in a tissue culture dish. How representative is this of what happens in vivo? Perhaps, the biggest barrier to better understanding of the biology of EBV is its strict human tropism. At the same time, almost all of us are infected, so the population we have available for study is huge. We need to make better and more imaginative use of it.

## References

Allday MJ, Crawford DH (1988) Role of epithelium in EBV persistence and pathogenesis of B cell tumours. Lancet 1:855–856

Atanasiu D, Whitbeck JC, Cairns TM, Reilly B, Cohen GH, Eisenberg RJ (2007) Bimolecular complementation reveals that glycoproteins gB and gH/gL of herpes simplex virus interact with each other during cell fusion. Proc Natl Acad Sci USA 104:18718–18723

Avitabile E, Forghieri C, Campadelli-Fiume G (2007) Complexes between herpes simplex virus glycoproteins gD, gB, and gH detected in cells by complementation of split enhanced green fluorescent protein. J Virol 81:11532–11537

Backovic M, Jardetzky TS (2009) Class III viral membrane fusion proteins. Curr Opin Struct Biol 19:189–196

Backovic M, Jardetzky TS, Longnecker R (2007a) Hydrophobic residues that form putative fusion loops of Epstein-Barr virus glycoprotein B are critical for fusion activity. J Virol 81:9596–9600

Backovic M, Leser GP, Lamb RA, Longnecker R, Jardetzky TS (2007b) Characterization of EBV gB indicates properties of both class I and class II fusion proteins. Virology 368:102–103

Backovic M, Longnecker R, Jardetzky TS (2009) Structure of a trimeric variant of the Epstein-Barr virus glycoprotein B. Proc Natl Acad Sci USA 106:2880–2885

Beisel C, Tanner J, Matsuo T, Thorley-Lawson D, Kezdy F, Kieff E (1985) Two major outer envelope glycoproteins of Epstein-Barr virus are encoded by the same gene. J Virol 54:665–674

Borza C, Hutt-Fletcher LM (1998) Epstein-Barr virus recombinant lacking expression of glycoprotein gp150 infects B cells normally but is enhanced for infection of the epithelial line SVKCR2. J Virol 72:7577–7582

Borza CM, Hutt-Fletcher LM (2002) Alternate replication in B cells and epithelial cells switches tropism of Epstein-Barr virus. Nature Med 8:594–599

Borza CM, Morgan AJ, Turk SM, Hutt-Fletcher LM (2004) Use of gHgL for attachment of Epstein-Barr virus to epithelial cells compromises infection. J Virol 78:5007–5014

Cairns TM, Whitbeck JC, Lou H, Heldwein EE, Chowdary TK, Eisenberg RJ, Cohen GH (2011) Capturing the herpes simplex core fusion complex (gB-gH/gL) in an acidic environment. J Virol 85:6175–6184

Chen J, Rowe CL, Jardetzky TS, Longnecker R (2012) The KGD motif of Epstein-Barr virus gH/gL is bifunctional, orchestrating infection of B cells and epithelial cells. MBio. doi:10.1128/mBio.00290-11

Chen J, Jardetzky TS, Longnecker R (2013) The large groove found in the gH/gL structure is an important functional domain for Epstein-Barr virus fusion. J Virol 87:3620–3627

Chesnokova LS, Hutt-Fletcher LM (2011) Fusion of EBV with epithelial cells can be triggered by αvβ5 in addition to αvβ6 and αvβ8 and integrin binding triggers a conformational change in gHgL. J Virol 85:13214–13223
Chesnokova LS, Nishimura S, Hutt-Fletcher L (2009) Fusion of epithelial cells by Epstein-Barr virus proteins is triggered by binding of viral proteins gHgL to integrins αvβ6 or αvβ8. Proc Natl Acad Sci USA 106:20464–20469
Chesnokova LS, Ahuja MK, Hutt-Fletcher LM (2014) Epstein-Barr virus glycoproteins gB and gHgL can mediate fusion and entry in trans; heat can act as a partial surrogate for gHgL and trigger a conformational change. J Virol 88 (in press)
D'Addario M, Libermann TA, Xu J, Ahmad A, Menezes J (2001) Epstein-Barr virus and its glycoprotein-350 upregulate IL-6 in human B cells via CD21, involving activation of NF-κB and different signaling pathways. J Mol Biol 308:501–514
Fearon DT, Carter RH (1995) The CD19/CR2/TAPA-1 complex of B lymphocytes: linking natural to acquired immunity. Ann Rev Immunol 13:127–149
Fingeroth JD, Weis JJ, Tedder TF, Strominger JL, Biro PA, Fearon DT (1984) Epstein-Barr virus receptor of human B lymphocytes is the C3d complement CR2. Proc Natl Acad Sci USA 81:4510–4516
Fingeroth JD, Diamond ME, Sage DR, Hayman J, Yates JL (1999) CD-21 dependent infection of an epithelial cell line, 293, by Epstein-Barr virus. J Virol 73:2115–2125
Gan Y, Chodosh J, Morgan A, Sixbey JW (1997) Epithelial cell polarization is a determinant in the infectious outcome of immunoglobulin A-mediated entry by Epstein-Barr virus. J Virol 71:519–526
Gill MB, Roecklein-Canfield J, Sage DR, Zambela-Soediono M, Longtine N, Uknis M, Fingeroth JD (2004) EBV attachment stimulates FHOS/FHOD1 redistribution and co-aggregation with CD21:formin interactions with the cytoplasmic domain of human CD21. J Cell Sci 117:2709–2720
Gill MB, Edgar R, May JS, Stevenson PG (2008) A gamma-herpesvirus glycoprotein complex manipulates actin to promote viral spread. PLoS ONE 3:e1808
Gore M, Hutt-Fletcher L (2008) The BDLF2 protein of Epstein-Barr virus is a type II glycosylated envelope protein whose processing is dependent on coexpression with the BMRF2 protein. Virology 383:162–167
Haan KM, Kwok WW, Longnecker R, Speck P (2000) Epstein-Barr virus entry utilizing HLA-DP or HLA-DQ as a coreceptor. J Virol 74:2451–2454
Hadinoto V, Shapiro M, Sun CC, Thorley-Lawson DA (2009) The dynamics of EBV shedding implicate a central role for epithelial cells in amplifying viral output. PLoS Pathog 7:e10000496
Haque T, Crawford DH (1997) PCR amplification is more sensitive than tissue culture methods for Epstein-Barr virus detection in clinical material. J Gen Virol 78:3357–3360
Heldwein EE, Lou H, Bender FC, Cohen GH, Eisenberg RJ, Harrison SC (2006) Crystal structure of glycoprotein B from herpes simplex virus 1. Science 313:217–220
Hutt-Fletcher LM (2007) Epstein-Barr virus entry. J Virol 81:7825–7832
Hutt-Fletcher LM, Chesnokova LS (2010) Integrins as triggers of Epstein-Barr virus fusion and epithelial cell infection. Virulence 1:395–398
Jiang R, Scott RS, Hutt-Fletcher LM (2006) Epstein-Barr virus shed in saliva is high in B cell tropic gp42. J Virol 80:7281–7283
Jiang R, Gu X, Nathan C, Hutt-Fletcher L (2008) Laser-capture microdissection of oropharyngeal epithelium indicates restriction of Epstein-Barr virus receptor/CD21 mRNA to tonsil epithelial cells. J Oral Path Med 37:626–633
Jiang R, Gu X, Moore-Medlin TN, Nathan C-A, Hutt-Fletcher LM (2012) Oral dysplasia and squamous cell carcinoma: correlation between increased expression of CD21, Epstein-Barr virus and CK19. Oral Oncol 48:836–841
Kirschner AN, Lowrey AS, Longnecker R, Jardetzky TS (2007) Binding site interactions between Epstein-Barr virus fusion proteins gp42 and gH/gL reveal a peptide that inhibits both epithelial and B-cell membrane fusion. J Virol 81:9216–9229

Kirschner AN, Sorem J, Longnecker R, Jardetzky TS (2009) Structure of Epstein-Barr virus glycoprotein gp42 suggests a mechanism for triggering receptor-activated virus entry. Structure 17:223–233

Lemon SM, Hutt LM, Shaw JE, Li J-LH, Pagano JS (1977) Replication of EBV in epithelial cells during infectious mononucleosis. Nature 268:268–270

Levine J, Pflugfelder SC, Yen M, Crouse CA, Atherton SS (1990) Detection of the complement (CD21)/Epstein-Barr virus receptor in human lacrimal gland and ocular surface epithelia. Reg Immunol 3:164–170

Li QX, Young LS, Niedobitek G, Dawson CW, Birkenbach M, Wang F, Rickinson AB (1992) Epstein-Barr virus infection and replication in a human epithelial system. Nature 356:347–350

Li QX, Turk SM, Hutt-Fletcher LM (1995) The Epstein-Barr virus (EBV) BZLF2 gene product associates with the gH and gL homologs of EBV and carries an epitope critical to infection of B cells but not of epithelial cells. J Virol 69:3987–3994

Li QX, Spriggs MK, Kovats S, Turk SM, Comeau MR, Nepom B, Hutt-Fletcher LM (1997) Epstein-Barr virus uses HLA class II as a cofactor for infection of B lymphocytes. J Virol 71:4657–4662

Lo KW, Chung GT, To KF (2012) Deciphering the molecular genetic basis of NPC through molecular, cytogenetic, and epigenetic approaches. Semin Cancer Biol 22:79–86

Martin DR, Yuryev A, Kalli KR, Fearon DT, Ahearn JM (1991) Determination of the structural basis for selective binding of Epstein-Barr virus to human complement receptor type 2. J Exp Med 174:1299–1311

Matsuura H, Kirschner AN, Longnecker R, Jardetzky TS (2010) Crystal structure of the Epstein-Barr virus (EBV) glycoprotein H/glycoprotein L (gH/gL) complex. Proc Natl Acad Sci USA 107:22641–22646

McShane MP, Longnecker R (2004) Cell-surface expression of a mutated Epstein-Barr virus glycoprotein B allows fusion independent of other viral proteins. Proc Natl Acad Sci USA 101:17474–17479

Miller N, Hutt-Fletcher LM (1992) Epstein-Barr virus enters B cells and epithelial cells by different routes. J Virol 66:3409–3414

Moore MD, DiScipio RG, Cooper NR, Nemerow GR (1989) Hydrodynamic, electron microscopic and ligand binding analysis of the Epstein-Barr virus/C3dg receptor (CR2). J Biol Chem 34:20576–20582

Mullen MM, Haan KM, Longnecker R, Jardetzky TS (2002) Structure of the Epstein-Barr virus gp42 protein bound to the MHC class II receptor HLA-DR1. Mol Cell 9:375–385

Nemerow GR, Cooper NR (1984) Early events in the infection of human B lymphocytes by Epstein-Barr virus. Virology 132:186–198

Nemerow GR, Mold C, Schwend VK, Tollefson V, Cooper NR (1987) Identification of gp350 as the viral glycoprotein mediating attachment of Epstein-Barr virus (EBV) to the EBV/C3d receptor of B cells: sequence homology of gp350 and C3 complement fragment C3d. J Virol 61:1416–1420

Nemerow GR, Houghton RA, Moore MD, Cooper NR (1989) Identification of the epitope in the major envelope proteins of Epstein-Barr virus that mediates viral binding to the B lymphocyte EBV receptor (CR2). Cell 56:369–377

Niedobitek G, Herbst H, Stein H (1989) Epstein-Barr virus/complement receptor and epithelial cells. Lancet 2:110

Niedobitek G, Young LS, Lau R, Brooks L, Greenspan D, Greenspan JS, Rickinson AB (1991) Epstein-Barr virus infection in oral hairy leukoplakia: virus replication in the absence of a detectable latent phase. J Gen Virol 72:3035–3146

Ogembo JG, Kannan L, Ghiran I, Nicholson-Weller A, Finberg R, Fingeroth JD (2013) Human complement receptor type1/CD35 is an Epstein-Barr virus receptor. Cell Rep. 3:1–15

Reem A-D, Mooney N, Charron D (2004) MHC class II signaling in antigen-presenting cells. Curr Opin Immunol 16:108–113

Roche S, Albertini AAV, Lepault S, Bressanelli S, Gaudin Y (2008) Structures of vesicular stomatitis virus glycoprotein: membrane fusion revisited. Cell Mol Life Sci 65:1716–1728

Sathiyamoorthy K, Jiang J, Hu YX, Rowe CL, Mohl BS, Chen J, Jiang W, Mellins ED, Longnecker R, Zhou ZH, Jardetzky TS (2014) Assembly and architecture of the EBV B cell entry triggering complex. PLoS Pathog 10:e1004309

Shannon-Lowe C, Rowe M (2011) Epstein-Barr virus infection of polarized epithelial cells via the basolateral surface by memory B cell-mediated transfer infection. PLoS Pathog 7:e1001338

Shannon-Lowe C, Baldwin G, Feederle R, Bell AI, Rickinson A, Delecluse H-J (2005) Epstein-Barr virus induced B-cell transformation: quantitating events from virus binding to cell outgrowth. J Gen Virol 86:3009–3019

Shannon-Lowe CD, Neuhierl B, Baldwin G, Rickinson AB, Delecluse H-J (2006) Resting B cells as a transfer vehicle for Epstein-Barr virus infection of epithelial cells. Proc Natl Acad Sci USA 103:7065–7070

Shannon-Lowe C, Adland A, Bell AI, Delecluse HJ, Rickinson AB, Rowe M (2009) Features distinguishing Epstein-Barr virus infections of epithelial cells and B cells: viral genome expression, genome maintenance, and genome amplification. J Virol 83:7749–7760

Silva AL, Omerovic J, Jardetzky TS, Longnecker R (2004) Mutational analysis of Epstein-Barr virus glycoprotein gp42 reveals functional domains not involved in receptor binding but required for membrane fusion. J Virol 78:5946–5956

Sinclair AJ, Farrell PJ (1995) Host cell requirements for efficient infection of quiescent primary B lymphocytes by Epstein-Barr virus. J Virol 69:5461–5468

Sixbey JW (1989) Epstein-Barr virus and epithelial cells. Adv Viral Oncol 8:187–202

Sixbey JW, Yao Q-Y (1992) Immunoglobulin A-induced shift of Epstein-barr virus tissue tropism. Science 255:1578–1580

Sixbey JW, Nedrud JG, Raab-Traub N, Hanes RA, Pagano JS (1984) Epstein-Barr virus replication in oropharyngeal epithelial cells. New Engl J Med 310:1225–1230

Sorem J, Jardetzky TS, Longnecker R (2009) Cleavage and secretion of Epstein-Barr virus glycoprotein gp42 promote membrane fusion with B lymphocytes. J Virol 83:6664–6672

Sugano N, Chen W, Roberts ML, Cooper NR (1997) Epstein-Barr virus binding to CD21 activates the initial viral promoter via NFκB induction. J Exp Med 186:731–737

Szakonyi G, Klein MG, Hannan JP, Young KA, Ma RZ, Asokan R, Holers VM, Chen XS (2006) Structure of the Epstein-Barr virus major envelope glycoprotein. Nat Struct Mol Biol 13:996–1001

Talacko AA, Teo CG, Griffin BE, Johnson NW (1991) Epstein-Barr virus receptors but not viral DNA are present in normal and malignant oral epithelium. J Oral Path Med 20:20–25

Tanner J, Weis J, Fearon D, Whang Y, Kieff E (1987) Epstein-Barr virus gp350/220 binding to the B lymphocyte C3d receptor mediates adsorption, capping and endocytosis. Cell 50:203–213

Tanner JE, Alfieri C, Chatila TA, Diaz-Mitoma F (1996) Induction of interleukin-6 after stimulation of human B-cell CD21 by Epstein-Barr virus glycoproteins gp350 and gp220. J Virol 70:570–575

Temple RM, Zhu J, Budgeon LR, Christensen ND, Meyers C, Sample CE (2014) Efficient replication of Epstein-Barr virus in stratified epithelium in vitro. Proc Natl Acad Sci USA 111:16544–16549

Thomas JA, Crawford DH (1989) Epstein-Barr virus/complement receptor and epithelial cells. Lancet 2:449–450

Tugizov SM, Berline JW, Palefsky JM (2003) Epstein-Barr virus infection of polarized tongue and nasopharyngeal epithelial cells. Nat Med 9:307–314

Tugizov SM, Herrera R, Veluppillai P, Greenspan J, Greenspan D, Palefsky JM (2007) Epstein-Barr virus (EBV)-infected monocytes facilitate dissemination of EBV within the oral mucosal epithelium. J Virol 81:5484–5496

Tugizov SM, Herrera R, Palefsky JM (2013) Epstein-Barr virus transcytosis through polarized oral epithelial cells. J Virol 87:8179–8194

Tuveson DA, Ahearn JM, Matsumoto AK, Fearon DT (1991) Molecular interactions of complement receptors on B lymphocytes: a CR1/CR2 complex distinct from CR2/CD19 complex. J Exp Med 173:1083–1089
Valencia SM, Hutt-Fletcher LM (2012) Important but differential roles for actin in trafficking of Epstein-Barr virus in B cells and epithelial cells. J Virol 86:2–10
Vanarsdall AL, Ryckman BJ, Chase MC, Johnson DC (2008) Human cytomegalovirus glycoproteins gB and gH/gL mediate epithelial cell-cell fusion when expressed either in cis or in trans. J Virol 82:11837–11850
Wang X, Kenyon WJ, Li QX, Mullberg J, Hutt-Fletcher LM (1998) Epstein-Barr virus uses different complexes of glycoproteins gH and gL to infect B lymphocytes and epithelial cells. J Virol 72:5552–5558
Wu L, Hutt-Fletcher LM (2007) Point mutations in EBV gH that abrogate or differentially affect B cell and epithelial cell fusion. Virology 363:148–155
Wu L, Borza CM, Hutt-Fletcher LM (2005) Mutations of Epstein-Barr virus gH that are differentially able to support fusion with B cells or epithelial cells. J Virol 79:10923–10930
Xiao J, Palefsky JM, Herrera R, Berline J, Tugizov SM (2008) The Epstein-Barr virus BMRF-2 protein facilitates attachment to oral epithelial cells. Virology 370:430–442
Xiao J, Palefsky JM, Herrera R, Berline J, Tugizov SM (2009) EBV BMRF-2 facilitates cell-to-cell spread within polarized oral epithelial cells. Virology 388:335–343
Young LS, Dawson CW, Brown KW, Rickinson AB (1989) Identification of a human epithelial cell surface protein sharing an epitope with the C3d/Epstein-Barr virus receptor molecule of B lymphocytes. Int J Cancer 43:786–794
Young LS, Lau R, Rowe M, Niedobitek G, Packham G, Shanahan F, Rowe DT, Greenspan D, Greenspan JS, Rickinson AB, Farrell PJ (1991) Differentiation-associated expression of the Epstein-Barr virus BZLF1 transactivator protein in oral hairy leukoplakia. J Virol 65:2868–2874
Zeng Y (1985) Seroepidemiological studies on nasopharyngeal carcinoma in China. Adv Cancer Res 44:121–138
Zeng Y, Zhang LG, Wu YC, Huang YS, Huang NQ, Li JY, Wang B, Jiang MK, Fang Z, Meng NN (1985) Prospective studies on nasopharyngeal carcinoma in Epstein-Barr virus IgA/VCA antibody-positive persons in Wuzhou City, China. Int J Cancer 36:545–547

# Epstein-Barr Virus Lytic Cycle Reactivation

**Jessica McKenzie and Ayman El-Guindy**

**Abstract** Epstein-Barr virus, which mainly infects B cells and epithelial cells, has two modes of infection: latent and lytic. Epstein-Barr virus infection is predominantly latent; however, lytic infection is detected in healthy seropositive individuals and becomes more prominent in certain pathological conditions. Lytic infection is divided into several stages: early gene expression, DNA replication, late gene expression, assembly, and egress. This chapter summarizes the most recent progress made toward understanding the molecular mechanisms that regulate the different lytic stages leading to production of viral progeny. In addition, the chapter highlights the potential role of lytic infection in disease development and current attempts to purposely induce lytic infection as a therapeutic approach.

## Contents



J. McKenzie · A. El-Guindy (✉)
Department of Pediatrics, Division of Infectious Diseases,
Yale University School of Medicine, New Haven, CT 06520, USA
e-mail: ayman.el-guindy@yale.edu



C. Münz (ed.), *Epstein Barr Virus Volume 2*, Current Topics in Microbiology and Immunology 391, DOI 10.1007/978-3-319-22834-1_8

## Abbreviations

| ATF | CAMP-dependent transcription factor |
|---|---|
| ATM | Ataxia telangiectasia-mutated |
| AP-1 | Activator protein-1 |
| BCR | B cell receptor |
| BLIMP1 | B lymphocyte-induced maturation protein-1 |
| bZIP | Basic leucine zipper |
| CBP | CREB-binding protein |
| C/EBP | CCAAT/enhancer-binding protein |
| CREB | CAMP response element-binding protein |
| EBF1 | Early B cell factor 1 |
| EBV | Epstein-Barr virus |
| HAT | Histone acetyltransferase |
| HDAC | Histone deacetylase |
| HIF | Hypoxia-inducible factor |
| IP3 | Inositol triphosphate |
| JDP2 | Jun dimerization protein |
| KLF | Kruppel-like factors |
| MEF2 | Myocyte enhancer factor 2 |
| MAPK | Mitogen-activated protein kinase(s) |
| MCAF1 | MBD1-containing chromatin-associated factor 1 |
| PKC | Protein kinase C |
| PI3K | Phosphoinositide 3-kinase |
| RNAP | RNA polymerase |
| SBE | SMAD-binding elements |
| SP | Specificity protein |
| TBP | TATA-binding protein |
| vIL-10 | Viral homolog of interleukin 10 |
| XBP1 | X-box-binding protein 1 |
| ZEB | Zinc finger E-box-binding proteins |
| ZEBRA | BamHI Z fragment Epstein-Barr virus Replication Activator |
| ZREs | ZEBRA response elements |

# 1 Introduction

Lytic infection is an integral component of the EBV life cycle. During the lytic cycle, most of the virally encoded genes are expressed. The products of lytic genes mediate amplification of the viral genome, synthesis of viral structural proteins, formation of the viral capsid, and production of new virus particles that spread through saliva to infect other individuals. Recent models suggest that virus released from B cells in Waldeyer's ring enters the epithelium tonsil where it becomes amplified, a process that is considered necessary to promote virus shedding into the saliva for more efficient transmission (Hadinoto et al. 2009).

The lytic cycle is not only required for horizontal transmission of EBV among hosts; lytic reactivation contributes to primary infection and establishment of latency. During primary infection, several reports suggest a model in which EBV particles infect epithelial cells in the oral cavity and undergo lytic replication resulting in production of new viral progeny. Released viral particles spread to oral submucosa to infect circulating B lymphocytes and establish primary infection (Lemon et al. 1978; Sixbey et al. 1983, 1984). Following entry into B cells, EBV undergoes a necessary transient phase of expression of a subset of lytic genes. Expression of this limited set of early lytic genes is critical to initiate and maintain latency in newly infected B lymphocytes (reviewed in Kalla and Hammerschmidt 2012; Altmann and Hammerschmidt 2005; Seto et al. 2010; Wen et al. 2007; Zeidler et al. 1997). Furthermore, spontaneous lytic reactivation from latency replenishes the pool of latently infected B cells that are constantly targeted by immune surveillance (Henle and Henle 1966; Miller et al. 1970).

# 2 Switch from the Latent to the Lytic Phase

Transition from the latent to the lytic state is triggered by expression of two virally encoded transcription factors, ZEBRA (also known as BZLF1, Zta, EB1, or Z) and Rta (also known as BRLF1 or R) (Miller et al. 2007). ZEBRA is the master regulator of the lytic cycle; expression of ZEBRA in latently infected cells is sufficient to initiate the lytic cycle and drive it to completion (Countryman and Miller 1985; Grogan et al. 1987). Regulation of ZEBRA expression is central to entry into the lytic cycle. In a limited number of EBV-positive cell lines, Rta also induces the lytic cycle by activating expression of ZEBRA (Ragoczy et al. 1998; Zalani et al. 1996). The promoters regulating expression of ZEBRA and Rta (Zp and Rp, respectively) are tightly repressed during latency.

## *2.1 Regulation of Zp*

Several cis-DNA regulatory elements were identified in Zp, numbered ZI to ZV, in addition to SMAD-binding elements (SBEs) and the HIF response elements (Fig. 1a).

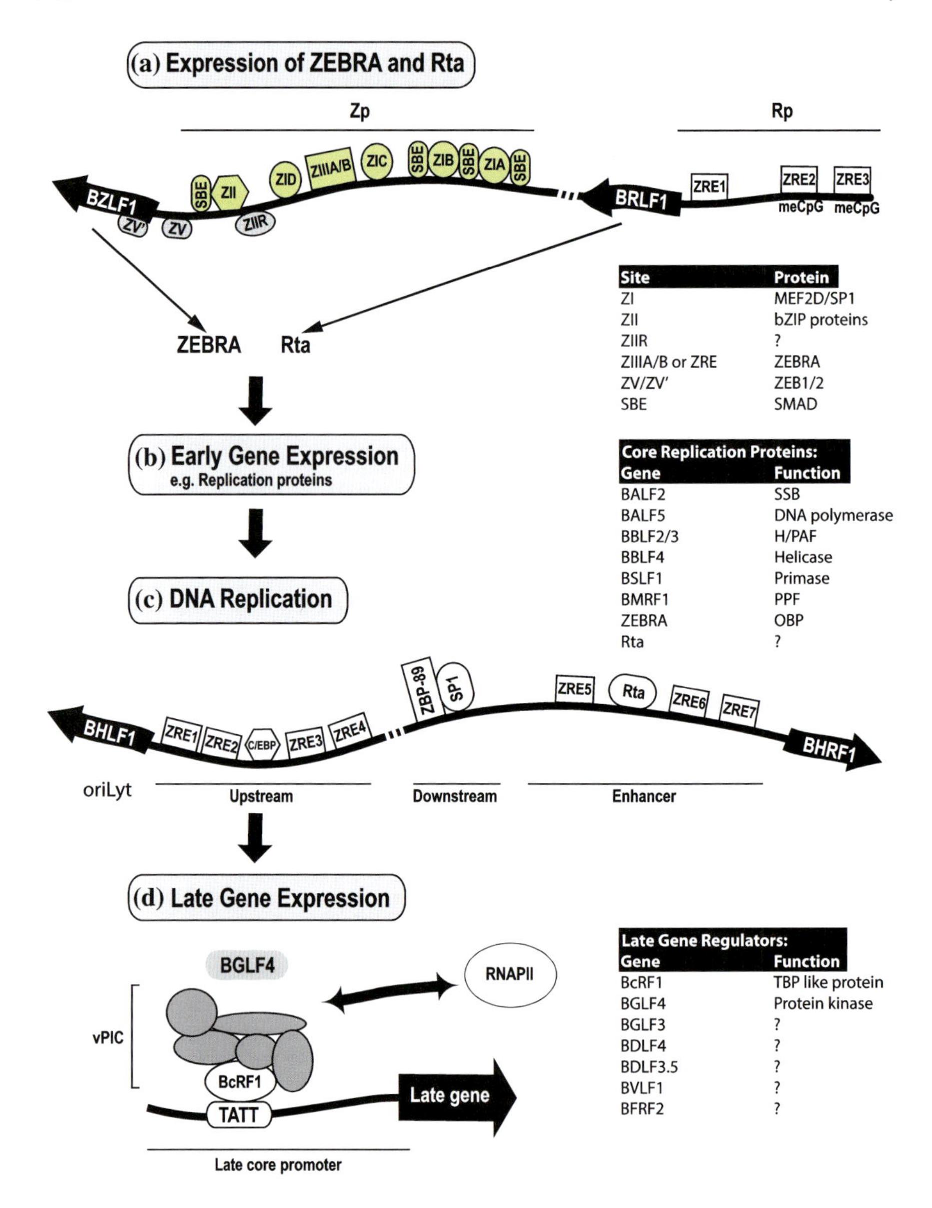
(a) Expression of ZEBRA and Rta
Zp
Rp
BZLF1
ZV'
ZV
SBE
ZII
ZIIR
ZID
ZIIIA/B
ZIC
SBE
ZIB
SBE
ZIA
SBE
BRLF1
ZRE1
ZRE2
ZRE3
meCpG
meCpG
ZEBRA
Rta
Site
Protein
ZI
MEF2D/SP1
ZII
bZIP proteins
ZIIR
?
ZIIIA/B or ZRE
ZEBRA
ZV/ZV'
ZEB1/2
SBE
SMAD
(b) Early Gene Expression
e.g. Replication proteins
Core Replication Proteins:
Gene
Function
BALF2
SSB
BALF5
DNA polymerase
BBLF2/3
H/PAF
BBLF4
Helicase
BSLF1
Primase
BMRF1
PPF
ZEBRA
OBP
Rta
?
(c) DNA Replication
BHLF1
ZRE1
ZRE2
C/EBP
ZRE3
ZRE4
ZBP-89
SP1
ZRE5
Rta
ZRE6
ZRE7
BHRF1
oriLyt
Upstream
Downstream
Enhancer
(d) Late Gene Expression
BGLF4
RNAPII
vPIC
BcRF1
TATT
Late gene
Late core promoter
Late Gene Regulators:
Gene
Function
BcRF1
TBP like protein
BGLF4
Protein kinase
BGLF3
?
BDLF4
?
BDLF3.5
?
BVLF1
?
BFRF2
?

**Fig. 1** Summary of the different stages of lytic reactivation. Upon induction of the lytic cycle, two essential transcription factors of EBV are expressed: ZEBRA and Rta. The ZEBRA promoter (Zp) is strictly regulated by a variety of proteins at its many regulatory motifs (*a*). Once expressed, ZEBRA activates Rta expression through interaction with its promoter (Rp) at ZEBRA response elements. Note that the BRLF1 gene extends into Zp (not depicted). Each of these proteins, ZEBRA and Rta, activate a subset of early genes on their own as well as synergistically (*b*). Many early genes code for proteins involved in viral DNA replication. ZEBRA, also a replication protein, interacts with the origin of lytic replication (oriLyt) along with other factors to initiate DNA amplification (*c*). Following viral DNA replication, late genes are expressed (*d*), a process which is mediated by the viral pre-initiation complex (vPIC) and the viral protein kinase BGLF4. vPIC is composed of six late gene regulators that presumably play a role in recruitment of cellular RNA polymerase II (RNAPII) to late promoters. A number of lytic genes then work in concert to package viral DNA into capsids and to create fully formed viral progeny, thus completing the lytic life cycle. SBE, SMAD-binding element; SSB, viral single-stranded DNA-binding protein; H/PAF, helicase/primase-associated factor; PPF, DNA polymerase processivity factor; OBP, origin-binding protein; TBP, TATA box-binding protein; and ZRE, ZEBRA response element. Zp-negative regulatory sites are in *gray*, while positive regulatory sites are in *yellow*

#### 2.1.1 Zp-Positive Regulatory Motifs

There are four ZI elements (ZIA through ZID) interspersed in Zp (Borras et al. 1996; Flemington and Speck 1990c). These elements are recognized by two families of transcription factors, SP1/KLF and MEF2 (Liu et al. 1997a, b). During lytic induction, phosphorylation of Sp1 and dephosphorylation of MEF2D disrupt their association with HDACs and promote association with HATs leading to activation of Zp (Bryant and Farrell 2002; Gruffat et al. 2002; Li et al. 2001; Liu et al. 1997b; Tsai et al. 2011). Importantly, insertion of mutations in the ZI motifs that disrupt binding of MEF2 or SP1/KLF abolishes expression of ZEBRA in response to lytic cycle inducing agents (Murata et al. 2013).

The ZII element, located near the TATA box, is essential for activation of Zp (Chatila et al. 1997; Daibata et al. 1994; Feng et al. 2004; Flemington and Speck 1990c; Murata et al. 2013). The ZII motif, which resembles binding sites of AP-1 and CREB, is recognized by several bZIP proteins (Flamand and Menezes 1996; Flemington and Speck 1990c; Huang et al. 2006; Liu et al. 1998; Murata et al. 2011; Sun and Thorley-Lawson 2007; Wang et al. 1997; Wu et al. 2004). The fact that several transcription factors can interact with ZII could be attributed to the inducing agent or cell type (Bhende et al. 2007; Huang et al. 2006; Murata et al. 2011; Shirley et al. 2011; Sun and Thorley-Lawson 2007; Wu et al. 2004).

The ZIIIA and ZIIIB domains are ZEBRA response elements (ZREs) (Flemington and Speck 1990a). Binding of ZEBRA to ZIIIA and ZIIIB auto-stimulates Zp to achieve full promoter activation (Flemington and Speck 1990a; Yin et al. 2004). ZEBRA also interacts with C/EBPα, and both proteins cooperatively activate Zp through direct binding to ZII and ZIIIB (Wu et al. 2004).

Zp contains five SBEs that regulate induction of the lytic cycle via the canonical TGF-β signaling pathway (Iempridee et al. 2011; Liang et al. 2002; Matsuzaki 2011). Mutation of any Zp-SBE compromises efficient activation of the promoter by TGF-β, suggesting that all five sites work in concert. Chromatin immunoprecipitation experiments demonstrated association of SMAD4 with Zp in Burkitt's lymphoma MutuI cells treated with TGF-β (Iempridee et al. 2011). Since SMADs interact with Sp1 and MEF2 proteins (Quinn et al. 2001; Zhang et al. 2000), it is possible that activation of Zp might also involve formation of SMAD/Sp1 or SMAD/MEF2 complexes on the promoter.

### 2.1.2 Zp-Negative Regulatory Motifs

ZIIR, ZV, and ZV' are three cis elements that negatively regulate Zp. Individual and concurrent mutations of these silencing elements resulted in viruses that are predominantly lytic and fail to transform primary B cells (Yu et al. 2007, 2012). The protein that binds to ZIIR and mediates repression of Zp is still elusive. ZV and ZV' are binding sites for ZEB1 and ZEB2, respectively, which are zinc finger E-box-binding proteins (Ellis et al. 2010; Kraus et al. 2003). Absence of ZEB1 expression in EBV-infected gastric carcinoma cells resulted in spontaneous lytic reactivation (Feng et al. 2007). Targeting of ZEB1 and ZEB2 mRNAs by cellular miRNAs 200b and 429 leads to upregulation of lytic gene expression (Ellis-Connell et al. 2010; Park et al. 2008; Yu et al. 2011). Therefore, the abundance of miRNA 200 or, inversely, the lack of ZEB1/2, in certain cell backgrounds, e.g., gastric carcinoma cells, might explain why lytic infection is favored (Ellis-Connell et al. 2010; Lin et al. 2010; Shinozaki et al. 2010).

## 2.2 *In Vitro Activation of the Lytic Cycle*

In tissue culture, several chemical and biological agents induce lytic reactivation. These agents include HDAC inhibitors (e.g., sodium butyrate), protein kinase C agonists (e.g., phorbol ester), DNA methyltransferase inhibitors (e.g., azacytidine), anti-immunoglobulin, and transforming growth factor beta (Ben-Sasson and Klein 1981; Fahmi et al. 2000; Luka et al. 1979; Takada 1984; zur Hausen et al. 1978). Data from reporter assays suggest that inducing stimuli converge on and activate Zp. The mechanism by which these inducing stimuli activate Zp is still under investigation.

Perhaps one of the more physiologically relevant stimuli of the lytic cycle is cross-linking of BCR with anti-immunoglobulin. The process mimics antigen-mediated activation of BCR and triggers a cascade of signal transduction that leads to activation of Zp (Packard and Cambier 2013; Takada and Ono 1989; Tovey et al. 1978). BCR engagement leads to activation of one or more of the phosphotyrosine kinases, Btk, Syk, and Lyn. The involvement of Btk and Syk in lytic reactivation is still ambiguous; inhibitors of these kinases had no effect on reactivation of the lytic program in Burkitt's lymphoma Akata cells (Goswami et al. 2012). Three additional downstream cellular kinases, PI3K, PKC, and MAPK, transduce the induction signal to Zp (Davies et al. 1991; Goswami et al. 2012; Iwakiri and Takada 2004; Satoh et al. 1999). Interestingly, EBV encodes a latent membrane protein, LMP2A, which mimics BCR and disrupts its association with downstream kinases (Fruehling and Longnecker 1997). Latently infected cells that express high levels of LMP2A are resistant to lytic reactivation via BCR engagement.

Silencing of Zp is mediated by both repressors of transcription (discussed earlier) as well as epigenetic histone modifications (Murata et al. 2012, 2013; Ramasubramanyan et al. 2012). Several HDAC inhibitors induce the lytic cycle partly by causing histone hyperacetylation of Zp (Chang and Liu 2000; Gruffat et al. 2002; Jenkins et al. 2000; Murata et al. 2011). However, recent reports indicate that histone hyperacetylation alone is not sufficient to disrupt latency. Treatment of EBV-positive Burkitt's lymphoma cells with sodium butyrate resulted in histone hyperacetylation of Zp and Rp in cells induced and refractory to lytic reactivation (Daigle et al. 2010). Moreover, valproic acid, an HDAC inhibitor, not only failed to reactivate the lytic cycle but also blocked the capacity of other inducing agents to do so in B lymphocytes (Countryman et al. 2008; Daigle et al. 2011). Additional factors contribute to the capacity of HDAC inhibitors to regulate expression of ZEBRA. These factors include synthesis of cellular immediate-early proteins upstream of ZEBRA and Rta (e.g., EGR1 and nuclear orphan receptors) (Ye et al. 2007), stimulation of the kinase activity of ATM and PKC-delta (Hagemeier et al. 2012; Tsai et al. 2011), and post-translational modifications of histones other than acetylation (Mansouri et al. 2013).

## 2.3 Cellular Events that Contribute to Disruption of Latency

Recent studies established a series of associations between terminal cell differentiation and lytic reactivation of EBV as well as other gammaherpesviruses (Lai et al. 2011; Liang et al. 2009). Induction of terminal differentiation in squamous epithelial cells augmented the activity of Zp (Karimi et al. 1995). Two cell differentiation factors, XBP1 and BLIMP1, mediate EBV reactivation in B lymphocytes and epithelial cells. In oral hairy leukoplakia, a condition observed in the tongue of HIV-infected patients, EBV replicates only in differentiated epithelial cells expressing BLIMP1 (Buettner et al. 2012). BLIMP1 induces the lytic cycle in epithelial cells by activating Zp and Rp (Reusch et al. 2014). In plasma-differentiated cells, activated XBP1 binds to the ZID/ZII elements in Zp and triggers lytic replication (Bhende et al. 2007; McDonald et al. 2010; Sun and Thorley-Lawson 2007). Several other cellular factors (e.g., EBF1 and Pax5) that play a role in maintaining the B cell lineage were described as repressors of lytic reactivation (Davies et al. 2010). EBF1, a transcription factor, regulates expression of Pax5. Loss of endogenous EBF1 or Pax5 expression in lymphoblastoid or Burkitt's lymphoma cell lines induces lytic viral reactivation (Arvey et al. 2012; Raver et al. 2013). Furthermore, induction of the lytic cycle using inducing agents reduces the level of Pax5 protein (Raver et al. 2013). In addition to cell differentiation, hypoxia and endoplasmic reticulum stress also have the potential to reactivate latent EBV (Jiang et al. 2006; Kenney and Mertz 2014; Taylor et al. 2011).

# 3 Early Gene Expression

## 3.1 The ZEBRA Protein

The BZLF1 gene codes for the ZEBRA protein (Countryman and Miller 1985; Grogan et al. 1987; Rooney et al. 1988). The capacity of ZEBRA to disrupt latency resides in its function as a transcription activator of lytic genes and as an essential origin-binding protein in the process of lytic DNA replication. ZEBRA belongs to the bZIP family of proteins and is composed of four functional regions: transactivation domain, regulatory domain, basic DNA-binding domain, and coiled-coil dimerization domain (El-Guindy et al. 2007; Flemington et al. 1992). ZEBRA activates transcription via its capacity to: bind to specific DNA motifs present in cellular and viral promoters, interact with general transcription factors (e.g., TFIID and TFIIA), and recruit histone-modifying enzymes to target promoters (Adamson and Kenney 1999; Chen et al. 2001; Chi et al. 1995; Deng et al. 2001; Lieberman and Berk 1991, 1994; Zerby et al. 1999). ZEBRA recognizes a broad range of DNA motifs that include AP-1 and methylated and unmethylated binding sites (Flemington and Speck 1990b; Glover and Harrison 1995; Lehman et al. 1998; Lieberman et al. 1990; Petosa et al. 2006; Farrell et al. 1989; Kouzarides et al. 1991).

Both transcription and replication functions of ZEBRA are regulated by post-translational modifications. ZEBRA is constitutively phosphorylated at sites clustered in the regulatory domain and the transactivation domain. Phosphorylation of residues in the regulatory domain is essential for its role as an origin-binding protein during lytic DNA replication and as a transcription repressor of a subclass of late genes during the early phase of the lytic cycle (El-Guindy and Miller 2004; El-Guindy et al. 2006, 2007). Oxidation of cysteine residues in the bZIP domain, particularly C189, regulates the function of ZEBRA as a replication protein (Wang et al. 2005). Furthermore, ZEBRA is sumoylated at K12, a modification that partly suppresses the transcriptional activity of the protein (Hagemeier et al. 2010).

## 3.2 Regulation of Rp and Other Early Promoters Through ZEBRA Binding to Methylated DNA

A primary role of ZEBRA during lytic reactivation is to stimulate the promoter of the BRLF1 gene (Rp), which codes for Rta. ZEBRA activates Rp via its capacity to bind to methylated ZREs (Bhende et al. 2004, 2005) (Fig. 1a). ZEBRA also binds to and activates the promoters of several lytic genes in a manner dependent on the presence of methylated CpGs. Six of these genes encode the different components of the viral replication machinery (Bergbauer et al. 2010) (Fig. 1). Discovering the capacity of ZEBRA to bind methylated DNA has transformed our understanding of the mechanisms that regulate viral gene expression during lytic infection. The finding revealed that the extent of viral DNA methylation is a determining factor in the outcome of lytic reactivation. During primary infection, ZEBRA is transiently expressed; however, most of the viral DNA is unmethylated (Kintner and Sugden 1981; Wen et al. 2007). Lack of viral DNA methylation impedes the capacity of ZEBRA to activate expression of genes encoding the lytic replication machinery. As a result, only few lytic genes are activated and no viral genome amplification takes place (Kalla et al. 2012). This outcome leads to a short phase of abortive lytic activation during primary infection that is considered essential for transformation of resting B cells (Kalla et al. 2010). In contrast, reactivation from the latent viral genome, which is heavily methylated, results in expression of all lytic genes and subsequent production of viral progeny. Therefore, methylation of viral DNA serves as a key determinant for the extent of lytic activation and its consequences (Bergbauer et al. 2010; Dickerson et al. 2009).

S186 is a key residue for the capacity of ZEBRA to bind to methylated DNA (Bhende et al. 2004, 2005). Sequence alignment of the basic domain of ZEBRA with other bZIP proteins revealed the uniqueness of S186 (Francis et al. 1997). Mutation of S186 to alanine induced a conformational change and disrupted ZEBRA's capacity to reactivate the lytic cycle (El-Guindy et al. 2002; Francis et al. 1999). AP-1 proteins with serine substitution of residues corresponding to S186 bound to methylated DNA and activated expression of Rta (Yu et al. 2013). The current crystal structure of ZEBRA employed a Z(S186A) mutant bound to a non-methylated AP-1 site (Petosa et al. 2006). These new findings highlight the need for a new crystal structure of ZEBRA in which S186 is intact and the protein is bound to a methylated ZRE.

### *3.3 Role of Rta in EBV Reactivation*

Rta is essential for disruption of EBV latency in infected B lymphocytes and epithelial cells (Feederle et al. 2000; Ragoczy et al. 1998; Zalani et al. 1996). All gammaherpesviruses encode a similar transcription factor; however, there are no cellular homologs of Rta. The protein has three functional domains, an amino-terminal DNA recognition domain, a dimerization domain, and a transcription activation domain (Manet et al. 1991). The DNA recognition function of Rta is regulated by an intrinsic 55 amino acid sequence termed the DNA-binding inhibitory sequence (DBIS). Mutations in DBIS enhance the DNA-binding activity of Rta in vitro (Chen et al. 2009). Rta is a phosphoprotein; however, the sites, the kinases, and the functional significance of phosphorylation have not yet been characterized (Zacny et al. 1998).

Genome-wide analyses demonstrated association of Rta with several lytic promoters (Heilmann et al. 2012). Rta binds to cognate DNA motifs termed Rta response elements (RREs) (Chen et al. 2005; Gruffat et al. 1990). In addition to direct binding, Rta is tethered to viral promoters, e.g., Zp, and the viral DNA polymerase (BALF5) promoter, through interaction with cellular transcription factors (Chang et al. 2005; Liu et al. 1996). Rta interacts with TBP, TFIIB, and CBP, which contributes to the recruitment of RNAPII and the opening of chromatin during the process of transcription activation (Manet et al. 1993; Swenson et al. 2001).

## 3.4 Early Events of Lytic Reactivation

While ZEBRA and Rta are the two main activators of the EBV lytic cycle, recent findings support a model in which expression of ZEBRA precedes that of Rta. ZEBRA activates expression of Rta and BRRF1 through its capacity to bind to methylated DNA (Dickerson et al. 2009). BRRF1 is a nuclear factor that synergizes with Rta to activate expression of lytic genes (Hong et al. 2004). In a positive feedback loop, Rta and BRRF1 cooperate together to activate Zp. In epithelial cells, expression of BRRF1 alone was sufficient to induce the lytic cycle, yet to a lesser extent relative to ZEBRA and Rta (Hagemeier et al. 2011). Once ZEBRA and Rta are expressed, each protein activates transcription of a separate class of EBV early lytic genes, and together the two proteins synergize to activate transcription of a third class of lytic genes (Ragoczy and Miller 1999) (Fig. 1b). The mechanism that regulates synergy is not fully understood and likely involves complex formation among ZEBRA, MCAF1, and Rta (Chang et al. 2010; Quinlivan et al. 1993).

## 3.5 Regulation of ZEBRA and Rta by Cellular and Viral Proteins

Several cellular and viral proteins modulate the function of ZEBRA and Rta in disruption of latency. Interaction of Rta with Oct1 enhances its capacity to bind to and activate lytic viral promoters (Robinson et al. 2011). The activity of three cellular protein kinases, PI3K, p38, and JNK, is required for Rta-mediated activation of Zp (Adamson et al. 2000; Darr et al. 2001). The EBV LF2 protein suppresses the function of Rta by retaining the protein at the extranuclear cytoskeleton (Heilmann et al. 2010). Cellular proteins counteract lytic reactivation by negatively regulating the function of ZEBRA. Pax5 and Oct2 interact with the bZIP domain of ZEBRA and inhibit its binding to lytic promoters (Raver et al. 2013; Robinson et al. 2012). These functional and physical interactions between the lytic cycle activators and viral or host proteins reflect the intricate events that regulate the switch from latent to lytic infection.

# 4 DNA Replication

Lytic viral DNA replication occurs independent of cellular DNA replication. EBV encodes proteins necessary for lytic DNA replication (Fixman et al. 1992, 1995). Two copies of duplicated core elements, termed *oriLyt* (origin of lytic replication), mediate replication of viral DNA during lytic infection resulting in several 100-fold genome amplification (Hammerschmidt and Sugden 1988). Because of the circularity of the EBV genome, lytic DNA replication is thought to follow a rolling circle model. However, other forms of bidirectional DNA replication might ensue to increase the number of circular viral templates prior to amplification (Pfuller and Hammerschmidt 1996). Lytic replication results in long concatemers of linear EBV genomes fused head to tail that are then cleaved and packaged into viral capsids (Bloss and Sugden 1994; Jacob and Roizman 1977).

## 4.1 oriLyt

A complete linear viral genome contains two copies of oriLyt located in the left and right duplicated sequences of the genome ($DS_L$ and $DS_R$) (Hammerschmidt and Sugden 1988). Naturally occurring EBV strains with deletions of one copy of either origin, such as the B95-8 and P3HR1 virus strains, still maintain the capacity to replicate the entire viral genome (Cho et al. 1984; Raab-Traub et al. 1980). Each origin of replication contains the DNA regulatory elements sufficient to replicate a surrogate oriLyt plasmid in cis. Three cis elements were mapped in the $DS_L$ origin, also known as BamHI-H oriLyt (Fig. 1c). These cis elements are the upstream and downstream elements, which are essential for genome amplification, and a dispensable enhancer element (Schepers et al. 1993). The detailed structure of oriLyt was recently reviewed (Hammerschmidt and Sugden 2013).

## 4.2 Assembly of the EBV Replication Machinery

The process of genome amplification requires core replication proteins encoded by EBV. These proteins include ZEBRA, which functions as the origin-binding protein, the single-stranded DNA-binding protein (BALF2), and five replication enzymes and co-enzymes, namely the helicase (BBLF4), primase (BSLF1), primase-associated factor (BBLF2/3), DNA polymerase (BALF5), and DNA polymerase processivity factor (BMRF1) (Fixman et al. 1992, 1995). Replication occurs in intranuclear sites termed replication compartments (Daikoku et al. 2005). One of the first events that take place during replication is origin recognition. ZEBRA binds to response elements present in oriLyt and recruits other replication proteins. Three replication proteins, BMRF1, BSLF1, and BALF2, augment binding of ZEBRA to oriLyt (El-Guindy et al. 2010). ZEBRA interacts with most replication proteins except BALF2 and is thought to nucleate the replication complex on oriLyt (Baumann et al. 1999; Gao et al. 1998; Liao et al. 2005; Zhang et al. 1996). Phosphorylation of S173 in the regulatory domain of ZEBRA is crucial for its capacity to bind oriLyt and to support viral DNA replication (El-Guindy et al. 2007). The Rta protein also plays an essential role in lytic DNA replication. The exact role of Rta in replication is unknown. Rta binds to oriLyt in vivo and localizes to replication compartments (El-Guindy et al. 2013; Heilmann et al. 2012). One potential role of Rta in replication is to activate oriLyt through synergy with ZEBRA. Both Rta and ZEBRA are necessary to activate transcription of the BHLF1 gene from the endogenous genome (El-Guindy et al. 2013), an event that is necessary for initiation of lytic DNA replication (Rennekamp and Lieberman 2011).

## 5 Late Gene Expression

EBV encodes around 36 late genes (Yuan et al. 2006) that code for viral structural proteins, glycoproteins, tegument proteins, and vIL-10. Functionally, products of late genes are crucial for many events in the life cycle of EBV including capsid assembly and maturation; DNA packaging in viral capsids; viral attachment, fusion, and internalization during de novo infection; and protection of lytically infected cells from immune responses.

Little is known about the mechanism of regulation of late gene expression. Several factors are necessary for expression of EBV late genes (Fig. 1d). One main factor is the onset of viral DNA replication. Disruption of viral DNA replication abolishes synthesis of late products (El-Guindy et al. 2010; Summers and Klein 1976). Several models were proposed to explain the link between DNA replication and late gene expression. For example, replicating the viral genome could induce: alterations in chromatin structure, demethylation of viral DNA, expression of transcription activators, and displacement of repressors bound to late promoters.

Recent studies demonstrated that viral DNA replication alone is not sufficient to activate expression of late genes. Studying the EBV transcriptome in the absence and presence of BGLF4, an EBV-encoded serine/threonine kinase, revealed that expression of almost all late genes is dependent on the function of BGLF4 even after replication of the viral genome (El-Guindy et al. 2014). Six additional factors have been identified as regulators of late gene expression; all six factors have homologs in both beta- and gammaherpesviruses (Aubry et al. 2014). Two factors, BGLF3 and BcRF1, interact with the largest subunit of RNA polymerase-II, RPB1, suggesting that these two proteins are part of the transcription machinery involved in transcribing late mRNAs (Aubry et al. 2014; El-Guindy et al. 2014). There are no known functional domains or cellular homologs of BGLF3. BcRF1, however, is structurally related to TBP, based on in silico structure prediction methods (Wyrwicz and Rychlewski 2007). BcRF1 binds to a non-canonical TATA element (TATT) present in many late promoters (Gruffat et al. 2012; Serio et al. 1998). Binding of BcRF1 to a late promoter is likely to recruit other late gene regulators through protein–protein interaction (Aubry et al. 2014).

Viral late promoters have a different structure compared to cellular promoters or viral promoters of other kinetic classes. Activation of late promoters is dependent on a core region encompassing the distinct TATT box element and a downstream sequence that spans the transcription start site (Amon et al. 2004; Serio et al. 1997, 1998) (Fig. 1d). Rta associates with at least two late promoters, BFRF3p and BLRF2p, even prior to amplification of the viral genome, suggesting a role in late gene expression (Heilmann et al. 2012). In the absence of ZEBRA, expression of Rta in EBV-positive cells bypasses the requirement for viral DNA replication and activates a subclass of late genes (Lu et al. 2006; Ragoczy and Miller 1999). ZEBRA phosphorylated at S173 suppresses the capacity of Rta to activate late genes in the absence of viral DNA replication (El-Guindy and Miller 2004). This finding suggests that ZEBRA plays a role in maintaining proper temporal gene expression during the early stage of lytic infection.

## 6 Role of Lytic Infection in Development of EBV-Associated Malignancies

Several observations support the conclusion that lytic infection contributes to the oncogenicity of EBV. Lytic infection precedes development of EBV-associated Hodgkin's lymphoma, nasopharyngeal carcinoma (NPC), Burkitt's lymphoma (BL), and post-transplant lymphoproliferative disease (PTLD) (Chan et al. 1991; Chen et al. 1985; Mueller et al. 1989, 1991). EB viruses isolated from NPC patients vary from other EBV isolates and have the tendency to spontaneously switch to lytic replication (Crawford et al. 1979; Tsai et al. 2013). High viral load, due to lytic replication, and elevated antibody titers against EBV lytic products are predictive markers for the onset of these diseases (Cao et al. 2011; Chien et al. 2001; Hong et al. 2005; Hsu et al. 2009; Katsumura et al. 2012; Lee et al. 2005; Lucas et al. 1998; Rasche et al. 2013; Rooney et al. 1995; Tsai et al. 2008; van Esser et al. 2001).

Exclusively lytic EBV infection is irreconcilable with B cell immortalization (Miller et al. 1974). Hence, it is more likely that lytic infection contributes to tumor development by expanding the population of latently infected cells, thus increasing the likelihood of a latently infected cell becoming neoplastic. Lytic infection also could contribute to EBV oncogenicity by inducing expression of proinflammatory cytokines, growth factors, and cellular signaling molecules that promote cell proliferation, genomic instability, and angiogenesis (Arvey et al. 2012; Bejarano and Masucci 1998; Hsu et al. 2008). In a humanized mouse model, induction of the lytic cycle was shown to be necessary for development of B cell lymphoma (Ma et al. 2011). Prophylaxis and preemptive antiviral therapy against lytic replication reduced the incidence of PTLD in patients with solid organ transplants

(Darenkov et al. 1997; Farmer et al. 2002; Green 2001; McDiarmid et al. 1998). Therefore, it is crucial to continue to study the mechanisms that regulate the different stages of lytic infection with a goal to develop new antiviral drugs.

## 7 Lytic Reactivation as a Therapeutic Approach

EBV is linked to the development of several lymphomas and carcinomas (reviewed in Kutok and Wang 2006). Most of the cells in EBV-associated tumors harbor latent virus. This prompted attempts to use reactivation of the lytic cycle as part of an oncolytic treatment repertoire. The approach involves administering two types of drugs: The first drug induces lytic gene expression in latently infected tumor cells, and the second drug, ganciclovir (GCV), is an inhibitor of cellular and viral DNA replication (Feng et al. 2002). Phosphorylation of GCV by the lytic BGLF4 kinase is necessary to change the prodrug into its active antiviral and cytotoxic form (Meng et al. 2010). Ideally, phosphorylation of GCV by BGLF4 will permit selective eradication of lytic tumor cells and block the release of infectious viral particles. The oncolytic therapeutic approach has been studied in tissue culture, in xenograft models, and clinically in patients with advanced EBV-associated lymphoma or nasopharyngeal carcinoma (Feng and Kenney 2006; Fu et al. 2008; Perrine et al. 2007; Wildeman et al. 2012). However, there are some limitations to this approach: (i) the limited proportion of latently infected cells that become induced into the lytic stage, (ii) the general toxicity of the inducing agent, and (iii) the poor efficacy of GCV in blocking EBV DNA replication in vivo; inefficient blockage of lytic replication bears the risk of releasing more virus particles, thus increasing the number of latently infected cells. Current studies aim at overcoming some of these hurdles. Small molecules related to tetrahydrocarboline were shown to exhibit lower toxicity but higher efficiency in activating the lytic cycle compared to HDAC inhibitors (Tikhmyanova et al. 2014). Furthermore, the ability to separate lytic from refractory cells provided the means to study the mechanism responsible for subduing lytic induction in refractory cells (Bhaduri-McIntosh and Miller 2006). STAT3 serves as a hallmark of refractory cells; STAT3 promotes the refractory state, likely via induction of cellular transcriptional repressors (Daigle et al. 2010; Hill et al. 2013). Targeting STAT3 or other regulators of the refractory state might increase the number of lytic reactivated cells in response to an inducing agent.

## 8 Conclusion

Lytic infection has been repeatedly shown to precede inception of several EBV-associated malignancies and is used as a means to monitor disease in patients undergoing treatment. The role of lytic infection in the development of these diseases is still poorly understood. Inhibition of lytic infection has been suggested as a prophylactic strategy in immunocompromised patients. However, none of the available drugs are competent to efficiently block lytic infection in patients. Discovering new anti-EBV drugs relies on understanding the different mechanisms employed by the virus to promote lytic infection. In recent years, significant advances were achieved in deciphering some of these mechanisms, which include the role of proteins involved in cell differentiation in disruption of latency, identification of cellular repressors of Zp, methylation-dependent activation of lytic promoters, and identification of several late gene regulators. Despite this important progress, the function of many lytic macromolecules remains elusive.

## References


Adamson AL, Kenney S (1999) The Epstein-Barr virus BZLF1 protein interacts physically and functionally with the histone acetylase CREB-binding protein. J Virol 73:6551–6558

Adamson AL, Darr D, Holley-Guthrie E, Johnson RA, Mauser A, Swenson J, Kenney S (2000) Epstein-Barr virus immediate-early proteins BZLF1 and BRLF1 activate the ATF2 transcription factor by increasing the levels of phosphorylated p38 and c-Jun N-terminal kinases. J Virol 74:1224–1233

Altmann M, Hammerschmidt W (2005) Epstein-Barr virus provides a new paradigm: a requirement for the immediate inhibition of apoptosis. PLoS Biol 3:e404

Amon W, Binne UK, Bryant H, Jenkins PJ, Karstegl CE, Farrell PJ (2004) Lytic cycle gene regulation of Epstein-Barr virus. J Virol 78:13460–13469

Arvey A, Tempera I, Tsai K, Chen HS, Tikhmyanova N, Klichinsky M, Leslie C, Lieberman PM (2012) An atlas of the Epstein-Barr virus transcriptome and epigenome reveals host-virus regulatory interactions. Cell Host Microbe 12:233–245

Aubry V, Mure F, Mariame B, Deschamps T, Wyrwicz LS, Manet E, Gruffat H (2014) Epstein-barr virus late gene transcription depends on the assembly of a virus-specific preinitiation complex. J Virol 88:12825–12838

Baumann M, Feederle R, Kremmer E, Hammerschmidt W (1999) Cellular transcription factors recruit viral replication proteins to activate the Epstein-Barr virus origin of lytic DNA replication, oriLyt. EMBO J 18:6095–6105 (published erratum appears in EMBO J 2000 Jan 17; 19(2):315)

Bejarano MT, Masucci MG (1998) Interleukin-10 abrogates the inhibition of Epstein-Barr virus-induced B-cell transformation by memory T-cell responses. Blood 92:4256–4262

Ben-Sasson SA, Klein G (1981) Activation of the Epstein-Barr virus genome by 5-aza-cytidine in latently infected human lymphoid lines. Int J Cancer 28:131–135

Bergbauer M, Kalla M, Schmeinck A, Gobel C, Rothbauer U, Eck S, Benet-Pages A, Strom TM, Hammerschmidt W (2010) CpG-methylation regulates a class of Epstein-Barr virus promoters. PLoS Pathog 6:e1001114

Bhaduri-McIntosh S, Miller G (2006) Cells lytically infected with Epstein-Barr virus are detected and separable by immunoglobulins from EBV-seropositive individuals. J Virol Methods 137:103–114

Bhende PM, Seaman WT, Delecluse HJ, Kenney SC (2004) The EBV lytic switch protein, Z, preferentially binds to and activates the methylated viral genome. Nat Genet 36:1099–1104

Bhende PM, Seaman WT, Delecluse HJ, Kenney SC (2005) BZLF1 activation of the methylated form of the BRLF1 immediate-early promoter is regulated by BZLF1 residue 186. J Virol 79:7338–7348

Bhende PM, Dickerson SJ, Sun X, Feng WH, Kenney SC (2007) X-box-binding protein 1 activates lytic Epstein-Barr virus gene expression in combination with protein kinase D. J Virol 81:7363–7370
Bloss TA, Sugden B (1994) Optimal lengths for DNAs encapsidated by Epstein-Barr virus. J Virol 68:8217–8222
Borras AM, Strominger JL, Speck SH (1996) Characterization of the ZI domains in the Epstein-Barr virus BZLF1 gene promoter: role in phorbol ester induction. J Virol 70:3894–3901
Bryant H, Farrell PJ (2002) Signal transduction and transcription factor modification during reactivation of Epstein-Barr virus from latency. J Virol 76:10290–10298
Buettner M, Lang A, Tudor CS, Meyer B, Cruchley A, Barros MH, Farrell PJ, Jack HM, Schuh W, Niedobitek G (2012) Lytic Epstein-Barr virus infection in epithelial cells but not in B-lymphocytes is dependent on Blimp1. J Gen Virol 93:1059–1064
Cao SM, Liu Z, Jia WH, Huang QH, Liu Q, Guo X, Huang TB, Ye W, Hong MH (2011) Fluctuations of epstein-barr virus serological antibodies and risk for nasopharyngeal carcinoma: a prospective screening study with a 20-year follow-up. PLoS ONE 6:e19100
Chan CK, Mueller N, Evans A, Harris NL, Comstock GW, Jellum E, Magnus K, Orentreich N, Polk BF, Vogelman J (1991) Epstein-Barr virus antibody patterns preceding the diagnosis of nasopharyngeal carcinoma. Cancer Causes Control 2:125–131
Chang LK, Liu ST (2000) Activation of the BRLF1 promoter and lytic cycle of Epstein-Barr virus by histone acetylation. Nucleic Acids Res 28:3918–3925
Chang LK, Chung JY, Hong YR, Ichimura T, Nakao M, Liu ST (2005) Activation of Sp1-mediated transcription by Rta of Epstein-Barr virus via an interaction with MCAF1. Nucleic Acids Res 33:6528–6539
Chang LK, Chuang JY, Nakao M, Liu ST (2010) MCAF1 and synergistic activation of the transcription of Epstein-Barr virus lytic genes by Rta and Zta. Nucleic Acids Res 38:4687–4700
Chatila T, Ho N, Liu P, Liu S, Mosialos G, Kieff E, Speck SH (1997) The Epstein-Barr virus-induced Ca2+/calmodulin-dependent kinase type IV/Gr promotes a Ca(2+)-dependent switch from latency to viral replication. J Virol 71:6560–6567
Chen JY, Hwang LY, Beasley RP, Chien CS, Yang CS (1985) Antibody response to Epstein-Barr-virus-specific DNase in 13 patients with nasopharyngeal carcinoma in Taiwan: a retrospective study. J Med Virol 16:99–105
Chen CJ, Deng Z, Kim AY, Blobel GA, Lieberman PM (2001) Stimulation of CREB binding protein nucleosomal histone acetyltransferase activity by a class of transcriptional activators. Mol Cell Biol 21:476–487
Chen LW, Chang PJ, Delecluse HJ, Miller G (2005) Marked variation in response of consensus binding elements for the Rta protein of Epstein-Barr virus. J Virol 79:9635–9650
Chen LW, Raghavan V, Chang PJ, Shedd D, Heston L, Delecluse HJ, Miller G (2009) Two phenylalanines in the C-terminus of Epstein-Barr virus Rta protein reciprocally modulate its DNA binding and transactivation function. Virology 386:448–461
Chi T, Lieberman P, Ellwood K, Carey M (1995) A general mechanism for transcriptional synergy by eukaryotic activators. Nature 377:254–257
Chien YC, Chen JY, Liu MY, Yang HI, Hsu MM, Chen CJ, Yang CS (2001) Serologic markers of Epstein-Barr virus infection and nasopharyngeal carcinoma in Taiwanese men. N Engl J Med 345:1877–1882
Cho MS, Bornkamm GW, zur Hausen H (1984) Structure of defective DNA molecules in Epstein-Barr virus preparations from P3HR-1 cells. J Virol 51:199–207
Countryman J, Miller G (1985) Activation of expression of latent Epstein-Barr herpesvirus after gene transfer with a small cloned subfragment of heterogeneous viral DNA. Proc Natl Acad Sci USA 82:4085–4089
Countryman JK, Gradoville L, Miller G (2008) Histone hyperacetylation occurs on promoters of lytic cycle regulatory genes in Epstein-Barr virus-infected cell lines which are refractory to disruption of latency by histone deacetylase inhibitors. J Virol 82:4706–4719

Crawford DH, Epstein MA, Bornkamm GW, Achong BG, Finerty S, Thompson JL (1979) Biological and biochemical observations on isolates of EB virus from the malignant epithelial cells of two nasopharyngeal carcinomas. Int J Cancer 24:294–302

Daibata M, Speck SH, Mulder C, Sairenji T (1994) Regulation of the BZLF1 promoter of Epstein-Barr virus by second messengers in anti-immunoglobulin-treated B cells. Virology 198:446–454

Daigle D, Megyola C, El-Guindy A, Gradoville L, Tuck D, Miller G, Bhaduri-McIntosh S (2010) Upregulation of STAT3 marks Burkitt lymphoma cells refractory to Epstein-Barr virus lytic cycle induction by HDAC inhibitors. J Virol 84:993–1004

Daigle D, Gradoville L, Tuck D, Schulz V, Wang'ondu R, Ye J, Gorres K, Miller G (2011) Valproic acid antagonizes the capacity of other histone deacetylase inhibitors to activate the Epstein-barr virus lytic cycle. J Virol 85:5628–5643

Daikoku T, Kudoh A, Fujita M, Sugaya Y, Isomura H, Shirata N, Tsurumi T (2005) Architecture of replication compartments formed during Epstein-Barr virus lytic replication. J Virol 79:3409–3418

Darenkov IA, Marcarelli MA, Basadonna GP, Friedman AL, Lorber KM, Howe JG, Crouch J, Bia MJ, Kliger AS, Lorber MI (1997) Reduced incidence of Epstein-Barr virus-associated posttransplant lymphoproliferative disorder using preemptive antiviral therapy. Transplantation 64:848–852

Darr CD, Mauser A, Kenney S (2001) Epstein-Barr virus immediate-early protein BRLF1 induces the lytic form of viral replication through a mechanism involving phosphatidylinositol- 3 kinase activation. J Virol 75:6135–6142

Davies AH, Grand RJ, Evans FJ, Rickinson AB (1991) Induction of Epstein-Barr virus lytic cycle by tumor-promoting and non-tumor-promoting phorbol esters requires active protein kinase C. J Virol 65:6838–6844

Davies ML, Xu S, Lyons-Weiler J, Rosendorff A, Webber SA, Wasil LR, Metes D, Rowe DT (2010) Cellular factors associated with latency and spontaneous Epstein-Barr virus reactivation in B-lymphoblastoid cell lines. Virology 400:53–67

Deng Z, Chen CJ, Zerby D, Delecluse HJ, Lieberman PM (2001) Identification of acidic and aromatic residues in the Zta activation domain essential for Epstein-Barr virus reactivation. J Virol 75:10334–10347

Dickerson SJ, Xing Y, Robinson AR, Seaman WT, Gruffat H, Kenney SC (2009) Methylation-dependent binding of the epstein-barr virus BZLF1 protein to viral promoters. PLoS Pathog 5:e1000356

El-Guindy AS, Miller G (2004) Phosphorylation of Epstein-Barr virus ZEBRA protein at its casein kinase 2 sites mediates its ability to repress activation of a viral lytic cycle late gene by Rta. J Virol 78:7634–7644

El-Guindy A, Heston L, Endo Y, Cho MS, Miller G (2002) Disruption of Epstein-Barr Virus Latency in the Absence of phosphorylation of ZEBRA by Protein Kinase C. J Virol 76:11199–11208

El-Guindy AS, Paek SY, Countryman J, Miller G (2006) Identification of constitutive phosphorylation sites on the Epstein-Barr virus ZEBRA protein. J Biol Chem 281:3085–3095

El-Guindy A, Heston L, Delecluse HJ, Miller G (2007) Phosphoacceptor site S173 in the regulatory domain of Epstein-Barr virus ZEBRA protein is required for lytic DNA replication but not for activation of viral early genes. J Virol 81:3303–3316

El-Guindy A, Heston L, Miller G (2010) A subset of replication proteins enhances origin recognition and lytic replication by the Epstein-Barr virus ZEBRA protein. PLoS Pathog 6

El-Guindy A, Ghiassi-Nejad M, Golden S, Delecluse HJ, Miller G (2013) Essential role of rta in lytic DNA replication of Epstein-Barr virus. J Virol 87:208–223

El-Guindy A, Lopez-Giraldez F, Delecluse HJ, McKenzie J, Miller G (2014) A locus encompassing the Epstein-Barr virus bglf4 kinase regulates expression of genes encoding viral structural proteins. PLoS Pathog 10:e1004307

Ellis AL, Wang Z, Yu X, Mertz JE (2010) Either ZEB1 or ZEB2/SIP1 can play a central role in regulating the Epstein-Barr virus latent-lytic switch in a cell-type-specific manner. J Virol 84:6139–6152
Ellis-Connell AL, Iempridee T, Xu I, Mertz JE (2010) Cellular microRNAs 200b and 429 regulate the Epstein-Barr virus switch between latency and lytic replication. J Virol 84:10329–10343
Fahmi H, Cochet C, Hmama Z, Opolon P, Joab I (2000) Transforming growth factor beta 1 stimulates expression of the Epstein-Barr virus BZLF1 immediate-early gene product ZEBRA by an indirect mechanism which requires the MAPK kinase pathway. J Virol 74:5810–5818
Farmer DG, McDiarmid SV, Winston D, Yersiz H, Cortina G, Dry S, Maxfield AJ, Vandenbogaart B, Correa M, Kroeber A, Geevarghese S, Busuttil RW (2002) Effectiveness of aggressive prophylatic and preemptive therapies targeted against cytomegaloviral and Epstein-Barr viral disease after human intestinal transplantation. Transp Proc 34:948–949
Farrell PJ, Rowe DT, Rooney CM, Kouzarides T (1989) Epstein-Barr virus BZLF1 trans-activator specifically binds to a consensus AP-1 site and is related to c-fos. EMBO J 8:127–132
Feederle R, Kost M, Baumann M, Janz A, Drouet E, Hammerschmidt W, Delecluse HJ (2000) The Epstein-Barr virus lytic program is controlled by the co-operative functions of two transactivators. EMBO J 19:3080–3089
Feng WH, Kenney SC (2006) Valproic acid enhances the efficacy of chemotherapy in EBV-positive tumors by increasing lytic viral gene expression. Cancer Res 66:8762–8769
Feng WH, Israel B, Raab-Traub N, Busson P, Kenney SC (2002) Chemotherapy induces lytic EBV replication and confers ganciclovir susceptibility to EBV-positive epithelial cell tumors. Cancer Res 62:1920–1926
Feng WH, Hong G, Delecluse HJ, Kenney SC (2004) Lytic induction therapy for Epstein-Barr virus-positive B-cell lymphomas. J Virol 78:1893–1902
Feng WH, Kraus RJ, Dickerson SJ, Lim HJ, Jones RJ, Yu X, Mertz JE, Kenney SC (2007) ZEB1 and c-Jun levels contribute to the establishment of highly lytic Epstein-Barr virus infection in gastric AGS cells. J Virol 81:10113–10122
Fixman ED, Hayward GS, Hayward SD (1992) trans-acting requirements for replication of Epstein-Barr virus ori-Lyt. J Virol 66:5030–5039
Fixman ED, Hayward GS, Hayward SD (1995) Replication of Epstein-Barr virus oriLyt: lack of a dedicated virally encoded origin-binding protein and dependence on Zta in cotransfection assays. J Virol 69:2998–3006
Flamand L, Menezes J (1996) Cyclic AMP-responsive element-dependent activation of Epstein-Barr virus zebra promoter by human herpesvirus 6. J Virol 70:1784–1791
Flemington E, Speck SH (1990a) Autoregulation of Epstein-Barr virus putative lytic switch gene BZLF1. J Virol 64:1227–1232
Flemington E, Speck SH (1990b) Epstein-Barr virus BZLF1 trans activator induces the promoter of a cellular cognate gene, c-fos. J Virol 64:4549–4552
Flemington E, Speck SH (1990c) Identification of phorbol ester response elements in the promoter of Epstein-Barr virus putative lytic switch gene BZLF1. J Virol 64:1217–1226
Flemington EK, Borras AM, Lytle JP, Speck SH (1992) Characterization of the Epstein-Barr virus BZLF1 protein transactivation domain. J Virol 66:922–929
Francis AL, Gradoville L, Miller G (1997) Alteration of a single serine in the basic domain of the Epstein-Barr virus ZEBRA protein separates its functions of transcriptional activation and disruption of latency. J Virol 71:3054–3061
Francis A, Ragoczy T, Gradoville L, El-Guindy A, Miller G (1999) Amino Acid substitutions reveal distinct functions of serine 186 of the ZEBRA protein in activation of lytic cycle genes and synergy with the EBV Rta transactivator. J Virol 73:4543–4551
Fruehling S, Longnecker R (1997) The immunoreceptor tyrosine-based activation motif of Epstein-Barr virus LMP2A is essential for blocking BCR-mediated signal transduction. Virology 235:241–251
Fu DX, Tanhehco Y, Chen J, Foss CA, Fox JJ, Chong JM, Hobbs RF, Fukayama M, Sgouros G, Kowalski J, Pomper MG, Ambinder RF (2008) Bortezomib-induced enzyme-targeted radiation therapy in herpesvirus-associated tumors. Nat Med 14:1118–1122

Gao Z, Krithivas A, Finan JE, Semmes OJ, Zhou S, Wang Y, Hayward SD (1998) The Epstein-Barr virus lytic transactivator Zta interacts with the helicase-primase replication proteins. J Virol 72:8559–8567
Glover JN, Harrison SC (1995) Crystal structure of the heterodimeric bZIP transcription factor c-Fos- c-Jun bound to DNA. Nature 373:257–261
Goswami R, Gershburg S, Satorius A, Gershburg E (2012) Protein kinase inhibitors that inhibit induction of lytic program and replication of Epstein-Barr virus. Antiviral Res 96:296–304
Green M (2001) Management of Epstein-Barr virus-induced post-transplant lymphoproliferative disease in recipients of solid organ transplantation. Am J Transp 1:103–108
Grogan EJ, Jenson J, Countryman J, Heston L, Gradoville L, Miller G (1987) Transfection of a rearranged viral DNA fragment WZhet, stably converts latent Epstein-Barr virus infection to productive infection in lymphoid cells. Proc Natl Acad Sci USA 84:1332–1336
Gruffat H, Manet E, Rigolet A, Sergeant A (1990) The enhancer factor R of Epstein-Barr virus (EBV) is a sequence- specific DNA binding protein. Nucleic Acids Res 18:6835–6843
Gruffat H, Manet E, Sergeant A (2002) MEF2-mediated recruitment of class II HDAC at the EBV immediate early gene BZLF1 links latency and chromatin remodeling. EMBO Rep 3:141–146
Gruffat H, Kadjouf F, Mariame B, Manet E (2012) The Epstein-Barr virus BcRF1 gene product is a TBP-like protein with an essential role in late gene expression. J Virol 86:6023–6032
Hadinoto V, Shapiro M, Sun CC, Thorley-Lawson DA (2009) The dynamics of EBV shedding implicate a central role for epithelial cells in amplifying viral output. PLoS Pathog 5:e1000496
Hagemeier SR, Dickerson SJ, Meng Q, Yu X, Mertz JE, Kenney SC (2010) Sumoylation of the Epstein-Barr virus BZLF1 protein inhibits its transcriptional activity and is regulated by the virus-encoded protein kinase. J Virol 84:4383–4394
Hagemeier SR, Barlow EA, Kleman AA, Kenney SC (2011) The Epstein-Barr virus BRRF1 protein, Na, induces lytic infection in a TRAF2- and p53-dependent manner. J Virol 85:4318–4329
Hagemeier SR, Barlow EA, Meng Q, Kenney SC (2012) The cellular ataxia telangiectasia-mutated kinase promotes Epstein-Barr virus lytic reactivation in response to multiple different types of lytic reactivation-inducing stimuli. J Virol 86:13360–13370
Hammerschmidt W, Sugden B (1988) Identification and characterization of oriLyt, a lytic origin of DNA replication of Epstein-Barr virus. Cell 55:427–433
Hammerschmidt W, Sugden B (2013) Replication of Epstein-Barr viral DNA. Cold Spring Harb Perspect Biol 5:a013029
Heilmann AM, Calderwood MA, Johannsen E (2010) Epstein-Barr virus LF2 protein regulates viral replication by altering Rta subcellular localization. J Virol 84:9920–9931
Heilmann AM, Calderwood MA, Portal D, Lu Y, Johannsen E (2012) Genome-wide analysis of Epstein-Barr virus Rta DNA binding. J Virol 86:5151–5164
Henle G, Henle W (1966) Immunofluorescence in cells derived from Burkitt's lymphoma. J Bacteriol 91:1248–1256
Hill ER, Koganti S, Zhi J, Megyola C, Freeman AF, Palendira U, Tangye SG, Farrell PJ, Bhaduri-McIntosh S (2013) Signal transducer and activator of transcription 3 limits Epstein-Barr virus lytic activation in B lymphocytes. J Virol 87:11438–11446
Hong GK, Delecluse HJ, Gruffat H, Morrison TE, Feng WH, Sergeant A, Kenney SC (2004) The BRRF1 early gene of Epstein-Barr virus encodes a transcription factor that enhances induction of lytic infection by BRLF1. J Virol 78:4983–4992
Hong GK, Gulley ML, Feng WH, Delecluse HJ, Holley-Guthrie E, Kenney SC (2005) Epstein-Barr virus lytic infection contributes to lymphoproliferative disease in a SCID mouse model. J Virol 79:13993–14003
Hsu M, Wu SY, Chang SS, Su IJ, Tsai CH, Lai SJ, Shiau AL, Takada K, Chang Y (2008) Epstein-Barr virus lytic transactivator Zta enhances chemotactic activity through induction of interleukin-8 in nasopharyngeal carcinoma cells. J Virol 82:3679–3688

Hsu WL, Chen JY, Chien YC, Liu MY, You SL, Hsu MM, Yang CS, Chen CJ (2009) Independent effect of EBV and cigarette smoking on nasopharyngeal carcinoma: a 20-year follow-up study on 9,622 males without family history in Taiwan. Cancer Epidemiol Biomark Prev 18:1218–1226

Huang J, Liao G, Chen H, Wu FY, Hutt-Fletcher L, Hayward GS, Hayward SD (2006) Contribution of C/EBP proteins to Epstein-Barr virus lytic gene expression and replication in epithelial cells. J Virol 80:1098–1109

Iempridee T, Das S, Xu I, Mertz JE (2011) Transforming growth factor beta-induced reactivation of Epstein-Barr virus involves multiple Smad-binding elements cooperatively activating expression of the latent-lytic switch BZLF1 gene. J Virol 85:7836–7848

Iwakiri D, Takada K (2004) Phosphatidylinositol 3-kinase is a determinant of responsiveness to B cell antigen receptor-mediated Epstein-Barr virus activation. J Immunol 172:1561–1566

Jacob RJ, Roizman B (1977) Anatomy of herpes simplex virus DNA VIII. Properties of the replicating DNA. J Virol 23:394–411

Jenkins PJ, Binne UK, Farrell PJ (2000) Histone acetylation and reactivation of Epstein-Barr virus from latency. J Virol 74:710–720

Jiang JH, Wang N, Li A, Liao WT, Pan ZG, Mai SJ, Li DJ, Zeng MS, Wen JM, Zeng YX (2006) Hypoxia can contribute to the induction of the Epstein-Barr virus (EBV) lytic cycle. J Clin Virol 37:98–103

Kalla M, Hammerschmidt W (2012) Human B cells on their route to latent infection–early but transient expression of lytic genes of Epstein-Barr virus. Eur J Cell Biol 91:65–69

Kalla M, Schmeinck A, Bergbauer M, Pich D, Hammerschmidt W (2010) AP-1 homolog BZLF1 of Epstein-Barr virus has two essential functions dependent on the epigenetic state of the viral genome. Proc Natl Acad Sci USA 107:850–855

Kalla M, Gobel C, Hammerschmidt W (2012) The lytic phase of Epstein-Barr virus requires a viral genome with 5-methylcytosine residues in CpG sites. J Virol 86:447–458

Karimi L, Crawford DH, Speck S, Nicholson LJ (1995) Identification of an epithelial cell differentiation responsive region within the BZLF1 promoter of the Epstein-Barr virus. J Gen Virol 76:759–765

Katsumura KR, Maruo S, Takada K (2012) EBV lytic infection enhances transformation of B-lymphocytes infected with EBV in the presence of T-lymphocytes. J Med Virol 84:504–510

Kenney SC, Mertz JE (2014) Regulation of the latent-lytic switch in Epstein-Barr virus. Semin Cancer Biol

Kintner C, Sugden B (1981) Conservation and progressive methylation of Epstein-Barr viral DNA sequences in transformed cells. J Virol 38:305–316

Kouzarides T, Packham G, Cook A, Farrell PJ (1991) The BZLF1 protein of EBV has a coiled coil dimerisation domain without a heptad leucine repeat but with homology to the C/EBP leucine zipper. Oncogene 6:195–204

Kraus RJ, Perrigoue JG, Mertz JE (2003) ZEB negatively regulates the lytic-switch BZLF1 gene promoter of Epstein-Barr virus. J Virol 77:199–207

Kutok JL, Wang F (2006) Spectrum of Epstein-Barr virus-associated diseases. Annu Rev Pathol 1:375–404

Lai IY, Farrell PJ, Kellam P (2011) X-box binding protein 1 induces the expression of the lytic cycle transactivator of Kaposi's sarcoma-associated herpesvirus but not Epstein-Barr virus in co-infected primary effusion lymphoma. J Gen Virol 92:421–431

Lee TC, Savoldo B, Rooney CM, Heslop HE, Gee AP, Caldwell Y, Barshes NR, Scott JD, Bristow LJ, O'Mahony CA, Goss JA (2005) Quantitative EBV viral loads and immunosuppression alterations can decrease PTLD incidence in pediatric liver transplant recipients. Am J Transp (Official Journal of The American Society of Transplantation and the American Society of Transplant Surgeons) 5:2222–2228

Lehman AM, Ellwood KB, Middleton BE, Carey M (1998) Compensatory energetic relationships between upstream activators and the RNA polymerase II general transcription machinery. J Biol Chem 273:932–939

Lemon SM, Hutt LM, Shaw JE, Li JL, Pagano JS (1978) Replication of Epstein-Barr virus DNA in epithelial cells in vivo. IARC Sci Publ 739–744

Li M, Linseman DA, Allen MP, Meintzer MK, Wang X, Laessig T, Wierman ME, Heidenreich KA (2001) Myocyte enhancer factor 2A and 2D undergo phosphorylation and caspase-mediated degradation during apoptosis of rat cerebellar granule neurons. J Neurosci 21:6544–6552

Liang CL, Chen JL, Hsu YP, Ou JT, Chang YS (2002) Epstein-Barr virus BZLF1 gene is activated by transforming growth factor-beta through cooperativity of Smads and c-Jun/c-Fos proteins. J Biol Chem 277:23345–23357

Liang X, Collins CM, Mendel JB, Iwakoshi NN, Speck SH (2009) Gammaherpesvirus-driven plasma cell differentiation regulates virus reactivation from latently infected B lymphocytes. PLoS Pathog 5:e1000677

Liao G, Huang J, Fixman ED, Hayward SD (2005) The Epstein-Barr virus replication protein BBLF2/3 provides an origin-tethering function through interaction with the zinc finger DNA binding protein ZBRK1 and the KAP-1 corepressor. J Virol 79:245–256

Lieberman PM, Berk AJ (1991) The Zta trans-activator protein stabilizes TFIID association with promoter DNA by direct protein-protein interaction. Genes Dev 5:2441–2454

Lieberman PM, Berk AJ (1994) A mechanism for TAFs in transcriptional activation: activation domain enhancement of TFIID-TFIIA–promoter DNA complex formation. Genes Dev 8:995–1006

Lieberman PM, Hardwick JM, Sample J, Hayward GS, Hayward SD (1990) The zta transactivator involved in induction of lytic cycle gene expression in Epstein-Barr virus-infected lymphocytes binds to both AP-1 and ZRE sites in target promoter and enhancer regions. J Virol 64:1143–1155

Lin Z, Wang X, Fewell C, Cameron J, Yin Q, Flemington EK (2010) Differential expression of the miR-200 family microRNAs in epithelial and B cells and regulation of Epstein-Barr virus reactivation by the miR-200 family member miR-429. J Virol 84:7892–7897

Liu C, Sista ND, Pagano JS (1996) Activation of the Epstein-Barr virus DNA polymerase promoter by the BRLF1 immediate-early protein is mediated through USF and E2F. J Virol 70:2545–2555

Liu S, Borras AM, Liu P, Suske G, Speck SH (1997a) Binding of the ubiquitous cellular transcription factors Sp1 and Sp3 to the ZI domains in the Epstein-Barr virus lytic switch BZLF1 gene promoter. Virology 228:11–18

Liu S, Liu P, Borras A, Chatila T, Speck SH (1997b) Cyclosporin A-sensitive induction of the Epstein-Barr virus lytic switch is mediated via a novel pathway involving a MEF2 family member. EMBO J 16:143–153

Liu P, Liu S, Speck SH (1998) Identification of a negative cis element within the ZII domain of the Epstein-Barr virus lytic switch BZLF1 gene promoter. J Virol 72:8230–8239

Lu CC, Jeng YY, Tsai CH, Liu MY, Yeh SW, Hsu TY, Chen MR (2006) Genome-wide transcription program and expression of the Rta responsive gene of Epstein-Barr virus. Virology 345:358–372

Lucas KG, Burton RL, Zimmerman SE, Wang J, Cornetta KG, Robertson KA, Lee CH, Emanuel DJ (1998) Semiquantitative Epstein-Barr virus (EBV) polymerase chain reaction for the determination of patients at risk for EBV-induced lymphoproliferative disease after stem cell transplantation. Blood 91:3654–3661

Luka J, Kallin B, Klein G (1979) Induction of the Epstein-Barr virus (EBV) cycle in latently infected cells by n-butyrate. Virology 94:228–231

Ma SD, Hegde S, Young KH, Sullivan R, Rajesh D, Zhou Y, Jankowska-Gan E, Burlingham WJ, Sun X, Gulley ML, Tang W, Gumperz JE, Kenney SC (2011) A new model of Epstein-Barr virus infection reveals an important role for early lytic viral protein expression in the development of lymphomas. J Virol 85:165–177

Manet E, Rigolet A, Gruffat H, Giot JF, Sergeant A (1991) Domains of the Epstein-Barr virus (EBV) transcription factor R required for dimerization, DNA binding and activation. Nucleic Acids Res 19:2661–2667

Manet E, Allera C, Gruffat H, Mikaelian I, Rigolet A, Sergeant A (1993) The acidic activation domain of the Epstein-Barr virus transcription factor R interacts in vitro with both TBP and TFIIB and is cell- specifically potentiated by a proline-rich region. Gene Expr 3:49–59

Mansouri S, Wang S, Frappier L (2013) A role for the nucleosome assembly proteins TAF-Ibeta and NAP1 in the activation of BZLF1 expression and Epstein-Barr virus reactivation. PLoS ONE 8:e63802

Matsuzaki K (2011) Smad phosphoisoform signaling specificity: the right place at the right time. Carcinogenesis 32:1578–1588

McDiarmid SV, Jordan S, Kim GS, Toyoda M, Goss JA, Vargas JH, Martin MG, Bahar R, Maxfield AL, Ament ME, Busuttil RW (1998) Prevention and preemptive therapy of postransplant lymphoproliferative disease in pediatric liver recipients. Transplantation 66:1604–1611

McDonald C, Karstegl CE, Kellam P, Farrell PJ (2010) Regulation of the Epstein-Barr virus Zp promoter in B lymphocytes during reactivation from latency. J Gen Virol 91:622–629

Meng Q, Hagemeier SR, Fingeroth JD, Gershburg E, Pagano JS, Kenney SC (2010) The Epstein-Barr virus (EBV)-encoded protein kinase, EBV-PK, but not the thymidine kinase (EBV-TK), is required for ganciclovir and acyclovir inhibition of lytic viral production. J Virol 84:4534–4542

Miller MH, Stitt D, Miller G (1970) Epstein-Barr viral antigen in single cell clones of two human leukocytic lines. J Virol 6:699–701

Miller G, Robinson J, Heston L, Lipman M (1974) Differences between laboratory strains of Epstein-Barr virus based on immortalization, abortive infection, and interference. Proc Natl Acad Sci USA 71:4006–4010

Miller G, El-Guindy A, Countryman J, Ye J, Gradoville L (2007) Lytic cycle switches of oncogenic human gammaherpesviruses. Adv Cancer Res 97:81–109

Mueller N, Evans A, Harris NL, Comstock GW, Jellum E, Magnus K, Orentreich N, Polk BF, Vogelman J (1989) Hodgkin's disease and Epstein-Barr virus. Altered antibody pattern before diagnosis. N Engl J Med 320:689–695

Mueller N, Mohar A, Evans A, Harris NL, Comstock GW, Jellum E, Magnus K, Orentreich N, Polk BF, Vogelman J (1991) Epstein-Barr virus antibody patterns preceding the diagnosis of non-Hodgkin's lymphoma. Int J Cancer 49:387–393

Murata T, Noda C, Saito S, Kawashima D, Sugimoto A, Isomura H, Kanda T, Yokoyama KK, Tsurumi T (2011) Involvement of Jun dimerization protein 2 (JDP2) in the maintenance of Epstein-Barr virus latency. J Biol Chem 286:22007–22016

Murata T, Kondo Y, Sugimoto A, Kawashima D, Saito S, Isomura H, Kanda T, Tsurumi T (2012) Epigenetic histone modification of Epstein-Barr virus BZLF1 promoter during latency and reactivation in Raji cells. J Virol 86:4752–4761

Murata T, Narita Y, Sugimoto A, Kawashima D, Kanda T, Tsurumi T (2013) Contribution of myocyte enhancer factor 2 family transcription factors to BZLF1 expression in Epstein-Barr virus reactivation from latency. J Virol 87:10148–10162

Packard TA, Cambier JC (2013) B lymphocyte antigen receptor signaling: initiation, amplification, and regulation. F1000Prime Rep 5:40

Park SM, Gaur AB, Lengyel E, Peter ME (2008) The miR-200 family determines the epithelial phenotype of cancer cells by targeting the E-cadherin repressors ZEB1 and ZEB2. Genes Dev 22:894–907

Perrine SP, Hermine O, Small T, Suarez F, O'Reilly R, Boulad F, Fingeroth J, Askin M, Levy A, Mentzer SJ, Di Nicola M, Gianni AM, Klein C, Horwitz S, Faller DV (2007) A phase 1/2 trial of arginine butyrate and ganciclovir in patients with Epstein-Barr virus-associated lymphoid malignancies. Blood 109:2571–2578

Petosa C, Morand P, Baudin F, Moulin M, Artero JB, Muller CW (2006) Structural basis of lytic cycle activation by the Epstein-Barr virus ZEBRA protein. Mol Cell 21:565–572

Pfuller R, Hammerschmidt W (1996) Plasmid-like replicative intermediates of the Epstein-Barr virus lytic origin of DNA replication. J Virol 70:3423–3431

Quinlivan EB, Holley-Guthrie EA, Norris M, Gutsch D, Bachenheimer SL, Kenney SC (1993) Direct BRLF1 binding is required for cooperative BZLF1/BRLF1 activation of the Epstein-Barr virus early promoter, BMRF1. Nucleic Acids Res 21:1999–2007 (corrected and republished with original paging, article originally printed in Nucleic Acids Res 1993 Apr 25; 21(8):1999–2007)
Quinn ZA, Yang CC, Wrana JL, McDermott JC (2001) Smad proteins function as co-modulators for MEF2 transcriptional regulatory proteins. Nucleic Acids Res 29:732–742
Raab-Traub N, Dambaugh T, Kieff E (1980) DNA of Epstein-Barr virus VIII: B95-8, the previous prototype, is an unusual deletion derivative. Cell 22:257–267
Ragoczy T, Miller G (1999) Role of the Epstein-Barr virus RTA protein in activation of distinct classes of viral lytic cycle genes. J Virol 73:9858–9866
Ragoczy T, Heston L, Miller G (1998) The Epstein-Barr virus Rta protein activates lytic cycle genes and can disrupt latency in B lymphocytes. J Virol 72:7978–7984
Ramasubramanyan S, Osborn K, Flower K, Sinclair AJ (2012) Dynamic chromatin environment of key lytic cycle regulatory regions of the Epstein-Barr virus genome. J Virol 86:1809–1819
Rasche L, Kapp M, Einsele H, Mielke S (2013) EBV-induced post transplant lymphoproliferative disorders: a persisting challenge in allogeneic hematopoetic SCT. Bone Marrow Transp
Raver RM, Panfil AR, Hagemeier SR, Kenney SC (2013) The B-cell-specific transcription factor and master regulator Pax5 promotes Epstein-Barr virus latency by negatively regulating the viral immediate early protein BZLF1. J Virol 87:8053–8063
Rennekamp AJ, Lieberman PM (2011) Initiation of Epstein-Barr virus lytic replication requires transcription and the formation of a stable RNA-DNA hybrid molecule at OriLyt. J Virol 85:2837–2850
Reusch JA, Nawandar DM, Wright KL, Kenney SC, Mertz JE (2014) Cellular differentiation regulator BLIMP1 induces Epstein-Barr virus lytic reactivation in epithelial and B cells by activating transcription from both the R and Z promoters. J Virol
Robinson AR, Kwek SS, Hagemeier SR, Wille CK, Kenney SC (2011) Cellular transcription factor Oct-1 interacts with the Epstein-Barr virus BRLF1 protein to promote disruption of viral latency. J Virol 85:8940–8953
Robinson AR, Kwek SS, Kenney SC (2012) The B-cell specific transcription factor, Oct-2, promotes Epstein-Barr virus latency by inhibiting the viral immediate-early protein, BZLF1. PLoS Pathog 8:e1002516
Rooney C, Taylor N, Countryman J, Jenson H, Kolman J, Miller G (1988) Genome rearrangements activate the Epstein-Barr virus gene whose product disrupts latency. Proc Natl Acad Sci USA 85:9801–9805
Rooney CM, Loftin SK, Holladay MS, Brenner MK, Krance RA, Heslop HE (1995) Early identification of Epstein-Barr virus-associated post-transplantation lymphoproliferative disease. Br J Haematol 89:98–103
Satoh T, Hoshikawa Y, Satoh Y, Kurata T, Sairenji T (1999) The interaction of mitogen-activated protein kinases to Epstein-Barr virus activation in Akata cells. Virus Genes 18:57–64
Schepers A, Pich D, Mankertz J, Hammerschmidt W (1993) cis-acting elements in the lytic origin of DNA replication of Epstein-Barr virus. J Virol 67:4237–4245
Serio TR, Kolman JL, Miller G (1997) Late gene expression from the Epstein-Barr virus BcLF1 and BFRF3 promoters does not require DNA replication in cis. J Virol 71:8726–8734
Serio TR, Cahill N, Prout ME, Miller G (1998) A functionally distinct TATA box required for late progression through the Epstein-Barr virus life cycle. J Virol 72:8338–8343
Seto E, Moosmann A, Gromminger S, Walz N, Grundhoff A, Hammerschmidt W (2010) Micro RNAs of Epstein-Barr virus promote cell cycle progression and prevent apoptosis of primary human B cells. PLoS Pathog 6:e1001063
Shinozaki A, Sakatani T, Ushiku T, Hino R, Isogai M, Ishikawa S, Uozaki H, Takada K, Fukayama M (2010) Downregulation of microRNA-200 in EBV-associated gastric carcinoma. Cancer Res 70:4719–4727

Shirley CM, Chen J, Shamay M, Li H, Zahnow CA, Hayward SD, Ambinder RF (2011) Bortezomib induction of C/EBPbeta mediates Epstein-Barr virus lytic activation in Burkitt lymphoma. Blood 117:6297–6303
Sixbey JW, Vesterinen EH, Nedrud JG, Raab-Traub N, Walton LA, Pagano JS (1983) Replication of Epstein-Barr virus in human epithelial cells infected in vitro. Nature 306:480–483
Sixbey JW, Nedrud JG, Raab-Traub N, Hanes RA, Pagano JS (1984) Epstein-Barr virus replication in oropharyngeal epithelial cells. N Engl J Med 310:1225–1230
Summers WC, Klein G (1976) Inhibition of Epstein-Barr virus DNA synthesis and late gene expression by phosphonoacetic acid. J Virol 18:151–155
Sun CC, Thorley-Lawson DA (2007) Plasma cell-specific transcription factor XBP-1s binds to and transactivates the Epstein-Barr virus BZLF1 promoter. J Virol 81:13566–13577
Swenson JJ, Holley-Guthrie E, Kenney SC (2001) Epstein-Barr virus immediate-early protein BRLF1 interacts with CBP, promoting enhanced BRLF1 transactivation. J Virol 75:6228–6234
Takada K (1984) Cross-linking of cell surface immunoglobulins induces Epstein-Barr virus in Burkitt lymphoma lines. Int J Cancer 33:27–32
Takada K, Ono Y (1989) Synchronous and sequential activation of latently infected Epstein-Barr virus genomes. J Virol 63:445–449
Taylor GM, Raghuwanshi SK, Rowe DT, Wadowsky RM, Rosendorff A (2011) Endoplasmic reticulum stress causes EBV lytic replication. Blood 118:5528–5539
Tikhmyanova N, Schultz DC, Lee T, Salvino JM, Lieberman PM (2014) Identification of a new class of small molecules that efficiently reactivate latent Epstein-Barr virus. ACS Chem Biol 9:785–795
Tovey MG, Lenoir G, Begon-Lours J (1978) Activation of latent Epstein-Barr virus by antibody to human IgM. Nature 276:270–272
Tsai DE, Douglas L, Andreadis C, Vogl DT, Arnoldi S, Kotloff R, Svoboda J, Bloom RD, Olthoff KM, Brozena SC, Schuster SJ, Stadtmauer EA, Robertson ES, Wasik MA, Ahya VN (2008) EBV PCR in the diagnosis and monitoring of posttransplant lymphoproliferative disorder: results of a two-arm prospective trial. Am J Transp 8:1016–1024 (Official Journal of The American Society of Transplantation and The American Society of Transplant Surgeons)
Tsai PF, Lin SJ, Weng PL, Tsai SC, Lin JH, Chou YC, Tsai CH (2011) Interplay between PKCdelta and Sp1 on histone deacetylase inhibitor-mediated Epstein-Barr virus reactivation. J Virol 85:2373–2385
Tsai MH, Raykova A, Klinke O, Bernhardt K, Gartner K, Leung CS, Geletneky K, Sertel S, Munz C, Feederle R, Delecluse HJ (2013) Spontaneous lytic replication and epitheliotropism define an Epstein-Barr virus strain found in carcinomas. Cell Rep 5:458–470
van Esser JW, van der Holt B, Meijer E, Niesters HG, Trenschel R, Thijsen SF, van Loon AM, Frassoni F, Bacigalupo A, Schaefer UW, Osterhaus AD, Gratama JW, Lowenberg B, Verdonck LF, Cornelissen JJ (2001) Epstein-Barr virus (EBV) reactivation is a frequent event after allogeneic stem cell transplantation (SCT) and quantitatively predicts EBV-lymphoproliferative disease following T-cell–depleted SCT. Blood 98:972–978
Wang YC, Huang JM, Montalvo EA (1997) Characterization of proteins binding to the ZII element in the Epstein-Barr virus BZLF1 promoter: transactivation by ATF1. Virology 227:323–330
Wang P, Day L, Dheekollu J, Lieberman PM (2005) A redox-sensitive cysteine in Zta is required for Epstein-Barr virus lytic cycle DNA replication. J Virol 79:13298–13309
Wen W, Iwakiri D, Yamamoto K, Maruo S, Kanda T, Takada K (2007) Epstein-Barr virus BZLF1 gene, a switch from latency to lytic infection, is expressed as an immediate-early gene after primary infection of B lymphocytes. J Virol 81:1037–1042
Wildeman MA, Novalic Z, Verkuijlen SA, Juwana H, Huitema AD, Tan IB, Middeldorp JM, de Boer JP, Greijer AE (2012) Cytolytic virus activation therapy for Epstein-Barr virus-driven tumors. Clin Cancer Res 18:5061–5070

Wu FY, Wang SE, Chen H, Wang L, Hayward SD, Hayward GS (2004) CCAAT/enhancer binding protein alpha binds to the Epstein-Barr virus (EBV) ZTA protein through oligomeric interactions and contributes to cooperative transcriptional activation of the ZTA promoter through direct binding to the ZII and ZIIIB motifs during induction of the EBV lytic cycle. J Virol 78:4847–4865

Wyrwicz LS, Rychlewski L (2007) Identification of Herpes TATT-binding protein. Antiviral Res 75:167–172

Ye J, Gradoville L, Daigle D, Miller G (2007) De novo protein synthesis is required for lytic cycle reactivation of Epstein-Barr virus, but not Kaposi's sarcoma-associated herpesvirus, in response to histone deacetylase inhibitors and protein kinase C agonists. J Virol 81:9279–9291

Yin Q, Jupiter K, Flemington EK (2004) The Epstein-Barr virus transactivator Zta binds to its own promoter and is required for full promoter activity during anti-Ig and TGF-beta1 mediated reactivation. Virology 327:134–143

Yu X, Wang Z, Mertz JE (2007) ZEB1 regulates the latent-lytic switch in infection by Epstein-Barr virus. PLoS Pathog 3:e194

Yu X, McCarthy PJ, Lim HJ, Iempridee T, Kraus RJ, Gorlen DA, Mertz JE (2011) The ZIIR element of the Epstein-Barr virus BZLF1 promoter plays a central role in establishment and maintenance of viral latency. J Virol 85:5081–5090

Yu X, McCarthy PJ, Wang Z, Gorlen DA, Mertz JE (2012) Shutoff of BZLF1 gene expression is necessary for immortalization of primary B cells by Epstein-Barr virus. J Virol 86:8086–8096

Yu KP, Heston L, Park R, Ding Z, Wang'ondu R, Delecluse HJ, Miller G (2013) Latency of Epstein-Barr virus is disrupted by gain-of-function mutant cellular AP-1 proteins that preferentially bind methylated DNA. Proc Natl Acad Sci USA 110:8176–8181

Yuan J, Cahir-McFarland E, Zhao B, Kieff E (2006) Virus and cell RNAs expressed during Epstein-Barr virus replication. J Virol 80:2548–2565

Zacny VL, Wilson J, Pagano JS (1998) The Epstein-Barr virus immediate-early gene product, BRLF1, interacts with the retinoblastoma protein during the viral lytic cycle. J Virol 72:8043–8051

Zalani S, Holley-Guthrie E, Kenney S (1996) Epstein-Barr viral latency is disrupted by the immediate-early BRLF1 protein through a cell-specific mechanism. Proc Natl Acad Sci USA 93:9194–9199

Zeidler R, Eissner G, Meissner P, Uebel S, Tampe R, Lazis S, Hammerschmidt W (1997) Downregulation of TAP1 in B lymphocytes by cellular and Epstein-Barr virus-encoded interleukin-10. Blood 90:2390–2397

Zerby D, Chen CJ, Poon E, Lee D, Shiekhattar R, Lieberman PM (1999) The amino-terminal C/H1 domain of CREB binding protein mediates zta transcriptional activation of latent Epstein-Barr virus. Mol Cell Biol 19:1617–1626

Zhang Q, Hong Y, Dorsky D, Holley-Guthrie E, Zalani S, Elshiekh NA, Kiehl A, Le T, Kenney S (1996) Functional and physical interactions between the Epstein-Barr virus (EBV) proteins BZLF1 and BMRF1: effects on EBV transcription and lytic replication. J Virol 70:5131–5142

Zhang W, Ou J, Inagaki Y, Greenwel P, Ramirez F (2000) Synergistic cooperation between Sp1 and Smad3/Smad4 mediates transforming growth factor beta1 stimulation of alpha 2(I)-collagen (COL1A2) transcription. J Biol Chem 275:39237–39245

zur Hausen H, O'Neil F, Freese U (1978) Persisting oncogenic herpesviruses induced by the tumor promoter TPA. Nature 272:373–375

# Part III
# Immune Responses to EBV

# Innate Immune Recognition of EBV

**Anna Lünemann, Martin Rowe and David Nadal**

**Abstract** The ability of Epstein–Barr virus (EBV) to establish latency despite specific immune responses and to successfully persist lifelong in the human host shows that EBV has developed powerful strategies and mechanisms to exploit, evade, abolish, or downsize otherwise effective immune responses to ensure its own survival. This chapter focuses on current knowledge on innate immune responses against EBV and its evasion strategies for own benefit and summarizes the questions that remain to be tackled. Innate immune reactions against EBV originate both from the main target cells of EBV and from nontarget cells, which are elements of the innate immune system. Thus, we structured our review accordingly but with a particular focus on the innate recognition of EBV in its two stages in its life cycle, latent state and lytic replication. Specifically, we discuss (I) innate sensing and resulting innate immune responses against EBV by its main target cells, focusing on (i) EBV transmission between epithelial cells and B cells and their life cycle stages; and (ii) elements of innate immunity in EBV's target cells. Further, we debate (II) the innate recognition and resulting innate immune responses against EBV by cells other than the main target cells, focusing on (iii) myeloid cells: dendritic cells, monocytes, macrophages, and neutrophil granulocytes; and (iv) natural killer cells. Finally, we address (III) how EBV counteracts or exploits innate immunity in its latent and lytic life cycle stages, concentrating on (v) TLRs; (vi) EBERs; and (vii) microRNAs.

A. Lünemann · D. Nadal (✉)
Division of Infectious Diseases and Hospital Epidemiology, University Children's Hospital of Zurich, Steinwiesstrasse 75, 8032 Zurich, Switzerland
e-mail: david.nadal@kispi.uzh.ch

A. Lünemann · D. Nadal
Children's Research Center, University Children's Hospital of Zurich, Zurich, Switzerland

M. Rowe
Centre for Human Virology, School of Cancer Sciences, College of Medical and Dental Sciences, The University of Birmingham, Birmingham, UK



C. Münz (ed.), *Epstein Barr Virus Volume 2*, Current Topics in Microbiology and Immunology 391, DOI 10.1007/978-3-319-22834-1_9

## Contents



# 1 Introduction

The human host has evolved various strategies and mechanisms to efficiently combat pathogens or malignant cells invading and emerging within its organism. Central elements of these are innate and adaptive immune responses. While innate responses are thought to be unspecific and fast, adaptive immunity can build long-lasting, efficient, and specific memory and thus protect against cognate antigens (Murphy et al. 2012).

According to textbook immunology, innate immune responses build the crucial but unspecific first line of defense. They provide rapid action if specific adaptive immunity has not been mounted yet. Additionally, there was thought to be no difference between primary and secondary encounters in innate responses. These dogmas have been recently challenged from several angles (Waggoner et al. 2012; Gasteiger and Rudensky 2014; O'Leary et al. 2006; Cooper et al. 2009; Paust et al. 2010). Innate immune responses differ substantially depending on the nature of the intruder (Murphy et al. 2012). Moreover, there have been descriptions of targeted memory-like functions independent of the adaptive immune system upon secondary encounters (Cooper et al. 2009; O'Leary et al. 2006; Paust et al. 2010; Iwasaki and Medzhitov 2004). In addition, innate immune cells have been shown to contribute to, modify, and regulate adaptive immune responses (Waggoner et al. 2012; White et al. 2014), and they have been found to be regulated directly or indirectly by and interact with adaptive immune cells (Gasteiger and Rudensky 2014; Horowitz et al. 2010; White et al. 2014). Thus, innate immunity not only provides the crucial first line of defense. Additionally, it paves the way for adaptive immunity upon a primary encounter of the culprit, since it closely interacts with adaptive immunity and can powerfully augment and complement its efficient protection on secondary encounters.

Epstein–Barr virus (EBV) infects more than 90 % of the population (Young and Rickinson 2004) and establishes lifelong persistence in the host despite immune responses. Innate and adaptive immunities, therefore, need to cope with primary and secondary encounters with EBV, in addition to its latency and reactivation. In a minority of infected individuals, EBV is strongly linked to a remarkable variety of nonmalignant and malignant diseases. It is associated with 2 % of all tumors in humans (Münz 2014). This successful lifelong persistence and, in particular, its ability to establish latency despite specific immune responses show that EBV has developed powerful strategies and mechanisms to exploit, evade, abolish, or downsize effective immune responses to ensure its own survival.

The evasion strategies against adaptive immune responses will be reviewed elsewhere in another chapter in this book by Emmanuel Wiertz. This chapter aims at reviewing current knowledge and summarizing open questions on innate immune responses against EBV and the evasion strategies of the virus for its own benefit. Innate immune reactions against EBV can be triggered both in its main target cells of EBV and in other cells of the innate immune system. Thus, we segregate our review accordingly but with a particular focus on the innate recognition of EBV. Since EBV as a gammaherpesvirus has two stages in its life cycle, latent state and lytic replication, which express distinct viral gene patterns, we address the innate immune recognition for all relevant cells in relation to the life cycle stages of EBV.

# 2 Innate Sensing and Resulting Innate Immune Responses Against EBV by Its Main Target Cells

## *2.1 EBV Transmission Between Epithelial Cells and B Cells and Their Life Cycle Stages*

The main target cells of EBV are B cells and oropharyngeal epithelial cells, the first innate barrier for infection being the oropharyngeal epithelium.

The understanding of how EBV gets access to B cells through the epithelial barrier is still debated and not clear-cut. One current model stipulates that EBV released from epithelial cells is B-cell tropic and that EBV released from B cells is epithelial cell tropic (reviewed in Shannon-Lowe and Rowe 2014). Recent in vitro studies have documented that apical to basolateral transcytosis of EBV in epithelial cells is possible, suggesting that this route might contribute to initial EBV penetration resulting in systemic infection (Tugizov et al. 2013). EBV's initial penetration, however, might happen also via transcytosis through M cells in tonsillar epithelium that function as antigen samplers (Fujimura 2000). Such pathways would be compatible with the fact that EBV in saliva is mainly B-cell tropic and has to be able to penetrate the basal membrane to access the B cells, but the physiologically relevant mechanisms remain to be determined.

Considerably, more studies have addressed the egress of EBV from the body via epithelial cells. In vitro studies have shown that cell-free EBV virions preferentially enter epithelial cells at the basolateral membrane and that highly efficient apical EBV cell-to-cell transmission can ocurr (Tugizov et al. 2003). In addition, other in vitro studies have revealed that direct cell–cell contact of B cells undergoing lytic virus replication greatly enhances the efficiency of epithelial cell infection (Imai et al. 1998) and that EBV bound to the surface of primary B cells initiates high-affinity binding to epithelial cells and its subsequent transfer (Shannon-Lowe et al. 2006). There is some evidence that oropharyngeal epithelium may act as an amplifier for EBV when the virus is shed into saliva (Hadinoto et al. 2009). Figure 1 shows current hypotheses of EBV's infection cycle between tonsils and peripheral blood in the main target cells. In summary, while it is clear that epithelial cells can be infected with EBV by different mechanisms, further studies are required to establish which mechanisms are relevant for virus entry to the host and which are relevant for virus egress.

In B cells, EBV establishes latency as default (Fig. 1a, b) and undergoes lytic replication rarely and only under certain tightly regulated conditions (Fig. 1a, c), one being their differentiation to plasma cells in tonsils (Laichalk and Thorley-Lawson 2005). It generally presumed that EBV replicates mainly in epithelial cells in the oropharynx (Hadinoto et al. 2009) where the default stage of infection seems to be lytic replication. Although EBV may establish a latent infection in normal epithelial cells, the episomal genome appears not to be amplified and subsequently the virus is lost in cells following cell division (Shannon-Lowe et al. 2009). Thus, EBV is thought to undergo replication to be efficiently transmitted and spread via saliva to other hosts (Fig. 1a, c).

This distinct behavior of EBV in the different target cells underlines that the innate immune recognition and subsequent responses by innate immune cells must combat distinct challenges. Furthermore, the interaction of EBV with its host cell might differ markedly. Indeed, this diversity is inherently reflected by their expression patterns of innate immune receptors of these cells.

## 2.2 Elements of Innate Immunity in EBV's Target Cells

Innate immunity receptors sense diverse pathogen-associated molecular patterns (PAMPs), and distinct PAMPs are recognized via a limited number of germline-encoded pattern-recognition receptors (PRRs). These include Toll-like receptors (TLRs), nucleotide-binding oligomerization-like receptors (NLRs), retinoic acid-inducible gene-like receptors (RIG), and C-type lectin-like receptors. The expression of PRRs varies considerably between subsets of immune cells and also between immune cells and other cell types (Iwasaki and Medzhitov 2004). In the recent years, PAMP triggering by EBV products has been found to result in

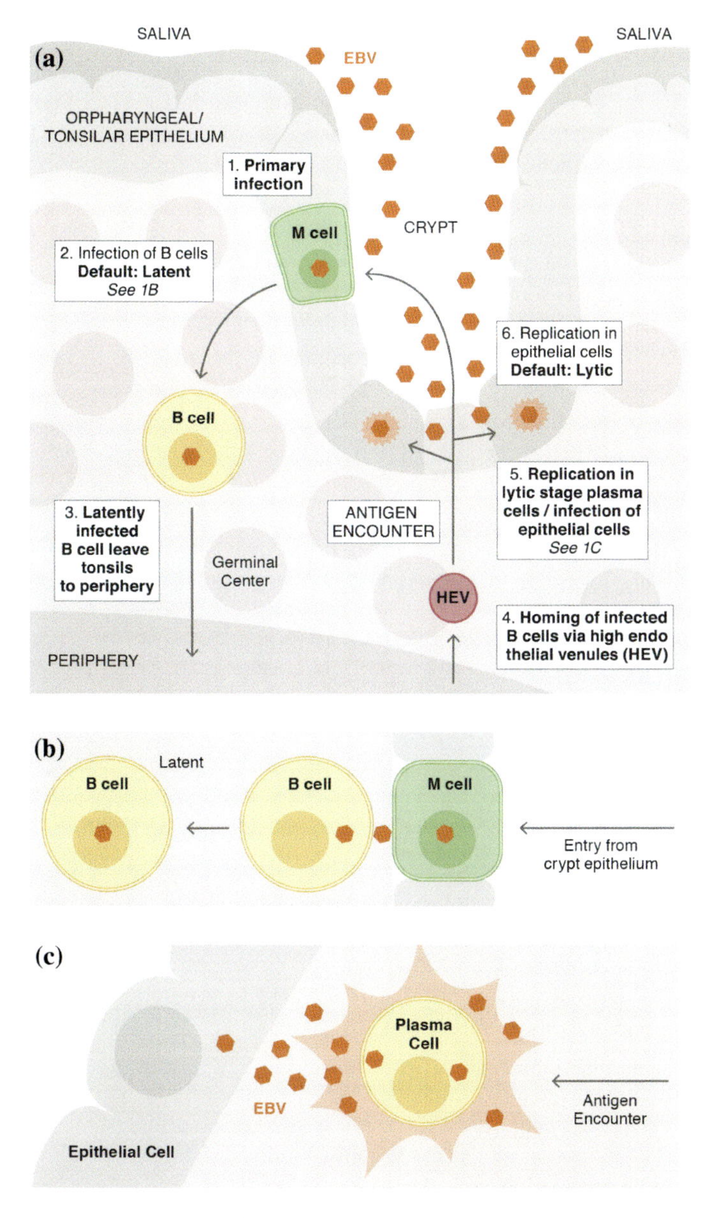


**Fig. 1** EBV infection cycle of main target cells. In B cells, EBV establishes latency as default. It undergoes lytic replication rarely and only under certain tightly regulated conditions, one being their differentiation to plasma cells in tonsils. In epithelial cells, the default stage for EBV seems to be lytic replication. **a** Infection cycle between tonsils and periphery according to current hypotheses; Step 1: infection of M cell(s); Step 2: infection of B cells via M cells (see **b**); Step 3: latently infected B cells leave tonsils via efferent lymph to periphery (default latent); Step 4: homing of infected B cells via high endothelial venules (HEVs); Step 5: replication in lytic stage plasma cells/infection of epithelial cells (see **c**); and Step 6: replication in epithelial cells (default lytic), resulting egression of EBV particles in saliva. **b** Step 2: infection of B cells via M cells, which enter from crypt epithelium, establishing the default latent infection for B cells. **c** Replication in lytic stage plasma cells/infection of epithelial cells

activation of the inflammasome in cancer cells, B cells, and epithelial cells (Ansari et al. 2013; Chen et al. 2012). This sensing of EBV products might induce triggering of the interferon (IFN)- or transforming growth factor (TGF)-pathways as well as release of the danger signal HMGB1. The role and regulation of inflammasomes in EBV infection and its innate immune modulation remain to be further elucidated.

So far, 10 TLRs are known in humans. TLRs 1–9 are specialized to respond to different PAMPs (Ishii et al. 2008). TLR10, which has been only very recently been suggested to play a role in influenza infection, remains without known ligand or antigen (Lee et al. 2014). Thus, we focus here on TLRs 1–9.

B cells are classically considered to be part of the adaptive immune system, although they do abundantly express *TLRs* 1, 2, 5, 6, 7, and 9 and can, if stimulated through them, function as innate-like effectors (Dorner et al. 2009). The second major host cell target for EBV infection, oropharyngeal epithelial cells, expresses *TLRs* 1, 3, 4, and 6, and inconsistently *TLRs* 2, 5, 7, and 9 (Lange et al. 2009; Rac J et al., unpublished). TLR 9 seems to be most important among the TLRs for EBV innate immune sensing, as it has been shown that EBV itself triggers TLR9 (van Gent et al. 2011) that is abundantly and preferentially expressed in B cells (Dorner et al. 2009).

The endosomally located TLR 9 senses unmethylated CpG dsDNA motifs (Ishii et al. 2008). EBV's dsDNA is unmethylated in virus particles and early after primary infection of the B cell (Kalla et al. 2010). Due to the compartmentalization of virus DNA to the nucleus, in the latent stage the virus might be efficiently protected. During lytic replication, the virus is protected by its envelope in the cytoplasm and TLR9 sensing will be limited to some strain-free unmethylated virus DNA. Nevertheless, although sensing of EBV by TLR9 has formally been demonstrated for monocytes and plasmacytoid dendritic cells only (Fiola et al. 2010), it is conceivable that EBV is sensed by TLR9 in B cells during their primary infection. Triggering of TLR9 eventually results in NFkB activation with subsequent transcription of pro-inflammatory cytokines and proliferation of B cells, and TLR9 agonists have been demonstrated to synergize with other B-cell proliferation and activation effects of EBV (Iskra et al. 2010). Also, triggering of TLR9 has been shown to inhibit reactivation and subsequent lytic replication of EBV by inducing chromatin remodeling, thereby impairing transcription of EBV's immediate-early lytic gene, the expression of which is sufficient to switch latent into lytic EBV (Zauner et al. 2010). Notably, similar inhibition of reactivation and subsequent lytic replication after TLR9 triggering of B cells have been observed also in infections with murine gamma-herpesvirus 68 (Haas et al. 2014), which is widely used to model infection biology of the human gamma-herpesviruses EBV and Kaposi's sarcoma herpesvirus (KSHV) (Speck and Ganem 2010). From the host's point of view, sensing of EBV by B-cellular TLR9 during EBV's primary infection of B cells has the advantageous potential to inhibit ensuing replication of EBV and to prevent subsequent lysis and death of the cell, thereby preserving important immune functions. It is likely that the sensing by and triggering of TLR9 by EBV will also occur during EBV reactivation.

By contrast, oropharyngeal epithelial cells can hardly sense EBV via TLR9 upon their primary infection by the virus as cells either do not express this receptor or do so only in very limited amounts (Rac J et al., unpublished). Thus, suppression of lytic EBV replication thereby promoting latent infection via TLR9 signaling as in B cells is lacking. Notably, latent EBV infection in epithelial cells has so far not been established in vitro, and lytic infection in these cells seems to be the default (Fig. 1a, c). Importantly, the mechanisms leading to latent EBV infection in epithelial cells, as observed in nasopharyngeal carcinoma (NPC), remain elusive. How and whether oropharyngeal epithelial cells, however, sense EBV and react to EBV still needs to be elucidated.

## 3 Innate Recognition and Resulting Innate Immune Responses Against EBV by Cells Other Than the Main Target Cells

EBV can also be sensed by other innate immune elements, which are not directly linked to the main target cells. Innate immune cells that are thought to be relevant for EBV recognition and its control are dendritic cells (DCs), monocytes/macrophages, neutrophils, and natural killer (NK) cells.

### *3.1 Myeloid Cells: Dendritic Cells, Monocytes, Macrophages, and Neutrophil Granulocytes*

DCs are rare but highly specialized cells. They can sense, phagocytize, process and present antigens, instruct adaptive immune responses and tolerance, and thus bridge innate and adaptive responses. They are divided in the two functional and developmentally distinct major subsets of the conventional DC (formerly myeloid DC, cDC) and the plasmacytoid DCs (pDCs)(Merad et al. 2013). The latter express TLR 9 abundantly and also have been reported to sense EBV via TLR9 (Fiola et al. 2010; Severa et al. 2013). TLR 9 senses unmethylated dsDNA and thus can sense EBV before and early after infection, but is probably not sensing viral DNA of dying latently infected B cells as these are heavily methylated (Münz 2014; Woellmer et al. 2012). This sensing by pDCs and monocytes through TLR9 thus will most likely be restricted to areas of lytic reactivation where dying cells and free virus will present unmethylated viral genomes, i.e., in the tonsils. It has been reported that the EBER could additionally stimulate pDCs via TLR7 and the resulting type I interferon production could be abrogated with a TLR 7 inhibitor, but not a TLR9 inhibitor (Quan et al. 2010). However, due to the limited number and characterization of samples and the lack of further data supporting this line of thought, these results in our opinion must at this point be considered with caution.

In contrast to pDCs, conventional (formerly myeloid) DCs (cDCs) in humans express no TLR9 but endosomally located TLR3 which recognizes dsRNA (Iwasaki and Medzhitov 2004). For EBV recognition by macrophages and conventional DCs, TLR2 and TLR3 have been implicated (Gaudreault et al. 2007; Ariza et al. 2009). The role of TLR2 remains so far enigmatic (Münz 2014).

Noncoding EBV-RNAs (EBERs) are abundantly present in all EBV-infected cells and are commonly used as the means to ascertain whether a cell is infected with EBV or not during latent infection in resting memory B cells when the virus does not express any of its coding genes (Kieff and Rickinson 2001). EBERs have been shown to be released by EBV-infected cells, and very recently found to be released in exosomes and to be sensed via TLR3 (Iwakiri et al. 2009; Ahmed et al. 2014). They are thought to form stem-loop structures resulting in RNA conformations that bind to TLR3. Sera from patients with the symptomatic form of primary EBV infection, i.e., infectious mononucleosis, with chronic active EBV infection, or EBV-associated hemophagocytic lymphocytosis were reported to contain high concentrations of EBERs (Iwakiri et al. 2009). EBERs may contribute to the pathology of these diseases hallmarked by the effects from high pro-inflammatory cytokine levels. Indeed, these sera induced TLR3-dependent cytokine production and DC maturation in vitro (Iwakiri et al. 2009). EBV-infected target cells release EBERs in high numbers in complex with the La protein (Iwakiri et al. 2009), a frequent auto-antigen in systemic lupus erythematosus or Sjögren's syndrome, which has been shown before to be able to bind EBERs (Lerner et al. 1981). In addition, the EBERs seem to be sensed in B cells by the intracellular PAMP receptor retinoic acid-inducible gene-1 (RIG-I) contributing to type I interferons (Samanta et al. 2006). The interaction of TLRs and RIG-I has been suggested to be central in antiviral responses in human DCs (Szabo and Rajnavolgyi 2013).

Both the triggering of TLR3 in cDCs and the TLR9 sensing of pDCs result in high type I interferon production and thus contribute to the innate and adaptive immune regulation. This will have multiple effects. Type I interferons have been shown to be able to restrict EBV-induced B-cell transformation within the first 24 h after infection (Lotz et al. 1985). In addition, DCs will hereby further activate and recruit other immune cells such as granulocytes and NK cells (Fig. 2), which have been shown to inhibit B-cell transformation and target lytic replicating cells very efficiently (Lünemann et al. 2013; Pappworth et al. 2007; Chijioke et al. 2013). An overview of current knowledge and hypotheses on cellular interactions during EBV infection is summarized in Fig. 2.

Monocytes and monocyte-derived macrophages have been described to sense EBV via TLR2 and TLR9 resulting in cytokines and chemokines secretion (Fiola et al. 2010). They have been described to spread EBV to other cells and facilitate epithelial infection (Tugizov et al. 2007). They also have been implied as amplifiers within tonsils and found to produce type I interferons and inflammatory cytokines upon encounter with EBV (Savard et al. 2000; Fiola et al. 2010). Encounter of EBV has, however, also been shown to promote apoptosis and restrict the development to DCs in vitro (Guerreiro-Cacais et al. 2004; Li et al. 2002). EBV-encoded secreted BARF1 (sBARF1) was found to inhibit M-CSF, thereby inhibiting monocyte differentiation and macrophage function; and it is produced in acute and chronic (lytic) infection and in NPC/EBVaGC tumors (Shim et al. 2012; Hoebe et al. 2012, 2013). The importance of this inhibition was further confirmed in an EBV-related herpesvirus infection in rhesus macaques (Ohashi et al. 2012). Taken together, further studies will be needed to understand the role of monocytes and macrophages in EBV infection.

The role of neutrophils in EBV infection and associated diseases remains enigmatic. IM can result in severe neutropenia, and this might indicate a role for neutrophils, which roles in antiviral defenses have recently been reviewed (Drescher and Bai 2013). Even though it was suggested that EBV would be able to infect neutrophils and viral DNA can be detected, neutrophils have been suggested to initiate programmed cell death in response, which might lead to the described neutropenia (Savard and Gosselin 2006; Gosselin et al. 2001). In HIV infection, neutrophils have recently been implicated to recognize the virus via TLR7/8 with a resulting host defense mechanism (Saitoh et al. 2012). Future studies will be needed to address the role of neutrophils for EBV and the interaction with its host, as they have been often overlooked in the past.

## *3.2 Natural Killer Cells*

NK cells are regarded the main innate effectors of cellular responses against EBV (Rickinson et al. 2014). NK cells were first named for their ability to kill target cells without preactivation (Kiessling et al. 1975a, b). Two main subsets have been described, $CD56^{bright}$ and $CD56^{dim}$ NK cells, which have distinct phenotypes and functions (reviewed by Lünemann et al. 2009). Precursors of human NK cells develop in the bone marrow and then are thought to migrate to secondary lymphoid organs, e.g., tonsils, where they mature before egressing to the peripheral blood to patrol the body for intruding pathogens and emerging cancer cells (Carrega and Ferlazzo 2012; Freud et al. 2014).

An important antiviral role of human NK against EBV was strongly implicated by the finding of patients with isolated NK cell deficiencies exhibiting a markedly increased susceptibility to herpesviruses (Orange 2013). In tonsils, which are the likely portal of entry for EBV, the most mature NK cells are the $CD56^{bright}$ cells (Freud et al. 2014). A distinct subset of these can restrict EBV-induced transformation in autologous B cells in vitro very efficiently, probably via IFN gamma (Lünemann et al. 2013; Strowig et al. 2008). This subset also has the potency to restrict EBV infection of autologous B cells, and this seems to happen within the first 4 days (Lünemann and Nadal, unpublished observations).

Observations in patients with the acute EBV infection, infectious mononucleosis (IM), also point toward the ability of NK cells to restrict EBV-infected cells also in vivo. These studies are all, however, hampered by the inaccessibility of samples, as they are undertaken in peripheral blood, where NK cell subsets differ distinctively from tonsilar subsets in phenotype, maturation stages, and functions

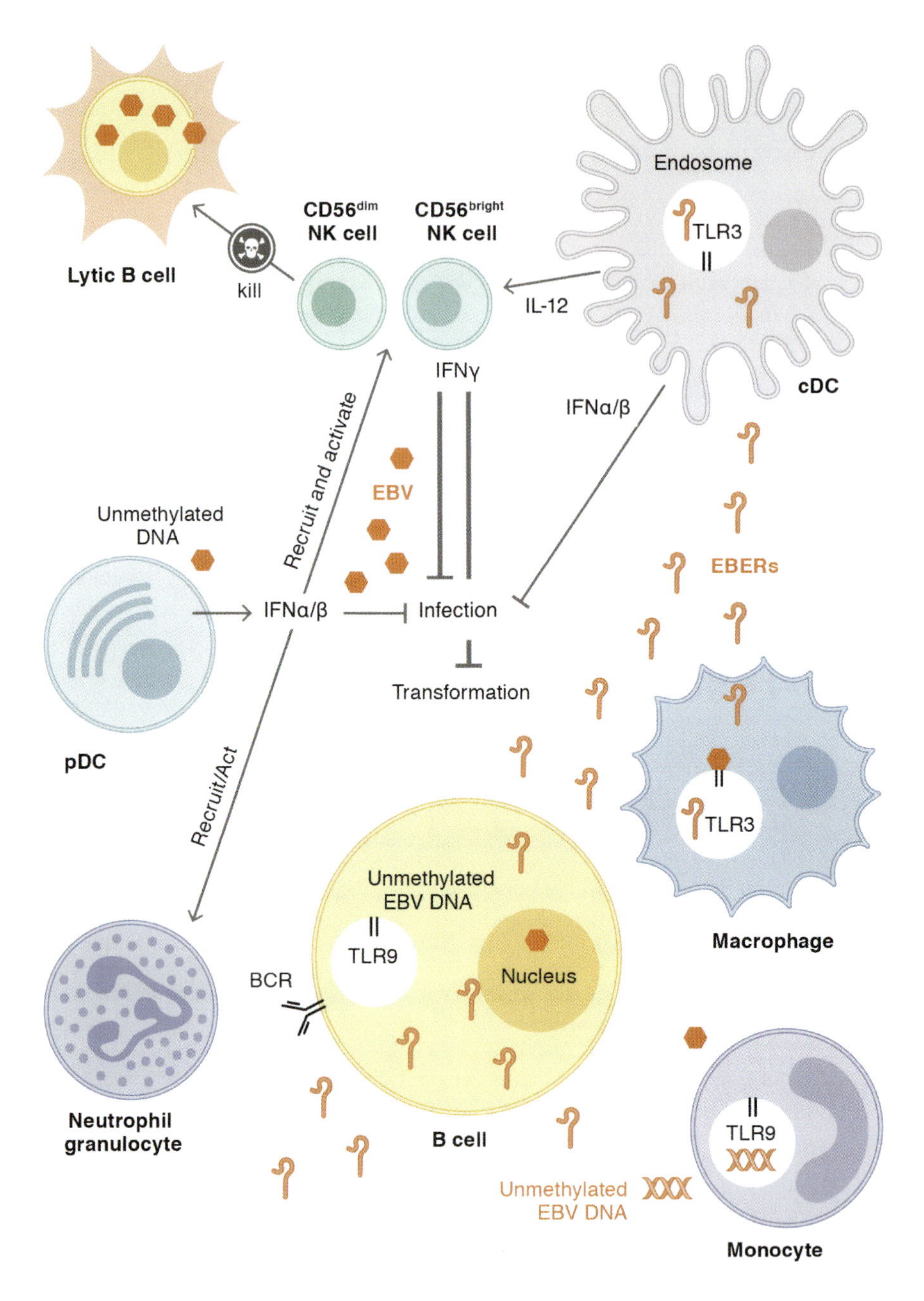
Lytic B cell
kill
CD56dim NK cell
CD56bright NK cell
Endosome
TLR3
IL-12
cDC
IFNγ
IFNα/β
Recruit and activate
EBV
EBERs
Unmethylated DNA
IFNα/β
Infection
Transformation
pDC
Recruit/Act
TLR3
Macrophage
Unmethylated EBV DNA
TLR9
Nucleus
BCR
Neutrophil granulocyte
B cell
TLR9
Unmethylated EBV DNA
Monocyte

◀ **Fig. 2** Current knowledge and hypotheses on cellular interactions during EBV infection. Classical DCs (cDCs) can produce IL-12, activating CD56$^{bright}$ NK cell subsets, and INF-α/β, which can directly inhibit infection and transformation within the first 24 h. All EBV-infected B cells produce and secrete EBERs. EBERs can be sensed by cDCS via TLR3, which can activate the cytokine production. Plasmacytoid DCs can also produce IFNa/b, and this can in addition recruit and activate NK cells and neutrophil granulocytes. The role of neutrophil granulocytes remains to be determined. Macrophages can sense EBV and thus EBERs via TLR3, but their role still remains enigmatic. Monocytes sense unmethylated EBV DNA via TLR9, as these cells are not main target cells, and the in vivo relevance of this remains to be determined and is still debated. NK cells have been shown to inhibit B-cell transformation and target lytic replicating cells very efficiently. This seems to be dependent on different distinct subsets

(Freud et al. 2014). In contrast to tonsils and other SLOs, in the peripheral blood, the least mature NK cells are CD56$^{bright}$. Absolute numbers and frequency of NK cells in peripheral blood from IM patients are increased (Williams et al. 2005; Balfour et al. 2013; Azzi et al. 2014).

There have been contradictory findings about the correlation of this expansion with disease severity and viral blood load. An expansion of CD56$^{bright}$ NK was associated with milder disease severity and a lower viral blood load in a smaller adult IM cohort (Williams et al. 2005). Apparently, contrary results were described in a large prospective study in college students (Balfour et al. 2013); unfortunately, this study did not distinguish between subsets of NK cells. In a recent longitudinal study on pediatric patients with IM, preferential proliferation of early differentiated CD56$^{dim}$ NKG2A+ KIR− NK cells was identified (Azzi et al. 2014). Moreover, this NK cell subset developed features of terminal differentiation and persisted at higher frequency over at least the first 6 months after acute IM. Also, this NK cell subset preferentially degranulated and proliferated upon exposure to EBV-infected B cells expressing lytic antigens, which fits with earlier descriptions of preferential targeting of lytically infected B cells by NK cells (Pappworth et al. 2007). The accumulation of the early differentiated NK cell subset that prevails especially in children under the age of 5 years (Azzi, Lünemann, and Nadal, unpublished observations) clearly contrasts with the expansion of the terminally differentiated NKG2C+ KIR+ subset observed in other viral infections, including those with cytomegalovirus, hantavirus, or chikungunya virus (Lopez-Verges et al. 2011; Bjorkstrom et al. 2011; Petitdemange et al. 2011).

In summary, these data implicate that early differentiated NK cells of the peripheral blood might play a key role in the immune control of primary infection with EBV and distinct subsets of NK cells are specialized for different viruses. This view is further supported by a humanized mouse study, which also finds an expansion of early differentiated NK cells in the blood, which preferentially target lytically replicating cells (Chijioke et al. 2013). Lytic replication, however, is confined to the tonsils, where initial infection occurs, and after amplification loops in epithelial cells, EBV throughout infection is released into the saliva. Hypothetically, the expansion of these NK subsets in the peripheral blood of IM patients might reflect a heightened maturation and release and perhaps overflow of mature tonsilar NK cells, where innate immune responses could clearly not sufficiently control and restrict EBV infection in situ. Thus, we hypothesize that younger children and potentially patients with a higher number of mature cytokine producing NK cells, which can be mobilized through maturation and/or migration to the tonsil in primary infection, might be able to sufficiently control primary EBV infection innately, so the adaptive immune system reaction (i.e., CD8 T cell expansion and massive cytokine release), and thus, IM is prevented. Confirmation of this speculative interpretation of the role of NK cells in primary EBV acquisition in IM and non-IM patients will require further elucidation of phenotypes, maturation stages, and functions in tonsils and the peripheral blood.

In addition to a role in IM, it has long been suggested that failing NK cell-mediated control of persisting EBV infection might be responsible for an outgrowth of EBV-associated tumors. Besides the described restriction of EBV-induced transformation in vitro (Lünemann et al. 2013), several case reports of patients with selective immunodeficiencies, including NK cells deficiencies and EBV-associated tumors, have documented the potential role of NK cells exerting antineoplastic effects (recently reviewed in Orange 2013; Rickinson et al. 2014). Clear causal relationships, however, remain elusive. The EBV-modified secreted exosomes containing galactin 1 and 9 have also been suggested to modulate NK cell control and could contribute to the virus' evasion of the innate immune system (Gandhi et al. 2007; Klibi et al. 2009; Golden-Mason et al. 2013). More work and new models, as well as detailed and thorough investigation in patient samples, will be needed to understand and potentially exploit these natural protectors for patient treatment.

## 4 EBV Counteracts or Exploits Innate Immunity in Its Latent and Lytic Life Cycle Stages

It is conceivable that EBV has adapted itself to the host's innate immunity during coevolution with man. The virus has been found to influence innate immunity responses through modulation of TLR expression, NFkB stimulation, and interaction with intracellular signaling. Since EBV persists and coexists lifelong in the host, the virus seemingly has no advantage from harming the host's cell. Thus, in general, it might be presumed that EBV subtly modulates innate immunity with a carefully preserved balance, rather than outright inhibition of signaling pathways and events (Fig. 3: visualized current knowledge). Recently emerging, different EBV-modified exosomes were described as potential vehicles for immune evasion and modulation (van der Grein and Nolte-'t Hoen 2014; Robbins and Morelli 2014; Wurdinger et al. 2012). Intriguingly, not only does the modulation of innate responses result in only partial immune evasion, but the virus seems to hijack some of the innate response pathways to facilitate its own growth-transforming functions.

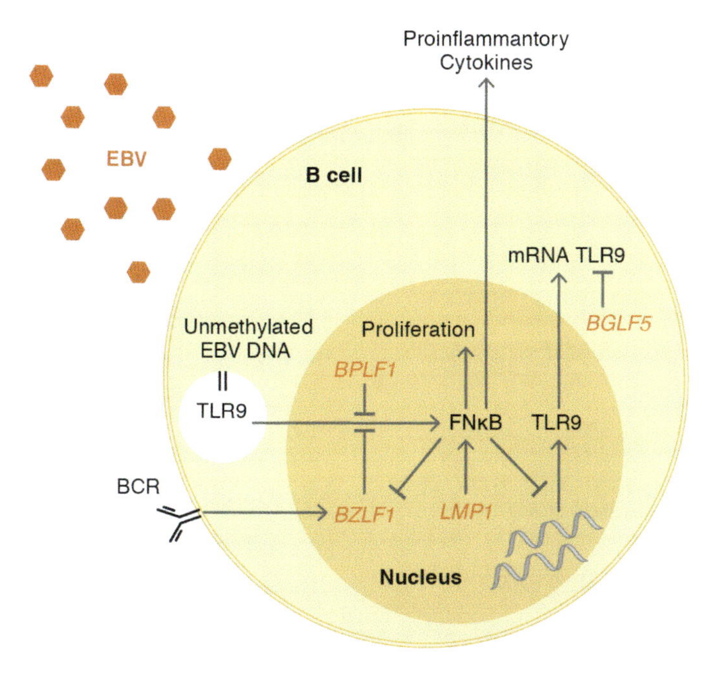


**Fig. 3** EBV seems to subtly modulate innate immunity with carefully preserved balance. An EBV-infected B cell can sense unmethylated EBV DNA via TLR9, and this will activate the NKkB pathway. This will lead to production of proinflammatory cytokines and to proliferation. EBV protein BPLF1 molecules inhibit this. BZLF1 (early lytic protein) expression is inhibited by TLR9 activation and by NFkB expression. It is activated, however, if the B-cell receptor is triggered. This might potentially lead to a balanced control when lytic cycle is activated. BGLF5 inhibits mRNA TLR9 transcription and thus inhibits this pathway. *Dark red* EBV proteins. *Dark blue* Cell proteins

## *4.1 TLRs*

As mentioned above, EBV has been described to activate TLR2, TLR3, TLR7, and TLR9. Thus, it might be of benefit for EBV if it were to evade innate immune activation by these TLRs. It was shown in BL cells that EBV is able to reduce mRNA expression of *TLR*s 1, 6, 9, and 10 during lytic reactivation, with *TLR9* mRNA undergoing a particularly strong inhibition. Notably, EBV reduces TLR9 expression during lytic reactivation via its early lytic protein BGLF5 and this protein was shown to degrade *TLR9* mRNA in vitro (van Gent et al. 2011). Furthermore, in naïve B cells, TLR9 was shown to be downregulated and TLR7 to be upregulated both by UV-inactivated and by untreated EBV (Martin et al. 2007). Downstream signaling of TLR7 was modified via induced expression of the constitutively nuclear splice variant V12 of interferon responsive factor (IRF) 5, which acts as a dominant-negative form of IRF-5 (Martin et al. 2007). Another group, however, did not observe modulation of TLR7 expression in primary B cells following

EBV infection leading to transformation to lymphoblastoid cell lines (LCLs), but recorded decreasing levels of *TLR9* mRNA and diminished TLR9 functionality that was linked to expression of EBV's LMP1 (Fathallah et al. 2010).

The latent EBV protein LMP1 is expressed in EBV latencies III and II that are observed in vivo in tonsillar naïve B cells and germinal center B cells, in post-transplant lymphoproliferative disease (PTLD) and classical Hodgkin's lymphoma (HL), and can also be expressed during EBV's lytic replication (Gottschalk et al. 2005). The suppression of TLR9 function by LMP1 could be seen as EBV's attempt to successfully evade innate immune recognition and resulting restriction on the way to establish latency and in addition during lytic replication. The partial diversion of TLR7 signaling might support the EBV reactivation process, because the recognition of newly synthesized EBV mRNAs via TLR7 could be downsized, thus resulting in lowered NF-κB activation and lessened suppression of lytic EBV. LMP1 is also secreted in exosomes, which can modulate innate immune responses (Dukers et al. 2000; Flanagan et al. 2003; Keryer-Bibens et al. 2006; Middeldorp and Pegtel 2008).

EBV was furthermore reported to induce a significant inhibition of the proliferative responses of PBMC or enriched B cells to TLR9 or TLR7/8 agonists (Younesi et al. 2010). The proliferation was described to be independent of the timepoint of EBV exposure before or after TLR triggering, suggesting that the suppressive effect was not due to EBV-induced downregulation of TLR protein expression. Given that the study was conducted using primary cells from a very restricted number of healthy donors with their EBV infection status not characterized and the number of infected B cells not explored, the results need to be interpreted with caution.

EBV was shown to additionally counteract the host's innate immune responses by interfering with the deubiquitinate signaling intermediates in the TLR cascade.

Exosomes containing dUTPase were described to influence innate responses (Ariza et al. 2013). Very recently, for one of the three putative EBV-encoded deubiquitinases, it was demonstrated that the conserved large herpesvirus tegument protein BPLF1 acts as a functional deubiquitinase in EBV-producing cells (van Gent et al. 2014). BPLF1 is an enzyme that is expressed in the late lytic phase. The protein's N-terminal part contains the catalytic site for deubiquitinase activity and suppresses TLR-mediated activation of NF-kB. The signaling intermediates targeted included TRAF6, NEMO, and IκBα (van Gent et al. 2014).

In summary, EBV seems to be able to counteract TLR signaling in host target cells by either degrading TLR mRNA or influencing downstream signaling of TLRs by inducing expression of dominant-negative acting forms of transcripts. The resulting suppression of TLR9 signaling during lytic EBV reactivation, for example, might be helpful for the virus, as the sensing of unmethylated EBV dsDNA to be packed in new virions would be reduced. Consequently, the replicative cycle could be completed and would evade innate immune control, and EBV particles can then exit the host B cell.

### 4.2 EBERs

Another example of how EBV escapes and exploits innate immunity for its own benefit without jeopardizing the life of the host target cell is illustrated by the non-coding EBV-RNAs (EBERs). Their secondary architecture features extensively base-paired structures and thus resembles that of dsRNA from adenoviruses. In general, for DNA viruses converging bidirectional transcription of the viral genome was suggested for double-stranded RNA expression (Jacobs and Langland 1996; Weber et al. 2006). Unlike cellular dsRNA transcripts, viral dsRNA consists of long and perfectly complementary strands, which might function as potent controllers of viral replication in latently infected cells as their strands are complementary to any viral RNA (Umbach and Cullen 2009). Notably, EBERs are expressed in every EBV-infected B cell (EBERs reviewed in Iwakiri 2014). Importantly, B cells do not express TLR3 (Dorner et al. 2009) that sense dsRNA (Ishii et al. 2008) including EBERs and, as described above, can trigger type I interferon production in cDCs and monocytes (Iwakiri 2014). Moreover, EBERs have been reported to inhibit interferon-stimulated gene activity by binding to PKR, a double-stranded RNA-dependent protein kinase, and inhibiting its activation (Nanbo et al. 2002). Taken together, EBV has therefore chosen strategies that allow the virus to successfully hide from antiviral innate (and adaptive) immunity when residing in resting memory B cells in latency 0, when no EBV proteins are expressed. Nevertheless, EBERs have been found to be capable to trigger cytosolic innate immune receptor RIG-I (Samanta et al. 2006). Although this triggering in B cells theoretically could lead to antiviral responses as induction of protective cellular genes, including type I IFNs and inflammatory cytokines, as has been suggested for DCs (Szabo and Rajnavolgyi 2013), EBERs in B cells may also lead to induction of the B-cell autocrine growth factor IL-10 that also exhibits anti-inflammatory properties (Iwakiri 2014). In addition, pronounced compartmentalization in EBV-positive B cells, where EBERs are predominantly nuclear and thus shielded from TLR sensing. In the cytoplasm, EBERs appear to be carefully shuttled to exit the cells, either complexed to RNA-binding proteins (e.g., La) or via a multivesicular body (MVB) into exosomes, thus evading RIG-I triggering (Ahmed et al. 2014; Zomer et al. 2010). The finding that the cytoplasmic RIG-I does not seem to be constantly activated in EBV-infected B cells despite high nuclear abundance of EBERs additionally supports this. Nevertheless, as detailed above upon delivery to monocytes or cDC's, EBER complexes can regardless of La or exosome context strongly activate IFN-I signaling via TLR3, TLR7, and RIG-I.

### 4.3 microRNAs

As with EBERs, microRNAs (miRNAs) encoded by EBV may exert important functions that ultimately safeguard the virus without compromising the host target cell. Although around 40 EBV-encoded miRNAs have been identified and characterized (reviewed in Klinke et al. 2014), only recently the first EBV miRNA was demonstrated to target a gene. BART 18-5p specifically targets the 3′UTR of the mRNA encoding the cellular signaling molecule MAP kinase kinase kinase 2 (MAP3K2). Given that MAP3K2 was also found to be an intermediary in the signaling pathways that can initiate lytic EBV replication, 18-5p by this interaction helps to block EBV replication. In general, miRNAs are believed to typically exert subtle inhibitory effects on many mRNAs embedded within intricate regulatory networks (Qiu and Thorley-Lawson 2014). Thus, as more targets of EBV miRNAs are identified, the extent to which the viral miRNAs protect EBV from innate (and adaptive) immunity will become apparent.

MicroRNAs are secreted actively through exosomes, which protect them from RNAses and remain functional in their target cells. One example for which this has been described is EBV miRNA BHRF1-3 and its effect on the confirmed EBV target gene CXCL-11 (Pegtel et al. 2010). The EBV miRNA BART2-5p was found to modulate MICB expression potentially to escape immune activation by this danger signal and direct NK killing (Nachmani et al. 2009). Exosomes containing the EBV miRNA BART 15 were identified to inhibit inflammasome formation (Haneklaus et al. 2012). Most recently, Callegari et al. (2014) reported on their investigations on the capacity of miRNAs encoded by Epstein–Barr virus (EBV) to interfere with the SUMO signaling network. From in silico analyses, they found that one or more EBV miRNA target a minimal set of 575 members of the SUMO interactome. Furthermore, upon upregulation of the BHRF1-encoded miRNAs in cells transduced with recombinant lentiviruses or entering the productive virus cycle, multiple components of the TGF-beta signaling pathway were inhibited. Thus, EBV's miRNAs seem to have the capacity to interfere with SUMO-regulated cellular functions that control key aspects of viral replication and pathogenesis. It can be speculated that additional interactions between EBV-encoded miRNAs and host intracellular signaling pathways including those of innate immunity will be discovered.

## 5 Conclusions

Innate immune recognition and sensing of EBV, and evasion strategies of the virus by both exploiting and counteracting innate immune elements and resulting immune responses in nonmalignant and malignant diseases, still need much further work to understand the underlying causal relations and mechanisms. Innate immune effects on and of EBV within its life cycles and host life preservation seem to have found a balanced relationship of "live and let live" as the virus persists lifelong usually without causing obvious harm to the host. However, in rare occasions, when for example associated malignant diseases develop, the seemingly sturdy balance and regulation of this persistence apparently fails. Even though many factors and players have been indicated and investigated, the overall picture of causal relations and interplay of the innate elements still remains enigmatic.

Open questions include but are not limited to causal interactions and subsequent regulation in acute and persistent EBV infection and also in EBV-associated cancers. This work will step-by-step not only teach us about EBV itself, but will also elucidate many more of the innate immune functions and how they counteract viral threats to the body, balancing and regulating persistent infections and monitoring and destroying emerging cancer cells. Ultimately we aim to efficiently harness our knowledge for treatment of EBV-associated diseases.

**Acknowledgements** AL and DN were supported by the Cancer League of the Canton of Zürich, the Swiss National Science Foundation (310030_135028), the Wolfermann Naegeli Stiftung, and the Foundation for Scientific Research of the University of Zurich. MR was supported by the Cancer Research UK, London (C5575/A15032). We thank all patients and healthy donors contributing to science, as well as all clinicians, which together by their contributions make research possible. We sincerely apologize to all colleagues whose work we could not cite in this review due to space restrictions.

## References

Ahmed W, Philip PS, Tariq S, Khan G (2014) Epstein-Barr virus-encoded small RNAs (EBERs) are present in fractions related to exosomes released by EBV-transformed cells. PLoS ONE 9(6):e99163. doi:10.1371/journal.pone.0099163, PONE-D-13-54217 [pii]

Ansari MA, Singh VV, Dutta S, Veettil MV, Dutta D, Chikoti L, Lu J, Everly D, Chandran B (2013) Constitutive interferon-inducible protein 16-inflammasome activation during Epstein-Barr virus latency I, II, and III in B and epithelial cells. J Virol 87(15):8606–8623. doi:10.1128/JVI.00805-13, JVI.00805-13 [pii]

Ariza ME, Glaser R, Kaumaya PT, Jones C, Williams MV (2009) The EBV-encoded dUTPase activates NF-kappa B through the TLR2 and MyD88-dependent signaling pathway. J Immunol 182(2):851–859. doi:182/2/851 [pii]

Ariza ME, Rivailler P, Glaser R, Chen M, Williams MV (2013) Epstein-Barr virus encoded dUTPase containing exosomes modulate innate and adaptive immune responses in human dendritic cells and peripheral blood mononuclear cells. PLoS ONE 8(7):e69827. doi:10.1371/journal.pone.0069827, PONE-D-13-19147 [pii]

Azzi T, Lunemann A, Murer A, Ueda S, Beziat V, Malmberg KJ, Staubli G, Gysin C, Berger C, Munz C, Chijioke O, Nadal D (2014) Role for early-differentiated natural killer cells in infectious mononucleosis. Blood 124(16):2533–2543. doi:10.1182/blood-2014-01-553024, blood-2014-01-553024 [pii]

Balfour HH Jr, Odumade OA, Schmeling DO, Mullan BD, Ed JA, Knight JA, Vezina HE, Thomas W, Hogquist KA (2013) Behavioral, virologic, and immunologic factors associated with acquisition and severity of primary Epstein-Barr virus infection in university students. J Infect Dis 207(1):80–88. doi:10.1093/infdis/jis646, jis646 [pii]

Bjorkstrom NK, Lindgren T, Stoltz M, Fauriat C, Braun M, Evander M, Michaelsson J, Malmberg KJ, Klingstrom J, Ahlm C, Ljunggren HG (2011) Rapid expansion and long-term persistence of elevated NK cell numbers in humans infected with hantavirus. J Exp Med 208(1):13–21. doi:10.1084/jem.20100762, jem.20100762 [pii]

Callegari S, Gastaldello S, Faridani OR, Masucci MG (2014) Epstein-Barr virus encoded microRNAs target SUMO-regulated cellular functions. FEBS J 281(21):4935–4950. doi:10.1111/febs.13040

Carrega P, Ferlazzo G (2012) Natural killer cell distribution and trafficking in human tissues. Front Immunol 3:347. doi:10.3389/fimmu.2012.00347

Chen LC, Wang LJ, Tsang NM, Ojcius DM, Chen CC, Ouyang CN, Hsueh C, Liang Y, Chang KP, Chang YS (2012) Tumour inflammasome-derived IL-1beta recruits neutrophils and improves local recurrence-free survival in EBV-induced nasopharyngeal carcinoma. EMBO Mol Med 4(12):1276–1293. doi:10.1002/emmm.201201569

Chijioke O, Muller A, Feederle R, Barros MH, Krieg C, Emmel V, Marcenaro E, Leung CS, Antsiferova O, Landtwing V, Bossart W, Moretta A, Hassan R, Boyman O, Niedobitek G, Delecluse HJ, Capaul R, Munz C (2013) Human natural killer cells prevent infectious mononucleosis features by targeting lytic Epstein-Barr virus infection. Cell Rep 5(6):1489–1498. doi:10.1016/j.celrep.2013.11.041, S2211-1247(13)00725-0 [pii]

Cooper MA, Elliott JM, Keyel PA, Yang L, Carrero JA, Yokoyama WM (2009) Cytokine-induced memory-like natural killer cells. Proc Natl Acad Sci USA 106(6):1915–1919. doi:10.1073/pnas.0813192106, 0813192106 [pii]

Dorner M, Brandt S, Tinguely M, Zucol F, Bourquin JP, Zauner L, Berger C, Bernasconi M, Speck RF, Nadal D (2009) Plasma cell toll-like receptor (TLR) expression differs from that of B cells, and plasma cell TLR triggering enhances immunoglobulin production. Immunology 128(4):573–579

Drescher B, Bai F (2013) Neutrophil in viral infections, friend or foe? Virus Res 171(1):1–7. doi:10.1016/j.virusres.2012.11.002, S0168-1702(12)00447-9 [pii]

Dukers DF, Meij P, Vervoort MB, Vos W, Scheper RJ, Meijer CJ, Bloemena E, Middeldorp JM (2000) Direct immunosuppressive effects of EBV-encoded latent membrane protein 1. J Immunol 165(2):663–670. doi:ji_v165n2p663 [pii]

Fathallah I, Parroche P, Gruffat H, Zannetti C, Johansson H, Yue J, Manet E, Tommasino M, Sylla BS, Hasan UA (2010) EBV latent membrane protein 1 is a negative regulator of TLR9. J Immunol 185(11):6439–6447. doi:10.4049/jimmunol.0903459, jimmunol.0903459 [pii]

Fiola S, Gosselin D, Takada K, Gosselin J (2010) TLR9 contributes to the recognition of EBV by primary monocytes and plasmacytoid dendritic cells. J Immunol 185(6):3620–3631. doi:10.4049/jimmunol.0903736, jimmunol.0903736 [pii]

Flanagan J, Middeldorp J, Sculley T (2003) Localization of the Epstein-Barr virus protein LMP 1 to exosomes. J Gen Virol 84(Pt 7):1871–1879

Freud AG, Yu J, Caligiuri MA (2014) Human natural killer cell development in secondary lymphoid tissues. Semin Immunol 26(2):132–137. doi:10.1016/j.smim.2014.02.008, S1044-5323(14)00021-9 [pii]

Fujimura Y (2000) Evidence of M cells as portals of entry for antigens in the nasopharyngeal lymphoid tissue of humans. Virchows Arch 436(6):560–566

Gandhi MK, Moll G, Smith C, Dua U, Lambley E, Ramuz O, Gill D, Marlton P, Seymour JF, Khanna R (2007) Galectin-1 mediated suppression of Epstein-Barr virus specific T-cell immunity in classic Hodgkin lymphoma. Blood 110(4):1326–1329. doi:10.1182/blood-2007-01-066100, blood-2007-01-066100 [pii]

Gasteiger G, Rudensky AY (2014) Interactions between innate and adaptive lymphocytes. Nat Rev Immunol. doi:10.1038/nri3726, nri3726 [pii]

Gaudreault E, Fiola S, Olivier M, Gosselin J (2007) Epstein-Barr virus induces MCP-1 secretion by human monocytes via TLR2. J Virol 81(15):8016–8024. doi:10.1128/JVI.00403-07, JVI.00403-07 [pii]

Golden-Mason L, McMahan RH, Strong M, Reisdorph R, Mahaffey S, Palmer BE, Cheng L, Kulesza C, Hirashima M, Niki T, Rosen HR (2013) Galectin-9 functionally impairs natural killer cells in humans and mice. J Virol 87(9):4835–4845. doi:10.1128/JVI.01085-12, JVI.01085-12 [pii]

Gosselin J, Savard M, Tardif M, Flamand L, Borgeat P (2001) Epstein-Barr virus primes human polymorphonuclear leucocytes for the biosynthesis of leukotriene B4. Clin Exp Immunol 126(3):494–502. doi:1687 [pii]

Gottschalk S, Heslop HE, Rooney CM (2005) Adoptive immunotherapy for EBV-associated malignancies. Leuk Lymphoma 46(1):1–10. doi:10.1080/10428190400002202, X6NJBGGNJEVVJBYN [pii]
Guerreiro-Cacais AO, Li L, Donati D, Bejarano MT, Morgan A, Masucci MG, Hutt-Fletcher L, Levitsky V (2004) Capacity of Epstein-Barr virus to infect monocytes and inhibit their development into dendritic cells is affected by the cell type supporting virus replication. J Gen Virol 85(Pt 10):2767–2778
Haas F, Yamauchi K, Murat M, Bernasconi M, Yamanaka N, Speck RF, Nadal D (2014) Activation of NF-kappaB via endosomal Toll-like receptor 7 (TLR7) or TLR9 suppresses murine herpesvirus 68 reactivation. J Virol 88(17):10002–10012. doi:10.1128/JVI.01486-14
Hadinoto V, Shapiro M, Sun CC, Thorley-Lawson DA (2009) The dynamics of EBV shedding implicate a central role for epithelial cells in amplifying viral output. PLoS Pathog 5(7):e1000496. doi:10.1371/journal.ppat.1000496
Haneklaus M, Gerlic M, Kurowska-Stolarska M, Rainey AA, Pich D, McInnes IB, Hammerschmidt W, O'Neill LA, Masters SL (2012) Cutting edge: miR-223 and EBV miR-BART15 regulate the NLRP3 inflammasome and IL-1beta production. J Immunol 189(8):3795–3799. doi:10.4049/jimmunol.1200312, jimmunol.1200312 [pii]
Hoebe EK, Le Large TY, Tarbouriech N, Oosterhoff D, De Gruijl TD, Middeldorp JM, Greijer AE (2012) Epstein-Barr virus-encoded BARF1 protein is a decoy receptor for macrophage colony stimulating factor and interferes with macrophage differentiation and activation. Viral Immunol 25(6):461–470. doi:10.1089/vim.2012.0034
Hoebe EK, Le Large TY, Greijer AE, Middeldorp JM (2013) BamHI-A rightward frame 1, an Epstein-Barr virus-encoded oncogene and immune modulator. Rev Med Virol 23(6):367–383. doi:10.1002/rmv.1758
Horowitz A, Newman KC, Evans JH, Korbel DS, Davis DM, Riley EM (2010) Cross-talk between T cells and NK cells generates rapid effector responses to Plasmodium falciparum-infected erythrocytes. J Immunol 184(11):6043–6052. doi:10.4049/jimmunol.1000106, jimmunol.1000106 [pii]
Imai S, Nishikawa J, Takada K (1998) Cell-to-cell contact as an efficient mode of Epstein-Barr virus infection of diverse human epithelial cells. J Virol 72(5):4371–4378
Ishii KJ, Koyama S, Nakagawa A, Coban C, Akira S (2008) Host innate immune receptors and beyond: making sense of microbial infections. Cell Host Microbe 3(6):352–363. doi:10.1016/j.chom.2008.05.003, S1931-3128(08)00151-0 [pii]
Iskra S, Kalla M, Delecluse HJ, Hammerschmidt W, Moosmann A (2010) Toll-like receptor agonists synergistically increase proliferation and activation of B cells by epstein-barr virus. J Virol 84(7):3612–3623. doi:10.1128/JVI.01400-09
Iwakiri D (2014) Epstein-Barr virus-encoded RNAs: key molecules in viral pathogenesis. Cancers (Basel) 6(3):1615–1630. doi:10.3390/cancers6031615, cancers6031615 [pii]
Iwakiri D, Zhou L, Samanta M, Matsumoto M, Ebihara T, Seya T, Imai S, Fujieda M, Kawa K, Takada K (2009) Epstein-Barr virus (EBV)-encoded small RNA is released from EBV-infected cells and activates signaling from Toll-like receptor 3. J Exp Med 206(10):2091–2099
Iwasaki A, Medzhitov R (2004) Toll-like receptor control of the adaptive immune responses. Nat Immunol 5(10):987–995. doi:10.1038/ni1112, ni1112 [pii]
Jacobs BL, Langland JO (1996) When two strands are better than one: the mediators and modulators of the cellular responses to double-stranded RNA. Virology 219(2):339–349. doi:10.1006/viro.1996.0259, S0042-6822(96)90259-7 [pii]
Kalla M, Schmeinck A, Bergbauer M, Pich D, Hammerschmidt W (2010) AP-1 homolog BZLF1 of Epstein-Barr virus has two essential functions dependent on the epigenetic state of the viral genome. In: Proceedings of the National Academy of Sciences of the United States of America 107(2):850–855. doi:10.1073/pnas.0911948107
Keryer-Bibens C, Pioche-Durieu C, Villemant C, Souquere S, Nishi N, Hirashima M, Middeldorp J, Busson P (2006) Exosomes released by EBV-infected nasopharyngeal

carcinoma cells convey the viral latent membrane protein 1 and the immunomodulatory protein galectin 9. BMC Cancer 6:283. doi:10.1186/1471-2407-6-283, 1471-2407-6-283 [pii]
Kieff E, Rickinson AB (2001) In: Knipe DM, Howley PM (eds) Fields Virology. Lippincott Williams and Wilkins, Philadelphia, pp 2511–2574
Kiessling R, Klein E, Pross H, Wigzell H (1975a) "Natural" killer cells in the mouse. II. Cytotoxic cells with specificity for mouse Moloney leukemia cells. Characteristics of the killer cell. Eur J Immunol 5(2):117–121. doi:10.1002/eji.1830050209
Kiessling R, Klein E, Wigzell H (1975b) "Natural" killer cells in the mouse. I. Cytotoxic cells with specificity for mouse Moloney leukemia cells. Specificity and distribution according to genotype. Eur J Immunol 5(2):112–117. doi:10.1002/eji.1830050208
Klibi J, Niki T, Riedel A, Pioche-Durieu C, Souquere S, Rubinstein E, Le Moulec S, Guigay J, Hirashima M, Guemira F, Adhikary D, Mautner J, Busson P (2009) Blood diffusion and Th1-suppressive effects of galectin-9-containing exosomes released by Epstein-Barr virus-infected nasopharyngeal carcinoma cells. Blood 113(9):1957–1966. doi:10.1182/blood-2008-02-142596, blood-2008-02-142596 [pii]
Klinke O, Feederle R, Delecluse HJ (2014) Genetics of Epstein-Barr virus microRNAs. Semin Cancer Biol 26:52–59. doi:10.1016/j.semcancer.2014.02.002, S1044-579X(14)00026-1 [pii]
Laichalk LL, Thorley-Lawson DA (2005) Terminal differentiation into plasma cells initiates the replicative cycle of Epstein-Barr virus in vivo. J Virol 79(2):1296–1307
Lange MJ, Lasiter JC, Misfeldt ML (2009) Toll-like receptors in tonsillar epithelial cells. Int J Pediatr Otorhinolaryngol 73(4):613–621. doi:10.1016/j.ijporl.2008.12.013
Lee SM, Kok KH, Jaume M, Cheung TK, Yip TF, Lai JC, Guan Y, Webster RG, Jin DY, Peiris JS (2014) Toll-like receptor 10 is involved in induction of innate immune responses to influenza virus infection. Proc Natl Acad Sci USA 111(10):3793–3798. doi:10.1073/pnas.1324266111, 1324266111 [pii]
Lerner MR, Andrews NC, Miller G, Steitz JA (1981) Two small RNAs encoded by Epstein-Barr virus and complexed with protein are precipitated by antibodies from patients with systemic lupus erythematosus. Proc Natl Acad Sci USA 78(2):805–809
Li L, Liu D, Hutt-Fletcher L, Morgan A, Masucci MG, Levitsky V (2002) Epstein-Barr virus inhibits the development of dendritic cells by promoting apoptosis of their monocyte precursors in the presence of granulocyte macrophage-colony-stimulating factor and interleukin-4. Blood 99(10):3725–3734
Lopez-Verges S, Milush JM, Schwartz BS, Pando MJ, Jarjoura J, York VA, Houchins JP, Miller S, Kang SM, Norris PJ, Nixon DF, Lanier LL (2011) Expansion of a unique CD57(+)NKG2Chi natural killer cell subset during acute human cytomegalovirus infection. Proc Natl Acad Sci USA 108(36):14725–14732. doi:10.1073/pnas.1110900108, 1110900108 [pii]
Lotz M, Tsoukas CD, Fong S, Carson DA, Vaughan JH (1985) Regulation of Epstein-Barr virus infection by recombinant interferons. Selected sensitivity to interferon-gamma. Eur J Immunol 15(5):520–525
Lünemann A, Lünemann JD, Munz C (2009) Regulatory NK cell functions in inflammation and autoimmunity. Mol Med. doi:10.2119/molmed.2009.00035
Lünemann A, Vanoaica LD, Azzi T, Nadal D, Münz C (2013) A distinct subpopulation of human NK cells restricts B cell transformation by EBV. J Immunol 191(10):4989–4995. doi:10.4049/jimmunol.1301046
Martin HJ, Lee JM, Walls D, Hayward SD (2007) Manipulation of the toll-like receptor 7 signaling pathway by Epstein-Barr virus. J Virol 81(18):9748–9758. doi:10.1128/JVI.01122-07, JVI.01122-07 [pii]
Merad M, Sathe P, Helft J, Miller J, Mortha A (2013) The dendritic cell lineage: ontogeny and function of dendritic cells and their subsets in the steady state and the inflamed setting. Annu Rev Immunol 31:563–604. doi:10.1146/annurev-immunol-020711-074950
Middeldorp JM, Pegtel DM (2008) Multiple roles of LMP1 in Epstein-Barr virus induced immune escape. Semin Cancer Biol 18(6):388–396. doi:10.1016/j.semcancer.2008.10.004, S1044-579X(08)00080-1 [pii]

Münz C (2014) Dendritic cells during Epstein Barr virus infection. Front Microbiol 5:308. doi: 10.3389/fmicb.2014.00308

Murphy KTP, Walport M, Janeway C (2012) Janeway's Immunobiology, 8th edn. Garland Science, New York

Nachmani D, Stern-Ginossar N, Sarid R, Mandelboim O (2009) Diverse herpesvirus microRNAs target the stress-induced immune ligand MICB to escape recognition by natural killer cells. Cell Host Microbe 5(4):376–385. doi:10.1016/j.chom.2009.03.003, S1931-3128(09)00094-8 [pii]

Nanbo A, Inoue K, Adachi-Takasawa K, Takada K (2002) Epstein-Barr virus RNA confers resistance to interferon-alpha-induced apoptosis in Burkitt's lymphoma. EMBO J 21(5):954–965. doi:10.1093/emboj/21.5.954

O'Leary JG, Goodarzi M, Drayton DL, von Andrian UH (2006) T cell- and B cell-independent adaptive immunity mediated by natural killer cells. Nat Immunol 7(5):507–516. doi:10.1038/ni1332, ni1332 [pii]

Ohashi M, Fogg MH, Orlova N, Quink C, Wang F (2012) An Epstein-Barr virus encoded inhibitor of colony stimulating factor-1 signaling is an important determinant for acute and persistent EBV infection. PLoS Pathog 8(12):e1003095. doi:10.1371/journal.ppat.1003095, PPATHOGENS-D-12-02098 [pii]

Orange JS (2013) Natural killer cell deficiency. J Allergy Clin Immunol 132(3):515–525; quiz 526. doi:10.1016/j.jaci.2013.07.020, S0091-6749(13)01123-8 [pii]

Pappworth IY, Wang EC, Rowe M (2007) The switch from latent to productive infection in epstein-barr virus-infected B cells is associated with sensitization to NK cell killing. J Virol 81(2):474–482. doi:10.1128/JVI.01777-06, JVI.01777-06 [pii]

Paust S, Gill HS, Wang BZ, Flynn MP, Moseman EA, Senman B, Szczepanik M, Telenti A, Askenase PW, Compans RW, von Andrian UH (2010) Critical role for the chemokine receptor CXCR6 in NK cell-mediated antigen-specific memory of haptens and viruses. Nat Immunol 11(12):1127–1135. doi:10.1038/ni.1953, ni.1953 [pii]

Pegtel DM, Cosmopoulos K, Thorley-Lawson DA, van Eijndhoven MA, Hopmans ES, Lindenberg JL, de Gruijl TD, Wurdinger T, Middeldorp JM (2010) Functional delivery of viral miRNAs via exosomes. Proc Natl Acad Sci USA 107(14):6328–6333. doi:10.1073/pnas.0914843107, 0914843107 [pii]

Petitdemange C, Becquart P, Wauquier N, Beziat V, Debre P, Leroy EM, Vieillard V (2011) Unconventional repertoire profile is imprinted during acute chikungunya infection for natural killer cells polarization toward cytotoxicity. PLoS Pathog 7(9):e1002268. doi:10.1371/journal.ppat.1002268, PPATHOGENS-D-11-00590 [pii]

Qiu J, Thorley-Lawson DA (2014) EBV microRNA BART 18-5p targets MAP3K2 to facilitate persistence in vivo by inhibiting viral replication in B cells. Proc Natl Acad Sci USA 111(30):11157–11162. doi:10.1073/pnas.1406136111, 1406136111 [pii]

Quan TE, Roman RM, Rudenga BJ, Holers VM, Craft JE (2010) Epstein-Barr virus promotes interferon-alpha production by plasmacytoid dendritic cells. Arthritis Rheum 62(6):1693–1701. doi:10.1002/art.27408

Rickinson AB, Long HM, Palendira U, Munz C, Hislop AD (2014) Cellular immune controls over Epstein-Barr virus infection: new lessons from the clinic and the laboratory. Trends Immunol 35(4):159–169. doi:10.1016/j.it.2014.01.003, S1471-4906(14)00016-7 [pii]

Robbins PD, Morelli AE (2014) Regulation of immune responses by extracellular vesicles. Nat Rev Immunol 14(3):195–208. doi:10.1038/nri3622, nri3622 [pii]

Saitoh T, Komano J, Saitoh Y, Misawa T, Takahama M, Kozaki T, Uehata T, Iwasaki H, Omori H, Yamaoka S, Yamamoto N, Akira S (2012) Neutrophil extracellular traps mediate a host defense response to human immunodeficiency virus-1. Cell Host Microbe 12(1):109–116. doi:10.1016/j.chom.2012.05.015, S1931-3128(12)00201-6 [pii]

Samanta M, Iwakiri D, Kanda T, Imaizumi T, Takada K (2006) EB virus-encoded RNAs are recognized by RIG-I and activate signaling to induce type I IFN. EMBO J 25(18):4207–4214. doi:10.1038/sj.emboj.7601314, 7601314 [pii]

Savard M, Gosselin J (2006) Epstein-Barr virus immunossuppression of innate immunity mediated by phagocytes. Virus Res 119(2):134–145. doi:10.1016/j.virusres.2006.02.008, S0168-1702(06)00069-4 [pii]

Savard M, Belanger C, Tardif M, Gourde P, Flamand L, Gosselin J (2000) Infection of primary human monocytes by Epstein-Barr virus. J Virol 74(6):2612–2619

Severa M, Giacomini E, Gafa V, Anastasiadou E, Rizzo F, Corazzari M, Romagnoli A, Trivedi P, Fimia GM, Coccia EM (2013) EBV stimulates TLR- and autophagy-dependent pathways and impairs maturation in plasmacytoid dendritic cells: implications for viral immune escape. Eur J Immunol 43(1):147–158. doi:10.1002/eji.201242552

Shannon-Lowe C, Rowe M (2014) Epstein Barr virus entry; kissing and conjugation. Curr Opin Virol 4:78–84. doi:10.1016/j.coviro.2013.12.001, S1879-6257(13)00204-6 [pii]

Shannon-Lowe CD, Neuhierl B, Baldwin G, Rickinson AB, Delecluse HJ (2006) Resting B cells as a transfer vehicle for Epstein-Barr virus infection of epithelial cells. Proc Natl Acad Sci USA 103(18):7065–7070. doi:10.1073/pnas.0510512103, 0510512103 [pii]

Shannon-Lowe C, Adland E, Bell AI, Delecluse HJ, Rickinson AB, Rowe M (2009) Features distinguishing Epstein-Barr virus infections of epithelial cells and B cells: viral genome expression, genome maintenance, and genome amplification. J Virol 83(15):7749–7760. doi: 10.1128/JVI.00108-09, JVI.00108-09 [pii]

Shim AH, Chang RA, Chen X, Longnecker R, He X (2012) Multipronged attenuation of macrophage-colony stimulating factor signaling by Epstein-Barr virus BARF1. Proc Natl Acad Sci U S A 109(32):12962–12967. doi:10.1073/pnas.1205309109, 1205309109 [pii]

Speck SH, Ganem D (2010) Viral latency and its regulation: lessons from the gamma-herpesviruses. Cell Host Microbe 8(1):100–115. doi:10.1016/j.chom.2010.06.014

Strowig T, Brilot F, Arrey F, Bougras G, Thomas D, Muller WA, Munz C (2008) Tonsilar NK cells restrict B cell transformation by the Epstein-Barr virus via IFN-gamma. PLoS Pathog 4(2):e27. doi:10.1371/journal.ppat.0040027, 07-PLPA-RA-0357 [pii]

Szabo A, Rajnavolgyi E (2013) Collaboration of Toll-like and RIG-I-like receptors in human dendritic cells: tRIGgering antiviral innate immune responses. Am J Clin Exp Immunol 2(3):195–207

Tugizov SM, Berline JW, Palefsky JM (2003) Epstein-Barr virus infection of polarized tongue and nasopharyngeal epithelial cells. Nat Med 9(3):307–314

Tugizov S, Herrera R, Veluppillai P, Greenspan J, Greenspan D, Palefsky JM (2007) Epstein-Barr virus (EBV)-infected monocytes facilitate dissemination of EBV within the oral mucosal epithelium. J Virol 81(11):5484–5496. doi:10.1128/JVI.00171-07, JVI.00171-07 [pii]

Tugizov SM, Herrera R, Palefsky JM (2013) Epstein-Barr virus transcytosis through polarized oral epithelial cells. J Virol 87(14):8179–8194. doi:10.1128/JVI.00443-13, JVI.00443-13 [pii]

Umbach JL, Cullen BR (2009) The role of RNAi and microRNAs in animal virus replication and antiviral immunity. Genes Dev 23(10):1151–1164. doi:10.1101/gad.1793309, 23/10/1151 [pii]

van der Grein SG, Nolte-'t Hoen EN (2014) "Small Talk" in the innate immune system via RNA-containing extracellular vesicles. Front Immunol 5:542. doi:10.3389/fimmu.2014.00542

van Gent M, Griffin BD, Berkhoff EG, van Leeuwen D, Boer IG, Buisson M, Hartgers FC, Burmeister WP, Wiertz EJ, Ressing ME (2011) EBV lytic-phase protein BGLF5 contributes to TLR9 downregulation during productive infection. J Immunol 186(3):1694–1702. doi:10.4049/jimmunol.0903120, jimmunol.0903120 [pii]

van Gent M, Braem SG, de Jong A, Delagic N, Peeters JG, Boer IG, Moynagh PN, Kremmer E, Wiertz EJ, Ovaa H, Griffin BD, Ressing ME (2014) Epstein-Barr virus large tegument protein BPLF1 contributes to innate immune evasion through interference with Toll-like receptor signaling. PLoS Pathog 10(2):e1003960. doi:10.1371/journal.ppat.1003960, PPATHOGENS-D-13-01814 [pii]

Waggoner SN, Cornberg M, Selin LK, Welsh RM (2012) Natural killer cells act as rheostats modulating antiviral T cells. Nature 481 (7381):394–398. doi:10.1038/nature10624, nature10624 [pii]

Weber F, Wagner V, Rasmussen SB, Hartmann R, Paludan SR (2006) Double-stranded RNA is produced by positive-strand RNA viruses and DNA viruses but not in detectable amounts by negative-strand RNA viruses. J Virol 80(10):5059–5064. doi:10.1128/JVI.80.10.5059-5064.2006, 80/10/5059 [pii]

White MJ, Nielsen CM, McGregor RH, Riley EH, Goodier MR (2014) Differential activation of CD57-defined natural killer cell subsets during recall responses to vaccine antigens. Immunology 142(1):140–150

Williams H, McAulay K, Macsween KF, Gallacher NJ, Higgins CD, Harrison N, Swerdlow AJ, Crawford DH (2005) The immune response to primary EBV infection: a role for natural killer cells. Br J Haematol 129(2):266–274

Woellmer A, Arteaga-Salas JM, Hammerschmidt W (2012) BZLF1 governs CpG-methylated chromatin of Epstein-Barr Virus reversing epigenetic repression. PLoS Pathog 8(9):e1002902. doi:10.1371/journal.ppat.1002902, PPATHOGENS-D-12-00492 [pii]

Wurdinger T, Gatson NN, Balaj L, Kaur B, Breakefield XO, Pegtel DM (2012) Extracellular vesicles and their convergence with viral pathways. Adv Virol 2012:767694. doi:10.1155/2012/767694

Younesi V, Nikzamir H, Yousefi M, Khoshnoodi J, Arjmand M, Rabbani H, Shokri F (2010) Epstein Barr virus inhibits the stimulatory effect of TLR7/8 and TLR9 agonists but not CD40 ligand in human B lymphocytes. Microbiol Immunol 54(9):534–541. doi:10.1111/j.1348-0421.2010.00248.x, MIM248 [pii]

Young LS, Rickinson AB (2004) Epstein-Barr virus: 40 years on. Nat Rev Cancer 4(10):757–768

Zauner L, Melroe GT, Sigrist JA, Rechsteiner MP, Dorner M, Arnold M, Berger C, Bernasconi M, Schaefer BW, Speck RF, Nadal D (2010) TLR9 triggering in Burkitt's lymphoma cell lines suppresses the EBV BZLF1 transcription via histone modification. Oncogene 29(32):4588–4598. doi:10.1038/onc.2010.203

Zomer A, Vendrig T, Hopmans ES, van Eijndhoven M, Middeldorp JM, Pegtel DM (2010) Exosomes: fit to deliver small RNA. Commun Integr Biol 3(5):447–450. doi:10.4161/cib.3.5.12339

# Epstein-Barr Virus-Specific Humoral Immune Responses in Health and Disease

**Jaap M. Middeldorp**

**Abstract** Epstein-Barr virus (EBV) is widely distributed in the world and associated with a still increasing number of acute, chronic, malignant and autoimmune disease syndromes. Humoral immune responses to EBV have been studied for diagnostic, pathogenic and protective (vaccine) purposes. These studies use a range of methodologies, from cell-based immunofluorescence testing to antibody-diversity analysis using immunoblot and epitope analysis using recombinant or synthetic peptide-scanning. First, the individual EBV antigen complexes (VCA, MA, EA(D), EA(R) and EBNA) are defined at cellular and molecular levels, providing a historic overview. The characteristic antibody responses to these complexes in health and disease are described, and differences are highlighted by clinical examples. Options for EBV vaccination are briefly addressed. For a selected number of immunodominant proteins, in particular EBNA1, the interaction with human antibodies is further detailed at the epitope level, revealing interesting insights for structure, function and immunological aspects, not considered previously. Humoral immune responses against EBV-encoded tumour antigens LMP1, LMP2 and BARF1 are addressed, which provide novel options for targeted immunotherapy. Finally, some considerations on EBV-linked autoimmune diseases are given, and mechanisms of antigen mimicry are briefly discussed. Further analysis of humoral immune responses against EBV in health and disease in carefully selected patient cohorts will open new options for understanding pathogenesis of individual EBV-linked diseases and developing targeted diagnostic and therapeutic approaches.

J.M. Middeldorp (✉)
VU Medical Center, Amsterdam, The Netherlands
e-mail: j.middeldorp@vumc.nl



C. Münz (ed.), *Epstein Barr Virus Volume 2*, Current Topics in Microbiology and Immunology 391, DOI 10.1007/978-3-319-22834-1_10

## Contents



Viral infections are countered by multiple levels of host defences, generally described as innate, and adaptive cellular and humoral (antibody-based) immune responses. The innate immune response can also be subdivided into cellular (e.g. natural killer and myeloid cells) and humoral (e.g. cytokines and chemokines) components, triggered by generic molecular pattern recognition, not to be addressed here (Chap. 9). Adaptive humoral immune responses are triggered by direct interaction of antigenic components (proteins and polypeptide fragments, protein–DNA/protein–RNA complexes, glycoproteins and glycolipids) with immunoglobulin receptors on naïve B-lymphocytes and require help from T-lymphocytes and defined cytokines for class switching and affinity maturation. Adaptive cellular immune responses are triggered by pathogen-specific peptide fragments of 8-15 amino acids originating from proteolytic digestion of viral proteins and presented in MHC-I or MHC-II molecules on the surface of nucleated host cells (Chap. 11). These antigenic fragments can be acquired either "in cis" on the infected cell itself or "in trans" via uptake and presentation of infected cell- or virion-derived antigenic components by professional antigen-presenting cells (APC). The activation of virus-specific adaptive immune responses is costimulated by cytokines and chemokines from early innate responses. Adaptive T-cell immune responses on their turn provide help for evolving humoral responses, i.e. antibody responses, either specifically by cell–cell contact (e.g. CD40L on T cells interacting with CD40 on activated B cells) or non-specifically via interleukin–cytokine release. This chapter will be dealing with the Epstein-Barr virus (EBV)-specific humoral immune responses in health and disease. Such responses can be considered as functionally relevant for eliminating infected cells and neutralising virus infectivity or merely as reflecting exposure to viral components released from infected cells. A focus is on the specific molecular recognition of EBV antigens by human antibodies and their importance for understanding viral persistence and pathogenesis, as well as their contribution to design of diagnostics and vaccines.

# 1 Historical Aspects of EBV Antigen Complex Discovery and Definition

During first exposure to EBV in young adults, a number of distinct antibody responses can be detected. Initially, as identified by Paul and Bunnell in 1932, a non-EBV-specific antibody response can be detected, referred to as the heterophile antibody (HA) response. HAs are largely of the IgM class and directed against glycolipid antigens on (non-self) red blood cells, causing such cells to aggregate ex vivo. These HAs are considered to arise from non-specific cytokine and EBV-stimulated polyreactive naïve B cells at early stages of primary EBV infection (Robinson 1982; Mockridge et al. 2004) and are detectable in about 80–90 % of juvenile IM cases. HAs are largely absent in childhood EBV infection, in adults and in other complications of EBV infection, and therefore, are of limited relevance to EBV biology and immunity. For assessing humoral immunity to EBV in health and disease, EBV-specific IgM and IgG (and IgA) serology testing is required. This provides the diagnostic tools and pathogenic insight into the role of EBV in patients presenting with fever, fatigue, enlarged lymph nodes and other non-specific symptoms associating with an ever increasing list of acute, chronic, autoimmune and malignant diseases linked to EBV.

Our understanding of EBV-specific humoral immune responses starts with the discovery in 1964 of EBV virion particles in cultured Burkitt lymphoma (BL) cells (Chap. 1). Using these first BL cells propagated in vitro, it was soon established that defined viral antigens are expressed and that these antigens interact with serum antibodies of most individuals on the globe (Henle et al. 1974, 1979). Further detailed analysis by the Henle's laboratory in Philadelphia, USA, of antibody responses to distinct EBV antigen complexes, as expressed under defined conditions in different BL cells, has established the basis for immunofluorescence-based EBV serology, which ruled diagnostic and epidemiological studies for some 40 years and is still used today (Fig. 1). Using indirect antibody-based immunofluorescence testing (IFT) and defined human serum samples, it was recognised that a small subset of cultured BL cells contain replicating virus, as confirmed by nuclear capsid structures and cytoplasmic virion particles visualised by electron microscopy. These cells are spontaneously present at 1–5 % in many BL cell lines, depending on cell culture conditions, and express an antigen complex referred to as virus capsid antigen (VCA). Some original BL cell lines proved to have more abundant VCA expression, like the P3-HR1 cell line, which were therefore selected for use in standardised serological studies. Most individuals proved to have stable levels of IgG antibodies to the VCA antigen complex, which defined past infection and lifelong virus carriership. In the mid-1980s, the group of George Miller at Yale University found that this spontaneous VCA expression in the P3HR1 line was related to the presence of a transposon-like element (WZhet), which carried the lytic switch protein Zebra (Z) located behind a latent EBV promoter element (Wp), allowing it to switch on lytic gene expression in latently infected cells (Miller et al. 1984). A P3HR1-derived cell line, called

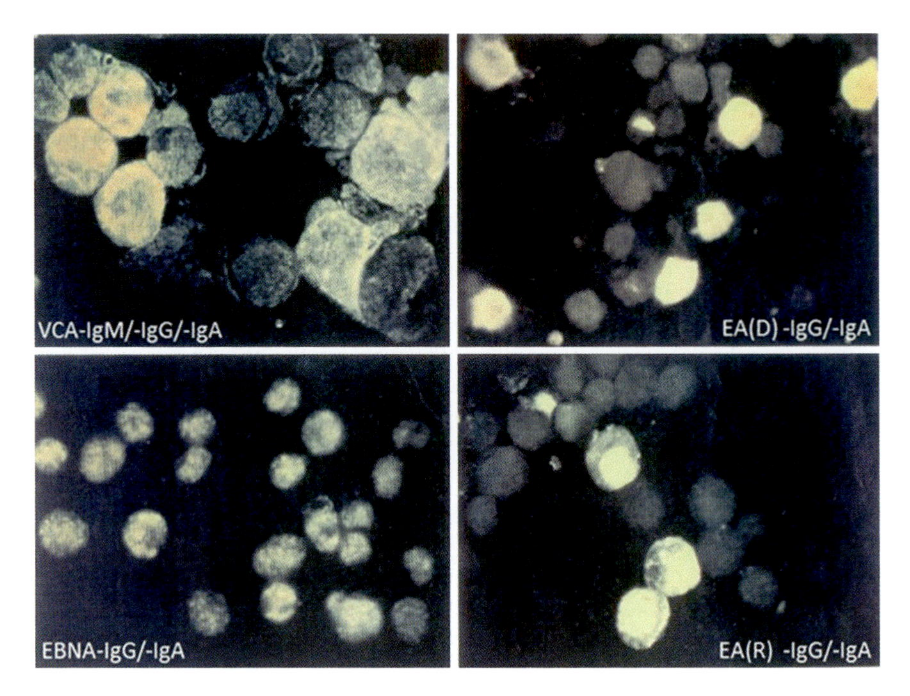


**Fig. 1** Indirect immunofluorescent antibody testing (IFT) using different Burkitt lymphoma-derived cell lines (from Henle, Henle and Lennette 1979). IFT serology requires serum titration and visual microscopic analysis of IFT staining patterns with different cell substrates by techniques that vary from laboratory to laboratory (poorly standardised). *Top-left* VCA-IgM/VCA-IgG staining as revealed in chemically induced P3HR1 cells, showing a diffuse or coarse nuclear and somewhat cytoplasmic staining; *Bottom-left* EBNA-IgG staining by anti-complement immunofluorescence (ACIF), showing a coarse speckled nuclear staining pattern as revealed in Raji cells; *Top-right* EA(D) IgG staining as revealed in chemically induced Raji cells as a fine and homogeneous nuclear staining pattern; *Bottom-left* EA(R) IgG staining with a coarse clustered cytoplasmic and nuclear staining as revealed in chemically induced Raji cells fixed with acetone only

HH514 clone-16, was created by the Miller laboratory, which was deprived of WZhet and carries a tightly latent EBV genome, but which can easily be induced for VCA expression to high levels by TPA/Butyrate treatment as will be detailed later (Rabson et al. 1983). Under defined cell culture conditions (low serum, low temperature and ageing), many EBV-transformed lymphoblastoid B cell lines (LCL) created in vitro also show some spontaneous lytic cycle VCA activity in >1 % of cells, including cell lines such as JY, X50/7, RN, but also the original marmoset cell line B95-8 (Jenson et al. 1998). These LCL cell lines all can be used for VCA-type EBV serology. The VCA antigen reactivity is operationally defined by antibody binding to nuclear and cytoplasmic antigens, as visualised by IFT staining in a limited number of cells. More abundant expression of VCA complex antigens was possible by treating BL cells for several days with sodium butyrate and/or phorbol esters, such as tetradecanoyl phorbol acetate (TPA). Later, it was found that part of the VCA complex antigens can also be detected by IFT and human antibodies on the surface of viable virus-producing BL cells and LCL, and this staining was referred to as membrane antigen (MA) complex (Klein et al. 1969; Pearson et al. 1970).

Characteristic for all herpesviruses, late antigen or VCA expression is preceded by the expression of so-called early antigens (EA), which mostly consist of virus-encoded enzymes that modulate and prepare the host cell for viral DNA replication and include the viral DNA polymerase (DNApol), encoded in the BALF5 gene. Inhibition of viral DNApol by certain antiviral agents, such as phosphonoformic acid (PFA; Foscarnet) and phosphonoacetic acid (PAA), effectively prevents the synthesis of new unmethylated viral DNA molecules that are required for expression of VCA gene products (Magalith et al. 1982). Treatment of chemically induced BL cells with PFA or PAA therefore blocks VCA expression and allows analysis of EA-specific antigens. Originally, the EA complex as defined by IFT was subdivided into EA(D) (diffuse) and EA(R) (restricted) based on sensitivity of EA-R staining to methanol fixation. EA(D) was defined as a homogeneous staining pattern restricted to the nucleus and EA(R) as a more coarse staining in both nucleus and cytoplasm. The BL Raji cell line was routinely used for EA-specific serology (Fig. 1, right side) because it lacks (inducible) VCA expression due to a deletion of defined EA genes in the BamHI-A region of the EBV genome, crucial for replication of the viral genome.

Another feature of the Raji BL cell line and linked to its lack of VCA expression was discovered by George Klein and Barbara Reedman (Reedman et al. 1974, 1975), who noted that all Raji cells, in the absence of VCA staining, contain a nuclear antigen weakly recognised by antibodies in all human sera, but which could be properly visualised by using an enhancement technique based on complement deposition around nuclear antibodies and detection by anti-complement immunofluorescence (ACIF). Using this ACIF technique, the Epstein-Barr nuclear antigen (EBNA) complex was defined as being present in all EBV-infected (tumour) cell lines in vitro, and in EBV-driven tumours in vivo as well. Raji cells proved most suitable substrate for ACIF serology among different BL lines. The Raji BL line was later shown to harbour a high copy number of EBV genomes per cell, which are maintained in dividing cells by the major EBNA-defining molecule, EBNA1 (see Chap. 1).

Although cell-based IFT/ACIF methods created the foundation for EBV serodiagnostic serology and allowed for the detection of distinct and aberrant antibody responses in defined diseases that were only later aetiologically linked to EBV, the IFT/ACIF methods are poorly standardised from laboratory to laboratory, requiring well-trained personnel to make cell slides and read the results, and the whole IFT/ACIF procedure is difficult to automate (Geser et al. 1974). Furthermore, the presence of autoantibodies in patients' sera may seriously hamper proper EBV serological read-out. Therefore, in the early 1980s, studies were initiated to define the molecular basis of the different EBV antigen complexes, which was greatly facilitated by newly developed molecular cloning techniques and not in the least by resolving the genomic DNA sequence of the EBV B95-8 strain in 1984 (Baer et al. 1984).

# 2 Molecular Definition of EBV Antigenic Complexes

## 2.1 The VCA and MA Complex

The virus capsid antigen (VCA) complex is operationally defined by IFT staining as the set of "structural" virion-associated protein antigens expressed in virus producing cells, such as chemically induced P3HR1 cells as described above (Fig. 1). Because virus-producing cells by definition also produce EA and IEA antigens, definition of the molecular identity of IFT-based VCA antigens became only possible with the production of VCA-specific (monoclonal) antibodies. As such, the major capsid protein (p160; BcLF1) (Vroman et al. 1985), the small capsid protein (VCA-p18, BFRF3) and the scaffold protein (VCA-p40, BdRF1) were first defined as the major immunoreactive VCA antigens (van Grunsven et al. 1993a, b). Subsequently, the tegument protein p23 (BLRF2) was added as immunoreactive component of the VCA complex (Hinderer et al. 1999). In addition, VCA reactivity defined by IFT also includes glycoprotein antigens residing in the nuclear and cytoplasmic membranes and the virion envelope, separately named membrane antigens (MA). Responses to these proteins can be studied using IFT and viable intact BL cells that are induced for VCA expression. In particular, gp125/110 (BALF4) and gp350/220 (BLLF1) are immunodominant determinants of the VCA complex (Kishishita et al. 1984). In part, these glycoproteins are incorporated in the virion envelope, mediating binding to and fusion with susceptible host cell membranes and thereby form a target for virus-neutralising antibody responses. On the other hand, these glycoproteins are also expressed on the nuclear membrane (gp125/110; Lee 1999) and plasma membrane of virus-producing cells (gp350/220; Thorley-Lawson and Geilinger 1980), thus functioning as target for cytolytic antibody responses capable of killing the target cell either through complement deposition/activation or through action of Fc-receptor-bearing killer cells (Khyatti et al 1994; Jilg et al. 1994).

## 2.2 The EA Complex

The diffuse nuclear EA(D) and restricted EA(R) early antigen complexes were originally defined based on sensitivity for methanol fixation and specifically detected in chemically induced or P3HR1 virus-superinfected Raji cells.

The EA(D) antigenic reactivity was initially located to the M-fragment of the viral genome by Glaser et al. (1983), subsequently defined as the BMRF1 47–54 KDa and BMLF1 60 KDa proteins (Wong and Levine 1989, Cho in Hayward group), but later EA(D) was found to consist of multiple nuclear proteins assembling in a nuclear DNA-binding scaffold preparing the cell for viral DNA replication and quantitatively dominated by the BMRF1-encoded DNA polymerase accessory protein (Cho et al. 1985; Zeng et al. 1997; Daikoku et al. 2005).

Additional immunoreactive EA(D) components were subsequently defined as the 138 KDa major DNA-binding protein (BALF2), DNA polymerase (BALF5), and DNAse (BGLF5), but also include enzymes as thymidine kinase (BXLF5) and protein kinase (BGLF4). Separately, it was demonstrated that antibodies in sera of NPC patients could neutralise a viral alkaline exonuclease (DNAse) function which had diagnostic relevance, but only when using native DNAse enzyme prepared from P3HR1 cells (Cheng et al. 1980; Stolzenberg et al. 1996; Middeldorp and Ooka unpublished data). This proved to be due to small amino acid variations between type 1 (B95-8 prototype) and type 2 (AG786, P3HR1 prototype) EBV strains altering the antigenic structure of the EBV DNAse, but not its function (Liu et al. 1998; Paramita et al. 2007; Horst et al. 2012). In 1983, Glaser et al., described first evidence that EA(R) was encoded by a distinct gene compared to EA(D), which subsequently is defined as the BHRF1 open reading frame, encoding the 17 KDa viral Bcl-2 homologue. An additional EA(R) marker was defined by Luka et al. (1986) as the large subunit of ribonucleotide reductase (BORF2), a cytoplasmic filamentous 85 KDa protein (Glaser et al. 1983; Luka et al. 1986). The EBV genomes in Raji cells carry a deletion in the BamHI-A region, are unable to express some major EA(D) proteins (BALF2–BALF5) and therefore cannot proceed from induced EA to the VCA stage of viral gene expression (Polack et al. 1984; Zhang et al. 1988).

As for all herpesviruses, the EA complex is regulated by a set of immediate early proteins (IEA), for EBV named ZEBRA (also Zta or Z) and Rta (R), which were first defined by the Miller group (Yale) and Kenney group (Chapel Hill) in the early 1980s (Grogan et al. 1987; Kenney et al. 1989a, b; Miller 1990). Although initially not classified as such, the Zta and Rta antigens operationally reside under the nuclear EA(D) group as defined by IFT staining in Raji cells.

## 2.3 The EBNA Complex

EBNA was originally defined by its speckled nuclear staining and biochemical association with chromosomes and purified DNA. Already in early studies it was suggested that the EBNA complex consisted of different molecules with different biochemical and DNA-binding properties and distinct serological reactivity. EBNA-specific enzyme-linked immunosorbent assays (ELISA) were developed by several groups, using defined and purified nuclear DNA-binding proteins from Raji cells, which avoided part of the autoantibody-related complications encountered by ACIF and IFT. In 1982, the Miller group at Yale and subsequently Hennessy and Kieff at Harvard (Boston, USA) first defined EBNA1 and EBNA2 as being encoded by distinct fragments on the EBV genome using molecular cloning techniques (Summers et al. 1982; Hennesy and kieff 1983). EBNA1 was shown to have a long glycine–alanine (Gly-Ala) repeat, which varied in size between different EBV genomes. In 1984, a synthetic Gly-Ala repeat peptide was identified as major antigenic domain of EBNA1, but later found to be cross-reactive with "self-proteins" and reactive with sera from acute CMV infection and some autoantibodies as well (Fox et al. 1986; Rhodes et al. 1990; Levine et al. 1994). In 1985, Milman and colleagues at John's Hopkins, Baltimore, USA, defined that the C-terminal domain of EBNA1 formed a reliable, stable and highly reactive EBNA reagent suitable for serological studies (Milman et al. 1985). In subsequent years, peptide technology and molecular cloning techniques revealed 6 distinct nuclear antigens that can be recognised by human antibodies, called EBNA1-6, or EBNA1, 2, 3a,3b,3c, and EBNA-LP (Falk et al. 1995a, b; Dillner et al. 1986; Finke et al. 1987).

The EBNA1 protein, encoded in the BKRF1 open reading frame on the viral genome, has now been firmly established as the major antigenic determinant of the serological EBNA complex and was further defined at the epitope level for interaction with human antibodies as detailed later in this chapter. At early times after primary infection, IgG antibodies to EBNA2 and higher EBNAs may be present, but these disappear or diminish with time postinfection. Aberrant (elevated) EBNA2 antibody responses may indicate risk of lymphomagenesis, in particular in immunosuppressed individuals and prior to development of Hodgkin's disease (Winkelspecht et al. 1996; Levin et al. 2012). Overall anti-EBNA1 responses dominate during lifelong latent virus carriership (Henle et al. 1987; Linde et al. 1990).

## 3 EBV Humoral Immune Responses as Defined by Classic Serology Testing

Using defined EBV antigen complexes in the "standard" EBV-infected cell substrates with serum titration and interpretation of IFT/ACIF staining patterns, the evolution of EBV-specific antibody reactivity during acute primary, convalescent and latent persistent stages of EBV infection in immune competent individuals, as well as altering responses upon viral reactivation and in chronic and malignant diseases, has been described by many studies and is widely used by commercial parties to illustrate use of EBV markers in diagnostic approaches (De Paschale and Clerici 2012; Gärtner et al. 2003; see Fig. 2).

The dynamics and antigen-specificity of anti-EBV humoral immune responses after primary infection reflects distinct stages of viral activity, including initial B-cell transformation, activation and proliferation, and (sub)epithelial cell lytic viral replication, virion shedding and mucosal/salivary spread. This process involves distinct phases of well-regulated and spatially separated viral gene (antigen) expression, triggering local and systemic immune responses. Complete understanding of the exact localisation and biology of early events during primary infection is currently still lacking, but the variability of anti-EBV antibody responses over time suggests this is not a uniform process in all individuals.

In general, de novo exposure to EBV structural antigens will trigger an early VCA-IgM reaction, which is the diagnostic hallmark for acute primary infection and usually detectable in the absence of EBNA1-IgG. A positive EA(D)-IgG

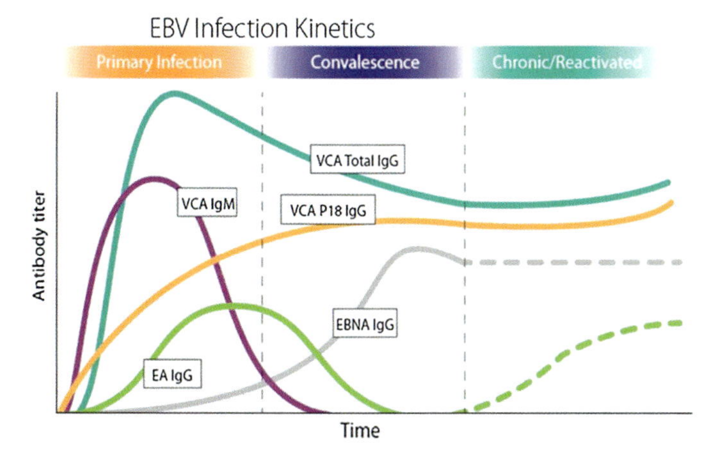


**Fig. 2** Dynamics of anti-EBV antibody responses over time during primary infection, convalescence, virus carriership and reactivation. Historically, the development of antibody responses to the different EBV antigen complexes VCA, EBNA and EA is defined by Burkitt cell-based IFT antibody titration. More recent serological techniques employ molecularly defined immunodominant markers representing the classic VCA, EBNA and EA complexes, thus maintaining the classic interpretation of EBV serology. Because different markers are used by individual diagnostic test manufacturers, universal standardisation of routine EBV serology is still lacking, although much has improved compared to the IFT methodology. The *dashed lines* for EBNA1 indicate possible alteration (loss or gain in reactivity) in seroreactivity when developing chronic (loss) or malignant (gain) disease, and *dashed line* for EA(D) indicates increase in seroreactivity relating to (chronic or malignant) reactivating infection (for details, see text)

reaction (triggered by exposure to released "particulate" early antigens) can be used as confirmation of acute active EBV infection, but may be missed occasionally when Raji cells are used as substrate (lacking the major EA(D) marker p138 encoded in BALF2). A positive and rising VCA-IgG response in follow-up samples, paralleled by a waning VCA-IgM and EA(D)-IgG response and a low but rising EBNA1-IgG response (triggered by exposure to released EBNA–DNA complexes from EBV-transformed cells), is indicative of convalescence and conversion of acute to latent persistent infection. Persistent elevated VCA-IgM may indicate problematic convalescence (insufficient T-cell help to provide Ig-class switching) and is usually paralleled by persistent positive EA(D)-IgG responses and delayed EBNA1-IgG seroconversion. Chronic (re)active EBV is characterised by elevated (high) VCA-IgG and persistently positive EA(D)-IgG levels, with low and occasionally absent EBNA1-IgG responses and negative VCA-IgM, reflecting persistent exposure to viral replicative antigens, and a possible defect in T-cell clearance of transformed B cells (Jones and Straus 1987; Miller et al. 1987). In general, individuals with aberrant (elevated) levels of EA(D) antibodies, are considered to (temporally) lack proper immunological control over endogenous latent EBV, resulting in increased viral reactivation and replication and re-expression of immunogenic early antigens, as observed in "stress" and "fatigue" cohort studies (Glaser et al. 2005; Lerner et al. 2012), but which are also observed in patients with certain autoimmune disorders, such as SLE and RA (Ballandraud et al. 2004; Lossius et al. 2012), and nasopharyngeal carcinoma (Paramita et al. 2007).

In EBV-related malignant diseases, virtually all patients are EBV seropositive (VCA and EBNA1-IgG detectable) and aberrant antibody responses prevail as defined by IFT serology, but no consistent pattern is observed. Burkitt lymphoma patients have high VCA-IgG levels, generally somewhat elevated EA(D)-IgG, but not elevated EBNA1 IgG and lack EBV IgM (Magrath and Henle 1975). NPC patients are characterised by high levels of IgG and especially IgA antibodies to all EBV antigen complexes, but lack EBV IgM (Cheng et al. 2002; Fachiroh et al. 2004; Ji et al. 2007). Elevated VCA-IgG levels were the first indication of a link between EBV and classic Hodgkin's disease (Mueller et al. 1989). This has proven to be a consistent finding, but cannot be used diagnostically because of too much overlap with other EBV syndromes (Meij et al. 2002; Levin et al. 2012).

In immunosuppressed transplant recipients, EBV serological responses may be dramatically different, delayed and reduced, but can also be compromised by (EBV seropositive or negative) blood transfusions or treatment with hyperimmune immunoglobulin for parallel HCMV infection (Middeldorp 2002; Verschuuren et al. 2003; Jaksch et al. 2013).

In HIV-infected patients, EBV-specific antibody responses are normal to elevated, with a specifically elevated IgA response, reflecting increased mucosal activity of the virus (Stevens et al. 2007). Interestingly, individuals that experience periods of physical or mental stress seem to reactivate EBV, as reflected by fluctuations in EBV DNA loads and anti-EBV antibody responses, particularly against early antigens (Glaser et al. 2005; Cacioppo et al. 2002). Patients with chronic fatigue syndrome also frequently have aberrant EBV serology, which may have therapeutic implications (Watt et al. 2012). Many patients with systemic autoimmune diseases have elevated antibody levels to VCA and frequently EA(D) as well, suggesting virus involvement in the pathogenesis, or merely loss of immunological control, which is confirmed by parallel presence of elevated levels of EBV DNA in the circulation (Balandraud et al. 2004; Poole et al. 2009; James and Robertson 2012; Fattal et al. 2014). Patients with multiple sclerosis tend to have an elevated IgG response to a defined domain in EBNA1, but not to other EBV antigens, which may suggest a pathogenic (mimicry) link (Munger et al. 2011; Jafari et al. 2010; Ruprecht et al. 2014).

The above illustrations of aberrant EBV serology in distinct clinical conditions suggest direct or indirect disease associations with EBV, which however remain difficult to prove without further molecular evidence of pathogenic involvement. The increasing realisation that EBV is involved in a wide spectrum of disease syndromes in both immunocompetent and immunocompromised hosts has led to confusion on how to interpret "classic" IFT-/ACIF-based EBV serology. Molecular characterisation of defined EBV antigens and epitopes underlying and forming the basis of the antibody responses visualised by IFT/ACIF serology became possible only recently, as will be described in detail here, and has contributed to our more fundamental understanding of the basis of anti-EBV humoral immune responses.

## 4 Molecular Details of Anti-EBV Antibody Responses During and After Acute Primary Infection

Using the Burkitt lymphoma cell line HH514.c16, created by single-cell cloning from the parent P3HR1 (Jijoye) line by the Miller group (Heston et al. 1982), which is ultrasensitive to chemical induction and expresses high levels of lytic-phase EBV proteins, it became possible to define the underlying molecular diversity and protein specificity of anti-EBV IgM, IgG and IgA antibody responses in great detail (Middeldorp and Herbrink 1988). Others have used "classic" EBV cell lines and specific immunoprecipitation and gene cloning techniques to define the exact molecular characteristics of immunologically and diagnostically relevant EBV antigens (Bayliss et al. 1983; Zhang et al. 1991).

As in most virus infections, primary EBV exposure is characterised by an early-onset IgM response. Interestingly, IgM responses to VCA are considered to be first, but include responses to EAd antigens as well (Fig. 3). In fact, the latter even can precede the true VCA response, as revealed by early immunoblot studies (Middeldorp and Herbrink 1988). IgM responses are initially triggered by particulate antigens carrying repetitive epitopes released from infected cells, which can directly activate naïve B-cell IgD/IgM immunoglobulin receptors with the appropriate antigen recognition specificity, resulting in pentameric IgM release (Hinton et al. 2008; Tolar et al. 2009). As such, virion particles and tegument-capsomeric VCA-related structures, but also protein–DNA/protein–RNA EAd-related complexes that are released from apoptotic virus-infected cells, form strong immunogens for triggering this early IgM response. Continued release of such multimeric antigen structures will lead to their uptake and digestion by antigen-presenting cells, which can present viral antigen peptide fragments to T cells, thereby triggering the antigen-specific adaptive immune responses and providing CD40-ligand- and cytokine-assisted maturation, differentiation and Ig-class switching of activated B cells. This is paralleled by enhancing affinity of virus-specific antibody responses through immunoglobulin variable domain hypermutation. This increase in affinity of anti-EBV IgG antibody responses over time can be diagnostically useful to differentiate acute-recent infection from past infection, i.e. by avidity testing (De Paschale and Clerici 2012).

The evolution and molecular characteristics (dynamics and diversity) of anti-EBV IgM and IgG responses over time in two acute IM patients are visualised in Fig. 3 using antigens extracted from VCA-induced HH514 cells (Middeldorp and Herbrink 1988). This profiling has been reproduced and confirmed for many IM cases.

Detailed immunoblot-based "antibody-diversity" analysis revealed that earliest IgM and IgG responses are directed against 45/54 KDa EA(D) components, representing different phosphorylated forms of the DNA polymerase accessory protein (BMRF1 gene) and 138 KDa EA(D), representing the major DNA-binding protein (BALF2). These proteins form the core of the EBV DNA replication complex (Zeng et al. 1997; Sugimoto et al. 2013) and are probably released as

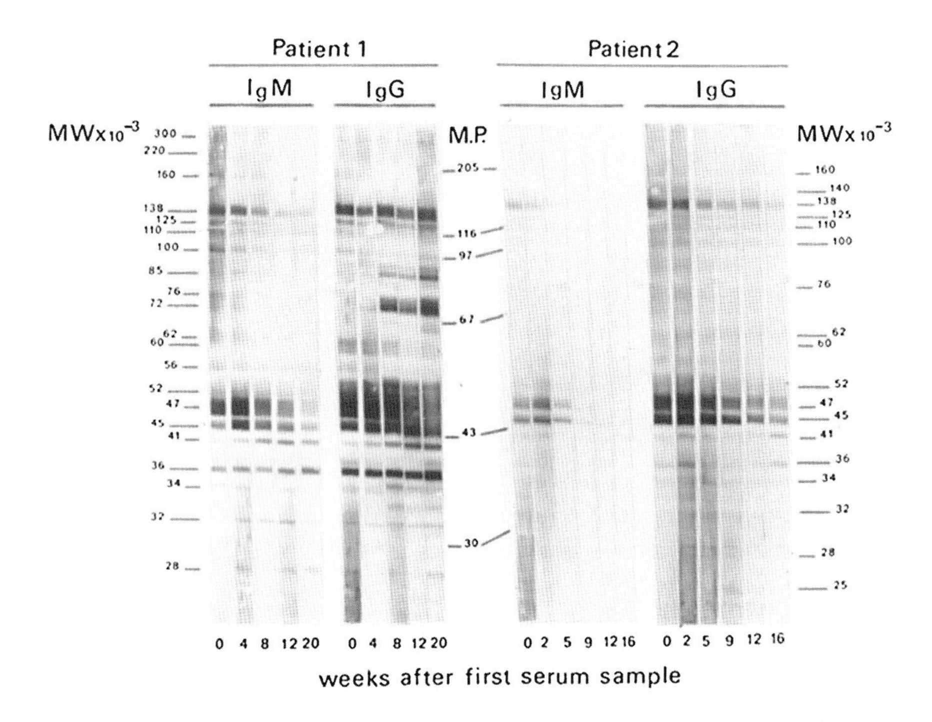


**Fig. 3** Dynamic antibody-diversity changes in EBV IgM and IgG responses in time after acute primary infection (mononucleosis). Standardised immunoblot strips were prepared from nuclear antigen extracts of VCA-induced HH514.c16 as described by Middeldorp and Herbrink (1988). Sequential serum samples were obtained from two patients with acute mononucleosis and probed for specific IgM and IgG reactivity. Detailed description of the individual EBV antigenic bands is given in the text

multimeric complexes from lytic-infected cells at early stage of infection in vivo. IgG responses against the EBV DNAse (BGLF5) at 55-58 kDa (doublet) are frequently detected at early times postinfection. Antibodies to the VCA-specific scaffold protein at 40 KDa (BdRF1, and processed from its precursor BVRF1 by the viral capsid protease) and the immediate early protein Zebra/Zta at 36-38 KDa (BZLF1) are induced at early times as well in most cases or follow shortly thereafter. These observations hold for both IgM and IgG responses. During early times postinfection, IgM and IgG antibodies to the virion structural VCA-p18, and VCA-p23 proteins are also detectable, but not visualised in Fig. 3 (van Grunsven et al. 1993a, b; Reishl et al. 1996; Hinderer et al. 1999). Importantly, early-onset VCA- and EA(D)-IgA responses, reflecting oropharyngeal mucosa activity of EBV, are frequently detected in acutely infected individuals (IM patients), but fade rapidly and have received very little attention in the past (Marklund et al. 1986). Usually weeks to months after primary infection and detectable VCA/EA(D) IgM and IgG responses, the anti-EBNA1 response (72 kDa) follows during convalescence, notably consisting primarily of IgG antibodies. In contrast to other markers, the EBNA1-IgG response is rarely preceded by EBNA1-IgM, with exception of the Gly-Ala repeat domain, which can be targeted by non-EBV-specific IgM responses as well (Rhodes et al. 1990). Humoral responses to the glycoproteins gp125/110 (perinuclear localisation and a major VCA component) and gp350/220 (cytoplasmic/membrane localisation) develop in parallel with IgM and IgG antibodies against structural VCA capsid proteins (Xu et al. 1998), whereas neutralising IgG and IgA antibody responses, predominantly directed against conformational epitopes on gp350/220, tend to develop with some delay, compared to responses against linear or denatured epitopes (Yao et al. 1991). Gp350/220 reactive IgA antibodies have been described but seem to be less effective in neutralising virus infectivity (Yao et al. 1991; Xu et al. 1998).

The identity of the EBV proteins originally identified by immunoblot and/or immunoprecipitation analysis has been verified in recent years by recombinant protein and monoclonal antibody technology, but is not detailed here further (reviewed in Hess 2003; De Paschale and Clerici 2012). EBV antibody profiling by immunoblot for serological confirmation and detection of acute, chronic or aberrant EBV infection has proven informative and has reached commercial platforms (Bauer 2001; Fachiroh et al. 2004; de Sanjosé et al. 2007; Paramita et al. 2008, de Ory et al. 2014).

Further detailed immunoblot studies revealed that following primary infection (either IM or non-symptomatic), healthy EBV carriers develop and maintain a remarkable similar anti-EBV IgG antibody-diversity pattern, recognising only a limited number of (immunodominant) antigens (Fig. 4). These antigens include EBNA1 (72 kDa), VCA-p40 (scaffold protein; BdRF1), Zebra (in 60 %) and the VCA-p18/VCA-p23 markers in nearly all healthy EBV carriers (Fig. 4b). The membrane-linked gp125 (BALF4) and gp350/220 (BLLF1) are also triggering abundant IgG responses in most individuals but are not visualised in the immunoblot approach. This persistent stable IgG response of limited diversity directed against "virion structural" VCA/MA components plus nuclear EBNA1 and occasional Zta reflects the well-balanced persistence of EBV in its human host. Surprisingly, though EA antigens precede VCA, are highly immunogenic and among the first to trigger IgM and IgG responses during acute primary infection, EA(D) antibodies are barely or not detectable in healthy EBV carriers worldwide. Patients with non-IM EBV-related syndromes, such as chronic active EBV syndrome or nasopharyngeal carcinoma (NPC), have distinctly different antibody-diversity profiles, frequently including responses to the EA(D) complex (Fig. 4). In fact, aberrant IgG and IgA antibody diversity is indicative of abnormal viral behaviour in the host as compared to normal latency (Stevens et al. 2002, 2007; Fachiroh et al. 2004; de Sanjosé et al. 2007).

EA(D)-IgG/EA(D)-IgA is an excellent marker for detecting active and reactivating EBV infection conditions (Glaser et al. 2005; Kimura 2006; Draborg et al. 2012). However, Zebra/Zta IgG responses are not limited to patients with acute chronic (malignant) infections, but are generally detectable in healthy EBV carriers, well beyond the acute phase. Aberrant EBV profiling has proven particular informative in chronic active EBV cases, which have very high IFT titres to VCA and EA(D), reflected in abundant IgG diversity on immunoblot, but may lack EBNA1 (Jones and Straus 1987; Miller et al. 1987; Kimura 2006; Fig. 4).

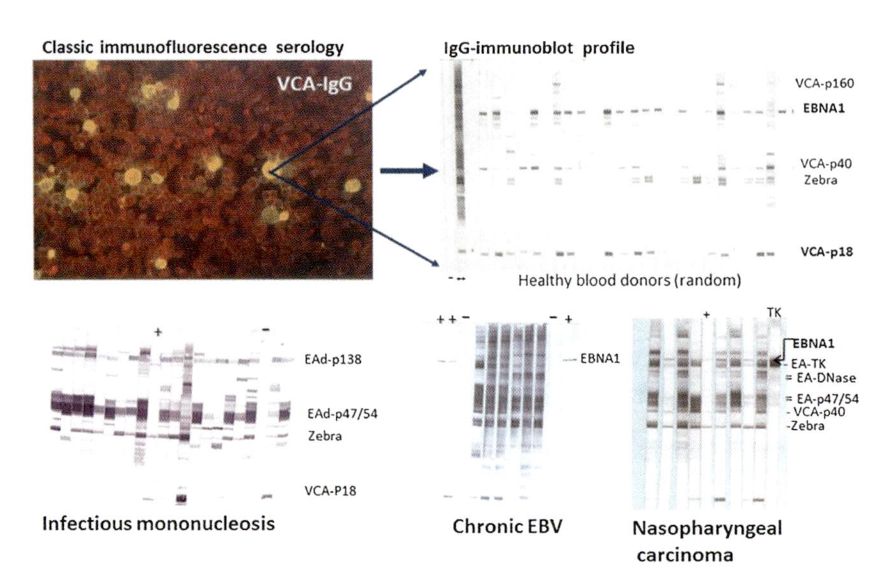


**Fig. 4** Diversity of IgG antibody profiles underlying the classic immunofluorescent antibody technique in different clinical situations. Overview of different IgG antibody profiles underlying the VCA reactivity, as observed by standardised analysis of serum samples from healthy EBV carriers and patients with various EBV-related diseases. No direct correlation is found between VCA-IFT or EBNA-ACIF titre and number of bands on the immunoblot. Healthy EBV carriers usually have a restricted IgG diversity pattern, irrespective of their VCA-IFT or EBNA-ACIF antibody titre. However, presence of EA(D) antibodies in various patient populations is reflected by a major increase in antibody diversity on the immunoblot (Meij et al. 1999; Fachiroh et al. 2004; de Sanjosé et al. 2007)

In NPC, clearly aberrant IgG diversity patterns are observed in virtually all cases, which is not the case in BL, HD nor in EBV-associated gastric cancer (data not shown; Kim et al. 2009). This aberrant IgG response was well known from IFT-based studies, but molecular details and diversity were revealed only recently. The spectrum of EBV antigens recognised by antibodies in sera from NPC patients appears to differ from chronic EBV and IM cases. Parallel analysis of IgG and IgA responses in large cohorts of NPC cases revealed that IgG and IgA antibody responses may not be targeting the same EBV proteins, an observation that was not possible by using IFT technology. Details are shown as example in Fig. 5.

This observation suggests that in NPC patients during carcinoma progression EBV antigens may be expressed differentially in endogenous lymphoid tissues, triggering predominantly IgG responses, and at nasopharyngeal mucosal surfaces, triggering NALT-associated IgA responses. This distinct triggering of IgG and IgA responses in NPC may relate to aetiological and pathogenic features in NPC currently not well understood, and has relevance for understanding EBV behaviour in vivo in mucosal and submucosal layers in the oropharyngeal (nasal)-associated lymphoid tissue (NALT). The fact that such aberrant anti-EBV antibody responses are not observed in EBV-associated gastric carcinoma (GC, Kim et al. 2009) supports the different (late) role of EBV infection in the pathogenesis of this malignancy, distinct from NPC (zur Hausen et al. 2004). In NPC, early-stage-enhanced EBV lytic replication in nasopharyngeal epithelia, possibly caused by food components such as nitrosamines and butyrates, is considered to play a pathogenic role which is reflected in early-onset aberrant antibody responses to EBV lytic genes (Ji et al. 2007). In contrast, EBV infection of gastric epithelia is considered a late event in the pathogenesis of GC, not involving abundant lytic replication (zur Hausen et al. 2004).

Overall, immunoblot analysis has revealed that "classic" EBV antibody profiles are more complex that previously considered from an immunological point of view, and are perhaps not fully appreciated thus far. The antibody profile, for either IgM, IgG or IgA, indirectly reflects viral activity and antigen exposure, being distinct in many clinical conditions. Further studies should be done, to detail serological abnormalities at the molecular level, which may reveal unexpected disease associations (de Sanjosé et al. 2007).

## 5 Immunodominance and Epitopes Within EBV Antigenic Complexes

Immunoblot studies made us understand that the humoral immune response to EBV as defined by IFT/ACIF-based serology is dominated by responses to a distinct and limited set of viral proteins (of the potentially 80–90 proteins encoded in the EBV genome), which are therefore considered as immunodominant. These markers are currently employed in commercial testing and facilitated the transfer of complex IFT/ACIF-based technology to well-standardised automated enzyme immune assays in many single and multiplex formats. How, from the immunological point of view, immune dominance is regulated in vivo among complex sets of viral proteins is largely unclear, but aspects such as multimeric structure and epitope repetitiveness and T-cell recognition are important (Tolar et al. 2009; Tolar 2014). The submolecular details of the interaction with human antibodies have been revealed for several immunodominant EBV proteins, using epitope mapping techniques, with either overlapping peptides (Middeldorp and Meloen 1988) or recombinant fragment analysis (Hinderer et al. 1999). These detailed studies have included the VCA markers p18 (BFRF3), p23 (BLRF2), p40 (BdRF1), p143 (BNRF1), p160 (BcLF1), gp125/110 (BALF4) and gp350/220 (BLLF1), the EA(D) markers p47/54 (BMRF1), p138 (BALF2), the IEA marker ZEBRA/Zta (BZLF1) and EBNA1 (BKRF1) (Middeldorp, largely unpublished data).

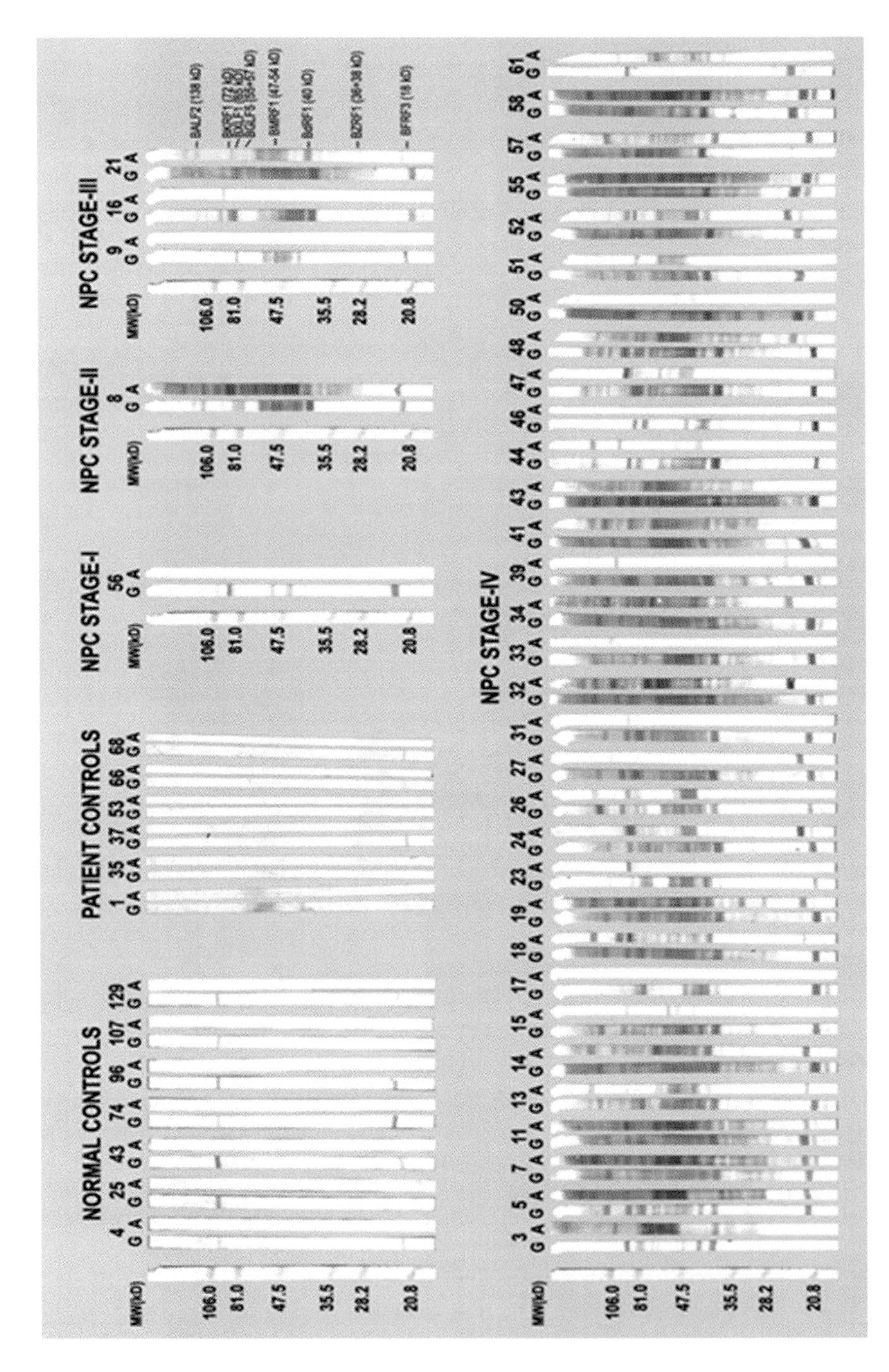
NORMAL CONTROLS
PATIENT CONTROLS
NPC STAGE-I
NPC STAGE-II
NPC STAGE-III
NPC STAGE-IV
MW(kD)
106.0
81.0
47.5
35.5
28.2
20.8

◀ **Fig. 5** Overview of IgG and IgA antibody diversity profiles by immunoblot in an endemic NPC population. Healthy Indonesian EBV carriers (blood bank donors) and non-EBV-related malignancies (hospital controls) have a "normal" restricted IgG antibody profile and negative IgA reactivity, with lower intensity in a standardised blot staining time. Already at early stage of NPC (rarely encountered in the clinic), increased staining intensity and aberrant antibody reactivity can be observed. In stages III–IV NPC (most commonly encountered at first presentation to the clinic in Indonesia), clearly abnormal immunoblot diversity patterns are observed in all patients. Of importance is the notion that IgG and IgA frequently show distinct diversity profiles, indicating independent antigen triggering of reactive B cells. This type of analysis reveals the molecular diversity of anti-EBV humoral immune responses, reflecting the pathogenic events underlying NPC development

## *5.1 VCA*

The major EBV VCA antigen for IgM, IgG and IgA responses proved to be a small capsid protein, called VCA-p18, which has a unique immunodominant and virus-specific antigenic domain in its C-terminus (AA 110-176). This domain contains several small peptide regions (epitopes) which can be combined to form a powerful diagnostic reagent for both VCA-IgM, and IgG and IgA antibody responses (van Grunsven et al. 1994; Fachiroh et al. 2006). The N-terminus (1–110) of VCA-p18 lacks reactivity with human serum antibodies and has strong homology with the N-terminus of its homologues in other herpesviruses, in particular KSHV (van Grunsven et al. 1994). This indicates that in vivo the C-terminus of VCA-p18 forms an exposed antigenic array on the surface of viral capsids, whereas the N-terminal domain is shielded from interacting with the immunoglobulin B-cell receptor due to its hidden position inside the virion capsid structure, as revealed by cryo-electron microscopy (Henson et al. 2009). The VCA-p40 protein forms the highly repetitive nuclear capsid scaffold and is reactive with IgG antibodies in most EBV carriers and IgA antibodies in NPC patients. However, detailed peptide epitope mapping for VCA-p40 failed to identify linear epitopes reactive with human serum antibodies (van Grunsven et al. 1993a, b; Middeldorp unpublished data), suggesting that antibody responses to this protein are triggered by exposure of B cells to intact capsomers yielding responses to conformational epitopes. A fusion protein between the VCA-p18 and VCA-p40 was recently described to have excellent diagnostic performance (Fachiroh et al. 2010). The VCA-p23 marker is a tegument protein, similar to VCA-p143 (BNRF1), located between the virion capsid and envelope. Both proteins require a largely intact polypeptide sequence to be recognised by human antibodies, suggesting involvement of conformational epitopes (Hinderer et al. 1999). The major capsid protein VCA-p160 was first identified as major VCA marker (Vroman et al. 1985), but contains only few small linear antigenic epitopes (Middeldorp and Meloen 1988). Overall IgG responses to this protein are less compared to the small VCA-p18, VCA-p23 and VCA-p40 proteins. In contrast to the VCA-p160 protein, the viral nuclear and virion envelop glycoprotein gp125/110 as well as the virion envelope protein gp350/220, which also lack short peptide domains for binding to human IgG (Middeldorp unpublished data), proved to be major VCA-specific components, reactive with IgG antibodies in virtually all EBV carriers (Kishishita et al. 1984; Luka et al. 1986). Although antibody responses to gp125/110 and gp350/220 are easily detected and diagnostically useful, their production and purification for diagnostics is cumbersome and requires proper glycosylation and folding (Hinderer et al. 1999; Hu et al. 2007). Some recent data indicate that the gp350/220 may be recognised rather diversely by individual EBV carriers, involving different glycopeptide domains (D'Arrigo et al. 2013). Most attention is focussed on the function of EBV glycoproteins as vaccine component to induce neutralising antibody responses, requiring the intact native antigenic structure (Pither et al. 1992; Cohen et al. 2013). This will not be detailed here.

## 5.2 EA(D)

The major immunodominant component of the EA(D) complex is p47/54 (BMRF1), a heavily phosphorylated component of the early nuclear EBV DNA replication machinery. It is one of the first antigens triggering EBV-specific IgM and IgG responses during acute primary infection (Fig. 3). Such antibodies initially interact with a defined set of peptide epitopes in the C-terminus (AA 300-380), but this response fades rapidly during convalescence and only reappears upon (chronic) reactivating EBV infection (Middeldorp, unpublished data; US Patent 5843405). Elevated EA(D) antibody responses are particularly strong in chronic active EBV infections and associate with the development of certain cancers, such as BL and NPC, reflecting the role of lytic virus activity in these diseases. Interestingly, whereas acute and chronic infection seems to trigger antibody responses to linear (peptide) epitopes, EA(D) responses in malignant disease such as NPC and BL are mainly directed against conformational epitopes on the native protein complex (Paramita et al. 2007). An explanation may be found in the lack of sufficient (early stage) T-cell help to assist in maturing anti-peptide EA(D) responses, and/or the chronicity of DNA-bound EA(D) complex exposure associating with malignant behaviour. Further study is required. The second dominant EA(D) component is the p138 protein (BALF2), present in the same nuclear complex (Zeng et al. 1997), which carries several linear epitopes that can be combined into a small immunoreactive component for diagnostics (Hinderer et al. 1999; Middeldorp 1993a). Patients with NPC develop characteristic IgG and IgA antibodies to conformational determinants of the EA(D) component known as thymidine kinase (BXLF1), DNA polymerase (BALF5) and viral DNAse (BGLF5), which are detectable by enzyme inhibition assays, but which are sensitive to denaturation and EBV strain specific (Ooka and Middeldorp unpublished). Such anti-TK, anti-DNApol and anti-DNAse antibodies can be detected in mononucleosis cases at low level and at much higher levels in chronic EBV infection as well as in NPC. Variable data have been reported on the ZEBRA-Zta antibody responses. Originally reported as being detectable in 60 % of EBV carriers, IgG-Zta seems to be a minor component of the EA(D) complex (van Grunsven et al. 1994). Some reports indicate good IgM responses in mononucleosis patients, whereas others do not (Bravo et al. 2009; Cameron et al. 2010). IgA responses to Zta can be readily detected in NPC patients (Joab et al. 1991; Dardari et al. 2008). Three main epitopes have been mapped as immunodominant (Dardari et al. 2008; Asito et al. 2010).

## *5.3 EBNA1*

The EBNA1 protein (BKRF1) is the major antigenic component of the EBNA complex reactive with human antibodies. It assembles into multimeric dimers tightly bound to DNA. EBNA1 binds with several cellular DNA-binding proteins to AT-rich regions in host chromosomes as well as to defined domains on the viral episome, known as "origin of plasmid replication" or "Ori-P" (Sears et al. 2004; Frappier 2012). The C-terminal domain (AA 460-614) of EBNA1 forms a tightly structured dimer that binds to Ori-P DNA at multiple positions via a flexible arm-like structure (AA 461-476) and interacting with other EBNA1 dimers via the flanking helix (AA 477-489) to form multimers (Bochkarev et al. 1998). The EBNA1 protein is further characterised by a nuclear localisation sequence (AA 379-386) and a long glycine–alanine (GA) repeat (AA 95-325), which prevents EBNA1 endogenous degradation by the proteasome. The GA repeat is flanked on both sides by glycine–arginine (GR) repeat regions (AA 33-53 and AA 327-377), which interact with host chromosomes (Fig. 6, top). A recent study used molecular modelling to assess the overall structure of full-length EBNA1 bound to DNA further addressing the 3D-folding, and DNA–protein interactions largely confirmed the above findings (Hussain et al. 2014).

Early biochemical studies had shown that EBNA1 is easily degraded, but that the C-terminal DNA-binding dimer (AA 450-614) is particularly stable and resistant to exogenous proteolytic degradation (Ambinder et al. 1991; Chen et al. 1993). It was found that the intact C-terminal domain of EBNA1 was highly reactive with human antibodies (Milman et al. 1985) and a good substrate for diagnostic testing. Subsequent detailed PEPSCAN peptide epitope mapping revealed that human IgG antibodies bind to EBNA1 via several immunoreactive domains, particularly in relatively non-structured regions, but that regions involved in dimer/multimer formation, internal folding and DNA interaction, were devoid of interaction with antibodies (Fig. 6). As noticed by others, the GA repeat is a major immunogenic domain in EBNA1, triggering abundant IgG antibody responses in all EBV carriers, as reflected by the high overall OD450 values. However, this region is also target for low-affinity IgM responses in other infections and target for cross-reacting (autoimmune) antibody responses (Rhodes et al. 1990). On the other hand, the small GR repeats on both sides of the large GA repeat appear to be non- or poorly immunogenic upon natural infection in humans, similar to the internally shielded dimerising and DNA-binding C-terminal sub-domains. Of note is the region at AA 390-450 which has clear reactivity with most human sera and can be mimicked by a synthetic peptide of 60 AAs having high EBNA1-specific reactivity and true EBNA1 characteristics in terms of dynamics of antibody responses (Middeldorp 1993b; Fachiroh

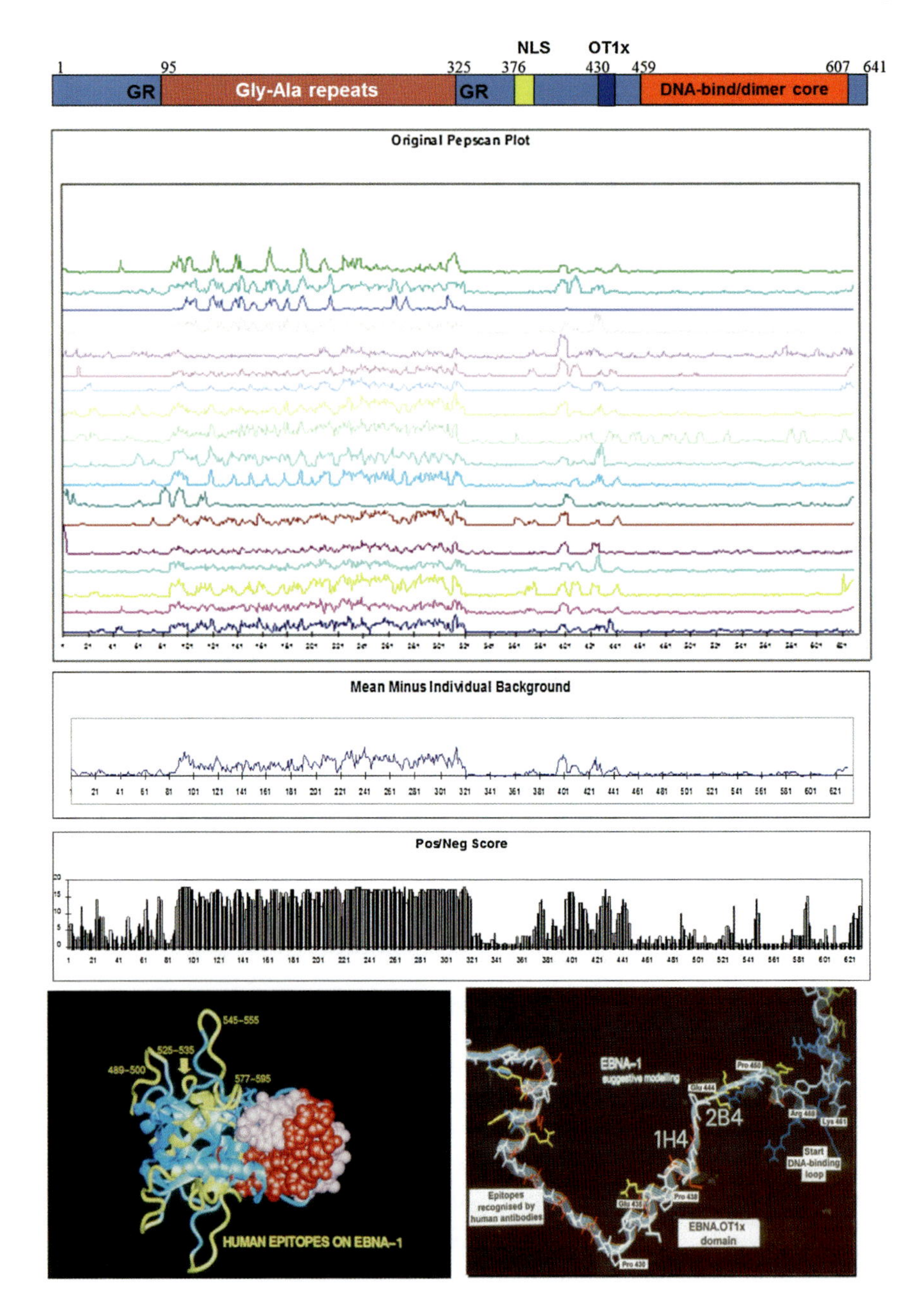
NLS
OT1x
1
95
325
376
430
459
607
641
GR
Gly-Ala repeats
GR
DNA-bind/dimer core
Original Pepscan Plot
Mean Minus Individual Background
Pos/Neg Score
545–555
525–535
489–500
577–595
HUMAN EPITOPES ON EBNA-1
EBNA-1
suggestive modelling
2B4
1H4
Start DNA-binding loop
Epitopes recognised by human antibodies
EBNA.OT1x domain

**Fig. 6** Detailed epitope mapping of polyclonal human humoral immune responses to EBNA1. *Top line* schematically illustrates the different functional domains in EBNA1 (AA 1-641). The large panel represents the OD450 ELISA values of PEPSCAN analysis with different human sera at 1:100 dilution, using 12-mer peptides overlapping by 11 AAs and covering the full EBNA1 sequence. The sera ($n = 18$) are collected from healthy EBV carriers via blood bank facilities in different countries such as Europe, the USA and Hong Kong. Details of this approach are described elsewhere (Middeldorp and Meloen 1988; van Grunsven et al. 1994; Middeldorp unpublished patent data). In the second panel, OD450 values were normalised for each serum sample against the cut-off value (COV) defined as the average OD450 value for each peptide with a set of EBV negative sera ($n = 5$). Normalised values for all sera were added-up for each peptide, revealing the most immune-reactive domains as peaks in the EBNA1 sequence, with the GA-repeat region as most reactive domain and the so-called OT1x domain at AA 390-460 as the second most reactive region. Alternatively, for each 12-mer peptide, a positive/negative score above COV (+1/0) was made for each serum and plotted for all sera, revealing all immunoreactive domains, even when having marginal OD450 values. Thus, additional regions of antibody binding (epitopes) were revealed in the GR repeats and the C-terminal domain. Interestingly, the human epitopes in the C-terminus overlap with sequences present at and extending from the surface of the EBNA1 dimer as revealed by crystal structure analysis (lower left picture). Further detailed epitope mapping of the "OT1x domain" revealed additional immunogenic domains as visualised in the lower right picture. Further details are described in the text

et al. 2006; Cameron et al. 2010). This region (Fig. 6, bottom right) is not involved in DNA binding and is readily accessible for antibody interaction, as tested in Ori-P DNA-based gel-shift assays (Chen et al. 1993; Greijer and Middeldorp unpublished data). Although the overall OD450 values for human antibody binding in the C-terminal domain seem very low, the listing of peptides that react just above the COV in this screening approach (pos/neg score) revealed small areas of antibody interaction, i.e. AA 489-500, 525-535, 545-555 and 577-590. When plotting these areas onto the crystal structure of EBNA1 C-terminus, it was revealed that these small reactive regions map exactly on the surface-exposed regions of the C-terminus and should be considered as conformational epitopes representing the strong antigenic properties of the 28 KDa C-terminal DNA-binding domain.

Most interesting about this overall assessment of EBNA1 epitope structure is the finding that all domains with known ligand interaction [i.e. Ori-P, dimerisation, flanking, host AT-rich DNA binding, nuclear localisation signal (NLS)] seem to be non-immunogenic, suggesting these domains are shielded for interaction with the Ig receptor on B cells (Hinton et al. 2008; Tolar et al. 2009). From these observations it can be speculated that EBNA1 is "seen" by the humoral immune system as an intact DNA-bound multimeric complex, probably released as such in a stable form from infected cells undergoing apoptosis and presented on the surface of APC via Fc- or complement receptors.

We further analysed the interaction of the EBNA1 unique immunodominant domain at AA 390-460 in more detail using well-known monoclonal antibodies (MoAbs; OT1x, 1H4, 2B4) and detailed epitope optimisation using PEPSCAN peptide length variation (i.e. analysing all overlapping peptides of 3-15 AA in length covering this region) and replacement net analysis (i.e. replacing every amino acid in the target peptide by any other amino acid at each position). These data were combined with computer-based minimal energy molecular modelling, to yield the proposed structure as shown in Fig. 6 bottom right panel). The rat Moabs 1H4 and 2B4 (Grässer et al. 1994) seem to interact with small linear epitopes requiring only 3-5 AA for antibody binding domain (AA 441-444 and 446-450) similar to human antibodies binding to this region. On the other hand, the so-called OT1x domain at AA 430-442 required at least 10-12 AAs for optimal binding to the OT1x MoAb and human sera (Chen et al. 1993). A replacement net analysis of this domain with human sera revealed $Pro^{431}$, $Ile^{434}$, $Glu^{435}$ and $Gly^{437}$ to be essential for proper folding of the epitope, whereas for the OT1x Moab, $Gly^{432}$, $Gly^{437}$ and $Pro^{438}$ proved to be essential. Combined, these data indicated that the EBNA1 430–442 domain has an alpha-helical structure, which was confirmed by minimal energy calculations as the preferred structure for this subdomain (Middeldorp and Grootenhuis unpublished data).

The EBNA1 region AA 441-448 has an important biological function in the infected cell by interacting with USP7/HAUSP, a key regulator of p53 and Mdm2 (Saridakis et al. 2005). This domain seems to be readily available for interaction with human antibodies, suggesting the USP7-EBNA1 interaction is resolved upon apoptotic release. Interestingly, the structural analysis of EBNA1 AA 441-448, as free peptide or bound to USP7, reveals a non-ordered "near linear" conformation, in agreement with the structure derived from our immunological and computational analysis (Fig. 6; Saridakis et al. 2005). Interestingly, antibodies recognising small linear epitopes have the potency of cross-reacting with other proteins containing the same sequence, as shown for the 2B4 antibody (Hennard et al. 2006). This may have implications for understanding the phenomenon of antigen mimicry, which is frequently observed between EBNA1 and "self" proteins in a variety of autoimmune diseases.

Overall, these data reveal how a single EBV protein (i.c. EBNA1) is recognised by human antibodies at the submolecular (epitope) level, reflecting its natural antigen presentation and processing. Further details of the position and relevance of T- and B-cell epitopes in healthy EBV carriers versus patients with various EBV-related (autoimmune) diseases may further deepen our insights into the pathogenic role of EBNA1 "in trans", i.e. outside the infected cell. It is important to notice that EBNA1 appears an important component in triggering antigen mimicry, as will be detailed below.

The EBNA1 C-terminal dimer region (AA 460-614) and the unique region AA 390-460 are widely used in commercial diagnostic tests and prove as excellent markers for EBNA-specific serology.

## 6 Antibody Responses to Tumour-Associated Antigens

In contrast to EBNA1, little has been described about humoral immune responses to other latent and tumour-associated EBV antigens, such as LMP1, LMP2 and BARF1. The overall impression is that these responses are non-existent or at best low marginal in comparison with EBNA1 and VCA responses, irrespective of the underlying EBV-driven diseases in the patients studied (Chen et al. 1993; Lennette et al. 1995; Meij et al. 1999, 2002; Paramita et al. 2011; Hoebe et al. 2011). This may indicate failure of these proteins to be presented to the BCR directly in an immunogenic form, possibly due to their hydrophobic nature and/or association with host proteins and exosomes (Dukers et al. 2000; Verweij et al. 2011). If detectable, the human antibody response seems to target intracellular cytoplasmic repeat or aggregation domains of LMP1 and LMP2, and particularly antibody responses to epitopes located in the putative extracellular membrane loops seem absent. Antibodies to BARF1 could possibly interfere with its function as scavenger of CSF-1 (M-CSF), but also these are barely detectable even in NPC patients with high VCA and EBNA1 antibody responses (Hoebe et al. 2011, 2013; Paramita et al. 2011). This apparent void in the anti-EBV humoral immune repertoire may open options for de novo immunisation with peptides derived from these extracellular domains, in order to fill this immunological gap and to create therapeutic antibody responses that potentially can eliminate EBV-driven tumours expressing LMP1 and LMP2. Experiments in animal models and selection of human phage-derived antibodies are in progress and hold promise to achieve such a goal (Middeldorp 2001; Delbende et al. 2009; Paramita et al. 2011; Chen et al. 2012).

## 7 Autoimmunity and Antigen Mimicry

EBV has been implicated in a wide variety of autoimmune diseases, including rheumatoid arthritis (RA), systemic lupus erythematosus (SLE), Sjögren's syndrome (SS), multiple sclerosis (MS) and most recently Parkinson's disease (Woulfe et al. 2014). Antigen mimicry, induced by aberrant EBV exposure or defective immune control, plays a central role in each disease and has been addressed in a number of recent reviews (Ballandraud et al. 2004; Pender 2011; Lossius et al. 2012; James and Robertson 2012; Füst 2013; Igoe and Scofield 2013). All affected individuals, including the juvenile cases, are EBV seropositive (i.e. EBV carriers) and show distinct serological abnormalities in response to a number of EBV antigens, suggestive of EBV reactivation. In many instances, such abnormalities are present at early stages of disease, prior to therapeutic intervention. However, it proves difficult to demonstrate the direct involvement of EBV in these diseases, as was recently discussed for EBV and MS (Lassmann et al. 2011). General consensus seems exist for a role of EBV as inflammatory trigger, possibly through the release of EBER molecules into the circulation, providing general innate and adaptive activation signals (Iwakiri 2014) and EBNA1–DNA complexes as a rich source of antigen sequences that may induce cross-reactive antibodies, depending on the host MHC-II background (Yadav et al. 2011). The release of complex and structured multi-epitope EBNA1–DNA complexes presented on APC during the cytokine-rich convalescent phase of mononucleosis together with innate signalling triggered by released EBER–protein/exosome

**Table 1** Overview of EBV-related epitopes involved in antigen mimicry in autoimmune diseases

| Disease (MHC-II) | Self-antigen | Self-antigen sequence | EBV antigen | EBV sequence |
|---|---|---|---|---|
| Systemic lupus erythematosus | Sm B/B' | PPPGMRPP | EBNA-1 | PPPGRRP |
| Sjögren's syndrome (HLA-DR2, 3) | Sm D1 (95–119) | RRPGGRGRGRGR GRGRGRGRGRGA | EBNA-1 (35–58) | GPAGPRGGGRGRGR GRGRGHNDGG |
| | Ro (169–180) | TKYKQRNGWSHK | EBNA-1 (58e72) | GGSGSGPRHRDGVRR |
| Rheumatoid arthritis (HLA-DRB1) (HLA-DR4, DQ8) | Cytokeratin | GGGYGGGFSSSSSS FGSFFGGGYGGGL | EBNA-1 | AGGGGGAGGAGA GGGAGGAGA |
| | Type II collagen | AGAAGA/AGA /AGA/AGA/ | EBNA-1 | AGAGGGAGGAGA GGGAGGAGA |
| | Citrullinated peptides CCP/ACPA | GRGRGRGRGR | EBNA1 (35–58) EBNA2 (338–358) | GGRGRGRGRGRGRG SRGRGRGRGRGRGKG |
| | HLA-DRB1*04101 | QK/RAA | gp110 | QKRAA |
| Multiple sclerosis (HLA-DR1501) | Myelin<br>beta-crystallin<br>Neurafilament LC<br>Myelin basic protein | RAKFQLL<br>PxxRRPFF…<br>495–519<br>QKRPS and PRHRD | BZLF1<br>EBNA1 (398–306)<br>EBNA1 (629–641)<br>EBNA1 | RAKFKQLL<br>PxxRRPFF…<br>Asp-Glu repeats<br>QKRPS and PRHRD |
| Parkinson | Alpha-synuclein | EDxxxDPDN | LMP-1 | $(DDxxxxDPDN)_4$ |

complexes may trigger autoantibody formation by crossing the tolerance threshold and extending the restricted repertoire to include autoreactive clones, a phenomenon called epitope spreading (Füst 2013; Cornaby et al. 2014). Evidence for such B-cell repertoire disturbance during mononucleosis has been presented and requires further analysis (Mockridge et al. 2004). In particular, EBNA1 is implicated in triggering antigenic mimicry, predominantly through epitopes in non-structured regions as detailed above and illustrated by examples in Table 1.

When comparing EBNA1 epitope recognition by antibodies versus T-cell receptors, it is interesting to note that these seem to be strictly separated. For CD4+ T cells, which abundantly respond to EBNA1 in most EBV carriers, the MHC-II-restricted epitopes are mostly located within the highly structured C-terminal DNA-binding dimer (AA 470-610), with virtually no responses to the GA repeats and non-structured regions (Munz et al. 2000; Leen et al. 2001; Long et al. 2005). In contrast, CD8+ CTL-targeted MHC-I recognition of EBNA1 is rather poor and mediated by a limited number of epitope combinations located in the N- (72-80), middle (378-386) and C-terminal domain (470-480) (Lee et al. 2004). This most likely reflects the in vivo release of intact DNA-bound EBNA1 complexes with subsequent processing and presentation of EBNA1 by APC to the cellular components of the immune system. Direct endogenous presentation of EBNA1 by the infected B cell is rather poor, possibly due to the strict nuclear localisation and immune evasive functions of the GA repeat (Levitskaya et al. 1985; Taylor et al. 2006; Mackay et al. 2009). Cross-presentation may be a dominant feature for T-cell responses to EBNA1.

EBNA1 plays an interesting role in the pathogenesis of MS, with CD4-T cells showing more predominant reactivity to EBNA1 at early stage of disease and potential reactivity with myelin "self" protein (Lünemann et al. 2008, 2010; Munger et al. 2011) whereas the anti-EBNA1 antibody response in MS seems to be elevated, particularly to regions in the AA 390-450 region (Jafari et al. 2010). Interestingly, antibodies recognising small linear epitopes have the potency of cross-reacting with other proteins containing the same sequence, as shown for the 2B4 antibody (Hennard et al. 2006, see above). Further illustrations are cross-reactive antibodies to Ro/SSA and La/SSB antigens (James and Robertson 2012) and dsDNA which can be induced by immunisation with EBNA1–DNA complexes and are primarily observed in patients with SLE and SS. Another recent illustration is the observation that post-translationally modified EBNA1- or EBNA2-derived GR-repeat sequences form the best antigen for the anti-citrullinated peptide response in early-stage RA (Cornillet et al. 2015). Similar situations may exist in other autoimmune diseases and require further attention (Table 1).

## 8 Concluding Remarks

Much progress has been made in defining the molecular basis of anti-EBV antibody responses in health and disease, originally assessed by IFT as macroscopic antigen complexes in BL cell lines and now slowly being elucidated at the submolecular (epitope) level. Although quite restricted in healthy carriers, humoral immune responses to EBV vary widely in different disease syndromes as revealed by immunoblot profiling, which was originally not appreciated as much using IFT serology. IgA and IgG responses may be triggered by distinct mechanisms, involving different B-cell repertoires. Detailed mapping of antigen–antibody interactions has been possible for a limited number of EBV proteins revealing insights in linear and conformational epitopes. New epitope-specific protein and peptide markers have resulted from this work that proved useful for diagnostic serology. Similar detailed analysis of EBV antigen–antibody interactions may provide insights into the molecular basis of antigen mimicry as pathogenic mechanism in EBV-related autoimmune diseases. The virus-neutralising response seems broad and universal, but is more complex than previously thought. Prophylactic EBV glycoprotein-based vaccine trials have been shown to be rather ineffective and require new initiatives. The low immunogenicity of (extracellular domains of) EBV-encoded tumour-associated proteins LMP1, LMP2 and BARF1 provides a natural window of opportunity for passive antibody-based immunotherapy and active peptide vaccination, which are still at infancy. The methods described in this chapter may be illustrative for further approaches in mapping additional details in the repertoire of anti-EBV humoral immune responses and will also provide insights into underlying natural and pathogenic interactions of EBV with its human host.

## References


Ambinder RF, Mullen MA, Chang YN, Hayward GS, Hayward SD (1991) Functional domains of Epstein-Barr virus nuclear antigen EBNA-1. J Virol 65(3):1466–1478

Asito AS, Piriou E, Odada PS, Fiore N, Middeldorp JM, Long C, Dutta S, Lanar DE, Jura WG, Ouma C, Otieno JA, Moormann AM, Rochford R (2010) Elevated anti-Zta IgG levels and EBV viral load are associated with site of tumor presentation in endemic Burkitt's lymphoma patients: a case control study. Infect Agent Cancer 5:13. doi:10.1186/1750-9378-5-13

Baer R, Bankier AT, Biggin MD, Deininger PL, Farrell PJ, Gibson TJ, Hatfull G, Hudson GS, Satchwell SC, Séguin C, Tuffnell PS, Barrell BG (1984) DNA sequence and expression of the B95-8 Epstein-Barr virus genome. Nature 310(5974):207–211

Balandraud N, Roudier J, Roudier C (2004) Epstein-Barr virus and rheumatoid arthritis. Autoimmun Rev 3(5):362–367

Bauer G (2001) Simplicity through complexity: immunoblot with recombinant antigens as the new gold standard in Epstein-Barr virus serology. Clin Lab 47(5–6):223–230

Bayliss GJ, Deby G, Wolf H (1983) An immunoprecipitation blocking assay for the analysis of EBV induced antigens. J Virol Methods 7(4):229–239

Bochkarev A, Bochkareva E, Frappier L, Edwards AM (1998) The 2.2 A structure of a permanganate-sensitive DNA site bound by the Epstein-Barr virus origin binding protein, EBNA1. J Mol Biol 284(5):1273–1278

Bravo D, Muñoz-Cobo B, Costa E, Clari MA, Tormo N, Navarro D (2009) Evaluation of an immunofiltration assay that detects immunoglobulin M antibodies against the ZEBRA protein for the diagnosis of Epstein-Barr virus infectious mononucleosis in immunocompetent patients. Clin Vaccine Immunol 16(6):885–888. doi:10.1128/CVI.00123-09

Cacioppo JT, Kiecolt-Glaser JK, Malarkey WB, Laskowski BF, Rozlog LA, Poehlmann KM, Burleson MH, Glaser R (2002) Autonomic and glucocorticoid associations with the steady-state expression of latent Epstein-Barr virus. Horm Behav 42(1):32–41

Cameron B, Flamand L, Juwana H, Middeldorp J, Naing Z, Rawlinson W, Ablashi D (2010) Lloyd A (2012) Serological and virological investigation of the role of the herpesviruses EBV, CMV and HHV-6 in post-infective fatigue syndrome. J Med Virol 82(10):1684–1688. doi:10.1002/jmv.21873
Chen MR, Middeldorp JM, Hayward SD (1993) Separation of the complex DNA binding domain of EBNA-1 into DNA recognition and dimerization subdomains of novel structure. J Virol 67(8):4875–4885
Chen R, Zhang D, Mao Y, Zhu J, Ming H, Wen J, Ma J, Cao Q, Lin H, Tang Q, Liang J, Feng Z (2012) A human Fab-based immunoconjugate specific for the LMP1 extracellular domain inhibits nasopharyngeal carcinoma growth in vitro and in vivo. Mol Cancer Ther 11(3):594–603. doi:10.1158/1535-7163.MCT-11-0725
Cheng YC, Chen JY, Glaser R, Henle W (1980) Frequency and levels of antibodies to Epstein-Barr virus-specific DNAs are elevated in patients with nasopharyngeal carcinoma. Proc Natl Acad Sci USA. 77(10):6162–6165
Cheng WM, Chan KH, Chen HL, Luo RX, Ng SP, Luk W, Zheng BJ, Ji MF, Liang JS, Sham JS, Wang DK, Zong YS, Ng MH (2002) Assessing the risk of nasopharyngeal carcinoma on the basis of EBV antibody spectrum. Int J Cancer 97(4):489–492
Cho MS, Milman G, Hayward SD (1985) A second Epstein-Barr virus early antigen gene in BamHI fragment M encodes a 48- to 50-kilodalton nuclear protein. J Virol 56(3):860–866
Cohen JI, Mocarski ES, Raab-Traub N, Corey L, Nabel GJ (2013) The need and challenges for development of an Epstein-Barr virus vaccine. Vaccine 31(Suppl 2):B194–B196. doi:10.1016/j.vaccine.2012.09.041
Cornaby C, Gibbons L, Mayhew V, Sloan CS, Welling A, Poole BD (2014) B cell epitope spreading: mechanisms and contribution to autoimmune diseases. Immunol Lett 163(1):56–68. doi:10.1016/j.imlet.2014.11.001
Cornillet M, Verrouil E, Cantagrel A, Serre G, Nogueira L (2015) In ACPA-positive RA patients, antibodies to EBNA35-58Cit, a citrullinated peptide from the Epstein-Barr nuclear antigen-1, strongly cross-react with the peptide β60-74Cit which bears the immunodominant epitope of citrullinated fibrin. Immunol Res 61(1–2):117–125. doi:10.1007/s12026-014-8584-2
Daikoku T, Kudoh A, Fujita M, Sugaya Y, Isomura H, Shirata N, Tsurumi T (2005) Architecture of replication compartments formed during Epstein-Barr virus lytic replication. J Virol 79(6):3409–3418
Dardari R, Menezes J, Drouet E, Joab I, Benider A, Bakkali H, Kanouni L, Jouhadi H, Benjaafar N, El Gueddari B, Hassar M, Khyatti M (2008) Analyses of the prognostic significance of the Epstein-Barr virus transactivator ZEBRA protein and diagnostic value of its two synthetic peptides in nasopharyngeal carcinoma. J Clin Virol 41(2):96–103
D'Arrigo I, Cló E, Bergström T, Olofsson S, Blixt O (2013) Diverse IgG serum response to novel glycopeptide epitopes detected within immunodominant stretches of Epstein-Barr virus glycoprotein 350/220: diagnostic potential of O-glycopeptide microarrays. Glycoconj J 30(7):633–640. doi:10.1007/s10719-012-9465-3
de Ory F, Guisasola E, Tarragó D, Sanz JC (2014) Application of a commercial immunoblot to define EBV IgG seroprofiles. J Clin Lab Anal. doi:10.1002/jcla.21726
De Paschale M, Clerici P (2012) Serological diagnosis of Epstein-Barr virus infection: problems and solutions. World J Virol. 1(1):31–43. doi:10.5501/wjv.v1.i1.31
de Sanjosé S, de Sanjosé S, Bosch R, Schouten T, Verkuijlen S, Nieters A, Foretova L, Maynadié M, Cocco PL, Staines A, Becker N, Brennan P, Benavente Y, Boffetta P, Meijer CJ, Middeldorp JM et al (2007) Epstein-Barr virus infection and risk of lymphoma: immunoblot analysis of antibody responses against EBV-related proteins in a large series of lymphoma subjects and matched controls. Int J Cancer 121(8):1806–1812
Delbende C, Verwaerde C, Mougel A, Tranchand Bunel D (2009) Induction of therapeutic antibodies by vaccination against external loops of tumor-associated viral latent membrane protein. J Virol 83(22):11734–11745. doi:10.1128/JVI.00578-09

Dillner J, Kallin B, Ehlin-Henriksson B, Rymo L, Henle W, Henle G, Klein G (1986) The Epstein-Barr virus determined nuclear antigen is composed of at least three different antigens. Int J Cancer 37(2):195–200

Draborg AH, Jørgensen JM, Müller H, Nielsen CT, Jacobsen S, Iversen LV, Theander E, Nielsen LP, Houen G, Duus K (2012) Epstein-Barr virus early antigen diffuse (EBV-EA/D)-directed immunoglobulin A antibodies in systemic lupus erythematosus patients. Scand J Rheumatol 41(4):280–289

Dukers DF, Meij P, Vervoort MB, Vos W, Scheper RJ, Meijer CJ, Bloemena E, Middeldorp JM (2000) Direct immunosuppressive effects of EBV-encoded latent membrane protein 1. J Immunol 165(2):663–670

Fachiroh J, Schouten T, Hariwiyanto B, Paramita DK, Harijadi A, Haryana SM, Ng MH, Middeldorp JM (2004) Molecular diversity of Epstein-Barr virus IgG and IgA antibody responses in nasopharyngeal carcinoma: a comparison of Indonesian, Chinese, and European subjects. J Infect Dis 190(1):53–62

Fachiroh J, Paramita DK, Hariwiyanto B, Harijadi A, Dahlia HL, Indrasari SR, Kusumo H, Zeng YS, Schouten T, Mubarika S, Middeldorp JM (2006) Single-assay combination of Epstein-Barr Virus (EBV) EBNA1- and viral capsid antigen-p18-derived synthetic peptides for measuring anti-EBV immunoglobulin G (IgG) and IgA antibody levels in sera from nasopharyngeal carcinoma patients: options for field screening. J Clin Microbiol 44(4):1459–1467

Fachiroh J, Stevens SJ, Haryana SM, Middeldorp JM (2010) Combination of Epstein-Barr virus scaffold (BdRF1/VCA-p40) and small capsid protein (BFRF3/VCA-p18) into a single molecule for improved serodiagnosis of acute and malignant EBV-driven disease. J Virol Methods 169(1):79–86. doi:10.1016/j.jviromet.2010.07.001

Falk K, Gratama JW, Rowe M, Zou JZ, Khanim F, Young LS, Oosterveer MA, Ernberg I (1995a) The role of repetitive DNA sequences in the size variation of Epstein-Barr virus (EBV) nuclear antigens, and the identification of different EBV isolates using RFLP and PCR analysis. J Gen Virol 76(Pt 4):779–790

Falk K, Linde A, Johnson D, Lennette E, Ernberg I, Lundkvist A (1995b) Synthetic peptides deduced from the amino acid sequence of Epstein-Barr virus nuclear antigen 6 (EBNA 6): antigenic properties, production of monoreactive reagents, and analysis of antibody responses in man. J Med Virol 46(4):349–357

Fattal I, Shental N, Molad Y, Gabrielli A, Pokroy-Shapira E, Oren S, Livneh A, Langevitz P, Pauzner R, Sarig O, Gafter U, Domany E, Cohen IR (2014) Epstein-Barr virus antibodies mark systemic lupus erythematosus and scleroderma patients negative for anti-DNA. Immunology 141(2):276–285. doi:10.1111/imm.12200

Finke J, Rowe M, Kallin B, Ernberg I, Rosén A, Dillner J, Klein G.(1987) Monoclonal and polyclonal antibodies against Epstein-Barr virus nuclear antigen 5 (EBNA-5) detect multiple protein species in Burkitt's lymphoma and lymphoblastoid cell lines. J Virol 61(12):3870–3878

Fox R, Sportsman R, Rhodes G, Luka J, Pearson G, Vaughan J (1986) Rheumatoid arthritis synovial membrane contains a 62,000-molecular-weight protein that shares an antigenic epitope with the Epstein-Barr virus-encoded associated nuclear antigen. J Clin Invest 77(5):1539–1547

Frappier L (2012) EBNA1 and host factors in Epstein-Barr virus latent DNA replication. Curr Opin Virol 733–739. doi:10.1016/j.coviro.2012.09.005

Füst G (2013) The role of the Epstein-Barr virus in the pathogenesis of some autoimmune disorders—Similarities and differences. Eur J Microbiol Immunol 1(4):267–278. doi:10.1556/EuJMI.1.2011.4.2

Gärtner BC, Hess RD, Bandt D, Kruse A, Rethwilm A, Roemer K, Mueller-Lantzsch N (2003) Evaluation of four commercially available Epstein-Barr virus enzyme immunoassays with an immunofluorescence assay as the reference method. Clin Diagn Lab Immunol 10(1):78–82

Geser A, Day NE, de-Thé GB, Chew BK, Freund RJ, Kwan HC, Lavoue MF, Simkovic D, Sohier R (1974) Bull World Health Organ 50(5):389–400

Glaser R, Boyd A, Stoerker J, Holliday J (1983) Functional mapping of the Epstein-Barr virus genome: identification of sites coding for the restricted early antigen, the diffuse early antigen, and the nuclear antigen. Virology 129(1):188–198

Glaser R, Padgett DA, Litsky ML, Baiocchi RA, Yang EV, Chen M, Yeh PE, Klimas NG, Marshall GD, Whiteside T, Herberman R, Kiecolt-Glaser J, Williams MV (2005) Stress-associated changes in the steady-state expression of latent Epstein-Barr virus: implications for chronic fatigue syndrome and cancer. Brain Behav Immunol 19(2):91–103

Grässer FA, Murray PG, Kremmer E, Klein K, Remberger K, Feiden W, Reynolds G, Niedobitek G, Young LS, Mueller-Lantzsch N (1994) Monoclonal antibodies directed against the Epstein-Barr virus-encoded nuclear antigen 1 (EBNA1): immunohistologic detection of EBNA1 in the malignant cells of Hodgkin's disease. Blood 84(11):3792–3798

Grogan E, Jenson H, Countryman J, Heston L, Gradoville L, Miller G (1987) Transfection of a rearranged viral DNA fragment, WZhet, stably converts latent Epstein-Barr viral infection to productive infection in lymphoid cells. Proc Natl Acad Sci USA 84(5):1332–1336

Henle W, Henle GE, Horwitz CA (1974) Epstein-Barr virus specific diagnostic tests in infectious mononucleosis. Hum Pathol 5(5):551–565

Henle W, Henle G, Lennette ET (1979) The Epstein-Barr virus. Sci Am 241(1):48–59

Henle W, Henle G, Andersson J, Ernberg I, Klein G, Horwitz CA, Marklund G, Rymo L, Wellinder C, Straus SE (1987) Antibody responses to Epstein-Barr virus-determined nuclear antigen (EBNA)-1 and EBNA-2 in acute and chronic Epstein-Barr virus infection. Proc Natl Acad Sci USA. 84(2):570–574

Hennard C, Pfuhl T, Buettner M, Becker KF, Knöfel T, Middeldorp J, Kremmer E, Niedobitek G, Grässer F (2006) The antibody 2B4 directed against the Epstein-Barr virus (EBV)-encoded nuclear antigen 1 (EBNA1) detects MAGE-4: implications for studies on the EBV association of human cancers. J Pathol 209(4):430–435

Hennesy K, Kieff E (1983) One of two Epstein-Barr virus nuclear antigens contains a glycine-alanine copolymer domain. Proc Natl Acad Sci USA 80(18):5665–5669

Henson BW, Perkins EM, Cothran JE, Desai P (2009) Self-assembly of Epstein-Barr virus capsids. J Virol 83(8):3877–3890. doi:10.1128/JVI.01733-08

Hess RD (2003) Routine Epstein-Barr virus diagnostics from the laboratory perspective: still challenging after 35 years. J Clin Microbiol 42(8):3381–3387

Heston L, Rabson M, Brown N, Miller G (1982) New Epstein-Barr virus variants from cellular subclones of P3 J-HR-1 Burkitt lymphoma. Nature 295(5845):160–163

Hinderer W, Lang D, Rothe M, Vornhagen R, Sonneborn HH, Wolf H (1999) Serodiagnosis of Epstein-Barr virus infection by using recombinant viral capsid antigen fragments and autologous gene fusion. J Clin Microbiol 37(10):3239–3244

Hinton HJ, Jegerlehner A, Bachmann MF (2008) Pattern recognition by B cells: the role of antigen repetitiveness versus Toll-like receptors. Curr Top Microbiol Immunol 319:1–15

Hoebe EK, Hutajulu SH, van Beek J, Stevens SJ, Paramita DK, Greijer AE, Middeldorp JM (2011) Purified hexameric Epstein-Barr virus-encoded BARF1 protein for measuring anti-BARF1 antibody responses in nasopharyngeal carcinoma patients. Clin Vaccine Immunol 18(2):298–304. doi:10.1128/CVI.00193-10

Hoebe EK, Le Large TY, Greijer AE, Middeldorp JM (2013) BamHI-A rightward frame 1, an Epstein-Barr virus-encoded oncogene and immune modulator. Rev Med Virol 23(6):367–383. doi:10.1002/rmv.1758

Horst D, Burmeister WP, Boer IG, van Leeuwen D, Buisson M, Gorbalenya AE, Wiertz EJ, Ressing ME (2012) The “Bridge” in the Epstein-Barr virus alkaline exonuclease protein BGLF5 contributes to shutoff activity during productive infection. J Virol 86(17):9175–9187. doi:10.1128/JVI.00309-12

Hu B, Hong G, Li Z, Xu J, Zhu Z, Li L (2007) Expression of VCA (viral capsid antigen) and EBNA1 (Epstein-Barr-virus-encoded nuclear antigen 1) genes of Epstein-Barr virus in Pichia pastoris and application of the products in a screening test for patients with nasopharyngeal carcinoma. Biotechnol Appl Biochem 47(Pt 1):59–69

Hussain M, Gatherer D, Wilson JB (2014) Modelling the structure of full-length Epstein-Barr virus nuclear antigen 1. Virus Genes 49(3):358–372. doi:10.1007/s11262-014-1101-9

Igoe A, Scofield RH (2013) Autoimmunity and infection in Sjögren's syndrome. Curr Opin Rheumatol 25(4):480–487. doi:10.1097/BOR.0b013e32836200d2

Iwakiri D (2014) Epstein-Barr virus-encoded RNAs: key molecules in viral pathogenesis. Cancers (Basel) 6(3):1615–1630. doi:10.3390/cancers6031615

Jafari N, van Nierop GP, Verjans GM, Osterhaus AD, Middeldorp JM, Hintzen RQ (2010) No evidence for intrathecal IgG synthesis to Epstein Barr virus nuclear antigen-1 in multiple sclerosis. J Clin Virol 49(1):26–31. doi:10.1016/j.jcv.2010.06.007

Jaksch P, Wiedemann D, Kocher A, Muraközy G, Augustin V, Klepetko W (2013) Effect of cytomegalovirus immunoglobulin on the incidence of lymphoproliferative disease after lung transplantation: single-center experience with 1157 patients. Transplantation 95(5):766–772. doi:10.1097/TP.0b013e31827df7a7

James JA, Robertson JM (2012) Lupus and Epstein-Barr. Curr Opin Rheumatol 24(4):383–388. doi:10.1097/BOR.0b013e3283535801

Jenson HB, Grant GM, Ench Y, Heard P, Thomas CA, Hilsenbeck SG, Moyer MP (1998) Immunofluorescence microscopy and flow cytometry characterization of chemical induction of latent Epstein-Barr virus. Clin Diagn Lab Immunol 5(1):91–97

Ji MF, Wang DK, Yu YL, Guo YQ, Liang JS, Cheng WM, Zong YS, Chan KH, Ng SP, Wei WI, Chua DT, Sham JS, Ng MH (2007) Sustained elevation of Epstein-Barr virus antibody levels preceding clinical onset of nasopharyngeal carcinoma. Br J Cancer 96(4):623–630

Jilg W, Bogedain C, Mairhofer H, Gu SY, Wolf H (1994) The Epstein-Barr virus-encoded glycoprotein gp 110 (BALF 4) can serve as a target for antibody-dependent cell-mediated cytotoxicity (ADCC). Virology 202(2):974–977

Joab I, Triki H, de Saint Martin J, Perricaudet M, Nicolas JC (1991) Detection of anti-Epstein-Barr virus trans-activator (ZEBRA) antibodies in sera from patients with human immunodeficiency virus. J Infect Dis 163(1):53–56

Jones JF, Straus SE (1987) Chronic Epstein-Barr virus infection. Annu Rev Med 38:195–209

Kenney S, Kamine J, Holley-Guthrie E, Lin JC, Mar EC, Pagano J (1989a) The Epstein-Barr virus (EBV) BZLF1 immediate-early gene product differentially affects latent versus productive EBV promoters. J Virol 63(4):1729–1736

Kenney S, Holley-Guthrie E, Mar EC, Smith M (1989b) The Epstein-Barr virus BMLF1 promoter contains an enhancer element that is responsive to the BZLF1 and BRLF1 transactivators. J Virol 63(9):3878–3883

Khyatti M, Stefanescu I, Blagdon M, Menezes J (1994) Epstein-Barr virus gp350-specific antibody titers and antibody-dependent cellular cytotoxic effector function in different groups of patients: a study using cloned gp350-expressing transfected human T cell targets. J Infect Dis 170(6):1439–1447

Kim Y, Shin A, Gwack J, Ko KP, Kim CS, Park SK, Hong YC, Kang D, Yoo KY (2009) Epstein-Barr virus antibody level and gastric cancer risk in Korea: a nested case-control study. Br J Cancer 101(3):526–529

Kimura H (2006) Pathogenesis of chronic active Epstein-Barr virus infection: is this an infectious disease, lymphoproliferative disorder, or immunodeficiency? Rev Med Virol 16(4):251–261

Kishishita M, Luka J, Vroman B, Poduslo JF, Pearson GR (1984) Production of monoclonal antibody to a late intracellular Epstein-Barr virus-induced antigen. Virology 133(2):363–375

Klein G, Clifford P, Henle G, Henle W, Geering G, Old LJ (1969) EBV-associated serological patterns in a Burkitt lymphoma patient during regression and recurrence. Int J Cancer 4(4):416–421

Lassmann H, Niedobitek G, Aloisi F, Middeldorp JM; NeuroproMiSe EBV Working Group (2011) Epstein-Barr virus in the multiple sclerosis brain: a controversial issue. Brain 134(Pt 9):2772–2786. doi:10.1093/brain/awr197

Lee SK (1999) Four consecutive arginine residues at positions 836-839 of EBV gp110 determine intracellular localization of gp110. Virology 264(2):350–358

Lee SP, Brooks JM, Al-Jarrah H, Thomas WA, Haigh TA, Taylor GS, Humme S, Schepers A, Hammerschmidt W, Yates JL, Rickinson AB, Blake NW (2004) CD8 T cell recognition of endogenously expressed Epstein-Barr virus nuclear antigen 1. J Exp Med 199(10):1409–1420

Leen A, Meij P, Redchenko I, Middeldorp J, Bloemena E, Rickinson A, Blake N (2001) Differential immunogenicity of Epstein-Barr virus latent-cycle proteins for human CD4(+) T-helper 1 responses. J Virol 75(18):8649–8659

Lennette ET, Winberg G, Yadav M, Enblad G, Klein G (1995) Antibodies to LMP2A/2B in EBV-carrying malignancies. Eur J Cancer 31A(11):1875–1878

Lerner AM, Ariza ME, Williams M, Jason L, Beqaj S, Fitzgerald JT, Lemeshow S (2012) Glaser R (2012) Antibody to Epstein-Barr virus deoxyuridine triphosphate nucleotidohydrolase and deoxyribonucleotide polymerase in a chronic fatigue syndrome subset. PLoS One 7(11):e47891. doi:10.1371/journal.pone.0047891

Levin LI, Chang ET, Ambinder RF, Lennette ET, Rubertone MV, Mann RB, Borowitz M, Weir EG, Abbondanzo SL, Mueller NE (2012) Atypical pre-diagnosis Epstein-Barr virus serology restricted to EBV-positive hodgkin lymphoma. Blood 120(18):3750–3755. doi:10.1182/blood-2011-12-390823

Levine D, Tilton RC, Parry MF, Klenk R, Morelli A, Hofreuter N (1994) False positive EBNA IgM and IgG antibody tests for infectious mononucleosis in children. Pediatrics 94(6):892–894

Levitskaya J, Coram M, Levitsky V, Imreh S, Steigerwald-Mullen PM, Klein G, Kurilla MG, Masucci MG (1985) Inhibition of antigen processing by the internal repeat region of the Epstein-Barr virus nuclear antigen-1. Nature 375(6533):685–688

Linde A, Kallin B, Dillner J, Andersson J, Jägdahl L, Lindvall A, Wahren B (1990) Evaluation of enzyme-linked immunosorbent assays with two synthetic peptides of Epstein-Barr virus for diagnosis of infectious mononucleosis. J Infect Dis 161(5):903–909

Liu MT, Hsu TY, Lin SF, Seow SV, Liu MY, Chen JY, Yang CS (1998) Distinct regions of EBV DNase are required for nuclease and DNA binding activities. Virology 242(1):6–13

Long HM, Haigh TA, Gudgeon NH, Leen AM, Tsang CW, Brooks J, Landais E, Houssaint E, Lee SP, Rickinson AB, Taylor GS (2005) CD4 + T-cell responses to Epstein-Barr virus (EBV) latent-cycle antigens and the recognition of EBV-transformed lymphoblastoid cell lines. J Virol 79(8):4896–4907

Lossius A, Johansen JN, Torkildsen Ø, Vartdal F, Holmøy T (2012) Epstein-Barr virus in systemic lupus erythematosus, rheumatoid arthritis and multiple sclerosis—association and causation. Viruses 4(12):3701–3730

Luka J, Miller G, Jörnvall H, Pearson GR (1986) Characterization of the restricted component of Epstein-Barr virus early antigens as a cytoplasmic filamentous protein. J Virol 58(3):748–756

Lünemann JD, Edwards N, Muraro PA, Hayashi S, Cohen JI, Münz C, Martin R (2008) Increased frequency and broadened specificity of latent EBV nuclear antigen-1-specific T cells in multiple sclerosis. Brain 129(Pt 6):1493–506

Lünemann JD, Tintoré M, Messmer B, Strowig T, Rovira A, Perkal H, Caballero E, Münz C, Montalban X, Comabella M (2010) Elevated Epstein-Barr virus-encoded nuclear antigen-1 immune responses predict conversion to multiple sclerosis. Ann Neurol 67(2):159–169. doi:10.1002/ana.21886

Mackay LK, Long HM, Brooks JM, Taylor GS, Leung CS, Chen A, Wang F, Rickinson AB (2009) T cell detection of a B-cell tropic virus infection: newly-synthesised versus mature viral proteins as antigen sources for CD4 and CD8 epitope display. PLoS Pathog 5(12):e1000699. doi:10.1371/journal.ppat.1000699

Magalith M, Manor D, Goldblum N (1982) Organophosphorous compounds as inhibitors of EBV infection and transformation. Dev Biol Stand 52:515–526
Magrath IT, Henle W (1975) Changes in antibodies to Epstein-Barr virus-associated antigens with the development of Burkitt's lymphoma. IARC Sci Publ 11(Pt 2):275–281
Marklund G, Henle W, Henle G, Ernberg I (1986) IgA antibodies to Epstein-Barr virus in infectious mononucleosis. Scand J Infect Dis 18(2):111–119
Meij P, Vervoort MB, Aarbiou J, van Dissel P, Brink A, Bloemena E, Meijer CJ, Middeldorp JM (1999) Restricted low-level human antibody responses against Epstein-Barr virus (EBV)-encoded latent membrane protein 1 in a subgroup of patients with EBV-associated diseases. J Infect Dis 179(5):1108–1115
Meij P, Vervoort MB, Bloemena E, Schouten TE, Schwartz C, Grufferman S, Ambinder RF, Middeldorp JM (2002) Antibody responses to Epstein-Barr virus-encoded latent membrane protein-1 (LMP1) and expression of LMP1 in juvenile Hodgkin's disease. J Med Virol 68(3):370–377
Middeldorp JM (1993a) Epstein-Barr virus peptides and antibodies against these peptides US Patent 5843405. http://www.google.com/patents/US5843405
Middeldorp JM (1993b) An epitope; for a diagnostic kit. US patent 5965353 http://www.google.nl/patents/US5965353
Middeldorp JM (2001) Method for the identification of extracellular domains of Epstein Barr virus (EBV) tumor-associated latent membrane proteins and for the selection of antibody reagents reactive therewith. US patent 20110064759, US 8241639; Extracellular domains of Epstein Barr Virus (EBV) tumor-associated latent membrane proteins, CN1526072A. http://www.google.com.ar/patents/US8241639
Middeldorp JM (2002) Molecular diagnosis of viral infections in renal transplant recipients. Curr Opin Nephrol Hypertens 11(6):665–672
Middeldorp JM, Herbrink P (1988) Epstein-Barr virus specific marker molecules for early diagnosis of infectious mononucleosis. J Virol Methods 21(1–4):133–146
Middeldorp JM, Meloen RH (1988) Epitope-mapping on the Epstein-Barr virus major capsid protein using systematic synthesis of overlapping oligopeptides. J Virol Methods 21(1–4):147–159
Miller G (1990) The switch between latency and replication of Epstein-Barr virus. J Infect Dis 161(5):833–844
Miller G, Rabson M, Heston L (1984) Epstein-Barr virus with heterogeneous DNA disrupts latency. J Virol 50(1):174–182
Miller G, Grogan E, Rowe D, Rooney C, Heston L, Eastman R, Andiman W, Niederman J, Lenoir G, Henle W, Henle G (1987) Selective lack of antibody to a component of EB nuclear antigen in patients with chronic active Epstein-Barr virus infection. J Infect Dis 156(1):26–35
Milman G, Scott AL, Cho MS, Hartman SC, Ades DK, Hayward GS, Ki PF, August JT, Hayward SD (1985) Carboxyl-terminal domain of the Epstein-Barr virus nuclear antigen is highly immunogenic in man. Proc Natl Acad Sci USA 82(18):6300–6304
Mockridge CI, Rahman A, Buchan S, Hamblin T, Isenberg DA, Stevenson FK, Potter KN (2004) Common patterns of B cell perturbation and expanded V4-34 immunoglobulin gene usage in autoimmunity and infection. Autoimmunity 37(1):9–15
Mueller N, Evans A, Harris NL, Comstock GW, Jellum E, Magnus K, Orentreich N, Polk BF, Vogelman J (1989) Hodgkin's disease and Epstein-Barr virus. Altered antibody pattern before diagnosis. N Engl J Med 320(11):689–695
Munger KL, Levin LI, O'Reilly EJ, Falk KI, Ascherio A (2011) Anti-Epstein-Barr virus antibodies as serological markers of multiple sclerosis: a prospective study among United States military personnel. Mult Scler 17(10):1185–1193. doi:10.1177/1352458511408991
Münz C, Bickham KL, Subklewe M, Tsang ML, Chahroudi A, Kurilla MG, Zhang D, O'Donnell M, Steinman RM (2000) Human CD4(+) T lymphocytes consistently respond to the latent Epstein-Barr virus nuclear antigen EBNA1. J Exp Med 191(10):1649–1660

Paramita DK, Fachiroh J, Artama WT, van Benthem E, Haryana SM, Middeldorp JM (2007) Native early antigen of Epstein-Barr virus, a promising antigen for diagnosis of nasopharyngeal carcinoma. J Med Virol 79(11):1710–1721

Paramita DK, Fachiroh J, Haryana SM, Middeldorp JM (2008) Evaluation of commercial EBV RecombLine assay for diagnosis of nasopharyngeal carcinoma. J Clin Virol 42(4):343–352. doi:10.1016/j.jcv.2008.03.006

Paramita DK, Fatmawati C, Juwana H, van Schaijk FG, Fachiroh J, Haryana SM, Middeldorp JM (2011) Humoral immune responses to Epstein-Barr virus encoded tumor associated proteins and their putative extracellular domains in nasopharyngeal carcinoma patients and regional controls. J Med Virol 83(4):665–678. doi:10.1002/jmv.21960

Pearson G, Dewey F, Klein G, Henle G, Henle W (1970) Relation between neutralization of Epstein-Barr virus and antibodies to cell-membrane antigens-induced by the virus. J Natl Cancer Inst 45(5):989–995

Pender MP (2011) The essential role of Epstein-Barr virus in the pathogenesis of multiple sclerosis. Neuroscientist 17(4):351–367. doi:10.1177/1073858410381531

Pither RJ, Zhang CX, Shiels C, Tarlton J, Finerty S, Morgan AJ (1992) Mapping of B-cell epitopes on the polypeptide chain of the Epstein-Barr virus major envelope glycoprotein and candidate vaccine molecule gp340. J Virol 66(2):1246–1251

Polack A, Delius H, Zimber U, Bornkamm GW (1984) Two deletions in the Epstein-Barr virus genome of the Burkitt lymphoma nonproducer line Raji. Virology 133(1):146–157

Poole BD, Schneider RI, Guthridge JM, Velte CA, Reichlin M, Harley JB, James JA (2009) Early targets of nuclear RNP humoral autoimmunity in human systemic lupus erythematosus. Arthritis Rheum 60(3):848–859. doi:10.1002/art.24306

Rabson M, Heston L, Miller G (1983) Identification of a rare Epstein-Barr virus variant that enhances early antigen expression in Raji cells. Proc Natl Acad Sci USA 80(9):2762–2766

Reedman BM, Klein G, Pope JH, Walters MK, Hilgers J, Singh S, Johansson B (1974) Epstein-Barr virus-associated complement-fixing and nuclear antigens in Burkitt lymphoma biopsies. Int J Cancer 13(6):755–763

Reedman BM, Hilgers J, Hilgers F, Klein G (1975) Immunofluorescence and anti-complement immunofluorescence absorption tests for quantitation of Epstein-Barr virus-associated antigens. Int J Cancer 15(4):566–571

Reischl U, Gerdes C, Motz M, Wolf H (1996) Expression and purification of an Epstein-Barr virus encoded 23-kDa protein and characterization of its immunological properties. J Virol Methods 57(1):71–85

Rhodes G, Smith RS, Rubin RE, Vaughan J, Horwitz CA (1990) Identical IgM antibodies recognizing a glycine-alanine epitope are induced during acute infection with Epstein-Barr virus and cytomegalovirus. J Clin Lab Anal 4(6):456–464

Robinson JE (1982) The biology of circulating B lymphocytes infected with Epstein-Barr virus during infectious mononucleosis. Yale J Biol Med 55(3–4):311–316

Ruprecht K, Wunderlich B, Gieß R, Meyer P, Loebel M, Lenz K, Hofmann J, Rosche B, Wengert O, Paul F, Reimer U, Scheibenbogen C (2014) Multiple sclerosis: the elevated antibody response to Epstein-Barr virus primarily targets, but is not confined to, the glycine-alanine repeat of Epstein-Barr nuclear antigen-1. J Neuroimmunol 272(1–2):56–61. doi:10.1016/j.jneuroim.2014.04.005

Saridakis V, Sheng Y, Sarkari F, Holowaty MN, Shire K, Nguyen T, Zhang RG, Liao J, Lee W, Edwards AM, Arrowsmith CH, Frappier L (2005) Structure of the p53 binding domain of HAUSP/USP7 bound to Epstein-Barr nuclear antigen 1 implications for EBV-mediated immortalization. Mol Cell 18(1):25–36

Sears J, Ujihara M, Wong S, Ott C, Middeldorp J, Aiyar A (2004) The amino terminus of Epstein-Barr Virus (EBV) nuclear antigen 1 contains AT hooks that facilitate the replication and partitioning of latent EBV genomes by tethering them to cellular chromosomes. J Virol 78(21):11487–11505

Stevens SJ, Blank BS, Smits PH, Meenhorst PL, Middeldorp JM (2002) High Epstein-Barr virus (EBV) DNA loads in HIV-infected patients: correlation with antiretroviral therapy and quantitative EBV serology. AIDS 16(7):993–1001

Stevens SJ, Smits PH, Verkuijlen SA, Rockx DA, van Gorp EC, Mulder JW, Middeldorp JM (2007) Aberrant Epstein-Barr virus persistence in HIV carriers is characterized by anti-Epstein-Barr virus IgA and high cellular viral loads with restricted transcription. AIDS 21(16):2141–2149

Stolzenberg MC, Debouze S, Ng M, Sham J, Choy D, Bouguermouh A, Chan KH, Ooka T (1996) Purified recombinant EBV desoxyribonuclease in serological diagnosis of nasopharyngeal carcinoma. Int J Cancer 66(3):337–341

Sugimoto A, Sato Y, Kanda T, Murata T, Narita Y, Kawashima D, Kimura H, Tsurumi T (2013) Different distributions of Epstein-Barr virus early and late gene transcripts within viral replication compartments. J Virol 87(12):6693–6699. doi:10.1128/JVI.00219-13

Summers WP, Grogan EA, Shedd D, Robert M, Liu CR, Miller G (1982) Stable expression in mouse cells of nuclear neoantigen after transfer of a 3.4-megadalton cloned fragment of Epstein-Barr virus DNA. Proc Natl Acad Sci USA 79(18):5688–5692

Taylor GS, Long HM, Haigh TA, Larsen M, Brooks J, Rickinson AB (2006) A role for intercellular antigen transfer in the recognition of EBV-transformed B cell lines by EBV nuclear antigen-specific CD4 + T cells. J Immunol 177(6):3746–3756

Thorley-Lawson DA, Geilinger K (1980) Monoclonal antibodies against the major glycoprotein (gp350/220) of Epstein-Barr virus neutralize infectivity. Proc Natl Acad Sci USA 77(9):5307–5311

Tolar P (2014) Spillane KM (2014) Force generation in B-cell synapses: mechanisms coupling B-cell receptor binding to antigen internalization and affinity discrimination. Adv Immunol 123:69–100. doi:10.1016/B978-0-12-800266-7.00002-9

Tolar P, Sohn HW, Liu W, Pierce SK (2009) The molecular assembly and organization of signaling active B-cell receptor oligomers. Immunol Rev 232(1):34–41. doi:10.1111/j.1600-065X.2009.00833

van Grunsven WM, Nabbe A, Middeldorp JM (1993a) Identification and molecular characterization of two diagnostically relevant marker proteins of the Epstein-Barr virus capsid antigen complex. J Med Virol 40(2):161–169

van Grunsven WM, van Heerde EC, de Haard HJ, Spaan WJ, Middeldorp JM (1993b) Gene mapping and expression of two immunodominant Epstein-Barr virus capsid proteins. J Virol 67(7):3908–3916

van Grunsven WM, Spaan WJ, Middeldorp JM (1994) Localization and diagnostic application of immunodominant domains of the BFRF3-encoded Epstein-Barr virus capsid protein. J Infect Dis 170(1):13–19

Verschuuren E, van der Bij W, de Boer W, Timens W, Middeldorp J, The TH (2003) Quantitative Epstein-Barr virus (EBV) serology in lung transplant recipients with primary EBV infection and/or post-transplant lymphoproliferative disease. J Med Virol 69(2):258–266

Verweij FJ, van Eijndhoven MA, Hopmans ES, Vendrig T, Wurdinger T, Cahir-McFarland E, Kieff E, Geerts D, van der Kant R, Neefjes J, Middeldorp JM, Pegtel DM (2011) LMP1 association with CD63 in endosomes and secretion via exosomes limits constitutive NF-κB activation. EMBO J 30(11):2115–2129. doi:10.1038/emboj.2011.123

Vroman B, Luka J, Rodriguez M, Pearson GR (1985) Characterization of a major protein with a molecular weight of 160,000 associated with the viral capsid of Epstein-Barr virus. J Virol 53(1):107–113

Watt T, Oberfoell S, Balise R, Lunn MR, Kar AK, Merrihew L, Bhangoo MS, Montoya JG (2012) Response to valganciclovir in chronic fatigue syndrome patients with human herpesvirus 6 and Epstein-Barr virus IgG antibody titers. J Med Virol 84(12):1967–1974. doi:10.1002/jmv.23411

Winkelspecht B, Grässer F, Pees HW, Mueller-Lantzsch N (1996) Anti-EBNA1/anti-EBNA2 ratio decreases significantly in patients with progression of HIV infection. Arch Virol 141(5):857–864

Wong KM, Levine AJ (1989) Characterization of proteins encoded by the Epstein-Barr virus transactivator gene BMLF1. Virology 168(1):101–111

Woulfe JM, Gray MT, Gray DA, Munoz DG, Middeldorp JM (2014) Hypothesis: a role for EBV-induced molecular mimicry in Parkinson's disease. Parkinsonism Relat Disord 20(7):685–694. doi:10.1016/j.parkreldis.2014.02.031

Xu J, Ahmad A, Blagdon M, D'Addario M, Jones JF, Dolcetti R, Vaccher E, Prasad U, Menezes J (1998) The Epstein-Barr virus (EBV) major envelope glycoprotein gp350/220-specific antibody reactivities in the sera of patients with different EBV-associated diseases. Int J Cancer 79(5):481–486

Yadav P, Tran H, Ebegbe R, Gottlieb P, Wei H, Lewis RH, Mumbey-Wafula A, Kaplan A, Kholdarova E, Spatz L (2011) Antibodies elicited in response to EBNA-1 may cross-react with dsDNA. PLoS One 6(1):e14488. doi:10.1371/journal.pone.0014488

Yao QY, Rowe M, Morgan AJ, Sam CK, Prasad U, Dang H, Zeng Y, Rickinson AB (1991) Salivary and serum IgA antibodies to the Epstein-Barr virus glycoprotein gp340: incidence and potential for virus neutralization. Int J Cancer 48(1):45–50

Zeng Y, Middeldorp JM, Madjar JJ, Ooka T (1997) A major DNA binding protein encoded by BALF2 open reading frame of Epstein-Barr virus (EBV) forms a complex with other EBV DNA-binding proteins: DNAase, EA-D, and DNA polymerase. Virology 239(2):285–295

Zhang CX, Decaussin G, Daillie J, Ooka T (1988) Altered expression of two Epstein-Barr virus early genes localized in BamHI-A in nonproducer Raji cells. J Virol 62(6):1862–1869

Zhang PF, Klutch M, Armstrong G, Qualtiere L, Pearson G, Marcus-Sekura CJ (1991) Mapping of the epitopes of Epstein-Barr virus gp350 using monoclonal antibodies and recombinant proteins expressed in Escherichia coli defines three antigenic determinants. J Gen Virol 72(Pt 11):2747–2755

Zur Hausen A, van Rees BP, van Beek J, Craanen ME, Bloemena E, Offerhaus GJ, Meijer CJ, van den Brule AJ (2004) Epstein-Barr virus in gastric carcinomas and gastric stump carcinomas: a late event in gastric carcinogenesis. J Clin Pathol 57(5):487–491

# T-Cell Responses to EBV

**Andrew D. Hislop and Graham S. Taylor**

**Abstract** Epstein–Barr virus (EBV) is arguably one of the most successful pathogens of humans, persistently infecting over ninety percent of the world's population. Despite this high frequency of carriage, the virus causes apparently few adverse effects in the vast majority of infected individuals. Nevertheless, the potent growth transforming ability of EBV means the virus has the potential to cause malignancies in infected individuals. Indeed, EBV is thought to cause 1 % of human malignancies, equating to 200,000 malignancies each year. A clear factor as to why virus-induced disease is relatively infrequent in healthy infected individuals is the presence of a potent immune response to EBV, in particular, that mediated by T cells. Thus, patient groups with immunodeficiencies or whose cellular immune response is suppressed have much higher frequencies of EBV-induced disease and, in at least some cases, these diseases can be controlled by restoration of the T-cell compartment. In this chapter, we will primarily review the role the αβ subset of T cells in the control of EBV in healthy and diseased individuals.

## Contents



A.D. Hislop (✉) · G.S. Taylor (✉)
School of Cancer Sciences, University of Birmingham, Edgbaston,
Birmingham B15 2TT, UK
e-mail: a.d.hislop@bham.ac.uk

G.S. Taylor
e-mail: g.s.taylor@bham.ac.uk



C. Münz (ed.), *Epstein Barr Virus Volume 2*, Current Topics in Microbiology and Immunology 391, DOI 10.1007/978-3-319-22834-1_11

## Abbreviations

| | |
|---|---|
| BL | Burkitt lymphoma |
| CMV | Cytomegalovirus |
| DLBCL | Diffuse large B-cell lymphoma |
| E | Early |
| EBNA | Epstein–Barr nuclear antigen |
| EBV | Epstein–Barr virus |
| ENKTL | Extranodal NK/T-cell lymphoma |
| GCa | Gastric carcinoma |
| HL | Hodgkin lymphoma |
| HLA | Human leucocyte antigen |
| IE | Immediate early |
| IM | Infectious mononucleosis |
| iNKT | Invariant natural killer T cell |
| L | Late |
| LCL | Lymphoblastoid cell line |
| LMP | Latent membrane protein |
| MHC | Major histocompatability complex |
| NK | Natural killer |
| NPC | Nasopharyngeal carcinoma |
| PTLD | Post-transplant lymphoproliferative disease |
| SAP | Signaling lymphocytic activation molecule-associated protein |
| TAP | Transporter associated with antigen processing |
| TCR | T-cell receptor |
| VCA | Viral capsid antigen |
| XIAP | X-linked inhibitor of apoptosis protein |
| XLP | X-linked lymphoproliferative disease |

# 1 Introduction

Like all herpesviruses, the EBV replication cycle oscillates between phases of either lytic or latent viral gene expression in infected cells. During lytic replication, up to 80 genes are expressed in a temporally regulated manner resulting in the production of new viral particles, the death of the infected cell and transmission of the virus to other cells or spread to new hosts. Latent gene expression, in the case of EBV, can induce growth transformation of infected cells; these genes include the six Epstein–Barr nuclear antigens (EBNAs) and the two latent membrane proteins (LMPs). Here, there is no virus production and ultimately the virus can enter a quiescent state where there is minimal, if any, viral gene expression allowing amplification of a reservoir of infected cells capable of reactivating virus in the future. These two patterns of gene expression give the immune system two potential sets of antigens that can be targeted to control the virus. Furthermore, in the normal biology of infection both sets of genes are likely expressed in different cellular backgrounds and different anatomical locations. Thus, latent antigens are expressed in B lymphocytes within lymphoid tissue, while lytic antigens are expressed also in B lymphocytes but likely more frequently expressed in epithelial cells of the oropharynx (Hadinoto et al. 2009). These different cellular backgrounds and anatomical locations likely impact on the virus-specific T-cell response in terms of the function and subset of antigen-specific T cells which traffic to these sites.

# 2 T-Cell Response in Primary Infection

Primary infection with EBV mostly occurs in childhood and in the main results in an infection with few if any symptoms of acute virus infection. However, if infection is delayed until adolescence, some 25–75 % of infected individuals will develop the primary infection syndrome infectious mononucleosis (IM): an acute self-limiting febrile illness characterised by the development of lymphadenopathy, sore throat and associated with the massive expansion of activated CD8+ T lymphocytes (Balfour et al. 2013; Crawford et al. 2006). Most of what we know about primary infection and the host response comes from the study of IM patients as the symptoms provide a convenient marker to identify people undergoing primary infection.

## *2.1 T-Cell Response in Infectious Mononucleosis*

People who develop infectious mononucleosis are thought to have been infected with the virus for up to six weeks before the symptoms became apparent. Why primary EBV infection can result in IM is presently unclear, but several proposals have been put forward. These include the development of T-cell specificities induced by non-EBV epitopes that cross-react with EBV epitopes and contribute to the expanded CD8+ T-cell response in IM (Clute et al. 2005; Selin et al. 2011). However, ex vivo analysis of such cross-reactive T-cell responses show they are low in frequency (Cornberg et al. 2010) and so their contribution to IM requires clarification. Others have proposed that IM may develop in part due to the reduced frequency of regulatory T cells found in IM patients compared to healthy donors (Wingate et al. 2009). Two recent epidemiological studies have suggested that a genetic component may be responsible for the development of IM. Thus, same sex twins have a higher incidence of IM compared to first-degree relatives (Rostgaard et al. 2014), while the concordance for development of IM was twice as frequent in monozygotic twins compared to dizygotic twins (Hwang et al. 2012). What form these genetic differences may take are unclear; however, polymorphisms or genetic markers associated with IM in immune response loci have been identified including polymorphisms in the promoters of IL-10 and IL-1 (Helminen et al. 1999; Hatta et al. 2007), polymorphisms in the TGF-β gene (Hatta et al. 2007) and presence of certain microsatellite markers in the HLA locus (McAulay et al. 2007).

At presentation, IM patients shed high levels of virus from the oropharynx and have high loads within their peripheral B-cell compartment. Few studies have been able to access samples from patients preceding the development of symptoms to determine what immune responses may be occurring at this time. Recent careful studies following EBV-negative university students who subsequently develop IM have shown that prior to the development of symptoms there is no obvious disturbance of the CD4+ or CD8+ lymphocyte compartment (Balfour et al. 2013). However, once symptoms develop in the patient there is a major disturbance, principally driven by the expansion of highly activated CD8+ T cells. Historically, the nature of these T cells was enigmatic, with initial studies analysing TCR usage indicating that these were monoclonal or polyclonal expansions suggestive of antigen-driven proliferation (Callan et al. 1996). Ex vivo analysis of these populations by either cytotoxicity or cytokine secretion assays showed that they contained EBV specificities (White et al. 1996; Steven et al. 1996, 1997; Hoshino et al. 1999). The development of MHC class I tetramer reagents that could identify EBV reactive CD8+ T cells clearly showed that significant proportions of the expanded CD8+ cell population were in fact EBV-specific (Callan et al. 1998). Further studies by numerous groups have now shown that in IM patients the expanded CD8+ population is mostly EBV-specific (Catalina et al. 2001; Hislop et al. 2002, 2005) although evidence has been presented for the activation of other virus specificities during acute infection (Odumade et al. 2012; Clute et al. 2005).

### 2.1.1 CD8+ T-Cell Response in Infectious Mononucleosis

Analysis of the specificities within the expanded CD8+ T-cell population has shown that it is dominated by T cells specific for epitopes derived from the immediate early (IE) gene products with lesser, yet still substantial, frequencies to some early (E) gene products and apparently few T cells specific for late (L) gene products (Pudney et al. 2005; Woodberry et al. 2005b). In some cases, individual lytic epitope-specific responses can constitute between 1 and 50 % of the expanded CD8+ repertoire in the peripheral circulation (Hislop et al. 2005). What drives the IE > E >> L immunodominance hierarchy is not clear. Two proposals have been suggested: firstly, as IE antigens are expressed in the first wave of viral protein synthesis IE-specific T cells may clear infected cells before other antigens are produced, reducing the supply of antigen to restimulate other later specificities (Pudney et al. 2005). Secondly, as immune evasion proteins are expressed predominantly in the E phase of lytic cycle, these reduce the presentation of epitopes on infected B cells that are derived from other E- and L-expressed proteins, limiting the ability of E- and L-phase epitope-specific T cells to be restimulated by infected cells (Hislop et al. 2007a).

Latent antigen-specific responses are also substantially expanded although generally not to the same degree as lytic epitope-specific T cells with individual epitope specificities representing from 0.1 to 5 % of the peripheral CD8+ T-cell population (Catalina et al. 2001; Hislop et al. 2002). In these cases, T cells are specific for epitopes predominantly drawn from the EBNA3 family of proteins and to a much lesser extent LMP2. EBNA1-specific responses are seen less frequently but in the context of certain HLA types strong responses have been observed (Blake et al. 2000) while CD8 specificities to LMP1 are rarely if ever seen (Hislop et al. 2002; Catalina et al. 2001; Woodberry et al. 2005b).

Although comprehensive studies of T-cell frequencies in the peripheral circulation of IM patients have demonstrated the high-frequency responses, a more relevant site to measure T-cell responses is the site of virus replication, namely the oropharyngeal lymphoid tissue making up Waldeyer's ring. Estimation of EBV-specific CD8+ T-cell frequencies in homogenised tonsillar preparations taken from IM patients showed substantial frequencies of EBV-specific CD8+ T cells in this anatomical compartment, with up to 25 % and 1.5 % of the CD8 population being specific for EBV lytic or latent epitopes, respectively (Hislop et al. 2005). However, comparing these frequencies to matched peripheral blood mononuclear cell (PBMC) preparations collected at the same time showed that these EBV-specific CD8+ T-cell frequencies were far lower in the tonsil than what was detected in periphery, particularly for lytic epitope-specific CD8+ T cells. This was despite very high virus genome loads being detected in these tonsillar preparations.

MHC class I tetramer-based analysis of the CD8+ T-cell response has yielded valuable information as to the phenotypic characteristics of EBV-specific T cells. These cells show evidence of being highly activated, expressing HLA-DR, CD38, CD69 and are in cycle as judged by the expression of ki-67(Callan et al. 1998). Historically, it has been known that PBMC from IM patients are highly susceptible to apoptosis when manipulated in vitro (Moss et al. 1985) and that this is likely a consequence of the low level of expression of the anti-apoptotic protein bcl-2 (Callan et al. 2000; Soares et al. 2004). Consistent with their highly activated status, these EBV-specific T cells express little in the way of lymphoid homing markers such as CCR7 or CD62-L (Catalina et al. 2002; Hislop et al. 2002). This lack of expression may give some clue as to why relatively lower frequencies of EBV-specific T cells are seen in the tonsils compared to the peripheral circulation. Entry of T cells into the tonsils requires expression of lymphoid homing markers, particularly since tonsils do not have afferent lymphatics and lymphocytes cannot therefore drain into this lymphoid tissue from peripheral sites. Nevertheless, EBV-specific T cells are found at this site and other mechanisms of recruitment may be relevant. Thus, activated T cells are known to express CXCR3, which binds the IFN-$\gamma$ inducible chemokines CXCL9 and CXCL10, which may direct effectors to this site. Conceivably, the inefficient recruitment of EBV-specific T cells to the tonsil during IM may explain why such high levels of virus are found at this location.

Within weeks, the symptoms of IM begin to resolve. During this time, there is a steep decline in virus genome loads found in the peripheral circulation although virus loads shed from the oropharynx remain high for several months (Fafi-Kremer et al. 2005; Hislop et al. 2005; Balfour et al. 2005). During this time, the EBV-specific T cells in the periphery are culled and the frequency and absolute number decline rapidly. Lytic epitope-specific populations are dramatically reduced, particularly those that are highly expanded (Catalina et al. 2001; Hislop et al. 2002) leading to a distribution of epitope specificities which does not necessarily reflect the frequencies seen during acute infection. Although frequencies of latent epitope-specific cells change less, with the contraction of the entire CD8 compartment numerically fewer latent epitope-specific cells will be present. Despite the dramatic impact, IM has on the T-cell compartment, when disease resolves there does not appear to be any attrition of pre-existing memory T cells, with absolute numbers of T cells specific to other viral epitopes broadly comparable before and after acute EBV infection (Odumade et al. 2012).

After resolution of disease, EBV-specific CD8+ T cells begin to return to a resting state, downregulate activation markers, come out of cycle and upregulate expression of anti-apoptotic proteins such as bcl-2 (Dunne et al. 2002). Latent, but not lytic, epitope-specific T cells begin to express CCR7 and CD62-L, and this is associated with recruitment of these T cells to the tonsil at a time when the control of growth transformation of B cells at this site is known to be controlled. Lytic epitope-specific cells remain poorly represented in this tissue, consistent with the continued high-level shedding of virus in saliva.

### 2.1.2 CD4+ T-Cell Response in Infectious Mononucleosis

Turning to the CD4+ T-cell response to EBV during IM, less is known about the response mediated by these cells due to the low frequency of specific responses and the fact that until very recently assays to measure CD4 T-cell responses were relatively insensitive. However, there is now a better appreciation of these cells as potential effectors against MHC class II targets, such as EBV-infected B cells. During IM there is little, if any, expansion of the global CD4 compartment (Balfour et al. 2013) and in contrast to the CD8 compartment, TCR analysis shows there is no evidence of antigen-driven monoclonal or oligoclonal expansions of CD4 + T cells (Maini et al. 2000). Initial attempts to study CD4+ T-cell responses used cytokine secretion assays to measure responses from CD4+ T cells stimulated with recombinant antigens or lysates of EBV-infected cells. These detected relatively weak responses to lytic and latent antigens with lytic antigens eliciting responses more frequently (Precopio et al. 2003; Amyes et al. 2003).

However, an increasing range of MHC class II tetramer reagents have recently become available and these have been employed to follow responses in IM patients from acute infection through resolution of disease (Long et al. 2013). With the epitope-specific reagents used in this work, particular latent responses appeared to dominate lytic responses with up to 1.5 % of CD4+ T cells specific for individual latent epitopes, compared to a maximum of 0.5 % of CD4+ T cells for lytic responses. These measurements were substantially higher than those estimated by cytokine secretion analysis in this and previous studies. The EBV-specific CD4 T cells, like CD8+ T cells, were highly activated and low frequencies expressed lymphoid homing markers.

With resolution of acute symptoms, tracking the CD4+ T-cell response shows that the frequencies of EBV-specific cells drops precipitously over a short period of time (Long et al. 2013; Precopio et al. 2003). In contrast to other latent epitope specificities, EBNA1-specific CD4+ T-cell responses are not readily detected during acute infection but emerge with delayed kinetics over several months. This delayed appearance has been attributed to the restricted release of EBNA1 from infected cells, thereby reducing antigen available for priming CD4+ T-cell responses (Long et al. 2013).

## 2.2 T-Cell Response in Asymptomatic Primary Infection

Although IM has taught us much about the immune responses made during primary EBV infection, these responses likely do not represent the situation in the majority of primary infections which occur asymptomatically in children. Determining the host response to the virus in this situation is extremely difficult since, by definition, the infection occurs without obvious evidence.

Nevertheless, careful methodical studies following EBV-seronegative individuals over time have allowed cases of asymptomatic EBV infections to be identified. Early studies of infants monitored monthly for seroconversion indicated that they showed none of the features seen in IM patients in terms of disruption to the lymphocyte compartment or febrile illness (Biggar et al. 1978; Fleisher et al. 1979). More contemporary studies have shown that young adults undergoing asymptomatic infection can have high virus loads in the peripheral circulation, equivalent to what is seen in IM patients; however, there is no lymphocytosis and unlike IM patients, most showed no disruption within the T-cell compartment of the clonality of their TCR Vβ repertoire (Silins et al. 2001). A recent study examining African children undergoing asymptomatic infection has suggested that, like the young adults in the earlier work, children can have genome loads equivalent to IM patients. Moreover, they can have substantial frequencies of activated EBV-specific CD8+ T cells as detected using MHC class I tetramers, up to 16 % of CD8+ T cells. Despite these high-frequency responses, there is no significant global expansion of CD8+ T-cell compartment (Jayasooriya et al. 2015).

Collectively these studies suggest that it is the global expansion of activated CD8+ T cells that is driving the pathology seen in IM. Although these studies are helpful for characterising the virus–host balance in asymptomatic infection, they do not explain the mechanism of control in the absence of pathology. Studies using the humanised mouse model of EBV infection have shed some light on this however, suggesting that specific subsets of natural killer cells may play a role in preventing the development of an IM like disease in this model (Chijioke et al. 2013) indicating that analysis of NK subsets in asymptomatic infected donors may be helpful.

# 3 T-Cell Response in Established Infection

The T-cell response seen in people with established EBV infections shows that there are substantial frequencies of EBV-specific CD8+ and CD4+ T cells present, although at a much reduced frequency and absolute number than what is seen in IM patients. In healthy donors, low frequencies of infected cells are detected in the memory B-cell compartment of the peripheral circulation (Babcock et al. 1998) and there is intermittent shedding of virus from the oropharynx (Fafi-Kremer et al. 2005) indicating that these responses are maintained in the presence of low-level antigen expression.

## *3.1 CD8+ T-Cell Response in Established Infection*

Individual lytic antigen-specific CD8 responses can account for up to 2 % of the CD8+ population, while latent antigen-specific responses are smaller constituting up to 1 % of the CD8 population. As seen in IM patients, CD8+ T cells specific for immediate early-expressed epitopes are the dominant specificities with lower responses to a subset of early-expressed epitopes and rare responses to late-expressed epitopes (Abbott et al. 2013). Although responses to late epitopes are of low frequency, a diverse range of late antigens are targeted and these responses are thought to increase with age (Orlova et al. 2011; Stowe et al. 2007). Latent responses are mostly made to epitopes derived from the EBNA3 family of proteins and to a lesser extent LMP2, EBNA1 and EBNA2, while infrequent responses are detected against EBNA-LP and LMP1. Interestingly, alterations to this hierarchy can be observed dependent on the HLA type of the donors. Thus, donors who are HLA B38 make strong responses to an epitope derived from EBNA2 (Chapman et al. 2001) while donors who are HLA-A*02.03 make a strong response to an EBNA-LP-derived epitope (unpublished observations).

Although there is evidence of chronic EBV shedding and thus antigen production, EBV-specific T cells circulating in the periphery show little evidence of activation. In contrast to the picture of IM, these cells appear mostly as resting antigen-experienced T cells expressing LFA-1 (Faint et al. 2001), they are not activated, are not in cycle and have relatively high levels of bcl2 (Callan et al. 2000; Hislop et al. 2001).

These T cells now also express variable levels of markers associated with homing to lymphoid tissues such as CCR7 and CD62L. The frequency of expression of these markers is higher on latent compared to lytic epitope-specific T cells for reasons that are unclear but perhaps related to the environment in which antigen is presented to the T cell. Analysis of the frequency of EBV-specific T cells in matched blood and lymph node specimens has, however, shown no obvious enrichment of EBV specificities in this latter compartment (Remmerswaal et al. 2012). Some enrichment of lytic but not latent epitope specificities in bone marrow specimens has been described although what is driving this enrichment of lytic specificities is unclear as there is no obvious increase in EBV genome loads in this compartment compared to blood (Palendira et al. 2008). However, an obvious enrichment of EBV-specific T cells is seen in the tonsil, where an approximate threefold and tenfold increase in lytic and latent specificities, respectively, is seen (Hislop et al. 2005; Woodberry et al. 2005a). In some tonsils then, at least 20 % of the CD8+ T-cell population is specific for EBV. Correlating with this enrichment is the surface expression of CD103 (αEβ7) by these EBV-specific T cells. This integrin binds to E-cadherin which is expressed by epithelial cells, thereby retaining the T cells at these sites. Furthermore, this molecule is now recognised as marker of resident memory T cells, a population of cells poised to reactivate at a site of previous antigen expression (Gebhardt et al. 2009).

In most healthy donors, the virus appears to establish a stable balance with the immune response although occasional fluctuations in the size of the response are seen, possibly due to subclinical reactivation (Crough et al. 2005). However, evidence for disruption of this balance and dysregulation of the T-cell response is emerging in older donors. Thus, elevated virus loads and CD8+ T-cell responses have been described in the elderly with some EBV-specific responses constituting up to 15 % of the CD8+ T-cell population in individuals over 60 years of age, although some loss of T-cell function was seen in these donors (Stowe et al. 2007; Khan et al. 2004). Interestingly, these expansions were not seen or were less marked in elderly donors co-infected with cytomegalovirus (CMV), suggesting the presence of CMV may suppress immunity to other viruses (Khan et al. 2004; Stowe et al. 2007; Colonna-Romano et al. 2007; Vescovini et al. 2004).

### 3.2 CD4+ T-Cell Response in Established Infection

Analysis of people with established EBV infections shows the EBV-specific CD4+ T-cell response differs from the CD8+ responses in several respects. First, the size of the memory CD4+ T-cell response to individual epitopes is much smaller (Leen et al. 2001; Amyes et al. 2003; Long et al. 2005, 2011, 2013). Second, the antigen targets of the CD4 T-cell response are also different, with latent reactivities tending to outnumber lytic reactivities (Long et al. 2013). Third, the proportion of central and effector memory CD4+ T cells is the same regardless of whether the cognate epitopes are from lytic or latent antigens (Long et al. 2013); CD8+ T-cell responses to the former are concentrated in the effector memory pool. Fourth, the CD4 T-cell response against lytic antigens is spread equally between IE, E and L viral proteins (Long et al. 2011), whereas the CD8 T-cell response is heavily skewed towards the former (Pudney et al. 2005).

The fact that EBV infects and persists in B cells, which constitutively express MHC class II, raises the possibility that CD4+ T cells may be able to act as direct effector cells in their own right. Reports from several groups clearly demonstrate that CD4+ T-cell clones against a wide range of EBV lytic and latent cycle antigens are able to recognise and kill newly infected B cells or established EBV-transformed lymphoblastoid cell lines (LCLs) (Adhikary et al. 2006; Kobayashi et al. 2008; Long et al. 2005; Munz et al. 2000; Sun et al. 2002b; Haigh et al. 2008; Khanna et al. 1997; Landais et al. 2004; Omiya et al. 2002; Demachi-Okamura et al. 2006; Rajnavolgyi et al. 2000). Although only a minority of cells in such lines are lytically infected, structural and non-structural lytic cycle proteins are efficiently transferred to neighbouring LCLs sensitising them to recognition by lytic antigen-specific CD4 T cells (Adhikary et al. 2006; Landais et al. 2004; Long et al. 2011). Receptor-mediated uptake of virions by LCLs likely explains the efficient transfer of structural proteins. The mechanism responsible for efficient uptake, processing and presentation of non-structural lytic proteins is currently unknown, but it appears to be part of a general phenomenon since the latent cycle proteins EBNA2, EBNA3A, EBNA3B and EBNA3C are also efficiently transferred from antigen-positive to antigen-negative B cells (Taylor et al. 2006; Mackay et al. 2009).

As described earlier, EBNA1 does not appear to be transferred between cells in culture (Leung et al. 2010; Long et al. 2013). Instead, EBNA1 is able to access the MHC-II pathway within the infected cell itself via macroautophagy (Paludan et al. 2005) a catabolic pathway in which cytoplasmic contents are enveloped by double-membrane vesicles that in turn fuse with lysosomes. However, EBNA1's normal nuclear localisation limits its processing by macroautophagy (Leung et al. 2010) and such 'nuclear shelter' from macroautophagy means only a subset of EBNA1 CD4+ T-cell epitopes are presented by LCLs (Paludan et al. 2005; Khanna et al. 1995; Mautner et al. 2004; Leung et al. 2010). These observations may resolve the paradox that although EBNA1 is essential for viral persistence (Humme et al. 2003), it nevertheless contains the largest number of CD4+ T-cell epitopes of any latent cycle protein and most EBV-infected individuals possess good CD4+ T-cell responses against the protein (Leen et al. 2001; Long et al. 2013; Munz et al. 2000).

### *3.3 Other T-Cell Subsets in Established Infection*

Turning to other subsets of T cells, little so far is known about the relevance of more specialised T cells such as Th9, Th17 or Th21 T cells in EBV infection. Some studies have suggested a role for invariant natural killer T cells (iNKT) in control of EBV. Thus, patients deficient in the SLAM-associated protein (SAP) encoded by *SH2D1A* have no NKT cells and are exquisitely sensitive to EBV infection and may develop a life-threatening lymphoproliferative disease upon EBV infection (Nichols et al. 2005). However, such mutations also affect conventional T-cell and natural killer (NK) cell function making it unclear whether this disease is solely due to lack of iNKT cells (Tangye 2014). Similarly, patients with mutations in the *BIRC4* gene, which encodes the X-linked inhibitor of apoptosis protein (XIAP), show sensitivity to EBV infection and have low numbers of iNKT cells (Rigaud et al. 2006). These patients have normal numbers of T cells; however, these are more sensitive to apoptotic stimuli, again making it unclear whether iNKT numbers are the solely responsible for controlling disease (Lopez-Granados et al. 2014). Other models and observations hint to a role of iNKT cells: patients with EBV-associated malignancies have lower circulating numbers of these cells, while iNKT cells adoptively transferred into immunodeficient mice then challenged with EBV-related malignant cells show reduced tumour formation (Yuling et al. 2009). Similarly, in vitro studies of resting B cells challenged with EBV showed higher frequencies of transformation when NKT cells are depleted from such cultures (Chung et al. 2013).

Some 1–10 % of the total T-cell population is comprised of γδ T cells that recognise a distinct range of antigenic targets and have a broad functional phenotype upon activation (Vantourout and Hayday 2013). The importance of these cells in controlling natural EBV infection is not known, but several observations suggest they could play a role. The Vδ1 subset of γδ T cells can directly recognise and lyse EBV-transformed LCLs in vitro (Hacker et al. 1992) and high frequencies of these cells have been described in transplant recipients who have previously experienced EBV reactivation (Fujishima et al. 2007; Farnault et al. 2013). The Vγ2 Vδ9 subset of γδ T cells can also recognise and lyse LCLs in vitro, but efficiency is low unless the cells are activated with pamidronate and then positively selected using anti-γδ-TCR-specific beads (thus delivering a TCR signal to the cells) (Xiang et al. 2014). However in mice reconstituted with human immune system components, pamidronate administration was sufficient to significantly reduce EBV-positive lymphoproliferative disease and this control was dependent upon Vγ2 Vδ2 T cells (Xiang et al. 2014). The antigens that allow selective recognition of EBV-infected LCLs by γδ T cells are currently unknown.

## 4 T-Cell Responses in Patients with EBV-Associated Malignancy

The EBV-specific T-cell response generated by natural infection is important to control the growth transforming activity of the virus for the lifetime of the host. Loss of this control, as occurs in patients receiving immunosuppression, can lead to post-transplant lymphoproliferative disease (PTLD). Tumours that occur in the first year of transplantation, when immunosuppression is greatest, typically express the full range of EBV latency proteins including the EBNA3A, EBNA3B and EBNA3C proteins that are immunodominant targets of CD8 T-cell immunity. Furthermore, these tumours display high levels of HLA class I and II molecules. Accordingly, restoring immunological control by adoptive transfer of EBV-specific T-cell preparations generated in vitro has been used by several groups for prophylaxis or treatment of PTLD following solid organ or haematopoetic stem cell transplantation (Rooney et al. 1995; Khanna et al. 1999; Sun et al. 2002a; Barker et al. 2010; Haque et al. 2007) with an excellent track record of safety and efficacy (Heslop et al. 2010).

Until recently, the T cells used to treat patients with PTLD were generated using EBV-transformed LCLs as the antigen source. The key T-cell effectors mediating clinical responses in patients were therefore thought to be the EBNA3A-, EBNA3B- and EBNA3C-specific CD8+ T cells that tend to dominate LCL-stimulated T-cell preparations. Undoubtedly, these T cells are important (Gottschalk et al. 2001), but recent data suggests that other specificities may also contribute to tumour control. Thus, LCL-stimulated T-cell preparations containing a higher level of CD4+ T cells were associated with better outcome in a multicentre phase II trial of adoptive therapy (Haque et al. 2007). The antigenic specificity and function of these CD4 T cells were not characterised, and it is possible that their importance reflects the provision of CD4-mediated T-cell help rather than that of a role as direct effectors. Nevertheless, this result is intriguing given the multiple reports of CD4 T-cell clones specific for a range of EBV latent and lytic cycle proteins directly recognising and killing LCLs (Adhikary et al. 2006; Kobayashi et al. 2008; Long et al. 2005; Munz et al. 2000; Sun et al. 2002b; Haigh et al. 2008; Khanna et al. 1997; Landais et al. 2004; Omiya et al. 2002; Demachi-Okamura et al. 2006; Rajnavolgyi et al. 2000). Not all T-cell clones are capable of such recognition and it is clear that the abundance of different CD4+ T-cell epitopes on the target cell surface can vary markedly, even for epitopes derived from the same protein (Long et al. 2005; Leung et al. 2010). CD4+ T cells incapable of direct recognition of EBV-positive cells could still be of value, however, by providing T-cell help to the overall immune response. Cultures of T cells prepared using LCLs as stimulators also include CD4+ T cells specific for non-viral antigens upregulated in B cells by EBV transformation (Gudgeon et al. 2005; Long et al. 2009). These cellular-antigen-specific CD4+ T cells can also control LCL outgrowth and may therefore enhance the anti-tumour effect, but do not appear to be essential since T-cell lines prepared without the use of LCLs, presumably

Table 1 EBV-associated malignancies and their expression of EBV antigens

| Tumour | Subtype | % EBV positive | EBV proteins expressed |
|---|---|---|---|
| Burkitt Lymphoma | Endemic<br>AIDS-related | 100<br>30–40 | EBNA1[a] |
| T/NK Lymphoma | Extranodal | 100 | EBNA1, LMP2B[b] |
| Diffuse large B-cell lymphoma | Late PT-DLBCL<br>Elderly DLBCL<br>AIDS-related | >50<br>>50<br>~50 | EBNA1, LMP1, LMP2[c] |
| Hodgkin lymphoma | Classical<br>AIDS-related | 30<br>100 | EBNA1, LMP1, LMP2 |
| Lympho-proliferative disease | Post-transplant, AIDS-related | 100 | EBNA1, EBNA2, EBNA3A, EBNA3B, EBNA3C, EBNA-LP, LMP1, LMP2 |
| Nasopharyngeal carcinoma | Undifferentiated | 100 | EBNA1, LMP1, LMP2[d] |
| Gastric carcinoma | | 5–15 | EBNA1, LMP2[d] |

[a]Some 10–15 % of endemic BLs express EBNA1, EBNA3A, EBNA3B EBNA3C, EBNA-LP and BHRF-1
[b] LMP2B is expressed from a novel mRNA transcript in the absence of LMP2A
[c]Reports of wider range of EBV latency genes reported in some cases
[d]BARF1 expression is reported in a proportion of these tumours

lacking such responses, yield clinical responses in transplant recipients with PTLD or undergoing EBV reactivation (Gerdemann et al. 2012; Icheva et al. 2013).

Most cases of EBV-associated malignancy, however, develop in people who are not iatrogenically immunosuppressed and express a smaller number EBV antigens (Table 1). These malignancies include lymphomas of B-cell origin such as Burkitt lymphoma (BL), Hodgkin lymphoma (HL) and diffuse large B-cell lymphoma and a smaller number of cases of non-B-cell origin such as extranodal NK/T-cell lymphoma (ENKTL). EBV is also linked to almost all cases of undifferentiated nasopharyngeal carcinoma (NPC) and a proportion of gastric carcinomas (GCa). The immunodominant EBNA3A, EBNA3B and EBNA3C proteins are absent in these tumours, but the EBV antigens that are expressed are still *bona fide* T-cell targets (Hislop et al. 2007b). Given that loss of EBV immune control underpins post-transplant lymphoma, an important question is whether the EBV-specific T-cell response is perturbed in patients with these other malignancies. This appears to be the case for endemic Burkitt lymphoma, which occurs in areas of holoendemic *Plasmodium falciparum* malaria. Recurrent malarial infection of young children adversely affects the EBV-specific T-cell response as measured using regression assays (Moss et al. 1983; Whittle et al. 1984). Studies of children in high- and low-incidence malaria areas show the former have high EBV viral loads (Moormann et al. 2005). Subsequent studies have shown that children living in endemic malarial areas have reduced CD8 T-cell responses to EBV lytic and latent antigens (Moormann et al. 2007) or phenotypic changes in these responses consistent with greater differentiation (Chattopadhyay et al. 2013); each could conceivably alter the virus–host balance to favour the development of Burkitt lymphoma.

An alterative, but not necessarily mutually exclusive explanation for the development of Burkitt lymphoma is that the tumours are able to develop because they escape immune control. The focus of attention here has been EBNA1 since this is the only EBV protein expressed in the majority of Burkitt lymphoma cases (Kelly et al. 2002). Although it is now known that the glycine/alanine repeat domain within EBNA1 does not afford complete protection from the MHC class I processing pathway (Tellam et al. 2004; Voo et al. 2004; Lee et al. 2004), Burkitt lymphoma cells show reduced expression of HLA class I molecules as well as the TAP-1 and TAP-2 proteins required for transport of antigenic peptides into the endoplasmic reticulum for HLA binding. Together, these defects contribute to a profound impairment of the ability of CD8+ T cells to recognise BL cells (Rowe et al. 1995; Khanna et al. 1994) even in cases when the tumours express strong immune targets (Kelly et al. 2002). Burkitt lymphoma cells do, however, express HLA class II and have normal HLA class II processing function (Khanna et al. 1997; Taylor et al. 2006). EBNA1-specific CD4+ T cells can recognise and lyse Burkitt lymphoma cells in vitro (Paludan et al. 2002) and can control tumours in a murine model (Fu et al. 2004). It is therefore interesting that fewer children with endemic Burkitt lymphoma had detectable EBNA1 T-cell responses in the one study published to date (Moormann et al. 2009).

The immunological situation in NPC and HL is quite different. In both cases, the tumours express a wider range of EBV proteins: EBNA1, LMP2 and, in a proportion of cases, LMP1 as well. Furthermore, cell lines derived from NPC (Lee et al. 2000; Khanna et al. 1998) and HL (Lee et al. 1998) have functional HLA class I processing capacity in vitro, and HLA class I, TAP-1 and TAP-2 are frequently detected in biopsies from patients with these diseases (Lee et al. 1998; Murray et al. 1998; Khanna et al. 1998; Yao et al. 2000). Indeed, compared to EBV-negative cases expression of HLA class I is more frequently detected and is present at higher levels in EBV-positive HL (Huang et al. 2010; Liu et al. 2013). HLA class II is also detected in over half of EBV-positive HL and NPC biopsies tested (Huang et al. 2010; Liu et al. 2013). These observations suggest that HL and NPC tumours could be susceptible to CD8 and CD4 T cells effectors of appropriate specificity and this certainly seems to be the case in the clinic with several groups reporting clinical responses in NPC and HL patients treated with EBV-specific T cells (Comoli et al. 2005; Louis et al. 2010; Chia et al. 2014; Smith et al. 2012; Bollard et al. 2014).

The fact that adoptively transferred EBV-specific T cells can control a proportion of NPC and HL cases raises the question whether T-cell responses are compromised in these patients in the first place. Screening of Chinese NPC patients and healthy donors using a panel of defined HLA class I and II epitope peptides has found that the T-cell response in patients is generally unimpaired apart from a single HLA-B*40.01 restricted LMP2 epitope that was absent in patients (Lin et al. 2008). This work, however, examined only a single EBNA1 CD8 T-cell epitope. A subsequent study focusing on the CD8+ T-cell response to EBNA1 reported that the frequency of such cells in patients was lower (Fogg et al. 2009). In some cases, T-cell responses could be rescued from patients by in vitro culture, suggesting that EBNA1-specific T cells may have become unresponsive rather than being lost. Similarly, the EBNA1-specific CD4 T-cell response is decreased in patients with Hodgkin lymphoma and AIDS–non-Hodgkin lymphoma (Heller et al. 2008; Piriou et al. 2005). The recent observation that adoptively transferred EBNA1-specific effectors can yield clinical benefit in patients with post-transplant lymphoma (Icheva et al. 2013) suggests that EBNA1-specific effectors could be candidates for the treatment of NPC and HL and that the above-described defects may therefore have clinical relevance.

In contrast to T-cell responses to EBNA1, LMP2-specific T cells appear less impaired in patients and CD8 T-cell responses are frequently detected in patients with HL and NPC (Lee et al. 2000; Fogg et al. 2009; Chapman et al. 2001; Lin et al. 2008) although the frequency of these cells may be lower in some cases (Gandhi et al. 2006). The increasing evidence linking HLA polymorphism with the risk of developing different EBV-associated malignancies is therefore intriguing. For Hodgkin lymphoma, HLA-A*01 increases and HLA-A*02 decreases the risk of developing EBV-positive but not EBV-negative disease (Niens et al. 2007). For NPC, a similar pattern is observed, although different HLA alleles are involved. Thus, a particular subtype of the HLA-A2 allele, A*02.07, increases disease risk, whereas the HLA-A*11.01 allele reduces risk (Su et al. 2013). Note that in most racial groups, the HLA-A*02.07 subtype is rare apart from the Chinese population, in whom NPC is a common malignancy and this subtype also increases the risk of developing HL in this population (Huang et al. 2012). For both diseases, it is notable that no EBV T-cell epitopes have yet to be identified as being presented by the HLA-A*01 or HLA-A*02.07 risk alleles, whereas the protective alleles can present multiple epitopes from a range of EBV proteins including LMP2 which is expressed in these malignancies. These observations are consistent with the hypothesis that a deficit in T-cell immunity may underpin the development of these malignancies (Niens et al. 2007; Brennan and Burrows 2008). However as noted earlier, patients with NPC and HL often possess detectable LMP2-specific T-cell responses at the time of their diagnosis. The in vivo situation is therefore likely to be more complex than a simple deficit in tumour surveillance.

Recent evidence suggests that some tumours may express additional viral antigens that could be exploited therapeutically. Although no LMP2 protein or mRNA could be detected in ENKTL cell lines, these cells were nevertheless efficiently recognised and killed by LMP2-specific CD8+ T-cell clones (Fox et al. 2010). This apparent paradox was resolved by the identification of a novel LMP2 mRNA transcript expressed from a different promoter that could not be detected by the standard molecular assays in use at the time but still contained the majority of T-cell epitopes. Although described as a lytic cycle protein, BARF1 is detected in many of the cases of the EBV-positive epithelial malignancies apparently in the absence of lytic replication (Decaussin et al. 2000; Seto et al. 2005; Stevens et al. 2006). Little is known about the immune response against BARF1. Several HLA-A2-restricted epitopes have been identified and T-cell responses are present at greater frequencies in NPC patients (Martorelli et al. 2008). The existence of other immune responses to BARF1 is possible since a systematic analysis of the immune response against the protein has not yet been performed.

Very little is known about the EBV-specific immune response in patients with other EBV-positive cancers such as GCa, ENKTL and DLBCL. EBV-positive DLBCL of the elderly is now recognised as a provisional entity in the World Health Organisation classification and is defined as a clonal B-cell lymphoid proliferation occurring in patients older than 50 years without immunodeficiency or prior lymphoma. The detection of EBNA3 expression in a proportion of cases (Nguyen-Van et al. 2011; Cohen et al. 2013) and the fact that the disease is associated with ageing has led to the suggestion that the disease arises in a background of lowered EBV-specific immunity caused by ageing (Dojcinov et al. 2011). However, paediatric EBV-positive DLBLCL cases have been reported and although these occur at higher frequency in immunocompromised children at least some cases arise in apparently immunocompetent children (Cohen et al. 2013). Careful analysis of the antigen-processing phenotype and pattern of EBV gene expression in the tumour and the corresponding EBV-specific immune response in the blood of the patient will be required to provide a complete picture of whether virus-specific immunity is compromised in patients with EBV-positive GCa, ENKTL or DLBCL.

## 5 Suppression of EBV-Specific T-Cell Responses in Patients with EBV-Associated Malignancy

Evading immune destruction is a recognised hallmark of cancer (Hanahan and Weinberg 2011). In this regard, several mechanisms are employed by EBV-associated malignancies to suppress T-cell responses. Although HL tumours are heavily infiltrated by immune cells, these infiltrates are dominated by CD4+ regulatory and CD4+ Th2 cells. Many reports do not differentiate between EBV-positive and EBV-negative cases, but this distinction is important to make because, although the two subtypes appear superficially similar, several important differences exist between them, namely the tumour immune microenvironment. Firstly, EBV-positive HL has a distinct gene signature with markers indicating cytotoxic and Th1 responses being increased (Chetaille et al. 2009; Barros et al. 2012) although markers of suppression such as LAG-3 and IL-10 are also raised (Morales et al. 2014). Secondly, while the frequency of regulatory T cells is increased in the blood and particularly the tumour infiltrates of HL patients (Marshall et al. 2004), the presence of EBV correlates with higher numbers of both natural and induced regulatory T cells (Assis et al. 2012; Morales et al. 2014). The increased numbers of the former in EBV-positive disease may stem from EBNA1-mediated upregulation of CCL20 in the malignant Hodgkin/Reed–Sternberg cells (Baumforth et al. 2008). The immunoregulatory molecule PD-L1 is also expressed by HRS cells and in the case of EBV-positive disease, this may result from LMP1- and LMP2-mediated upregulation of an AP-1-dependent pathway rather than an increase in PD-L1 gene dosage through chromosome 9p24.1 amplification (Yamamoto et al. 2008; Juszczynski et al. 2007; Green et al. 2012).

Less is known about the microenvironment of NPC. The most common subtype, undifferentiated NPC, is always EBV-positive and the tumours contain a sizeable infiltrate of lymphoid cells recruited to the tumour via CXCR6 and CCR5 (Parsonage et al. 2012). Regulatory CD4 + T-cell numbers are increased in the blood of some patients and these cells are also consistently detected in tumours (Lau et al. 2007; Yip et al. 2009). The presence in tumours of CD8+ FoxP3+ lymphocytes with suppressive function has also been reported (Li et al. 2011). In certain respects, HL and NPC tumour cells use similar strategies to evade immune responses. For example, an immunomodulatory galectin (galectin-1 in HL, galectin-9 in NPC) (Juszczynski et al. 2007; Gandhi et al. 2007; Klibi et al. 2009) is expressed by a proportion of cases of each disease as is Fas ligand, which may act as a tumour defence molecule (Dutton et al. 2004). However, important differences exist between HL and NPC. For example, few cases of EBV+ve HL express HLA-G, an inhibitor of T- and NK-cell function, whereas it is expressed by 80 % of NPC tumours with high expression predicting poor survival (Cai et al. 2012).

Expression of the immunoregulatory molecule PD-L1 was detected in 90 % of EBV-positive HL cases and NPC cases as well as a wide range of EBV-associated malignancies including extranodal NK/T-cell lymphoma, diffuse large B-cell lymphoma and PTLD (Chen et al. 2013). These diseases may therefore be amenable to immune checkpoint inhibitors that target the PD1/PD-L1 axis (Pardoll 2012). Several of these inhibitors are being tested in late-stage trials for melanoma and non-small cell lung cancer, and they have yielded impressive clinical outcomes. Whether PD1/PDL1 inhibition will be similarly effective in the context of EBV-associated malignancies is currently unknown. They may be effective when used as single agents or could be combined with existing adoptive T-cell therapy or therapeutic vaccination strategies (Smith et al. 2012; Chia et al. 2014; Bollard et al. 2014; Taylor et al. 2014; Hui et al. 2013) since they clearly have synergistic potential (Wolchok et al. 2013). Rational combination approaches may be of particular value in cases of advanced disease which currently represents a challenging clinical problem.

# 6 Future Directions

1. Defining the immunological factors influencing whether primary EBV infection is asymptomatic or leads to infectious mononucleosis.
2. Determining what sort of immunity is important in determining the viral load set point and how control over this is lost in elderly populations.
3. Understanding the role of innate immune cells in limiting primary EBV infection and whether such cells can be harnessed for therapy of EBV-associated malignancies.

4. Characterising the repertoire of immunomodulatory mechanisms operating in the different EBV-associated malignancies and whether perturbing those mechanisms can unleash EBV-specific T-cell immunity to attack the tumour.
5. Understanding the immunology of emerging EBV-associated malignancies such as EBV-positive gastric carcinoma and DLBCL, and how these causes of morbidity and mortality can be targeted immunologically.
6. Improving the efficacy of immunotherapies to treat EBV-associated malignancies, particularly in cases of advanced disease, and developing ways to apply immunotherapies to patients in low-resource countries where many cases of EBV-associated malignancy occur.

## References

Abbott RJ, Quinn LL, Leese AM, Scholes HM, Pachnio A, Rickinson AB (2013) CD8+ T cell responses to lytic EBV infection: late antigen specificities as subdominant components of the total response. J Immunol 191(11):5398–5409

Adhikary D, Behrends U, Moosmann A, Witter K, Bornkamm GW, Mautner J (2006) Control of Epstein-Barr virus infection in vitro by T helper cells specific for virion glycoproteins. J Exp Med 203(4):995–1006

Amyes E, Hatton C, Montamat-Sicotte D, Gudgeon N, Rickinson AB, McMichael AJ, Callan MF (2003) Characterization of the CD4 + T cell response to Epstein-Barr virus during primary and persistent infection. J Exp Med 198(6):903–911

Assis MC, Campos AH, Oliveira JS, Soares FA, Silva JM, Silva PB, Penna AD, Souza EM, Baiocchi OC (2012) Increased expression of CD4+ CD25+ FOXP3+ regulatory T cells correlates with Epstein-Barr virus and has no impact on survival in patients with classical Hodgkin lymphoma in Brazil. Med Oncol 29(5):3614–3619

Babcock GJ, Decker LL, Volk M, Thorley-Lawson DA (1998) EBV persistence in memory B cells in vivo. Immunity 9(3):395–404

Balfour HH Jr, Holman CJ, Hokanson KM, Lelonek MM, Giesbrecht JE, White DR, Schmeling DO, Webb CH, Cavert W, Wang DH, Brundage RC (2005) A prospective clinical study of Epstein-Barr virus and host interactions during acute infectious mononucleosis. J Infect Dis 192(9):1505–1512

Balfour HH Jr, Odumade OA, Schmeling DO, Mullan BD, Ed JA, Knight JA, Vezina HE, Thomas W, Hogquist KA (2013) Behavioral, virologic, and immunologic factors associated with acquisition and severity of primary Epstein-Barr virus infection in university students. J Infect Dis 207(1):80–88

Barker JN, Doubrovina E, Sauter C, Jaroscak JJ, Perales MA, Doubrovin M, Prockop SE, Koehne G, O'Reilly RJ (2010) Successful treatment of EBV-associated posttransplantation lymphoma after cord blood transplantation using third-party EBV-specific cytotoxic T lymphocytes. Blood 116(23):5045–5049

Barros MH, Vera-Lozada G, Soares FA, Niedobitek G, Hassan R (2012) Tumor microenvironment composition in pediatric classical Hodgkin lymphoma is modulated by age and Epstein-Barr virus infection. Int J Cancer 131(5):1142–1152

Baumforth KR, Birgersdotter A, Reynolds GM, Wei W, Kapatai G, Flavell JR, Kalk E, Piper K, Lee S, Machado L, Hadley K, Sundblad A, Sjoberg J, Bjorkholm M, Porwit AA, Yap LF, Teo S, Grundy RG, Young LS, Ernberg I, Woodman CB, Murray PG (2008) Expression of the Epstein-Barr virus-encoded Epstein-Barr virus nuclear antigen 1 in Hodgkin's lymphoma cells mediates up-regulation of CCL20 and the migration of regulatory T cells. Am J Pathol 173(1):195–204

Biggar RJ, Henle G, Bocker J, Lennette ET, Fleisher G, Henle W (1978) Primary Epstein-Barr virus infections in African infants. II. Clinical and serological observations during seroconversion. Int J Cancer 22(3):244–250
Blake N, Haigh T, Shaka'a G, Croom-Carter D, Rickinson A (2000) The importance of exogenous antigen in priming the human CD8+ T cell response: lessons from the EBV nuclear antigen EBNA1. J Immunol 165(12):7078–7087
Bollard CM, Gottschalk S, Torrano V, Diouf O, Ku S, Hazrat Y, Carrum G, Ramos C, Fayad L, Shpall EJ, Pro B, Liu H, Wu MF, Lee D, Sheehan AM, Zu Y, Gee AP, Brenner MK, Heslop HE, Rooney CM (2014) Sustained complete responses in patients with lymphoma receiving autologous cytotoxic T lymphocytes targeting Epstein-Barr virus latent membrane proteins. J Clin Oncol 32(8):798–808
Brennan RM, Burrows SR (2008) A mechanism for the HLA-A*01-associated risk for EBV+ Hodgkin lymphoma and infectious mononucleosis. Blood 112(6):2589–2590
Cai MB, Han HQ, Bei JX, Liu CC, Lei JJ, Cui Q, Feng QS, Wang HY, Zhang JX, Liang Y, Chen LZ, Kang TB, Shao JY, Zeng YX (2012) Expression of human leukocyte antigen G is associated with prognosis in nasopharyngeal carcinoma. Int J Biol Sci 8(6):891–900
Callan MF, Fazou C, Yang H, Rostron T, Poon K, Hatton C, McMichael AJ (2000) CD8(+) T-cell selection, function, and death in the primary immune response in vivo. J Clin Invest 106(10):1251–1261
Callan MF, Steven N, Krausa P, Wilson JD, Moss PA, Gillespie GM, Bell JI, Rickinson AB, McMichael AJ (1996) Large clonal expansions of CD8+ T cells in acute infectious mononucleosis. Nat Med 2(8):906–911
Callan MF, Tan L, Annels N, Ogg GS, Wilson JD, O'Callaghan CA, Steven N, McMichael AJ, Rickinson AB (1998) Direct visualization of antigen-specific CD8+ T cells during the primary immune response to Epstein-Barr virus in vivo. J Exp Med 187(9):1395–1402
Catalina MD, Sullivan JL, Bak KR, Luzuriaga K (2001) Differential evolution and stability of epitope-specific CD8(+) T cell responses in EBV infection. J Immunol 167(8):4450–4457
Catalina MD, Sullivan JL, Brody RM, Luzuriaga K (2002) Phenotypic and functional heterogeneity of EBV epitope-specific CD8+ T cells. J Immunol 168(8):4184–4191
Chapman AL, Rickinson AB, Thomas WA, Jarrett RF, Crocker J, Lee SP (2001) Epstein-Barr virus-specific cytotoxic T lymphocyte responses in the blood and tumor site of Hodgkin's disease patients: implications for a T-cell-based therapy. Cancer Res 61(16):6219–6226
Chattopadhyay PK, Chelimo K, Embury PB, Mulama DH, Sumba PO, Gostick E, Ladell K, Brodie TM, Vulule J, Roederer M, Moormann AM, Price DA (2013) Holoendemic malaria exposure is associated with altered Epstein-Barr virus-specific CD8(+) T-cell differentiation. J Virol 87(3):1779–1788
Chen BJ, Chapuy B, Ouyang J, Sun HH, Roemer MG, Xu ML, Yu H, Fletcher CD, Freeman GJ, Shipp MA, Rodig SJ (2013) PD-L1 expression is characteristic of a subset of aggressive B-cell lymphomas and virus-associated malignancies. Clin Cancer Res 19(13):3462–3473
Chetaille B, Bertucci F, Finetti P, Esterni B, Stamatoullas A, Picquenot JM, Copin MC, Morschhauser F, Casasnovas O, Petrella T, Molina T, Vekhoff A, Feugier P, Bouabdallah R, Birnbaum D, Olive D, Xerri L (2009) Molecular profiling of classical Hodgkin lymphoma tissues uncovers variations in the tumor microenvironment and correlations with EBV infection and outcome. Blood 113(12):2765–3775
Chia WK, Teo M, Wang WW, Lee B, Ang SF, Tai WM, Chee CL, Ng J, Kan R, Lim WT, Tan SH, Ong WS, Cheung YB, Tan EH, Connolly JE, Gottschalk S, Toh HC (2014) Adoptive T-cell transfer and chemotherapy in the first-line treatment of metastatic and/or locally recurrent nasopharyngeal carcinoma. Mol Ther 22(1):132–139
Chijioke O, Muller A, Feederle R, Barros MH, Krieg C, Emmel V, Marcenaro E, Leung CS, Antsiferova O, Landtwing V, Bossart W, Moretta A, Hassan R, Boyman O, Niedobitek G, Delecluse HJ, Capaul R, Munz C (2013) Human natural killer cells prevent infectious mononucleosis features by targeting lytic Epstein-Barr virus infection. Cell Rep 5(6):1489–1498

Chung BK, Tsai K, Allan LL, Zheng DJ, Nie JC, Biggs CM, Hasan MR, Kozak FK, van den Elzen P, Priatel JJ, Tan R (2013) Innate immune control of EBV-infected B cells by invariant natural killer T cells. Blood 122(15):2600–2608

Clute SC, Watkin LB, Cornberg M, Naumov YN, Sullivan JL, Luzuriaga K, Welsh RM, Selin LK (2005) Cross-reactive influenza virus-specific CD8+ T cells contribute to lymphoproliferation in Epstein-Barr virus-associated infectious mononucleosis. J Clin Invest 115(12):3602–3612

Cohen M, De Matteo E, Narbaitz M, Carreno FA, Preciado MV, Chabay PA (2013) Epstein-Barr virus presence in pediatric diffuse large B-cell lymphoma reveals a particular association and latency patterns: analysis of viral role in tumor microenvironment. Int J Cancer 132(7):1572–1580

Colonna-Romano G, Akbar AN, Aquino A, Bulati M, Candore G, Lio D, Ammatuna P, Fletcher JM, Caruso C, Pawelec G (2007) Impact of CMV and EBV seropositivity on CD8 T lymphocytes in an old population from West-Sicily. Exp Gerontol 42(10):995–1002

Comoli P, Pedrazzoli P, Maccario R, Basso S, Carminati O, Labirio M, Schiavo R, Secondino S, Frasson C, Perotti C, Moroni M, Locatelli F, Siena S (2005) Cell therapy of stage IV nasopharyngeal carcinoma with autologous Epstein-Barr virus-targeted cytotoxic T lymphocytes. J Clin Oncol 23(35):8942–8949

Cornberg M, Clute SC, Watkin LB, Saccoccio FM, Kim SK, Naumov YN, Brehm MA, Aslan N, Welsh RM, Selin LK (2010) CD8 T cell cross-reactivity networks mediate heterologous immunity in human EBV and murine vaccinia virus infections. J Immunol 184(6):2825–2838

Crawford DH, Macsween KF, Higgins CD, Thomas R, McAulay K, Williams H, Harrison N, Reid S, Conacher M, Douglas J, Swerdlow AJ (2006) A cohort study among university students: identification of risk factors for Epstein-Barr virus seroconversion and infectious mononucleosis. Clin Infect Dis 43(3):276–282

Crough T, Burrows JM, Fazou C, Walker S, Davenport MP, Khanna R (2005) Contemporaneous fluctuations in T cell responses to persistent herpes virus infections. Eur J Immunol 35(1):139–149

Decaussin G, Sbih-Lammali F, de Turenne-Tessier M, Bouguermouh A, Ooka T (2000) Expression of BARF1 gene encoded by Epstein-Barr virus in nasopharyngeal carcinoma biopsies. Cancer Res 60(19):5584–5588

Demachi-Okamura A, Ito Y, Akatsuka Y, Tsujimura K, Morishima Y, Takahashi T, Kuzushima K (2006) Epstein-Barr virus (EBV) latent membrane protein-1-specific cytotoxic T lymphocytes targeting EBV-carrying natural killer cell malignancies. Eur J Immunol 36(3):593–602

Dojcinov SD, Venkataraman G, Pittaluga S, Wlodarska I, Schrager JA, Raffeld M, Hills RK, Jaffe ES (2011) Age-related EBV-associated lymphoproliferative disorders in the Western population: a spectrum of reactive lymphoid hyperplasia and lymphoma. Blood 117(18):4726–4735

Dunne PJ, Faint JM, Gudgeon NH, Fletcher JM, Plunkett FJ, Soares MV, Hislop AD, Annels NE, Rickinson AB, Salmon M, Akbar AN (2002) Epstein-Barr virus-specific CD8(+) T cells that re-express CD45RA are apoptosis-resistant memory cells that retain replicative potential. Blood 100(3):933–940

Dutton A, O'Neil JD, Milner AE, Reynolds GM, Starczynski J, Crocker J, Young LS, Murray PG (2004) Expression of the cellular FLICE-inhibitory protein (c-FLIP) protects Hodgkin's lymphoma cells from autonomous Fas-mediated death. Proc Natl Acad Sci USA 101(17):6611–6616

Fafi-Kremer S, Morand P, Brion JP, Pavese P, Baccard M, Germi R, Genoulaz O, Nicod S, Jolivet M, Ruigrok RW, Stahl JP, Seigneurin JM (2005) Long-term shedding of infectious epstein-barr virus after infectious mononucleosis. J Infect Dis 191(6):985–989

Faint JM, Annels NE, Curnow SJ, Shields P, Pilling D, Hislop AD, Wu L, Akbar AN, Buckley CD, Moss PA, Adams DH, Rickinson AB, Salmon M (2001) Memory T cells constitute a subset of the human CD8+ CD45RA+ pool with distinct phenotypic and migratory characteristics. Immunol 167(1):212–220

Farnault L, Gertner-Dardenne J, Gondois-Rey F, Michel G, Chambost H, Hirsch I, Olive D (2013) Clinical evidence implicating gamma-delta T cells in EBV control following cord blood transplantation. Bone Marrow Transplant 48(11):1478–1479

Fleisher G, Henle W, Henle G, Lennette ET, Biggar RJ (1979) Primary infection with Epstein-Barr virus in infants in the United States: clinical and serologic observations. J Infect Dis 139(5):553–558

Fogg MH, Wirth LJ, Posner M, Wang F (2009) Decreased EBNA-1-specific CD8+ T cells in patients with Epstein-Barr virus-associated nasopharyngeal carcinoma. Proc Natl Acad Sci USA 106(9):3318–3323

Fox CP, Haigh TA, Taylor GS, Long HM, Lee SP, Shannon-Lowe C, O'Connor S, Bollard CM, Iqbal J, Chan WC, Rickinson AB, Bell AI, Rowe M (2010) A novel latent membrane 2 transcript expressed in Epstein-Barr virus-positive NK- and T-cell lymphoproliferative disease encodes a target for cellular immunotherapy. Blood 116(19):3695–3704

Fu T, Voo KS, Wang RF (2004) Critical role of EBNA1-specific CD4+ T cells in the control of mouse Burkitt lymphoma in vivo. J Clin Invest 114(4):542–550

Fujishima N, Hirokawa M, Fujishima M, Yamashita J, Saitoh H, Ichikawa Y, Horiuchi T, Kawabata Y, Sawada KI (2007) Skewed T cell receptor repertoire of Vdelta1(+) gammadelta T lymphocytes after human allogeneic haematopoietic stem cell transplantation and the potential role for Epstein-Barr virus-infected B cells in clonal restriction. Clin Exp Immunol 149(1):70–79

Gandhi MK, Lambley E, Duraiswamy J, Dua U, Smith C, Elliott S, Gill D, Marlton P, Seymour J, Khanna R (2006) Expression of LAG-3 by tumor-infiltrating lymphocytes is coincident with the suppression of latent membrane antigen-specific CD8+ T-cell function in Hodgkin lymphoma patients. Blood 108(7):2280–2289

Gandhi MK, Moll G, Smith C, Dua U, Lambley E, Ramuz O, Gill D, Marlton P, Seymour JF, Khanna R (2007) Galectin-1 mediated suppression of Epstein-Barr virus specific T-cell immunity in classic Hodgkin lymphoma. Blood 110(4):1326–1329

Gebhardt T, Wakim LM, Eidsmo L, Reading PC, Heath WR, Carbone FR (2009) Memory T cells in nonlymphoid tissue that provide enhanced local immunity during infection with herpes simplex virus. Nat Immunol 10(5):524–530

Gerdemann U, Keirnan JM, Katari UL, Yanagisawa R, Christin AS, Huye LE, Perna SK, Ennamuri S, Gottschalk S, Brenner MK, Heslop HE, Rooney CM, Leen AM (2012) Rapidly generated multivirus-specific cytotoxic T lymphocytes for the prophylaxis and treatment of viral infections. Mol Ther 20(8):1622–1632

Gottschalk S, Ng CY, Perez M, Smith CA, Sample C, Brenner MK, Heslop HE, Rooney CM (2001) An Epstein-Barr virus deletion mutant associated with fatal lymphoproliferative disease unresponsive to therapy with virus-specific CTLs. Blood 97(4):835–843

Green MR, Rodig S, Juszczynski P, Ouyang J, Sinha P, O'Donnell E, Neuberg D, Shipp MA (2012) Constitutive AP-1 activity and EBV infection induce PD-L1 in Hodgkin lymphomas and posttransplant lymphoproliferative disorders: implications for targeted therapy. Clin Cancer Res 18(6):1611–1618

Gudgeon NH, Taylor GS, Long HM, Haigh TA, Rickinson AB (2005) Regression of Epstein-Barr virus-induced B-cell transformation in vitro involves virus-specific CD8+ T cells as the principal effectors and a novel CD4+ T-cell reactivity. J Virol 79(9):5477–5488

Hacker G, Kromer S, Falk M, Heeg K, Wagner H, Pfeffer K (1992) V delta 1 + subset of human gamma delta T cells responds to ligands expressed by EBV-infected Burkitt lymphoma cells and transformed B lymphocytes. J Immunol 149(12):3984–3989

Hadinoto V, Shapiro M, Sun CC, Thorley-Lawson DA (2009) The dynamics of EBV shedding implicate a central role for epithelial cells in amplifying viral output. PLoS Pathog 5(7):e1000496

Haigh TA, Lin X, Jia H, Hui EP, Chan AT, Rickinson AB, Taylor GS (2008) EBV latent membrane proteins (LMPs) 1 and 2 as immunotherapeutic targets: LMP-specific CD4+ cytotoxic T cell recognition of EBV-transformed B cell lines. J Immunol 180(3):1643–1654

Hanahan D, Weinberg RA (2011) Hallmarks of cancer: the next generation. Cell 144(5):646–674

Haque T, Wilkie GM, Jones MM, Higgins CD, Urquhart G, Wingate P, Burns D, McAulay K, Turner M, Bellamy C, Amlot PL, Kelly D, MacGilchrist A, Gandhi MK, Swerdlow AJ, Crawford DH (2007) Allogeneic cytotoxic T-cell therapy for EBV-positive posttransplantation lymphoproliferative disease: results of a phase 2 multicenter clinical trial. Blood 110(4):1123–1131

Hatta K, Morimoto A, Ishii E, Kimura H, Ueda I, Hibi S, Todo S, Sugimoto T, Imashuku S (2007) Association of transforming growth factor-beta1 gene polymorphism in the development of Epstein-Barr virus-related hematologic diseases. Haematologica 92(11):1470–1474

Heller KN, Arrey F, Steinherz P, Portlock C, Chadburn A, Kelly K, Munz C (2008) Patients with Epstein Barr virus-positive lymphomas have decreased CD4(+) T-cell responses to the viral nuclear antigen 1. Int J Cancer 123(12):2824–2831

Helminen M, Lahdenpohja N, Hurme M (1999) Polymorphism of the interleukin-10 gene is associated with susceptibility to Epstein-Barr virus infection. J Infect Dis 180(2):496–499

Heslop HE, Slobod KS, Pule MA, Hale GA, Rousseau A, Smith CA, Bollard CM, Liu H, Wu MF, Rochester RJ, Amrolia PJ, Hurwitz JL, Brenner MK, Rooney CM (2010) Long-term outcome of EBV-specific T-cell infusions to prevent or treat EBV-related lymphoproliferative disease in transplant recipients. Blood 115(5):925–935

Hislop AD, Annels NE, Gudgeon NH, Leese AM, Rickinson AB (2002) Epitope-specific evolution of human CD8(+) T cell responses from primary to persistent phases of Epstein-Barr virus infection. J Exp Med 195(7):893–905

Hislop AD, Gudgeon NH, Callan MF, Fazou C, Hasegawa H, Salmon M, Rickinson AB (2001) EBV-specific CD8+ T cell memory: relationships between epitope specificity, cell phenotype, and immediate effector function. J Immunol 167(4):2019–2029

Hislop AD, Kuo M, Drake-Lee AB, Akbar AN, Bergler W, Hammerschmitt N, Khan N, Palendira U, Leese AM, Timms JM, Bell AI, Buckley CD, Rickinson AB (2005) Tonsillar homing of Epstein-Barr virus-specific CD8+ T cells and the virus-host balance. J Clin Invest 115(9):2546–2555

Hislop AD, Ressing ME, van Leeuwen D, Pudney VA, Horst D, Koppers-Lalic D, Croft NP, Neefjes JJ, Rickinson AB, Wiertz EJ (2007a) A CD8+ T cell immune evasion protein specific to Epstein-Barr virus and its close relatives in Old World primates. J Exp Med 204(8):1863–1873

Hislop AD, Taylor GS, Sauce D, Rickinson AB (2007b) Cellular responses to viral infection in humans: lessons from Epstein-Barr virus. Annu Rev Immunol 25:587–617

Hoshino Y, Morishima T, Kimura H, Nishikawa K, Tsurumi T, Kuzushima K (1999) Antigen-driven expansion and contraction of CD8+ -activated T cells in primary EBV infection. J Immunol 163(10):5735–5740

Huang X, Hepkema B, Nolte I, Kushekhar K, Jongsma T, Veenstra R, Poppema S, Gao Z, Visser L, Diepstra A, van den Berg A (2012) HLA-A*02:07 is a protective allele for EBV negative and a susceptibility allele for EBV positive classical Hodgkin lymphoma in China. PLoS ONE 7(2):e31865

Huang X, van den Berg A, Gao Z, Visser L, Nolte I, Vos H, Hepkema B, Kooistra W, Poppema S, Diepstra A (2010) Expression of HLA class I and HLA class II by tumor cells in Chinese classical Hodgkin lymphoma patients. PLoS ONE 5(5):e10865

Hui EP, Taylor GS, Jia H, Ma BB, Chan SL, Ho R, Wong WL, Wilson S, Johnson BF, Edwards C, Stocken DD, Rickinson AB, Steven NM, Chan AT (2013) Phase I trial of recombinant modified vaccinia ankara encoding Epstein-Barr viral tumor antigens in nasopharyngeal carcinoma patients. Cancer Res 73(6):1676–1688

Humme S, Reisbach G, Feederle R, Delecluse HJ, Bousset K, Hammerschmidt W, Schepers A (2003) The EBV nuclear antigen 1 (EBNA1) enhances B cell immortalization several thousandfold. Proc Natl Acad Sci USA 100(19):10989–10994

Hwang AE, Hamilton AS, Cockburn MG, Ambinder R, Zadnick J, Brown EE, Mack TM, Cozen W (2012) Evidence of genetic susceptibility to infectious mononucleosis: a twin study. Epidemiol Infect 140(11):2089–2095

Icheva V, Kayser S, Wolff D, Tuve S, Kyzirakos C, Bethge W, Greil J, Albert MH, Schwinger W, Nathrath M, Schumm M, Stevanovic S, Handgretinger R, Lang P, Feuchtinger T (2013) Adoptive transfer of epstein-barr virus (EBV) nuclear antigen 1-specific t cells as treatment for EBV reactivation and lymphoproliferative disorders after allogeneic stem-cell transplantation. J Clin Oncol 31(1):39–48

Jayasooriya S, de Silva TI, Njie-Jobe J, Sanyang C, Leese AM, Bell AI, McAulay KA, Yanchun P, Long HM, Dong T, Whittle HC, Rickinson AB, Rowland-Jones SL, Hislop AD, Flanagan KL (2015) Early virological and immunological events in asymptomatic Epstein-Barr virus infection in African children. PLoS Pathog *(in press)*

Juszczynski P, Ouyang J, Monti S, Rodig SJ, Takeyama K, Abramson J, Chen W, Kutok JL, Rabinovich GA, Shipp MA (2007) The AP1-dependent secretion of galectin-1 by Reed Sternberg cells fosters immune privilege in classical Hodgkin lymphoma. Proc Natl Acad Sci USA 104(32):13134–13139

Kelly G, Bell A, Rickinson A (2002) Epstein-Barr virus-associated Burkitt lymphomagenesis selects for downregulation of the nuclear antigen EBNA2. Nat Med 8(10):1098–1104

Khan N, Hislop A, Gudgeon N, Cobbold M, Khanna R, Nayak L, Rickinson AB, Moss PA (2004) Herpesvirus-specific CD8 T cell immunity in old age: cytomegalovirus impairs the response to a coresident EBV infection. J Immunol 173(12):7481–7489

Khanna R, Bell S, Sherritt M, Galbraith A, Burrows SR, Rafter L, Clarke B, Slaughter R, Falk MC, Douglass J, Williams T, Elliott SL, Moss DJ (1999) Activation and adoptive transfer of Epstein-Barr virus-specific cytotoxic T cells in solid organ transplant patients with post-transplant lymphoproliferative disease. Proc Natl Acad Sci USA 96(18):10391–10396

Khanna R, Burrows SR, Argaet V, Moss DJ (1994) Endoplasmic reticulum signal sequence facilitated transport of peptide epitopes restores immunogenicity of an antigen processing defective tumour cell line. Int Immunol 6(4):639–645

Khanna R, Burrows SR, Steigerwald-Mullen PM, Thomson SA, Kurilla MG, Moss DJ (1995) Isolation of cytotoxic T lymphocytes from healthy seropositive individuals specific for peptide epitopes from Epstein-Barr virus nuclear antigen 1: implications for viral persistence and tumor surveillance. Virology 214(2):633–637

Khanna R, Burrows SR, Thomson SA, Moss DJ, Cresswell P, Poulsen LM, Cooper L (1997) Class I processing-defective Burkitt's lymphoma cells are recognized efficiently by CD4+ EBV-specific CTLs. J Immunol 158(8):3619–3625

Khanna R, Busson P, Burrows SR, Raffoux C, Moss DJ, Nicholls JM, Cooper L (1998) Molecular characterization of antigen-processing function in nasopharyngeal carcinoma (NPC): evidence for efficient presentation of Epstein-Barr virus cytotoxic T-cell epitopes by NPC cells. Cancer Res 58(2):310–314

Klibi J, Niki T, Riedel A, Pioche-Durieu C, Souquere S, Rubinstein E, Le Moulec S, Guigay J, Hirashima M, Guemira F, Adhikary D, Mautner J, Busson P (2009) Blood diffusion and Th1-suppressive effects of galectin-9-containing exosomes released by Epstein-Barr virus-infected nasopharyngeal carcinoma cells. Blood 113(9):1957–1966

Kobayashi H, Nagato T, Takahara M, Sato K, Kimura S, Aoki N, Azumi M, Tateno M, Harabuchi Y, Celis E (2008) Induction of EBV-latent membrane protein 1-specific MHC class II-restricted T-cell responses against natural killer lymphoma cells. Cancer Res 68(3):901–908

Landais E, Saulquin X, Scotet E, Trautmann L, Peyrat MA, Yates JL, Kwok WW, Bonneville M, Houssaint E (2004) Direct killing of Epstein-Barr virus (EBV)-infected B cells by CD4 T cells directed against the EBV lytic protein BHRF1. Blood 103(4):1408–1416

Lau KM, Cheng SH, Lo KW, Lee SA, Woo JK, van Hasselt CA, Lee SP, Rickinson AB, Ng MH (2007) Increase in circulating Foxp3+ CD4+ CD25(high) regulatory T cells in nasopharyngeal carcinoma patients. Br J Cancer 96(4):617–622

Lee SP, Brooks JM, Al-Jarrah H, Thomas WA, Haigh TA, Taylor GS, Humme S, Schepers A, Hammerschmidt W, Yates JL, Rickinson AB, Blake NW (2004) CD8 T cell recognition of endogenously expressed epstein-barr virus nuclear antigen 1. J Exp Med 199(10):1409–1420

Lee SP, Chan AT, Cheung ST, Thomas WA, CroomCarter D, Dawson CW, Tsai CH, Leung SF, Johnson PJ, Huang DP (2000) CTL control of EBV in nasopharyngeal carcinoma (NPC): EBV-specific CTL responses in the blood and tumors of NPC patients and the antigen-processing function of the tumor cells. J Immunol 165(1):573–582

Lee SP, Constandinou CM, Thomas WA, Croom-Carter D, Blake NW, Murray PG, Crocker J, Rickinson AB (1998) Antigen presenting phenotype of Hodgkin Reed-Sternberg cells: analysis of the HLA class I processing pathway and the effects of interleukin-10 on Epstein-Barr virus-specific cytotoxic T-cell recognition. Blood 92(3):1020–1030

Leen A, Meij P, Redchenko I, Middeldorp J, Bloemena E, Rickinson A, Blake N (2001) Differential immunogenicity of Epstein-Barr virus latent-cycle proteins for human CD4(+) T-helper 1 responses. J Virol 75(18):8649–8659

Leung CS, Haigh TA, Mackay LK, Rickinson AB, Taylor GS (2010) Nuclear location of an endogenously expressed antigen, EBNA1, restricts access to macroautophagy and the range of CD4 epitope display. Proc Natl Acad Sci USA 107(5):2165–2170

Li J, Huang ZF, Xiong G, Mo HY, Qiu F, Mai HQ, Chen QY, He J, Chen SP, Zheng LM, Qian CN, Zeng YX (2011) Distribution, characterization, and induction of CD8+ regulatory T cells and IL-17-producing CD8+ T cells in nasopharyngeal carcinoma. J Transl Med 9:189

Lin X, Gudgeon NH, Hui EP, Jia H, Qun X, Taylor GS, Barnardo MC, Lin CK, Rickinson AB, Chan AT (2008) CD4 and CD8 T cell responses to tumour-associated Epstein-Barr virus antigens in nasopharyngeal carcinoma patients. Cancer Immunol Immunother 57(7):963–975

Liu Y, van den Berg A, Veenstra R, Rutgers B, Nolte I, van Imhoff G, Visser L, Diepstra A (2013) PML nuclear bodies and SATB1 are associated with HLA class I expression in EBV+ Hodgkin lymphoma. PLoS ONE 8(8):e72930

Long HM, Chagoury OL, Leese AM, Ryan GB, James E, Morton LT, Abbott RJ, Sabbah S, Kwok W, Rickinson AB (2013) MHC II tetramers visualize human CD4+ T cell responses to Epstein-Barr virus infection and demonstrate atypical kinetics of the nuclear antigen EBNA1 response. J Exp Med 210(5):933–949

Long HM, Haigh TA, Gudgeon NH, Leen AM, Tsang CW, Brooks J, Landais E, Houssaint E, Lee SP, Rickinson AB, Taylor GS (2005) CD4+ T-cell responses to Epstein-Barr virus (EBV) latent-cycle antigens and the recognition of EBV-transformed lymphoblastoid cell lines. J Virol 79(8):4896–4907

Long HM, Leese AM, Chagoury OL, Connerty SR, Quarcoopome J, Quinn LL, Shannon-Lowe C, Rickinson AB (2011) Cytotoxic CD4+ T cell responses to EBV contrast with CD8 responses in breadth of lytic cycle antigen choice and in lytic cycle recognition. J Immunol 187(1):92–101

Long HM, Zuo J, Leese AM, Gudgeon NH, Jia H, Taylor GS, Rickinson AB (2009) CD4+ T-cell clones recognizing human lymphoma-associated antigens: generation by in vitro stimulation with autologous Epstein-Barr virus-transformed B cells. Blood 114(4):807–815

Lopez-Granados E, Stacey M, Kienzler AK, Sierro S, Willberg CB, Fox CP, Rigaud S, Long HM, Hislop AD, Rickinson AB, Patel S, Latour S, Klenerman P, Chapel H (2014) A mutation in XIAP (G466X) leads to memory inflation of EBV-specific T cells. Clin Exp Immunol 178:470

Louis CU, Straathof K, Bollard CM, Ennamuri S, Gerken C, Lopez TT, Huls MH, Sheehan A, Wu MF, Liu H, Gee A, Brenner MK, Rooney CM, Heslop HE, Gottschalk S (2010) Adoptive transfer of EBV-specific T cells results in sustained clinical responses in patients with locoregional nasopharyngeal carcinoma. J Immunother 33(9):983–990

Mackay LK, Long HM, Brooks JM, Taylor GS, Leung CS, Chen A, Wang F, Rickinson AB (2009) T cell detection of a B-cell tropic virus infection: newly-synthesised versus mature viral proteins as antigen sources for CD4 and CD8 epitope display. PLoS Pathog 5(12):e1000699

Maini MK, Gudgeon N, Wedderburn LR, Rickinson AB, Beverley PC (2000) Clonal expansions in acute EBV infection are detectable in the CD8 and not the CD4 subset and persist with a variable CD45 phenotype. J Immunol 165(10):5729–5737

Marshall NA, Christie LE, Munro LR, Culligan DJ, Johnston PW, Barker RN, Vickers MA (2004) Immunosuppressive regulatory T cells are abundant in the reactive lymphocytes of Hodgkin lymphoma. Blood 103(5):1755–1762

Martorelli D, Houali K, Caggiari L, Vaccher E, Barzan L, Franchin G, Gloghini A, Pavan A, Da Ponte A, Tedeschi RM, De Re V, Carbone A, Ooka T, De Paoli P, Dolcetti R (2008) Spontaneous T cell responses to Epstein-Barr virus-encoded BARF1 protein and derived peptides in patients with nasopharyngeal carcinoma: bases for improved immunotherapy. Int J Cancer 123(5):1100–1107

Mautner J, Pich D, Nimmerjahn F, Milosevic S, Adhikary D, Christoph H, Witter K, Bornkamm GW, Hammerschmidt W, Behrends U (2004) Epstein-Barr virus nuclear antigen 1 evades direct immune recognition by CD4+ T helper cells. Eur J Immunol 34(9):2500–2509

McAulay KA, Higgins CD, Macsween KF, Lake A, Jarrett RF, Robertson FL, Williams H, Crawford DH (2007) HLA class I polymorphisms are associated with development of infectious mononucleosis upon primary EBV infection. J Clin Invest 117(10):3042–3048

Moormann AM, Chelimo K, Sumba OP, Lutzke ML, Ploutz-Snyder R, Newton D, Kazura J, Rochford R (2005) Exposure to holoendemic malaria results in elevated Epstein-Barr virus loads in children. J Infect Dis 191(8):1233–1238

Moormann AM, Chelimo K, Sumba PO, Tisch DJ, Rochford R, Kazura JW (2007) Exposure to holoendemic malaria results in suppression of Epstein-Barr virus-specific T cell immunosurveillance in Kenyan children. J Infect Dis 195(6):799–808

Moormann AM, Heller KN, Chelimo K, Embury P, Ploutz-Snyder R, Otieno JA, Oduor M, Munz C, Rochford R (2009) Children with endemic Burkitt lymphoma are deficient in EBNA1-specific IFN-gamma T cell responses. Int J Cancer 124(7):1721–1726

Morales O, Mrizak D, Francois V, Mustapha R, Miroux C, Depil S, Decouvelaere AV, Lionne-Huyghe P, Auriault C, de Launoit Y, Pancre V, Delhem N (2014) Epstein-Barr virus infection induces an increase of T regulatory type 1 cells in Hodgkin lymphoma patients. Br J Haematol 166(6):875–890

Moss DJ, Bishop CJ, Burrows SR, Ryan JM (1985) T lymphocytes in infectious mononucleosis. I. T cell death in vitro. Clin Exp Immunol 60(1):61–69

Moss DJ, Burrows SR, Castelino DJ, Kane RG, Pope JH, Rickinson AB, Alpers MP, Heywood PF (1983) A comparison of Epstein-Barr virus-specific T-cell immunity in malaria-endemic and -nonendemic regions of Papua New Guinea. Int J Cancer 31(6):727–732

Munz C, Bickham KL, Subklewe M, Tsang ML, Chahroudi A, Kurilla MG, Zhang D, O'Donnell M, Steinman RM (2000) Human CD4(+) T lymphocytes consistently respond to the latent Epstein-Barr virus nuclear antigen EBNA1. J Exp Med 191(10):1649–1660

Murray PG, Constandinou CM, Crocker J, Young LS, Ambinder RF (1998) Analysis of major histocompatibility complex class I, TAP expression, and LMP2 epitope sequence in Epstein-Barr virus-positive Hodgkin's disease. Blood 92(7):2477–2483

Nguyen-Van D, Keane C, Han E, Jones K, Nourse JP, Vari F, Ross N, Crooks P, Ramuz O, Green M, Griffith L, Trappe R, Grigg A, Mollee P, Gandhi MK (2011) Epstein-Barr virus-positive diffuse large B-cell lymphoma of the elderly expresses EBNA3A with conserved CD8 T-cell epitopes. Am J Blood Res 1(2):146–159

Nichols KE, Hom J, Gong SY, Ganguly A, Ma CS, Cannons JL, Tangye SG, Schwartzberg PL, Koretzky GA, Stein PL (2005) Regulation of NKT cell development by SAP, the protein defective in XLP. Nat Med 11(3):340–345

Niens M, Jarrett RF, Hepkema B, Nolte IM, Diepstra A, Platteel M, Kouprie N, Delury CP, Gallagher A, Visser L, Poppema S, te Meerman GJ, van den Berg A (2007) HLA-A*02 is associated with a reduced risk and HLA-A*01 with an increased risk of developing EBV+ Hodgkin lymphoma. Blood 110(9):3310–3315

Odumade OA, Knight JA, Schmeling DO, Masopust D, Balfour HH Jr, Hogquist KA (2012) Primary Epstein-Barr virus infection does not erode preexisting CD8(+) T cell memory in humans. J Exp Med 209(3):471–478

Omiya R, Buteau C, Kobayashi H, Paya CV, Celis E (2002) Inhibition of EBV-induced lymphoproliferation by CD4(+) T cells specific for an MHC class II promiscuous epitope. J Immunol 169(4):2172–2179

Orlova N, Wang F, Fogg MH (2011) Persistent infection drives the development of CD8+ T cells specific for late lytic infection antigens in lymphocryptovirus-infected macaques and Epstein-Barr virus-infected humans. J Virol 85(23):12821–12824

Palendira U, Chinn R, Raza W, Piper K, Pratt G, Machado L, Bell A, Khan N, Hislop AD, Steyn R, Rickinson AB, Buckley CD, Moss P (2008) Selective accumulation of virus-specific CD8+ T cells with unique homing phenotype within the human bone marrow. Blood 112(8):3293–3302

Paludan C, Bickham K, Nikiforow S, Tsang ML, Goodman K, Hanekom WA, Fonteneau JF, Stevanovic S, Munz C (2002) Epstein-Barr nuclear antigen 1-specific CD4(+) Th1 cells kill Burkitt's lymphoma cells. J Immunol 169(3):1593–1603

Paludan C, Schmid D, Landthaler M, Vockerodt M, Kube D, Tuschl T, Munz C (2005) Endogenous MHC class II processing of a viral nuclear antigen after autophagy. Science 307(5709):593–596

Pardoll DM (2012) The blockade of immune checkpoints in cancer immunotherapy. Nat Rev Cancer 12(4):252–264

Parsonage G, Machado LR, Hui JW, McLarnon A, Schmaler T, Balasothy M, To KF, Vlantis AC, van Hasselt CA, Lo KW, Wong WL, Hui EP, Chan AT, Lee SP (2012) CXCR6 and CCR5 localize T lymphocyte subsets in nasopharyngeal carcinoma. Am J Pathol 180(3):1215–1222

Piriou E, van Dort K, Nanlohy NM, van Oers MH, Miedema F, van Baarle D (2005) Loss of EBNA1-specific memory CD4+ and CD8+ T cells in HIV-infected patients progressing to AIDS-related non-Hodgkin lymphoma. Blood 106(9):3166–3174

Precopio ML, Sullivan JL, Willard C, Somasundaran M, Luzuriaga K (2003) Differential kinetics and specificity of EBV-specific CD4+ and CD8+ T cells during primary infection. J Immunol 170(5):2590–2598

Pudney VA, Leese AM, Rickinson AB, Hislop AD (2005) CD8+ immunodominance among Epstein-Barr virus lytic cycle antigens directly reflects the efficiency of antigen presentation in lytically infected cells. J Exp Med 201(3):349–360

Rajnavolgyi E, Nagy N, Thuresson B, Dosztanyi Z, Simon A, Simon I, Karr RW, Ernberg I, Klein E, Falk KI (2000) A repetitive sequence of Epstein-Barr virus nuclear antigen 6 comprises overlapping T cell epitopes which induce HLA-DR-restricted CD4(+) T lymphocytes. Int Immunol 12(3):281–293

Remmerswaal EB, Havenith SH, Idu MM, van Leeuwen EM, van Donselaar KA, Ten Brinke A, van der Bom-Baylon N, Bemelman FJ, van Lier RA, Ten Berge IJ (2012) Human virus-specific effector-type T cells accumulate in blood but not in lymph nodes. Blood 119(7):1702–1712

Rigaud S, Fondaneche MC, Lambert N, Pasquier B, Mateo V, Soulas P, Galicier L, Le Deist F, Rieux-Laucat F, Revy P, Fischer A, de Saint Basile G, Latour S (2006) XIAP deficiency in humans causes an X-linked lymphoproliferative syndrome. Nature 444(7115):110–114

Rooney CM, Smith CA, Ng CY, Loftin S, Li C, Krance RA, Brenner MK, Heslop HE (1995) Use of gene-modified virus-specific T lymphocytes to control Epstein-Barr-virus-related lymphoproliferation. Lancet 345(8941):9–13

Rostgaard K, Wohlfahrt J, Hjalgrim H (2014) A genetic basis for infectious mononucleosis: evidence from a family study of hospitalized cases in Denmark. Clin Infect Dis 58(12):1684–1689

Rowe M, Khanna R, Jacob CA, Argaet V, Kelly A, Powis S, Belich M, Croom-Carter D, Lee S, Burrows SR et al (1995) Restoration of endogenous antigen processing in Burkitt's lymphoma cells by Epstein-Barr virus latent membrane protein-1: coordinate up-regulation of peptide transporters and HLA-class I antigen expression. Eur J Immunol 25(5):1374–1384

Selin LK, Wlodarczyk MF, Kraft AR, Nie S, Kenney LL, Puzone R, Celada F (2011) Heterologous immunity: immunopathology, autoimmunity and protection during viral infections. Autoimmunity 44(4):328–347

Seto E, Yang L, Middeldorp J, Sheen TS, Chen JY, Fukayama M, Eizuru Y, Ooka T, Takada K (2005) Epstein-Barr virus (EBV)-encoded BARF1 gene is expressed in nasopharyngeal carcinoma and EBV-associated gastric carcinoma tissues in the absence of lytic gene expression. J Med Virol 76(1):82–88

Silins SL, Sherritt MA, Silleri JM, Cross SM, Elliott SL, Bharadwaj M, Le TT, Morrison LE, Khanna R, Moss DJ, Suhrbier A, Misko IS (2001) Asymptomatic primary Epstein-Barr virus infection occurs in the absence of blood T-cell repertoire perturbations despite high levels of systemic viral load. Blood 98(13):3739–3744

Smith C, Tsang J, Beagley L, Chua D, Lee V, Li V, Moss DJ, Coman W, Chan KH, Nicholls J, Kwong D, Khanna R (2012) Effective treatment of metastatic forms of Epstein-Barr virus-associated nasopharyngeal carcinoma with a novel adenovirus-based adoptive immunotherapy. Cancer Res 72(5):1116–1125

Soares MV, Plunkett FJ, Verbeke CS, Cook JE, Faint JM, Belaramani LL, Fletcher JM, Hammerschmitt N, Rustin M, Bergler W, Beverley PC, Salmon M, Akbar AN (2004) Integration of apoptosis and telomere erosion in virus-specific CD8+ T cells from blood and tonsils during primary infection. Blood 103(1):162–167

Steven NM, Annels NE, Kumar A, Leese AM, Kurilla MG, Rickinson AB (1997) Immediate early and early lytic cycle proteins are frequent targets of the Epstein-Barr virus-induced cytotoxic T cell response. J Exp Med 185(9):1605–1617

Steven NM, Leese AM, Annels NE, Lee SP, Rickinson AB (1996) Epitope focusing in the primary cytotoxic T cell response to Epstein-Barr virus and its relationship to T cell memory. J Exp Med 184(5):1801–1813

Stevens SJ, Verkuijlen SA, Hariwiyanto B, Harijadi Paramita DK, Fachiroh J, Adham M, Tan IB, Haryana SM, Middeldorp JM (2006) Noninvasive diagnosis of nasopharyngeal carcinoma: nasopharyngeal brushings reveal high Epstein-Barr virus DNA load and carcinoma-specific viral BARF1 mRNA. Int J Cancer 119(3):608–614

Stowe RP, Kozlova EV, Yetman DL, Walling DM, Goodwin JS, Glaser R (2007) Chronic herpesvirus reactivation occurs in aging. Exp Gerontol 42(6):563–570

Su WH, Hildesheim A, Chang YS (2013) Human leukocyte antigens and epstein-barr virus-associated nasopharyngeal carcinoma: old associations offer new clues into the role of immunity in infection-associated cancers. Front Oncol 3:299

Sun Q, Burton R, Reddy V, Lucas KG (2002a) Safety of allogeneic Epstein-Barr virus (EBV)-specific cytotoxic T lymphocytes for patients with refractory EBV-related lymphoma. Br J Haematol 118(3):799–808

Sun Q, Burton RL, Lucas KG (2002b) Cytokine production and cytolytic mechanism of CD4(+) cytotoxic T lymphocytes in ex vivo expanded therapeutic Epstein-Barr virus-specific T-cell cultures. Blood 99(9):3302–3309

Tangye SG (2014) XLP: clinical features and molecular etiology due to mutations in SH2D1A encoding SAP. J Clin Immunol. doi:10.1007/s10875-014-0083-7

Taylor GS, Jia H, Harrington K, Lee LW, Turner J, Ladell K, Price DA, Tanday M, Matthews J, Roberts C, Edwards C, McGuigan L, Hartley A, Wilson S, Hui EP, Chan AT, Rickinson AB, Steven NM (2014) A recombinant modified vaccinia ankara vaccine encoding Epstein-Barr Virus (EBV) target antigens: a phase I trial in UK patients with EBV-positive cancer. Clin Cancer Res 20(19):5009–5022

Taylor GS, Long HM, Haigh TA, Larsen M, Brooks J, Rickinson AB (2006) A role for intercellular antigen transfer in the recognition of EBV-transformed B cell lines by EBV nuclear antigen-specific CD4+ T cells. J Immunol 177(6):3746–3756

Tellam J, Connolly G, Green KJ, Miles JJ, Moss DJ, Burrows SR, Khanna R (2004) Endogenous presentation of CD8+ T cell epitopes from Epstein-Barr virus-encoded nuclear antigen 1. J Exp Med 199(10):1421–1431

Vantourout P, Hayday A (2013) A six-of-the-best: unique contributions of gammadelta T cells to immunology. Nat Rev Immunol 13(2):88–100

Vescovini R, Telera A, Fagnoni FF, Biasini C, Medici MC, Valcavi P, di Pede P, Lucchini G, Zanlari L, Passeri G, Zanni F, Chezzi C, Franceschi C, Sansoni P (2004) Different contribution of EBV and CMV infections in very long-term carriers to age-related alterations of CD8+ T cells. Exp Gerontol 39(8):1233–1243

Voo KS, Fu T, Wang HY, Tellam J, Heslop HE, Brenner MK, Rooney CM, Wang RF (2004) Evidence for the presentation of major histocompatibility complex class I-restricted Epstein-Barr virus nuclear antigen 1 peptides to CD8+ T lymphocytes. J Exp Med 199(4):459–470

White CA, Cross SM, Kurilla MG, Kerr BM, Schmidt C, Misko IS, Khanna R, Moss DJ (1996) Recruitment during infectious mononucleosis of CD3+ CD4+ CD8+ virus-specific cytotoxic T cells which recognise Epstein-Barr virus lytic antigen BHRF1. Virology 219(2):489–492

Whittle HC, Brown J, Marsh K, Greenwood BM, Seidelin P, Tighe H, Wedderburn L (1984) T-cell control of Epstein-Barr virus-infected B cells is lost during *P. falciparum* malaria. Nature 312(5993):449–450

Wingate PJ, McAulay KA, Anthony IC, Crawford DH (2009) Regulatory T cell activity in primary and persistent Epstein-Barr virus infection. J Med Virol 81(5):870–877

Wolchok JD, Kluger H, Callahan MK, Postow MA, Rizvi NA, Lesokhin AM, Segal NH, Ariyan CE, Gordon RA, Reed K, Burke MM, Caldwell A, Kronenberg SA, Agunwamba BU, Zhang X, Lowy I, Inzunza HD, Feely W, Horak CE, Hong Q, Korman AJ, Wigginton JM, Gupta A, Sznol M (2013) Nivolumab plus ipilimumab in advanced melanoma. N Engl J Med 369(2):122–133

Woodberry T, Suscovich TJ, Henry LM, August M, Waring MT, Kaur A, Hess C, Kutok JL, Aster JC, Wang F, Scadden DT, Brander C (2005a) Alpha E beta 7 (CD103) expression identifies a highly active, tonsil-resident effector-memory CTL population. J Immunol 175(7):4355–4362

Woodberry T, Suscovich TJ, Henry LM, Davis JK, Frahm N, Walker BD, Scadden DT, Wang F, Brander C (2005b) Differential targeting and shifts in the immunodominance of Epstein-Barr virus–specific CD8 and CD4 T cell responses during acute and persistent infection. J Infect Dis 192(9):1513–1524

Xiang Z, Liu Y, Zheng J, Liu M, Lv A, Gao Y, Hu H, Lam KT, Chan GC, Yang Y, Chen H, Tsao GS, Bonneville M, Lau YL, Tu W (2014) Targeted activation of human Vgamma9Vdelta2-T cells controls Epstein-Barr virus-induced B cell lymphoproliferative disease. Cancer Cell 26(4):565–576

Yamamoto R, Nishikori M, Kitawaki T, Sakai T, Hishizawa M, Tashima M, Kondo T, Ohmori K, Kurata M, Hayashi T, Uchiyama T (2008) PD-1–PD-1 ligand interaction contributes to immunosuppressive microenvironment of Hodgkin lymphoma. Blood 111(6):3220–3224

Yao Y, Minter HA, Chen X, Reynolds GM, Bromley M, Arrand JR (2000) Heterogeneity of HLA and EBER expression in Epstein-Barr virus-associated nasopharyngeal carcinoma. Int J Cancer 88(6):949–955

Yip WK, Abdullah MA, Yusoff SM, Seow HF (2009) Increase in tumour-infiltrating lymphocytes with regulatory T cell immunophenotypes and reduced zeta-chain expression in nasopharyngeal carcinoma patients. Clin Exp Immunol 155(3):412–422

Yuling H, Ruijing X, Li L, Xiang J, Rui Z, Yujuan W, Lijun Z, Chunxian D, Xinti T, Wei X, Lang C, Yanping J, Tao X, Mengjun W, Jie X, Youxin J, Jinquan T (2009) EBV-induced human CD8+ NKT cells suppress tumorigenesis by EBV-associated malignancies. Cancer Res 69(20):7935–7944

# Immune Evasion by Epstein-Barr Virus

**Maaike E. Ressing, Michiel van Gent, Anna M. Gram, Marjolein J.G. Hooykaas, Sytse J. Piersma and Emmanuel J.H.J. Wiertz**

**Abstract** Epstein-Bar virus (EBV) is widespread within the human population with over 90 % of adults being infected. In response to primary EBV infection, the host mounts an antiviral immune response comprising both innate and adaptive effector functions. Although the immune system can control EBV infection to a large extent, the virus is not cleared. Instead, EBV establishes a latent infection in B lymphocytes characterized by limited viral gene expression. For the production of new viral progeny, EBV reactivates from these latently infected cells. During the productive phase of infection, a repertoire of over 80 EBV gene products is expressed, presenting a vast number of viral antigens to the primed immune system. In particular the EBV-specific $CD4^+$ and $CD8^+$ memory T lymphocytes can respond within hours, potentially destroying the virus-producing cells before viral replication is completed and viral particles have been released. Preceding the adaptive immune response, potent innate immune mechanisms provide a first line of defense during primary and recurrent infections. In spite of this broad range of antiviral immune effector mechanisms, EBV persists for life and continues to replicate. Studies performed over the past decades have revealed a wide array of viral gene products interfering with both innate and adaptive immunity. These include EBV-encoded proteins as well as small noncoding RNAs with immune-evasive properties. The current review presents an overview of the evasion strategies that are employed by EBV to facilitate immune escape during latency and productive infection. These evasion mechanisms may also compromise the elimination of EBV-transformed cells, and thus contribute to malignancies associated with EBV infection.

M.E. Ressing · M. van Gent · A.M. Gram · M.J.G. Hooykaas · S.J. Piersma · E.J.H.J. Wiertz (✉)
Department of Medical Microbiology, University Medical Center Utrecht,
Utrecht, The Netherlands
e-mail: ewiertz@umcutrecht.nl

M.E. Ressing · A.M. Gram
Department of Molecular Cell Biology, Leiden University Medical Center,
Leiden, The Netherlands



C. Münz (ed.), *Epstein Barr Virus Volume 2*, Current Topics in Microbiology
and Immunology 391, DOI 10.1007/978-3-319-22834-1_12

## Contents



## Abbreviations

| | |
|---|---|
| APC | Antigen-presenting cell |
| ATP | Adenosine triphosphate |
| BART | BamHI fragment A rightward transcript |
| BL | Burkitt's lymphoma |
| CIITA | Class II, major histocompatibility complex, transactivator |
| CSF-1 | Colony-stimulating factor-1 |
| E | Early |
| EBER | EBV-encoded RNA |
| EBNA | Epstein-Barr nuclear antigen |
| EBV | Epstein-Barr virus |
| ER | Endoplasmic reticulum |
| GPCR | G-protein-coupled receptor |
| HCMV | Human cytomegalovirus |
| HIV | Human immunodeficiency virus |
| HLA | Human leukocyte antigen |
| HSV | Herpes Simplex virus |
| IE | Immediate-early |
| IFI16 | Interferon inducible protein 16 |
| IFN | Interferon |
| IKK | Inhibitor of NF-κB kinase |
| IL | Interleukin |
| iNKT | Invariant natural killer T cells |
| IRF | Interferon-regulatory factor |
| ISG | Interferon-stimulated gene |
| JAK | Janus-kinase |
| KSHV | Kaposi's sarcoma-associated herpesvirus |

| | |
|---|---|
| L | Late |
| LCL | Lymphoblastoid cell line |
| LMP | Latent membrane protein |
| MAPK | Mitogen-activated protein kinase |
| MHV68 | Murine herpesvirus 68 |
| MICB | MHC class I polypeptide-related sequence B |
| miRNA | MicroRNA |
| NF-κB | Nuclear factor-κB |
| NK cells | Natural killer cells |
| NLRP3 | NLR family, pyrin domain containing 3 |
| ORF | Open reading frame |
| PAMP | Pathogen-associated molecular pattern |
| PI3 K | Phosphatidylinositide 3-kinase |
| PKR | Protein kinase RNA-activated |
| PML-bodies | Promyelocytic leukemia bodies |
| PRR | Pattern-recognition receptor |
| qPCR | Quantitative PCR |
| RISC | RNA-induced silencing complex |
| RLR | RIG-I like receptor |
| shRNA | Short hairpin RNA |
| SOCS | Suppressor of cytokine signaling |
| STAT | Signal transducer and activator of transcription |
| TAP | Transporter associated with antigen processing |
| TCR | T-cell receptor |
| TGF | Transforming growth factor |
| TLR | Toll-like receptor |
| TNF | Tumor necrosis factor |
| vhs | Virion host shutoff |

# 1 Introduction

Like other herpesviruses, Epstein-Barr virus (EBV) persists in infected immunocompetent individuals by maintaining a delicate balance between viral replication and host antiviral immunity. During primary infection, EBV is detected by innate immune sensors that initiate direct antiviral responses and also orchestrate ensuing adaptive immunity. Innate and adaptive immune responses toward EBV have been described extensively in recent reviews (CTMI 2015 in press).

In contrast to the subfamilies of *Alpha-* and *Betaherpesvirinae*, the *Gammaherpesvirinae*, of which EBV is the prototype, have growth-transforming properties. EBV establishes a lifelong, latent infection within the memory B-cell compartment. EBV genomes are propagated during division of the transformed, latently infected B cells. During the latent phase, the number of viral gene products is limited, thereby reducing the amount of viral targets available for immune detection. With the development of model systems to study EBV infection of B cell and epithelial cell, it has become apparent that EBV encodes a wide range of gene products that are nonessential for replication in vitro, but contribute to creating an in vivo environment that is beneficial to the maintenance of viral episome and occasional reactivation of EBV. Following the identification of multiple EBV immune evasion molecules and elucidation of their mechanisms of action, it is estimated that over half of the viral gene products is dedicated to functions that modulate antiviral responses of the host. Furthermore, the incorporation of EBV mRNAs and noncoding miRNAs into viral particles provides a means to immediately express immune modulatory gene products in newly infected cells (Jochum et al. 2012b).

## 2 Immune Evasion by Lytic EBV Gene Products

Upon reactivation from latency in B cells, over 80 EBV-encoded proteins are expressed in a regulated fashion. The immediate-early BZLF1 and BRLF1 molecules act as transactivators to induce expression of over 30 early lytic gene products involved in replication of the EBV genome. Finally, over 30 late genes encode structural proteins for the formation of new viral particles. Epithelial cells are thought to support only productive EBV infection.

### *2.1 Evasion of Innate Immunity*

Cells targeted by EBV express a variety of pattern recognition receptors (PRRs) involved in initial sensing of viral infection. Examples of PRRs are cell surface and endosomal Toll-like receptors (TLRs) and cytoplasmic DNA and RNA sensors (for instance, IFI16, cGAS, and retinoic acid-inducible gene (RIG)-I-like receptors (RLRs)). These receptors detect pathogen-associated molecular patterns (PAMPs) derived from viruses or infected cells and initiate signaling cascades that culminate in activation of the transcription factors interferon-regulatory factors (IRF) 3 and 7 or NF-κB. Nuclear translocation of IRF3/7 induces production of type I interferons (IFN I) and NF-κB-induced transcription activates antiapoptotic and inflammatory processes, thereby creating an environment hostile to viral replication (Takeuchi and Akira 2009). Various PRR signaling pathways are activated during EBV infection. To allow establishment of infection and persistence, EBV has adopted strategies to modulate these signaling pathways at different levels to minimize their antiviral activity, while taking advantage of their growth-promoting effects (Fig. 1 and Table 1).

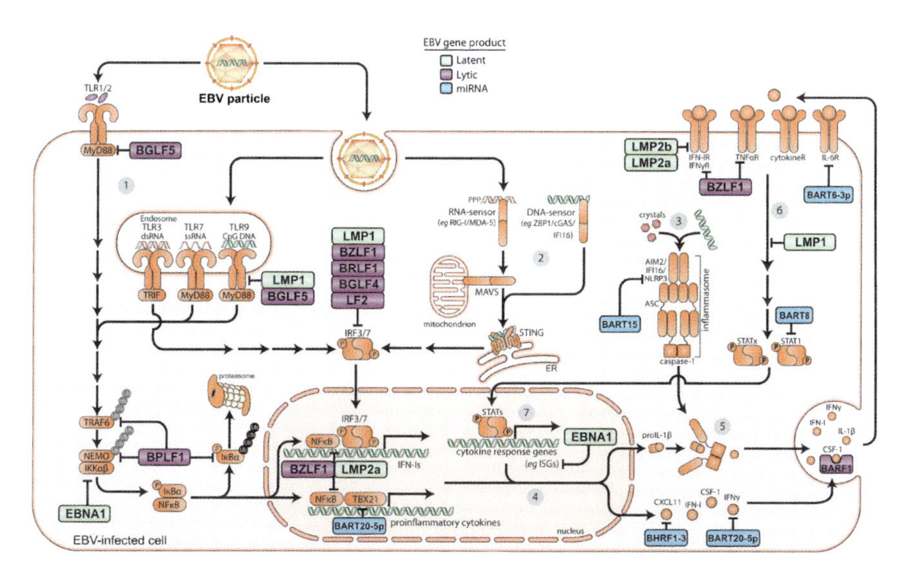


**Fig. 1** Overview of innate signaling pathways subjected to EBV modulation. EBV particles can be sensed by a number of pathogen recognition receptors (PRRs). At the cell surface and in endosomes, EBV is detected by Toll-like receptors (TLRs) (1). Within the cytosol, virus-derived or virus-induced components can be detected by RNA and DNA sensors (2) as well as by inflammasomes (3). Recognition by TLRs, RNA, and DNA sensors induces a cascade of intracellular signaling events resulting in activation of interferon-regulatory factors (IRFs) and NF-κB, and, consequently, in gene transcription leading to the production of cytokines and type I interferons (IFN I) (4). Detection by inflammasomes results in the generation of active caspase 1, which processes IL-1β and IL-18 precursors (5). Secreted cytokines are recognized by surface cytokine receptors that signal through signal transducer and activator of transcription molecules (STATs) (6). Nuclear translocation of STAT dimers results in the transcription of cytokine response genes, such as interferon-stimulated genes (ISGs) whose products can exert direct antiviral functions. Various steps in these PRR signaling pathways are targeted by latent (*green*) and/or lytic (*purple*) EBV proteins or by EBV miRNAs (*blue*), as has been appreciated more recently

### 2.1.1 Reduction of Toll-like Receptor Expression

Like other γ-herpesviruses, EBV inhibits cellular protein synthesis in productively infected cells through global mRNA destabilization. This process, termed shutoff, is mediated by the EBV DNase (alkaline exonuclease) BGLF5, which is expressed with early kinetics during the productive phase of infection (Zuo et al. 2008). BGLF5's additional RNase function utilizes the same catalytic site as its DNase activity, yet the substrate-binding site appears only partly shared by DNA and RNA substrates (Horst et al. 2012). The promiscuous RNA degradation induced by EBV BGLF5 can affect immunologically relevant proteins, including TLR2 and TLR9 that are capable of sensing EBV infection (Gaudreault et al. 2007; van Gent et al. 2011, 2015). The observation that BGLF5 did not downregulate TLR4, a TLR not reported to contribute to EBV detection, suggests some

**Table 1** Lytic EBV gene products interfering with innate or adaptive immunity

| Gene product | Functional mechanism(s) | References |
|---|---|---|
| BGLF5 | Alkaline exonuclease involved in host shutoff, downregulates TLR2 and TLR9, reduces HLA I and HLA II expression through mRNA degradation | van Gent et al. (2011, 2014), Horst et al. (2012), Rowe et al. (2007a, b), Zuo et al. (2008), Quinn et al. (2014) |
| BZLF1 | Inhibits IRF7 transcriptional activity on IFNα4 and IFNβ promoters, inhibits NF-κB expression, reduces expression of TNFα and IFNγ receptor leading to, e.g., reduced HLA II surface expression, induces SOCS3 expression to favor type I IFN irresponsiveness, induces immunosuppressive TGFβ expression, disrupts formation of PML-bodies, interferes with the function of the invariant chain to reduced HLA II levels | Hahn et al. (2005), Dreyfus et al. (1999), Gutsch et al. (1994), Morrison and Kenney (2004), Bristol et al. (2010), Morrison et al. (2001, 2004), Michaud et al. (2010), Cayrol and Flemington (1995), Adamson and Kenney (2001), Zuo et al. (2011) |
| BRLF1 | Reduces expression of IRF3 and IRF7 | Bentz et al. (2010) |
| LF2 (BILF4) | Inhibits IRF7 to suppress IFNα production | Wu et al. (2009) |
| BGLF4 | Inhibits IRF3 activity and reduces IFNβ production in response to poly(I:C), suppresses NF-κB activity by inhibiting UXT coactivator | Wang et al. (2009), Chang et al. (2012) |
| BPLF1 | Interferes with TLR- and LMP1-mediated NF-κB activation | van Gent et al. (2014), Saito et al. (2013) |
| BARF1 | Soluble form of CSF-1 receptor that neutralizes effects of CSF-1 | Cohen and Lekstrom (1999), Strockbine et al. (1998) |
| dUTPase (BLLF3) | Compromises lymphocyte responses and proliferation, induces NF-κB activity through TLR2, induces production of proinflammatory cytokines and IL-10 | Padgett et al. (2004), Ariza et al. (2009, 2013), Waldman et al. (2008), Glaser et al. (2006), Brooks et al. (2006) |
| BNLF2a | Depletes peptides from the ER through inhibiting peptide transport by TAP by blocking both TAP and ATP binding | Horst et al. (2011, 2009), Hislop et al. (2007), Croft et al. (2009), Quinn et al. (2014) |
| BILF1 | Constitutively active GPCR that reduces HLA I surface levels by interfering with HLA I transport from the trans-Golgi network and increasing turnover from the cells surface, affects most HLA I haplotypes, except HLA C | Zuo et al. (2009), Griffin et al. (2013), Quinn et al. (2014) |

(continued)

**Table 1** (continued)

| Gene product | Functional mechanism(s) | References |
|---|---|---|
| gp42 (BZLF2) | Entry receptor for EBV, binds to HLA II and interferes with $CD8^{+}$ T-cell activation and may block cross-presentation, cooperates with gH and gL | Ressing et al. (2003, 2005); manuscript in preparation |
| vIL-10 (BCRF1) | Inhibits antiviral $CD4^{+}$ T-cell responses, inhibits co-stimulatory molecules on human monocytes | Salek-Ardakani et al. (2002) |

selectivity in shutoff (Gaudreault et al. 2007; van Gent et al. 2015). By analogy to herpes simplex virus (HSV)-2 infection of epithelial cells, where shutoff induced by the virion host shutoff (vhs) protein contributes to downregulation of the PRRs TLR2, TLR3, RIG-I, and MDA-5 (Yao and Rosenthal 2011), the effects of BGLF5 may well extend to additional immune components. Indeed, many TLRs are differentially expressed in lytically EBV-infected B cells by mechanisms that remain elusive and may involve shutoff effects (van Gent et al. 2011). In vivo studies on HSV and the murine γ-herpesvirus MHV68 indicate that shutoff-induced reductions in protein levels mainly prevent production of newly synthesized effector molecules rather than reducing the levels of existing PRRs (Murphy et al. 2003; Pasieka et al. 2008; Sheridan et al. 2014). It remains to be assessed to what extent EBV BGLF5 affects PRR expression levels, and subsequent signal transduction, in vivo.

### 2.1.2 Modulation of IRF Signaling and Type I Interferon Production

A number of lytic EBV proteins interfere with the actions of host IRFs, the transcription factors that induce type I IFN production. The immediate-early EBV transactivator BZLF1 interacts with IRF7 and inhibits its transcriptional activity on the IFNα4 and IFNβ promoters to prevent induction of an antiviral environment (Hahn et al. 2005). The other immediate-early EBV transactivator, BRLF1, reduces expression of both IRF3 and IRF7, thereby inhibiting production of IFNβ (Bentz et al. 2010). The EBV tegument protein LF2, which is present in virions and gains access to the cell immediately upon virus entry, targets IRF7 and suppresses IFNα production (Wu et al. 2009). Finally, the EBV protein kinase BGLF4 phosphorylates and inhibits IRF3 transcriptional activity, thereby reducing IFNβ expression in response to treatment with the TLR3 agonist poly(I:C) (Wang et al. 2009). Recently, Dunmire et al. (2014) reported that a clear systemic IFN response is observed during acute EBV infection, but this response lacks some key components compared to observations for other viruses. This may illustrate the successful actions of the immune evasion mechanisms employed by EBV to repress secretion of interferon responsive genes.

### 2.1.3 Interference with NF-κB and Inflammatory Pathways

In general, productive EBV infection is associated with a reduction in NF-κB-dependent gene expression (Keating et al. 2002). Viral BZLF1 and cellular NF-κB reciprocally inhibit each other's expression and, as a consequence, higher levels of NF-κB in the absence of BZLF1 favor EBV latency, whereas increased expression of BZLF1 upon lytic cycle induction overwhelms the limiting amount of NF-κB (Dreyfus et al. 1999; Gutsch et al. 1994; Morrison and Kenney 2004). While NF-κB is still translocated to the nucleus, its transcriptional activity is suppressed by BZLF1, preventing induction of antiviral immune effector mechanisms (Morrison and Kenney 2004).

TLR signaling pathways leading to NF-κB activation are tightly controlled by post-translational modifications, such as phosphorylation and ubiquitination (Deribe et al. 2010; Wertz and Dixit 2010). EBV encodes lytic proteins that interfere with these modifications. For example, phosphorylation of the NF-κB coactivator UXT is targeted by BGLF4 to suppress NF-κB activity (Chang et al. 2012). BPLF1, the EBV homolog of a conserved herpesvirus deubiquitinase, reverses ubiquitination of several TLR signaling intermediates (van Gent et al. 2014). This inhibits NF-κB activation and cytokine production following TLR stimulation (Saito et al. 2013; van Gent et al. 2014), and promotes viral genome replication (Saito et al. 2013). Being a component of the EBV tegument, BPLF1 could act both in productively as well as in newly infected cells (Johannsen et al. 2004; van Gent et al. 2014).

### 2.1.4 Interference with Innate Effector Molecules

The abovementioned transcription factors induced upon PRR engagement greatly alter cellular gene expression to effectuate diverse effector mechanisms. Among these is the secretion of effector molecules, such as proinflammatory cytokines and interferons, which act in autocrine and paracrine ways. A number of EBV gene products interfere with the function of these innate effector molecules.

EBV counteracts the pleiotropic host cytokine colony-stimulating factor 1 (CSF-1), which stimulates macrophage differentiation and IFNα secretion. To this end, EBV encodes a soluble form of the CSF-1 receptor, BARF1, that neutralizes the effects of host CSF-1 in vitro, leading to reduced IFNα secretion by EBV-infected mononuclear cells (Cohen and Lekstrom 1999; Strockbine et al. 1998). Mutating the BARF1 homolog in a related rhesus macaque lymphocryptovirus decreases viral load during primary infection and leads to a lower persistence set point in vivo (Ohashi et al. 2012).

EBV BZLF1 counteracts innate effector molecules in several ways. First, BZLF1 downregulates the receptors for TNFα and IFNγ to reduce cellular responsiveness to these cytokines (Bristol et al. 2010; Morrison et al. 2001, 2004). Second, BZLF1 induces the suppressor of cytokine signaling SOCS3, which inhibits JAK/STAT signaling and thereby favors a state of type I IFN-irresponsiveness

(Michaud et al. 2010). Additionally, SOCS3 reduces IFNα production by monocytes. Third, BZLF1 causes expression of the immunosuppressive cytokine TGFβ (Cayrol and Flemington 1995) and disrupts the formation of PML-bodies (Adamson and Kenney 2001), which can have antiviral activity (Saffert and Kalejta 2008).

The early-expressed EBV-encoded dUTPase (encoded by the BLLF3 gene) also modulates cytokine-induced responses. In a mouse model, dUTPase compromises lymphocyte responses, e.g., secretion of IFNγ (Padgett et al. 2004). In human cells, EBV dUTPase has seemingly opposing effects: It induces NF-κB activation in a TLR2/MyD88-dependent way (Ariza et al. 2009, 2013; Waldman et al. 2008); it inhibits lymphocyte proliferation; and it induces production of both proinflammatory cytokines as well as IL-10 (Brooks et al. 2006; Glaser et al. 2006). Following this strategy, EBV appears to exploit the advantageous effects of NF-κB activation, while limiting ensuing antiviral T-cell responses.

## *2.2 Evasion of Adaptive Immunity*

Adaptive antiviral immunity relies on a powerful, memory-based response of virus-specific B- and T lymphocytes. Once EBV infection has been established, the virus resides intracellularly most of the time. Detection and elimination of EBV-infected cells depends to a large extent on T cells, which recognize virus-derived peptides in the context of surface HLA molecules. Primary EBV infection induces strong, virus-specific T-cell responses targeting both lytic and latent EBV-derived epitopes, described in detail by Hislop and Taylor (CTMI 2015 in press). EBV compromises activation of both $CD8^+$ and $CD4^+$ T cells by interfering at various stages of the HLA class I and class II antigen presentation pathways, in particular during the productive phase of infection (Fig. 2 and Table 1).

### 2.2.1 Evasion of $CD8^+$ T Lymphocytes

As nucleated human cells all express HLA class I molecules (HLA I), also EBV-infected cells will be targeted by antiviral $CD8^+$ T cells. To prevent recognition by EBV-specific (memory) T cells, EBV encodes at least three proteins that independently interfere with antigen presentation through downregulating the surface expression of HLA I in distinct ways. These three viral proteins are expressed early during the replicative cycle of EBV and act in concert with prevent recognition by $CD8^+$ T cells (Quinn et al. 2014; Ressing et al. 2008b).

General interference by BGLF5

Degradation of HLA I-encoding mRNAs by EBV BGLF5 reduces peptide presentation at the cell surface and, in turn, inhibits T-cell recognition (Rowe et al. 2007a; Zuo et al. 2008). When BGLF5 was expressed in isolation, T-cell activation

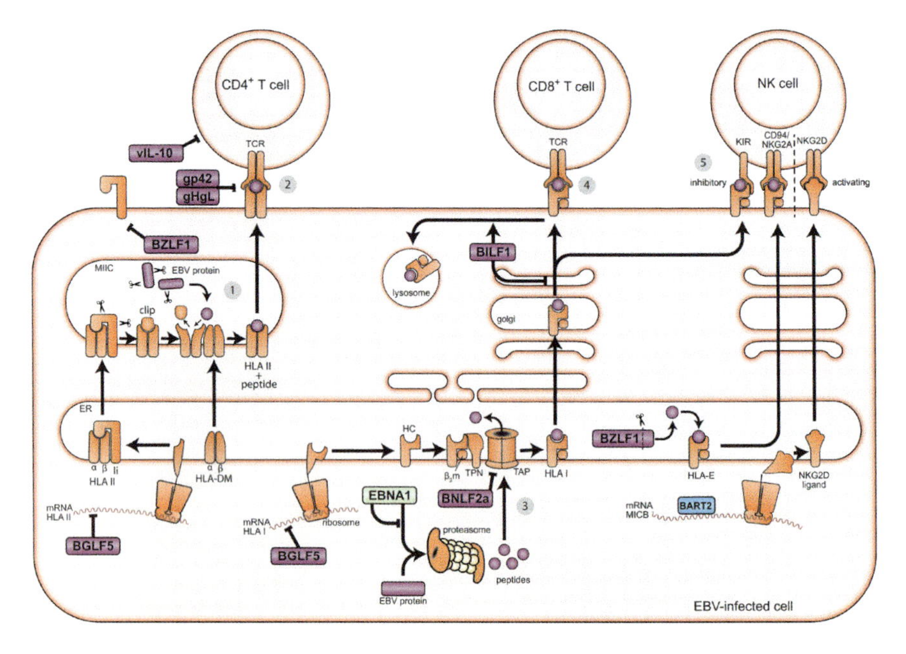


**Fig. 2** Cellular immune responses that are targeted by EBV. Multiple EBV-derived epitopes are presented in the context of HLA molecules to be scrutinized by specific T cells. Presentation to CD4$^+$ T cells (2) occurs by HLA II$^+$ antigen-presenting cells, where EBV proteins are degraded in MHC class-II-loading compartments (MIIC) and the resulting peptides are loaded onto HLA II molecules (1). For antigen presentation to CD8$^+$ T cells (4), cytosolic EBV proteins, or fragments thereof, are degraded by the proteasome into peptides that are transported across the ER membrane by the transporter associated with antigen presentation (TAP) and are subsequently loaded onto newly synthesized HLA I molecules (3); mature HLA I/peptide complexes travel through the Golgi compartments to the cell surface. In addition to their role in antigen presentation to CD8$^+$ cells, surface HLA I molecules can provide an inhibitory signal to natural killer (NK) cells (5). NK cells respond to a combination of inhibitory signals (e.g., through HLA-E) and activation signals (e.g., through NKG2D ligands). Latent (*green*), lytic (*purple*) EBV proteins, and EBV miRNAs (*blue*) interfere at various steps with activation of the cellular immune response

was decreased by 90 %. During the lytic cycle, silencing of BGLF5 expression by 75 % using shRNAs reduced, but not completely blocked, HLA I expression (van Gent et al. 2014), indicating that other EBV gene products contribute to HLA I downregulation. This role is fulfilled by BNLF2a and BILF1.

### Inhibition of TAP by BNLF2a

BNLF2a is a gene product unique to lymphocryptoviruses of Old World primates. BNLF2a was found to be strictly membrane-associated, although the protein lacks an N-terminal signal sequence. Through post-translational membrane insertion of its hydrophobic C-terminus, BNLF2a has acquired an alternative way of integrating itself into ER membranes (Horst et al. 2011). Expression of BNLF2a in isolation or in the context of EBV infection results in reduced CD8$^+$ T-cell recognition.

BNLF2a appears to deplete peptides from the ER (the HLA I loading compartment) through inhibition of peptide import by the transporter associated with antigen presentation (TAP) (Hislop et al. 2007; Horst et al. 2009). In vitro infection with BNLF2a-deleted recombinant EBV restores T-cell recognition of peptides expressed by these cells early after viral reactivation (Croft et al. 2009). The mechanism of action of BNLF2a is exceptional among viral TAP inhibitors known to date. BNLF2a corrupts the binding of both peptides and ATP to the TAP complex, thereby blocking its transporter function and, ultimately, surface display of peptide/HLA complexes.

Downregulation of surface HLA class I by BILF1

The third molecule of EBV interfering with HLA I function is BILF1, encoding a constitutively active G-protein-coupled receptor (GPCR). The GPCR-signaling function is not required for downregulation of HLA I (Zuo et al. 2009). The underlying mechanism involves reduced transport of HLA I from the trans-Golgi network, as well as an increased turnover from the cell surface and, subsequently, enhanced degradation via lysosomal proteases (Zuo et al. 2009). This molecular mechanism is distinct from the ones identified for other viruses that induce degradation of HLA I. The cytoplasmic C-termini of both BILF1 and its targets are critical for HLA I downregulation. Most HLA I haplotypes are downregulated by BILF1, yet HLA-C alleles appear resistant (Griffin et al. 2013); the latter could deviate NK cells.

Temporal expression of HLA I evasion proteins

The relative contribution of these three EBV proteins to HLA I downregulation differs during the IE, E, and L phases of the EBV lytic cycle (Quinn et al. 2014). Knockdown of BNLF2a in donor LCLs primarily results in reduced activation of $CD8^+$ T cells specific for IE and E antigens, while BILF1 knockdown increases recognition of E and especially L antigens. Contrary to observations in overexpression studies, reducing BGLF5 expression displays limited effects on antigen recognition in any of the phases (Quinn et al. 2014). Timing of expression partially explains these differences, although some synergy between BNLF2a and BILF1 is also observed in reducing late antigen recognition.

### 2.2.2 Evasion of $CD4^+$ T Lymphocytes

EBV interference with HLA class-II-restricted antigen presentation

HLA class-II-restricted $CD4^+$ T cells are an essential component of the adaptive immune response against EBV, especially since EBV infects B lymphocytes—cells that express high levels of HLA II. EBV-specific memory T-helper responses can eliminate virus-producing cells either directly through their cytotoxic capacity or indirectly through activation of $CD8^+$ cytotoxic T cells and B cells. Indeed, following lytic EBV infection, $CD4^+$ T cells against lytic antigens are readily detectable in the peripheral blood of infected individuals (Mackay et al. 2009). These responses may be generated by direct recognition of EBV-infected B cells, but can also be primed via cross-presentation of exogenous antigens taken up and presented by professional APCs such as dendritic cells. The importance of CD4$^+$ T-cell responses in controlling EBV infection is well illustrated by the observation that increased frequencies of EBV-associated malignancies occur in individuals with a defective CD4$^+$ T-cell compartment, such as HIV-infected individuals (Petrara et al. 2013). In response, EBV adopted a number of immune evasion strategies that interfere with CD4$^+$ T-cell immunity at multiple levels.

Interference with HLA class-II-restricted antigen presentation by gp42/gH/gL

Gp42 has initially been described as an entry receptor for EBV, binding to HLA class II molecules present on B cells. Additionally, gp42 acts as an immune evasion molecule. Its association with HLA class II/peptide complexes blocks T-cell receptor (TCR)—class II interactions and precludes activation of CD4$^+$ T cells (Ressing et al. 2003). EBV gp42 occurs in two forms: a full-length type II membrane protein and a truncated soluble form, generated upon proteolytic cleavage. Both forms are capable of providing immune-evasive properties, suggesting that secreted gp42 may block cross-presentation in addition to endogenous antigen presentation (Ressing et al. 2005). Whereas downregulation of HLA II occurs during productive EBV infection, this effect is not observed upon expression of gp42 in isolation (Ressing et al. 2005), indicating that EBV employs additional HLA II evasion strategies. We recently found that the viral interaction partners of gp42, gH, and gL cooperate to increase HLA II evasion (our unpublished observation). In line with this, T-cell activation was further diminished by additional inclusion of gH and gL in the gp42-HLA II complexes. Mechanistically, the major effect of gH/gL appeared to be stabilization and increased expression of gp42 (our unpublished observation).

Downregulation of HLA class II by EBV

In addition to the late protein gp42/gH/gL, the early host shutoff protein BGLF5 decreases cell-surface HLA II by degradation of HLA II mRNAs (Rowe et al. 2007b). In addition to these ORFs that directly impair HLA II recognition, other EBV gene products indirectly interfere with CD4$^+$ T-cell immunity. The immediate-early protein BZLF1 has been reported to impair IFN$\gamma$-signaling, thereby inhibiting CIITA promoter activity and, as a result, decreasing HLA II surface levels (Morrison et al. 2001). More recently, BZLF1 has been shown to impair HLA II presentation post-transcriptionally by interfering with the function of the invariant chain (Zuo et al. 2011). In this study, expression of BZLF1 in isolation resulted in approximately 50 % reduction in CD4$^+$ T-cell recognition. EBV also encodes a viral IL-10 homolog (BCRF1). IL-10 serves as an anti-inflammatory cytokine that is able to inhibit and modulate CD4$^+$ T-cell priming and effector functions; BCRF1 has been suggested to inhibit antiviral CD4$^+$ T-cell responses, similar to (host) IL-10 (Brooks et al. 2006). Moreover, BCRF1 has been shown to inhibit co-stimulatory molecules on human monocytes, which potentially results in inefficient priming and expansion of CD4$^+$ T cells (Salek-Ardakani et al. 2002).

#### 2.2.3 Interplay of Evasion Strategies Targeting the Adaptive Immune System

In conclusion, EBV has evolved multiple layers of immune evasion that interfere with the recognition of infected cells by $CD8^+$ and $CD4^+$ T cells. This wide range of evasion mechanisms explains how EBV can replicate and establish a lifelong infection of its host, despite the existence of strong $CD8^+$ and $CD4^+$ T-cell immunity against a broad repertoire of EBV antigens. Notably, the discovery of multiple EBV lytic cycle genes that cooperate to interfere with HLA class I and class II antigen processing underscores the need for EBV to evade $CD8^+$ and $CD4^+$ T-cell responses during replication, a time at which a large number of potential viral targets are expressed (Ressing et al. 2008a). Together, these immune evasion strategies ensure a window for undetected replication of EBV. Additionally, these evasion mechanisms facilitate the establishment and maintenance of a lifelong infection of the host.

## 3 Immune Evasion During Latency

EBV severely limits viral protein expression during latent infection to avoid recognition by the host immune system. Different forms of latency reflect the various stages leading from primary infection of the naive B cell to growth transformation. Thus, in latency III cells, EBNA1-6 and LMP1 and 2 are expressed. During latency II, expression is restricted to EBNA1 and LMP1 and 2. Latency I involves expression of EBNA1 only, and latency 0 exists without any EBV protein expression. These latency stages are represented in the various forms of cancer associated with EBV, namely post-transplant lymphomas (latency III), Hodgkin's disease and nasopharyngeal carcinoma (both latency II), and Burkitt's lymphoma (latency I). Immunomodulatory functions have been ascribed to several individual EBV proteins expressed during the various latency stages (Figs. 1 and 2; Table 2).

### *3.1 EBNA1*

EBNA1, expressed during all latency stages, contains a long glycine–alanine repeat that inhibits translation as well as proteasomal degradation of EBNA1 through interference with processing by the 19S proteasomal subunit (Apcher et al. 2009; Blake et al. 1997; Levitskaya et al. 1995, 1997; Yin et al. 2003). This strategy ensures sufficient EBNA1 levels to maintain the viral genome (Hochberg et al. 2004), while decreasing protein turnover to minimize viral antigen presentation to $CD8^+$ T cells. Initially, EBNA1-specific $CD8^+$ T-cell responses were indeed not observed in vitro (Blake et al. 1997). However, later studies did report EBNA1-specific T-cell responses initiated by endogenously presented

**Table 2** Latent EBV gene products interfering with host immunity

| Gene product | Functional mechanism(s) | References |
|---|---|---|
| EBNA1 | Inhibits its own translation and proteasomal degradation, inhibits canonical NF-κB pathway by interfering with IKK phosphorylation, modulates STAT1 and TGFβ signaling pathways | Apcher et al. (2009); Blake et al. (1997); Levitskaya et al. (1995); Yin et al. (2003); Hochberg et al. (2004); Valentine et al. (2010); Wood et al. (2007) |
| EBNA2 | Induces low level of IFNβ production, inhibits production of selective ISGs, enhances STAT3 activity, upregulates IL-18 receptor | Aman and von Gabain (1990); Kanda et al. (1992, 1999); Muromoto et al. (2009); Pagès et al. (2004) |
| LMP1 | Promotes B-cell growth and survival by mimicking CD40 signaling, induces type I IFN production, induces STAT1 and STAT2 activity, upregulates IRF7, reduces TLR9 expression | Kieser et al. (2007), Middeldorp and Pegtel (2008), Najjar et al. (2005), Richardson et al. (2003), Xu et al. (2006), Geiger and Martin (2006), Leslie et al. (2007), Ning et al. (2008), Zhang et al. (2001, 2004), Zhang and Pagano (2000), Zhao et al. (2010), Fathallah et al. (2010), Cahir-McFarland et al. (2000), Izumi and Kieff (1997), Gires et al. (1999) |
| LMP2a | Inhibits NF-κB activity, IL-6 production and subsequent JAK/STAT signaling in epithelial cells, induces NF-κB activity to induce anti-apoptotic Bcl-2, accelerates turnover of IFN receptors | Stewart et al. (2004), Swanson-Mungerson et al. (2010), Shah et al. (2009) |
| LMP2b | Accelerates turnover of IFN receptors | Shah et al. (2009) |

EBNA1-derived antigens (Lee et al. 2004; Tellam et al. 2004; Voo et al. 2004). Potential sources of these antigens include defective ribosomal products that lack the glycine–alanine repeat, or cross-presented exogenous antigens released by EBV-infected cells.

Other immune-evasive actions of EBNA1 include inhibition of the canonical NF-κB pathway by interfering with phosphorylation of the IKK complex signaling intermediate (Valentine et al. 2010) and modulation of the STAT1 and TGFβ signaling pathways (Wood et al. 2007).

## 3.2 EBNA2

EBNA2 applies a double-edged strategy by inducing low-level IFNβ production that leads to interferon-stimulated gene (ISG) production in BL cell lines (Kanda et al. 1999), whereas anti-proliferative effects are neutralized by EBNA2-mediated inhibition of selected ISGs (Aman and von Gabain 1990; Kanda et al. 1992) and enhanced transcriptional activity of STAT3 (Muromoto et al. 2009) following IFNα production. STAT3 modulates IFN-induced immune responses through STAT1 and suppresses production of inflammatory mediators (Ho and Ivashkiv 2006). In addition, EBNA2 upregulates the IL-18 receptor on BL cells (Pagès et al. 2004). IL-18 plays a role in regulating innate and adaptive immune responses and is elevated in certain EBV-associated malignancies (van de Veerdonk et al. 2012).

## 3.3 LMP1

LMP1 promotes B-cell growth and survival by mimicking constitutive CD40 signaling to activate NF-κB, JNK, MAPK, JAK/STAT, and PI3K signaling pathways (Kieser 2007). These pathways affect many immunological processes and allow LMP1 to steer the host immune response (reviewed in (Middeldorp and Pegtel 2008)). LMP1-mediated NF-κB activation in EBV-immortalized B cells results in type I IFN production that stimulates STAT1 expression in autocrine and paracrine fashion (Najjar et al. 2005; Richardson et al. 2003; Xu et al. 2006). STAT2 activity is inhibited by LMP1 (Geiger and Martin 2006). LMP1-mediated upregulation of IRF7 benefits EBV by promoting cell growth, while at the same time an inhibitory IRF7 splice variant is induced to repress the adverse effects of type I IFN production (Leslie et al. 2007; Ning et al. 2008; Zhang et al. 2001; Zhang and Pagano 2000; Zhao et al. 2010). Furthermore, LMP1-mediated induction of JAK/STAT signaling pathways may be advantageous to EBV as the antiviral activities of ISGs prevent superinfection and facilitate establishment of latency (Gires et al. 1999; Zhang et al. 2004). Finally, LMP-1-mediated NF-κB activation reduces TLR9 surface expression (Fathallah et al. 2010) and supplies growth benefits to infected cells (Cahir-McFarland et al. 2000; Izumi and Kieff 1997).

## 3.4 LMP2a + 2b

LMP2a inhibits NF-κB activity, IL-6 production, and subsequent JAK/STAT signaling pathways in carcinoma cell lines (Stewart et al. 2004). In contrast, LMP2a induces NF-κB activation in B cells and uses the subsequently increased levels of anti-apoptotic Bcl-2 to protect infected cells from apoptosis in a transgenic mouse model (Swanson-Mungerson et al. 2010). Furthermore, LMP2a and LMP2b accelerate turnover of IFN receptors, resulting in decreased responsiveness of epithelial cells to IFNα and IFNγ (Shah et al. 2009).

**Table 3** EBV noncoding RNAs interfering with host immunity

| miRNA | Functional mechanism(s) | References |
|---|---|---|
| EBERs | Inhibit PKR activity | Lerner et al. (1981), Clarke et al. (1991), Komano et al. (1999), Nanbo et al. (2002, 2005), Ruf et al. (2005) |
| miR-BHRF1–3 | Downregulates T cell attracting chemokine CXCL-11 | Xia et al. (2008) |
| miR-BART2-5p | Inhibition of this miRNA results in increased NK-cell killing in vitro | Nachmani et al. (2009) |
| miR-BART15 | Downregulates NLRP3 | Haneklaus et al. (2012) |
| miR-BART20-5p | Downregulates IFNγ transcriptional regulator T-bet and IFNγ | Huang and Lin (2014), Lin et al. (2013) |
| miR-BART8 | Inhibits STAT1 expression | Huang and Lin (2014) |
| miR-BART6-3p | Downregulation of IL-6 receptor chains | Ambrosio et al. (2014) |

# 4 EBV Noncoding RNAs

EBV encodes various forms of noncoding RNAs, including two EBV-encoded small RNAs (EBERs) and over forty miRNAs. The functions of these noncoding RNAs are now beginning to be elucidated and several have been implicated in immune evasion strategies (Figs. 1 and 2; Table 3).

## *4.1 EBERs*

Two approximately 170 nucleotide long RNA molecules called EBER-1 and EBER-2 are expressed during each of the latency stages (Lerner et al. 1981). The EBERs bind dsRNA-dependent protein kinase R (PKR) in vitro (Clarke et al. 1991) and inhibit apoptosis, but it is still a subject of debate whether suppression of IFNα-induced apoptosis by the EBERs is due to inhibition of PKR (Komano et al. 1999; Nanbo et al. 2002, 2005; Ruf et al. 2005).

## *4.2 EBV miRNAs*

EBV expresses a number of miRNAs, named the BART and BHRF1 miRNAs (reviewed by Cullen et al., CTMI 2015, in press) (Cai et al. 2006; Grundhoff et al. 2006; Pfeffer et al. 2004; Zhu et al. 2009). These miRNAs are organized in clusters and are expressed during all latency stages (and thus in EBV-associated tumors), although their expression levels differ between these stages (Amoroso et al. 2011; Motsch et al. 2012; Qiu et al. 2011). miRNAs have been detected in EBV virions, allowing their release in newly infected cells (Jochum et al. 2012a). Moreover, EBV miRNAs can be transferred via exosomes from EBV-infected cells to uninfected recipient cells in vitro and may in this way regulate uninfected cells and cell types not typically infected by EBV (Meckes et al. 2010; Pegtel et al. 2010). Although our knowledge on the function of the EBV miRNAs is limited, knockout EBV strains have indicated a role for the BHRF1 miRNAs and, to a lesser extent, for the BART miRNAs in the early phase of B-cell transformation (Feederle et al. 2011; Seto et al. 2010; Vereide et al. 2014). Experiments in humanized mice showed that the BHRF1 miRNAs were not crucial for infection, but knockout virus did show delayed kinetics compared to the wild-type virus (Wahl et al. 2013). In a xenograft mouse model, expression of the BART miRNAs appeared to provide a tumor cell growth advantage (Qiu et al. 2015). Apart from these growth-transforming and anti-apoptotic functions, EBV miRNAs target several host genes involved in antiviral immunity.

Among the first EBV miRNA targets identified was CXCL-11, a T cell attracting chemokine downregulated by EBV miRNAs BHRF1-3 (Xia et al. 2008). Stress-induced NK-cell ligands have been specifically investigated as potential viral miRNA targets. Initially, MICB was identified as a target of the HCMV-encoded miR-UL112, and later, it became apparent that also miRNAs encoded by KSHV and EBV downregulate MICB expression. Inhibiting EBV miRNA BART2-5p results in increased NK-cell killing in vitro (Nachmani et al. 2009). Inflammasomes are induced by various cytoplasmic and nuclear sensors (e.g., NLRP3 and IFI16) and lead to production of the inflammatory cytokines IL-1β and IL-18 (Rathinam and Fitzgerald 2010). Although EBV has so far only been observed to activate inflammasomes through IFI16 (Ansari et al. 2013), EBV miRNA BART15 downregulates the alternative inflammasome-activating sensor NLRP3 (Haneklaus et al. 2012). Co-culture of monocytic recipient cells with $EBV^+$ B cells secreting BART15-containing exosomes results in decreased IL-1β production. Additionally, EBV miRNAs regulate the IFNγ-STAT1 pathway in $EBV^+$ NK cells by downregulating IFNγ transcriptional regulator T-bet (BART20-5p), IFNγ (BART20-5p), and STAT1 (BART8) (Huang and Lin 2014; Lin et al. 2013). Inhibition of BART6-3p in a BL cell line caused upregulation of the IL-6 receptor chains (p80 and gp130) at the mRNA and protein level, indicating that BART6-3p may affect IL-6 signaling (Ambrosio et al. 2014).

Technological developments have greatly aided the search for miRNA targets. A number of RISC immunoprecipitation screens on EBV-infected B cells have generated a long list of putative miRNA targets that await functional validation, including C-type lectin receptors and PML body components (Dolken et al. 2010; Gottwein et al. 2011; Kuzembayeva et al. 2012; Riley et al. 2012; Skalsky et al. 2012; Vereide et al. 2014).

## 5 Conclusion and Future Perspectives

To date, several EBV-encoded proteins have been identified that interfere with innate and adaptive immune responses during the lytic and latent phase of infection. More recently, also EBV miRNAs have been implicated in modifying expression levels of genes relevant to antiviral immunity. These findings provide interesting, new insights into the biology of this important human pathogen.

The elusion of $CD4^+$ and $CD8^+$ T-cell immunity is a well-studied evasion strategy that is appreciated as an important mechanism successfully used by EBV to establish a lifelong infection of its host. Additionally, immune cells like iNKT cells, a specialized subset of T cells, and NK cells have been implicated in the control of EBV infection (Azzi et al. 2014; Chijioke et al. 2013; Pappworth et al. 2007; Pasquier et al. 2005; Rigaud et al. 2006).

Absence or reduced numbers of iNKT cells are associated with fatal EBV infection, while patients with selective defects in NK cells frequently have recurrent EBV-associated disease (Munz et al. 2000; CTMI 2015 in press). The iNKT cells can be activated via the non-classical MHC molecule CD1d presenting lipid antigens. Interestingly, CD1d expression is downregulated in B cells that have been infected with EBV in vitro (Chung 2013). KSHV and HSV-1 downregulate CD1d expression at the cell surface by different mechanisms during lytic infection (Rao et al. 2011; Sanchez et al. 2005). A recent study by van Gent et al. (2015) indicates that EBV BGLF5 reduces the expression of CD1d in the context of lytic replication of EBV in vitro.

Activation of NK cells is determined by the balance between inhibitory and activating signals. NKG2D is an activating receptor that has a role in controlling chronic EBV infection (Chaigne-Delalande et al. 2013). Induction of the EBV lytic cycle increases NKG2D-dependent NK-cell reactivity (Pappworth et al. 2007). These findings, however, do not preclude that EBV also encodes NK evasion molecules to protect lytic EBV-infected cells from an NK-cell attack. For example, BZLF1 contains an HLA-E-binding peptide that increases surface expression of HLA-E (Ulbrecht et al. 1998), which is a ligand for the inhibitory receptor NKG2A. Therefore, increased HLA-E surface expression may restrict NK-cell reactivity.

Several other herpesviruses have been found to interfere with additional immune signaling and effector pathways, including complement activation and inflammasome-dependent responses. As of yet, no such mechanisms have been identified for EBV, but it seems plausible that EBV employs such strategies as well. None of the EBV proteins shares homology to complement inhibitors identified for other herpesviruses. Still, EBV particles have been found to interfere with complement activation (Mold et al. 1988). Hence, identification of the EBV-encoded complement inhibitor(s) might reveal a novel complement evasion strategy.

Recent discoveries in the field of innate immunity, such as the identification of new cytoplasmic and nuclear DNA-sensing molecules, shed light on the detection of EBV infection. Upon stimulation, the DNA-sensing molecule IFI16 can induce the formation of inflammasomes or can trigger an IFN I response. In latently EBV-infected cells, IFI16 is constitutively active, causing IL-1β production (Ansari et al. 2013). Currently, it is unknown whether EBV triggers other inflammasome-forming or DNA-sensing molecules, or whether EBV interferes with inflammasome function, as has been observed for other herpesviruses.

Another important aspect not extensively explored yet is the in vivo relevance and relative contribution of the immune evasion strategies discussed here. Studying in vivo EBV infection is hampered by the strict host-specificity of this human pathogen, especially with respect to interactions with the immune system, and the concomitant lack of suitable animal models. Valuable information on the immune response to EBV has been gained from studies on related murine or rhesus macaque-specific herpesviruses in their respective hosts (Stevenson et al. 2009; Wang 2013). Mice with humanized immune system components (HIS mice) provide a valuable alternative, recapitulating several important aspects of EBV-specific immunity (Chatterjee et al. 2014; Munz 2014; Strowig et al. 2009; Traggiai et al. 2004). Future infection studies with mutant EBV strains in these model systems will provide important information on the contribution of immune evasion strategies to viral infection and pathogenesis.

In summary, this overview of immune evasion mechanisms of EBV shows that this virus employs a wide range of strategies to compromise innate and adaptive immunity during the latent and replicative phases of its life cycle. These evasion mechanisms are likely to contribute to the development of EBV-associated malignancies. A profound understanding of the immune-evasive maneuvres of EBV will aid in the development of novel strategies to combat EBV infections and associated diseases, especially the malignancies caused by this oncogenic human herpesvirus.

## References

Adamson AL, Kenney S (2001) Epstein-Barr virus immediate-early protein BZLF1 Is SUMO-1 modified and disrupts promyelocytic leukemia bodies. J Virol 75:2388–2399

Aman P, von Gabain A (1990) An Epstein-Barr virus immortalization associated gene segment interferes specifically with the IFN-induced anti-proliferative response in human B-lymphoid cell lines. The EMBO J 9:147–152

Ambrosio MR, Navari M, Di LL, Leon EA, Onnis A, Gazaneo S, Mundo L, Ulivieri C, Gomez G et al (2014).The Epstein Barr-encoded BART-6-3p microRNA affects regulation of cell growth and immuno response in Burkitt lymphoma. Infect Agent Cancer 9:12

Amoroso R, Fitzsimmons L, Thomas WA, Kelly GL, Rowe M, Bell AI (2011) Quantitative studies of Epstein-Barr virus-encoded microRNAs provide novel insights into their regulation. J Virol 85:996–1010

Ansari MA, Singh VV, Dutta S, Veettil MV, Dutta D, Chikoti L, Lu J, Everly D, Chandran B (2013) Constitutive interferon-inducible protein 16-Inflammasome activation during Epstein-Barr virus latency I, II, and III in B and epithelial cells. J Virol 87:8606–8623

Apcher S, Komarova A, Daskalogianni C, Yin Y, Malbert-Colas L, Fahraeus R (2009) mRNA translation regulation by the Gly-Ala repeat of Epstein-Barr virus nuclear antigen 1. J Virol 83:1289–1298

Ariza ME, Glaser R, Kaumaya PTP, Jones C, Williams MV (2009) The EBV-encoded dUTPase activates NF-kB through the TLR2 and MyD88-dependent signaling pathway. J Immunol 182:851–859

Ariza ME, Rivailler P, Glaser R, Chen M, Williams MV (2013) Epstein-Barr virus encoded dUTPase containing exosomes modulate innate and adaptive immune responses in human dendritic cells and peripheral blood mononuclear cells. PLoS ONE 8:e69827

Azzi T, Lunemann A, Murer A, Ueda S, Béziat V, Malmberg KJ, Staubli G, Gysin C, Berger C et al (2014) Role for early-differentiated natural killer cells in infectious mononucleosis. Blood 124:2533–2543

Bentz GL, Liu R, Hahn AM, Shackelford J, Pagano JS (2010) Epstein-Barr virus BRLF1 inhibits transcription of IRF3 and IRF7 and suppresses induction of interferon-ß. Virology 402:121–128

Blake N, Lee S, Redchenko I, Thomas W, Steven N, Leese A, Steigerwald-Mullen P, Kurilla MG, Frappier L et al (1997) Human CD8+ T Cell responses to EBV EBNA1: HLA class I presentation of the (Gly-Ala)-containing protein requires exogenous processing. Immunity 7:791–802

Bristol JA, Robinson AR, Barlow EA, Kenney SC (2010) The Epstein-Barr virus BZLF1 protein inhibits tumor necrosis factor receptor 1 expression through effects on cellular C/EBP proteins. J Virol 84:12362–12374

Brooks DG, Trifilo MJ, Edelmann KH, Teyton L, McGavern DB, Oldstone MBA (2006) Interleukin-10 determines viral clearance or persistence in vivo. Nat Med 12:1301–1309

Cahir-McFarland ED, Davidson DM, Schauer SL, Duong J, Kieff E (2000) NF-kB inhibition causes spontaneous apoptosis in Epstein-Barr virus-transformed lymphoblastoid cells. Proc Natl Acad Sci 97:6055–6060

Cai X, Schafer A, Lu S, Bilello JP, Desrosiers RC, Edwards R, Raab-Traub N, Cullen BR (2006) Epstein-Barr virus microRNAs are evolutionarily conserved and differentially expressed. PLoS Pathog 2:e23

Cayrol C, Flemington EK (1995) Identification of cellular target genes of the Epstein-Barr virus transactivator Zta: activation of transforming growth factor beta igh3 (TGF-beta igh3) and TGF-beta 1. J Virol 69:4206–4212

Chaigne-Delalande B, Li FY, O'Connor GM, Lukacs MJ, Jiang P, Zheng L, Shatzer A, Biancalana M, Pittaluga S et al (2013) $Mg2^{+}$ regulates cytotoxic functions of NK and CD8 T cells in chronic EBV infection through NKG2D. Science 341:186–191

Chang LS, Wang JT, Doong SL, Lee CP, Chang CW, Tsai CH, Yeh SW, Hsieh CY, Chen MR (2012) Epstein-Barr virus BGLF4 kinase downregulates NF-kB transactivation through phosphorylation of coactivator UXT. J Virol 86:12176–12186

Chatterjee B, Leung CS, Münz C (2014) Animal models of Epstein Barr virus infection. J Immunol Methods 410:80–87

Chijioke O, Muller A, Feederle R, Barros M, Krieg C, Emmel V, Marcenaro E, Leung CS, Antsiferova O et al (2013) Human natural killer cells prevent infectious mononucleosis features by targeting lytic Epstein-Barr virus infection. Cell Rep 5:1489–1498

Chung BKT (2013) Innate immune control of EBV-infected B cells by invariant natural killer T cells. Blood 122:2600–2608

Clarke PA, Schwemmle M, Schickinger J, Hilse K, Clemens MJ (1991) Binding of Epstein-Barr virus small RNA EBER-1 to the double-stranded RNA-activated protein kinase DAI. Nucleic Acids Res 19:243–248

Cohen JI, Lekstrom K (1999) Epstein-Barr virus BARF1 protein is dispensable for B-cell transformation and inhibits alpha interferon secretion from mononuclear cells. J Virol 73:7627–7632

Croft NP, Shannon-Lowe C, Bell AI, Horst D, Kremmer E, Ressing ME, Wiertz EJHJ, Middeldorp JM, Rowe M et al (2009) Stage-specific inhibition of MHC class I presentation by the Epstein-Barr virus BNLF2a protein during virus lytic cycle. PLoS Pathog 5:e1000490

Deribe YL, Pawson T, Dikic I (2010) Post-translational modifications in signal integration. Nat Struct Mol Biol 17:666–672

Dolken L, Malterer G, Erhard F, Kothe S, Friedel CC, Suffert G, Marcinowski L, Motsch N, Barth S et al (2010) Systematic analysis of viral and cellular microRNA targets in cells latently infected with human gamma-herpesviruses by RISC immunoprecipitation assay. Cell Host Microbe 7:324–334

Dreyfus DH, Nagasawa M, Pratt JC, Kelleher CA, Gelfand EW (1999) Inactivation of NF-kB by EBV BZLF-1-encoded ZEBRA protein in human T cells. J Immunol 163:6261–6268

Dunmire SK, Odumade OA, Porter JL, Reyes-Genere J, Schmeling DO, Bilgic H, Fan D, Baechler EC, Balfour HH et al (2014) Primary EBV infection induces an expression profile distinct from other viruses but similar to hemophagocytic syndromes. PLoS One 9:e85422

Fathallah I, Parroche P, Gruffat H, Zannetti C, Johansson H, Yue J, Manet E, Tommasino M, Sylla BS et al (2010) EBV latent membrane protein 1 is a negative regulator of TLR9. J Immunol 185:6439–6447

Feederle R, Linnstaedt SD, Bannert H, Lips H, Bencun M, Cullen BR, Delecluse HJ (2011) A viral microRNA cluster strongly potentiates the transforming properties of a human herpesvirus. PLoS Pathog 7:e1001294

Gaudreault E, Fiola S, Olivier M, Gosselin J (2007) Epstein-Barr virus induces MCP-1 secretion by human monocytes via TLR2. J Virol 81:8016–8024

Geiger TR, Martin JM (2006) The Epstein-Barr virus-encoded LMP-1 oncoprotein negatively affects Tyk2 phosphorylation and interferon signaling in human B cells. J Virol 80:11638–11650

Gires O, Kohlhuber F, Kilger E, Baumann M, Kieser A, Kaiser C, Zeidler R, Scheffer B, Ueffing M et al (1999) Latent membrane protein 1 of Epstein-Barr virus interacts with JAK3 and activates STAT proteins. EMBO J 18:3064–3073

Glaser R, Litsky ML, Padgett DA, Baiocchi RA, Yang EV, Chen M, Yeh PE, Green-Church KB, Caligiuri MA et al (2006) EBV-encoded dUTPase induces immune dysregulation: implications for the pathophysiology of EBV-associated disease. Virology 346:205–218

Gottwein E, Corcoran DL, Mukherjee N, Skalsky RL, Hafner M, Nusbaum JD, Shamulailatpam P, Love CL, Dave SS et al (2011) Viral microRNA targetome of KSHV-infected primary effusion lymphoma cell lines. Cell Host Microbe 10:515–526

Griffin BD, Gram AM, Mulder A, van Leeuwen D, Claas FHJ, Wang F, Ressing ME, Wiertz E (2013) EBV BILF1 evolved to downregulate cell surface display of a wide range of HLA class I molecules through their cytoplasmic tail. J Immunol 190:1672–1684

Grundhoff A, Sullivan CS, Ganem D (2006) A combined computational and microarray-based approach identifies novel microRNAs encoded by human gamma-herpesviruses. RNA 12:733–750

Gutsch DE, Holley-Guthrie EA, Zhang Q, Stein B, Blanar MA, Baldwin AS, Kenney SC (1994) The bZIP transactivator of Epstein-Barr virus, BZLF1, functionally and physically interacts with the p65 subunit of NF-kappa B. Mol Cell Biol 14:1939–1948

Hahn AM, Huye LE, Ning S, Webster-Cyriaque J, Pagano JS (2005) Interferon regulatory factor 7 is negatively regulated by the Epstein-Barr virus immediate-early gene, BZLF-1. J Virol 79:10040–10052

Haneklaus M, Gerlic M, Kurowska-Stolarska M, Rainey AA, Pich D, McInnes IB, Hammerschmidt W, O'Neill LA, Masters SL (2012) Cutting edge: miR-223 and EBV miR-BART15 regulate the NLRP3 inflammasome and IL-1beta production. J Immunol 189:3795–3799

Hislop AD, Ressing ME, van Leeuwen D, Pudney VA, Horst D, Koppers-Lalic D, Croft NP, Neefjes JJ, Rickinson AB et al (2007) A CD8+ T cell immune evasion protein specific to Epstein-Barr virus and its close relatives in old world primates. J Exp Med 204:1863–1873

Ho HH, Ivashkiv LB (2006) Role of STAT3 in type I interferon responses: negative regulation of STAT1-dependent inflammatory gene activation. J Biol Chem 281:14111–14118

Hochberg D, Middeldorp JM, Catalina M, Sullivan JL, Luzuriaga K, Thorley-Lawson DA (2004) Demonstration of the Burkitt's lymphoma Epstein-Barr virus phenotype in dividing latently infected memory cells in vivo. Proc Natl Acad Sci 101:239–244

Horst D, van Leeuwen D, Croft NP, Garstka MA, Hislop AD, Kremmer E, Rickinson AB, Wiertz EJHJ, Ressing ME (2009) Specific targeting of the EBV lytic phase protein BNLF2a to the transporter associated with antigen processing results in impairment of HLA class I-restricted antigen presentation. J Immunol 182:2313–2324

Horst D, Favaloro V, Vilardi F, van Leeuwen HC, Garstka MA, Hislop AD, Rabu C, Kremmer E, Rickinson AB et al (2011) EBV protein BNLF2a exploits host tail-anchored protein integration machinery to inhibit TAP. J Immunol 186:3594–3605

Horst D, Burmeister WP, Boer IGJ, van Leeuwen D, Buisson M, Gorbalenya AE, Wiertz EJHJ, Ressing ME (2012) The bridge in the Epstein-Barr virus alkaline exonuclease protein BGLF5 contributes to shutoff activity during productive infection. J Virol 86:9175–9187

Huang WT, Lin CW (2014) EBV-encoded miR-BART20-5p and miR-BART8 inhibit the IFN-gamma-STAT1 pathway associated with disease progression in nasal NK-cell lymphoma. Am J Pathol 184:1185–1197

Izumi KM, Kieff ED (1997) The Epstein-Barr virus oncogene product latent membrane protein 1 engages the tumor necrosis factor receptor-associated death domain protein to mediate B lymphocyte growth transformation and activate NF-kB. Proc Natl Acad Sci 94:12592–12597

Jochum S, Ruiss R, Moosmann A, Hammerschmidt W, Zeidler R (2012a) RNAs in Epstein-Barr virions control early steps of infection. Proc Natl Acad Sci USA 109:E1396–E1404

Jochum S, Moosmann A, Lang S, Hammerschmidt W, Zeidler R (2012b) The EBV immunoevasins vIL-10 and BNLF2a protect newly infected B cells from immune recognition and elimination. PLoS Pathog 8:e1002704

Johannsen E, Luftig M, Chase MR, Weicksel S, Cahir-McFarland E, Illanes D, Sarracino D, Kieff E (2004) Proteins of purified Epstein-Barr virus. Proc Natl Acad Sci USA 101:16286–16291

Kanda K, Decker T, Aman P, Wahlström M, von Gabain A, Kallin B (1992) The EBNA2-related resistance towards alpha interferon (IFN-alpha) in Burkitt's lymphoma cells effects induction of IFN-induced genes but not the activation of transcription factor ISGF-3. Mol Cell Biol 12:4930–4936

Kanda K, Kempkes B, Bornkamm GW, Gabain AV, Decker T (1999) The Epstein-Barr virus nuclear antigen 2 (EBNA2), a protein required for B lymphocyte immortalization, induces the synthesis of type I interferon in burkitts lymphoma cell lines. In: Biological chemistry, vol 380, p. 213

Keating S, Prince S, Jones M, Rowe M (2002) The lytic cycle of Epstein-Barr virus is associated with decreased expression of cell surface major histocompatibility complex class I and class II molecules. J Virol 76:8179–8188

Kieser A (2007) Signal transduction by the Epstein-Barr virus oncogene latent membrane protein 1 (LMP1). Sig Transduct 7:20–33

Komano J, Maruo S, Kurozumi K, Oda T, Takada K (1999) Oncogenic role of Epstein-Barr virus-encoded RNAs in Burkitt's lymphoma cell line Akata. J Virol 73:9827–9831

Kuzembayeva M, Chiu YF, Sugden B (2012) Comparing proteomics and RISC immunoprecipitations to identify targets of Epstein-Barr viral miRNAs. PLoS One 7:e47409

Lee SP, Brooks JM, Al-Jarrah H, Thomas WA, Haigh TA, Taylor GS, Humme S, Schepers A, Hammerschmidt W et al (2004) CD8 T cell recognition of endogenously expressed Epstein-Barr virus nuclear antigen 1. J Exp Med 199:1409–1420

Lerner MR, Andrews NC, Miller G, Steitz JA (1981) Two small RNAs encoded by Epstein-Barr virus and complexed with protein are precipitated by antibodies from patients with systemic lupus erythematosus. Proc Natl Acad Sci 78:805–809

Leslie E, Ning S, Kelliher M, Pagano JS (2007) Interferon regulatory factor 7 is activated by a viral oncoprotein through RIP-dependent ubiquitination. Mol Cell Biol 27:2910–2918

Levitskaya J, Coram M, Levitsky V, Imreh S, Steigerwald-Mullen PM, Klein G, Kurilla MG, Masucci MG (1995) Inhibition of antigen processing by the internal repeat region of the Epstein-Barr virus nuclear antigen-1. Nature 375:685–688

Levitskaya J, Sharipo A, Leonchiks A, Ciechanover A, Masucci MG (1997) Inhibition of ubiquitin/proteasome-dependent protein degradation by the Gly-Ala repeat domain of the Epstein-Barr virus nuclear antigen 1. Proc Natl Acad Sci 94:12616–12621

Lin TC, Liu TY, Hsu SM, Lin CW (2013) Epstein-Barr virus-encoded miR-BART20-5p inhibits T-bet translation with secondary suppression of p53 in invasive nasal NK/T-cell lymphoma. Am J Pathol 182:1865–1875

Mackay LK, Long HM, Brooks JM, Taylor GS, Leung CS, Chen A, Wang F, Rickinson AB (2009) T cell detection of a B-cell tropic virus infection: newly-synthesised versus mature viral proteins as antigen sources for CD4 and CD8 epitope display. PLoS Pathog 5:e1000699

Meckes DG Jr, Shair KH, Marquitz AR, Kung CP, Edwards RH, Raab-Traub N (2010) Human tumor virus utilizes exosomes for intercellular communication. Proc Natl Acad Sci USA 107:20370–20375

Michaud F, Coulombe F, Gaudreault E, Paquet-Bouchard C, Rola-Pleszczynski M, Gosselin J (2010) Epstein-Barr virus interferes with the amplification of IFNa secretion by activating suppressor of cytokine signaling 3 in primary human monocytes. PLoS One 5:e11908

Middeldorp JM, Pegtel DM (2008) Multiple roles of LMP1 in Epstein-Barr virus induced immune escape. Semin Cancer Biol 18:388–396

Mold C, Bradt BM, Nemerow GR, Cooper NR (1988) Epstein-Barr virus regulates activation and processing of the third component of complement. J Exp Med 168:949–969

Morrison TE, Kenney SC (2004) BZLF1, an Epstein-Barr virus immediate-early protein, induces p65 nuclear translocation while inhibiting p65 transcriptional function. Virology 328:219–232

Morrison TE, Mauser A, Wong A, Ting JPY, Kenney SC (2001) Inhibition of IFN-y signaling by an Epstein-Barr virus immediate-early protein. Immunity 15:787–799

Morrison TE, Mauser A, Klingelhutz A, Kenney SC (2004) Epstein-Barr virus immediate-early protein BZLF1 inhibits tumor necrosis factor alpha-induced signaling and apoptosis by downregulating tumor necrosis factor receptor 1. J Virol 78:544–549

Motsch N, Alles J, Imig J, Zhu J, Barth S, Reineke T, Tinguely M, Cogliatti S, Dueck A et al (2012) MicroRNA profiling of Epstein-Barr virus-associated NK/T-cell lymphomas by deep sequencing. PLoS One 7:e42193

Munz C (2014) Viral infections in mice with reconstituted human immune system components. Immunol Lett 161:118–124

Munz C, Bickham KL, Subklewe M, Tsang ML, Chahroudi A, Kurilla MG, Zhang D, O'Donnell M, Steinman RM (2000) Human CD4(+) T lymphocytes consistently respond to the latent Epstein-Barr virus nuclear antigen EBNA1. J Exp Med 191(10):1649–1660

Muromoto R, Ikeda O, Okabe K, Togi S, Kamitani S, Fujimuro M, Harada S, Oritani K, Matsuda T (2009) Epstein-Barr virus-derived EBNA2 regulates STAT3 activation. Biochem Biophys Res Commun 378:439–443

Murphy JA, Duerst RJ, Smith TJ, Morrison LA (2003) Herpes simplex virus type 2 virion host shutoff protein regulates alpha/beta interferon but not adaptive immune responses during primary infection in vivo. J Virol 77:9337–9345

Nachmani D, Stern-Ginossar N, Sarid R, Mandelboim O (2009) Diverse herpesvirus microRNAs target the stress-induced immune ligand MICB to escape recognition by natural killer cells. Cell Host Microbe 5:376–385

Najjar I, Baran-Marszak F, Le Clorennec C, Laguillier C, Schischmanoff O, Youlyouz-Marfak I, Schlee M, Bornkamm GW, Raphaël M et al (2005) Latent membrane protein 1 regulates STAT1 through NF-kB-dependent interferon secretion in Epstein-Barr virus-immortalized B cells. J Virol 79:4936–4943

Nanbo A, Inoue K, Adachi-Takasawa K, Takada K (2002) Epstein-Barr virus RNA confers resistance to interferon-alpha-induced apoptosis in Burkitt's lymphoma. EMBO J 21:954–965

Nanbo A, Yoshiyama H, Takada K (2005) Epstein-Barr virus-encoded poly(A)- RNA confers resistance to apoptosis mediated through Fas by blocking the PKR pathway in human epithelial intestine 407 cells. J Virol 79:12280–12285

Ning S, Campos AD, Darnay BG, Bentz GL, Pagano JS (2008) TRAF6 and the three C-terminal lysine sites on IRF7 are required for its ubiquitination-mediated activation by the tumor necrosis factor receptor family member latent membrane protein 1. Mol Cell Biol 28:6536–6546

Ohashi M, Fogg MH, Orlova N, Quink C, Wang F (2012) An Epstein-Barr virus encoded inhibitor of colony stimulating factor-1 signaling is an important determinant for acute and persistent EBV infection. PLoS Pathog 8:e1003095

Padgett DA, Hotchkiss AK, Pyter LM, Nelson RJ, Yang E, Yeh PE, Litsky M, Williams M, Glaser R (2004) Epstein-Barr virus-encoded dUTPase modulates immune function and induces sickness behavior in mice. J Med Virol 74:442–448

Pagès F, Galon J, Karaschuk G, Dudziak D, Camus M, Lazar V, Camilleri-Broet S, Lagorce-Pagès C, Lebel-Binay S et al (2004) Epstein-Barr virus nuclear antigen 2 induces interleukin-18 receptor expression in B cells. Blood 105:1632–1639

Pappworth IY, Wang EC, Rowe M (2007) The switch from latent to productive infection in Epstein-Barr virus-infected B cells is associated with sensitization to NK cell killing. J Virol 81:474–482

Pasieka TJ, Lu B, Crosby SD, Wylie KM, Morrison LA, Alexander DE, Menachery VD, Leib DA (2008) Herpes simplex virus virion host shutoff attenuates establishment of the Antiviral State. J Virol 82:5527–5535

Pasquier B, Yin L, Fondanèche MC, Relouzat F, Bloch-Queyrat C, Lambert N, Fischer A, de Saint-Basile G, Latour S (2005) Defective NKT cell development in mice and humans lacking the adapter SAP, the X-linked lymphoproliferative syndrome gene product. J Exp Med 201:695–701

Pegtel DM, Cosmopoulos K, Thorley-Lawson DA, van Eijndhoven MAJ, Hopmans ES, Lindenberg JL, de Gruijl TD, Würdinger T, Middeldorp JM (2010) Functional delivery of viral miRNAs via exosomes. Proc Natl Acad Sci USA 107:6328–6333

Petrara MR, Freguja R, Gianesin K, Zanchetta M, De Rossi A (2013) Epstein-Barr virus-driven lymphomagenesis in the context of human immunodeficiency virus type 1 infection. Front Microbiol 4

Pfeffer S, Zavolan M, Grasser FA, Chien M, Russo JJ, Ju J, John B, Enright AJ, Marks D et al (2004) Identification of virus-encoded micrornas. Science 304:734–736

Qiu J, Cosmopoulos K, Pegtel M, Hopmans E, Murray P, Middeldorp J, Shapiro M, Thorley-Lawson DA (2011) A novel persistence associated EBV miRNA expression profile is disrupted in neoplasia. PLoS Pathog 7:e1002193

Qiu J, Smith P, Leahy L, Thorley-Lawson DA (2015) The Epstein-Barr virus encoded bart mirnas potentiate tumor growth in vivo. PLoS Pathog 11:e1004561

Quinn LL, Zuo J, Abbott RJM, Shannon-Lowe C, Tierney RJ, Hislop AD, Rowe M (2014) Cooperation between Epstein-Barr virus immune evasion proteins spreads protection from CD8+ T cell recognition across all three phases of the Lytic cycle. PLoS Pathog 10:e1004322

Rao P, Pham HT, Kulkarni A, Yang Y, Liu X, Knipe DM, Cresswell P, Yuan W (2011) Herpes simplex virus 1 glycoprotein B and US3 collaborate to inhibit CD1d antigen presentation and NKT cell function. J Virol 85:8093–8104

Rathinam V, Fitzgerald K (2010) Inflammasomes and anti-viral immunity. J Clin Immunol 30:632–637

Ressing ME, van Leeuwen D, Verreck FA, Gomez R, Heemskerk B, Toebes M, Mullen MM, Jardetzky TS, Longnecker R et al (2003) Interference with T cell receptor-HLA-DR interactions by Epstein-Barr virus gp42 results in reduced T helper cell recognition. Proc Natl Acad Sci USA 100:11583–11588

Ressing ME, van Leeuwen D, Verreck FA, Keating S, Gomez R, Franken KL, Ottenhoff TH, Spriggs M, Schumacher TN et al (2005) Epstein-Barr virus gp42 is posttranslationally modified to produce soluble gp42 that mediates HLA class II immune evasion. J Virol 79:841–852

Ressing ME, Horst D, Griffin BD, Tellam J, Zuo J, Khanna R, Rowe M, Wiertz EJ (2008a) Epstein-Barr virus evasion of CD8(+) and CD4(+) T cell immunity via concerted actions of multiple gene products. Semin Cancer Biol 18:397–408

Ressing ME, Horst D, Griffin BD, Tellam J, Zuo J, Khanna R, Rowe M, Wiertz EJHJ (2008b) Epstein-Barr virus evasion of CD8+ and CD4+ T cell immunity via concerted actions of multiple gene products. Semin Cancer Biol 18:397–408

Richardson C, Fielding C, Rowe M, Brennan P (2003) Epstein-Barr virus regulates STAT1 through latent membrane protein 1. J Virol 77:4439–4443

Rigaud S, Fondaneche MC, Lambert N, Pasquier B, Mateo V, Soulas P, Galicier L, Le Deist F, Rieux-Laucat F et al (2006) XIAP deficiency in humans causes an X-linked lymphoproliferative syndrome. Nature 444:110–114

Riley KJ, Rabinowitz GS, Yario TA, Luna JM, Darnell RB, Steitz JA (2012) EBV and human microRNAs co-target oncogenic and apoptotic viral and human genes during latency. EMBO J 31:2207–2221

Rowe M, Glaunsinger B, van LD, Zuo J, Sweetman D, Ganem D, Middeldorp J, Wiertz EJ, Ressing ME (2007a) Host shutoff during productive Epstein-Barr virus infection is mediated by BGLF5 and may contribute to immune evasion. Proc Natl Acad Sci USA 104:3366–3371

Rowe M, Glaunsinger B, van LD, Zuo J, Sweetman D, Ganem D, Middeldorp J, Wiertz EJ, Ressing ME (2007b) Host shutoff during productive Epstein-Barr virus infection is mediated by BGLF5 and may contribute to immune evasion. Proc Natl Acad Sci USA 104:3366–3371

Ruf IK, Lackey KA, Warudkar S, Sample JT (2005) Protection from interferon-induced apoptosis by Epstein-Barr virus small RNAs is not mediated by inhibition of PKR. J Virol 79:14562–14569

Saffert RT, Kalejta RF (2008) Promyelocytic leukemia-nuclear body proteins: herpesvirus enemies, accomplices, or both? Future Virol 3:265–277

Saito S, Murata T, Kanda T, Isomura H, Narita Y, Sugimoto A, Kawashima D, Tsurumi T (2013) Epstein-Barr virus deubiquitinase downregulates TRAF6-mediated NF-kB signaling during productive replication. J Virol 87:4060–4070

Salek-Ardakani S, Arrand JR, Mackett M (2002) Epstein-Barr virus encoded interleukin-10 inhibits HLA-class I, ICAM-1, and B7 expression on human monocytes: implications for immune evasion by EBV. Virology 304:342–351

Sanchez DJ, Gumperz JE, Ganem D (2005) Regulation of CD1d expression and function by a herpesvirus infection. J Clin Invest 115:1369–1378

Seto E, Moosmann A, Gromminger S, Walz N, Grundhoff A, Hammerschmidt W (2010) Micro RNAs of Epstein-Barr virus promote cell cycle progression and prevent apoptosis of primary human B cells. PLoS Pathog 6:e1001063

Shah KM, Stewart SE, Wei W, Woodman CBJ, O'Neil JD, Dawson CW, Young LS (2009) The EBV-encoded latent membrane proteins, LMP2A and LMP2B, limit the actions of interferon by targeting interferon receptors for degradation. Oncogene 28:3903–3914

Sheridan V, Polychronopoulos L, Dutia BM, Ebrahimi B (2014) A shutoff and exonuclease mutant of murine gammaherpesvirus-68 yields infectious virus and causes RNA loss in type I interferon receptor knock-out cells. J Gen Virol 95(5):1135–1143

Skalsky RL, Corcoran DL, Gottwein E, Frank CL, Kang D, Hafner M, Nusbaum JD, Feederle R, Delecluse HJ et al (2012) The viral and cellular microRNA targetome in lymphoblastoid cell lines. PLoS Pathog 8:e1002484

Stevenson PG, Simas JP, Efstathiou S (2009) Immune control of mammalian gamma-herpesviruses: lessons from murid herpesvirus-4. J Gen Virol 90:2317–2330

Stewart S, Dawson CW, Takada K, Curnow J, Moody CA, Sixbey JW, Young LS (2004) Epstein-Barr virus-encoded LMP2A regulates viral and cellular gene expression by modulation of the NF-kB transcription factor pathway. Proc Natl Acad Sci USA 101:15730–15735

Strockbine LD, Cohen JI, Farrah T, Lyman SD, Wagener F, DuBose RF, Armitage RJ, Spriggs MK (1998) The Epstein-Barr virus BARF1 gene encodes a novel, soluble colony-stimulating factor-1 receptor. J Virol 72:4015–4021
Strowig T, Gurer C, Ploss A, Liu YF, Arrey F, Sashihara J, Koo G, Rice CM, Young JW et al (2009) Priming of protective T cell responses against virus-induced tumors in mice with human immune system components. J Exp Med 206:1423–1434
Swanson-Mungerson M, Bultema R, Longnecker R (2010) Epstein-Barr virus LMP2A imposes sensitivity to apoptosis. J Gen Virol 91:2197–2202
Takeuchi O, Akira S (2009) Innate immunity to virus infection. Immunol Rev 227:75–86
Tellam J, Connolly G, Green KJ, Miles JJ, Moss DJ, Burrows SR, Khanna R (2004) Endogenous presentation of CD8+ T cell epitopes from Epstein-Barr virus-encoded nuclear antigen 1. J Exp Med 199:1421–1431
Traggiai E, Chicha L, Mazzucchelli L, Bronz L, Piffaretti JC, Lanzavecchia A, Manz MG (2004) Development of a human adaptive immune system in cord blood cell-transplanted mice. Science 304:104–107
Ulbrecht M, Modrow S, Srivastava R, Peterson PA, Weiss EH (1998) Interaction of HLA-E with peptides and the peptide transporter in vitro: implications for its function in antigen presentation. J Immunol 160:4375–4385
Valentine R, Dawson C, Hu C, Shah K, Owen T, Date K, Maia S, Shao J, Arrand J et al (2010) Epstein-Barr virus-encoded EBNA1 inhibits the canonical NF-kappaB pathway in carcinoma cells by inhibiting IKK phosphorylation. Mol Cancer 9:1
van de Veerdonk FL, Wever PC, Hermans MHA, Fijnheer R, Joosten LAB, van der Meer JWM, Netea MG, Schneeberger PM (2012) IL-18 serum concentration is markedly elevated in acute EBV infection and can serve as a marker for disease severity. J Infect Dis 206:197–201
van Gent M, Griffin BD, Berkhoff EG, van Leeuwen D, Boer IGJ, Buisson M, Hartgers FC, Burmeister WP, Wiertz EJ et al (2011) EBV lytic-phase protein BGLF5 contributes to TLR9 downregulation during productive infection. J Immunol 186:1694–1702
van Gent M, Braem SGE, de Jong A, Delagic N, Peeters JGC, Boer IGJ, Moynagh PN, Kremmer E, Wiertz EJ et al (2014) Epstein-Barr virus large tegument protein BPLF1 contributes to innate immune evasion through interference with toll-like receptor signaling. PLoS Pathog 10:e1003960
van Gent M, Gram AM, Boer IGJ, Geerdink RJ, Lindenbergh MFS, Lebbink RJ, Wiertz EJ, Ressing ME (2015) Silencing the shutoff protein of Epstein-Barr virus in productively infected B cells points to (innate) targets for immune evasion. J Gen Virol 96(4):858–865
Vereide DT, Seto E, Chiu YF, Hayes M, Tagawa T, Grundhoff A, Hammerschmidt W, Sugden B (2014) Epstein-Barr virus maintains lymphomas via its miRNAs. Oncogene 33:1258–1264
Voo KS, Fu T, Wang HY, Tellam J, Heslop HE, Brenner MK, Rooney CM, Wang RF (2004) Evidence for the presentation of major histocompatibility complex class I-restricted Epstein-Barr virus nuclear antigen 1 peptides to CD8+ T lymphocytes. J Exp Med 199:459–470
Wahl A, Linnstaedt SD, Esoda C, Krisko JF, Martinez-Torres F, Delecluse HJ, Cullen BR, Garcia JV (2013) A cluster of virus-encoded microRNAs accelerates acute systemic Epstein-Barr virus infection but does not significantly enhance virus-induced oncogenesis in vivo. J Virol 87:5437–5446
Waldman WJ, Williams J, Lemeshow S, Binkley P, Guttridge D, Kiecolt-Glaser JK, Knight DA, Ladner KJ, Glaser R (2008) Epstein-Barr virus-encoded dUTPase enhances proinflammatory cytokine production by macrophages in contact with endothelial cells: Evidence for depression-induced atherosclerotic risk. Brain Behav Immun 22:215–223
Wang F (2013) Nonhuman primate models for Epstein-Barr virus infection. Curr Opin Virol 3:233–237
Wang JT, Doong SL, Teng SC, Lee CP, Tsai CH, Chen MR (2009) Epstein-Barr virus BGLF4 kinase suppresses the interferon regulatory factor 3 signaling pathway. J Virol 83:1856–1869

Wertz IE, Dixit VM (2010) Signaling to NF-kB: regulation by ubiquitination. Cold Spring Harb Perspect Biol 2:1–19

Wood VHJ, O'Neil JD, Wei W, Stewart SE, Dawson CW, Young LS (2007) Epstein-Barr virus-encoded EBNA1 regulates cellular gene transcription and modulates the STAT1 and TGF[beta] signaling pathways. Oncogene 26:4135–4147

Wu L, Fossum E, Joo CH, Inn KS, Shin YC, Johannsen E, Hutt-Fletcher LM, Hass J, Jung JU (2009) Epstein-Barr virus LF2: an antagonist to type I interferon. J Virol 83:1140–1146

Xia T, O'Hara A, Araujo I, Barreto J, Carvalho E, Sapucaia JB, Ramos JC, Luz E, Pedroso C et al (2008) EBV microRNAs in primary lymphomas and targeting of CXCL-11 by ebv-mir-BHRF1-3. Cancer Res 68:1436–1442

Xu D, Brumm K, Zhang L (2006) The latent membrane protein 1 of Epstein-Barr virus (EBV) primes EBV latency cells for type I interferon production. J Biol Chem 281:9163–9169

Yao XD, Rosenthal KL (2011) Herpes simplex virus type 2 virion host shutoff protein suppresses innate dsRNA antiviral pathways in human vaginal epithelial cells. J Gen Virol 92:1981–1993

Yin Y, Manoury B, Fahraeus R (2003) Self-inhibition of synthesis and antigen presentation by Epstein-Barr virus-encoded EBNA1. Science 5:1371–1374

Zhang L, Pagano JS (2000) Interferon regulatory factor 7 is induced by Epstein-Barr virus latent membrane protein 1. J Virol 74:1061–1068

Zhang L, Wu L, Hong K, Pagano JS (2001) Intracellular signaling molecules activated by Epstein-Barr virus for induction of interferon regulatory factor 7. J Virol 75:12393–12401

Zhang J, Das SC, Kotalik C, Pattnaik AK, Zhang L (2004) The latent membrane protein 1 of Epstein-Barr virus establishes an Antiviral State via induction of interferon-stimulated genes. J Biol Chem 279:46335–46342

Zhao Y, Xu D, Jiang Y, Zhang L (2010) Dual functions of interferon regulatory factors 7C in Epstein-Barr virus-mediated transformation of human B lymphocytes. PLoS One 5:e9459

Zhu JY, Pfuhl T, Motsch N, Barth S, Nicholls J, Grasser F, Meister G (2009) Identification of novel Epstein-Barr virus microRNA genes from nasopharyngeal carcinomas. J Virol 83:3333–3341

Zuo J, Thomas W, van Leeuwen D, Middeldorp JM, Wiertz EJHJ, Ressing ME, Rowe M (2008) The DNase of gammaherpesviruses impairs recognition by virus-specific CD8+ T cells through an additional host shutoff function. J Virol 82:2385–2393

Zuo J, Currin A, Griffin BD, Shannon-Lowe C, Thomas WA, Ressing ME, Wiertz EJHJ, Rowe M (2009) The Epstein-Barr virus G-protein-coupled receptor contributes to immune evasion by targeting MHC class I molecules for degradation. PLoS Pathog 5:e1000255

Zuo J, Thomas WA, Haigh TA, Fitzsimmons L, Long HM, Hislop AD, Taylor GS, Rowe M (2011) Epstein-Barr virus evades CD4+ T cell responses in lytic cycle through BZLF1-mediated downregulation of CD74 and the cooperation of vBcl-2. PLoS Pathog 7:e1002455

# Part IV
# Animal Models of EBV Infection

# Non-human Primate Lymphocryptoviruses: Past, Present, and Future

**Janine Mühe and Fred Wang**

**Abstract** Epstein-Barr virus (EBV) orthologues from non-human primates (NHPs) have been studied for nearly as long as EBV itself. Cross-reactive sera and DNA hybridization studies provided the first glimpses of the closely related herpesviruses that belonged to the same gamma-1 herpesvirus, or lymphocryptovirus, genus, as EBV. Over the years, detailed molecular and sequence analyses of LCVs that infect humans and other NHPs revealed similar colinear genome structures and homologous viral proteins expressed during latent and lytic infection. Despite these similarities, experimental infection of NHPs with EBV did not result in acute symptoms or persistent infection as observed in humans, suggesting some degree of host species restriction. Genome sequencing and a molecular clone of an LCV isolate from naturally infected rhesus macaques combined with domestic colonies of LCV-naïve rhesus macaques have opened the door to a unique experimental animal model that accurately reproduces the normal transmission, acute viremia, lifelong persistence, and immune responses found in EBV-infected humans. This chapter will summarize the advances made over the last 50 years in our understanding of LCVs that naturally infect both Old and New World NHPs, the recent, groundbreaking developments in the use of rhesus macaques as an animal model for EBV infection, and how NHP LCVs and the rhLCV animal model can advance future EBV research and the development of an EBV vaccine.

J. Mühe · F. Wang
Department of Medicine, Harvard Medical School, Brigham and Women's Hospital, 181 Longwood Ave, Boston, MA 02115

J. Mühe · F. Wang
Department of Molecular Biology and Immunobiology, Harvard Medical School, Brigham and Women's Hospital, 181 Longwood Ave, Boston, MA 02115

F. Wang (✉)
Infectious Diseases Division, Brigham and Women's Hospital, 181 Longwood Avenue, Boston, MA, USA
e-mail: fwang@research.bwh.harvard.edu



C. Münz (ed.), *Epstein Barr Virus Volume 2*, Current Topics in Microbiology and Immunology 391, DOI 10.1007/978-3-319-22834-1_13

## Contents



## Abbreviations

| | |
|---|---|
| baLCV | Baboon lymphocryptovirus |
| BARF1 | BamHI A rightward open reading frame 1 |
| chLCV | Chimpanzee lymphocryptovirus |
| CSF1-R | Colony-stimulating factor 1 receptor |
| cyLCV | Cynomolgus lymphocryptovirus |
| EBERs | Epstein-Barr virus-encoded small RNAs |
| EBNA | Epstein-Barr virus-associated nuclear antigen |
| EBV | Epstein-Barr virus |
| goLCV | Gorilla lymphocryptovirus |
| gp340 | Glycoprotein with a molecular mass of 340 kDa |
| IM | Infectious mononucleosis |
| LCL | Lymphoblastoid cell line |
| LCV | Lymphocryptovirus |
| LMP | Latent membrane protein |
| maLCV | Marmoset lymphocryptovirus |
| MHC-I | Major histocompatibility complex I |
| NHP | Non-human primate |
| ORF | Open reading frame |
| orLCV | Orangutan lymphocryptovirus |
| PBMC | Peripheral blood mononuclear cells |
| rhEBERs | Rhesus lymphocryptovirus homologue of Epstein-Barr virus-encoded small RNAs |
| rhLCV | Rhesus lymphocryptovirus |
| SIV | Simian immunodeficiency virus |
| SHIV | Simian–human immunodeficiency virus |
| spf | Specific pathogen free |
| TR | Terminal repeat |
| VCA | Viral capsid antigen |

# 1 Discovery of Lymphocryptoviruses in Non-human Primates

The discovery of herpes-like particles in cell cultures derived from Burkitt lymphoma cells initiated a search for similar viruses in non-human primates (NHPs) (Epstein et al. 1964). Gerber and Birch reported the presence of antibodies reactive to Epstein-Barr virus (EBV) in the sera of chimpanzees, baboons, cynomolgus and rhesus macaques, as well as African green monkeys (Gerber and Birch 1967). Seroprevalence of EBV cross-reactivity was usually as high in NHPs as in humans; 100 % of chimpanzees and cynomolgus macaques, and 67 and 90 % of rhesus macaques and African green monkeys, respectively, tested positive for anti-EBV antibodies (Gerber and Birch 1967). In contrast to Old World NHPs, other mammals, ranging from guinea pigs to cows, had no detectable EBV cross-reactive antibodies (Gerber and Birch 1967). Various New World monkeys (howlers, squirrel, and owl monkeys) and prosimians, such as the Galago (bush baby), were also reported to lack cross-reactive antibodies to EBV (Dunkel et al. 1972; Kalter et al. 1972). These observations led to the hypothesis either that Old World monkeys were susceptible to EBV infection or that they harbored an EBV-related herpesvirus not present in New World monkeys or lower primates.

In 1968, the first lymphoblastoid cell line (LCL) from an NHP was recovered (Landon et al. 1968). The cells grew spontaneously from culture of chimpanzee peripheral blood mononuclear cells (PBMCs), were characterized as B lymphocytes, and contained herpesvirus particles (Landon et al. 1968). In the following decade, spontaneous LCLs were derived from baboons, orangutans, and gorillas, all of which contained herpesvirus particles, and the respective viruses were referred to as *Herpesvirus pan*, *Herpesvirus papio*, *Herpesvirus pongo*, and *Herpesvirus gorilla* (Falk et al. 1976; Gerber et al. 1976; Neubauer et al. 1979; Rasheed et al. 1977). Later, spontaneous LCLs from cynomolgus and rhesus macaques were isolated as well (see Table 1 for summary) (Bocker 1980; Fujimoto et al. 1990; Heberling et al. 1981; Rangan et al. 1986; Rivadeneira et al. 1999). Since the biologic and genetic similarities of these viruses place them in the same lymphocryptovirus genus as EBV, we refer to the EBV-related viruses from NHPs as chimpanzee, baboon, orangutan, gorilla, cynomolgus, or rhesus lymphocryptovirus (chLCV, baLCV, orLCV, goLCV, cyLCV, or rhLCV, respectively), to provide clarity, accuracy, and simplicity.

The initial characterization of these newly isolated NHP LCVs included analyses of specific viral antigens detected by sera from humans and NHPs. The viral capsid antigen (VCA) expressed during lytic replication turned out to be a frequent target for naturally occurring antibodies. Sera from humans, chimpanzees, gorillas, orangutans, baboons, and rhesus monkeys readily detected EBV and baLCV VCA (Falk et al. 1976; Falk et al. 1977). Conversely, human and baboon sera detected VCA of EBV, ch-, ba-, or-, and rhLCV (Gerber et al. 1977; Rangan et al. 1986; Rasheed et al. 1977).

**Table 1** Overview of lymphocryptoviruses detected to date

| | | | Primate host[a] | LCV isolates propagated in vitro[b] | LCV-PCR positive hosts (number of species)[c] | LCV genome sequence |
|---|---|---|---|---|---|---|
| Catarrhini (Old World monkeys, great apes, gibbons, and human) | Hominoidae | Hominidae | *Homo* (humans) | Multiple (*H. sapiens*) | 1 | B95-8[e] |
| | | | *Pan* (chimpanzee) | Ch888 (*P. troglodytes*)[j] | 2 | *P. paniscus*[d] |
| | | | *Gorilla* (gorilla) | Machi (*G. gorilla*)[k] | 1 | |
| | | | *Pongo* (orangutan) | CP-81 (*P. pygmaeus*)[n] | 1 | |
| | | Hylobatidae | *Hylobates* (gibbon) | | 3 | |
| | | | *Nomascus* | | | |
| | | | *Symphalangus* | | 1 | |
| | Cercopithecidae | | *Lophocebus* (crested mangabey) | | 2 | |
| | | | *Papio* (baboon) | BA-65 (*P. cynocephalus*)[i] S594 (*P. hamadryas*)[l] | 3 | |
| | | | *Theropithecus* (gelada) | | | |
| | | | *Mandrillus* (mandrill) | | 1 | |
| | | | *Cercocebus* (white-eyelid managabey) | | 2 | |
| | | | *Macaca* (rhesus macaque) | Ts-B6 (*M. fascicularis*)[h] LCL8664 (*M. mulatta*)[q] HVMNE (*M. nemestina*)[o] | 6 | LCL8664[q] |
| | | | *Erythrocebus* (patas) | | 1 | |
| | | | *Chlorocebus* (vervet) | | | |

(continued)

**Table 1** (continued)

| | | Primate host[a] | LCV isolates propagated in vitro[b] | LCV-PCR positive hosts (number of species)[c] | LCV genome sequence |
|---|---|---|---|---|---|
| | | *Allenopithecus* (Allen's swamp monkey) | | | |
| | | *Miopithecus* (talapoin) | | 1 | |
| | | *Cercopithecus* (African green monkey) | AGM-2206 (*C. aethiops*)[f] | 4 | |
| | | *Pygathrix* (douc) | | | |
| | | *Nasalis* (proboscis monkey) | | | |
| | | *Rhinopithecus* (snub-nosed monkey) | | | |
| | | *Semnopithecus* (langur) | | 1 | |
| | | *Trachypithecus* (lutung) | | | |
| | | *Presbytis* (suruli) | | | |
| | | *Piliocolobus* (colobus) | | 1 | |
| | | *Colobus* (colobus) | | 2 | |
| Platyrrhini (New World monkeys) | Cebidae | *Mico* (marmoset/tamarin) | | | |
| | | *Cabuella* (marmoset) | | | |
| | | *Callithrix* (marmoset/tamarin) | Cj0149 (*C. jaccus*)[m] | 2 | CalHV-3[g, p] |
| | | *Callimico* (marmoset) | | | |
| | | *Leontopithecus* (lion tamarin) | | 1 | |
| | | *Saguinus* (gold-handed tamarin) | | 1 | |
| | | *Aotus* (owl monkey) | | | |
| | | *Saimirí* (squirrel monkey) | | 1 | |
| | | *Cebus* (capuchin) | | 1 | |

(continued)

**Table 1** (continued)

| | | Primate host[a] | LCV isolates propagated in vitro[b] | LCV-PCR positive hosts (number of species)[c] | LCV genome sequence |
|---|---|---|---|---|---|
| | Atelidae | *Lagothrix* (woolly monkey) | | | |
| | | *Brachyteles* (muriqui) | | | |
| | | *Ateles* (spider monkey) | | 1 | |
| | | *Alouatta* (howler) | | | |
| | Pitheciidae | *Cacajao* (uakari) | | | |
| | | *Chiropotes* (bearded saki) | | | |
| | | *Pithecia* (saki) | | 1 | |
| | | *Callicebus* (titi) | | | |

[a]Primate phylogenetic order according to Perelman and colleagues (Perelman et al. 2011)
[b]Examples of lymphoblastoid cell lines from NHP species established and used to propagate LCV in vitro
[c]Number of primate species with LCV detected by DNA-Pol PCR using degenerate primers (de Thoisy et al. 2003; Ehlers et al. 2003; Ehlers et al. 2010; Ramer et al. 2000)
[d]Partial genome, 78 kb only (Aswad and Katzourakis 2014); [e]Baer et al. 1984; [f]Bocker et al. 1980; [g]Cho et al. 2001; [h]Fujimoto et al. 1990; [i]Gerber et al. 1977; [j]Gerber et al. 1976; [k]Neubauer et al. 1979; [l]Rabin et al. 1978; [m]Ramer et al. 2000; [n]Rasheed et al. 1977; [o]Rivadeneira et al. 1999; [p]Rivailler et al. 2002a; [q]Rivailler et al. 2002b

The latent infection nuclear antigens (NAs) were not as frequently detected by cross-reacting sera, especially by sera from more distantly related species. Epstein-Barr virus-associated nuclear antigens (EBNAs) were only detected by human sera and not by any tested NHP sera (Falk et al. 1977; Gerber et al. 1976; Neubauer et al. 1979; Rangan et al. 1986). The sera of humans, chimpanzees, and orangutans detected NAs of both chLCV and orLCV (Gerber et al. 1976; Rasheed et al. 1977). Gorilla serum only stained goLCV-NA, but goLCV-NA was also detected by sera from humans, orangutans, and gibbons (Neubauer et al. 1979). RhLCV-NA was detected by sera from rhesus and cynomolgus macaques, as well as chimpanzees (Rangan et al. 1986). The less reliable cross-detection of NA versus VCA may have been due to the lower sensitivity of NA assays, as well as the more recently recognized sequence diversity between latent infection proteins, e.g., NAs, compared to lytic proteins, e.g., VCA.

The first genetic analyses of NHP LCVs were performed by DNA–DNA reassociation studies that compared viral DNA from human LCLs to NHP LCLs. These studies showed that EBV DNA had the highest homology to chLCV DNA (35–45 %), with only slightly more distant homology to baLCV (40 %), orLCV (30–40 %), and goLCV (30–40 %) (Falk et al. 1976; Gerber et al. 1976; Neubauer et al. 1979; Rabin et al. 1978). Once the EBV genome was cloned, more detailed cross-hybridization studies showed that almost all EBV BamHI DNA fragments cross-hybridized with baLCV DNA in a colinear fashion (Heller and Kieff 1981). The only fragments that failed to cross-hybridize were the BamHI E and Nhet fragments, DNA fragments that were later found to encode the EBNA-3 and LMP open reading frames (ORFs). These findings foretold the marked diversity of latent infection proteins among human and NHP LCV revealed by subsequent nucleotide sequencing.

The rhLCV genome was the first non-human LCV genome sequenced, and since it included homologues for LF1, LF2, and LF3 which are deleted from the B95-8 strain of EBV, it was also the first prototypical LCV genome fully sequenced (Baer et al. 1984; Rivailler et al. 2002). Comparative analyses of the rhLCV and EBV genomes revealed (i) an identical repertoire of lytic and latent genes, (ii) colinear organization of the ORFs (see Fig. 1 for latent infection ORFs), (iii) well-conserved amino acid sequences for the lytic infection proteins, and (iv) more distant amino acid homology for the latent infection proteins (Rivailler et al. 2002). Subsequently, at least 36 pre-miRNAs were characterized in the rhLCV genome, the largest number of miRNAs encoded by a single virus found to date (Cai et al. 2006; Riley et al. 2010; Walz et al. 2009).

Although LCVs were not believed to naturally infect New World primates, Ramer and colleagues found DNA evidence for an undiscovered LCV in lymphomas arising in common marmosets (*Callithrix jaccus*) using degenerate PCR primers capable of amplifying the DNA polymerase and terminase from all herpesviruses (Ramer et al. 2000). A cell line (Cj0149), producing B cell immortalizing virus, was derived from the tumor, and the maLCV became the third LCV genome fully sequenced (Cho et al. 2001; Rivailler et al. 2002). MaLCV has a similar colinear genome organization as EBV and rhLCV, and most ORFs share

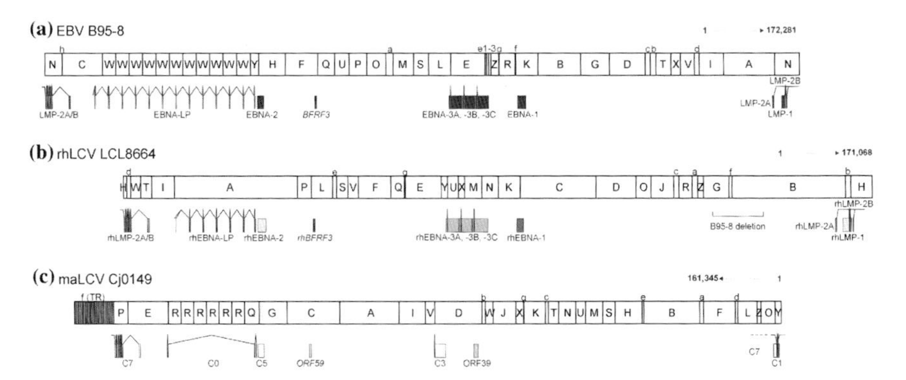


**Fig. 1** Genome organization of prototypical lymphocryptoviruses. BamHI restriction maps for the (A) EBV B95-8 strain, (B) rhLCV LCL8664 strain, and (C) maLCV Cj0149 strain are shown to scale. The maLCV genome is inverted relative to its nucleotide numbering in order to show it in the same colinear direction as EBV and rhLCV. The BamHI fragments are identified starting with a capital A for the largest sized fragment and proceeding in order of size with lower case used for fragments smaller than Z. Multiple copies of the maLCV BamHI f fragment are present in the maLCV terminal repeat (TR), and the actual size of the maLCV TR is underrepresented in the figure. The region of the EBV genome containing LF1, LF2, and LF3 deleted from the B95-8 strain is present in the rhLCV genome as shown. Relative positions in the genome of coding sequences for latent infection proteins are shown as boxes with the associated splice sites (e.g., EBNA-1, rhEBNA-1, and the maLCV EBNA-1 orthologue, ORF39). The relative amino acid homology to the EBV sequence is shown by shades of gray, i.e., *black* highest homology and *white* no homology. The positions of the sVCA orthologues in EBV (BFRF3), rhLCV (rhBFRF3), and maLCV (ORF59) are shown. The transcriptional start site for maLCV C7 has not been identified as indicated by the *dotted line*

positional and amino acid homology to an EBV ORF, but several features distinguished the prototypic New World LCV from the prototypic Old World LCV. The maLCV is notable for (i) the absence of sequence or positional orthologues for eleven genes conserved in EBV and Old World LCVs, including EBERs, BARF1, and vIL-10, (ii) the presence of positional orthologues that have little or no sequence homology to the EBV or rhLCV latent infection proteins (see Fig. 1), (iii) a single positional EBNA-3 orthologue (C3, Fig. 1), as opposed to the related EBNA-3A, -3B, and -3C ORFs, (iv) the absence of tandem repeats in ori-p, and (v) a very large GC-rich terminal repeat region (Cho et al. 2001; Rivailler et al. 2002). Some features were reminiscent of gamma-2 herpesviruses, e.g., the TR and N-terminal location of SH2 domains in the LMP2 homologue (C7), suggesting an evolutionary link of LCV with rhadinoviruses (Rivailler et al. 2002). Serologic assays using maLCV-specific antigens showed seroprevalence in multiple domestic marmoset colonies, as well as marmosets captured from the wild, indicating that this LCV is a natural pathogen for marmosets (Fogg et al. 2005a). The historic inability to detect serologic cross-reactivity to EBV antigens in New World NHP was probably due to a combination of the low homology between maLCV and EBV antigens and less sensitive detection of New World monkey serum immunoglobulins when anti-human Ig-specific secondary reagents were used. PCR detected similar LCV-like DNA sequences in the blood of squirrel monkeys, suggesting that LCV in New World NHP were as ubiquitous as in Old World NHP since species from two major families of New World NHP were infected with LCV (Cho et al. 2001).

Two different nomenclature approaches were taken to identify open reading frames in the Old and New World LCV prototypes. Due to an identical set of ORFs in rhLCV and EBV, rhLCV ORFs were named using the EBV acronym, e.g., BFRF3, with a "rh" prefix (e.g., rhBFRF3, to identify the homologous ORF encoding the small viral capsid protein (sVCA) in rhLCV). This approach was taken to more easily identify the rhLCV orthologues and to reflect the strong similarities between EBV and rhLCV ORFs. However, this convention is somewhat of a misnomer because BFRF3 is an acronym for the EBV **B**amHI **F** DNA fragment's third **R**ightward **ORF** and is not an accurate description for the location of the homologous rhLCV ORF which is not in the BamHI F DNA fragment of rhLCV. Technically, the rhsVCA is encoded by the first leftward reading frame in the rhLCV BamHI L fragment (see Fig. 1b), but accuracy of the genomic position in rhLCV was sacrificed for easier recognition of functional and sequence homology to EBV ORFs.

ORF nomenclature in the maLCV genome provided several different challenges. The homologous ORFs were much less well conserved, several orthologues for EBV ORFs were missing completely, and many of the presumptive orthologues for the latent infection proteins based on position had no sequence homology to the EBV proteins. Thus, a numbering approach similar to that used for rhadinoviruses was used for maLCV, i.e., ORFs with amino acid homology to ORFs present in other herpesviruses were numbered sequentially by location from the beginning of the genome, e.g., ORF1, 2. Unique ORFs with no sequence homology to other viral ORFs were numbered sequentially with a C for Callitrichine herpesvirus 3, e.g., C1, C2. Since the first report of the maLCV genome described approximately two-thirds of the sequence beginning from the right end of the genome when compared to EBV, maLCV ORFs were numbered in opposite orientation from the traditional EBV representation (see Fig. 1c). However, this maLCV orientation does align with that typically used for gamma-2 herpesviruses, making this convention appropriate given the evidence for a closer relationship of maLCV versus EBV or Old World LCV to rhadinoviruses. Based on their position and sequence homology, the sVCA and EBNA-1 orthologues in maLCV could be identified and were labeled as ORF59 and ORF39 (see Fig. 1). A positional and functional orthologue for LMP1 in maLCV with no sequence homology to EBV or Old World LCV was named C1, as the first unique maLCV ORF from the start of the genome (Cho et al. 2001).

The ubiquitous distribution of Old and New World LCVs has been confirmed more recently by PCR amplification of viral DNA fragments from the peripheral blood of many NHP species (see Table 1 for overview) (Ehlers et al. 2003; Ehlers et al. 2010). However, as of now, even this method has not uncovered any evidence for LCV infection of prosimians and lower mammals. Despite the high prevalence of LCVs detectable in Old and New World NHP by serology and PCR, viral isolates and full genome sequences have been recovered in only a small fraction of species (Table 1).

## 2 LCV Host Range

All LCV are able to immortalize B cells and infect hosts from their own primate species, but LCV are not able to immortalize B cells from any other mammals besides primates. Thus, LCV host range is restricted to primates, and given the high degree of viral genome and protein homology, as well as similarities among human and NHP hosts, one might expect that LCV could also immortalize B cells and infect hosts from other primate species (i.e., cross-species infection). This hypothesis is only partially true based on results in the literature where a limited number of NHP LCVs have been tested for cross-immortalization of B cells from a limited number of primate species.

Observations with EBV and several Old World LCVs suggest that LCV can immortalize B cells from closely related species, but not from more distantly related species. The prototypic EBV strain derived from the B95-8 cell line has been shown to transform cells from humans, chimpanzees, and gibbons (black boxes and references in column 1 of Fig. 2), but not cells from baboons, or various macaque species in vitro (gray boxes and references in column 1 of Fig. 2) (Gerber et al. 1976, 1977; Rabin et al. 1978). Only a few experiments have addressed go- and orLCV host range (columns 3 and 4 in Fig. 2). Both viruses were shown to immortalize gibbon cells, but neither human, nor baboon or rhesus cells were transformed by orLCV (Neubauer et al. 1979; Rabin et al. 1978; Rasheed et al. 1977). Rh- and cyLCV have also been extensively studied and were shown to only transform B cells from macaque species, but not from humans or other primates including New World monkeys (columns 6 and 7 in Fig. 2) (Heberling et al. 1981; Moghaddam et al. 1998; Rangan et al. 1986). Interestingly, EBV immortalizes B cells from multiple species of New World monkeys, including cotton-top tamarins that constitute the cellular background of the B95-8 cell line (Falk et al. 1974; Miller et al. 1972). Results with New World B cells seem somewhat idiosyncratic with a relatively unique susceptibility to EBV cross-immortalization, but not rh- or cyLCV. New World maLCV does not cross-immortalize rhesus or human B cells [unpublished observations].

Other published observations do not fit the model that LCV can immortalize B cells from closely related, but not more distant, species, but the interpretations of these experiments are complicated. ChLCV was reported to immortalize B cells from human, gibbon, and macaque species (Gerber et al. 1977). However, virus for these studies was produced from an LCL derived by infecting baboon PBMCs with throat swab material from chimpanzees, not from an LCL derived by infecting, or spontaneously arising from, chimpanzee PBMC (Gerber et al. 1977). Although the baboon was reported to be EBV seronegative, an adult baboon is likely to be naturally baLCV infected with a false-negative cross-reactive serology rather than to be truly naïve for LCV infection, since baLCV infection is ubiquitous in baboons and EBV cross-reactive serologic tests are unreliable due to lower sensitivity. Molecular studies were not available at the time to rule out the possibility for spontaneous outgrowth of a baboon B cell with endogenous baLCV

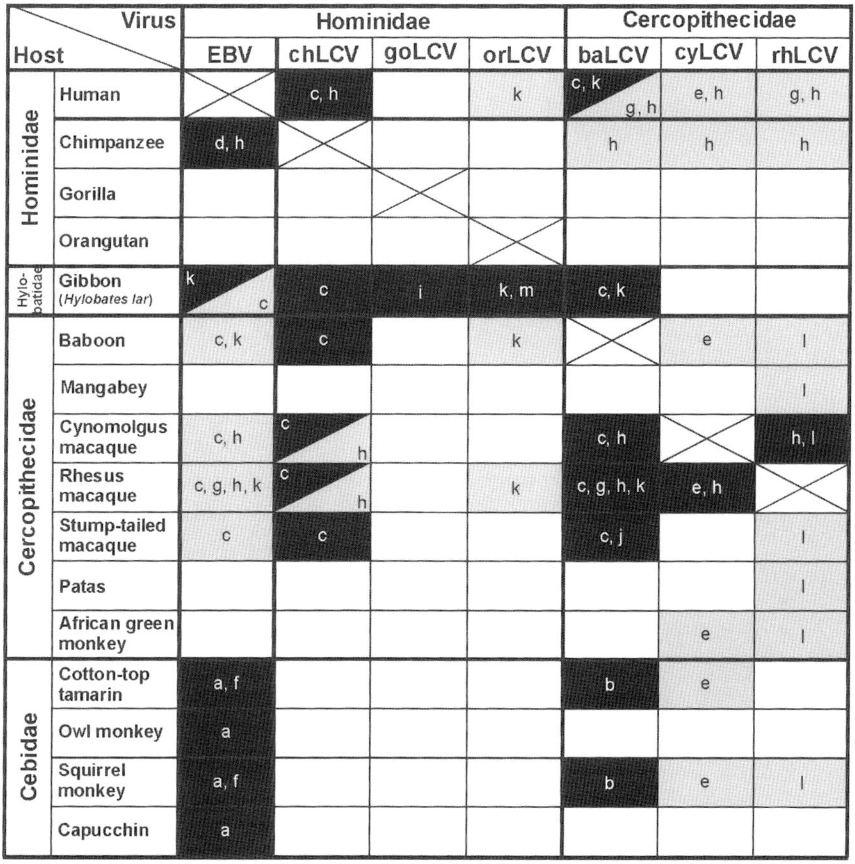

| Host | | Hominidae: EBV | Hominidae: chLCV | Hominidae: goLCV | Hominidae: orLCV | Cercopithecidae: baLCV | Cercopithecidae: cyLCV | Cercopithecidae: rhLCV |
|---|---|---|---|---|---|---|---|---|
| Hominidae | Human | X | c, h | | k | c, k / g, h | e, h | g, h |
| | Chimpanzee | d, h | X | | | h | h | h |
| | Gorilla | | | X | | | | |
| | Orangutan | | | | X | | | |
| Hylobatidae | Gibbon (*Hylobates lar*) | k / c | c | i | k, m | c, k | | |
| Cercopithecidae | Baboon | c, k | c | | k | X | e | l |
| | Mangabey | | | | | | | l |
| | Cynomolgus macaque | c, h | c / h | | | c, h | X | h, l |
| | Rhesus macaque | c, g, h, k | c / h | | k | c, g, h, k | e, h | X |
| | Stump-tailed macaque | c | c | | | c, j | | l |
| | Patas | | | | | | | l |
| | African green monkey | | | | | | e | l |
| Cebidae | Cotton-top tamarin | a, f | | | | b | e | |
| | Owl monkey | a | | | | | | |
| | Squirrel monkey | a, f | | | | b | e | l |
| | Capucchin | a | | | | | | |

**Fig. 2** Cross-species immortalization of primate PBMCs by LCVs. Viruses (*columns*) were used to infect PBMCs from various human and non-human primates (*rows*). Reports of positive LCV-induced immortalization of primate B cells are indicated by a *black box* with the associated publication reference. Reports of experiments with no LCV-induced immortalization are shown as *gray boxes*. Multiple reports with opposing results are shown as half boxes of *black* and *gray*. Results of LCV infection of PBMCs from the native species are not indicated and shown as an X. *White boxes* indicate that published reports for experiments with that combination of virus and host PBMC were not found. [a]Falk et al. 1974; [b]Falk et al. 1977; [c]Gerber et al. 1977; [d]Gerber et al. 1976; [e]Heberling et al. 1981; [f]Miller et al. 1972; [g]Moghaddam et al. 1998; [h]Mühe and Wang 2015[i]; Neubauer et al. 1979; [j]Rabin et al. 1977; [k]Rabin et al. 1978; [l]Rangan et al. 1986; [m]Rasheed et al. 1977

infection, as opposed to in vitro transformation with chLCV. Therefore, the cell line assumed to be producing chLCV could have actually been producing baLCV (Gerber et al. 1977).

Coincidentally, a baLCV isolate was reported to have a similarly wide host range capable of transforming B cells from baboon, gibbon, macaques, New World primates, as well as human cord blood B cells (Falk et al. 1976, 1977; Gerber et al. 1977; Rabin et al. 1977, 1978) (see Fig. 2). However, in our hands, baLCV could immortalize rhesus, but not human B cells (Moghaddam et al. 1998).

Although infection of human PBMC with baLCV did result in LCL outgrowth, viral DNA analyses showed that the immortalized human B cells were coinfected with both EBV and baLCV in all cases, i.e., baLCV was recovered only in endogenously EBV-infected B cells superinfected with baLCV. When virus replication was induced from these coinfected cells and viral supernatants were used to infect either human or rhesus B cells, the viruses could be separated. Only EBV was recovered in immortalized human B cells, and only baLCV was recovered in immortalized rhesus B cells (Moghaddam et al. 1998). Thus, baLCV is capable of infecting, persisting, and replicating in human B cells, but cannot immortalize human B cells without complementation by EBV infection. Our experiments demonstrated the potential complication from spontaneous outgrowth of endogenous LCV and the importance of characterizing the viral genome in cell lines when interpreting results of experimental LCV host range determination across primate species.

Data regarding cross-infection of primate hosts with LCV from a different species are even more limited. Rhesus macaques (*Macaca mulatta*) were experimentally inoculated with EBV, but there was no definitive evidence for infection (Levine et al. 1980; Shope et al. 1973). BaLCV inoculation of rhesus macaques was reported to induce VCA antibodies, but no virus excretion in the saliva or tumor development was observed (Gerber et al. 1977), and thus, one could not exclude the possibility that VCA antibodies were simply induced from the antigenic stimulus provided by viral inoculation. ChLCV inoculation of baboons (*Papio cynocephalus*) and rhesus monkeys resulted in seroconversion in all animals and virus excretion from baboons, but the nucleic acid identity of the shed virus was not identified (Gerber et al. 1977). The chLCV isolate used for these in vivo studies was produced from the same LCL derived from baboon PBMCs described above, and the baboon hosts inoculated experimentally may not have been truly baLCV naïve (Gerber et al. 1977). No tumors were observed in either NHP species inoculated with chLCV (Gerber et al. 1977). Thus, inoculation of animals with cell-free virus, or LCV-infected cells (not reviewed here), did not provide conclusive evidence that LCV can infect Old World NHP hosts from a different species.

The lack of EBV cross-reactive antibodies in New World monkeys and the ability to immortalize New World monkey B cells with EBV in vitro led investigators to inoculate cotton-top tamarins (*Saguinus oedipus*) with EBV to determine whether they could be experimentally infected (Miller et al. 1972). Inoculation of tamarins with cell-free EBV was shown to induce lymphomas, and the incidence rate increased when the animals were immunosuppressed (Shope et al. 1973). Infected New World monkeys also developed antibodies to EBV VCA (Miller et al. 1977). Experimental EBV inoculation of marmosets (*Callithrix jacchus*) and owl monkeys (*Aotus trivirgatus*) was also shown to result in low-level seroconversion and lymphoproliferative disease (Epstein 1976; Falk et al. 1976). Experimental chLCV inoculation of cotton-top tamarins was reported to induce seroconversion, whereas baLCV inoculation did not (Falk et al. 1977; Gerber et al. 1977). In both instances, there was no evidence of viral shedding or lymphoproliferative disease.

The ability to experimentally inoculate tamarins and detect EBV infection by lymphoma induction was used as an animal model to develop potential EBV vaccines. Purified gp340 was tested as a subunit vaccine and induced a strong antibody response in tamarins with serum-neutralizing activity detectable by in vitro transformation assays (Epstein 1986; Epstein et al. 1985). Gp340 vaccination was also able to prevent EBV-induced lymphoproliferation in cotton-top tamarins providing preclinical support for testing gp340 as a vaccine in humans (Epstein 1986; Epstein et al. 1985). Other researchers showed recombinant adenovirus expressing gp340 could induce gp340-specific antibodies in the tamarin model, but in vitro tests were unable to detect serum-neutralizing activity to EBV infection (Ragot et al. 1993). Nevertheless, animals immunized with Adgp340 were also protected against EBV-induced lymphomas after challenge (Ragot et al. 1993).

# 3 RhLCV Infection of Rhesus Macaques as a Model System for Studying EBV Infection

While experimental inoculation of New World monkeys provided an animal model for EBV infection, this system bypassed normal transmission through the oral mucosa. Percutaneous virus inoculation was either cleared or resulted in uncontrolled lymphomagenesis, and lifelong persistent viral infection in the peripheral blood B cell compartment was not established. The balance between virus and host may be disrupted in favor of the virus by incorporation of more immune evasion strategies in EBV than New World LCV and a more limited host immune response, e.g., low diversity of the major histocompatibility complex I (MHC-I) observed in cotton-top tamarins (Gyllensten et al. 1994). Thus, LCV infection in Old World monkeys may provide a more accurate model for the virus–host interaction during EBV infection in humans. Attention focused on rhesus macaques because their use in biomedical research has been well established, sequencing of the rhLCV genome revealed an identical gene repertoire as EBV, and investigations of natural rhLCV infection in rhesus macaques showed a remarkable similarity to EBV infection in humans.

Newborn macaques are seropositive for LCV infection due to placental transfer of maternal antibodies, which disappear around 6 months of age, similar to human newborns. LCV infection is spread very rapidly in captive colonies, presumably through oral transmission, so that virtually all animals become seropositive by one year of age [unpublished observations]. RhLCV persists asymptomatically in peripheral blood B cells for life, and detailed studies show that the repertoire and magnitude of cellular and humoral immune responses against lytic and latent infection proteins in rhLCV-infected macaques important for controlling LCV infection are remarkably similar to those in EBV-infected humans (Fogg et al. 2005b, 2006; Leskowitz et al. 2013; Orlova et al. 2011a, b; Rao et al. 2000).

Immunosuppression in macaques, whether drug-induced or secondary to simian immunodeficiency virus (SIV) infection, can result in LCV-induced lymphomas (Feichtinger et al. 1992; Habis et al. 2000; Rivailler et al. 2004). Virus-associated oral hairy leukoplakia also arises in SIV-infected macaques, showing that pathogenesis of LCV infection in epithelial cells is similar in NHP and humans (Baskin et al. 1995; Kutok et al. 2004). Other LCV-associated malignancies, such as nasopharyngeal carcinoma, gastric carcinoma, and Hodgkin's lymphoma, have not been observed in macaques, but given the nature of these malignancies in the human population, it is unlikely that these diseases would be observed in the limited number of captive macaques observed for relatively short durations.

Perhaps the most intriguing biologic observation in NHP was the finding of a second type of rhLCV similar to that of the type 2 EBV (Cho et al. 1999). The 208-95 rhLCV strain was isolated from a lymphoma arising in a SIV-infected macaque and was notable for a rhEBNA-2 protein that migrated dramatically different from the prototypic LCL8664 rhLCV strain. Sequencing showed a 208-95 rhEBNA-2 with only 41 % amino acid similarity to the LCL8664 rhEBNA-2 with most of the differences located centrally in the divergent region where most of the differences between the type 1 and type 2 EBV EBNA-2s are found. Sequencing showed that the 208-95 rhLMP1 ORF was nearly identical to that of LCL8664 rhLMP1, suggesting two similar rhLCV strains defined by their differences in rhEBNA-2, as occurring with EBV. Thus, the same biologic pressure that has selected for the evolution of two EBV strains in humans also selected for two LCV strains in macaques. Strain-specific PCR studies detected LCL8664 rhLCV in oral washes from 6 of 20 macaques and 208-95 rhLCV in 6 of 20 macaques, indicating that, as in humans, both strains are prevalent and some hosts are coinfected with both strains (Cho et al. 1999).

The formation of extended specific pathogen-free (spf) rhesus macaque colonies opened the door to experimental inoculations of rhLCV-naïve macaques as an animal model for EBV infection. Initially, spf colonies were derived to enhance the health and safety of the colonies by isolating and breeding animals free of herpes B and other specific pathogens. Animals hand-reared from birth are also likely to remain free of other herpesviruses, and many of these animals were found to be free of rhesus cytomegalovirus, rhesus rhadinovirus, and rhLCV infection. By vigilant screening for these extended pathogens, colonies free of rhCMV, RRV, and rhLCV can be raised into self-sustaining, spf colonies for experimental studies.

Experimental oral inoculation of rhLCV-naïve macaques with LCL8664 rhLCV reproduced both the acute and persistent phases of EBV infection seen in humans (Moghaddam et al. 1997). Infection of the peripheral blood following penetration of the oral mucosa was detected within 2–3 weeks by DNA PCR and as soon as 7 days post-inoculation by more sensitive rhEBERs RT-PCR (Ohashi et al. 2012; Rivailler et al. 2004). Acute viremia in immunocompetent macaques, detectable by DNA PCR positivity in PBMC, is typically seen between 3 and 10 weeks post-inoculation (Rivailler et al. 2004). Atypical lymphocytosis and lymphadenopathy were detected in a minority of animals between 3 and 15 weeks post-inoculation (Moghaddam et al. 1997). Thus, the kinetics of experimental rhLCV infection shows that penetration of the oral mucosa and invasion of the peripheral blood occur rapidly after oral inoculation, and viral amplification during acute infection precedes the onset of acute infectious mononucleosis symptoms in humans which has been temporally linked in sailors at sea to viral inoculation at least 6 weeks earlier during shore leave (Hoagland 1955).

Asymptomatic persistent infection was established in rhesus macaques after experimental oral inoculation as evidenced by the isolation of LCV-positive LCL arising spontaneously from in vitro culture of PBMC or by detection of rhEBERs by RT-PCR of PBMC RNA (Moghaddam et al. 1997; Ohashi et al. 2012). Persistent rhLCV infection was documented for more than 14 years after oral inoculation in one animal, suggesting that experimental infection is durable and lifelong, as in humans [unpublished observations]. Importantly, the asymptomatic nature of experimental infection in healthy animals is clearly due to immune control, and not simply an attenuated viral strain, since experimental inoculation of LCL8664 rhLCV can induce lymphomagenesis in immunosuppressed macaques (Rivailler et al. 2004).

Recently, the rhLCV genome has been cloned as a bacterial artificial chromosome (BAC) allowing site-specific manipulation of the rhLCV genome and both in vitro and in vivo functional analysis of viral genes for the first time (Ohashi et al. 2010). The first cloned rhLCV (clone 16) had an insertion of the BAC vector sequences in rhBARF1, a putative LCV immune evasion gene which encodes a secreted homologue of the colony-stimulating factor 1 receptor (CSF1-R) (Ohashi et al. 2010). The genetic manipulation resulted in expression of a truncated rhBARF1 that was neither secreted nor capable of binding CSF-1. As expected from EBV studies, the loss of functional rhBARF1 had no detectable effect on either rhLCV-induced B cell immortalization or rhLCV lytic replication, i.e., rhBARF1 was not essential for latent or lytic infection in vitro (Ohashi et al. 2010). However, experimental inoculation of naïve rhesus macaques showed that mutation of this lytic replication protein resulted in decreased viral load in acute infection and a 100-fold reduction in the number of infected B cells, or viral setpoint, during persistent infection (Ohashi et al. 2012). This abnormal phenotype in acute and persistent infection could be rescued either by immunosuppressing the host with simian–human immunodeficiency virus (SHIV) infection or by repairing the rhBARF1 mutation, showing that the defects in vivo were due specifically to the immune evasive properties of rhBARF1 (Ohashi et al. 2012). The reduction in acute viremia can be explained by the loss of rhBARF1 function during lytic replication, increased susceptibility of virus to immune control, and decreased acute viral amplification. However, the marked reduction in viral setpoint during persistent infection was unexpected since the classic paradigm suggests that rhBARF1 is not expressed in latently infected B cells. This finding may indicate a link between acute viremia and establishment of persistent viral setpoints, i.e., lower acute viremia resulting in seeding of fewer infected B cells and establishment of lower persistent viral setpoints. Alternatively, rhBARF1 may be expressed in LCV-infected peripheral blood B cells and provide an important function during persistent infection.

# 4 Future Directions for the Rhesus Macaque Animal Model

The first 50 years of EBV research have seen a phenomenal growth not only in our understanding of EBV, but also in our understanding of EBV-related LCVs in NHPs. From the initial observations of cross-reactive antibodies to EBV in a few NHP species, we now recognize the prevalence of EBV-related viruses not only in virtually all Old World NHPs, but also in New World NHPs. We have full-genome sequences for both Old and New World LCV prototypes that paint a dynamic picture for the evolution of these fascinating viruses. LCV-naïve rhesus macaques can be raised and experimentally infected with laboratory rhLCV to successfully reproduce natural transmission, acute infection, asymptomatic persistent infection, and lymphomagenesis. More recently, recombinant viral genetics has been incorporated into the rhesus LCV animal model to link the immune evasion functions of a single LCV lytic infection protein to defects in both viral amplification during acute infection and viral setpoint establishment during persistent infection.

The power of the rhLCV animal model system is clearly the strong similarities to EBV infection in humans. No other animal model for EBV infection reproduces the natural route of transmission through the oral mucosa, viral amplification and invasion of the peripheral blood during acute infection, establishment of asymptomatic persistent infection as the normal outcome, and the potential for lymphomagenesis as a result of abnormal perturbations of immune control. Even though EBV cannot be used to infect rhesus macaques, the rhLCV genome has an identical repertoire of viral genes, the molecular functions of the viral proteins are strongly conserved, and the host immune responses to rhLCV proteins are remarkably similar to the EBV-specific immune responses in humans. Thus, the rhesus LCV animal model provides a unique opportunity to experimentally interrogate EBV infection in ways not possible in humans or other animal model systems.

An obvious limitation of the system is the cost of animals and limited availability of rhLCV-naïve animals. The cost makes large studies powered to detect small biologic differences more difficult, and the finite number of naïve animals puts a priority on well-conceived experiments to test specific questions with the highest impact on our understanding of EBV biology, pathogenesis, and therapy. Also, NHP cannot be genetically modified as readily as rodents to manipulate host factors, but monoclonal antibodies that deplete specific cell populations, e.g., CD8+ T cells, in vivo can be used as an alternative experimental approach (Schmitz et al. 1999).

Conceptually, the rhLCV animal model is most powerful for studying those aspects of the virus–host interaction which cannot be readily reproduced in tissue culture or other small animal model systems. While it would be fruitful to simply delete various viral genes and study the effect of the mutant LCV in vivo, we believe that the rhesus animal model provides a unique opportunity to investigate three broad areas of EBV biology that are difficult to study in tissue culture, other animal models, or humans: (i) acute infection, especially penetration of the mucosal surface and invasion of the peripheral blood, (ii) persistent infection, especially how viral setpoints are established, and (iii) therapy, especially development of an EBV vaccine to reduce or prevent EBV-associated diseases.

In acute infection, the rhesus macaque model is valuable for being able to control the timing of viral inoculation, thereby enabling investigation of the earliest events in EBV infection. Studies of epithelial cell infection and viral shedding in infectious mononucleosis patients, who were presumably inoculated 6 weeks prior to symptoms, probably reflect more about how the virus gets out of the host, as opposed to studying how the virus gets into the host during the first few days after inoculation. When we studied rhesus macaque tissues collected from various parts of the oral cavity within 7 days of oral inoculation, we found it difficult to find infected cells, suggesting that widespread mucosal epithelial cell infection is not present even early in infection and penetration of the oral mucosa may be limited to more discrete foci [unpublished observations]. Thus, key questions that can be asked in the rhLCV animal model are as follows: How does virus penetrate the mucosal epithelium? Is there a role for epithelial cell infection? How and where is virus amplified, e.g., lytic replication in epithelial cells or lytic/latent infection of B cells? What are the immune modalities critical for interrupting these early events in acute infection? Understanding these fundamental steps in primary EBV infection is important for the rational development of vaccines to prevent EBV-associated infectious mononucleosis.

Viral setpoints during persistent infection have been extensively studied in humans with the number of EBV-infected B cells falling from high levels in acute IM to extremely low levels that persist for life (Hochberg et al. 2004). Adoptive transfer of EBV-specific T cells can reduce the viral setpoints in persistently infected hosts, and while there is consensus for persistent latent infection in memory B cells, the effect of antiviral therapy and viral lytic replication in the maintenance of viral setpoints remains uncertain (Babcock et al. 1998; Hoshino et al. 2009; Yao et al. 1989). Little is known about how viral setpoints are established, e.g., are they related to viral inoculum levels of acute viral replication? Indeed, viral setpoints appear to be much higher in naturally infected rhesus macaques than in humans, highlighting how little is known about how LCV viral setpoints are established in natural hosts (Ohashi et al. 2012). It is not known whether the frequency or absolute numbers of virally infected cells are the important factor, whether higher initial viral inoculums or repeated viral exposures may play a role in higher viral setpoints in NHP, or whether intrinsic differences in the immune response may determine viral setpoints. Many important questions about persistent infection and viral setpoints are difficult to address in humans and remain to be answered. Can vaccines targeting acute viral infection also lower viral setpoints during persistent infection? Will lower viral setpoints reduce the risk of EBV-associated malignancies in life? It is difficult to model EBV-associated malignancies such as HL and NPC in rhesus macaques since only a small fraction of infected hosts develop disease after long periods of time. However, viral setpoints may be an important surrogate marker for the risk of EBV-associated malignancies and understanding how viral setpoints are established and how they can be altered can be readily investigated in macaques.

EBV vaccine development is a rich target for the rhesus macaque animal model. The immune responses to LCV infection are well conserved, and protection can be tested in animals with viral infection beginning at the oral mucosal surface. Empiric testing of vaccine candidates is a resource-intensive approach; therefore, using rhesus macaques to better understand the biology underlying acute and persistent infection may provide the foundation for a more rational approach to vaccine development, e.g., should epithelial cell infection be targeted, can viral setpoints be reduced by vaccination?

The genetic and biologic similarities between EBV and rhLCV provide a powerful animal model for investigation of EBV infection, pathogenesis, and therapy. These similarities are due to the recent evolution of LCV and strong biologic selection for lifelong B cell persistence in primates. At the same time, nature has provided subtle, but biologically significant, differences that limit LCV-induced B cell immortalization and host range within primate species. Dissecting the mechanisms underlying the restriction of these closely related viruses and hosts will identify pathways that have driven LCV to evolve specifically for their native species, i.e., pathways critical for LCV-induced B cell immortalization in vitro and persistent infection in vivo. Thus, ongoing scientific investigations using the biological similarities, as well as molecular differences, of LCV infection in NHPs will continue to provide unique insights for the next 50 years of EBV research.

## References


Aswad A, Katzourakis A (2014) The first endogenous herpesvirus, identified in the tarsier genome, and novel sequences from primate rhadinoviruses and lymphocryptoviruses. PLoS Genet 10:e1004332

Babcock GJ, Decker LL, Volk M, Thorley-Lawson DA (1998) EBV persistence in memory B cells in vivo. Immunity 9:395–404

Baer R, Bankier AT, Biggin MD, Deininger PL, Farrell PJ, Gibson TJ, Hatfull G, Hudson GS, Satchwell SC, Seguin C et al (1984) DNA sequence and expression of the B95-8 Epstein-Barr virus genome. Nature 310:207–211

Baskin GB, Roberts ED, Kuebler D, Martin LN, Blauw B, Heeney J, Zurcher C (1995) Squamous epithelial proliferative lesions associated with rhesus Epstein-Barr virus in simian immunodeficiency virus-infected rhesus monkeys. J Infect Dis 172:535–539

Bocker, J. F., K. H. Tiedemann, G. W. Bornkamm, and H. zur Hausen. 1980. Characterization of an EBV-like virus from African green monkey lymphoblasts. Virology 101:291–5

Cai X, Schafer A, Lu S, Bilello JP, Desrosiers RC, Edwards R, Raab-Traub N, Cullen BR (2006) Epstein-Barr virus microRNAs are evolutionarily conserved and differentially expressed. PLoS Pathog 2:e23

Cho Y, Ramer J, Rivailler P, Quink C, Garber RL, Beier DR, Wang F (2001) An Epstein-Barr-related herpesvirus from marmoset lymphomas. Proc Natl Acad Sci USA 98:1224–1229

Cho YG, Gordadze AV, Ling PD, Wang F (1999) Evolution of two types of rhesus lymphocryptovirus similar to type 1 and type 2 Epstein-Barr virus. J Virol 73:9206–9212

de Thoisy B, Pouliquen JF, Lacoste V, Gessain A, Kazanji M (2003) Novel gamma-1 herpesviruses identified in free-ranging new world monkeys (golden-handed tamarin [*Saguinus midas*], squirrel monkey [*Saimiri sciureus*], and white-faced saki [*Pithecia pithecia*]) in French Guiana. J Virol 77:9099–9105

Dunkel VC, Pry TW, Henle G, Henle W (1972) Immunofluorescence tests for antibodies to Epstein-Barr virus with sera of lower primates. J Natl Cancer Inst 49:435–440
Ehlers B, Ochs A, Leendertz F, Goltz M, Boesch C, Matz-Rensing K (2003) Novel simian homologues of Epstein-Barr virus. J Virol 77:10695–10699
Ehlers B, Spiess K, Leendertz F, Peeters M, Boesch C, Gatherer D, McGeoch DJ (2010) Lymphocryptovirus phylogeny and the origins of Epstein-Barr virus. J Gen Virol 91:630–642
Epstein MA (1976) EB virus in the owl monkey (*Aotus trivirgatus*). Lab Anim Sci 26:1127–1130
Epstein MA (1986) Vaccination against Epstein-Barr virus: current progress and future strategies. Lancet 1:1425–1427
Epstein MA, Achong BG, Barr YM (1964) Virus particles in cultured lymphoblasts from Burkitt's lymphoma. Lancet 1:702–703
Epstein MA, Morgan AJ, Finerty S, Randle BJ, Kirkwood JK (1985) Protection of cotton top tamarins against Epstein-Barr virus-induced malignant lymphoma by a prototype subunit vaccine. Nature 318:287–289
Falk L, Deinhardt F, Nonoyama M, Wolfe LG, Bergholz C (1976) Properties of a baboon lymphotropic herpesvirus related to Epstein-Barr virus. Int J Cancer 18:798–807
Falk L, Wolfe L, Deinhardt F, Paciga J, Dombos L, Klein G, Henle W, Henle G (1974) Epstein-Barr virus: transformation of non-human primate lymphocytes in vitro. Int J Cancer 13:363–376
Falk LA, Henle G, Henle W, Deinhardt F, Schudel A (1977) Transformation of lymphocytes by Herpesvirus papio. Int J Cancer 20:219–226
Feichtinger H, Li SL, Kaaya E, Putkonen P, Grunewald K, Weyrer K, Bottiger D, Ernberg I, Linde A, Biberfeld G et al (1992) A monkey model for Epstein Barr virus-associated lymphomagenesis in human acquired immunodeficiency syndrome. J Exp Med 176:281–286
Fogg MH, Carville A, Cameron J, Quink C, Wang F (2005a) Reduced prevalence of Epstein-Barr virus-related lymphocryptovirus infection in sera from a new world primate. J Virol 79:10069–10072
Fogg MH, Garry D, Awad A, Wang F, Kaur A (2006) The BZLF1 homolog of an Epstein-Barr-related gamma-herpesvirus is a frequent target of the CTL response in persistently infected rhesus macaques. J Immunol 176:3391–3401
Fogg MH, Kaur A, Cho YG, Wang F (2005b) The CD8+ T-cell response to an Epstein-Barr virus-related gammaherpesvirus infecting rhesus macaques provides evidence for immune evasion by the EBNA-1 homologue. J Virol 79:12681–12691
Fujimoto K, Terato K, Miyamoto J, Ishiko H, Fujisaki M, Cho F, Honjo S (1990) Establishment of a B-lymphoblastoid cell line infected with Epstein-Barr-related virus from a cynomolgus monkey (Macaca fascicularis). J Med Primatol 19:21–30
Gerber P, Birch SM (1967) Complement-fixing antibodies in sera of human and nonhuman primates to viral antigens derived from Burkitt's lymphoma cells. Proc Natl Acad Sci USA 58:478–484
Gerber P, Kalter SS, Schidlovsky G, Peterson WD Jr, Daniel MD (1977) Biologic and antigenic characteristics of Epstein-Barr virus-related Herpesviruses of chimpanzees and baboons. Int J Cancer 20:448–459
Gerber P, Pritchett RF, Kieff ED (1976) Antigens and DNA of a chimpanzee agent related to Epstein-Barr virus. J Virol 19:1090–1099
Gyllensten U, Bergstrom T, Josefsson A, Sundvall M, Savage A, Blumer ES, Giraldo LH, Soto LH, Watkins DI (1994) The cotton-top tamarin revisited: Mhc class I polymorphism of wild tamarins, and polymorphism and allelic diversity of the class II DQA1, DQB1, and DRB loci. Immunogenetics 40:167–176
Habis A, Baskin G, Simpson L, Fortgang I, Murphey-Corb M, Levy LS (2000) Rhesus lymphocryptovirus infection during the progression of SAIDS and SAIDS-associated lymphoma in the rhesus macaque. AIDS Res Hum Retroviruses 16:163–171
Heberling RL, Bieber CP, Kalter SS (1981) Establishment of a lymphoblastoid cell line from a lymphomous cynomolgus monkey. In: Yohn DS, Blakeslee JR (eds) Advances in comparative leukemia research. Elsevier, Amsterdam, pp 385–386

Heller M, Kieff E (1981) Colinearity between the DNAs of Epstein-Barr virus and herpesvirus papio. J Virol 37:821–826

Hoagland RJ (1955) The transmission of infectious mononucleosis. Am J Med Sci 229:262–272

Hochberg D, Souza T, Catalina M, Sullivan JL, Luzuriaga K, Thorley-Lawson DA (2004) Acute infection with Epstein-Barr virus targets and overwhelms the peripheral memory B-cell compartment with resting, latently infected cells. J Virol 78:5194–5204

Hoshino Y, Katano H, Zou P, Hohman P, Marques A, Tyring SK, Follmann D, Cohen JI (2009) Long-term administration of valacyclovir reduces the number of Epstein-Barr virus (EBV)-infected B cells but not the number of EBV DNA copies per B cell in healthy volunteers. J Virol 83:11857–11861

Kalter SS, Heberling RL, Ratner JJ (1972) EBV antibody in sera of non-human primates. Nature 238:353–354

Kutok JL, Klumpp S, Simon M, MacKey JJ, Nguyen V, Middeldorp JM, Aster JC, Wang F (2004) Molecular evidence for rhesus lymphocryptovirus infection of epithelial cells in immunosuppressed rhesus macaques. J Virol 78:3455–3461

Landon JC, Ellis LB, Zeve VH, Fabrizio DP (1968) Herpes-type virus in cultured leukocytes from chimpanzees. J Natl Cancer Inst 40:181–192

Levine PH, Leiseca SA, Hewetson JF, Traul KA, Andrese AP, Granlund DJ, Fabrizio P, Stevens DA (1980) Infection of rhesus monkeys and chimpanzees with Epstein-Barr virus. Arch Virol 66:341–51

Leskowitz RM, Zhou XY, Villinger F, Fogg MH, Kaur A, Lieberman PM, Wang F, Ertl HC (2013) CD4+ and CD8+ T-cell responses to latent antigen EBNA-1 and lytic antigen BZLF-1 during persistent lymphocryptovirus infection of rhesus macaques. J Virol 87:8351–8362

Miller G, Shope T, Coope D, Waters L, Pagano J, Bornkamn G, Henle W (1977) Lymphoma in cotton-top marmosets after inoculation with Epstein-Barr virus: tumor incidence, histologic spectrum antibody responses, demonstration of viral DNA, and characterization of viruses. J Exp Med 145:948–967

Miller G, Shope T, Lisco H, Stitt D, Lipman M (1972) Epstein-Barr virus: transformation, cytopathic changes, and viral antigens in squirrel monkey and marmoset leukocytes. Proc Natl Acad Sci USA 69:383–387

Moghaddam A, Koch J, Annis B, Wang F (1998) Infection of human B lymphocytes with lymphocryptoviruses related to Epstein-Barr virus. J Virol 72:3205–3212

Moghaddam A, Rosenzweig M, Lee-Parritz D, Annis B, Johnson RP, Wang F (1997) An animal model for acute and persistent Epstein-Barr virus infection. Science 276:2030–2033

Mühe J, Wang F (2015) Host range restriction of Epstein-Barr virus and related lymphocryptoviruses. J Virol 89:9133–9136

Neubauer RH, Rabin H, Strnad BC, Nonoyama M, Nelson-Rees WA (1979) Establishment of a lymphoblastoid cell line and isolation of an Epstein-Barr-related virus of gorilla origin. J Virol 31:845–848

Ohashi M, Fogg MH, Orlova N, Quink C, Wang F (2012) An Epstein-Barr virus encoded inhibitor of Colony Stimulating Factor-1 signaling is an important determinant for acute and persistent EBV infection. PLoS Pathog 8:e1003095

Ohashi M, Orlova N, Quink C, Wang F (2010) Cloning of the Epstein-Barr virus-related rhesus lymphocryptovirus as a bacterial artificial chromosome: a loss-of-function mutation of the rhBARF1 immune evasion gene. J Virol 85:1330–1339

Orlova N, Fogg MH, Carville A, Wang F (2011a) Antibodies to lytic infection proteins in lymphocryptovirus-infected rhesus macaques: a model for humoral immune responses to epstein-barr virus infection. Clin Vaccine Immunol 18:1427–1434

Orlova N, Wang F, Fogg MH (2011b) Persistent infection drives the development of CD8+ T cells specific for late lytic infection antigens in lymphocryptovirus-infected macaques and Epstein-Barr virus-infected humans. J Virol 85:12821–12824

Perelman P, Johnson WE, Roos C, Seuanez HN, Horvath JE, Moreira MA, Kessing B, Pontius J, Roelke M, Rumpler Y, Schneider MP, Silva A, O'Brien SJ, Pecon-Slattery J (2011) A molecular phylogeny of living primates. PLoS Genet 7:e1001342

Rabin H, Neubauer RH, Hopkins RF 3rd, Dzhikidze EK, Shevtsova ZV, Lapin BA (1977) Transforming activity and antigenicity of an Epstein-Barr-like virus from lymphoblastoid cell lines of baboons with lymphoid disease. Intervirology 8:240–249
Rabin H, Neubauer RH, Hopkins RF 3rd, Nonoyama M (1978) Further characterization of a herpesvirus-positive orang-utan cell line and comparative aspects of in vitro transformation with lymphotropic old world primate herpesviruses. Int J Cancer 21:762–767
Ragot T, Finerty S, Watkins PE, Perricaudet M, Morgan AJ (1993) Replication-defective recombinant adenovirus expressing the Epstein-Barr virus (EBV) envelope glycoprotein gp340/220 induces protective immunity against EBV-induced lymphomas in the cottontop tamarin. J Gen Virol 74(Pt 3):501–507
Ramer JC, Garber RL, Steele KE, Boyson JF, O'Rourke C, Thomson JA (2000) Fatal lymphoproliferative disease associated with a novel gammaherpesvirus in a captive population of common marmosets. Comp Med 50:59–68
Rangan SR, Martin LN, Bozelka BE, Wang N, Gormus BJ (1986) Epstein-Barr virus-related herpesvirus from a rhesus monkey (*Macaca mulatta*) with malignant lymphoma. Int J Cancer 38:425–432
Rao P, Jiang H, Wang F (2000) Cloning of the rhesus lymphocryptovirus viral capsid antigen and Epstein-Barr virus-encoded small RNA homologues and use in diagnosis of acute and persistent infections. J Clin Microbiol 38:3219–3225
Rasheed S, Rongey RW, Bruszweski J, Nelson-Rees WA, Rabin H, Neubauer RH, Esra G, Gardner MB (1977) Establishment of a cell line with associated Epstein-Barr-like virus from a leukemic orangutan. Science 198:407–409
Riley KJ, Rabinowitz GS, Steitz JA (2010) Comprehensive analysis of Rhesus lymphocryptovirus microRNA expression. J Virol 84:5148–5157
Rivadeneira ED, Ferrari MG, Jarrett RF, Armstrong AA, Markham P, Birkebak T, Takemoto S, Johnson-Delaney C, Pecon-Slattery J, Clark EA, Franchini G (1999) A novel Epstein-Barr virus-like virus, HV(MNE), in a Macaca nemestrina with mycosis fungoides. Blood 94:2090–2101
Rivailler P, Carville A, Kaur A, Rao P, Quink C, Kutok JL, Westmoreland S, Klumpp S, Simon M, Aster JC, Wang F (2004) Experimental rhesus lymphocryptovirus infection in immunosuppressed macaques: an animal model for Epstein-Barr virus pathogenesis in the immunosuppressed host. Blood 104:1482–1489
Rivailler P, Cho YG, Wang F (2002a) Complete genomic sequence of an Epstein-Barr virus-related herpesvirus naturally infecting a new world primate: a defining point in the evolution of oncogenic lymphocryptoviruses. J Virol 76:12055–12068
Rivailler P, Jiang H, Cho YG, Quink C, Wang F (2002b) Complete nucleotide sequence of the rhesus lymphocryptovirus: genetic validation for an Epstein-Barr virus animal model. J Virol 76:421–426
Schmitz JE, Simon MA, Kuroda MJ, Lifton MA, Ollert MW, Vogel CW, Racz P, Tenner-Racz K, Scallon BJ, Dalesandro M, Ghrayeb J, Rieber EP, Sasseville VG, Reimann KA (1999) A nonhuman primate model for the selective elimination of CD8+ lymphocytes using a mouse-human chimeric monoclonal antibody. Am J Pathol 154:1923–1932
Shope T, Dechairo D, Miller G (1973) Malignant lymphoma in cottontop marmosets after inoculation with Epstein-Barr virus. Proc Natl Acad Sci U S A 70:2487–2491
Walz N, Christalla T, Tessmer U, Grundhoff A (2009) A global analysis of evolutionary conservation among known and predicted gammaherpesvirus microRNAs. J Virol 84:716–728
Yao QY, Ogan P, Rowe M, Wood M, Rickinson AB (1989) Epstein-Barr virus-infected B cells persist in the circulation of acyclovir-treated virus carriers. Int J Cancer 43:67–71

# EBV Infection of Mice with Reconstituted Human Immune System Components

**Christian Münz**

**Abstract** Epstein-Barr virus (EBV) was discovered 50 years ago as the first candidate human tumor virus. Since then, we have realized that this human γ-herpesvirus establishes persistent infection in the majority of adult humans, but fortunately causes EBV-associated diseases only in few individuals. This is an incredible success story of the human immune system, which controls EBV infection and its transforming capacity for decades. A better understanding of this immune control would not only benefit patients with EBV-associated malignancies, but could also provide clues how to establish such a potent, mostly cell-mediated immune control against other pathogens and tumors. However, the functional relevance of EBV-specific immune responses can only be addressed in vivo, and mice with reconstituted human immune system components (huMice) constitute a small animal model to interrogate the protective value of immune compartments during EBV infection, but also might provide a platform to test EBV-specific vaccines. This chapter will summarize the insights into EBV immunobiology that have already been gained in these models and provide an outlook into promising future avenues to develop this in vivo model of EBV infection and human immune responses further.

## Contents



C. Münz (✉)
Viral Immunobiology, Institute of Experimental Immunology,
University of Zürich, Winterthurerstrasse 190, 8057 Zurich, Switzerland
e-mail: christian.muenz@uzh.ch



C. Münz (ed.), *Epstein Barr Virus Volume 2*, Current Topics in Microbiology and Immunology 391, DOI 10.1007/978-3-319-22834-1_14

## Abbreviations

| | |
|---|---|
| BLT | Bone marrow, liver, thymus |
| CD | Cluster of differentiation |
| DNAM | DNAX accessory molecule |
| DLBCL | Diffuse large B cell lymphoma |
| EBNA | Epstein-Barr nuclear antigen |
| EBV | Epstein-Barr virus |
| HLA | Human leukocyte antigen |
| HLH | Hemophagocytic lymphohistiocytosis |
| HPC | Hematopoietic progenitor cell |
| HuMice | Mice with reconstituted human immune system components |
| IFN | Interferon |
| Ig | Immunoglobulin |
| IL | Interleukin |
| IM | Infectious mononucleosis |
| LMP | Latent membrane protein |
| MHC | Major histocompatibility complex |
| NK | Natural killer |
| NKG2D | Natural killer group 2, member D |
| NOD | Non-obese diabetic |
| NPC | Nasopharyngeal carcinoma |
| PAMP | Pathogen-associated molecular pattern |
| PBMC | Peripheral blood mononuclear cell |
| DC | Dendritic cell |
| Rag | Recombinase-activating gene |
| SCID | Severe combined immunodeficiency |
| SIRP | Signal regulatory protein |
| TCR | T cell receptor |
| TLR | Toll-like receptor |

## 1 Introduction

Epstein-Barr virus (EBV) is a ubiquitous $\gamma_1$-herpesvirus of humans with more than 95 % of adults being persistently infected (Rickinson et al. 2014). While $\gamma$-herpesviruses are estimated to have developed as a separate lineage from other herpesviruses 200 million years ago (McGeoch et al. 2000), $\gamma_1$-herpesviruses, also called lymphocryptoviruses, seem to have diverged from $\gamma_2$-herpesviruses, so-called rhadinoviruses, around 80 million years ago (McGeoch 2001). Even so this predates separation of mouse and man, which is estimated to have occurred 60–70 million years ago (Mestas and Hughes 2004), no rodent $\gamma_1$-herpesvirus has been identified so far and this $\gamma$-herpesvirus subgroup has so far only been identified in monkeys. Moreover, the cardinal feature of EBV, its tumorigenic potential, which led to its discovery 50 years ago (Epstein et al. 1964a, b), is linked to a distinct set of EBV proteins expressed during latent infection, which are only conserved with homologues in lymphocryptoviruses of old world monkeys (Rivailler et al. 2002). These species restrictions obviously complicate studying EBV and its closely related old world monkey virus cousins in vivo, but they also suggest that EBV has coevolved with humans over a significant period of time and might therefore have been one of the major training pathogens of the human immune system during evolution.

This should allow us to probe characteristic features of the human immune system by studying immune control of EBV. In order to do so in vivo, mice with reconstituted human immune system components (huMice) have been explored in the recent past (Chatterjee et al. 2014). For these models, mouse strains that lack mouse lymphocytes and tolerate human cells with their myeloid cell compartments are currently preferentially used (Rongvaux et al. 2013). These include NOD–*scid* $\gamma_c^{null}$, NOD Rag1$^{-/-}$ $\gamma_c^{null}$, BALB/c Rag2$^{-/-}$ $\gamma_c^{null}$ human or NOD SIRPα tg, C57BL/6 Rag2$^{-/-}$ $\gamma_c^{-/-}$ CD47$^{-/-}$, and C57BL/6 RAG$^{-/-}$ $\gamma_c^{-/-}$ C5$^{-/-}$ β2m$^{-/-}$ I-Aβ$^{-/-}$ HLA-DR1$^{+}$ HLA-A2$^{+}$ c-fms-p-hSIPRα tg mice (Ishikawa et al. 2005; Lavender et al. 2013; Legrand et al. 2011; Serra-Hassoun et al. 2014; Shultz et al. 2005; Strowig et al. 2011). In these mouse strains, the *scid* mutation or absence of one of the two recombinase-activating genes (Rag1 or 2) compromises B and T cell receptor somatic recombination, thus abolishing adaptive lymphocyte development. The $\gamma_c$ deficiency blocks interleukin (IL) -2, -4, -7, -9, -15, and -21 signaling, which compromises the development of innate lymphoid cell precursors (IL-7 dependent) and differentiation into the natural killer (NK) cell lineage (IL-15 dependent). Therefore, innate lymphocytes are also absent in these mice. Apart from lymphocytes, which could attack transplanted human cells, these could also be phagocytosed by murine myeloid cells. In order to avoid this, the above-mentioned mouse strains modulate the interaction between the signal regulatory protein α (SIRPα), an inhibitory receptor on myeloid cells, and its ligand CD47. NOD SIRPα cross-reacts with human CD47 (Takenaka et al. 2007). Therefore, transgenic introduction of human or mouse SIRPα and human immune system component reconstitution in the NOD background prevents myeloid restriction of human cells (Legrand et al. 2011; Strowig et al. 2011; Yamauchi et al. 2013). In addition, it was recently reported that the absence of CD47 might down-modulate mouse myeloid cell reactivity against CD47-negative cells (Wang et al. 2007) and seems to also enhance human cell acceptance in CD47-deficient mice (Lavender et al. 2013).

These mouse strains are then either neonatally injected with human CD34$^{+}$ hematopoietic progenitor cells (HPCs) or transplanted with a human fetal liver and thymus organoid under the kidney capsule followed by CD34$^{+}$ HPC injection into adult mice (BLT mice) (Rongvaux et al. 2013). In both instances, human immune

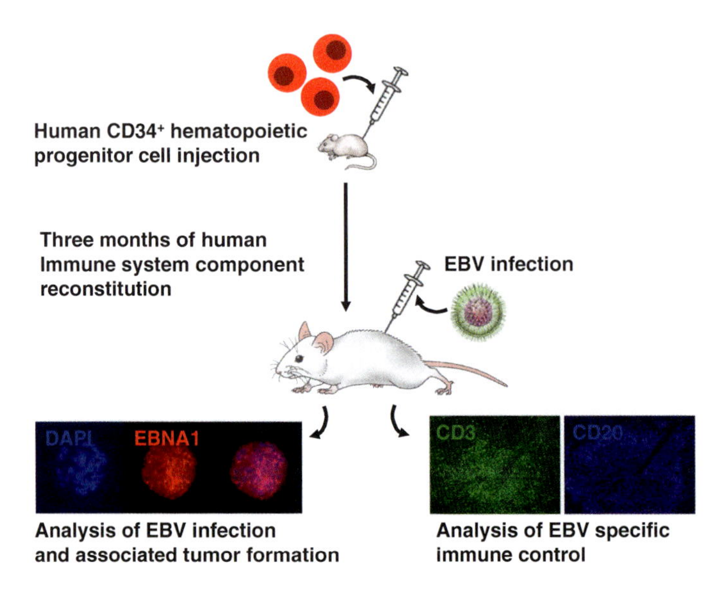


**Fig. 1** Mice with reconstituted human immune system components in EBV research. Lymphocyte-deficient mice get reconstituted with human $CD34^+$ hematopoietic progenitor cells via intrahepatic injection (huMice). The reconstitution efficiency is characterized after three months before these huMice are infected with EBV by intraperitoneal infection. Over the next one to three months, parameters of EBV infection and associated tumor formation as well as the developing immune control in response to this infection can be characterized

system components develop in the mouse host and are, therefore, tolerized against this environment, unlike peripheral blood mononuclear cell (PBMC) transfer, which readily causes graft-versus-host disease. After neonatal $CD34^+$ HPC transfer, human T cell education occurs in the mouse thymus (Watanabe et al. 2009), while in BLT mice, this happens in the transplanted human organoid. After three months, these models allow reconstitution of most human immune system components with on average around 60 % of the huMouse PBMCs being of human origin (Ramer et al. 2011). At this time point, the B and T cells are reconstituted at nearly equal frequencies (45 %:45 %), the majority of T cells are $CD4^+$, and 3 % monocytes, 2 % dendritic cells, and 5 % NK cells can be found in peripheral blood. These huMice can sustain EBV infection and raise cell-mediated immune control of this virus, which will be discussed next (Fig. 1).

## 2 Epstein-Barr Virus Infection in Mice with Reconstituted Human Immune System Components

Similar to other herpesviruses, EBV can elicit latent or lytic infection after oromucosal transmission via saliva exchange (Cesarman 2014). During latent infection, EBV replicates via the proliferation of its host cell, and during lytic infection, viral particles are formed and released for maximal viral genome amplification. Both modes of EBV infection occur in B cells of huMice (Strowig et al. 2009), but since they are tightly linked to human B cell differentiation, the extent of human B cell immunobiology in huMice also defines the frequency and quality of these programs of EBV infection. In huMice, we observe primarily transitional and naïve B cells after reconstitution (Salguero et al. 2014; Watanabe et al. 2009), not unlike the B cell phenotype in cord blood. Since infection of naïve B cells is associated with complete latent EBV antigen expression in humans (Babcock et al. 2000), it is maybe not too surprising that this latency III pattern (all latent nuclear [EBNAs] and membrane [LMPs] proteins expressed) is also primarily observed in huMice (Strowig et al. 2009). In contrast, germinal center reactions, class-switch recombination, and somatic hypermutation are poorly developed in huMice (Rongvaux et al. 2013), and therefore, lower latencies, which are found in centroblasts and centrocytes (latency II, only EBNA1, LMP1, and LMP2 proteins expressed) or memory B cells (latency I and 0, only EBNA1 or no EBV proteins expressed), are rarely observed (Cocco et al. 2008; Islas-Ohlmayer et al. 2004; Ma et al. 2011). However, this low level of latency I- and latency II-expressing cells might depend on $CD4^{+}$ T cell help and rudimentary germinal center formation (Heuts et al. 2014), which could be strengthened by inflammation induced secondary and tertiary lymphoid tissue induction (Salguero et al. 2014). In addition to memory B cell differentiation, plasma cell differentiation and plasmablast differentiation are only observed at low frequency in huMice and most of these models have with steady-state human IgG levels below 1 μg/ml more than 1000-fold less human IgG in their serum than adult humans (Billerbeck et al. 2013; Salguero et al. 2014). Since plasma cell differentiation is associated with lytic EBV replication in healthy human virus carriers (Laichalk and Thorley-Lawson 2005), the low level of this differentiation in huMice also supports only a low level of lytic EBV replication (Chijioke et al. 2013; Ma et al. 2011; Strowig et al. 2009). In addition, EBV infects epithelial cells in humans, presumably mainly causing lytic replication as a final amplification loop for shedding into saliva and further transmission (Hutt-Fletcher 2007). However, this, probably final step in the EBV life cycle in human hosts, cannot be modeled in huMice, because they lack human epithelial cells. Thus, due to the immaturity of the human B cell compartment in huMice, primarily latency III infection can be achieved with minor contributions of lytic EBV replication and maybe lower latency programs.

This pattern of EBV infection has, however, been primarily established with the B95-8 strain of EBV which was originally isolated from an American patients with symptomatic primary EBV infection, called infectious mononucleosis (IM) (Miller and Lipman 1973). Other viral isolates might behave completely differently. Along these lines, the M81 virus strain isolated from a Chinese nasopharyngeal carcinoma (NPC) patient was recently reported (Tsai et al. 2013). This virus displayed a much higher levels of lytic EBV replication in huMice and upon human B cell infection in vitro (Tsai et al. 2013), suggesting that some viral strains might switch to lytic EBV replication without plasma cell differentiation. This was especially apparent in the analysis of late lytic EBV antigen, like gp350, expression, which is rarely observed with B95-8 EBV infection, but labeled a high frequency of M81 infected cells in huMice (Tsai et al. 2013). Thus, increased plasma cell differentiation in both immunization (Salguero et al. 2014) and coinfection settings, and EBV isolates with a predisposition for lytic EBV replication might augment lytic EBV replication in huMice and should allow more comprehensive studies on this EBV infection program and its immune control.

## 3 EBV-Associated Tumorigenesis in HuMice

The bias for latency III of latent EBV infection in huMice is also seen in the tumors that emerge in this in vivo model system. A subset of mice (20–30 % after 4 weeks of infection with $10^5$ rB95-8 EBV infectious particles in huMice) develop B cell lymphoproliferative lesions (Antsiferova et al. 2014; Chijioke et al. 2013). Tumor incidence increases over time with more than 50 % of the animals developing tumors after 6–8 weeks (Antsiferova et al. 2014; Ma et al. 2011). This coincides with the highest viral load after infection, which then declines in the few mice that survive this peak of the acute infection (Yajima et al. 2008). The lymphoproliferative lesions are primarily composed of latency III-infected B cells (Strowig et al. 2009; Yajima et al. 2008), which are polyclonal in nature (Chijioke et al. 2013) and heavily infiltrated with human immune cells, including T cells, NK cells, neutrophils, and macrophages (Antsiferova et al. 2014; White et al. 2012). In line with the growth promoting latency III program of EBV-infected B cells in these lymphoproliferative lesions, EBV-transformed B cells can be grown out from the observed tumors (Strowig et al. 2009; White et al. 2012). However, in addition to the latency III program in the tumor cells, the low level of lytic EBV replication seems also to contribute to tumor formation (Ma et al. 2011), especially at extralymphoid sites (Antsiferova et al. 2014). It has been suggested that the lytic EBV program could stimulate vascularization at these sites (Hong et al. 2005). As we will see in the discussion below, the inflammatory leukocyte infiltrate in these EBV-driven lymphoproliferative lesions clearly controls tumor formation. This also becomes apparent in EBV infections with an EBNA3B-deficient virus (White et al. 2012). The absence of EBNA3B leads to increased tumor formation with around half of the mice developing tumors even at an EBV dose of $10^4$ infectious particles after 4 weeks. The resulting tumors are devoid of most inflammatory infiltrates and have, therefore, rather the appearance of diffuse large B cell lymphomas (DLBCL). Interestingly, a subset of human DLBCL tumors harbor also loss-of-expression mutations in EBNA3B (White et al. 2012), and these were very similar to the tumor cell lines growing out from EBNA3B-deficient EBV-infected huMice. At least partially responsible for the absence of inflammatory infiltrates was the observation that EBNA3B-deficient EBV-transformed B cell lines do not produce chemokines (CXCL9 and 10) that attract via CXCR3 human immune cells into their microenvironment, and transgenic expression of CXCL10 rendered these tumor cells again more susceptible to T cell-mediated immune control. Therefore, inflammatory infiltrates in EBV-induced lymphoproliferative lesions seem to be important in controlling EBV-associated tumorigenesis in huMice.

## 4 Modeling Other EBV-Associated Pathologies in HuMice

In addition to tumor formation, uncontrolled primary EBV infection can also be associated with hemophagocytic lymphohistiocytosis (HLH) and can predispose for the autoimmune disease multiple sclerosis (Rouphael et al. 2007; Thacker et al. 2006). Therefore, EBV infection of huMice was investigated for evidence of HLH and autoimmune diseases. Indeed, high-dose EBV infection for prolonged periods of time (10 weeks) was reported to result in signs of HLH, such as normocytic anemia and thrombocytopenia (Sato et al. 2011). Viral loads, however, persisted in the affected animals at $10^6$–$10^7$ copies per ml of blood for more than a month, while with a $10^5$ infectious dose, usually only $10^4$–$10^5$ viral copies per ml are reached. Presumably due to this difference, HLH symptoms have not been reported in other studies on EBV infections of huMice. These high viral loads could originate from a higher infectious dose or viral strain differences, because the Akata EBV strain was used in this study. Nevertheless, even under these extreme conditions of EBV infection, the virus remained restricted to the B cell populations. Irrespective of these considerations, prolonged high viral loads can model EBV-associated HLH in huMice.

With respect to autoimmune diseases, erosive arthritis was observed after EBV infection of huMice (Kuwana et al. 2011). Also in this study, the Akata EBV strain was used for infection. While synovial membrane proliferation, pannus formation, and bone marrow edema were found by histology of joints and adjacent bone marrow, autoimmune characteristics of, for example, rheumatoid arthritis, like antibodies against citrullinated self-peptides, were not observed. Therefore, it remains unclear whether true autoimmunity, rather than tissue damage by hyperinflammation, can be elicited via EBV infection of huMice. Nevertheless, apart from tumorigenesis, additional EBV-associated pathologies can be modeled in huMice, but the association of EBV with these diseases needs to be better understood in order to judge whether the modeling in mice reflects the mechanisms of disease association in humans.

## 5 Innate Immune Control of EBV Infection in HuMice

The risk for both EBV-associated tumorigenesis and autoimmune diseases is in some cases determined by the course of primary EBV infection with IM predisposing for EBV-associated Hodgkin's lymphoma and multiple sclerosis (Hjalgrim et al. 2003; Thacker et al. 2006). While primary EBV infection occurs usually asymptomatically early in life, delay into adolescence is frequently associated with IM and accompanied by elevated lytic EBV replication and a lymphocytosis of $CD8^+$ T cells, which are mostly specific for lytic EBV antigens (Luzuriaga and Sullivan 2010; Rickinson et al. 2014). This course of primary infection might be influenced by a variety of factors, including initial infectious dose, susceptibility of the B cell compartment in adults versus children, the quality of the primed T cell response, and the ability of the innate immune response to limit virus replication initially. In this respect, particularly two innate immune compartments, dendritic cells (DCs) and natural killer (NK) cells, are of interest. Particularly, plasmacytoid DCs are known for their superior ability to produce the anti-viral cytokines collectively termed type I interferons (IFNs), consisting of IFN-β and IFN-α subtypes (Reizis et al. 2011). They limit B cell transformation by EBV in the first 12 h of EBV encounter in vitro (Lotz et al. 1985). Plasmacytoid DCs have been found to produce copious amounts of type I IFNs in response to EBV (Fiola et al. 2010; Severa et al. 2013). Presumably, the unmethylated viral DNA of EBV particles activates the pathogen-associated molecular pattern (PAMP) receptor and toll-like receptor (TLR) 9 of plasmacytoid DCs (Fiola et al. 2010; Lim et al. 2006). This recognition limits EBV infection in a PBMC transfer model, because PBMCs without plasmacytoid DCs developed EBV-associated lymphoproliferative disease more rapidly and plasmacytoid DC transfer in addition to PBMCs increased the resistance to lymphoproliferative disease after EBV superinfection (Lim et al. 2006). While this plasmacytoid DC recognition of EBV might limit infection initially, no evidence could be found that plasmacytoid DCs can also prime protective T cell responses against this virus (Severa et al. 2013). Therefore, other DC subpopulations might perform this task, as has been shown for inflammatory monocyte-derived DCs in vitro (Bickham et al. 2003). Even so TLR2 and 3 have been implicated in EBV recognition (Ariza et al. 2009; Gaudreault et al. 2007; Iwakiri et al. 2009), it remains unclear by which mechanism conventional and inflammatory DCs might be activated by EBV.

Both monokine (IL-12, IL-18, and IL-15) as well as type I IFN production by DCs can also activate other innate immune responses. Along these lines, NK, NKT, and γδ T cells have been implicated in innate immune control of EBV (Bhaduri-McIntosh et al. 2008; Chijioke et al. 2013; Chung et al. 2013; Yuling et al. 2009). These different innate lymphocyte compartments might target different EBV infection programs. NKT cells have been shown to limit EBV-associated lymphoproliferative disease in huMice (Yuling et al. 2009). $CD8^+$ NKT cells were sufficient and necessary to mediate this effect in vivo (Yuling et al. 2009), and CD1d-restricted EBV-transformed lymphoblastoid cell line recognition led to IFN-γ production as well as cytotoxicity by NKT cells in vitro (Chung et al. 2013). Thus, NKT cells might directly recognize a CD1d restricted ligand on latently EBV-infected B cells, but the nature of this ligand remains unclear. In contrast to the recognition of latently EBV-infected B cells, NK cells seem to preferentially react to lytic EBV infection (Chijioke et al. 2013; Pappworth et al. 2007). In particular, early differentiated NK cells without killer immunoglobulin-like receptors (KIRs) expanded after EBV infection of huMice (Chijioke et al.

2013). This NK cell population composes half of the peripheral NK cells shortly after birth and is then successively replaced by KIR-positive NK cells during the first decade of life (Sundstrom et al. 2007). Depletion of NK cells in huMice led to higher viral loads and enhanced tumorigenesis (Chijioke et al. 2013). Only wild-type but not lytic replication-deficient BLZF1 knockout EBV infection was affected by NK cell depletion, suggesting that early differentiated NK cells control lytic EBV infection (Chijioke et al. 2013). This control might be mediated by direct recognition and killing of lytically EBV-replicating B cells via the activating NK cell receptors NKG2D and DNAM1, as has been shown for one model cell line in vitro so far (Pappworth et al. 2007). Therefore, early differentiated NK cells could efficiently control lytic EBV replication early in life and therefore prevent uncontrolled lytic replication, leading to IM symptoms and the increased risks for Hodgkin's lymphoma and multiple sclerosis.

# 6 Adaptive Immune Control of EBV Infection in HuMice

Adaptive anti-viral immune responses rely on neutralization or opsonization of viral particles and infected cells by antibodies as well as modifying infected cells by cytokines or killing them directly by T cells. These two arms of adaptive immunity are called humoral and cell-mediated adaptive immune responses and are mediated by B and T cells, respectively. Similar to newborn human immune compartments, huMice have difficulties in raising antibody responses during EBV infection (Yajima et al. 2008). Most of the reconstituted B cells are transitional or naïve, and the serum of huMice contains usually only low levels of antibodies. In particular, the concentration of class-switched and affinity-matured antibodies is more than 1000-fold lower than in human serum (Ishikawa et al. 2005; Shultz et al. 2005; Traggiai et al. 2004). This inability to mature antibody responses does probably not only result from the immaturity of the reconstituted human B cell compartment, but also result from the poor development of secondary lymphoid tissue structure with poorly developed germinal centers, in which both antibody isotype switching and somatic hypermutation for affinity maturation occur. Accordingly, only EBV-specific IgM antibodies have been observed after infection and EBV-specific vaccination (Gurer et al. 2008; Yajima et al. 2008). Therefore, adaptive humoral immune responses to EBV might be difficult to model in huMice, and the protective value of EBV-specific antibodies might have to be investigated initially by passive immunization.

In contrast, huMice raise considerable $CD4^+$ and $CD8^+$ T cell responses to EBV infection (Strowig et al. 2009; Traggiai et al. 2004). Overall, the T cell repertoire of huMice is polyclonal, but due to the contribution of mouse, MHC class I molecules in its thymic selection expected to be different from humans (Marodon et al. 2009; Watanabe et al. 2009). Nevertheless, huMice massively expand their $CD8^+$ T cell compartment after EBV infection, peaking at 5–6 weeks after infection. At that time, the expanded $CD8^+$ T cells recognize autologous EBV-transformed B cells by cytokine production (Melkus et al. 2006; Strowig et al. 2009; Yajima et al. 2008). In HLA class I transgenic huMice, some of them can also be stained by MHC class I/peptide tetramers, carrying both lytic and latent EBV antigen-derived and in humans immunodominant epitopes, but the intensity of the respective staining is low (Antsiferova et al. 2014; Shultz et al. 2010; Strowig et al. 2009), indicating maybe T cell receptor (TCR) down-regulation due to the high antigenic load in the animals at these time points. Nevertheless, cloning of these expanding $CD8^+$ T cells by limiting dilution results in clones that can kill autologous EBV-transformed B cells in vitro (Strowig et al. 2009). Depleting of $CD8^+$ T cells during EBV infection leads to higher viral loads and increased incidence of EBV-associated lymphoproliferative disease at 4–6 weeks after huMice infection (Chijioke et al. 2013; Strowig et al. 2009; Yajima et al. 2009). Adoptive transfer of $CD8^+$ T cell clones against lytic EBV antigens only affects viral load prior to that, curbing high viral loads at 3 weeks after infection and eliminating BZLF1 positive B cells from spleen sections (Antsiferova et al. 2014). Therefore, EBV infection primes lytic and latent EBV antigen-specific $CD8^+$ T cells in huMice, which control viral replication and associated tumorigenesis.

The role of $CD4^+$ T cells during EBV infection of huMice is less clear. They do not expand strongly after EBV infection and EBV-specific vaccination (Antsiferova et al. 2014; Chijioke et al. 2013; Gurer et al. 2008; Meixlsperger et al. 2013; Strowig et al. 2009). However, when cloned by limiting dilution, half of them are able to kill autologous EBV-transformed B cell lines and recognize these physiological targets by cytokine production (Meixlsperger et al. 2013; Strowig et al. 2009). Their depletion does also increase EBV viral loads 4 weeks after infection, but seems to only minimally influence EBV-associated tumorigenesis (Strowig et al. 2009). Thus, the recognized antigen breath and functional relevance of EBV-specific $CD4^+$ T cells in infected huMice require further investigations.

## 7 HuMice as a Preclinical Testing Platform for EBV-Specific Vaccines

One method to assess the protective value of distinct EBV-derived antigen specificities is by vaccinating with these antigens and challenging the induced immune responses with EBV infection. However, EBV challenge experiments after vaccination have not yet been performed in huMice. In addition to the targeted antigen, also an adjuvant, usually a PAMP binding to TLRs, is needed for such a vaccination in order to stimulate antigen-presenting cells to up-regulate costimulation and cytokine production for T cell priming. Indeed, huMice have a well-reconstituted human DC compartment, which is able to sense such adjuvants and up-regulate molecules that are essential for T cell priming (Ding et al. 2014;

Meixlsperger et al. 2013). Among different TLR agonists, the double-stranded RNA that mimics polyI:C binding to TLR3 potently activates and matures conventional DCs (Meixlsperger et al. 2013). Moreover, it elicits potent IL-12 and type I IFN production, reaching concentrations that have also been observed in healthy human volunteers injected with this adjuvant formulation (Caskey et al. 2011; Meixlsperger et al. 2013). This adjuvant was combined with the EBV antigen EBNA1, which is expressed in all EBV latencies and even during lytic replication (Lear et al. 1992; Rowe et al. 1992). Moreover, EBNA1 is the most consistently recognized latent $CD4^+$ T cell antigen in healthy EBV carriers (Leen et al. 2001; Münz et al. 2000), and EBNA1-specific T cells were successfully used to treat EBV-associated lymphoproliferative disease after bone marrow transplantation (Icheva et al. 2013). For vaccination, EBNA1 was targeted to DCs with a hybrid antibody specific for the endocytic receptor DEC-205 on DCs (Trumpfheller et al. 2012) and combined with polyI:C (Gurer et al. 2008; Meixlsperger et al. 2013). This vaccination elicited a low frequency of EBNA1-specific $CD4^+$ T cell responses and EBNA1-specific IgM antibodies in the huMice with the highest EBNA1-specific $CD4^+$ T cell responses (Gurer et al. 2008; Meixlsperger et al. 2013). Limiting dilution cloning of these responses revealed that a broad, multiple epitopes recognizing $CD4^+$ T cell response could be primed and that the individual EBNA1-specific $CD4^+$ T cell clones recognized autologous EBV-transformed B cells in a HLA class II-restricted fashion by IFN-$\gamma$ production and degranulation of their cytotoxic vesicles (Meixlsperger et al. 2013). However, in addition to DCs, also other human leukocyte populations were targeted, and indeed, DEC-205 targeting leads also to efficient MHC class II antigen presentation on B cells, which augment the primed $CD4^+$ T cell responses by restimulation (Leung et al. 2013). Moreover, no $CD8^+$ T cell responses were obtained, and therefore, this vaccination strategy most likely needs to be combined with antigen formulations that lead to efficient antigen presentation on MHC class I molecules. Along these lines, recombinant viral vectors encoding EBNA1 can be explored (Duraiswamy et al. 2004; Hui et al. 2013; Taylor et al. 2004).

## 8 Limitations and Future Challenges for Human Immune System Reconstitution in Vivo

Both with respect to the development of humoral immunity and EBV infection stages that are dependent on B cell differentiation, the poor development of germinal centers and secondary lymphoid tissues at mucosal sites are a major impediment for studying EBV immunobiology in huMice. Mucosal reconstitution of secondary lymphoid organ is probably also required for EBV infection via the oropharyngeal physiological route. This deficiency seems to be particularly pronounced in mouse strains without common gamma chain expression ($\gamma_c^{null}$), because IL-7 signaling is essential for innate lymphoid precursor development, which is a prerequisite for the differentiation of lymphoid tissue inducer (LTi) cells (Spits et al. 2013). Indeed, BLT mice that are generated on the NOD–*scid* background have mucosa-associated lymphoid tissues, while BLT mice on the NOD–*scid* $\gamma_c^{null}$ background do not (Nochi et al. 2013). However, even in these, NOD–*scid*-based BLT mice antibody responses develop poorly (Melkus et al. 2006). Another possibility to overcome this shortcoming is to induce secondary lymphoid structures by proinflammatory DC injection (Salguero et al. 2014). The resulting secondary lymphoid tissues then seem to support also more robust antibody responses, but the steady-state physiology of the immune system is altered upon continuously cytokine-producing DC injection, and adjuvant effects of vaccine candidates might be difficult to interpret afterward. The effects of these manipulations on latent EBV infection have not been investigated so far. Nevertheless, overcoming the block of germinal center development is probably the biggest challenge in huMice.

In addition, the analysis of T cell responses in huMice could greatly benefit from improving the tracing of distinct T cell specificities. The low staining intensity of MHC class I/peptide tetramers on T cells of EBV-infected mice makes it difficult to correctly enumerate the respective T cell populations and positively identify low frequencies. Therefore, the introduction of transgenic, easily traceable TCRs into the reconstituting T cell compartment could greatly facilitate investigations of longitudinal T cell differentiation and expansion during EBV infection for both latent and lytic EBV antigen-specific T cell responses. Lessons that could be learned with such possibly lentiviral TCR transfer would then also later on enable us to manipulate other surface receptors and transcription factor of reconstituting human immune system compartments, which could reveal the molecular basis of EBV-specific immune control.

## 9 Important Future Questions in EBV Immunobiology that Can Be Addressed in HuMice

Both virological and immunological questions for EBV can be further addressed in huMice. Mutant viral strains and new viral isolates can be tested for their behavior in vivo (Antsiferova et al. 2014; Tsai et al. 2013; White et al. 2012). We still have to clarify which latency programs can be assessed with the current infection protocols and how we might be able to alter these to access latencies 0, I, and II and their respective malignancies. With respect to EBV-specific immune control, we can now start to interrogate distinct receptors and their role in it, both by overexpression, but also down-regulation in the reconstituting human immune system compartments. These might include T cell receptors with distinct antigen specificities and activating receptors on NK cells. HuMice hold the promise that we might be able to perform sophisticated mechanistic studies, so far only possible in regular mouse models, with the human immune system and real human pathogens like EBV in vivo.

**Acknowledgements** Work in my laboratory is supported by Cancer Research Switzerland (KFS-3234-08-2013), the Association for International Cancer Research (14-1033), KFSP$^{MS}$ and KFSP$^{HLD}$ of the University of Zurich, the Baugarten Foundation, the Sobek Foundation, Fondation Acteria, the Wellcome Trust, the Leukaemia and Lymphoma Research, the Medical Research Council, and the Swiss National Science Foundation (310030_143979 and CRSII3_136241).

## References

Antsiferova O, Müller A, Rämer P, Chijioke O, Chatterjee B, Raykova A, Planas R, Sospedra M, Shumilov A, Tsai MH et al (2014) Adoptive transfer of EBV specific CD8$^+$ T cell clones can transiently control EBV infection in humanized mice. PLoS Pathog 10(8):e1004333

Ariza ME, Glaser R, Kaumaya PT, Jones C, Williams MV (2009) The EBV-encoded dUT-Pase activates NF-kappa B through the TLR2 and MyD88-dependent signaling pathway. J Immunol 182:851–859

Babcock JG, Hochberg D, Thorley-Lawson AD (2000) The expression pattern of Epstein-Barr virus latent genes in vivo is dependent upon the differentiation stage of the infected B cell. Immunity 13:497–506

Bhaduri-McIntosh S, Rotenberg MJ, Gardner B, Robert M, Miller G (2008) Repertoire and frequency of immune cells reactive to Epstein-Barr virus-derived autologous lymphoblastoid cell lines. Blood 111:1334–1343

Bickham K, Goodman K, Paludan C, Nikiforow S, Tsang ML, Steinman RM, Münz C (2003) Dendritic cells initiate immune control of Epstein-Barr virus transformation of B lymphocytes in vitro. J Exp Med 198:1653–1663

Billerbeck E, Horwitz JA, Labitt RN, Donovan BM, Vega K, Budell WC, Koo GC, Rice CM, Ploss A (2013) Characterization of human antiviral adaptive immune responses during hepatotropic virus infection in HLA-transgenic human immune system mice. J Immunol 191:1753–1764

Caskey M, Lefebvre F, Filali-Mouhim A, Cameron MJ, Goulet JP, Haddad EK, Breton G, Trumpfheller C, Pollak S, Shimeliovich I et al (2011) Synthetic double-stranded RNA induces innate immune responses similar to a live viral vaccine in humans. J Exp Med 208:2357–2366

Cesarman E (2014) Gammaherpesviruses and lymphoproliferative disorders. Annu Rev Pathol 9:349–372

Chatterjee B, Leung CS, Münz C (2014) Animal models of Epstein Barr virus infection. J Immunol Methods 410:80–87

Chijioke O, Muller A, Feederle R, Barros MH, Krieg C, Emmel V, Marcenaro E, Leung CS, Antsiferova O, Landtwing V et al (2013) Human natural killer cells prevent infectious mononucleosis features by targeting lytic Epstein-Barr virus infection. Cell Rep 5:1489–1498

Chung BK, Tsai K, Allan LL, Zheng DJ, Nie JC, Biggs CM, Hasan MR, Kozak FK, van den Elzen P, Priatel JJ, Tan R (2013) Innate immune control of EBV-infected B cells by invariant natural killer T cells. Blood 122:2600–2608

Cocco M, Bellan C, Tussiwand R, Corti D, Traggiai E, Lazzi S, Mannucci S, Bronz L, Palummo N, Ginanneschi C et al (2008) CD34$^+$ cord blood cell-transplanted Rag2$^{-/-}$ gamma c$^{-/-}$ mice as a model for Epstein-Barr virus infection. Am J Pathol 173:1369–1378

Ding Y, Wilkinson A, Idris A, Fancke B, O'Keeffe M, Khalil D, Ju X, Lahoud MH, Caminschi I, Shortman K et al (2014) FLT3-ligand treatment of humanized mice results in the generation of large numbers of CD141$^+$ and CD1c$^+$ dendritic cells in vivo. J Immunol 192:1982–1989

Duraiswamy J, Bharadwaj M, Tellam J, Connolly G, Cooper L, Moss D, Thomson S, Yotnda P, Khanna R (2004) Induction of therapeutic T-cell responses to subdominant tumor-associated viral oncogene after immunization with replication-incompetent polyepitope adenovirus vaccine. Cancer Res 64:1483–1489

Epstein MA, Achong BG, Barr YM (1964a) Virus particles in cultured lymphoblasts from Burkitt's lymphoma. Lancet 1:702–703
Epstein MA, Henle G, Achong BG, Barr YM (1964b) Morphological and biological studies on a virus in cultured lymphoblasts from Burkitt's lymphoma. J Exp Med 121:761–770
Fiola S, Gosselin D, Takada K, Gosselin J (2010) TLR9 contributes to the recognition of EBV by primary monocytes and plasmacytoid dendritic cells. J Immunol 185:3620–3631
Gaudreault E, Fiola S, Olivier M, Gosselin J (2007) Barr virus induces MCP-1 secretion by human monocytes via TLR2. J Virol 81:8016–8024
Gurer C, Strowig T, Brilot F, Pack M, Trumpfheller C, Arrey F, Park CG, Steinman RM, Münz C (2008) Targeting the nuclear antigen 1 of Epstein Barr virus to the human endocytic receptor DEC-205 stimulates protective T-cell responses. Blood 112:1231–1239
Heuts F, Rottenberg ME, Salamon D, Rasul E, Adori M, Klein G, Klein E, Nagy N (2014) T cells modulate Epstein-Barr virus latency phenotypes during infection of humanized mice. J Virol 88:3235–3245
Hjalgrim H, Askling J, Rostgaard K, Hamilton-Dutoit S, Frisch M, Zhang JS, Madsen M, Rosdahl N, Konradsen HB, Storm HH, Melbye M (2003) Characteristics of Hodgkin's lymphoma after infectious mononucleosis. N Engl J Med 349:1324–1332
Hong GK, Gulley ML, Feng WH, Delecluse HJ, Holley-Guthrie E, Kenney SC (2005) Epstein-Barr virus lytic infection contributes to lymphoproliferative disease in a SCID mouse model. J Virol 79:13993–14003
Hui EP, Taylor GS, Jia H, Ma BB, Chan SL, Ho R, Wong WL, Wilson S, Johnson BF, Edwards C et al (2013) Phase I trial of recombinant modified vaccinia ankara encoding Epstein-Barr viral tumor antigens in nasopharyngeal carcinoma patients. Cancer Res 73:1676–1688
Hutt-Fletcher LM (2007) Epstein-Barr virus entry. J Virol 81:7825–7832
Icheva V, Kayser S, Wolff D, Tuve S, Kyzirakos C, Bethge W, Greil J, Albert MH, Schwinger W, Nathrath M et al (2013) Adoptive transfer of epstein-barr virus (EBV) nuclear antigen 1-specific t cells as treatment for EBV reactivation and lymphoproliferative disorders after allogeneic stem-cell transplantation. J Clin Oncol 31:39–48
Ishikawa F, Yasukawa M, Lyons B, Yoshida S, Miyamoto T, Yoshimoto G, Watanabe T, Akashi K, Shultz LD, Harada M (2005) Development of functional human blood and immune systems in NOD/SCID/IL2 receptor gamma chain(null) mice. Blood 106:1565–1573
Islas-Ohlmayer M, Padgett-Thomas A, Domiati-Saad R, Melkus MW, Cravens PD, Martin Mdel P, Netto G, Garcia JV (2004) Experimental infection of NOD/SCID mice reconstituted with human $CD34^+$ cells with Epstein-Barr virus. J Virol 78:13891–13900
Iwakiri D, Zhou L, Samanta M, Matsumoto M, Ebihara T, Seya T, Imai S, Fujieda M, Kawa K, Takada K (2009) Epstein-Barr virus (EBV)-encoded small RNA is released from EBV-infected cells and activates signaling from Toll-like receptor 3. J Exp Med 206:2091–2099
Kuwana Y, Takei M, Yajima M, Imadome K, Inomata H, Shiozaki M, Ikumi N, Nozaki T, Shiraiwa H, Kitamura N et al (2011) Epstein-Barr virus induces erosive arthritis in humanized mice. PLoS ONE 6:e26630
Laichalk LL, Thorley-Lawson DA (2005) Terminal differentiation into plasma cells initiates the replicative cycle of Epstein-Barr virus in vivo. J Virol 79:1296–1307
Lavender KJ, Pang WW, Messer RJ, Duley AK, Race B, Phillips K, Scott D, Peterson KE, Chan CK, Dittmer U et al (2013) BLT-humanized C57BL/6 $Rag2^{-/-}$gamma $c^{-/-}CD47^{-/-}$ mice are resistant to GVHD and develop B- and T-cell immunity to HIV infection. Blood 122:4013–4020
Lear AL, Rowe M, Kurilla MG, Lee S, Henderson S, Kieff E, Rickinson AB (1992) The Epstein-Barr virus (EBV) nuclear antigen 1 BamHI F promoter is activated on entry of EBV-transformed B cells into the lytic cycle. J Virol 66:7461–7468
Leen A, Meij P, Redchenko I, Middeldorp J, Bloemena E, Rickinson A, Blake N (2001) Differential immunogenicity of Epstein-Barr virus latent-cycle proteins for human $CD4^+$ T-helper 1 responses. J Virol 75:8649–8659

Legrand N, Huntington ND, Nagasawa M, Bakker AQ, Schotte R, Strick-Marchand H, de Geus SJ, Pouw SM, Bohne M, Voordouw A et al (2011) Functional CD47/signal regulatory protein alpha (SIRP(alpha)) interaction is required for optimal human T- and natural killer- (NK) cell homeostasis in vivo. Proc Natl Acad Sci USA 108:13224–13229

Leung CS, Maurer MA, Meixlsperger S, Lippmann A, Cheong C, Zuo J, Haigh TA, Taylor GS, Münz C (2013) Robust T-cell stimulation by Epstein-Barr virus-transformed B cells after antigen targeting to DEC-205. Blood 121:1584–1594

Lim WH, Kireta S, Russ GR, Coates PT (2006) Human plasmacytoid dendritic cells regulate immune responses to Epstein-Barr virus (EBV) infection and delay EBV-related mortality in humanized NOD-SCID mice. Blood 109:1043–1050

Lotz M, Tsoukas CD, Fong S, Carson DA, Vaughan JH (1985) Regulation of Epstein-Barr virus infection by recombinant interferons. Selected sensitivity to interferon-gamma. Eur J Immunol 15:520–525

Luzuriaga K, Sullivan JL (2010) Infectious mononucleosis. N Engl J Med 362:1993–2000

Ma SD, Hegde S, Young KH, Sullivan R, Rajesh D, Zhou Y, Jankowska-Gan E, Burlingham WJ, Sun X, Gulley ML et al (2011) A new model of Epstein-Barr virus infection reveals an important role for early lytic viral protein expression in the development of lymphomas. J Virol 85:165–177

Marodon G, Desjardins D, Mercey L, Baillou C, Parent P, Manuel M, Caux C, Bellier B, Pasqual N, Klatzmann D (2009) High diversity of the immune repertoire in humanized NOD.SCID. gamma $c^{-/-}$ mice. Eur J Immunol 39:2136–2145

McGeoch DJ (2001) Molecular evolution of the gamma-Herpesvirinae. Philos Trans R Soc Lond B Biol Sci 356:421–435

McGeoch DJ, Dolan A, Ralph AC (2000) Toward a comprehensive phylogeny for mammalian and avian herpesviruses. J Virol 74:10401–10406

Meixlsperger S, Leung CS, Ramer PC, Pack M, Vanoaica LD, Breton G, Pascolo S, Salazar AM, Dzionek A, Schmitz J et al (2013) $CD141^{+}$ dendritic cells produce prominent amounts of IFN-alpha after dsRNA recognition and can be targeted via DEC-205 in humanized mice. Blood 121:5034–5044

Melkus MW, Estes JD, Padgett-Thomas A, Gatlin J, Denton PW, Othieno FA, Wege AK, Haase AT, Garcia JV (2006) Humanized mice mount specific adaptive and innate immune responses to EBV and TSST-1. Nat Med 12:1316–1322

Mestas J, Hughes CC (2004) Of mice and not men: differences between mouse and human immunology. J Immunol 172:2731–2738

Miller G, Lipman M (1973) Comparison of the yield of infectious virus from clones of human and simian lymphoblastoid lines transformed by Epstein-Barr virus. J Exp Med 138:1398–1412

Münz C, Bickham KL, Subklewe M, Tsang ML, Chahroudi A, Kurilla MG, Zhang D, O'Donnell M, Steinman RM (2000) Human $CD4^{+}$ T lymphocytes consistently respond to the latent Epstein-Barr virus nuclear antigen EBNA1. J Exp Med 191:1649–1660

Nochi T, Denton PW, Wahl A, Garcia JV (2013) Cryptopatches are essential for the development of human GALT. Cell Rep 3:1874–1884

Pappworth IY, Wang EC, Rowe M (2007) The switch from latent to productive infection in epstein-barr virus-infected B cells is associated with sensitization to NK cell killing. J Virol 81:474–482

Ramer PC, Chijioke O, Meixlsperger S, Leung CS, Münz C (2011) Mice with human immune system components as in vivo models for infections with human pathogens. Immunol Cell Biol 89:408–416

Reizis B, Bunin A, Ghosh HS, Lewis KL, Sisirak V (2011) Plasmacytoid dendritic cells: recent progress and open questions. Annu Rev Immunol 29:163–183

Rickinson AB, Long HM, Palendira U, Münz C, Hislop A (2014) Cellular immune controls over Epstein-Barr virus infection: new lessons from the clinic and the laboratory. Trends Immunol 35:159–169

Rivailler P, Cho YG, Wang F (2002) Complete genomic sequence of an Epstein-Barr virus-related herpesvirus naturally infecting a new world primate: a defining point in the evolution of oncogenic lymphocryptoviruses. J Virol 76:12055–12068

Rongvaux A, Takizawa H, Strowig T, Willinger T, Eynon EE, Flavell RA, Manz MG (2013) Human hemato-lymphoid system mice: current use and future potential for medicine. Annu Rev Immunol 31:635–674

Rouphael NG, Talati NJ, Vaughan C, Cunningham K, Moreira R, Gould C (2007) Infections associated with haemophagocytic syndrome. Lancet Infect Dis 7:814–822

Rowe M, Lear AL, Croom-Carter D, Davies AH, Rickinson AB (1992) Three pathways of Epstein-Barr virus gene activation from EBNA1- positive latency in B lymphocytes. J Virol 66:122–131

Salguero G, Daenthanasanmak A, Münz C, Raykova A, Guzman CA, Riese P, Figueiredo C, Länger F, Schneider A, Macke L et al (2014) Dendritic cell-mediated immune humanization of mice: implications for allogeneic and xenogeneic stem cell transplantation. J Immunol 192:4636–4647

Sato K, Misawa N, Nie C, Satou Y, Iwakiri D, Matsuoka M, Takahashi R, Kuzushima K, Ito M, Takada K, Koyanagi Y (2011) A novel animal model of Epstein-Barr virus-associated hemophagocytic lymphohistiocytosis in humanized mice. Blood 117:5663–5673

Serra-Hassoun M, Bourgine M, Boniotto M, Berges J, Langa F, Michel ML, Freitas AA, Garcia S (2014) Human hematopoietic reconstitution and HLA-restricted responses in nonpermissive alymphoid mice. J Immunol 193:1504–1511

Severa M, Giacomini E, Gafa V, Anastasiadou E, Rizzo F, Corazzari M, Romagnoli A, Trivedi P, Fimia GM, Coccia EM (2013) EBV stimulates TLR- and autophagy-dependent pathways and impairs maturation in plasmacytoid dendritic cells: implications for viral immune escape. Eur J Immunol 43:147–158

Shultz LD, Lyons BL, Burzenski LM, Gott B, Chen X, Chaleff S, Kotb M, Gillies SD, King M, Mangada J et al (2005) Human lymphoid and myeloid cell development in NOD/LtSz-scid IL2R gamma null mice engrafted with mobilized human hemopoietic stem cells. J Immunol 174:6477–6489

Shultz LD, Saito Y, Najima Y, Tanaka S, Ochi T, Tomizawa M, Doi T, Sone A, Suzuki N, Fujiwara H et al (2010) Generation of functional human T-cell subsets with HLA-restricted immune responses in HLA class I expressing NOD/SCID/IL2r gamma(null) humanized mice. Proc Natl Acad Sci USA 107:13022–13027

Spits H, Artis D, Colonna M, Diefenbach A, Di Santo JP, Eberl G, Koyasu S, Locksley RM, McKenzie AN, Mebius RE et al (2013) Innate lymphoid cells–a proposal for uniform nomenclature. Nat Rev Immunol 13:145–149

Strowig T, Gurer C, Ploss A, Liu YF, Arrey F, Sashihara J, Koo G, Rice CM, Young JW, Chadburn A et al (2009) Priming of protective T cell responses against virus-induced tumors in mice with human immune system components. J Exp Med 206:1423–1434

Strowig T, Rongvaux A, Rathinam C, Takizawa H, Borsotti C, Philbrick W, Eynon EE, Manz MG, Flavell RA (2011) Transgenic expression of human signal regulatory protein alpha in Rag2$^{-/-}$gamma c$^{-/-}$ mice improves engraftment of human hematopoietic cells in humanized mice. Proc Natl Acad Sci USA 108:13218–13223

Sundstrom Y, Nilsson C, Lilja G, Karre K, Troye-Blomberg M, Berg L (2007) The expression of human natural killer cell receptors in early life. Scand J Immunol 66:335–344

Takenaka K, Prasolava TK, Wang JC, Mortin-Toth SM, Khalouei S, Gan OI, Dick JE, Danska JS (2007) Polymorphism in Sirpa modulates engraftment of human hematopoietic stem cells. Nat Immunol 8:1313–1323

Taylor GS, Haigh TA, Gudgeon NH, Phelps RJ, Lee SP, Steven NM, Rickinson AB (2004) Dual stimulation of Epstein-Barr Virus (EBV)-specific CD4$^{+}$- and CD8$^{+}$-T-cell responses by a chimeric antigen construct: potential therapeutic vaccine for EBV-positive nasopharyngeal carcinoma. J Virol 78:768–778

Thacker EL, Mirzaei F, Ascherio A (2006) Infectious mononucleosis and risk for multiple sclerosis: a meta-analysis. Ann Neurol 59:499–503

Traggiai E, Chicha L, Mazzucchelli L, Bronz L, Piffaretti JC, Lanzavecchia A, Manz MG (2004) Development of a human adaptive immune system in cord blood cell-transplanted mice. Science 304:104–107

Trumpfheller C, Longhi MP, Caskey M, Idoyaga J, Bozzacco L, Keler T, Schlesinger SJ, Steinman RM (2012) Dendritic cell-targeted protein vaccines: a novel approach to induce T-cell immunity. J Intern Med 271:183–192

Tsai MH, Raykova A, Klinke O, Bernhardt K, Gartner K, Leung CS, Geletneky K, Sertel S, Münz C, Feederle R, Delecluse HJ (2013) Spontaneous lytic replication and epitheliotropism define an Epstein-Barr virus strain found in carcinomas. Cell Rep 5:458–470

Wang H, Madariaga ML, Wang S, Van Rooijen N, Oldenborg PA, Yang YG (2007) Lack of CD47 on nonhematopoietic cells induces split macrophage tolerance to CD47null cells. Proc Natl Acad Sci USA 104:13744–13749

Watanabe Y, Takahashi T, Okajima A, Shiokawa M, Ishii N, Katano I, Ito R, Ito M, Minegishi M, Minegishi N et al (2009) The analysis of the functions of human B and T cells in humanized NOD/shi-scid/gammac$^{null}$ (NOG) mice (hu-HSC NOG mice). Int Immunol 21:843–858

White RE, Ramer PC, Naresh KN, Meixlsperger S, Pinaud L, Rooney C, Savoldo B, Coutinho R, Bodor C, Gribben J et al (2012) EBNA3B-deficient EBV promotes B cell lymphomagenesis in humanized mice and is found in human tumors. J Clin Invest 122:1487–1502

Yajima M, Imadome K, Nakagawa A, Watanabe S, Terashima K, Nakamura H, Ito M, Shimizu N, Honda M, Yamamoto N, Fujiwara S (2008) A new humanized mouse model of Epstein-Barr virus infection that reproduces persistent infection, lymphoproliferative disorder, and cell-mediated and humoral immune responses. J Infect Dis 198:673–682

Yajima M, Imadome K, Nakagawa A, Watanabe S, Terashima K, Nakamura H, Ito M, Shimizu N, Yamamoto N, Fujiwara S (2009) T cell-mediated control of Epstein-Barr virus infection in humanized mice. J Infect Dis 200:1611–1615

Yamauchi T, Takenaka K, Urata S et al (2013) Polymorphic SIRPalpha is the genetic determinant for NOD-based mouse lines to achieve efficient human cell engraftment. Blood 121:1316–1325

Yuling H, Ruijing X, Li L, Xiang J, Rui Z, Yujuan W, Lijun Z, Chunxian D, Xinti T, Wei X et al (2009) EBV-induced human CD8$^{+}$ NKT cells suppress tumorigenesis by EBV-associated malignancies. Cancer Res 69:7935–7944

# Part V
# Therapy of EBV Associated Diseases

# Adoptive T-Cell Immunotherapy

**Stephen Gottschalk and Cliona M. Rooney**

**Abstract** Epstein-Barr virus (EBV) is associated with a range of malignancies involving B cells, T cells, natural killer (NK) cells, epithelial cells, and smooth muscle. All of these are associated with the latent life cycles of EBV, but the pattern of latency-associated viral antigens expressed in tumor cells depends on the type of tumor. EBV-specific T cells (EBVSTs) have been explored as prophylaxis and therapy for EBV-associated malignancies for more than two decades. EBVSTs have been most successful as prophylaxis and therapy for post-transplant lymphoproliferative disease (PTLD), which expresses the full array of latent EBV antigens (type 3 latency), in hematopoietic stem-cell transplant (HSCT) recipients. While less effective, clinical studies have also demonstrated their therapeutic potential for PTLD post-solid organ transplant and for EBV-associated malignancies such as Hodgkin's lymphoma, non-Hodgkin's lymphoma, and nasopharyngeal carcinoma (NPC) that express a limited array of latent EBV antigens (type 2 latency). Several approaches are actively being pursued to improve the antitumor activity of EBVSTs including activation and expansion of T cells specific for the EBV antigens expressed in type 2 latency, genetic approaches to render EBVSTs resistant to the immunosuppressive tumor environment, and combination approaches with other immune-modulating modalities. Given the recent advances and renewed interest in cell therapy, we hope that EBVSTs will become an integral part of our treatment armamentarium against EBV-positive malignancies in the near-future.

S. Gottschalk (✉) · C.M. Rooney (✉)
Center for Cell and Gene Therapy, Texas Children's Hospital,
Houston Methodist Hospital, Baylor College of Medicine,
1102 Bates Street, Suite 1770, Houston, TX 77030, USA
e-mail: smg@bcm.edu

C.M. Rooney
e-mail: crooney@bcm.edu



C. Münz (ed.), *Epstein Barr Virus Volume 2*, Current Topics in Microbiology and Immunology 391, DOI 10.1007/978-3-319-22834-1_15

## Contents



## 1 Introduction

Epstein-Barr virus (EBV) is associated with a range of malignancies involving B cells, T cells, natural killer (NK) cells, epithelial cells, and smooth muscle. All of these are associated with the latent life cycles of EBV, but the pattern of latency-associated viral antigens expressed in tumor cells depends on the type of tumor. True latency (no expression of viral antigens) is found only in normal memory B cells and never in EBV-associated malignancies.

The viral antigens expressed in EBV-positive tumors provide target antigens for immune-based therapies and T cells specific for each of the latency-associated antigens have been detected in patients with malignancies, as well as in healthy individuals (Fig. 1). Therefore, even tumors such as Burkitt's lymphoma (BL) and gastric carcinoma (GC) that express only EBNA1 and BARF1 (type 1 latency) can, in principal, be targeted by T cells. Malignancies such as B-, T-, and NK-cell lymphomas and nasopharyngeal carcinoma (NPC) express additional, more immunogenic target antigens, LMP1 and LMP2, a pattern termed type 2 latency. Type 3 latency involves the expression of all latency-associated antigens and adds EBNA's -2, -3a, -3b, -3c, and -LP to the range of viral antigens that can be targeted. This highly immunogenic form of latency is observed only in patients who are severely immunosuppressed for example by stem-cell or solid organ transplantation, congenital immunodeficiency, or HIV infection. All healthy seropositive

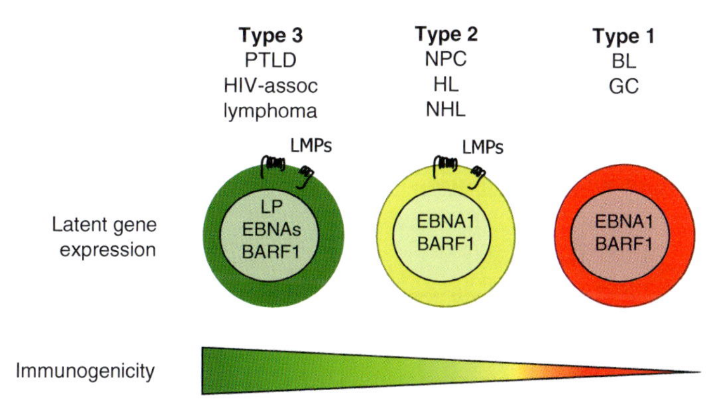


**Fig. 1** Immunogenicity of EBV-positive tumors according to latency. For details see text. EBNAs: EBNAs -2, -3a, -3b, and -3c; LMPs: LMP1 and LMP2

individuals and most patients carry a broad repertoire of T cells specific for a range of EBV latency antigens that can be reactivated and expanded ex vivo for therapeutic use. The frequency of T cells specific for EBV early lytic cycle antigens is usually higher than for the latency antigens (Steven et al. 1997; Pudney et al. 2005), and while these T cells likely control virus spread by killing lytically infected cells before they can release infectious, their role, if any, in the control of malignancies is unknown.

EBV-specific T cells (EBVSTs) have had outstanding success for the treatment of immunogenic type 3 latency, and infusion of donor-derived EBVSTs in hematopoietic stem-cell transplant (HSCT) recipients rapidly restores EBV-specific immunity. EBVSTs are less effective in type 2 malignancies that develop in immune-competent hosts because these have developed sophisticated immune evasion strategies. However, EBVSTs have produced CRs in patients with locoregional NPC (Louis et al. 2010) and prolonged overall survival in a larger group of patients with more extensive disease (Chia et al. 2014). Responses in type 2 latency lymphoma were achieved by only focusing T cells on the type 2 latency antigens, but such T cells produce tumor responses in over 70 % of patients and complete responses (CRs) in over 50 % (Bollard et al. 2004, 2007, 2014). However, to ensure clinical efficacy in all patients, additional strategies will be required to overcome tumor immune evasion strategies and enable T-cell expansion and continued antitumor function after infusion (Poppema 2005). Gene modifications of EBVSTs may be used to provide intrinsic resistance to inhibitory molecules, to express growth-promoting genes, or to provide additional specificity for stromal cells. Alternatively, EBVSTs may be combined with other immunomodulatory agents, such as checkpoint inhibitors or vaccines.

There are many advantages to the use of EBVSTs for the treatment of EBV-associated malignancies, not least of which their lack of short- or long-term toxicities demonstrated in hundreds of patients who continue with their normal lives during and after therapy. Further, a single infusion of a small dose of T cells can proliferate exponentially in the patient, eliminate tumors, enter the memory compartment, and provide life-long antitumor immunity. While previously thought to be a boutique therapy available only in institutions with specialized cell culture facilities, recent successes of gene-modified T cells in more common malignancies (Kalos et al. 2011; Grupp et al. 2013; Brentjens et al. 2013; Rosenberg and Dudley 2009) have captured the interest of the pharmaceutical industry which is now bringing T-cell therapies to a wider patient population.

Other chapters have discussed the different malignancies associated with EBV (Chapters in book section '**Viral Associated Diseases**'), their complex patterns of gene expression, and the functions of the latency-associated genes (Chapters in book section '**EBV Latency**'). For the purpose of this chapter, we will discuss the protein products of latency genes predominantly as targets for adoptive immunotherapy with EBVSTs in malignancies with different patterns of latent gene expression.

## 2 T Cells for Type 3 Malignancies

### *2.1 Adoptive Transfer of EBVSTs for the Prevention and Treatment of EBV-Associated Lymphoproliferative Disease (PTLD) After HSCT*

Prior to the advent of B-cell-depleting monoclonal antibody (MAb) to CD20, Rituxan®, and FDA approval of the CD52-specific MAb Campath® in the USA in 2001 allowing in vitro or in vivo depletion of T cells and B cells, EBV-associated post-transplant lymphoproliferative disease (PTLD) was a significant problem in recipients of T-cell-depleted stem cells from HLA-mismatched or unrelated donors. While T-cell depletion reduces the incidence of graft versus host disease (GVHD), it leaves patients more vulnerable to infections with viruses including EBV, allowing investigators to evaluate the safety and efficacy of T cells for the treatment of malignancy. The first studies used unmanipulated donor lymphocyte infusions (DLIs) that had proved effective for the treatment of leukemic relapse after allogeneic HSCT (Fig. 2; Table 1) (Porter et al. 1994). DLIs were also effective for PTLD, but were associated with significant toxicity due to GVHD (Chia et al. 2014).

Our group therefore specifically activated EBVSTs from stem-cell donors using autologous EBV-transformed B-lymphoblastoid cell lines (LCLs) as antigen-presenting cells (APCs) in the hope that the incidence of GVHD would be reduced (Wallace et al. 1982; Rooney et al. 1995). T cells reactivated using LCLs are ideal for the treatment of type 3 malignancies since they express the same range of viral antigens and stimulate CD4- and CD8-positive T cells with specificity for a broad range of viral and non-viral tumor antigens (Cobbold et al. 2013). LCLs are outstanding APCs, since they can be prepared from most donors, provide an

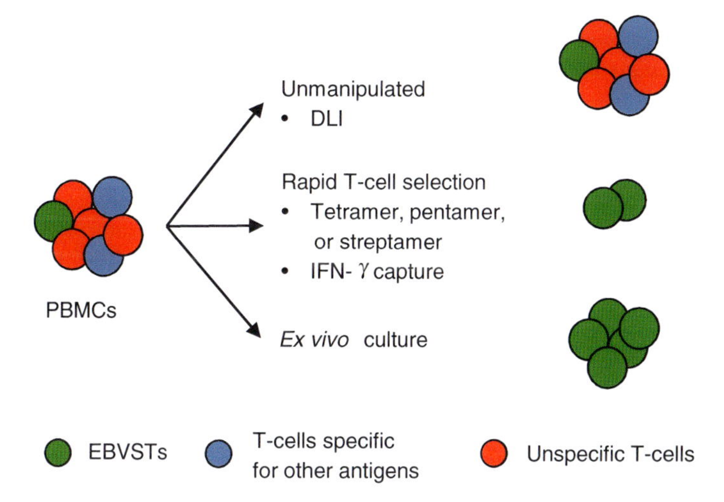


**Fig. 2** T-cell products for EBV-positive malignancies. For details see text. *DLI*—donor lymphocyte infusion

unlimited source of autologous professional APCs, and present a full gamut of EBV antigens. 101 patients received EBVSTs as prophylaxis and 13 as treatment for active PTLD (Heslop et al. 2010). After infusion, T cells proliferated extensively, restored immunity to EBV, and persisted for up to 10 years as demonstrated by gene-marking in the first 26 patients (Rooney et al. 1995, 1998; Heslop et al. 1996). No patient who received EBVSTs as prophylaxis developed PTLD, by comparison with 12 % of historical controls and high virus loads were reduced. Of the 13 patients who received EBVSTs as treatment for active disease, 11 patients had complete and permanent tumor responses. Importantly, there were few short- and no long-term toxicities associated with treatment. There was no incidence of de novo GVHD and no patients suffered cytokine storm, even during the elimination of bulky tumors. This contrasts with responses of B-cell malignancies to adoptively transferred CD3- and CD28-activated T cells modified with chimeric antigen receptors (CARs) for CD19 (Brentjens et al. 2013; Porter et al. 2011).

EBVSTs produce similar clinical efficacy when infused as a component of multivirus-specific T cells (Leen et al. 2006, 2009). Peripheral blood mononuclear cells (PBMCs) stimulated with monocytes and LCLs transduced with a recombinant adenovirus vector expressing pp65 of cytomegalovirus (CMV) demonstrated robust specificity for EBV and CMV pp65 as well as to hexon and penton of adenovirus that are processed and presented from the virion. As few as $2 \times 10^7$ total T cells were effective at clearing all three viruses, and since EBVSTs comprised only a minor fraction of the total, these studies showed that very small numbers of EBVSTs could reconstitute EBV-specific immunity in the HSCT setting (Leen et al. 2006).

**Table 1** Selected T-cell therapy clinical studies for type 3 malignancy—PTLD post-HSCT

| T-cell product | Production method | Comment |
|---|---|---|
| *Allogeneic—donor derived* | | |
| DLI | n/a | *MSKCC experience*: Thr (n = 30): 22 CR or PR; Incidence of GVHD: 17 % (Doubrovina et al. 2012)<br>Smaller case series have reported similar response rates, but a higher incidence of GHVD (up to 40 %) |
| EBV-specific Ts | LCL | *SJCRH and BCM experience*: Pro (n = 101): no PTLD development; Thr (n = 13): 11 CR (Heslop et al. 2010)<br>*MSKCC experience*: Thr (n = 19): 13 CR or PR (Doubrovina et al. 2012)<br>Smaller case series have reported similar response rates |
| | IFN-γ capture post-stim with peptides derived from 11 EBV antigens | Thr (n = 6): 3 CR (Moosmann et al. 2010) |
| | IFN-γ capture post-stim with EBNA1 antigen | Thr (n = 10): 7 CR or PR (Icheva et al. 2013) |
| | Pentamer selection | Thr (n = 1): 1 CR (Uhlin et al. 2012) |
| EBV-, Adv-specific Ts | LCL modified with Ad5f35 | Pro (n = 13): no PTLD development<br>Thr (n = 1): 1 CR (Leen et al. 2009) |
| EBV-,CMV-, Adv-specific Ts | LCL modified with Ad5f35.CMVpp65 | Pro (n = 20): no PTLD development<br>Thr (n = 6): 6 CR (Leen et al. 2006) |
| | DC/plasmid E1ΔGA-LMP2-I-BZLF1 | Pro (n=9): no PTLD development<br>Thr (n=2): 2 CR (Gerdemann et al. 2013) |
| EBV-,CMV-, Adv-, BKV-, HHV6-specific Ts | Monocytes loaded with pepmixes for E1ΔGA, LMP2, BZLF1 | Thr (n = 5): 5 CR (Papadopoulou et al. 2014) |
| *Allogeneic—3rd party* | | |
| EBV-specific Ts | LCL | *Edinburgh experience*: Thr (n=3): 2 CR (Haque et al. 2002)<br>*MSKCC experience*: Thr (n=2): 2 CR (Barker et al. 2010) |
| EBV-,CMV-, Adv-specific Ts | LCL modified with Ad5f35.CMVpp65 | Thr (n = 9): 6 CR or PR (Haque et al. 2007) |

*DCs* Dendritic cells; *CR* Complete response; *PR* Partial Response; *Pro* Prophylaxis; *Thr* Therapy

## 2.2 Adoptive Transfer of EBVSTs for the Prevention and Treatment of PTLD After Solid Organ Transplant (SOT)

PTLDs occurring after SOT are usually of recipient origin, and therefore, EBVSTs are ideally autologous. EBVSTs are readily generated from SOT recipients, even those with active PTLD, despite their in vivo suppression with drugs such as steroids, calcineurin inhibitors, or mTor inhibitors, suggesting that EBVSTs are present, but are unable to respond adequately to virus reactivation (Savoldo et al. 2001). SOT recipients receiving EBVSTs must remain on immunosuppressants to prevent graft rejection and unlike HSCT recipients are not lymphodepleted. Nevertheless, Haque et al. showed that autologous EBVSTs infused in 3 escalating doses one month apart, expanded after infusion into 3 SOT recipients, and reduced virus load for 3 months without causing graft rejection (Haque et al. 1998). Comoli et al. (1997, 2002) generated EBVSTs from 23 SOT recipients deemed at high risk for PTLD based on a high virus load. Seven of these were safely infused with up to 5 doses of EBVSTs with a decrease of virus load in 5 patients and increases in the EBV-specific T cell precursor frequency in those tested (Table 2). The demonstration of safety in this patient population is important, but the ability of the adoptively transferred cells to persist and function long term remains unclear (Savoldo et al. 2006; Khanna et al. 1999; Sherritt et al. 2003). To enable adoptively transferred EBVSTs to expand and function in patients on immunosuppressant drugs to prevent graft rejection, two groups rendered T-cell resistant to specific immunosuppressive drugs (Brewin et al. 2009; De Angelis et al. 2009; Huye et al. 2011). These will be discussed in the Sect. 5.1.

**Table 2** Selected T-cell therapy clinical studies for Type 3 malignancy—PTLD post-SOT

| T-cell product | Production method | Comment |
|---|---|---|
| *Autologous* | | |
| EBV-specific Ts | LCL | *Edinburgh experience*: Pro (n = 3): decrease in viral load; no PTLD development (Haque et al. 1998)<br>*Pavia experience*: Pro (n = 7): decrease in viral load in 5/7 patients; no PTLD development (Comoli et al. 2002)<br>*BCM experience*: Pro (n=12): variable effects on viral load; no PTLD development (Savoldo et al. 2006)<br>*Two case reports*: Thr (n=2): 1 CR or 1 PR (Khanna et al. 1999; Sherritt et al. 2003) |
| *Allogeneic*—3rd *party* | | |
| EBV-specific Ts | LCL | Thr (n = 38): 23 CR or PR (Haque et al. 2002, 2007) |

*CR* Complete response; *PR* Partial Response; *Pro* Prophylaxis; *Thr* Therapy

## 2.3 *Rapid Selection of EBVSTs from Donor Blood for Adoptive Transfer*

The activation and expansion of EBVSTs using LCLs is a lengthy process, requiring six weeks to establish LCLs, then at least 4 weeks to expand EBVSTs followed by two weeks for quality-control testing (Fig. 3). Since PTLD is rapidly progressive, EBVSTs must be made in advance to be of clinical benefit and since the incidence of PTLD is low, many lines would never be infused. Therefore, unless all patients are treated prophylactically, this type of manufacturing is not practical for wider use. Hence, investigators evaluated strategies to isolate virus-specific T cells (VSTs) directly from donor peripheral blood. This strategy was first evaluated for the control of CMV reactivation. CMV-specific T cells (CMVSTs) were selected either using HLA-peptide multimers or streptamers, or by magnetic isolation of T cells that secrete interferon (IFN)-γ in response to antigen stimulation (gamma capture) (Cobbold et al. 2005; Peggs et al. 2011; Feuchtinger et al. 2004). Even using large starting blood volumes, the VST numbers recovered are generally small allowing only small doses of cells to be infused (Cobbold et al. 2005; Peggs et al. 2011; Uhlin et al. 2012). The silver lining to this drawback was finding that a very few VSTs could expand exponentially in HSCT recipients and control disease. Less than $10^4$ tetramer-selected CMVSTs per kg of patient body weight were able to expand in patients and eliminate CMV viremia

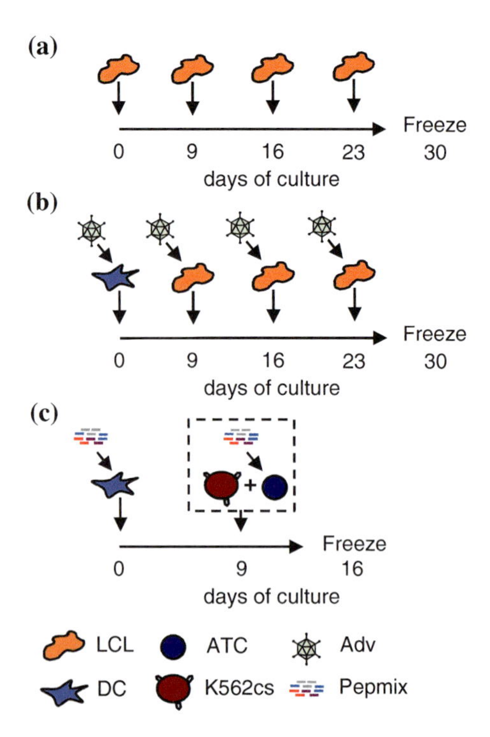


**Fig. 3** Ex vivo generation of EBVSTs for clinical studies. **a** For most clinical studies, EBVSTs have been generated using LCLs as antigen-presenting cells (APCs). **b** In an effort to increase the frequency of T cells for type 2 latency antigens, DCs and LCLs have been used as APCs that are modified with recombinant adenoviruses expressing LMP2 or LMP1 and LMP2. **c** EBVST generation not requiring LCLs or recombinant adenoviruses. For additional details see text

(Cobbold et al. 2005). Similarly, a mean of 21 × $10^3$/kg CMV pp65-specific gamma-captured T cells eliminated CMV viremia in 83 % of patients (Feuchtinger et al. 2010).

One problem with tetramer/streptamers selection is that these are limited to single epitopes presented by single HLA alleles, so that multiple clinical grade reagents would be required for the treatment of all patients and a lack of reagents for the selection of CD4-positive T cells. This may be overcome as more reagents including class II multimers become available. By contrast, the 'gamma capture strategy' is not HLA-dependent and can use whole antigen to activate polyclonal $CD^{4+}$ and $CD^{8+}$ T cells. This is ideally suited to CMV and adenovirus, since all seropositive individuals recognize pp65 and immediate early (IE) of CMV and hexon and penton of adenoviruses and T cells with these specificities have proved protective. However, the antigen specificity of EBVSTs is broad and highly HLA-dependent, and there are no universally dominant EBV antigens, so that an even wider range of clinical grade tetramers/streptamers would be required to cover all individuals. Further, it is not clear which antigen or antigens are required to induce protective T cells, and T cells elicited by LCLs recognize a range of latency antigens, early lytic cycle antigens, and phosphoproteins and studies using LCL-activated T cells did not identify critical target antigens for tumor recognition (Cobbold et al. 2013; Linnerbauer et al. 2014). However, two recent studies used donor-derived IFN-γ-captured EBVSTs in HSCT recipients with PTLD or viremia (Table 1) (Moosmann et al. 2010; Icheva et al. 2013). In one of these studies, patients received EBNA1-specific T cells, which expanded in 8 of 10 patients between 3 and 45 days after transfer and produced clinical and anti-viral responses in seven (Icheva et al. 2013). This was an important finding, since EBNA1 is poorly presented by MHC class I and hence has not been considered an ideal target antigen (Steigerwald-Mullen et al. 1995; Levitskaya et al. 1997).

## *2.4 Rapid Expansion of EBVSTs from Donor Blood for Adoptive Transfer*

Rapid selection strategies require large amounts of donor blood that cannot always be obtained from unrelated donors and in some donors the frequency of VSTs is limiting. Gerdemann et al. (2013) developed a rapid expansion strategy in which small numbers of donor PBMCs were stimulated for 10 days with autologous dendritic cells transfected with DNA plasmids expressing EBNA1, LMP2, and the immediate early lytic cycle antigen, BZLF1 as well as pp65 and IE of CMV and hexon and penton of adenovirus in the presence of interleukin (IL)-4 and IL-7. The total manufacturing time including 7 days for dendritic cell (DC) manufacture was 17 days plus 7 days for quality-control testing (Gerdemann et al. 2009, 2012). PBMCs expanded by about 1.5 logs in 9–11 days, so that starting with 15 × $10^6$ PBMCs, a median of 212.5 × $10^6$ VSTs was obtained, more than sufficient for the infusion of 5–20 × $10^6$ VSTs per $m^2$. 10 patients were treated for 12 viral infections, including four patients with EBV reactivations. 80 % of patients had CRs and a single patient whose EBV load did not respond to T-cell infusion did not develop PTLD.

This rapid expansion strategy was shortened by using overlapping peptide libraries (pepmixes) as a source of antigen instead of plasmids. These could be pulsed directly onto PBMCs eliminating the requirement for DCs and 7 days of culture. Pepmixes comprising 15-mers overlapping by 11 amino acids that span the entire protein sequence of the antigen of interest contain all possible class-I-restricted epitopes and many class-II-restricted epitopes (Kern et al. 2000). Papadopolou infused pentavirus-specific T cells targeting EBV, CMV, adenoviruses, BKV, and human herpes virus (HHV) 6 into 8 patients with 18 viral infections including 5 with PTLD or EBV viremia. CRs were observed in 80 % of patients, including all 5 with EBV (Papadopoulou et al. 2014). The EBV pepmixes used spanned EBNA1, LMP2, and BZLF1. Although EBNA1 and LMP2 are not the strongest antigens, they are recognized by most individuals and are expressed in most EBV-associated malignancies and EBNA1 strongly induces CD4-positive T cells. BZLF1 was included as a strong antigen recognized by most donors that may mediate elimination of productively infected cells before they can release infectious virus and therefore should help control virus spread after HSCT. Together, these antigens induced T cells able to eliminate bulky PTLD (Papadopoulou et al. 2014).

## 2.5 *Off the Shelf, Third-Party T Cells*

Even 10 days of culture may be too long to wait to treat a patient with rapidly progressive PTLD, and in the case of HSCT recipients, the stem-cell donor is not always available or willing to provide additional blood and the donor may be seronegative or cord blood. Further, patients developing PTLD after solid organ transplant do not have a healthy donor. Haque et al. therefore established a bank of 60 LCL-activated EBVST lines and used them to treat 33 transplant recipients with PTLD occurring after SOT (31) or HSCT (2) (Haque et al. 2002, 2007). All patients had failed standard therapies, and EBVST lines were selected based on best HLA match and ability to kill patient LCLs if available. 64 % of patients showed tumor responses at 5-week post-infusion and 14 (42 %) had CRs. Responses correlated with the number of HLA matches and the presence of $CD4^{+}$ T cells. The major anticipated toxicity was graft rejection or GVHD, but no adverse events were observed. TCR spectratyping in 5 patients revealed the presence of donor T cells for up to 7 days, but major T-cell expansion and persistence, as observed with donor T cells in the post-HSCT setting, was not observed. Subsequently, other third-party banks have been established at Memorial Sloan-Kettering Cancer Center (MSKCC) (Barker et al. 2010; Doubrovina et al. 2012) and Baylor College of Medicine (BCM), (Leen et al. 2013) and a new bank consisting of EBVSTs from New Zealanders has been established and used by the Edinburgh group to avoid potential transmission of Creutzfeld-Jacob disease (Vickers et al. 2014). Our (BCM) bank comprised 32 lines with specificity not only for EBV, but also for CMV and adenovirus. Eighteen lines, selected for their ability to recognize the culprit virus through the shared HLA allele(s), were administered to 50 patients (23 for CMV, 18 for adenovirus, and 9 for EBV) in a multicenter trial. Responses to all viruses were observed in around 75 % of recipients at 6 weeks of post-infusion and of the responders; only 4 patients recurred or progressed. Despite the HLA disparities between donor and recipient, de novo GVHD occurred in only two patients. Together, these results support the development of third-party banks in the transplant setting, since the cells are rapidly available, safe, and have high efficacy. The mechanism of action of third-party T cells is however mysterious since they do not expand in the peripheral blood like EBVSTs given to HSCT recipients. It is possible T cells may remain active at tumor sites or may create an inflammatory response that induces endogenous tumor-specific T cells specific for non-viral antigens.

# 3 T Cells for Type 2 Malignancies

## *3.1 LCL-Activated EBVSTs for the Treatment of Type 2 Latency Malignancies*

NPC and the T- and B-cell lymphomas occurring outside the transplant setting express only EBNA1, LMP1, LMP2, and the BARF1 gene products at the protein level, while EBERs and miRNAs are also expressed but not translated (type 2 malignancy) (Gilligan et al. 1991; Brooks et al. 1992; Decaussin et al. 2000; Chiang et al. 1996; Herbst et al. 1991; Pallesen et al. 1991). LCL-activated EBVSTs are usually dominated by T cells specific for early lytic cycle antigens and EBNA's 3A, 3B, and 3C, with unpredictable activity toward type 2 latency antigens (Steven et al. 1997; Pudney et al. 2005). This problem is exacerbated in patients with malignancies, since T cells specific for tumor antigens may be suppressed or anergized by the tumor microenvironment (Fogg et al. 2009). Nevertheless, LCL-activated EBVSTs were evaluated in patients with Hodgkin's lymphoma (HL) and NPC (Louis et al. 2010; Chia et al. 2014; Bollard et al. 2004; Chua et al. 2001; Lucas et al. 2004; Comoli et al. 2005; Secondino et al. 2012; Straathof et al. 2005a). In 14 patients with multiple-relapsed HL, EBVSTs were able to control B-symptoms and reduce peripheral blood EBV load (Bollard et al. 2004). Two patients with minimal disease had CRs, one had a partial response and five had stable disease (Table 3).

Our group also infused EBVSTs into twenty-three patients with recurrent/refractory NPC (Louis et al. 2010; Straathof et al. 2005a). Three of 4 patients with locoregional disease had CRs. By contrast, only 1 CR was observed in 11 patients with metastatic disease (Louis et al. 2010). Comoli et al. reported control

**Table 3** Selected T-cell therapy clinical studies for type 2 malignancy

| T-cell product | Production method | Comment |
|---|---|---|
| *HL and NHL* | | |
| Auto EBV-specific Ts | LCL | Thr (n = 14): 2 CR, 1 PR, 5 SD (Bollard et al. 2004) |
| | LCL/Ad5f35.LMP2 or Ad5f35.ΔLMP1-I-LMP2 | Thr (n = 21): 11 CR, 2 PR (Bollard et al. 2007, 2014) |
| | DCs and ATCs/K562cs loaded with pepmixes for E1ΔGA, LMP1, LMP2, BARF1 | Clinical study in progress |
| Allo EBV-specific Ts | LCL | T cells (n = 3): 3 PR; Chemo + T cells (n=3): 1 SD, 2 PR (Lucas et al. 2004) |
| *NPC* | | |
| Auto EBV-specific Ts | LCL | *Brisbane experience* (n = 4): 4 NR (Chua et al. 2001)<br>*Pavia experience*: T cells (n = 10): 2 PR, 4 SD; Chemo + T cells (n = 11): 6 SD (Comoli et al. 2005; Secondino et al. 2012)<br>*BCM experience* (n = 15): 3 CR, 2 CRu, 2 PR, 3 SD (Louis et al. 2010; Straathof et al. 2005a)<br>*Singapore experience* (n = 35): Chemo + T cells; 3 CR, 22 PR; 3-year OS: 37.1 % (Chia et al. 2014) |
| | Monocytes modified with Ad5f35.E1ΔGA-LMPpoly | As adjuvant (n = 16); prolonged OS in comparison with patients (n = 8) who did not receive T cells (Smith et al. 2012) |

*Allo* Allogeneic; *Auto* Autologous; *ATCs* Activated T cells; *Chemo* chemotherapy; *CR* Complete response; *CRu* Complete response undefined; *DCs* Dendritic cells; *NR* no response; *OS* Overall survival; *PR* Partial Response; *Pro* Prophylaxis; *Thr* Therapy

of disease in 6 of 10 patients with stage 4 NPC and similar results in a later study that used lymphodepleting chemotherapy prior to EBVST infusion in 11 patients (Comoli et al. 2005; Secondino et al. 2012). In a larger study in Singapore, 35 patients received up to six doses of EBVSTs after four cycles of gemcitabine and carboplatin producing a response rate of 71.4 % with 3 CRs and 22 partial responses (Chia et al. 2014). The 2-year and 3-year overall survival (OS) rates were 62.9 and 37.1 %, respectively. Tumor responses correlated with the presence of LMP2-specific T cells in the infused cell line. Tessa Therapeutics has started a phase III trial to compare the efficacy of this strategy with chemotherapy alone.

## 3.2 Targeting T Cells to Type 2 Latency Antigens

To increase the frequency of T cells specific for type 2 latency antigens, our group used APCs overexpressing LMP1 and/or LMP2 from recombinant adenovirus (Ad) vectors encoding either LMP2 alone (Ad5f35.LMP2) or LMP2 and a non-toxic, truncated from of LMP1 (Ad5f35.ΔLMP1-I-LMP2) to stimulate PBMCs (Gahn et al. 2001; Gottschalk et al. 2003). DCs in the first stimulation proved essential to ensure the activation of anergic LMP1- and LMP2-specific T cells from patient PBMCs, differing from healthy donors in this respect. Subsequent stimulations used LCLs transduced with the same Ad vectors to produce sufficient antigen-specific T cells for infusion. In our hands, this protocol has evolved over time and our ability to reactivate and expand LMP-specific T cells from patients has been improved by the incorporation of cytokines and superior media formulations. Nevertheless, LMP-specific T cells could not always be detected, although LCL killing was almost always observed (Bollard et al. 2014).

We performed two clinical trials using these LMP-directed T cells in patients with EBV-associated HL and non-Hodgkin lymphoma (NHL), including NKT lymphoma; the first trial targeted LMP2 alone, and the second targeted LMP1 and LMP2 (Bollard et al. 2007, 2014). When the two trials are considered together, 50 patients received LMP-targeted T-cell lines. Twenty-nine patients received LMP-targeted EBVSTs as adjuvant therapy after autologous HSCT or chemotherapy and 28 remained in remission for a median of 3.1 years and although 8 patients died during this time, deaths resulted from complications of prior chemotherapy. Twenty-one patients received T-cell therapy for active disease and of 13 patients with objective responses, 11 were complete. EBVSTs could be generated in 91 % of patients and LMP1 and/or LMP2 specificity was detected in about 66 %. After infusion, increases in LMP-specific activity could be detected in the majority of responders and epitope spreading to non-viral tumor antigens was detected only in responders (Bollard et al. 2014).

In an Australian study, EBVSTs enriched in LMP- and EBNA1-specific T cells generated by stimulation of PBMCs with monocytes transduced with an Ad vector encoding HLA-class-I-restricted LMP1 and LMP2 epitopes and the n-terminus of EBNA1 (Ad5f35.E1ΔGA-LMPpoly) were given as an adjuvant to sixteen patients with NPC, and these patients had a prolonged median OS in comparison with patients (n = 8), who did not receive cells (523 vs. 220 days) (Smith et al. 2012).

### *3.3 Overcoming Problems with Manufacturing Using LCLs as APCs*

The generation of LCLs for use as APCs adds at least 6 weeks to the EBVST manufacturing time with an additional four weeks to expand T cells and 2 weeks for quality-control (QC) testing. During this time, patients may progress and become ineligible for infusion. Further, EBV-LCLs cannot be generated from patients who have received the B-cell depleting monoclonal antibody Rituxan®, and while this was not a problem in early studies, this drug has become standard of care for patients with B-cell malignancies. Finally, the presence of live EBV is a regulatory hurdle, particularly in European and some Asian countries even though it has not presented a problem in hundreds of patients who have received LCL-activated T cells.

Pepmixes can also be used to generate EBVSTs for the adoptive immunotherapy of type 2 latency malignancies. Again, in our hands, DCs were required for the first stimulation of PBMCs with pepmixes from most patients (Ngo et al. 2014), and since DCs are limiting, especially in patients, for the second stimulation we developed an antigen-presenting complex comprising pepmix-pulsed, autologous activated T cells (ATCs), and HLA-negative K562 costimulatory cells (K562cs) from a master cell bank to replace autologous LCLs (Ngo et al. 2014). ATCs upregulate HLA class II molecules and therefore can present peptide epitopes in association with HLA class I and class II, while HLA class-I- and class-II-negative K562cs cells have been gene-modified to express the costimulatory molecules CD80, CD86, CD83, and 4-1BB ligand (Suhoski et al. 2007). Using this antigen-presenting complex, we have generated T-cell lines with specificity for LMP1, LMP2, EBNA1, and BARF1 from patients with lymphoma and NPC and are currently evaluating them clinically in patients with lymphoma. Of note, the resultant T-cell lines, even from healthy donors, rarely recognize all four antigens, but T cells specific for at least one and up to 4 antigens can be generated from most patients.

Improving media formulations, the addition of cytokines and the use of gas-permeable GRex culture vessels have further reduced the manufacturing time by increasing the rate of T-cell expansion, so that pepmix-activated EBVST for the treatment of patients with type 2 malignancies can now be manufactured and released in about one month: seven days for DC differentiation and maturation, 16 days for T-cell expansion, and 7–14 days for QC (Table 4).

**Table 4** Release criteria for pepmix-activated EBVSTs[a]

| Test | Method | Release criterion | Specification |
|---|---|---|---|
| Antigen-specific function | Elispot | No | Recognition of type 2 latency antigens |
| Antigen-specific function | Chromium release | No | Killing of LCLs and pepmix-pulsed targets |
| Phenotype subset and memory markers | Flow cytometry | No | None |
| Killing of autologous targets | Chromium release | Yes | <10 % killing at E:T ratio of 20:1 in a 4 h assay |
| Phenotype | Flow cytometry | Yes | <0.1 % K562 cells |
| Identity | HLA PCR | Yes | HLA identical with blood |
| Mycoplasma | MycoAlert PCR | Yes | Negative |
| Endotoxin | Endosafe PCR | Yes | <5.0 EU/ml |
| Fungus and bacteria | Bactec and CFR sterility | Yes | Negative at 7–21 days |

[a]T cells are fully characterized for research purposes and to develop potency and purity criteria for later phase studies

While the pepmix strategy provides a more specific T-cell product, caveats may be that T cells specific for non-viral phosphoproteins that are reactivated by LCLs (Cobbold et al. 2013), and may be important clinical target antigens (Linnerbauer et al. 2014). Thus, EBV-LCL-activated EBVSTs have been shown to recognize uninfected B-cell blasts, but not other autologous cells (Linnerbauer et al. 2014). Further, T cells specific for early lytic antigens is present in LCL-activated T-cell lines (Pudney et al. 2005). These T cells may play an important role in the control of malignancies that support even low levels of virus replication, since lytic cycle antigens are abundant relative to latent cycle antigens and may be cross-presented by other tumor cells after death of even a minority of lytically infected cells.

# 4 T Cells for Type 1 Malignancies

BL and GC express type 1 latency, thus provide only EBNA1 and BARF1 as target antigens for T cells (Rowe et al. 1987; Xue et al. 2002; Seto et al. 2005; Fiorini and Ooka 2008). EBNA1 contains a glycine alanine repeat (GAr) that was shown to prevent EBNA1 transfer to the endoplasmic reticulum for processing and presentation on HLA class I molecules; transfer of the GAr domain to other immunogenic proteins conferred resistance to processing and presentation, and deletion of the GAr from EBNA1 increased its immune recognition in LCLs (Levitskaya et al. 1995; Lee et al. 2004). Later it was suggested that the purine-rich mRNA sequence encoding the GAR reduced the rate of EBNA1 translation initiation and the availability of protein for processing and that it was the nucleotide sequence, not the amino acid sequence that reduced EBNA1 presentation (Yin et al. 2003; Apcher et al. 2010; Tellam et al. 2012). Regardless, it was assumed for a long time that EBNA1 would not be a good target for CD8-positive cytotoxic VSTs. More recently, several groups showed that EBNA1 could be presented on specific HLA class I alleles, for example HLA B35, perhaps due to processing from defective ribosomal products (DRiPs) (Lee et al. 2004; Voo et al. 2004; Munz 2004). Further, EBNA1 contains numerous HLA-class-II-restricted epitopes (Munz 2004; Munz et al. 2000; Leen et al. 2001) and CD4-positive T cells were shown to kill BL cell lines in vitro (Munz et al. 2000; Paludan et al. 2002). EBNA1-specific T-cell clones were also able to inhibit the outgrowth of EBV-infected B cells in vitro (Nikiforow et al. 2001), so that EBNA1 may in fact be an ideal target antigen, being expressed in all EBV-positive malignancies and inducing CD4-positive helper/killer T cells and some CD8-positive VSTs (Munz 2004). However, as yet, no clinical trials have evaluated T-cell therapy for either EBV-positive BL or GC.

# 5 Genetic Modifications to Improve Antitumor Activity of T Cells

While the adoptive transfer of EBVSTs is an effective therapy for PTLD post-HSCT, EBVSTs have been less effective for PTLD in SOT recipients and for type 2 latency malignancies for a number of reasons. Most SOT recipients receive immunosuppressive medications that inhibit T-cell function, and are not lymphodepleted, a prerequisite for significant in vivo expansion of adoptively transferred T cells. For type 2 malignancies, the limited array of EBV antigens expressed, lack of lymphoid space to expand, and the immunosuppressive tumor microenvironment are key factors that may limit EBVST efficacy. Several genetic modifications have been evaluated as potential countermeasures to these road-blocks (Table 5) and are discussed in this section.

## *5.1 Rendering T-Cell Resistant to Immunosuppressive Medications*

Most SOT recipients receive immunosuppressive medications such as FK506, rapamycin, or mycophenolate mofetil (MMF) to prevent graft rejection. Ricciardelli et al. showed that overexpression of a calcineurin A mutant in

**Table 5** Genetic modifications to improve antitumor activity of T cells

| Goal | Genetic modification |
|---|---|
| Rendering T-cell resistant to immunosuppressive Thr | *Molecules conferring resistance*: Calcineurin A mutant (Brewin et al. 2009; Ricciardelli et al. 2014); rapa-resistant mTOR (Huye et al. 2011); mutant IMDH II (Jonnalagadda et al. 2013)<br>*Gene silencing*: FKBP12 (De et al. 2009) |
| Enhancing T-cell expansion | *Cytokines*: IL-15 (Perna et al. 2013)<br>*Cytokine receptors*: IL-7Rα (Vera et al. 2009; Perna et al. 2014) |
| Rendering T-cell resistant to immunosuppressive tumor environment | *Cytokines*: IL-12 (Wagner et al. 2004; Pegram et al. 2012; Chinnasamy et al. 2012), IL-15 (Perna et al. 2013)<br>*Dominant-negative receptors (DNR)*: TGFβ RII DNR (Bollard et al. 2002; Foster et al. 2008)<br>*Chimeric cytokine receptors*: TGFβ RII/TLR4 (Watanabe et al. 2013); IL-4/IL-7 (Leen et al. 2014); IL-4/IL-2 (Wilkie et al. 2010)<br>*Silencing negative regulators*: FAS (Dotti et al. 2005) |
| Enhancing T-cell homing to tumor sites | *Chemokine receptors*: CCR4 (Di Stasi et al. 2009) |
| Redirecting T cells to non-EBV antigens | *CARs*: CD30 (Savoldo et al. 2007), CD70 (Shaffer et al. 2011), GD2 (Pule et al. 2008; Louis et al. 2011) |

*CAR* Chimeric antigen receptor; *FKBP* FK-binding protein; *IMDH* inosine monophosphate dehydrogenase; *mTOR* mammalian target of rapamycin; *Rapa* rapamycin

EBVSTs provided resistance to the calcineurin inhibitor FK506 and restored their ability to eliminate established LCLs in NOD/SCID/IL2r$\gamma^{null}$ mice in the presence of FK506 (Ricciardelli et al. 2014). This strategy is under evaluation clinically. De Angelis et al. produced similar results by silencing the expression of FK-binding protein 12 in EBVSTs (De et al. 2009). Huye et al. (2011) expressed a rapamycin-resistant mTOR in CD19-specific chimeric antigen receptor (CAR)-modified T cells that synergized with rapamycin in the elimination of B-cell lymphoma, a strategy that could be adapted to EBVSTs for SOT recipients receiving this drug (Huye et al. 2011). Lastly, investigators have rendered T-cell resistant to MMF by expressing a mutant inosine monophosphate dehydrogenase II in T cells (Jonnalagadda et al. 2013). Any of these strategies should improve the ability of T cells to function in patients receiving immunosuppressive drugs to prevent graft rejection and are at elevated risk for viral infections.

## *5.2 Enhancing T-Cell Expansion in Vivo*

While administration of lymphodepleting chemotherapy such as cyclophosphamide and/or fludarabine prior to T-cell infusion enhances the in vivo expansion of adoptively transferred T cells (Dudley et al. 2008), this approach lacks specificity and carries the risk of serious adverse events. Vaccination post-T-cell transfer enhances the expansion of adoptively transferred T cells in preclinical models (Song et al. 2010); however, there is no commercially available EBV vaccine to adapt this approach to cell therapy with EBVSTs. To test this concept in humans, we are currently evaluating the potency of commercially available varicella zoster virus vaccines to boost adoptively transferred varicella-specific T cells in vivo. Systemic administration of IL-2 has shown promise in patients with melanoma to promote T-cell expansion in vivo. However, systemic administration of IL-2 is associated with serious toxicities such as capillary leak (Rosenstein et al. 1986). In addition, IL-2 potentially expands inhibitory, regulatory T cells (Tregs) that could potentially inhibit the antitumor activity of infused EBVSTs. To overcome this limitation, investigators have explored the use of IL-15. While IL-15 shares the growth-promoting effects of IL-2 on effector T cells and does not preferentially promote Treg expansion, its systemic administration is also toxic, preventing its wider use (Conlon et al. 2014). Therefore, investigators have genetically modified EBVSTs to express IL-15. Genetically modified EBVSTs expressing IL-15 at low levels produced autocrine expansion in an antigen-dependent fashion (Perna et al. 2013) and were resistant to Treg-mediated immunosuppression (Perna et al. 2013). Transgenic expression of cytokine receptors that sensitize EBVSTs to cytokines with low systemic toxicity is another potential strategy to enhance T-cell expansion. For example, expressing IL-7R$\alpha$ on EBVSTs restores their sensitivity to IL-7, a cytokine that has an encouraging safety profile in humans (Vera et al. 2009; Perna et al. 2014; Sportes et al. 2008).

### 5.3 Rendering T-Cell Resistant to the Immunosuppressive Tumor Microenvironment

Hodgkin Reed Sternberg (HRS) cells, NHL, and NPC cells have developed—like other malignant cells—an intricate system to suppress the immune system (Leen et al. 2007; Rabinovich et al. 2007; Cruz-Merino et al. 2012; Fogg et al. 2013; Mrizak et al. 2015; Chen et al. 2007). For example, they (1) secrete immunosuppressive cytokines such as transforming growth factor (TGFβ) or IL-10; (2) recruit immuonsuppressive cells such Tregs- or myeloid-derived suppressor cells; (3) express molecules on the cell surface that suppress immune cells including FAS ligand (FAS-L) and PD-L1; and (4) create a metabolic environment (e.g., high lactate, low tryptophan) that is immunosuppressive.

TGFβ expression is employed by many tumors as an immune evasion strategy since it promotes tumor growth, limits T-cell effector function, and activates Tregs (Yang et al. 2010). These detrimental effects of TGFβ can be overcome by modifying T cells to express a dominant-negative TGFβ receptor type II (DNR), which lacks its intracellular signaling domain (Bollard et al. 2002; Foster et al. 2008). DNR expression blocks TGFβ-signaling and restores T-cell effector function in the presence of TGFβ. A clinical study evaluating this strategy is in progress for patients with EBV-positive NHL and preliminary results indicate that DNR-modified EBVSTs benefit patients who failed therapy with unmodified EBVSTs (Bollard et al. 2012). Preclinical studies have further shown that not only is it possible to render T-cell resistant to the detrimental effects of TGFβ, but also to convert the 'negative' TGFβ signal into a 'positive' signal by expressing a chimeric cytokine receptor in T cells, which consists of the extracellular domain of the TGFβ receptor type II and the endodomain of toll-like receptor (TLR) 4. Transgenic expression of this chimeric cytokine induced T-cell activation and expansion in the presence of TGFβ and in the absence of growth-promoting cytokines (Watanabe et al. 2013). Similar approaches have been developed to 'convert' the inhibitory effects of IL-4 on T cells (Leen et al. 2014; Wilkie et al. 2010).

Silencing negative regulators or transgenic expression of cytokines is other strategy to render EBVSTs resistant to the immunosuppressive microenvironment. For example, HRS cells as well as NPC cells express FAS-L, and preclinical studies have shown that silencing FAS-expression in EBVSTs renders T-cell resistant to the FAS-mediated apoptosis (Dotti et al. 2005). As mentioned in the Sect. 5.2, transgenic expression of IL-15 renders EBVSTs resistant to Tregs (Perna et al. 2013). Transgenic expression of another cytokine, IL-12, is also actively being explored to render T cells resistant to the immunosuppressive tumor microenvironment (or to convert the environment to one more conducive to T-cell growth) (Wagner et al. 2004; Pegram et al. 2012; Chinnasamy et al. 2012).

### 5.4 Enhancing T-Cell Homing to Tumor Sites

T-cell homing to tumor sites depends on the secretion of chemokines by the tumor and expression of the corresponding chemokine receptors on T cells. Often, there is a 'chemokine/chemokine receptor mismatch' limiting T-cell homing to tumor sites. HL cells secrete thymus- and activation-regulated chemokine (TARC) and macrophage-derived chemokine (MDC), which both preferentially attract inhibitory T cells such as Tregs and T helper (TH)2 cells that express the corresponding chemokine receptor CCR4, but not CCR4 negative, CD8-positive cytotoxic T cells. In a preclinical HL xenograft model, transgenic expression of CCR4 on effector T cells that were specific for the HL antigen CD30 showed improved migration toward tumor sites and antitumor activity, indicating that this approach could potentially improve the antitumor activity of EBVSTs (Di Stasi et al. 2009). Of note however, EBV-positive lymphomas often express RANTES that is induced by LMP1 and recruits CCR5-expressing effector T cells (Buettner et al. 2007).

### 5.5 Redirecting T Cells to Non-EBV Antigens

Redirecting EBVSTs to non-EBV antigens is actively being explored not only to enhance their antitumor activity against EBV-positive malignancies, but also to broaden their utility for malignancies that are not virus-associated and lack strong viral antigens. CARs, which consist of an antigen-binding domain, most commonly derived from a MAb, a transmembrane domain, and intracellular domain signaling domains (Dotti et al. 2014) can redirect EBVSTs to tumor antigens, since CARs do not interfere with the native T-cell receptor expressed on EBVSTs. In addition, CARs recognize cell surface antigens in an MHC-independent fashion, rendering the CAR/antigen recognition immune to commonly used immune evasion strategies such as downregulation of MHC class I expression or defects in the antigen processing machinery of tumor cells. Antigens that have been explored for type 2 malignancies include CD30 and CD70, which are expressed on HRS or NPC cells, respectively (Savoldo et al. 2007; Shaffer et al. 2011). A clinical study with CD30-specific CAR T cells is in progress.

Since latently infected memory B cells maintain infused EBVSTs in EBV-seropositive patients (Heslop et al. 2010), our group expressed a CAR directed to GD2 (a disialoganglioside) on EBVSTs and gave them to 11 children with advanced neuroblastoma (Pule et al. 2008; Louis et al. 2011). Three of them had complete responses while an additional two with bulky tumors showed substantial tumor necrosis. This approach has been adapted to prevent and treat recurrent CD19-positive B-cell malignancies with CD19-specific CAR-modified VSTs post-HSCT (Cruz et al. 2013).

In summary, gene transfer is an attractive approach to enhance the antitumor activity and broaden the scope of EBVSTs. While the majority of approaches have, as yet, only been evaluated in preclinical studies, the encouraging results obtained so far warrant further active exploration of this modified approach to immunotherapy with EBVSTs. While gene transfer has resulted in malignant transformation of hematopoietic stem cells (Hacein-Bey-Abina et al. 2010; Cavazzana-Calvo et al. 2012), no such event has been observed in more than 500 patients, who received T cells that were genetically modified with retroviral or lentiviral vectors (Bonini et al. 2011). Nevertheless, some of the discussed genetic modifications such as transgenic expression of cytokines might increase the risk of autonomous T-cell growth and/or transformation. Thus, depending on the inserted transgene, inclusion of an inducible suicide gene is advisable so that selective T-cell death can be induced in the event of unwanted toxicities. In this regard, several inducible suicide systems have been developed and evaluated in preclinical or clinical studies. For example, T cells that express an inducible, modified caspase 9 gene can be effectively ablated in preclinical models and in patients by administration of a 'chemical inducer of dimerization' (Straathof et al. 2005b; Di Stasi et al. 2011). Other strategies include the expression of a cell surface antigen on T cells such as truncated CD20 or EGFR, which allows the elimination of T cells with FDA-approved MAbs (Vogler et al. 2010; Wang et al. 2011).

# 6 Combinatorial T-Cell Therapy

Combining EBVSTs with other therapies is an attractive approach to increase their antitumor activity. For example, type 2 and type 1 malignancies can potentially be rendered more sensitive to EBVST-mediated killing by inducing the expression of immunodominant, lytic EBV antigens. In preclinical studies, several agents including chemotherapy, histone deacetylase inhibitors, or proteasome inhibitors have shown promise in inducing the expression of lytic cycle antigens (Feng and Kenney 2006; Feng et al. 2004; Shirley et al. 2011). However, clinical experience indicate so far that the induction of EBV lytic antigens with currently available agents in type 2 malignancies is either limited in humans (Chan et al. 2004; Wildeman et al. 2012) or requires their continued infusion to be effective (Perrine et al. 2007).

Combining EBVSTs with MAbs that block immune-cell-intrinsic checkpoints is another strategy to enhance their antitumor activity. In this regard, MAbs that block the inhibitory receptor CTLA-4 on T cells or the interaction between the inhibitory receptor PD-1 and its ligand (PD-L1) have shown promising antitumor activity as single agents in early Phase clinical studies for patients with solid tumors and HL (Hodi et al. 2010; Topalian et al. 2012; Brahmer et al. 2012; Moskwa et al. 2014). While CTLA-4 blockade or PD-1/PD-L1 blockade has not been combined with the adoptive transfer of T cells including EBVSTs in humans, these MAbs enhance the antitumor activity of adoptive T-cell therapies in preclinical models (John et al. 2013).

# 7 Conclusions

EBVSTs have been explored as prophylaxis and therapy for EBV-associated malignancies for more than two decades. While most successful for PTLD post-HSCT, clinical studies have also demonstrated their therapeutic potential for PTLD post-SOT and type 2 malignancies including HL, NHL, and NPC. Advances in the production technology of EBVSTs and renewed interest of biotech companies should facilitate later phase clinical studies. Lastly, gene transfer and approaches to combine EBVSTs with other targeted therapies hold the promise to further improve their antitumor activity. Giving the recent advances in the field, we hope that EBVSTs will become an integral part of our treatment armamentarium against EBV-positive malignancies in the near-future.

**Acknowledgements** We are grateful for the long-term support from the NIH (NCI, NHLBI), The Department of Defense, and The Leukemia and Lymphoma Society. We are also grateful to the foundations that have supported our research focused on EBV-associated malignancies including Alex's Lemonade Stand Foundation, and the V Foundation. We thank the supporting faculty and staff of the Stem Cell Transplant Unit, Clinical Research Unit, and Cell Processing Laboratory at Texas Children's Hospital and Houston Methodist Hospital. We also thank the patients who participated in our studies, and the parents who entrusted the care of their children to us.

# References

Apcher S, Daskalogianni C, Manoury B et al (2010) Epstein Barr virus-encoded EBNA1 interference with MHC class I antigen presentation reveals a close correlation between mRNA translation initiation and antigen presentation. PLoS Pathog 6(10):e1001151

Barker JN, Doubrovina E, Sauter C et al (2010) Successful treatment of EBV-associated post-transplantation lymphoma after cord blood transplantation using third-party EBV-specific cytotoxic T lymphocytes. Blood 116(23):5045–5049

Bollard CM, Rossig C, Calonge MJ et al (2002) Adapting a transforming growth factor beta-related tumor protection strategy to enhance antitumor immunity. Blood 99(9):3179–3187

Bollard CM, Aguilar L, Straathof KC et al (2004) Cytotoxic T lymphocyte therapy for Epstein-Barr virus+ Hodgkin's disease. J Exp Med 200(12):1623–1633

Bollard CM, Gottschalk S, Leen AM et al (2007) Complete responses of relapsed lymphoma following genetic modification of tumor-antigen presenting cells and T-lymphocyte transfer. Blood 110(8):2838–2845

Bollard CM, Dotti G, Gottschalk S et al (2012) Administration of TGF-beta resistant tumor-specific CTL to patienst with EBV-associated HL and NHL. Mol Ther 20(Supplement 1):S22

Bollard CM, Gottschalk S, Torrano V et al (2014) Sustained complete responses in patients with lymphoma receiving autologous cytotoxic T lymphocytes targeting Epstein-Barr virus latent membrane proteins. J Clin Oncol 32(8):798–808

Bonini C, Brenner MK, Heslop HE et al (2011) Genetic modification of T cells. Biol Blood Marrow Transplant 17(1 Suppl):S15–S20

Brahmer JR, Tykodi SS, Chow LQ et al (2012) Safety and activity of anti-PD-L1 antibody in patients with advanced cancer. N Engl J Med 366(26):2455–2465

Brentjens RJ, Davila ML, Riviere I et al (2013) CD19-targeted T cells rapidly induce molecular remissions in adults with chemotherapy-refractory acute lymphoblastic leukemia. Sci Transl Med 5(177):177ra38

Brewin J, Mancao C, Straathof K et al (2009) Generation of EBV-specific cytotoxic T cells that are resistant to calcineurin inhibitors for the treatment of posttransplantation lymphoproliferative disease. Blood 114(23):4792–4803

Brooks L, Yao QY, Rickinson AB et al (1992) Epstein-Barr virus latent gene transcription in nasopharyngeal carcinoma cells: coexpression of EBNA1, LMP1, and LMP2 transcripts. J Virol 66(5):2689–2697

Buettner M, Meyer B, Schreck S et al (2007) Expression of RANTES and MCP-1 in epithelial cells is regulated via LMP1 and CD40. Int J Cancer 121(12):2703–2710

Cavazzana-Calvo M, Fischer A, Hacein-Bey-Abina S et al (2012) Gene therapy for primary immunodeficiencies: Part 1. Curr Opin Immunol 24(5):580–584

Chan AT, Tao Q, Robertson KD et al (2004) Azacitidine induces demethylation of the Epstein-Barr virus genome in tumors. J Clin Oncol 22(8):1373–1381

Chen L, Huang TG, Meseck M et al (2007) Rejection of metastatic 4T1 breast cancer by attenuation of Treg cells in combination with immune stimulation. Mol Ther 15(12):2194–2202

Chia WK, Teo M, Wang WW et al (2014) Adoptive T-cell transfer and chemotherapy in the first-line treatment of metastatic and/or locally recurrent nasopharyngeal carcinoma. Mol Ther 22(1):132–139

Chiang AK, Tao Q, Srivastava G et al (1996) Nasal NK- and T-cell lymphomas share the same type of Epstein-Barr virus latency as nasopharyngeal carcinoma and Hodgkin's disease. Int J Cancer 68(3):285–290

Chinnasamy D, Yu Z, Kerkar SP et al (2012) Local delivery of interleukin-12 using T cells targeting VEGF receptor-2 eradicates multiple vascularized tumors in mice. Clin Cancer Res 18(6):1672–1683

Chua D, Huang J, Zheng B et al (2001) Adoptive transfer of autologous Epstein-Barr virus-specific cytotoxic T cells for nasopharyngeal carcinoma. Int J Cancer 94(1):73–80

Cobbold M, Khan N, Pourgheysari B et al (2005) Adoptive transfer of cytomegalovirus-specific CTL to stem cell transplant patients after selection by HLA-peptide tetramers. J Exp Med 202(3):379–386

Cobbold M, De La Pena H, Norris A et al (2013) MHC class I-associated phosphopeptides are the targets of memory-like immunity in leukemia. Sci Transl Med 5(203):203ra125

Comoli P, Locatelli F, Gerna G et al (1997) Autologous EBV-specific cytotoxic T cells to treat EBV-associated post-transplant lymphoproliferative disease (PTLD). Blood 90(10):249a

Comoli P, Labirio M, Basso S et al (2002) Infusion of autologous Epstein-Barr virus (EBV)-specific cytotoxic T cells for prevention of EBV-related lymphoproliferative disorder in solid organ transplant recipients with evidence of active virus replication. Blood 99(7):2592–2598

Comoli P, Pedrazzoli P, Maccario R et al (2005) Cell therapy of stage IV nasopharyngeal carcinoma with autologous Epstein-Barr virus-targeted cytotoxic T lymphocytes. J Clin Oncol 23(35):8942–8949

Conlon KC, Lugli E, Welles HC et al (2014) Redistribution, hyperproliferation, activation of natural killer cells and CD8 T cells, and cytokine production during first-in-human clinical trial of recombinant human interleukin-15 in patients with cancer. J Clin Oncol 33:74–84

Cruz CR, Micklethwaite KP, Savoldo B et al (2013) Infusion of donor-derived CD19-redirected virus-specific T cells for B-cell malignancies relapsed after allogeneic stem cell transplant: a phase 1 study. Blood 122(17):2965–2973

Cruz-Merino L, Lejeune M, Nogales FE et al (2012) Role of immune escape mechanisms in Hodgkin's lymphoma development and progression: a whole new world with therapeutic implications. Clin Dev Immunol 2012:756353

De Angelis B, Dotti G, Quintarelli C et al (2009) Generation of virus-specific cytotoxic T lymphocytes (Ctls) resistant to the immunosuppressive drug tacrolimus (Fk506). Biol Blood Marrow Transplant 15(2):377

De AB, Dotti G, Quintarelli C et al (2009) Generation of Epstein-Barr virus-specific cytotoxic T lymphocytes resistant to the immunosuppressive drug tacrolimus (FK506). Blood 114(23):4784–4791

Decaussin G, Sbih-Lammali F, de Turenne-Tessier M et al (2000) Expression of BARF1 gene encoded by Epstein-Barr virus in nasopharyngeal carcinoma biopsies. Cancer Res 60(19):5584–5588

Di Stasi A, De Angelis B, Rooney CM et al (2009) T lymphocytes coexpressing CCR4 and a chimeric antigen receptor targeting CD30 have improved homing and antitumor activity in a Hodgkin tumor model. Blood 113(25):6392–6402

Di Stasi A, Tey SK, Dotti G et al (2011) Inducible apoptosis as a safety switch for adoptive cell therapy. N Engl J Med 365(18):1673–1683

Dotti G, Savoldo B, Pule M et al (2005) Human cytotoxic T lymphocytes with reduced sensitivity to Fas-induced apoptosis. Blood 105(12):4677–4684

Dotti G, Gottschalk S, Savoldo B et al (2014) Design and development of therapies using chimeric antigen receptor-expressing T cells. Immunol Rev 257(1):107–126

Doubrovina E, Oflaz-Sozmen B, Prockop SE et al (2012) Adoptive immunotherapy with unselected or EBV-specific T cells for biopsy-proven EBV+ lymphomas after allogeneic hematopoietic cell transplantation. Blood 119(11):2644–2656

Dudley ME, Yang JC, Sherry R et al (2008) Adoptive cell therapy for patients with metastatic melanoma: evaluation of intensive myeloablative chemoradiation preparative regimens. J Clin Oncol 26(32):5233–5239

Feng WH, Kenney SC (2006) Valproic acid enhances the efficacy of chemotherapy in EBV-positive tumors by increasing lytic viral gene expression. Cancer Res 66(17):8762–8769

Feng WH, Hong G, Delecluse HJ et al (2004) Lytic induction therapy for Epstein-Barr virus-positive B-cell lymphomas. J Virol 78(4):1893–1902

Feuchtinger T, Lang P, Hamprecht K et al (2004) Isolation and expansion of human adenovirus-specific $CD4^+$ and $CD8^+$ T cells according to IFN-gamma secretion for adjuvant immunotherapy. Exp Hematol 32(3):282–289

Feuchtinger T, Opherk K, Bethge WA et al (2010) Adoptive transfer of pp65-specific T cells for the treatment of chemorefractory cytomegalovirus disease or reactivation after haploidentical and matched unrelated stem cell transplantation. Blood 116(20):4360–4367

Fiorini S, Ooka T (2008) Secretion of Epstein-Barr virus-encoded BARF1 oncoprotein from latently infected B cells. Virol J 5:70

Fogg MH, Wirth LJ, Posner M et al (2009) Decreased EBNA-1-specific $CD8^+$ T cells in patients with Epstein-Barr virus-associated nasopharyngeal carcinoma. Proc Natl Acad Sci USA 106(9):3318–3323

Fogg M, Murphy JR, Lorch J et al (2013) Therapeutic targeting of regulatory T cells enhances tumor-specific $CD8^+$ T cell responses in Epstein-Barr virus associated nasopharyngeal carcinoma. Virology 441(2):107–113

Foster AE, Dotti G, Lu A et al (2008) Antitumor activity of EBV-specific T lymphocytes transduced with a dominant negative TGF-beta receptor. J Immunother 31(5):500–505

Gahn B, Siller-Lopez F, Pirooz AD et al (2001) Adenoviral gene transfer into dendritic cells efficiently amplifies the immune response to the LMP2A-antigen: a potential treatment strategy for Epstein-Barr virus-positive Hodgkin's lymphoma. Int J Cancer 93(5):706–713

Gerdemann U, Christin AS, Vera JF et al (2009) Nucleofection of DCs to generate Multivirus-specific T cells for prevention or treatment of viral infections in the immunocompromised host. Mol Ther 17(9):1616–1625

Gerdemann U, Keirnan JM, Katari UL et al (2012) Rapidly generated multivirus-specific cytotoxic T lymphocytes for the prophylaxis and treatment of viral infections. Mol Ther 20(8):1622–1632

Gerdemann U, Katari UL, Papadopoulou A et al (2013) Safety and clinical efficacy of rapidly-generated trivirus-directed T cells as treatment for adenovirus, EBV, and CMV infections after allogeneic hematopoietic stem cell transplant. Mol Ther 21(11):2113–2121

Gilligan KJ, Rajadurai P, Lin JC et al (1991) Expression of the Epstein-Barr virus BamHI a fragment in nasopharyngeal carcinoma: evidence for a viral protein expressed in vivo. J Virol 65(11):6252–6259

Gottschalk S, Edwards OL, Sili U et al (2003) Generating CTL against the subdominant Epstein-Barr virus LMP1 antigen for the adoptive immunotherapy of EBV-associated malignancies. Blood 101(5):1905–1912

Grupp SA, Kalos M, Barrett D et al (2013) Chimeric antigen receptor-modified T cells for acute lymphoid leukemia. N Engl J Med 368(16):1509–1518

Hacein-Bey-Abina S, Hauer J, Lim A et al (2010) Efficacy of gene therapy for X-linked severe combined immunodeficiency. N Engl J Med 363(4):355–364

Haque T, Amlot PL, Helling N et al (1998) Reconstitution of EBV-specific T cell immunity in solid organ transplant recipients. J Immunol 160(12):6204–6209

Haque T, Wilkie GM, Taylor C et al (2002) Treatment of Epstein-Barr-virus-positive post-transplantation lymphoproliferative disease with partly HLA-matched allogeneic cytotoxic T cells. Lancet 360(9331):436–442

Haque T, Wilkie GM, Jones MM et al (2007) Allogeneic cytotoxic T-cell therapy for EBV-positive posttransplantation lymphoproliferative disease: results of a phase 2 multicenter clinical trial. Blood 110(4):1123–1131

Herbst H, Dallenbach F, Hummel M et al (1991) Epstein-Barr virus latent membrane protein expression in Hodgkin and Reed-Sternberg cells. Proc Natl Acad Sci USA 88(11):4766–4770

Heslop HE, Ng CYC, Li C et al (1996) Long-term restoration of immunity against Epstein-Barr virus infection by adoptive transfer of gene-modified virus-specific T lymphocytes. Nat Med 2:551–555

Heslop HE, Slobod KS, Pule MA et al (2010) Long-term outcome of EBV-specific T-cell infusions to prevent or treat EBV-related lymphoproliferative disease in transplant recipients. Blood 115(5):925–935

Hodi FS, O'Day SJ, McDermott DF et al (2010) Improved survival with ipilimumab in patients with metastatic melanoma. N Engl J Med 363(8):711–723

Huye LE, Nakazawa Y, Patel MP et al (2011) Combining mTor inhibitors with rapamycin-resistant T cells: a two-pronged approach to tumor elimination. Mol Ther 19(12):2239–2248

Icheva V, Kayser S, Wolff D et al (2013) Adoptive transfer of epstein-barr virus (EBV) nuclear antigen 1-specific T cells as treatment for EBV reactivation and lymphoproliferative disorders after allogeneic stem-cell transplantation. J Clin Oncol 31(1):39–48

John LB, Devaud C, Duong CP et al (2013) Anti-PD-1 antibody therapy potently enhances the eradication of established tumors by gene-modified T cells. Clin Cancer Res 19(20):5636–5646

Jonnalagadda M, Brown CE, Chang WC et al (2013) Engineering human T cells for resistance to methotrexate and mycophenolate mofetil as an in vivo cell selection strategy. PLoS ONE 8(6):e65519

Kalos M, Levine BL, Porter DL et al (2011) T cells with chimeric antigen receptors have potent antitumor effects and can establish memory in patients with advanced leukemia. Sci Transl Med 3(95):95ra73

Kern F, Faulhaber N, Frommel C et al (2000) Analysis of CD8 T cell reactivity to cytomegalovirus using protein-spanning pools of overlapping pentadecapeptides. Eur J Immunol 30(6):1676–1682

Khanna R, Bell S, Sherritt M et al (1999) Activation and adoptive transfer of Epstein-Barr virus-specific cytotoxic T cells in solid organ transplant patients with posttransplant lymphoproliferative disease. Proc Natl Acad Sci USA 96(18):10391–10396

Lee SP, Brooks JM, Al-Jarrah H et al (2004) CD8 T cell recognition of endogenously expressed epstein-barr virus nuclear antigen 1. J Exp Med 199(10):1409–1420

Leen A, Meij P, Redchenko I et al (2001) Differential immunogenicity of Epstein-Barr virus latent-cycle proteins for human CD4(+) T-helper 1 responses. J Virol 75(18):8649–8659

Leen AM, Myers GD, Sili U et al (2006) Monoculture-derived T lymphocytes specific for multiple viruses expand and produce clinically relevant effects in immunocompromised individuals. Nat Med 12(10):1160–1166

Leen AM, Rooney CM, Foster AE (2007) Improving T cell therapy for cancer. Annu Rev Immunol 25:243–265
Leen AM, Christin A, Myers GD et al (2009) Cytotoxic T lymphocyte therapy with donor T cells prevents and treats adenovirus and Epstein-Barr virus infections after haploidentical and matched unrelated stem cell transplantation. Blood 114(19):4283–4292
Leen AM, Bollard CM, Mendizabal AM et al (2013) Multicenter study of banked third-party virus-specific T cells to treat severe viral infections after hematopoietic stem cell transplantation. Blood 121(26):5113–5123
Leen AM, Sukumaran S, Watanabe N et al (2014) Reversal of tumor immune inhibition using a chimeric cytokine receptor. Mol Ther 22(6):1211–1220
Levitskaya J, Coram M, Levitsky V et al (1995) Inhibition of antigen processing by the internal repeat region of the Epstein-Barr virus nuclear antigen-1. Nature 375(6533):685–688
Levitskaya J, Sharipo A, Leonchiks A et al (1997) Inhibition of ubiquitin/proteasome-dependent protein degradation by the Gly-Ala repeat domain of the Epstein-Barr virus nuclear antigen 1. Proc Natl Acad Sci 94(23):12616–12621
Linnerbauer S, Behrends U, Adhikary D et al (2014) Virus and autoantigen-specific $CD4^+$ T cells are key effectors in a SCID mouse model of EBV-associated post-transplant lymphoproliferative disorders. PLoS Pathog 10(5):e1004068
Louis CU, Straathof K, Bollard CM et al (2010) Adoptive transfer of EBV-specific T cells results in sustained clinical responses in patients with locoregional nasopharyngeal carcinoma. J Immunother 33(9):983–990
Louis CU, Savoldo B, Dotti G et al (2011) Antitumor activity and long-term fate of chimeric antigen receptor-positive T cells in patients with neuroblastoma. Blood 118(23):6050–6056
Lucas KG, Salzman D, Garcia A et al (2004) Adoptive immunotherapy with allogeneic Epstein-Barr virus (EBV)-specific cytotoxic T-lymphocytes for recurrent, EBV-positive Hodgkin disease. Cancer 100(9):1892–1901
Moosmann A, Bigalke I, Tischer J et al (2010) Effective and long-term control of EBV PTLD after transfer of peptide-selected T cells. Blood 115(14):2960–2970
Moskwa M, Ribrag V, Michot J-M et al (2014) PD-1 blockade with the monoclonal antibody pembrolizumab (MK-3475) in patients with classical Hodgkin lymphoma after brentuximab vedotin failure: preliminary results from a phase 1b study. Blood 24(21):290
Mrizak D, Martin N, Barjon C et al (2015) Effect of nasopharyngeal carcinoma-derived exosomes on human regulatory T cells. J Natl Cancer Inst 107(1):363
Munz C (2004) Epstein-barr virus nuclear antigen 1: from immunologically invisible to a promising T cell target. J Exp Med 199(10):1301–1304
Munz C, Bickham KL, Subklewe M et al (2000) Human CD4(+) T lymphocytes consistently respond to the latent Epstein-Barr virus nuclear antigen EBNA1. J Exp Med 191(10):1649–1660
Ngo MC, Ando J, Leen AM et al (2014) Complementation of antigen-presenting cells to generate T lymphocytes with broad target specificity. J Immunother 37(4):193–203
Nikiforow S, Bottomly K, Miller G (2001) $CD4^+$ T-cell effectors inhibit Epstein-Barr virus-induced B-cell proliferation. J Virol 75(8):3740–3752
Pallesen G, Hamilton-Dutoit SJ, Rowe M et al (1991) Expression of Epstein-Barr virus latent gene products in tumour cells of Hodgkin's disease. Lancet 337:320–322
Paludan C, Bickham K, Nikiforow S et al (2002) Epstein-Barr nuclear antigen 1-specific CD4(+) Th1 cells kill Burkitt's lymphoma cells. J Immunol 169(3):1593–1603
Papadopoulou A, Gerdemann U, Katari UL et al (2014) Activity of broad-spectrum T cells as treatment for AdV, EBV, CMV, BKV, and HHV6 infections after HSCT. Sci Transl Med 6(242):242ra83
Peggs KS, Thomson K, Samuel E et al (2011) Directly selected cytomegalovirus-reactive donor T cells confer rapid and safe systemic reconstitution of virus-specific immunity following stem cell transplantation. Clin Infect Dis 52(1):49–57
Pegram HJ, Lee JC, Hayman EG et al (2012) Tumor-targeted T cells modified to secrete IL-12 eradicate systemic tumors without need for prior conditioning. Blood 119(18):4133–4141

Perna SK, De AB, Pagliara D et al (2013) Interleukin 15 provides relief to CTLs from regulatory T cell-mediated inhibition: implications for adoptive T cell-based therapies for lymphoma. Clin Cancer Res 19(1):106–117
Perna SK, Pagliara D, Mahendravada A et al (2014) Interleukin-7 mediates selective expansion of tumor-redirected cytotoxic T lymphocytes (CTLs) without enhancement of regulatory T-cell inhibition. Clin Cancer Res 20(1):131–139
Perrine SP, Hermine O, Small T et al (2007) A phase 1/2 trial of arginine butyrate and ganciclovir in patients with Epstein-Barr virus-associated lymphoid malignancies. Blood 109(6):2571–2578
Poppema S (2005) Immunobiology and pathophysiology of hodgkin lymphomas. Hematol Am Soc Hematol Educ Program 2005:231–238
Porter DL, Roth MS, McGarigle C et al (1994) Induction of graft-versus-host disease as immunotherapy for relapsed chronic myeloid leukemia. N Engl J Med 330(2):100–106
Porter DL, Levine BL, Kalos M et al (2011) Chimeric antigen receptor-modified T cells in chronic lymphoid leukemia. N Engl J Med 365(8):725–733
Pudney VA, Leese AM, Rickinson AB et al (2005) $CD8^+$ immunodominance among Epstein-Barr virus lytic cycle antigens directly reflects the efficiency of antigen presentation in lytically infected cells. J Exp Med 201(3):349–360
Pule MA, Savoldo B, Myers GD et al (2008) Virus-specific T cells engineered to coexpress tumor-specific receptors: persistence and antitumor activity in individuals with neuroblastoma. Nat Med 14(11):1264–1270
Rabinovich GA, Gabrilovich D, Sotomayor EM (2007) Immunosuppressive strategies that are mediated by tumor cells. Annu Rev Immunol 25:267–296
Ricciardelli I, Blundell MP, Brewin J et al (2014) Towards gene therapy for EBV-associated posttransplant lymphoma with genetically modified EBV-specific cytotoxic T cells. Blood 124(16):2514–2522
Rooney CM, Smith CA, Ng C et al (1995) Use of gene-modified virus-specific T lymphocytes to control Epstein-Barr virus-related lymphoproliferation. Lancet 345:9–13
Rooney CM, Smith CA, Ng CYC et al (1998) Infusion of cytotoxic T cells for the prevention and treatment of Epstein-Barr virus-induced lymphoma in allogeneic transplant recipients. Blood 92(5):1549–1555
Rosenberg SA, Dudley ME (2009) Adoptive cell therapy for the treatment of patients with metastatic melanoma. Curr Opin Immunol 21(2):233–240
Rosenstein M, Ettinghausen SE, Rosenberg SA (1986) Extravasation of intravascular fluid mediated by the systemic administration of recombinant interleukin 2. J Immunol 137(5):1735–1742
Rowe M, Rowe DT, Gregory CD et al (1987) Differences in B cell growth phenotype reflect novel patterns of Epstein-Barr virus latent gene expression in Burkitt's lymphoma cells. EMBO J 6:2743–2751
Savoldo B, Goss J, Liu Z et al (2001) Generation of autologous Epstein Barr virus (EBV)-specific cytotoxic T cells (CTL) for adoptive immunotherapy in solid organ transplant recipients. Transplantation 72(6):1078–1086
Savoldo B, Goss JA, Hammer MM et al (2006) Treatment of solid organ transplant recipients with autologous Epstein Barr virus-specific cytotoxic T lymphocytes (CTLs). Blood 108(9):2942–2949
Savoldo B, Rooney CM, Di Stasi A et al (2007) Epstein Barr virus specific cytotoxic T lymphocytes expressing the anti-CD30zeta artificial chimeric T-cell receptor for immunotherapy of Hodgkin disease. Blood 110(7):2620–2630
Secondino S, Zecca M, Licitra L et al (2012) T-cell therapy for EBV-associated nasopharyngeal carcinoma: preparative lymphodepleting chemotherapy does not improve clinical results. Ann Oncol 23(2):435–441
Seto E, Yang L, Middeldorp J et al (2005) Epstein-Barr virus (EBV)-encoded BARF1 gene is expressed in nasopharyngeal carcinoma and EBV-associated gastric carcinoma tissues in the absence of lytic gene expression. J Med Virol 76(1):82–88

Shaffer DR, Savoldo B, Yi Z et al (2011) T cells redirected against CD70 for the immunotherapy of CD70-positive malignancies. Blood 117(16):4304–4314
Sherritt MA, Bharadwaj M, Burrows JM et al (2003) Reconstitution of the latent T-lymphocyte response to Epstein-Barr virus is coincident with long-term recovery from posttransplant lymphoma after adoptive immunotherapy. Transplantation 75(9):1556–1560
Shirley CM, Chen J, Shamay M et al (2011) Bortezomib induction of C/EBPbeta mediates Epstein-Barr virus lytic activation in Burkitt lymphoma. Blood 117(23):6297–6303
Smith C, Tsang J, Beagley L et al (2012) Effective treatment of metastatic forms of Epstein-Barr virus-associated nasopharyngeal carcinoma with a novel adenovirus-based adoptive immunotherapy. Cancer Res 72(5):1116–1125
Song XT, Turnis M, Zhou X et al (2010) A Th1-inducing adenoviral vaccine for boosting adoptively transferred T cells. Mol Ther 19(1):211–217
Sportes C, Hakim FT, Memon SA et al (2008) Administration of rhIL-7 in humans increases in vivo TCR repertoire diversity by preferential expansion of naive T cell subsets. J Exp Med 205(7):1701–1714
Steigerwald-Mullen PM, Klein G, Kurilla MG et al (1995) Inhibition of antigen processing by the internal repeat region of the Epstein-Barr virus nuclear antigen-1. Nature 375(6533):685–688
Steven NM, Annels NE, Kumar A et al (1997) Immediate early and early lytic cycle proteins are frequent targets of the Epstein-Barr virus-induced cytotoxic T cell response. J Exp Med 185(9):1605–1617
Straathof KC, Bollard CM, Popat U et al (2005a) Treatment of nasopharyngeal carcinoma with Epstein-Barr virus-specific T lymphocytes. Blood 105:1898–1904
Straathof KC, Pule MA, Yotnda P et al (2005b) An inducible caspase 9 safety switch for T-cell therapy. Blood 105(11):4247–4254
Suhoski MM, Golovina TN, Aqui NA et al (2007) Engineering artificial antigen-presenting cells to express a diverse array of co-stimulatory molecules. Mol Ther 15(5):981–988
Tellam JT, Lekieffre L, Zhong J et al (2012) Messenger RNA sequence rather than protein sequence determines the level of self-synthesis and antigen presentation of the EBV-encoded antigen, EBNA1. PLoS Pathog 8(12):e1003112
Topalian SL, Hodi FS, Brahmer JR et al (2012) Safety, activity, and immune correlates of anti-PD-1 antibody in cancer. N Engl J Med 366(26):2443–2454
Uhlin M, Gertow J, Uzunel M et al (2012) Rapid salvage treatment with virus-specific T cells for therapy-resistant disease. Clin Infect Dis 55(8):1064–1073
Vera JF, Hoyos V, Savoldo B et al (2009) Genetic manipulation of tumor-specific cytotoxic T lymphocytes to restore responsiveness to IL-7. Mol Ther 17(5):880–888
Vickers MA, Wilkie GM, Robinson N et al (2014) Establishment and operation of a good manufacturing practice-compliant allogeneic Epstein-Barr virus (EBV)-specific cytotoxic cell bank for the treatment of EBV-associated lymphoproliferative disease. Br J Haematol 167(3):402–410
Vogler I, Newrzela S, Hartmann S et al (2010) An improved bicistronic CD20/tCD34 vector for efficient purification and in vivo depletion of gene-modified T cells for adoptive immunotherapy. Mol Ther 18(7):1330–1338
Voo KS, Fu T, Wang HY et al (2004) Evidence for the presentation of major histocompatibility complex class I-restricted Epstein-Barr virus nuclear antigen 1 peptides to $CD8^+$ T lymphocytes. J Exp Med 199(4):459–470
Wagner HJ, Bollard CM, Vigouroux S et al (2004) A strategy for treatment of Epstein-Barr virus-positive Hodgkin's disease by targeting interleukin 12 to the tumor environment using tumor antigen-specific T cells. Cancer Gene Ther 11(2):81–91
Wallace LE, Rickinson AB, Rowe M et al (1982) Stimulation of human lymphocytes with irradiated cells of the autologous Epstein-Barr virus-transformed cell line. I. Virus-specific and nonspecific components of the cytotoxic response. Cell Immunol 67(1):129–140
Wang X, Chang WC, Wong CW et al (2011) A transgene-encoded cell surface polypeptide for selection, in vivo tracking, and ablation of engineered cells. Blood 118(5):1255–1263

Watanabe N, Anurathapan U, Brenner M et al (2013) Transgenic expression of a novel immunosuppressive signal converter on T Cells. Mol Ther 22(S1):S153

Wildeman MA, Novalic Z, Verkuijlen SA et al (2012) Cytolytic virus activation therapy for epstein-barr virus-driven tumors. Clin Cancer Res 18(18):5061–5070

Wilkie S, Burbridge SE, Chiapero-Stanke L et al (2010) Selective expansion of chimeric antigen receptor-targeted T-cells with potent effector function using interleukin-4. J Biol Chem 285(33):25538–25544

Xue SA, Labrecque LG, Lu QL et al (2002) Promiscuous expression of Epstein-Barr virus genes in Burkitt's lymphoma from the central African country Malawi. Int J Cancer 99(5):635–643

Yang L, Pang Y, Moses HL (2010) TGF-beta and immune cells: an important regulatory axis in the tumor microenvironment and progression. Trends Immunol 31(6):220–227

Yin Y, Manoury B, Fahraeus R (2003) Self-inhibition of synthesis and antigen presentation by Epstein-Barr virus-encoded EBNA1. Science 301(5638):1371–1374

# The Development of Prophylactic and Therapeutic EBV Vaccines

**Corey Smith and Rajiv Khanna**

**Abstract** Over the last century, the development of effective vaccine approaches to treat a number of viral infections has provided the impetus for the continual development of vaccine platforms for other viral infections, including Epstein–Barr virus (EBV). The clinical manifestations associated with EBV infection occur either following primary infection, such as infectious mononucleosis, or following an extended period of latency, primarily the EBV-associated malignancies and potentially including a number of autoimmune disorders, such as multiple sclerosis. As a consequence, two independent vaccine approaches are under development to prevent or control EBV-associated diseases. The first approach, which has been widely successful against other viral infections, is aimed at inducing a viral neutralisation antibody response to prevent primary infection. The second approach focuses upon the induction of cell-mediated immunity to control latent infected cells in persistently infected individuals. Early clinical studies have offered some insight into the potential efficacy of both of these approaches.

C. Smith · R. Khanna
QIMR Centre for Immunotherapy and Vaccine Development and Department of Immunology, QIMR Berghofer Medical Research Institute, Brisbane, QLD 4029, Australia

R. Khanna (✉)
Tumour Immunology Laboratory, Department of Immunology, QIMR Berghofer Medical Research Institute, 300 Herston Rd, Brisbane 4006, Australia
e-mail: rajiv.khanna@qimrberghofer.edu.au



C. Münz (ed.), *Epstein Barr Virus Volume 2*, Current Topics in Microbiology and Immunology 391, DOI 10.1007/978-3-319-22834-1_16

## Contents



## Abbreviations

| | |
|---|---|
| EBV | Epstein–Barr virus |
| IM | Infectious mononucleosis |
| HL | Hodgkin's lymphoma |
| MS | Multiple sclerosis |
| PTLD | Post-transplant lymphoproliferative disease |
| DLBCL | Diffuse large B cell lymphoma |
| NK/TL | NK/T cell lymphoma |
| NPC | Nasopharyngeal carcinoma |
| GC | Gastric carcinoma |
| VLP | Virus-like particle |
| DC | Dendritic cell |
| EBNA | EBV nuclear antigen |
| LMP | Latent membrane protein |
| DTH | Delayed-type hypersensitivity |
| MVA | Modified Vaccinia Ankara |
| APC | Antigen-presenting cell |
| LCV | Lymphocryptoviruses |

## 1 Introduction

Research over the past five decades has provided significant insight into Epstein–Barr virus (EBV) in the context of disease and aspects of both humoral and cellular immunity that will likely be critical mediators for effective vaccine development (see Fig. 1 for a timeline of important events in EBV vaccine

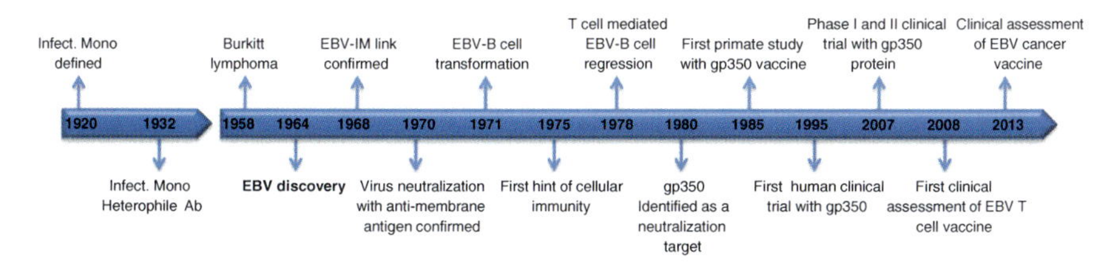


**Fig. 1** A timeline of EBV discovery and vaccine development

development). The clinical manifestations associated with EBV infection indicate that the development of an EBV vaccine could have a potentially significant impact upon a wide range of diseases. Primary symptomatic EBV infection is associated with 500 cases per 100,000 individuals per year in the USA and typically arises in adolescents and in young adults (Luzuriaga and Sullivan 2010). While most of these cases are self-limiting, many patients will endure infectious mononucleosis (IM)-associated symptoms for up to 6 months post-diagnosis and severe complications will develop in some individuals. IM is also associated with a threefold increased risk of developing EBV-associated Hodgkin's lymphoma (HL) (Hjalgrim et al. 2003), and potentially a twofold increase risk of developing multiple sclerosis (MS) (Handel et al. 2010). In developed countries including the USA, an effective prophylactic vaccine delivered during adolescence to seronegative individuals could therefore have a profound effect on not only IM, but also on the risk of developing other EBV-associated complications, including HL and MS, and other potentially EBV-associated autoimmune diseases (Hanlon et al. 2014).

EBV infection is also associated with a range of malignancies of both B cell and epithelial cell origin, which often arise many years after primary infection. In addition to EBV-associated HL, which affects 33,000 worldwide annually (de Martel et al. 2012), EBV is associated with rarer B cell malignancies including post-transplant lymphoproliferative disease (PTLD) and diffuse large B cell lymphoma (DLBCL), and with NK/T cell lymphoma (NK/TL), endemic in regions of Asia. EBV is also associated with 100 % of undifferentiated nasopharyngeal carcinoma (NPC) also endemic in south-east Asia and to a lesser extent in Northern Africa, with an annual incidence of 72,000 cases (de Martel et al. 2012). Although less definitively shown to be associated with EBV, estimates suggest that 7–10 % of all gastric carcinoma (GC) cases worldwide are EBV-associated (Murphy et al. 2009; Fukayama 2010). The prevalence of NPC and GC suggest there is significant potential for the development of vaccines to prevent these malignancies. EBV seropositivity reaches close to 100 % in childhood in regions where EBV malignancies are endemic (Xiong et al. 2014), suggesting that prophylactic vaccination to prevent EBV infection in these areas will require early childhood delivery and the establishment of lifelong immunity. However, given our comprehensive understanding of EBV latency in the different malignancies, vaccine approaches that directly target the induction of cellular immunity to control latently infected malignant cells offer an attractive alternative option to treat these diseases. This chapter will discuss the distinct prophylactic and therapeutic approaches to EBV vaccine development, their clinical assessment and strategies aimed at improving the efficacy of these approaches.

## 2 Prophylactic Vaccination to Prevent EBV Infection

As with all infectious agents, the most effective mechanism to prevent EBV-associated diseases would be the development of a vaccination approach to prevent infection and induce sterilising immunity (see Fig. 2 for a summary of the current and future vaccine strategies being explored for EBV-associated diseases). This would offer the dual benefit of preventing disease associated with primary lytic infection, such as infectious mononucleosis, and by preventing the establishment of latent infection, would also likely impact upon the development of EBV-associated malignancies. Despite the development of a number of vaccine approaches, one of which reached phase II clinical studies, it is as yet unclear how sterilising immunity against EBV infection could be achieved. This is primarily a

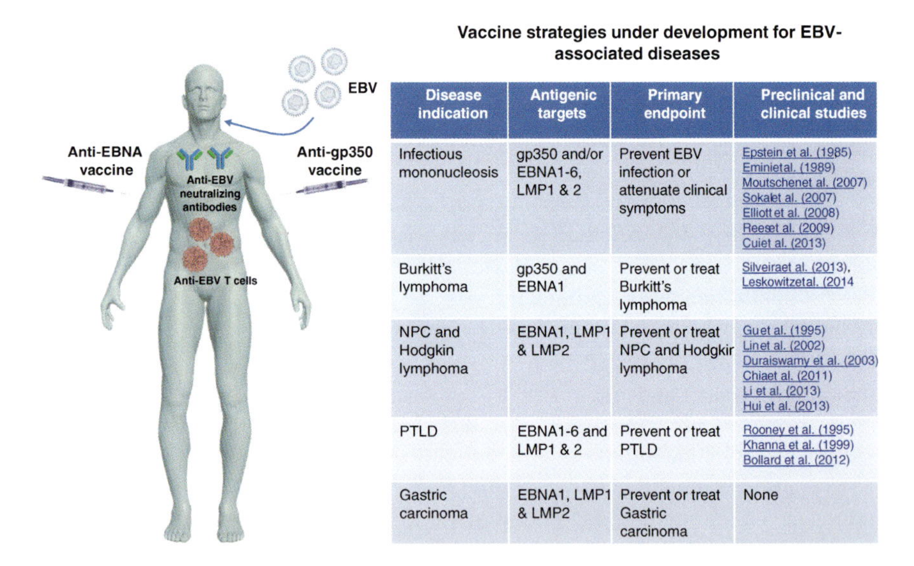


Vaccine strategies under development for EBV-associated diseases

| Disease indication | Antigenic targets | Primary endpoint | Preclinical and clinical studies |
|---|---|---|---|
| Infectious mononucleosis | gp350 and/or EBNA1-6, LMP1 & 2 | Prevent EBV infection or attenuate clinical symptoms | Epstein et al. (1985) Eminietal. (1989) Moutschenet al. (2007) Sokalet al. (2007) Elliott et al. (2008) Reeset al. (2009) Cuiet al. (2013) |
| Burkitt's lymphoma | gp350 and EBNA1 | Prevent or treat Burkitt's lymphoma | Silveiraet al. (2013), Leskowitzetal. (2014 |
| NPC and Hodgkin lymphoma | EBNA1, LMP1 & LMP2 | Prevent or treat NPC and Hodgkir lymphoma | Guet al. (1995) Linet al. (2002) Duraiswamy et al. (2003) Chiaet al. (2011) Li et al. (2013) Hui et al. (2013) |
| PTLD | EBNA1-6 and LMP1 & 2 | Prevent or treat PTLD | Rooney et al. (1995) Khanna et al. (1999) Bollard et al. (2012) |
| Gastric carcinoma | EBNA1, LMP1 & LMP2 | Prevent or treat Gastric carcinoma | None |

**Fig. 2** EBV vaccine approaches currently under development. The life cycle of EBV and the distinct disease manifestations associated with primary and persistent infection have promoted the development of two platforms in EBV vaccine research. The key role of neutralising antibodies in restricting viral spread during primary infection has promoted the development of vaccine approaches that target the major membrane glycoprotein, gp350. The critical role of T cells in preventing the uncontrollable spread of latently infected cells has led to the development of vaccine approaches for EBV-associated malignancies that promote the induction of cellular immunity to latent viral antigens relevant to each malignancy

consequence of our lack of a definitive understanding of the route and mechanism of primary infection. Despite this, recent developments have suggested that while current vaccination approaches may be unable to prevent asymptomatic infection with EBV, they do have the capacity to control viral infection and prevent disease associated with primary infection.

## *2.1 Clinical Assessment of gp350 as a Vaccine Candidate*

Following the identification of EBV in Burkitt's lymphoma (Epstein et al. 1964), it was evident very quickly that infection of primary B cells could be neutralised by antibodies from immune individuals targeting cell membrane antigens (Pearson et al. 1970). The subsequent isolation of the EBV membrane antigen gp350 (Hoffman et al. 1980) lead the first likely vaccine antigen candidate. Since its initial identification, the focus of the majority of clinical EBV vaccine studies has been on the induction of a gp350 neutralisation antibody response. As a major component of the outer membrane of EBV, gp350 binds to the CD21 molecule on the surface of B cells, promoting B cell infection. Although, gp350-deficient virus can still infect B cells, infection is less efficient, and neutralisation of B cell infection is predominantly mediated by gp350-specific antibodies. The first clinical study involving gp350 was performed in China using a recombinant vaccinia strain encoding gp350 in a small cohort of children (Gu et al. 1995). A gp350-speciifc antibody response was detectable in seronegative individuals and could be boosted in seropositive children. Despite this, no further studies were undertaken with this approach, likely due to safety concern with the vaccinia vector. Subsequent gp350 vaccine approaches have employed recombinant protein expressed in Chinese hamster ovary cells (Jackman et al. 1999). Formulated in combination with the AS04 adjuvant system, initial phase I/II studies demonstrated the immunogenicity and safety of gp350 vaccination. Three doses of 50 μg of gp350 in AS04 delivered at 0, 1 and 5 months were sufficient to induce gp350 neutralising titres similar to that seen in seropositive individuals (Moutschen et al. 2007). In a subsequent placebo-controlled double-blinded phase II study in 181 seronegative adults, 98.7 % of vaccinees displayed gp350 seroconversion by 6 months (Sokal et al. 2007). This gp350 seroconversion correlated with a reduced incidence of IM compared to placebo controls. While 8 of 90 individuals receiving the placebo developed IM, only 2 of 86 individuals receiving gp350/AS04 developed IM, a demonstrable efficacy rate of 78 %. Despite these encouraging results, no evidence of protection against asymptomatic infection was detected, although the overall number of total EBV infections in the gp350/AS04 group was reduced compared to placebo, and no phase III studies are currently planned to test the efficacy of the gp350/AS04 formulation in a larger cohort.

Another recent phase I clinical study investigated the use of a gp350 vaccine in a cohort of children awaiting kidney transplant (Rees et al. 2009). The rationale of this approach is that pre-existing gp350-specific immunity in seronegative children will reduce viral burden upon primary infection following transplantation and the subsequent risk of developing PTLD due to immunosuppression. In this study, EBV-seronegative children received three to four doses of either 12.5 or 25 μg of gp350 formulated in 0.2 % alhydrogel. The vaccine was well tolerated, and all evaluated patients in both dose ranges generated gp350-specific antibody responses. However, only 4 of 13 patients generated a neutralising antibody response following gp350, and the vaccination did not appear to have an impact on viral titres post-transplantation. These observations suggest that while this gp350/alhydrogel formulation was safe, the poor immunogenicity, as measure by the induction of EBV-neutralisation, suggests that an improved formulation is needed to see any potential efficacy in this cohort of patients.

The observations from the clinical evaluations of gp350 suggest that it is safe and well tolerated. One clear limitation of the current vaccine formulations, particularly in alhydrogel, is the effectiveness of neutralising antibody induction against gp350. While the AS04 formulation was capable of generating a neutralising antibody response in all recipients, protective efficacy was only evident after administration of the third dose (Sokal et al. 2007), suggesting that efficient induction of a neutralising antibody response required administration of all three doses. New vaccine formulations that increase the titre and rate of gp350-specific neutralising antibody induction may therefore offer a mechanism to improve the efficacy of the current vaccine formulations. This may require improved adjuvant combinations or antigenic modifications to improve the immunogenicity of gp350. One recent novel gp350 vaccine formulation that improved antibody titres in pre-clinical studies used tetrameric gp350 in alum or alum and CpG (Cui et al. 2013). In both settings, tetrameric gp350 induced significantly higher antibody titres than the monomeric gp350, boosted gp350-specific T cell immunity and improved viral neutralisation titres.

## *2.2 Are There Other Viral Neutralisation Targets?*

As with all members of the Herpesviridae family, EBV has a number of surface glycoproteins that play a role in viral attachment and entry, and function differentially to promote infection of different cell types, particularly epithelial cells. However, unlike the well-established in vitro model of B cell infection, in vitro models of EBV epithelial infection are not as well developed, and it still has not been definitively proven that the oropharygeal epithelial cells are the site of primary infection, with some models still proposing that locally infiltrating B cells may also be the primary cell infected (Rickinson et al. 2014). Nevertheless, it has been shown that infection of epithelial cell is not dependent upon gp350 interaction with CD21, but is dependent upon other surface glycoproteins including the gHgL complex and gB (Wang et al. 1998; Haan et al. 2001; Tugizov et al. 2003; Tsao et al. 2012), which may offer alternative targets in vaccine development. However, work investigating the immunogenicity of these other potential antibody targets in the context of vaccine development has only recently commenced.

One approach that may offer more neutralising antibody targets, in addition to gp350, is an EBV virus-like particle (VLP) (Ruiss et al. 2011). VLPs have the potential advantage over recombinant antigens because they can mimic the natural structure of the virion, while still containing no viral DNA. Their main potential limitation is the more complex manufacturing requirements compared to single recombinant antigens. To generate an EBV-VLP, Ruiss and colleagues produced a dedicated cell line that provides all of the essential viral genes required for the production and release of the VLPs, but without packaging of the viral DNA. Vaccination of animals with the EBV-VLP induced high EBV-specific antibody titres that were capable of neutralising EBV infection of human B cells in vitro. More recent studies by Pavlova and colleagues developed a series of EBV mutants in order to optimise VLP production, demonstrating that deletion of the BLFL1 and BRRF1A genes provided optimal for the production of DNA-free EBV-VLPs (Pavlova et al. 2013).

## 3 T Cell-Mediated Immunity as a Target of EBV Vaccines

The role of T cell-mediated immunity in controlling EBV-infected cells was first identified by Moss and Rickinson in the 1970s. Their observations demonstrated that the outgrowth of in vitro transformed B cells by EBV was restricted in EBV-seropositive individuals and was dependent upon cytotoxic T cells (Moss et al. 1978; Rickinson et al. 1979). Since these early observations, a significant amount of research has focused upon understanding the role of T cell immunity in controlling latent EBV infection and has lead to the definition of a significant number of $CD4^+$ and $CD8^+$ T cell determinants associated with both the lytic and latent stages of EBV infection (Burrows et al. 2011; Khanna et al. 1999). While the dramatic expansion of EBV-specific $CD8^+$ T cells during IM provides co-incidental evidence that T cell immunity plays an important role in restricting viral growth during primary exposure (Callan et al. 1996; Hislop et al. 2002), the association of EBV infection with post-transplant lymphoma's (PTLD) has provided the most compelling evidence that T cell immunity is critical for the effective control of EBV infection (Smets et al. 2002; Sherritt et al. 2003; Sebelin-Wulf et al. 2007).

### *3.1 A Prophylactic T Cell-Mediated Vaccine for EBV*

Following primary EBV exposure in transplant patients, active immunosuppression can lead to a failure to prime T cell immunity to latently expressed antigens, likely resulting in the uncontrolled proliferation of EBV-infected B cells and the development of PTLD. Effective immune control can be instigated by the adoptive transfer of T cells specific for latent antigen (Bollard et al. 2012; Rooney et al. 1995; Khanna et al. 1999), providing compelling evidence that T cells are critical mediators of immune control of latently infected cells. While it is unlikely that pre-existing T cell immunity induced following vaccination will have the capacity to prevent the establishment of latent infection, it is likely that used in the right context such an approach could have a significant impact upon primary symptomatic infectious mononucleosis and in preventing PTLD in seronegative transplant recipients, predominantly children, who receive a seropositive graft. The only study to specifically investigate the induction of a cell-mediated immune response following vaccination used a single EBV nuclear antigen (EBNA) 3-encoded HLA-B8-restricted epitope, FLRGRAYGL, formulated with tetanus toxoid and Montanide ISA 720 (Elliott et al. 2008). The vaccine was well tolerated, and epitope-specific T cell responses could be detected in 8 of 9 vaccine recipients. While a similar rate of seroconversion following exposure to EBV was evident in both the immunised and the placebo control group, all four vaccinated individuals who seroconverted in the vaccine group were asymptomatically infected, whereas one of two individuals in the placebo control group who seroconverted had primary symptomatic infection. Although this study was too small to conclusively determine whether the induction of T cell immunity prevented primary symptomatic infection, it does demonstrate the feasibility of inducing cell-mediated immunity in healthy vaccine recipients by targeting antigens associated with EBV latency. Such an approach would likely require the inclusion of multiple viral determinants to provide more broad coverage in ethnically distinct populations. In the context of a prophylactic vaccine for EBV-associated PTLD, particularly following solid organ transplantation, the inclusion of a T cell component will likely be critical due to primary infection of recipient B cells occurring following transplantation of infected cells within the organ, bypassing the mucosal surfaces whereby pre-existing humoral immunity plays its most critical role in restricting viral entry. Once infection is established, antibody-mediated neutralisation has also likely a very limited role in controlling the spread of the latently infected B cells that drive PTLD.

## *3.2 Latency Programs in Malignancies Associated with Immunocompetent Individuals*

While an efficient vaccination approach to boost immunity in transplant patients to prevent PTLD would likely succeed due to the atypical nature of EBV T cell immunity in transplant patients, and the prophylactic success demonstrated following adoptive therapy of EBV-specific T cells into transplant recipients, the malignancies that arise in otherwise immunocompetent individuals are much more prevalent, accounting for >90 % of all EBV-associated malignancies (de Martel et al. 2012). Unlike PTLD whereby the EBV latency III gene expression profile includes expression of the full array of latent genes, including EBNA1-6 and latent membrane protein (LMP) 1 and 2, the latency II type malignancies which include NPC and GC, as well as EBV-associated HL and EBV-associated NK/TL, express a limited array of latent antigens, restricted to EBNA1 and LMP1&2. This limited gene expression pattern provides unique challenges to the development of therapeutic vaccine approaches targeting the latency II malignancies, predominantly due to the well-established poor immunogenicity of these antigens.

### *3.3 Immunisation of NPC Patients with Dendritic Cell-based Vaccines*

The first clinical study evaluating the targeting of either the LMP antigens or EBNA1 was a dendritic cell (DC)-based vaccine trial that incorporated defined epitopes from LMP2 (Lin et al. 2002). Sixteen NPC patients received four injections of peptide-pulsed monocyte-derived DC at weekly intervals. A boost in the LMP2-specific T cell response was evident in 9 of 16 patients following treatment, which was associated with a partial clinical response in 2 of these patients. A similar recent study using LMP2-pulsed DC also showed an increase in the LMP2 response, while this did not correlate with a reduction in the peripheral EBV load, there was evidence that the induction of a delayed-type hypersensitivity (DTH) response did correlate to a reduction in EBV load (Li et al. 2013). Another approach using DC involved transduction with an adenoviral vector encoding a truncated LMP1 (ΔLMP1) and full-length LMP2 (Chia et al. 2011). In this phase II study, 9 of 12 NPC patients generated a DTH response. Although no change in frequency of peripheral LMP-specific T cells was detected, three patients did show a clinical response. These observations suggest that the induction of a LMP-specific T cell response is possible following DC vaccination. However, it is unlikely that the broad application of DC vaccination for EBV-associated malignancies is practical, particularly given the cost associated with personalised DC preparation, which is likely to be prohibitively expensive in a number of countries where NPC is endemic. A more appropriate approach is likely to incorporate the delivery of immunological determinant from LMP1&2 and EBNA1 in a viral vector or formulated with adjuvants already licensed for human use.

### *3.4 Viral Vectors for the Delivery of LMP1&2 and EBNA1*

The most advanced viral vector vaccine candidate to date targeting EBNA1 and the LMP antigens is the Modified Vaccinia Ankara (MVA) vector developed by Taylor and Colleagues. The MVA vector is an attenuated strain of the vaccinia virus and has been used clinically in a number of vaccine trials targeting other pathogens. It has also been shown to have a good safety record with the capacity to induce T cell and B cell responses in human subjects (Parrino et al. 2007). The MVA construct, termed MVA-EL, has been designed to encode a fusion protein of the 3' half of the EBNA1 gene and the full-length LMP2 gene. Pre-clinical observations with this vector demonstrated that it could efficiently expand both EBNA1- and LMP2-specific $CD4^+$ and $CD8^+$ T cells from the peripheral blood lymphocytes of EBV-seropositive individuals (Taylor et al. 2004). The MVA-EL vector has now been evaluated for safety in two Phase I clinical trials. The first study was a dose-escalation trial in Hong Kong, using NPC patients in remission at least 12 weeks post-primary therapy (Hui et al. 2013). In all, 18 patients were treated with three doses of MVA-EL over a period of 9 weeks. While some adverse events were recorded following vaccination, these were primarily grade 1 in most patients and grade 2–3 in only 3 patients. In addition to the assessment of safety, the amplification of LMP2- and EBNA1-specific T cell responses was also assessed in these donors. These observations revealed that 12 of 18 patients and 9 of 18 patients demonstrated at least a twofold increase in IFN-$\gamma$ ELISPOT response to EBNA1 and LMP2, respectively, following the full three cycles of vaccination. Response rates were also higher in those patients who received the highest dose of $5 \times 10^8$ pfu. In some individuals, these increase responses were shown to correspond to defined $CD4^+$ and $CD8^+$ epitopes in EBNA1 and LMP2 and were maintained for more than 14 weeks post the final dose. The MVA-EL vector has also been assessed in $EBV^+$ NPC patients in the UK (Taylor et al. 2014).

One potential limitation of using full-length LMP1&2 and EBNA1 antigens is the poor natural immunogenicity of these antigens. As a consequence, most constructs under development, including MVA-EL, encode an EBNA1 construct deficient in the glycine–alanine repeat (Taylor et al. 2004; Smith et al. 2006). Similarly, both LMP1 and to a lesser extent LMP2A are known to be presented inefficiently in the context of EBV-infected cells, potentially limiting the efficacy of any vaccine candidate encoding the full-length antigens (Smith et al. 2009a, b; Gavioli et al. 2002). The potential to promote oncogenic transformation may also limit the potential use of full-length LMP proteins in any EBV vaccine (Dawson et al. 2012). As a consequence, vaccine strategies are underdevelopment that not only limit this potential oncogenicity, but also improve immunogenicity. The ΔLMP1 construct used in the DC vaccine approach was designed to improve the immunogenicity of LMP1 by removing the first transmembrane domain preventing LMP1 aggregation and boosting presentation via the MHC class I pathway (Gottschalk et al. 2003). Although this approach was largely ineffectual at inducing significant T cells responses following DC vaccination, the adenoviral vector encoding the ΔLMP1 has been shown in vitro to be efficient at inducing the expansion of LMP1 T cells responses. Furthermore, considering the efficacy adenoviral vaccine vectors have shown in other systems, a direct vaccination approach rather than infection of DC in vitro would likely offer a better outcome for the induction of LMP1 and 2 responses using the Ad-ΔLMP1/LMP2A vector.

The approach our group has undertaken to avoid the poor immunogenicity and any oncogenic potential of LMP1&2 has been to use defined $CD8^+$ T cell epitopes from LMP1&2 encoded in a single recombinant polyepitope construct. Work nearly two decades ago demonstrated that when encoded end to end without any intervening flanking residues, $CD8^+$ T cell epitopes could be efficiently processed and presented via the MHC class I pathway to $CD8^+$ T cells (Thomson et al. 1995). Many subsequent pre-clinical animals studies have shown that polyepitope constructs can be used effectively to induce T cell responses following delivering in a number of different vector platforms, including recombinant vaccinia vectors, adenoviral vectors and DNA-based vaccines (Smith et al. 2006; Thomson et al. 1995, 1996, 1998). Our initial studies demonstrated that immunisation with a recombinant vaccinia polyepitope construct comprising LMP1 HLA A2-restricted epitopes could induce $CD8^+$ T cell responses in HLA A2 transgenic mice which were protective against challenge in a model tumour setting (Duraiswamy et al. 2003). Our subsequent studies have focused upon developing a LMP1&2 polyepitope encoding T cell epitopes that would provide coverage across broad HLA-types, including those prevalent in regions where EBV-associated malignancies are endemic. Encoded in a modified adenoviral vector, the LMP1&2 polyepitope is capable of efficiently inducing the expansion of $CD8^+$ T cells from a wide range of HLA type (Duraiswamy et al. 2004). This vector has since been modified to encode the EBNA1 gene with a deletion in the glycine–alanine repeat, generating the AdE1-LMPpoly vector (Smith et al. 2006). The AdE1-LMPpoly can efficiently induce the activation of both LMP1&2 and EBNA1 T cells from healthy individuals and patients with EBV-associated malignancies. Our pre-clinical animal studies in HLA A2 transgenic mice have demonstrated that the AdE1-LMPpoly vector is very efficient at inducing both $CD8^+$ and $CD4^+$ T cell responses including in settings of immunodeficiency (Smith et al. 2008), and we are optimistic that this approach would be effective at inducing strong T cell immunity in patients with EBV-associated malignancies given our recent success at generating T cells from both NPC and lymphoma patients using the AdE1-LMPpoly vector (Smith et al. 2006, 2012).

### *3.5 Recombinant EBV Antigens and in Vivo Targeting of Antigen-Presenting Cells*

Another vaccine approach potentially applicable to the induction of cellular immune responses is the use of recombinant EBV antigens. Subunit vaccines based upon recombinant proteins offer improved safety profiles compared to vaccine vectors; however, the successful use of recombinant antigens will likely be dependent upon the efficient delivery of antigen to professional antibody-presenting cells (APCs). One approach under development to target antigen to APCs is the use of recombinant antibody-antigen fusion proteins that target surface receptors on professional APCs. Observations by Gurer et al. (2008) demonstrated that an anti-DEC205-EBNA1 fusion protein promoted DC-activation of EBNA1-specific $CD4^+$ and $CD8^+$ T cell responses that could restrict the outgrowth of EBV-transformed B cells. They also demonstrated in a humanised mouse model the induction of EBNA1-specific IFN-$\gamma$ production following vaccination and have more recently shown that this targeting of DCs likely occurs via $CD141^+$

DCs that are thought to function as the critical cross-presenting cell in the activation of human T cell responses (Meixlsperger et al. 2013). Recent observations have shown that the DEC205 is expressed on the surface of EBV-infected cells, and the EBNA1 fusion protein also has the capacity to directly target EBV-transformed B cells in vitro, inducing the proliferation of EBNA1-specific T cells, offering the potential to directly target EBV-infected cells in vivo, potentially boosting immune recognition of transformed B cells (Leung et al. 2013). The EBV-VLP discussed earlier may also have the capacity to target EBV antigens to EBV-infected B cells in vivo via gp350 (Pavlova et al. 2013). This approach has been shown to induce the activation of EBV-specific CD4$^{+}$ T cells following in vitro loading of EBV-transformed B cells with VLP.

### *3.6 Challenges in the Development of a Vaccine for EBV-Related Malignancies*

A major challenge to the efficacy of any vaccine targeting LMP1&2 and EBNA1 will be the capacity of the T cell response generated to recognise the EBV-infected malignant cells. As outlined earlier, LMP1&2 and EBNA1 efficiently avoid immune recognition by cytotoxic T cells. Immune recognition of EBNA1 is restricted via self-regulation of protein translation by the glycine–alanine repeat (Tellam et al. 2004, 2008), while restriction of LMP1 processing and presentation appears to be associated with its ability to form large self-aggregates and through the destruction of T cell epitopes by the immunoproteasome (Smith et al. 2009; Gavioli et al. 2002). Although the mechanisms by which LMP2 avoid immune recognition are not full-delineated, there is evidence that LMP2 also avoids recognition by T cells (Smith et al. 2009). These mechanisms likely play a critical role in the efficient establishment of EBV infection, promoting the survival of latently infected cells and restricting recognition of EBV-transformed B cells by cytotoxic T lymphocytes in particular. These evasion mechanisms are also likely critical mediators that enable the survival of malignant cells expressing these foreign antigens. Despite these in vitro observations, it remains to be determined what impact this restricted recognition has upon the immune surveillance of LMP1&2 and EBNA1 expressing cells in vivo. Additionally, the majority of these studies have been performed on EBV-infected B cells, with very little analysis performed on infected epithelial lines or with EBV-bearing malignant cells. Another consequence of the poor natural immunogenicity, in addition to a low frequency of LMP1&2 and EBNA1 memory T cells, is the functional/phenotypic inferiority of these cells relative to T cells recognising the EBNA3 antigens (Smith et al. 2009). These phenotypic differences may also rendered LMP1&2 and EBNA1-specific T cells more susceptible to tumour-mediated immune regulation (Smith et al. 2009; Gandhi et al. 2006, 2007). Therefore, a vaccine approach that improves the frequency of LMP1&2 and EBNA1 may also improve the functionality of these cells, increasing their capacity to potential traffic to sites of malignancy, promoting effector functions that can kill malignant cells and potentially enhancing resistance to some tumour-mediated immune evasion mechanisms. Larger clinical studies are therefore required to definitively prove if a vaccine approach targeting LMP1&2 and EBNA1 can both improve the functionality of specific T cells and consequentially lead to better tumour control.

# 4 Animal Models for EBV Vaccine Development

Effective vaccine development in most settings has been dependent upon the ability to test vaccine formulations in animal challenge settings. This has remained a potential obstacle in EBV research due to the host tropism of EBV. Animal models used to assess direct efficacy of vaccines against EBV have primarily been based upon the use of other primate species found to be permissive to EBV infection, including both cotton-top tamarins and common marmosets, which were used in early gp350 vaccine studies (Epstein et al. 1985; Emini et al. 1989). These studies demonstrated that vaccination with different gp350 formulations induced neutralising antibodies and were efficacious against EBV challenge. Despite these promising observations, the cotton-top tamarin is no longer an ethically viable model primarily due to their critically endangered status; and both models require a high dose of viral challenge via a non-physiological route to establish infection.

Recent observations have also suggested that rabbits can be infected with EBV, offering another potential model to test the efficacy of EBV vaccines (Kanai et al. 2010; Takashima et al. 2008). Rabbits are a well-established model for vaccine development and offer obvious advantages over the use of the previously described primate models of EBV infection. However, the establishment of infection in rabbits required intravenous inoculation, and it seems unlikely that the pathogenesis of EBV infection in rabbits will closely resemble that of humans, potentially restricting their potential use in the development of any EBV vaccine, particularly those designed to treat EBV-related lymphomas. The humanised mouse offers another potential model system to assess the therapeutic efficacy of EBV vaccines. Severely immunocompromised mice can be reconstitution with human haematopoietic cell lineages following transfer of $CD34^+$ progenitor cells (Traggiai et al. 2004). The efficient reconstitution of human B cells provides a source for EBV infection in humanised mice (Traggiai et al. 2004), which can replicate some of the hallmarks of primary human EBV infection and the development of EBV-associated lymphoproliferative disease (Yajima et al. 2008). EBV-infected humanised mice have also been shown to develop symptoms similar to those seen hemophagocytic lymphohistiocytosis patients (Sato et al. 2011) and can develop EBV-specific cell-mediated and humoral immune response (Yajima et al. 2008; Strowig et al. 2009). Humanised mice may therefore offer a robust system to test the use of vaccines in a therapeutic setting following the establishment of infection. It remains to be determined whether the reconstitution of the immune components, including T and B cells and dendritic cells populations, that are critical mediators in the induction of adaptive immune responses, will be sufficient to promote effective immunity following vaccination in humanised mouse models. It is also evident, given the lack of human epithelial cells and requirement for infection via the intraperitoneal route, that current humanised mouse models may not be appropriate for testing current and future prophylactic vaccines targeting the induction of neutralising antibody responses and the establishment of mucosal immunity.

Other potentially more appropriate models to develop EBV vaccines rely on the use of comparative studies with related lymphocryptoviruses (LCVs) in their host primate species. While not offering a direct mechanism to test EBV-specific vaccination, these models offer a potential platform to investigate the role of different immunological parameters in promoting protection against EBV following immunisation. The most developed model involves the use of the rhesus LCV, whose ORFs share approximately 75 % homology with EBV, and encodes genes homologous with all ORFs in EBV (Moghaddam et al. 1997; Rivailler et al. 2002). Rhesus macaques can be challenged orally with rhesus LCV, and persistent infection can be detected in B cells and in the oropharyngeal compartment (Moghaddam et al. 1997). Despite the similarities with EBV, rhesus LCV does not induce the immortalisation of B cells in vitro, although it has been associated with lymphoma development in immunocompromised macaques. Studies by Cohen and colleagues have investigated the impact of both antibody and T cell-based vaccines on rhesus LCV infection (Sashihara et al. 2011). Immunisation with rhesus gp350 was shown to be protective against viral challenge, reducing viral load following infection and seroconversion to LCV viral capsid antigen. This study also demonstrated the induction of T cell responses against EBNA3A and EBNA3C. Studies by others have also demonstrated that vaccination of rhesus macaques can boost the endogenous response to the EBNA1 homologue following vaccination (Silveira et al. 2013; Leskowitz et al. 2014), a potential model for the development of therapeutic vaccines to treat EBV-associated malignancies. The rhesus model therefore offers the potential to further develop correlates of protection potentially relevant to EBV infection.

## 5 Future Developments in EBV Vaccine Research

It is apparent that while a great magnitude of research over the last five decades has significantly advanced our understanding of EBV virology, immunology and related diseases, the challenges in the development of EBV-targeted vaccines remain. As outlined in this chapter and summarised in Fig. 2, a number of vaccine approaches are underdevelopment for the prevention/treatment of most EBV-associated diseases. An effective vaccination program could potentially have significant benefit for the treatment and prevention of a range of EBV-associated diseases, and early clinical studies have provided evidence that vaccination may offer an effective measure to prevent EBV-associated diseases (Cohen 2015; Balfour 2014; Kanekiyo et al. 2015). Further studies into the immunological components critical to the establishment of EBV immunity, particularly the role of B cell immunity during primary infection (Panikkar et al. 2015a; Hagn et al. 2015), and the implementation of more appropriate animal models will also hopefully provide further insight into the immune mechanisms and target antigens that are critical to prevent/control infection (Panikkar et al. 2015b). However, it is also apparent that the advancement of any potential further EBV vaccine strategies will require the translation of these findings into more clinical studies.

## References


Balfour HH Jr (2014) Progress, prospects, and problems in Epstein-barr virus vaccine development. Curr Opin Virol 6:1–5

Bollard CM, Rooney CM, Heslop HE (2012) T-cell therapy in the treatment of post-transplant lymphoproliferative disease. Nat Rev Clin Oncol 9(9):510–519

Burrows SR, Moss DJ, Khanna R (2011) Understanding human T-cell-mediated immunoregulation through herpesviruses. Immunol Cell Biol 89(3):352–358

Callan MF, Steven N, Krausa P et al (1996) Large clonal expansions of $CD8^+$ T cells in acute infectious mononucleosis. Nat Med 2(8):906–911

Chia WK, Wang WW, Teo M, et al (2011) A phase II study evaluating the safety and efficacy of an adenovirus-{Delta}LMP1-LMP2 transduced dendritic cell vaccine in patients with advanced metastatic nasopharyngeal carcinoma. *Ann Oncol*

Cohen JI (2015) Epstein–barr virus vaccines. Clin Transl Immunol 4:e32

Cui X, Cao Z, Sen G et al (2013) A novel tetrameric gp350 1-470 as a potential Epstein-barr virus vaccine. Vaccine 31(30):3039–3045

Dawson CW, Port RJ, Young LS (2012) The role of the EBV-encoded latent membrane proteins LMP1 and LMP2 in the pathogenesis of nasopharyngeal carcinoma (NPC). Semin Cancer Biol 22(2):144–153

de Martel C, Ferlay J, Franceschi S et al (2012) Global burden of cancers attributable to infections in 2008: a review and synthetic analysis. Lancet Oncol 13(6):607–615

Duraiswamy J, Sherritt M, Thomson S et al (2003) Therapeutic LMP1 polyepitope vaccine for EBV-associated Hodgkin disease and nasopharyngeal carcinoma. Blood 101(8):3150–3156

Duraiswamy J, Bharadwaj M, Tellam J et al (2004) Induction of therapeutic T-cell responses to subdominant tumor-associated viral oncogene after immunization with replication-incompetent polyepitope adenovirus vaccine. Cancer Res 64(4):1483–1489

Elliott SL, Suhrbier A, Miles JJ et al (2008) Phase I trial of a $CD8^+$ T-cell peptide epitope-based vaccine for infectious mononucleosis. J Virol 82(3):1448–1457

Emini EA, Schleif WA, Silberklang M, Lehman D, Ellis RW (1989) Vero cell-expressed Epstein-barr virus (EBV) gp350/220 protects marmosets from EBV challenge. J Med Virol 27(2):120–123

Epstein MA, Achong BG, Barr YM (1964) Virus particles in cultured lymphoblasts from Burkitt's lymphoma. Lancet 1(7335):702–703

Epstein MA, Morgan AJ, Finerty S, Randle BJ, Kirkwood JK (1985) Protection of cottontop tamarins against Epstein-barr virus-induced malignant lymphoma by a prototype subunit vaccine. Nature 318(6043):287–289

Fukayama M (2010) Epstein-Barr virus and gastric carcinoma. Pathol Int 60(5):337–350

Gandhi MK, Lambley E, Duraiswamy J et al (2006) Expression of LAG-3 by tumor-infiltrating lymphocytes is coincident with the suppression of latent membrane antigen-specific $CD8^+$ T-cell function in Hodgkin lymphoma patients. Blood 108(7):2280–2289

Gandhi MK, Moll G, Smith C et al (2007) Galectin-1 mediated suppression of Epstein-barr virus specific T-cell immunity in classic Hodgkin lymphoma. Blood 110(4):1326–1329

Gavioli R, Vertuani S, Masucci MG (2002) Proteasome inhibitors reconstitute the presentation of cytotoxic T-cell epitopes in Epstein-barr virus-associated tumors. Int J Cancer 101(6):532–538

Gottschalk S, Edwards OL, Sili U et al (2003) Generating CTLs against the subdominant Epstein-barr virus LMP1 antigen for the adoptive immunotherapy of EBV-associated malignancies. Blood 101(5):1905–1912

Gu SY, Huang TM, Ruan L et al (1995) First EBV vaccine trial in humans using recombinant vaccinia virus expressing the major membrane antigen. Dev Biol Stand 84:171–177

Gurer C, Strowig T, Brilot F et al (2008) Targeting the nuclear antigen 1 of Epstein-barr virus to the human endocytic receptor DEC-205 stimulates protective T-cell responses. Blood 112(4):1231–1239

Haan KM, Lee SK, Longnecker R (2001) Different functional domains in the cytoplasmic tail of glycoprotein B are involved in Epstein-barr virus-induced membrane fusion. Virology 290(1):106–114

Hagn M, Panikkar A, Smith C, Balfour HH Jr, Khanna R, Voskoboinik I, Trapani JA (2015) B cell-derived circulating granzyme B is afeature of acute infectious mononucleosis. Clin Transl Immunology. 4(6):e38. doi:10.1038/cti.2015.10

Handel AE, Williamson AJ, Disanto G, Handunnetthi L, Giovannoni G, Ramagopalan SV (2010) An updated meta-analysis of risk of multiple sclerosis following infectious mononucleosis. *PLoS ONE* 5(9)

Hanlon P, Avenell A, Aucott L, Vickers MA (2014) Systematic review and meta-analysis of the sero-epidemiological association between Epstein-Barr virus and systemic lupus erythematosus. Arthritis Res Ther 16(1):R3

Hislop AD, Annels NE, Gudgeon NH, Leese AM, Rickinson AB (2002) Epitope-specific evolution of human CD8(+) T cell responses from primary to persistent phases of Epstein-barr virus infection. J Exp Med 195(7):893–905

Hjalgrim H, Askling J, Rostgaard K et al (2003) Characteristics of Hodgkin's lymphoma after infectious mononucleosis. N Engl J Med 349(14):1324–1332

Hoffman GJ, Lazarowitz SG, Hayward SD (1980) Monoclonal antibody against a 250,000-dalton glycoprotein of Epstein-barr virus identifies a membrane antigen and a neutralizing antigen. Proc Natl Acad Sci USA 77(5):2979–2983

Hui EP, Taylor GS, Jia H et al (2013) Phase I trial of recombinant modified vaccinia ankara encoding Epstein-Barr viral tumor antigens in nasopharyngeal carcinoma patients. Cancer Res 73(6):1676–1688

Jackman WT, Mann KA, Hoffmann HJ, Spaete RR (1999) Expression of Epstein-Barr virus gp350 as a single chain glycoprotein for an EBV subunit vaccine. Vaccine 17(7–8):660–668

Kanai K, Takashima K, Okuno K et al (2010) Lifelong persistent EBV infection of rabbits with EBER1-positive lymphocyte infiltration and mild sublethal hemophagocytosis. Virus Res 153(1):172–178

Kanekiyo M, Bu W, Joyce MG, Meng G, Whittle JR, Baxa U, Yamamoto T, Narpala S, Todd JP, Rao SS, McDermott AB, Koup RA, Rossmann MG, Mascola JR, Graham BS, Cohen JI, Nabel GJ (2015) Rational design of an epstein-barr virus vaccine targeting the receptor-binding site. Cell 12. doi:10.1016/j.cell.2015.07.043. [Epub ahead of print]

Khanna R, Moss DJ, Burrows SR (1999) Vaccine strategies against Epstein-barr virus-associated diseases: lessons from studies on cytotoxic T-cell-mediated immune regulation. *Immunol Rev* 170:49–64

Khanna R, Bell S, Sherritt M et al (1999b) Activation and adoptive transfer of Epstein-barr virus-specific cytotoxic T cells in solid organ transplant patients with posttransplant lymphoproliferative disease. Proc Natl Acad Sci USA 96(18):10391–10396

Leskowitz R, Fogg MH, Zhou XY et al (2014) Adenovirus-based vaccines against rhesus lymphocryptovirus EBNA-1 induce expansion of specific $CD8^{+}$ and $CD4^{+}$ T cells in persistently infected rhesus macaques. J Virol 88(9):4721–4735

Leung CS, Maurer MA, Meixlsperger S et al (2013) Robust T-cell stimulation by Epstein-barr virus-transformed B cells after antigen targeting to DEC-205. Blood 121(9):1584–1594

Li F, Song D, Lu Y, Zhu H, Chen Z, He X (2013) Delayed-type hypersensitivity (DTH) Immune Response related with EBV-DNA in nasopharyngeal carcinoma treated with autologous dendritic cell vaccination after radiotherapy. J Immunother 36(3):208–214

Lin CL, Lo WF, Lee TH et al (2002) Immunization with Epstein-Barr Virus (EBV) peptide-pulsed dendritic cells induces functional $CD8^+$ T-cell immunity and may lead to tumor regression in patients with EBV-positive nasopharyngeal carcinoma. Cancer Res 62(23):6952–6958

Luzuriaga K, Sullivan JL (2010) Infectious mononucleosis. N Engl J Med 362(21):1993–2000

Meixlsperger S, Leung CS, Ramer PC et al (2013) CD141+ dendritic cells produce prominent amounts of IFN-alpha after dsRNA recognition and can be targeted via DEC-205 in humanized mice. Blood 121(25):5034–5044

Moghaddam A, Rosenzweig M, Lee-Parritz D, Annis B, Johnson RP, Wang F (1997) An animal model for acute and persistent Epstein-barr virus infection. Science 276(5321):2030–2033

Moss DJ, Rickinson AB, Pope JH (1978) Long-term T-cell-mediated immunity to Epstein-barr virus in man. I. Complete regression of virus-induced transformation in cultures of seropositive donor leukocytes. Int J Cancer 22(6):662–668

Moutschen M, Leonard P, Sokal EM et al (2007) Phase I/II studies to evaluate safety and immunogenicity of a recombinant gp350 Epstein-barr virus vaccine in healthy adults. Vaccine 25(24):4697–4705

Murphy G, Pfeiffer R, Camargo MC, Rabkin CS (2009) Meta-analysis shows that prevalence of Epstein-Barr virus-positive gastric cancer differs based on sex and anatomic location. Gastroenterology 137(3):824–833

Panikkar A, Smith C, Hislop A, Tellam N, Dasari V, Hogquist KA, Wykes M, Moss DJ, Rickinson A, Balfour HH Jr, Khanna R (2015a) Impaired Epstein-Barr Virus-specific neutralizing antibody response during acute infectious mononucleosis is coincident with global B-Cell dysfunction. J Virol 89(17):9137–41

Panikkar A, Smith C, Hislop A, Tellam N, Dasari V, Hogquist KA, Wykes M, Moss DJ, Rickinson A, Balfour HH Jr, Khanna R (2015b) Cytokine-mediated loss of blood dendritic cells during Epstein-Barr Virus-associated acute infectious mononucleosis: Implication for immune dysregulation. J Infect Dis. jiv340 [Epub ahead of print]

Parrino J, McCurdy LH, Larkin BD et al (2007) Safety, immunogenicity and efficacy of modified vaccinia Ankara (MVA) against Dryvax challenge in vaccinia-naive and vaccinia-immune individuals. Vaccine 25(8):1513–1525

Pavlova S, Feederle R, Gartner K, Fuchs W, Granzow H, Delecluse HJ (2013) An Epstein-barr virus mutant produces immunogenic defective particles devoid of viral DNA. J Virol 87(4):2011–2022

Pearson G, Dewey F, Klein G, Henle G, Henle W (1970) Relation between neutralization of Epstein-barr virus and antibodies to cell-membrane antigens-induced by the virus. J Natl Cancer Inst 45(5):989–995

Rees L, Tizard EJ, Morgan AJ et al (2009) A phase I trial of epstein-barr virus gp350 vaccine for children with chronic kidney disease awaiting transplantation. Transplantation 88(8):1025–1029

Rickinson AB, Moss DJ, Pope JH (1979) Long-term C-cell-mediated immunity to Epstein-barr virus in man. II. Components necessary for regression in virus-infected leukocyte cultures. Int J Cancer 23(5):610–617

Rickinson AB, Long HM, Palendira U, Munz C, Hislop AD (2014) Cellular immune controls over Epstein-barr virus infection: new lessons from the clinic and the laboratory. Trends Immunol 35(4):159–169

Rivailler P, Jiang H, Cho YG, Quink C, Wang F (2002) Complete nucleotide sequence of the rhesus lymphocryptovirus: genetic validation for an Epstein-barr virus animal model. J Virol 76(1):421–426

Rooney CM, Smith CA, Ng CY et al (1995) Use of gene-modified virus-specific T lymphocytes to control Epstein-Barr-virus-related lymphoproliferation. Lancet 345(8941):9–13

Ruiss R, Jochum S, Wanner G, Reisbach G, Hammerschmidt W, Zeidler R (2011) A virus-like particle-based Epstein-barr virus vaccine. J Virol 85(24):13105–13113

Sashihara J, Hoshino Y, Bowman JJ et al (2011) Soluble rhesus lymphocryptovirus gp350 protects against infection and reduces viral loads in animals that become infected with virus after challenge. PLoS Pathog 7(10):e1002308

Sato K, Misawa N, Nie C et al (2011) A novel animal model of Epstein-barr virus-associated hemophagocytic lymphohistiocytosis in humanized mice. Blood 117(21):5663–5673

Sebelin-Wulf K, Nguyen TD, Oertel S et al (2007) Quantitative analysis of EBV-specific CD4/CD8 T cell numbers, absolute CD4/CD8 T cell numbers and EBV load in solid organ transplant recipients with PLTD. Transpl Immunol 17(3):203–210

Sherritt MA, Bharadwaj M, Burrows JM et al (2003) Reconstitution of the latent T-lymphocyte response to Epstein-barr virus is coincident with long-term recovery from posttransplant lymphoma after adoptive immunotherapy. Transplantation 75(9):1556–1560

Silveira EL, Fogg MH, Leskowitz RM et al (2013) Therapeutic vaccination against the rhesus lymphocryptovirus EBNA-1 homologue, rhEBNA-1, elicits T cell responses to novel epitopes in rhesus macaques. J Virol 87(24):13904–13910

Smets F, Latinne D, Bazin H et al (2002) Ratio between Epstein-barr viral load and anti-Epstein-barr virus specific T-cell response as a predictive marker of posttransplant lymphoproliferative disease. Transplantation 73(10):1603–1610

Smith C, Cooper L, Burgess M et al (2006) Functional reversion of antigen-specific $CD8^+$ T cells from patients with Hodgkin lymphoma following in vitro stimulation with recombinant polyepitope. J Immunol 177(7):4897–4906

Smith C, Martinez M, Cooper L, Rist M, Zhong J, Khanna R (2008) Generating functional $CD8^+$ T cell memory response under transient $CD4^+$ T cell deficiency: implications for vaccination of immunocompromised individuals. Eur J Immunol 38(7):1857–1866

Smith C, Beagley L, Khanna R (2009a) Acquisition of polyfunctionality by Epstein-Barr virus-specific $CD8^+$ T cells correlates with increased resistance to galectin-1-mediated suppression. J Virol 83(12):6192–6198

Smith C, Wakisaka N, Crough T et al (2009b) Discerning regulation of cis- and trans-presentation of $CD8^+$ T-cell epitopes by EBV-encoded oncogene LMP-1 through self-aggregation. Blood 113(24):6148–6152

Smith C, Tsang J, Beagley L et al (2012) Effective treatment of metastatic forms of Epstein-barr virus-associated nasopharyngeal carcinoma with a novel adenovirus-based adoptive immunotherapy. Cancer Res 72(5):1116–1125

Sokal EM, Hoppenbrouwers K, Vandermeulen C et al (2007) Recombinant gp350 vaccine for infectious mononucleosis: a phase 2, randomized, double-blind, placebo-controlled trial to evaluate the safety, immunogenicity, and efficacy of an Epstein-barr virus vaccine in healthy young adults. J Infect Dis 196(12):1749–1753

Strowig T, Gurer C, Ploss A et al (2009) Priming of protective T cell responses against virus-induced tumors in mice with human immune system components. J Exp Med 206(6):1423–1434

Takashima K, Ohashi M, Kitamura Y et al (2008) A new animal model for primary and persistent Epstein-barr virus infection: human EBV-infected rabbit characteristics determined using sequential imaging and pathological analysis. J Med Virol 80(3):455–466

Taylor GS, Haigh TA, Gudgeon NH et al (2004) Dual stimulation of Epstein-barr virus (EBV)-specific $CD4^+$- and $CD8^+$-T-cell responses by a chimeric antigen construct: potential therapeutic vaccine for EBV-positive nasopharyngeal carcinoma. J Virol 78(2):768–778

Taylor GS, Jia H, Harrington K, Lee LW, Turner J, Ladell K, Price DA, Tanday M, Matthews J, Roberts C, Edwards C, McGuigan L, Hartley A, Wilson S, Hui EP, Chan AT, Rickinson AB, Steven NM (2014) A recombinant modified vaccinia ankara vaccine encoding Epstein-Barr Virus (EBV) target antigens: a phase I trial in UK patients with EBV-positive cancer. Clin Cancer Res 20(19):5009–22

Tellam J, Connolly G, Green KJ et al (2004) Endogenous presentation of $CD8^+$ T cell epitopes from Epstein-barr virus-encoded nuclear antigen 1. J Exp Med 199(10):1421–1431
Tellam J, Smith C, Rist M et al (2008) Regulation of protein translation through mRNA structure influences MHC class I loading and T cell recognition. Proc Natl Acad Sci USA 105(27):9319–9324
Thomson SA, Khanna R, Gardner J et al (1995) Minimal epitopes expressed in a recombinant polyepitope protein are processed and presented to $CD8^+$ cytotoxic T cells: implications for vaccine design. Proc Natl Acad Sci USA 92(13):5845–5849
Thomson SA, Elliott SL, Sherritt MA et al (1996) Recombinant polyepitope vaccines for the delivery of multiple CD8 cytotoxic T cell epitopes. J Immunol 157(2):822–826
Thomson SA, Sherritt MA, Medveczky J et al (1998) Delivery of multiple CD8 cytotoxic T cell epitopes by DNA vaccination. J Immunol 160(4):1717–1723
Traggiai E, Chicha L, Mazzucchelli L et al (2004) Development of a human adaptive immune system in cord blood cell-transplanted mice. Science 304(5667):104–107
Tsao SW, Tsang CM, Pang PS, Zhang G, Chen H, Lo KW (2012) The biology of EBV infection in human epithelial cells. Semin Cancer Biol 22(2):137–143
Tugizov SM, Berline JW, Palefsky JM (2003) Epstein-barr virus infection of polarized tongue and nasopharyngeal epithelial cells. Nat Med 9(3):307–314
Wang X, Kenyon WJ, Li Q, Mullberg J, Hutt-Fletcher LM (1998) Epstein-barr virus uses different complexes of glycoproteins gH and gL to infect B lymphocytes and epithelial cells. J Virol 72(7):5552–5558
Xiong G, Zhang B, Huang MY et al (2014) Epstein-Barr virus (EBV) infection in Chinese children: a retrospective study of age-specific prevalence. PLoS ONE 9(6):e99857
Yajima M, Imadome K, Nakagawa A et al (2008) A new humanized mouse model of Epstein-barr virus infection that reproduces persistent infection, lymphoproliferative disorder, and cell-mediated and humoral immune responses. J Infect Dis 198(5):673–682

# The Biology and Clinical Utility of EBV Monitoring in Blood

Jennifer Kanakry and Richard Ambinder

**Abstract** Epstein-Barr virus (EBV) DNA in blood can be quantified in peripheral blood mononuclear cells, in circulating cell-free (CCF) DNA specimens, or in whole blood. CCF viral DNA may be actively released or extruded from viable cells, packaged in virions or passively shed from cells during apoptosis or necrosis. In infectious mononucleosis, viral DNA is detected in each of these kinds of specimens, although it is only transiently detected in CCF specimens. In nasopharyngeal carcinoma, CCF EBV DNA is an established tumor marker. In EBV-associated Hodgkin lymphoma and in EBV-associated extranodal NK-/T-cell lymphoma, there is growing evidence for the utility of CCF DNA as a tumor marker.

## Contents



J. Kanakry · R. Ambinder (✉)
Department of Oncology, Johns Hopkins School of Medicine, 389 CRB1 1650 Orleans, Baltimore, MD 21287, USA
e-mail: rambind1@jhmi.edu



C. Münz (ed.), *Epstein Barr Virus Volume 2*, Current Topics in Microbiology and Immunology 391, DOI 10.1007/978-3-319-22834-1_17

## Abbreviations

| | |
|---|---|
| EBV | Epstein-Barr virus |
| PBMC | Peripheral blood mononuclear cells |
| CCF | Circulating cell-free |
| NK | Natural killer |
| HIV | Human immunodeficiency virus |
| IM | Infectious mononucleosis |
| CAEBV | Chronic active Epstein-Barr virus |
| IL-6 | Interleukin-6 |
| TNF-α | Tumor necrosis factor alpha |
| HAART | Highly active antiretroviral therapy |
| MGUS | Monoclonal gammopathy of uncertain significance |
| PTLD | Post-transplantation lymphoproliferative disorder |
| NPV | Negative predictive value |
| NPC | Nasopharyngeal carcinoma |
| PPV | Positive predictive value |
| PFS | Progression-free survival |
| DFS | Disease-free survival |
| OS | Overall survival |
| CR | Complete response |
| PET/CT | Positron emission tomography/computed tomography |
| HSCT | Hematopoietic stem cell transplantation |
| HLH | Hemophagocytic lymphohistiocytosis |
| DLBCL | Diffuse large B-cell lymphoma |
| ENKTL | Extranodal NK-/T-cell lymphoma |
| PTCL | Peripheral T-cell lymphoma |
| PCNSL | Primary CNS lymphoma |
| LDH | Lactate dehydrogenase |
| HL | Hodgkin lymphoma |
| CSF | Cerebrospinal fluid |
| BAL | Bronchoalveolar lavage |

## 1 Introduction

Measurement of viral nucleic acids in blood now plays an important role in the diagnosis and management of a variety of viral diseases including HIV and hepatitis B and C among others. For EBV DNA, there are three approaches to quantitation that have been investigated: measurements in whole blood, measurements in peripheral blood mononuclear cells (PBMC), and measurements in plasma or serum. DNA in plasma or serum falls into the broader category of circulating cell-free (CCF) DNA. CCF DNA has been increasingly appreciated as providing a window on cellular compartments that are distinct from cells circulating in the blood. It is now clear that there are situations in which measurements of viral sequences in CCF DNA yield very different information than measurements of viral DNA in PBMC.

This chapter reviews the literature with regard to EBV measurements in various blood specimens in health and disease. It also touches on viral DNA in other body fluid specimens. The chapter concludes with some thoughts about the interpretation of quantitative EBV DNA measurements in various settings.

## *1.1 Replication of EBV DNA*

EBV DNA replicates by two distinct mechanisms (Hammerschmidt and Sugden 2013). In latently infected cells, EBV DNA is generally present as a closed circular nuclear plasmid. Replication proceeds in synchrony with cell cycle and requires only a single viral protein, EBNA1. Replicated plasmids segregate to daughter cells with mitosis. Latency viral replication is not inhibited by antiviral agents such as acyclovir, ganciclovir, or their congeners. Lytic viral replication proceeds through double-stranded multigenome length linear concatemers that are cleaved to give rise to genomes that can be packaged in virions. The process requires many viral proteins including the viral DNA polymerase. Lytic replication is inhibited by antiviral agents that inhibit the viral DNA polymerase including acyclovir and ganciclovir, although lytic EBV replication is less sensitive to inhibition than herpes simplex or varicella zoster virus (Coen et al. 2014).

## *1.2 CCF DNA*

CCF DNA is present in healthy individuals but increases with tissue injury such as stroke, myocardial infarction, surgery, or inflammation, and also with tissue growth and development as accompanies normal pregnancy or neoplasia (Diaz and Bardelli 2014). Important differences in the character of DNA associated with circulating cells versus CCF DNA are illustrated by studies of noninvasive prenatal testing. In plasma, the ratio of fetal to maternal DNA is nearly 1000-fold greater than the ratio of fetal to maternal cells in the blood (Bischoff et al. 2005). The CCF fetal DNA predominantly derives from DNA released from the fetus rather than from fetal cells in the maternal circulation. Furthermore, cells detected in the maternal circulation may reflect past pregnancies, whereas CCF DNA is short-lived and reflects the active gestation. Monitoring CCF DNA after 9- to 10-weeks of gestation has proven more sensitive and specific than invasive screening tests (Lo et al. 2014).

Processes of apoptosis and necrosis in tumors also result in cellular debris and release of CCF tumor DNA (Sausen et al. 2014; Diaz and Bardelli 2014). The amount of circulating DNA derived from tumor is a function of tumor burden, vascularity, location, cell turnover, and the efficiency with which cellular debris is cleared by infiltrating phagocytes. In some instances, more than 90 % of CCF DNA is tumor derived and tumor DNA is present in excess of 100,000 DNA copies per mL of plasma. Although CCF tumor DNA can be detected in early-stage cancers, it is more abundant in late-stage cancers. CCF tumor DNA can be detected in >75 % of patients with a variety of advanced cancers (Bettegowda et al. 2014). In situations where cancer cells or cancer-related cells sharing a particular genetic or epigenetic pattern also circulate in the blood, their detection may indicate tumor, but in some instances reflects the presence of long-lived biologically inactive cells. Like fetal cells from a previous pregnancy, these long-lived but possibly inert cells do not indicate the presence of a growing malignancy. In contrast, the short half-life of CCF DNA including tumor DNA (approximately 2 h) means that the tumor cells from which the CCF tumor DNA derives are turning over. Long-lived cells that are not cycling cannot account for CCF DNA.

# 2 Healthy EBV Seropositives and Infectious Mononucleosis

## *2.1 Healthy EBV Seropositives*

In healthy EBV-seropositive individuals, viral DNA is present as nuclear double-stranded DNA plasmids in resting memory B lymphocytes (Decker et al. 1996; Thorley-Lawson et al. 2013). There are typically approximately 1–50 EBV-infected cells per 1000,000 B lymphocytes (Khan et al. 1996). While there is variation in the number of infected B lymphocytes among individuals, these numbers appear to be stable over time (Khan et al. 1996; Stevens et al. 2007). EBV DNA levels in mouthwash samples from EBV-seropositive children and adults correlate with levels in PBMCs in some but not all series (Yao et al. 1991; Hug et al. 2010). CCF EBV DNA is detected in only a minority (0–4 %) and, if present, is typically at low levels (Wagner et al. 2001, 2002; Pajand et al. 2011).

## *2.2 Infectious Mononucleosis*

During acute infectious mononucleosis (IM), EBV DNA is readily detected in PBMC and is almost exclusively found within B lymphocyte fractions (Calattini et al. 2010; Fafi-Kremer et al. 2004; Cheng et al. 2007; Balfour and Verghese 2013; Fafi-Kremer et al. 2005; Hadinoto et al. 2008). The EBV DNA in PBMCs has been shown to decrease between day 0 and day 30, but rises again in the majority of patients at day 60 and/or day 90, when most are asymptomatic. EBV

DNA is also often detected in plasma of IM patients, but is very transient and becomes undetectable in most people within a couple of weeks of symptom onset (Wagner et al. 2001; Cheng et al. 2007; Pitetti et al. 2003; Yamamoto et al. 1995; Teramura et al. 2002). In one series, EBV DNA was detected in the plasma of 95 % of IM patients at diagnosis, decreased by day 3 in most patients, and became undetectable in all patients by day 15 (Fafi-Kremer et al. 2005). In another series, IM patients had no EBV DNA detected in plasma by day 7 (Fafi-Kremer et al. 2004). A third study demonstrated that 100 % of IM patients had EBV DNA in plasma during the acute phase of illness (first 14 days), but only 44 % had EBV DNA detected in plasma during the convalescent phase (days 15–40) (Yamamoto et al. 1995). By contrast, all patients had EBV DNA detected in PBMCs during both phases. The highest copy number was seen in plasma specimens collected within 7 days of symptom onset.

The effects of acyclovir or its prodrug valacyclovir have been studied in IM patients in a series of trials. A meta-analysis of 5 randomized controlled trials found a significant reduction in the rate of oropharyngeal EBV shedding at the end of the therapy but no difference in EBV shedding 3 weeks after discontinuation of therapy (Torre and Tambini 1999). In a more recent study, in people with IM treated with valacyclovir, EBV DNA copy number in saliva decreased during therapy and rebounded after treatment, while EBV DNA copy number in whole blood was stable (Vezina et al. 2010). Taken together, these results suggest that in IM patients, EBV DNA in the saliva is largely virion DNA, while EBV DNA in the blood is predominantly latent viral DNA.

In most patients, IM is a self-limited illness. In rare patients, IM is fatal. In one report, patients with fatal IM had 100-fold higher copy number in plasma as compared to those who survived (Yamamoto et al. 1995).

## 3 Immunocompromised Patients

In immunocompromised patients, increased numbers of latently infected lymphocytes are detected in the circulation (Babcock et al. 1999; Wagner et al. 2002; Fafi-Kremer et al. 2004; Yang et al. 2000; Calattini et al. 2010; Gotoh et al. 2010). The number of EBV genome copies in each infected B lymphocyte varies from individual to individual and in relation to underlying immune function. As in healthy seropositives, in transplant recipients and in HIV patients EBV DNA in the cellular fraction is predominantly harbored by CD19(+) resting B lymphocytes (Calattini et al. 2010; Gotoh et al. 2010; Babcock et al. 1999). In chronic active EBV (CAEBV), viral DNA can also be detected in plasma cells/plasmablasts, monocytes, or in the T cells (Calattini et al. 2010).

In HIV patients, EBV DNA is present in whole blood at levels higher than in healthy, HIV-seronegative patients (Stevens et al. 2002, 2007; Petrara et al. 2012). EBV DNA levels do not correlate with CD4 count or, in some series, HIV viral load. In one study, EBV DNA copy number in PBMCs was higher in patients with detectable HIV viremia and corresponded to higher levels of pro-inflammatory cytokines such as IL-6 and TNF-α and higher numbers of activated B cells (Petrara et al. 2012). HIV patients with no EBV DNA detected at one time point were unlikely to have EBV DNA detected at time points years later while those who had elevated EBV DNA in PBMCs had detectable levels on follow-up specimens, indicating some stability in levels over time (Stevens et al. 2007). When compared to banked specimens collected during HIV monotherapy, highly active retroviral therapy (HAART) does not seem to decrease the copy number of EBV DNA in whole blood (Stevens et al. 2002). In a study of HIV patients treated with HAART, EBV DNA copy number in PBMCs was noted to stay stable or increase as CD4 counts improved (Righetti et al. 2002). For those who had an increase in EBV DNA copy number in PBMCs with CD4 count recovery, a rise in IgG levels was also observed (Righetti et al. 2002). In a separate study, EBV DNA copy number in B cells was found to be several fold higher in HIV patients with persistent monoclonal gammopathies (MGUS) as compared to HIV patients with transient MGUS and HIV patients without MGUS (Ouedraogo et al. 2013). In hospitalized AIDS patients without lymphoma or with EBV(−) lymphomas, EBV DNA copy number in plasma is typically low if detected, with no clear relationship between EBV DNA copy number and HIV viral load or CD4 count (Fan et al. 2005).

In cancer patients without EBV-associated tumors, CCF EBV DNA is detected more frequently in patients receiving chemotherapy, particularly T-cell-depleting agents or in those with opportunistic infections (Martelius et al. 2010; Ogata et al. 2011). In healthy individuals, CCF EBV DNA is more frequently detected in the elderly perhaps reflecting age-related immune senescence (Stowe et al. 2007).

# 4 EBV-Associated Tumors

## *4.1 Nasopharyngeal Carcinoma (NPC)*

EBV DNA quantification in PBMCs has no clinical utility as a tumor marker in NPC (Shao et al. 2004). By contrast, CCF EBV DNA is an established tumor marker in undifferentiated NPC (Leung et al. 2014; Lin et al. 2004, 2007; Kalpoe et al. 2006; Shao et al. 2004). More than 90 % of untreated NPC patients have EBV DNA detectable in plasma compared to only a small percentage of healthy controls. As a screening tool for NPC, plasma EBV DNA quantification is highly sensitive and specific, with both high PPV and NPV (O et al. 2007). Plasma EBV DNA copy number has been shown to be positively correlated with tumor stage (Ma et al. 2006; Sun et al. 2014; Wang et al. 2013; Lin et al. 2007; Hou et al. 2011; Ferrari et al. 2012). In studies that have evaluated plasma and serum EBV DNA levels, both appear to be sensitive and specific for NPC and correlate well with each other, although there are discrepancies between plasma and serum levels (Jones et al. 2012). In meta-analyses of plasma and serum EBV DNA assessment for the diagnosis of NPC and its utility in distinguishing those with disease from healthy individuals, plasma has been shown to perform better than serum (Liu et al. 2011; Han et al. 2012).

### 4.1.1 Screening

In Southern China, EBV DNA in CCF DNA has been investigated for screening for NPC (Ji et al. 2014). With a cutoff of 0 copies/mL plasma, EBV DNA was 87 % sensitive for detecting NPC within a one year, with a PPV of 30 % and NPV of 99.3 %, although sensitivity was lower for early-stage disease. In Hong Kong in an evaluation of prospective screening, individuals with EBV DNA detected in plasma at enrollment and at two-week follow-up were referred for further evaluation with nasal endoscopy (Chan et al. 2013). NPC was diagnosed in 15 % of these patients.

### 4.1.2 Prognosis

CCF EBV DNA is detectable in the plasma in NPC patients at the time of diagnosis or relapse, but is rarely detectable in NPC patients during periods of remission (Fan et al. 2004). In several series, median plasma EBV DNA copy number for untreated NPC patients with stage II–IV disease ranges from a few hundred to a few thousand copies/mL (Lin et al. 2007; Leung et al. 2014; Tan et al. 2006). Pre-treatment plasma EBV DNA levels of ≥1500 copies/mL have been associated with inferior progression-free survival (PFS), disease-free survival (DFS), and overall survival (OS) in many studies including prospective cohorts (Wei et al. 2014; Wang et al. 2013; Lin et al. 2004, 2007). The association between high pre-treatment plasma EBV DNA levels and inferior OS has also been demonstrated in NPC cohorts using other cutoffs (Chai et al. 2012). Pre-treatment plasma EBV DNA levels have been shown to be a prognostic marker for treatment response and distant metastasis-free survival (Li et al. 2013; An et al. 2011; Hsu et al. 2012; Wang et al. 2010; Leung et al. 2003). In one study, the probability of distant failure in NPC patients was significantly higher in those with pre-treatment EBV DNA copy number of >4000 copies/mL plasma (Leung et al. 2003). In patients with relapsed NPC, preoperative plasma EBV DNA copy number corresponded to tumor burden, positive surgical margins, and subsequent systemic metastasis (Chan and Wong 2014).

### 4.1.3 Detecting Residual or Relapsed Disease After Therapy

EBV DNA in plasma after local radiotherapy is associated with the presence of distant metastatic disease (Lin et al. 2004, 2007; Twu et al. 2007; Hou et al. 2011; An et al. 2011; Chan et al. 2002). Persistent plasma EBV DNA one week after therapy was associated with inferior OS and DFS compared to those with undetectable EBV DNA in their plasma after therapy (Lin et al. 2004). In another study, post-treatment levels of >500 copies/mL plasma were associated with inferior PFS and OS (Chan et al. 2002). Return of plasma EBV DNA positivity after therapy has been associated with subsequent relapse, and patients who develop distant metastatic disease have been shown to have corresponding increases in EBV DNA copy number in plasma on serial evaluation (Ferrari et al. 2012; Kalpoe et al. 2006; Chan et al. 2004b). At the end of therapy, those achieving a complete response (CR) had EBV DNA copies remain below 500 copies/mL, while those in apparent CR who later relapsed had EBV DNA copy number increase from <500 copies/mL at the end of therapy to >500 copies/mL 2–16 months prior to clinically detectable relapse (Chan et al. 2004a). PET/CT has been shown to be valuable in detecting distant metastatic disease in patients with low degree of spread to regional lymph nodes but high EBV DNA copy number in plasma (Tang et al. 2013). In treated NPC patients in remission who were prospectively monitored for relapse with serial plasma EBV DNA measurements, all patients with EBV DNA in plasma on follow-up were found to have disease recurrence on PET, whereas no recurrences were detected in patients with undetectable EBV DNA in the plasma, even if there were symptoms or radiographic findings suggestive of potential relapse (Wang et al. 2011).

### 4.1.4 EBV DNA Copy Number Kinetics

Intraoperative plasma EBV DNA levels were checked one hour after NPC resection and were found to be undetectable in most patients and significantly lower than preoperative levels in remaining patients (Chan and Wong 2014). One week post-op, all patients had undetectable EBV DNA in plasma. This finding reinforces the conclusion that in NPC patients, viral DNA detect in CCF DNA is not derived from circulating tumor cells nor from virions released from lymphocytes. Since the viral DNA disappears with excision of the tumor, the tumor itself must be the source of the viral DNA.

Molecular investigations have investigated whether the CCF EBV DNA was encapsidated (virion) DNA or was DNA released from cells. Three techniques have been used to make this differentiation. Ultracentrifugation will pellet virions but not free DNA. Virions but not free DNA will be relatively protected from DNase digestion. Finally, DNA released from apoptotic cells is exposed to nucleases that clip DNA not protected by nucleosomes. The result is DNA fragment lengths that are 180–200 bps or multiples thereof forming a characteristic "ladder" on sizing gels. All three approaches to distinguishing viral sequences released from latently infected cells and virion DNA have shown that very little if any of the DNA detected is virion DNA.

The clearance rate of plasma EBV DNA with treatment has been found to be prognostic for response and overall survival (Hsu et al. 2012; Wang et al. 2010). Plasma EBV DNA half-life at a cutoff of >7 or 8 days also predicted treatment response and OS in patients with metastatic or relapsed disease (Hsu et al. 2012; Wang et al. 2010). On multivariate analysis, pre-treatment plasma EBV DNA levels in combination with assessment of half-life added additional prognostic information for OS (Hsu et al. 2012). Serial assessment of EBV DNA copy number in plasma showed a decline to undetectable levels within three weeks of starting therapy in responding patients (Kalpoe et al. 2006). In another study, most patients had undetectable EBV DNA in plasma after four weeks of therapy and the presence of EBV DNA in plasma at this mid-treatment time point was associated with inferior PFS, OS, and distant failure (Leung et al. 2014). In patients with metastatic NPC, those whose EBV DNA in plasma became undetectable after one cycle of chemotherapy had better survival than the others (An et al. 2011).

## *4.2 EBV Post-Transplant Lymphoproliferative Disease (EBV-PTLD)*

In lung transplant patients, EBV DNA copy number in whole blood was associated with lower rates of graft rejection (Ahya et al. 2007). Similarly, in heart transplant patients, detectable EBV DNA levels in whole blood were associated with higher drug levels of calcineurin inhibitors (Doesch et al. 2008). Thus, some have used EBV DNA copy number in blood as a marker of adequate immunosuppression in transplant recipients.

A rise in EBV DNA in PBMCs can be observed weeks prior to the onset of clinical symptoms or signs suggestive of EBV-PTLD providing the rationale for the monitoring strategies leading to preemptive interventions employed by some transplant centers (Meerbach et al. 2008). However, detection of EBV DNA in PBMCs of transplant patients is not associated with EBV-PTLD in most cases. EBV DNA copy number thresholds that should lead to the initiation of preemptive interventions are not well established. The organ transplanted, the particular immunosuppressive regimen, and host factors may all be important (Tsai et al. 2008; Meerbach et al. 2008; Ono et al. 2008; Wagner et al. 2002). While EBV DNA in the PBMCs of transplant patients is not diagnostic or highly predictive of EBV-PTLD, the absence of EBV DNA in PBMCs has a high NPV (Tsai et al. 2008).

CCF EBV DNA is not routinely detected in transplant recipients. However, it may be detected after intensified pharmacologic immunosuppression for graft rejection, antithymocyte globulin, or T-cell-depleted stem cell transplant (Barkholt et al. 2005; van Esser et al. 2001; Haque et al. 2011). A progressive rise to high levels is usually indicative of EBV-PTLD (Loginov et al. 2006). In comparison with PBMCs, EBV detection in plasma has superior specificity, PPV, and NPV and comparable sensitivity in detecting EBV-PTLD (Tsai et al. 2008; Ruf et al. 2012; van Esser et al. 2001).

When monitored after hematopoietic stem cell transplant (HSCT), cell-free EBV DNA is typically not detected until ~60 days after transplant, occurring after engraftment and with the recovery of lymphocyte counts (Clave et al. 2004; van Esser et al. 2001). In pediatric transplant patients with chronically elevated EBV DNA copies in whole blood (>5000 copies/mL for >60 months), none developed PTLD and the majority did not have EBV DNA detected in plasma (Gotoh et al. 2010). In adult transplant patients, EBV DNA was detected in 24 % of whole blood specimens but only 6 % of plasma specimens (Wada et al. 2007). In another study of transplant patients monitored for PTLD, EBV DNA was commonly detected in the whole blood and was often persistent, but was not associated with EBV-PTLD unless EBV DNA was also detected in the plasma (Tsai et al. 2008). Several studies report that among transplant patients who develop systemic EBV-PTLD, all have EBV DNA detected in plasma at diagnosis (Wada et al. 2007; Ishihara et al. 2011; Tsai et al. 2008; Meerbach et al. 2008). Conversely, EBV DNA is rarely detected in the plasma of transplant patients without EBV-PTLD, with EBV(−) PTLD, or with central nervous system (CNS)-only EBV-PTLD (Tsai et al. 2008). In screening for EBV-PTLD, the NPV associated with a plasma EBV DNA copy number <1000 copies/mL can be as high as 100 % (van Esser et al. 2001; Ruf et al. 2012). At diagnosis, patients often have cell-free EBV DNA copy numbers upward of 10,000 copies/mL (Wagner et al. 2001; Haque et al. 2011; van Esser et al. 2001). CCF EBV DNA can be detected weeks in advance of the development of clinical signs and symptoms of EBV-PTLD (van Esser et al. 2001). Furthermore, patients with a log increase in plasma EBV DNA copy number are at increased risk for developing EBV-PTLD, speaking to the potential value of serial measurements. However, while the copy number is predictive, CCF EBV DNA quantification alone is inadequate for diagnosis.

Whereas EBV DNA levels can often be elevated in whole blood, PBMCs, and B lymphocyte fractions even during PTLD-free episodes, clinically meaningful fluctuations in EBV DNA levels that correspond to PTLD disease activity and treatment response are more reliably observed in plasma (Ruf et al. 2012; Wagner et al. 2001). With treatment of EBV-PTLD, plasma EBV DNA has been shown to decline or become undetectable in responding patients (Tsai et al. 2008; Savoldo et al. 2006). In solid organ transplant patients at high risk for EBV-PTLD or with PTLD who were treated with autologous EBV-specific cytotoxic T cells, EBV DNA copy number in plasma was detectable prior to therapy, transiently increased early in therapy, and then declined in line with clinical response (Savoldo et al. 2006). In HSCT patients treated preemptively for EBV-PTLD with a single dose of rituximab if they had EBV DNA >1000 copies/mL plasma, EBV DNA became undetectable in plasma at a median of eight days in responding patients (van Esser et al. 2002).

It is worth considering the detection of cytomegalovirus DNA versus EBV DNA in CCF in the post-transplant setting. Detection of cytomegalovirus DNA is always indicative of active lytic replication. In the absence of resistance mutations, cytomegalovirus DNA will always clear with inhibitors of lytic viral replication. In contrast, latently infected EBV lymphocytes may increase in number without lytic replication. Cell turnover alone such as occurs in neoplasia will result in release of viral DNA fragments into CCF blood. Although these DNA fragments may be indicative of EBV-associated pathology, they are typically not responsive to inhibition with antivirals. Furthermore, the viral DNA need not reflect the presence of infectious virions at all and whereas it is entirely appropriate to refer to cytomegalovirus "viral load," referring to the EBV DNA debris from proliferating cells as "viral load" is misleading and should be avoided. The term "EBV copy number" is less likely to be misinterpreted as an indication of infectious virions in the blood.

## *4.3 EBV(+) Lymphomas and Lymphoproliferative Disorders*

CCF EBV DNA has been investigated as a potential tumor marker in patients with EBV(+) lymphoma and other EBV-associated diseases. Patients with untreated EBV(+) lymphomas or EBV-associated hemophagocytic lymphohistiocytosis (EBV-HLH) consistently have EBV DNA detected in plasma, often at high copy number (Martelius et al. 2010; Lei et al. 2001, 2002; Kanakry et al. 2013; Musacchio et al. 2006; Donati et al. 2006; Gallagher et al. 1999; Elazary et al. 2007; Teramura et al. 2002; Morishima et al. 2014; Suwiwat et al. 2007; Beutel et al. 2009). Across EBV(+) lymphomas of B-, T-, and NK-cell lineage, EBV DNA has been demonstrated to be detectable in the plasma prior to therapy in the majority of patients and to be markedly elevated in many (Machado et al. 2010; Au et al. 2004; Kanakry et al. 2013). By contrast, immunocompetent patients with EBV(−) lymphomas rarely have EBV DNA detected in plasma (Au et al. 2004; Machado et al. 2010; Kanakry et al. 2013).

In diffuse large B-cell lymphoma (DLBCL), EBV DNA is detected in the plasma of the majority of patients with EBV(+) tumors but not in patients with EBV(−) tumors or in healthy controls (Morishima et al. 2014). In EBV(+) extranodal NK-/T-cell lymphoma (ENKTL), more patients had EBV DNA detected in plasma than in PBMCs at diagnosis (Suzuki et al. 2011). In patients with EBV(+) peripheral T-cell lymphoma (PTCL) or EBV(+) T-cell proliferative diseases, EBV DNA is commonly detected in plasma, at significantly higher levels than seen in patients with EBV(−) PTCL or lymphoproliferative diseases (Suwiwat et al. 2007). On DNase I digestion, EBV DNA in plasma became undetectable or very low in all patients studied, suggesting that the circulating viral DNA is not encapsidated and is likely tumor derived (Suwiwat et al. 2007). In patients with HIV-associated EBV(+) lymphomas, EBV DNA was detected in plasma or serum in all patients in two series, but was undetectable in the plasma of HIV(+) controls matched for CD4 count and HIV viral load (Fan et al. 2005; Ouedraogo et al. 2013). However, in EBV(+) primary CNS lymphoma (PCNSL), EBV DNA may be detected in the plasma in only a minority of patients and at low copy number (Bossolasco et al. 2002, 2006; Fan et al. 2005), perhaps reflecting a "brain–blood" barrier for tumor DNA.

In many EBV(+) lymphomas, CCF EBV DNA has been shown to fall to undetectable levels for those achieving remission, to remain elevated in those with refractory disease, and to rise prior to clinically detectable relapse (Au et al. 2004; Machado et al. 2010; Lei et al. 2001; Martelius et al. 2010; Jones et al. 2012). By contrast, changes in cellular EBV DNA over time were not associated with response in patients with EBV(+) lymphomas (Jones et al. 2012). In patients with B-cell malignancies and high EBV DNA copy number in plasma, a rapid decrease in copy number has been shown to occur with rituximab treatment (Martelius et al. 2010). As rituximab targets CD20(+) B lymphocytes, it should be noted that a fall in EBV DNA copy number in PBMCs after rituximab may not as accurately reflect tumor response (Yang et al. 2000). In patients with HIV, EBV DNA was detected in plasma in all patients with untreated EBV(+) lymphomas and fell to undetectable in the majority of responding patients, some as quickly as three weeks into therapy (Fan et al. 2005).

In ENKTL, CCF EBV DNA copy number at diagnosis correlated with lactate dehydrogenase (LDH) and disease stage, with high copy number found to be associated with inferior treatment response rates and DFS (Au et al. 2004; Lei et al. 2002; Kwong et al. 2014; Ito et al. 2012). In patients with early-stage ENKTL treated with radiotherapy, high EBV copy number in pre-treatment plasma (> 500/mL) was associated with B symptoms, high LDH values, and inferior OS (Wang et al. 2012a). In another study, plasma EBV DNA copy number of 0, <1000 copies/mL, and ≥1000 copies/mL stratified patients into three prognostic groups for OS, with high copy number patients having the worst outcomes (Suzuki et al. 2011). CCF EBV DNA has also been shown to be a marker of disease status in patients with ENKTL where changes in plasma EBV DNA levels on serial assessment corresponded to degree of treatment response (Suzuki et al. 2011; Lei et al. 2002). Plasma EBV DNA positivity appears to be an early indicator of relapse in ENKTL, as patients in apparent clinical remission but with elevated EBV DNA in plasma have been observed to subsequently relapse (Lei et al. 2002). Furthermore, patients with detectable EBV DNA in plasma after therapy had inferior PFS and OS outcomes, where undetectable EBV DNA in plasma after chemotherapy was the best predictor of good OS (Kwong et al. 2014; Wang et al. 2012b). While plasma specimens have been more frequently studied in ENKTL, whole blood and plasma EBV DNA measurements have been shown to be highly correlated (Ito et al. 2012). However, pre-treatment plasma EBV DNA appears to be a better indicator of clinical stage, B symptoms, performance status, and prognosis than PBMC measurements in these patients (Suzuki et al. 2011).

In patients with classical Hodgkin lymphoma (HL), detection of EBV DNA in plasma is highly specific for EBV(+) disease and appears promising as a prognostic marker and indicator of treatment response. EBV DNA copy number in plasma was higher in HL patients with advanced disease, higher prognostic scores, and B symptoms (Hohaus et al. 2011). In one study, EBV DNA was detectable in the plasma of all untreated patients with EBV(+) HL and was undetectable in all responding EBV(+) HL patients after therapy, as well as patients with EBV(−) HL (Gandhi et al. 2006). By contrast, there was no association between EBV

DNA copy number in PBMCs and EBV(+) HL disease activity (Gandhi et al. 2006). In a study of children with EBV(+) HL, 85 % had EBV DNA detectable in plasma prior to therapy and those with post-treatment plasma specimens and complete response had no EBV DNA detectable after therapy (Sinha et al. 2013). In adults with HL, plasma EBV DNA status was closely concordant with tumor EBV status by tissue-based techniques and pre-treatment plasma EBV DNA positivity was associated with inferior PFS on multivariate analysis (Kanakry et al. 2013). At month six of therapy, patients who were plasma EBV DNA-positive had significantly inferior PFS compared to those who were plasma EBV DNA-negative (Kanakry et al. 2013). Other studies have also shown that declines in plasma EBV DNA copy number are associated with treatment response, while increases in plasma EBV DNA copy number precede disease relapse (Spacek et al. 2011). Plasma EBV DNA copy number has also been associated with higher numbers of tumor-associated macrophages and soluble CD163 levels, both of which may also have prognostic significance in HL (Jones et al. 2013; Hohaus et al. 2011). There is evidence to suggest that very little of the EBV CCF DNA detected in Hodgkin lymphoma is virion DNA (Ryan et al. 2004) and we (Kanakry, Ambinder) have unpublished data to corroborate these findings.

Aggressive NK leukemia is EBV(+), and in one series, all patients had EBV DNA detected in their serum at diagnosis (Zhang et al. 2013). Those who responded to chemotherapy had a decrease in serum EBV DNA copy number post-therapy, with undetectable post-treatment levels only observed in patients with a clinical complete remission (Zhang et al. 2013). Serum EBV DNA levels were noted to rise one to two weeks prior to clinically detectable relapse (Zhang et al. 2013).

In CAEBV patients undergoing HSCT, plasma EBV DNA copy number at diagnosis of CAEBV was significantly higher in those who died after transplant as compared to survivors (Gotoh et al. 2008). Patients who had a disappearance of clinical symptoms of CAEBV after HSCT had an accompanying decrease in EBV DNA in the plasma, whereas patients who had relapsed/refractory disease did not have a decrease in EBV DNA in the plasma (Gotoh et al. 2008). In EBV-HLH patients, EBV DNA was no longer detected in the plasma at four months in patients responding to therapy, although EBV DNA copy number prior to therapy or at two months did not distinguish responders from non-responders (Teramura et al. 2002). Taken together, CCF EBV DNA is a promising potential biomarker of EBV-associated lymphomas and lymphoproliferative disorders which may have diagnostic and prognostic value akin to its clinical utility in NPC, although further studies are still needed.

# 5 Other Patient Populations and Other Body Fluids

## *5.1 Critical Illness*

CCF EBV DNA is less often detected in the absence of an EBV-associated cancer or lymphoproliferative disorder (Fafi-Kremer et al. 2004; Hakim et al. 2007). In hospitalized non-transplant patients, EBV DNA is rarely detected in plasma and, when detected, is typically observed in patients who are critically ill or immunocompromised by HIV, immunosuppressive drugs, or malignancy (Martelius et al. 2010). In one prospective series, one-third of patients with sepsis had EBV DNA detectable in plasma, whereas in non-septic critically ill patients and healthy controls, EBV DNA was detected in plasma in 5 % and 0.6 % of cases, respectively (Walton et al. 2014). Septic patients who were plasma EBV DNA-positive were more likely than plasma EBV DNA-negative septic patients to have fungal infections (Walton et al. 2014). Detection of EBV DNA in plasma correlated with longer intensive care unit stay, but did not correlate with mortality risk scores.

## *5.2 Malaria*

The relationship between endemic Burkitt lymphoma and malaria remains poorly understood but has generated interest in possible interplay between the parasitic illness and the viral infection. In children with acute *Plasmodium falciparum* infection, EBV DNA is detected more frequently and at higher copy number than in plasma from children without malaria (Donati et al. 2006). Treatment of malaria has been reported to be associated with decreases in EBV DNA copy number in plasma (Chene et al. 2011; Donati et al. 2006). In whole blood, EBV DNA is more frequently detected among those with a history of severe malaria infection compared to those with a history of mild malaria infection and whole blood EBV DNA levels have been shown to be correlated with the number of malaria attacks (Yone et al. 2006). However, whether malaria is associated with more virions in the blood or the release of viral DNA from latently infected cells has not yet been determined.

## *5.3 Other Body Fluids*

It is quite common to periodically detect EBV DNA in the saliva of EBV-seropositive individuals (Ling et al. 2003). Seasonal variation of EBV DNA shedding has been demonstrated, with higher frequency of shedding in spring and fall (Ling et al. 2003). In college students followed longitudinally and diagnosed with primary EBV infection, EBV DNA was undetectable in the saliva prior to the onset of symptoms but became detectable in the oral cell pellet at diagnosis, remaining positive for a median of 175 days (Balfour and Verghese 2013). In the oral supernatant of these IM patients, EBV DNA copy number was lower and detected more transient (Balfour and Verghese 2013). In patients with HIV, EBV DNA is very commonly detected in the saliva, even among those on effective antiretroviral therapy (Jacobson et al. 2009; Griffin et al. 2008). However, patients with lower HIV RNA viral loads or those on HAART have been shown to have EBV DNA less frequently detected in saliva and, if detected, present at lower copy number (Ling et al. 2003; Griffin et al. 2008). In some studies, EBV DNA is frequently detected in saliva regardless of CD4 count (Jacobson et al. 2009), while other studies have shown HIV patients with higher CD4 counts to be less likely to shed EBV DNA in saliva (Griffin et al. 2008). Patients with endemic BL have been shown to have high EBV DNA copy number in saliva, with a frequency of detection of 100 % in one series (Donati et al. 2006).

In HIV-PCNSL, EBV DNA is detected in cerebral spinal fluid (CSF) in over 75 % of patients, but does not correlate with EBV DNA copy number in plasma (Bossolasco et al. 2002, 2006). In patients with imaging studies showing central nervous system mass lesions, the presence of viral DNA in the CSF is viewed by many as adequate to establish a diagnosis of PCNSL. EBV DNA can also be detected in the CSF of patients with central nervous system involvement by an EBV(+) tumor, such as leptomeningeal involvement by NPC (Ma et al. 2008).

In NPC patients, EBV DNA was detected in the urine of 56 % of untreated patients and patients with EBV DNA in urine had significantly higher plasma EBV DNA copy numbers (Chan et al. 2008). Urine and plasma values were positively correlated and it is presumed that small DNA fragments from plasma are filtered through the glomerulus into the urine.

In lung transplant patients, EBV DNA was detected in 44 % of patients' bronchoalveolar lavage fluid (Bauer et al. 2007). By contrast, EBV DNA was detected in 5 % of BAL fluid specimens from healthy controls. Over 33 % of lung transplant patients had both EBV and CMV detected in BAL fluid, and EBV and CMV DNA copy number in BAL fluid were positively correlated. The clinical implications of EBV DNA in BAL fluid of transplant patients, however, are not clear.

## 6 Summary

EBV is associated with many diseases but most often not associated with any disease at all. EBV DNA can be measured in PBMC, serum or plasma, or in whole blood. As a tool for basic investigation, measuring EBV DNA in PBMC and in other cellular fractions has provided important insights into the biology of persistence and the role of the resting memory B cell. In the most sensitive assays, viral DNA can be detected in PBMC in almost all subjects who have been infected. Among organ transplant recipients, higher copy number of viral DNA in PBMC may be indicative of the level of immunosuppression achieved and is perhaps useful in guiding pharmacologic immunosuppression. However, present evidence suggests that most of the viral DNA even in immunocompromised patients is not in lymphoblastoid-like immortalized cells but is in resting memory B cells. In any case, absent or very low viral DNA copy number in PBMC makes a diagnosis of EBV-PTLD unlikely. Although high copy number should raise suspicion for EBV-PTLD and may be useful in guiding adjustment of immunosuppression in the organ transplant setting, high copy number alone is not adequate for diagnosis. Typically, the addition of acyclovir or ganciclovir has little if any impact on viral copy number in PBMC insofar as at least a large fraction of the viral DNA is latent. In patients with EBV-PTLD, the administration of rituximab or other anti-B-cell therapy typically eliminates measurable copy number in PBMC even when the EBV-PTLD continues to progress and thus further measurements in PBMC have little value. In other settings such as patients with HIV infection and lymphoma, in patients with NPC, or in patients with HL, EBV copy number in PBMC does not seem promising with regard to either tumor diagnosis or monitoring. Measurement of viral sequences in CCF DNA is a very different measurement insofar as viral DNA is not detected in most seropositives in the absence of disease. However, the viral sequences detected in CCF DNA may be either virion DNA or may be DNA released from cells. As with detection of viral DNA in PBMC, inhibition of viral lytic replication does not impact on viral DNA sequences released in association with turnover of latently infected cells. It is perhaps worth noting that some virion DNA is present in patients with acute IM and some in patients with CAEBV, although the relative contributions of EBV from latently infected cells versus that from virion DNA have not yet been well defined. In NPC and HL, present evidence suggests that in most instances, the viral CCF DNA derives from tumor cells and is not packaged as virions. Persistence of CCF DNA appears to correlate closely with the presence of residual tumor. In NPC, assays of CCF DNA have become fairly standard in tumor monitoring. This is not yet the case in HL but seems a promising path for further investigation. EBV DNA is assayed in whole blood in many settings. This has the virtue of identifying high copy number in either cells or CCF DNA and of involving minimal processing. And on the other hand, assay of whole blood obscures the differences between compartments. In particular circumstances, assay of body fluids, particularly cerebrospinal fluid in patients with HIV and central nervous system lesions

**Table 1** EBV DNA in blood compartments

| **PBMC** |
|---|
| Latent infection in lymphocytes |
| Lytic infection in lymphocytes including cell-associated virions |
| **Plasma or Serum** |
| Virions |
| Lytically replicated DNA released from cells |
| Latently replicated DNA released from lymphocytes by processes of apoptosis, necrosis, and secretion |

**Table 2** DNA in blood compartments

| **Cellular** |
|---|
| Normal blood cells, endothelial cells, fetal cells (that may circulate and persist long after parturition), circulating tumor cells. The half-life of the cells is a function of the particular cell type but may be measured in days, months, or years. |
| **Circulating cell-free DNA** |
| Derived from cells in association with cell death (apoptosis, necrosis), secretory processes, or lytic viral replication. The source of the DNA includes but is not limited to circulating cells. The half-life of CCF DNA is measured in hours. |

on imaging, may be useful to establish a diagnosis of primary central nervous system lymphoma (see Tables 1 and 2).

## References

Ahya VN, Douglas LP, Andreadis C, Arnoldi S, Svoboda J, Kotloff RM, Hadjiliadis D, Sager JS, Woo YJ, Pochettino A, Schuster SJ, Stadtmauer EA, Tsai DE (2007) Association between elevated whole blood Epstein-Barr virus (EBV)-encoded RNA EBV polymerase chain reaction and reduced incidence of acute lung allograft rejection. J Heart Lung Transpl Official Publ Int Soc Heart Transpl 26(8):839–844. doi:10.1016/j.healun.2007.05.009

An X, Wang FH, Ding PR, Deng L, Jiang WQ, Zhang L, Shao JY, Li YH (2011) Plasma Epstein-Barr virus DNA level strongly predicts survival in metastatic/recurrent nasopharyngeal carcinoma treated with palliative chemotherapy. Cancer 117(16):3750–3757. doi:10.1002/cncr.25932

Au WY, Pang A, Choy C, Chim CS, Kwong YL (2004) Quantification of circulating Epstein-Barr virus (EBV) DNA in the diagnosis and monitoring of natural killer cell and EBV-positive lymphomas in immunocompetent patients. Blood 104(1):243–249. doi:10.1182/blood-2003-12-4197

Babcock GJ, Decker LL, Freeman RB, Thorley-Lawson DA (1999) Epstein-barr virus-infected resting memory B cells, not proliferating lymphoblasts, accumulate in the peripheral blood of immunosuppressed patients. J Exp Med 190(4):567–576

Balfour HH Jr, Verghese P (2013) Primary Epstein-Barr virus infection: impact of age at acquisition, coinfection, and viral load. J Infect Dis 207(12):1787–1789. doi:10.1093/infdis/jit096

Barkholt L, Linde A, Falk KI (2005) OKT3 and ganciclovir treatments are possibly related to the presence of Epstein-Barr virus in serum after liver transplantation. Transpl Int Official J Eur Soc Organ Transpl 18(7):835–843. doi:10.1111/j.1432-2277.2005.00145.x

Bauer CC, Jaksch P, Aberle SW, Haber H, Lang G, Klepetko W, Hofmann H, Puchhammer-Stockl E (2007) Relationship between cytomegalovirus DNA load in epithelial lining fluid and plasma of lung transplant recipients and analysis of coinfection with Epstein-Barr virus and human herpesvirus 6 in the lung compartment. J Clin Microbiol 45(2):324–328. doi:10.1128/JCM.01173-06

Bettegowda C, Sausen M, Leary RJ, Kinde I, Wang Y, Agrawal N, Bartlett BR, Wang H, Luber B, Alani RM, Antonarakis ES, Azad NS, Bardelli A, Brem H, Cameron JL, Lee CC, Fecher LA, Gallia GL, Gibbs P, Le D, Giuntoli RL, Goggins M, Hogarty MD, Holdhoff M, Hong SM, Jiao Y, Juhl HH, Kim JJ, Siravegna G, Laheru DA, Lauricella C, Lim M, Lipson EJ, Marie SK, Netto GJ, Oliner KS, Olivi A, Olsson L, Riggins GJ, Sartore-Bianchi A, Schmidt K, Shih l M, Oba-Shinjo SM, Siena S, Theodorescu D, Tie J, Harkins TT, Veronese S, Wang TL, Weingart JD, Wolfgang CL, Wood LD, Xing D, Hruban RH, Wu J, Allen PJ, Schmidt CM, Choti MA, Velculescu VE, Kinzler KW, Vogelstein B, Papadopoulos N, Diaz LA, Jr.

(2014) Detection of circulating tumor DNA in early- and late-stage human malignancies. Sci Transl Med 6(224):224ra224. doi:10.1126/scitranslmed.3007094

Beutel K, Gross-Wieltsch U, Wiesel T, Stadt UZ, Janka G, Wagner HJ (2009) Infection of T lymphocytes in Epstein-Barr virus-associated hemophagocytic lymphohistiocytosis in children of non-Asian origin. Pediatr Blood Cancer 53(2):184–190. doi:10.1002/pbc.22037

Bischoff FZ, Lewis DE, Simpson JL (2005) Cell-free fetal DNA in maternal blood: kinetics, source and structure. Human Reprod Update 11(1):59–67. doi:10.1093/humupd/dmh053

Bossolasco S, Cinque P, Ponzoni M, Vigano MG, Lazzarin A, Linde A, Falk KI (2002) Epstein-Barr virus DNA load in cerebrospinal fluid and plasma of patients with AIDS-related lymphoma. J Neurovirol 8(5):432–438. doi:10.1080/13550280260422730

Bossolasco S, Falk KI, Ponzoni M, Ceserani N, Crippa F, Lazzarin A, Linde A, Cinque P (2006) Ganciclovir is associated with low or undetectable Epstein-Barr virus DNA load in cerebrospinal fluid of patients with HIV-related primary central nervous system lymphoma. Clin Infect Dis Official Publ Infect Dis Soc Am 42(4):e21–e25. doi:10.1086/499956

Calattini S, Sereti I, Scheinberg P, Kimura H, Childs RW, Cohen JI (2010) Detection of EBV genomes in plasmablasts/plasma cells and non-B cells in the blood of most patients with EBV lymphoproliferative disorders by using Immuno-FISH. Blood 116(22):4546–4559. doi:10.1182/blood-2010-05-285452

Chai SJ, Pua KC, Saleh A, Yap YY, Lim PV, Subramaniam SK, Lum CL, Krishnan G, Mahiyuddin WR, Malaysian NPCSG, Teo SH, Khoo AS, Yap LF (2012) Clinical significance of plasma Epstein-Barr Virus DNA loads in a large cohort of Malaysian patients with nasopharyngeal carcinoma. J Clin Virol Official Publ Pan Am Soc Clin Virol 55(1):34–39. doi:10.1016/j.jcv.2012.05.017

Chan AT, Lo YM, Zee B, Chan LY, Ma BB, Leung SF, Mo F, Lai M, Ho S, Huang DP, Johnson PJ (2002) Plasma Epstein-Barr virus DNA and residual disease after radiotherapy for undifferentiated nasopharyngeal carcinoma. J Natl Cancer Inst 94(21):1614–1619

Chan AT, Ma BB, Lo YM, Leung SF, Kwan WH, Hui EP, Mok TS, Kam M, Chan LS, Chiu SK, Yu KH, Cheung KY, Lai K, Lai M, Mo F, Yeo W, King A, Johnson PJ, Teo PM, Zee B (2004a) Phase II study of neoadjuvant carboplatin and paclitaxel followed by radiotherapy and concurrent cisplatin in patients with locoregionally advanced nasopharyngeal carcinoma: therapeutic monitoring with plasma Epstein-Barr virus DNA. J Clin Oncol Official J Am Soc Clin Oncol 22(15):3053–3060. doi:10.1200/JCO.2004.05.178

Chan AT, Tao Q, Robertson KD, Flinn IW, Mann RB, Klencke B, Kwan WH, Leung TW, Johnson PJ, Ambinder RF (2004b) Azacitidine induces demethylation of the Epstein-Barr virus genome in tumors. J Clin Oncol Official J Am Soc Clin Oncol 22(8):1373–1381. doi:10.1200/JCO.2004.04.185

Chan JY, Wong ST (2014) The role of plasma Epstein-Barr virus DNA in the management of recurrent nasopharyngeal carcinoma. Laryngoscope 124(1):126–130. doi:10.1002/lary.24193

Chan KC, Hung EC, Woo JK, Chan PK, Leung SF, Lai FP, Cheng AS, Yeung SW, Chan YW, Tsui TK, Kwok JS, King AD, Chan AT, van Hasselt AC, Lo YM (2013) Early detection of nasopharyngeal carcinoma by plasma Epstein-Barr virus DNA analysis in a surveillance program. Cancer 119(10):1838–1844. doi:10.1002/cncr.28001

Chan KC, Leung SF, Yeung SW, Chan AT, Lo YM (2008) Quantitative analysis of the transrenal excretion of circulating EBV DNA in nasopharyngeal carcinoma patients. Clin Cancer Res Official J Am Assoc Cancer Res 14(15):4809–4813. doi:10.1158/1078-0432.CCR-08-1112

Chene A, Nylen S, Donati D, Bejarano MT, Kironde F, Wahlgren M, Falk KI (2011) Effect of acute Plasmodium falciparum malaria on reactivation and shedding of the eight human herpes viruses. PLoS ONE 6(10):e26266. doi:10.1371/journal.pone.0026266

Cheng CC, Chang LY, Shao PL, Lee PI, Chen JM, Lu CY, Lee CY, Huang LM (2007) Clinical manifestations and quantitative analysis of virus load in Taiwanese children with Epstein-Barr virus-associated infectious mononucleosis. J Microbiol Immunol Infect (Wei mian yu gan ran za zhi) 40(3):216–221

Clave E, Agbalika F, Bajzik V, Peffault de Latour R, Trillard M, Rabian C, Scieux C, Devergie A, Socie G, Ribaud P, Ades L, Ferry C, Gluckman E, Charron D, Esperou H, Toubert A, Moins-Teisserenc H (2004) Epstein-Barr virus (EBV) reactivation in allogeneic stem-cell transplantation: relationship between viral load, EBV-specific T-cell reconstitution and rituximab therapy. Transplantation 77(1):76–84. doi:10.1097/01.TP.0000093997.83754.2B
Coen N, Duraffour S, Topalis D, Snoeck R, Andrei G (2014) Spectrum of activity and mechanisms of resistance of various nucleoside derivatives against gammaherpesviruses. Antimicrob Agents Chemother 58(12):7312–7323. doi:10.1128/AAC.03957-14
Decker LL, Klaman LD, Thorley-Lawson DA (1996) Detection of the latent form of Epstein-Barr virus DNA in the peripheral blood of healthy individuals. J Virol 70(5):3286–3289
Diaz LA Jr, Bardelli A (2014) Liquid biopsies: genotyping circulating tumor DNA. J Clin Oncol Official J Am Soc Clin Oncol 32(6):579–586. doi:10.1200/JCO.2012.45.2011
Doesch AO, Konstandin M, Celik S, Kristen A, Frankenstein L, Sack FU, Schnabel P, Schnitzler P, Katus HA, Dengler TJ (2008) Epstein-Barr virus load in whole blood is associated with immunosuppression, but not with post-transplant lymphoproliferative disease in stable adult heart transplant patients. Transpl Int Official J Eur Soc Organ Transpl 21(10):963–971. doi:10.1111/j.1432-2277.2008.00709.x
Donati D, Espmark E, Kironde F, Mbidde EK, Kamya M, Lundkvist A, Wahlgren M, Bejarano MT, Falk KI (2006) Clearance of circulating Epstein-Barr virus DNA in children with acute malaria after antimalaria treatment. J Infect Dis 193(7):971–977. doi:10.1086/500839
Elazary AS, Wolf DG, Amir G, Avni B, Rund D, Yehuda DB, Sviri S (2007) Severe Epstein-Barr virus-associated hemophagocytic syndrome in six adult patients. J Clin Virol Official Publ Pan Am Soc Clin Virol 40(2):156–159. doi:10.1016/j.jcv.2007.06.014
Fafi-Kremer S, Brengel-Pesce K, Bargues G, Bourgeat MJ, Genoulaz O, Seigneurin JM, Morand P (2004) Assessment of automated DNA extraction coupled with real-time PCR for measuring Epstein-Barr virus load in whole blood, peripheral mononuclear cells and plasma. J Clin Virol Official Publ Pan Am Soc Clin Virol 30(2):157–164. doi:10.1016/j.jcv.2003.10.002
Fafi-Kremer S, Morand P, Brion JP, Pavese P, Baccard M, Germi R, Genoulaz O, Nicod S, Jolivet M, Ruigrok RW, Stahl JP, Seigneurin JM (2005) Long-term shedding of infectious epstein-barr virus after infectious mononucleosis. J Infect Dis 191(6):985–989. doi:10.1086/428097
Fan H, Kim SC, Chima CO, Israel BF, Lawless KM, Eagan PA, Elmore S, Moore DT, Schichman SA, Swinnen LJ, Gulley ML (2005) Epstein-Barr viral load as a marker of lymphoma in AIDS patients. J Med Virol 75(1):59–69. doi:10.1002/jmv.20238
Fan H, Nicholls J, Chua D, Chan KH, Sham J, Lee S, Gulley ML (2004) Laboratory markers of tumor burden in nasopharyngeal carcinoma: a comparison of viral load and serologic tests for Epstein-Barr virus. Int J Cancer J Int du Cancer 112(6):1036–1041. doi:10.1002/ijc.20520
Ferrari D, Codeca C, Bertuzzi C, Broggio F, Crepaldi F, Luciani A, Floriani I, Ansarin M, Chiesa F, Alterio D, Foa P (2012) Role of plasma EBV DNA levels in predicting recurrence of nasopharyngeal carcinoma in a Western population. BMC Cancer 12:208. doi:10.1186/1471-2407-12-208
Gallagher A, Armstrong AA, MacKenzie J, Shield L, Khan G, Lake A, Proctor S, Taylor P, Clements GB, Jarrett RF (1999) Detection of Epstein-Barr virus (EBV) genomes in the serum of patients with EBV-associated Hodgkin's disease. Int J Cancer J Int du Cancer 84(4):442–448
Gandhi MK, Lambley E, Burrows J, Dua U, Elliott S, Shaw PJ, Prince HM, Wolf M, Clarke K, Underhill C, Mills T, Mollee P, Gill D, Marlton P, Seymour JF, Khanna R (2006) Plasma Epstein-Barr virus (EBV) DNA is a biomarker for EBV-positive Hodgkin's lymphoma. Clin Cancer Res Official J Am Assoc Cancer Res 12(2):460–464. doi:10.1158/1078-0432.CCR-05-2008
Gotoh K, Ito Y, Ohta R, Iwata S, Nishiyama Y, Nakamura T, Kaneko K, Kiuchi T, Ando H, Kimura H (2010) Immunologic and virologic analyses in pediatric liver transplant

recipients with chronic high Epstein-Barr virus loads. J Infect Dis 202(3):461–469. doi:10.1086/653737

Gotoh K, Ito Y, Shibata-Watanabe Y, Kawada J, Takahashi Y, Yagasaki H, Kojima S, Nishiyama Y, Kimura H (2008) Clinical and virological characteristics of 15 patients with chronic active Epstein-Barr virus infection treated with hematopoietic stem cell transplantation. Clin Infect Dis Official Publ Infect Dis Soc Am 46(10):1525–1534. doi:10.1086/587671

Griffin E, Krantz E, Selke S, Huang ML, Wald A (2008) Oral mucosal reactivation rates of herpesviruses among HIV-1 seropositive persons. J Med Virol 80(7):1153–1159. doi:10.1002/jmv.21214

Hadinoto V, Shapiro M, Greenough TC, Sullivan JL, Luzuriaga K, Thorley-Lawson DA (2008) On the dynamics of acute EBV infection and the pathogenesis of infectious mononucleosis. Blood 111(3):1420–1427. doi:10.1182/blood-2007-06-093278

Hakim H, Gibson C, Pan J, Srivastava K, Gu Z, Bankowski MJ, Hayden RT (2007) Comparison of various blood compartments and reporting units for the detection and quantification of Epstein-Barr virus in peripheral blood. J Clin Microbiol 45(7):2151–2155. doi:10.1128/JCM.02308-06

Hammerschmidt W, Sugden B (2013) Replication of Epstein-Barr viral DNA. Cold Spring Harb Perspect Biol 5(1):a013029. doi:10.1101/cshperspect.a013029

Han BL, Xu XY, Zhang CZ, Wu JJ, Han CF, Wang H, Wang X, Wang GS, Yang SJ, Xie Y (2012) Systematic review on Epstein-Barr virus (EBV) DNA in diagnosis of nasopharyngeal carcinoma in Asian populations. Asian Pacific J Cancer Prevention APJCP 13(6):2577–2581

Haque T, Chaggar T, Schafers J, Atkinson C, McAulay KA, Crawford DH (2011) Soluble CD30: a serum marker for Epstein-Barr virus-associated lymphoproliferative diseases. J Med Virol 83(2):311–316. doi:10.1002/jmv.21953

Hohaus S, Santangelo R, Giachelia M, Vannata B, Massini G, Cuccaro A, Martini M, Cesarini V, Cenci T, D'Alo F, Voso MT, Fadda G, Leone G, Larocca LM (2011) The viral load of Epstein-Barr virus (EBV) DNA in peripheral blood predicts for biological and clinical characteristics in Hodgkin lymphoma. Clin Cancer Res Off J Am Assoc Cancer Res 17(9):2885–2892. doi:10.1158/1078-0432.CCR-10-3327

Hou X, Zhao C, Guo Y, Han F, Lu LX, Wu SX, Li S, Huang PY, Huang H, Zhang L (2011) Different clinical significance of pre- and post-treatment plasma Epstein-Barr virus DNA load in nasopharyngeal carcinoma treated with radiotherapy. Clin Oncol 23(2):128–133. doi:10.1016/j.clon.2010.09.001

Hsu CL, Chang KP, Lin CY, Chang HK, Wang CH, Lin TL, Liao CT, Tsang NM, Lee LY, Chan SC, Ng SH, Li HP, Chang YS, Wang HM (2012) Plasma Epstein-Barr virus DNA concentration and clearance rate as novel prognostic factors for metastatic nasopharyngeal carcinoma. Head Neck 34(8):1064–1070. doi:10.1002/hed.21890

Hug M, Dorner M, Frohlich FZ, Gysin C, Neuhaus D, Nadal D, Berger C (2010) Pediatric Epstein-Barr virus carriers with or without tonsillar enlargement may substantially contribute to spreading of the virus. J Infect Dis 202(8):1192–1199. doi:10.1086/656335

Ishihara M, Tanaka E, Sato T, Chikamoto H, Hisano M, Akioka Y, Dohno S, Maeda A, Hattori M, Wakiguchi H, Fujieda M (2011) Epstein-Barr virus load for early detection of lymphoproliferative disorder in pediatric renal transplant recipients. Clin Nephrol 76(1):40–48

Ito Y, Kimura H, Maeda Y, Hashimoto C, Ishida F, Izutsu K, Fukushima N, Isobe Y, Takizawa J, Hasegawa Y, Kobayashi H, Okamura S, Kobayashi H, Yamaguchi M, Suzumiya J, Hyo R, Nakamura S, Kawa K, Oshimi K, Suzuki R (2012) Pretreatment EBV-DNA copy number is predictive of response and toxicities to SMILE chemotherapy for extranodal NK/T-cell lymphoma, nasal type. Clin Cancer Res Official J Am Assoc Cancer Res 18(15):4183–4190. doi:10.1158/1078-0432.CCR-12-1064

Jacobson MA, Ditmer DP, Sinclair E, Martin JN, Deeks SG, Hunt P, Mocarski ES, Shiboski C (2009) Human herpesvirus replication and abnormal CD8+ T cell activation and low CD4+ T cell counts in antiretroviral-suppressed HIV-infected patients. PLoS ONE 4(4):e5277. doi:10.1371/journal.pone.0005277

Ji MF, Huang QH, Yu X, Liu Z, Li X, Zhang LF, Wang P, Xie SH, Rao HL, Fang F, Guo X, Liu Q, Hong MH, Ye W, Zeng YX, Cao SM (2014) Evaluation of plasma Epstein-Barr virus DNA load to distinguish nasopharyngeal carcinoma patients from healthy high-risk populations in Southern China. Cancer 120(9):1353–1360. doi:10.1002/cncr.28564

Jones K, Nourse JP, Keane C, Crooks P, Gottlieb D, Ritchie DS, Gill D, Gandhi MK (2012) Tumor-specific but not nonspecific cell-free circulating DNA can be used to monitor disease response in lymphoma. Am J Hematol 87(3):258–265. doi:10.1002/ajh.22252

Jones K, Vari F, Keane C, Crooks P, Nourse JP, Seymour LA, Gottlieb D, Ritchie D, Gill D, Gandhi MK (2013) Serum CD163 and TARC as disease response biomarkers in classical Hodgkin lymphoma. Clin Cancer Res Official J Am Assoc Cancer Res 19(3):731–742. doi:10.1158/1078-0432.CCR-12-2693

Kalpoe JS, Dekker PB, van Krieken JH, Baatenburg de Jong RJ, Kroes AC (2006) Role of Epstein-Barr virus DNA measurement in plasma in the clinical management of nasopharyngeal carcinoma in a low risk area. J Clin Pathol 59(5):537–541. doi:10.1136/jcp.2005.030544

Kanakry JA, Li H, Gellert LL, Lemas MV, Hsieh WS, Hong F, Tan KL, Gascoyne RD, Gordon LI, Fisher RI, Bartlett NL, Stiff P, Cheson BD, Advani R, Miller TP, Kahl BS, Horning SJ, Ambinder RF (2013) Plasma Epstein-Barr virus DNA predicts outcome in advanced Hodgkin lymphoma: correlative analysis from a large North American cooperative group trial. Blood 121(18):3547–3553. doi:10.1182/blood-2012-09-454694

Khan G, Miyashita EM, Yang B, Babcock GJ, Thorley-Lawson DA (1996) Is EBV persistence in vivo a model for B cell homeostasis? Immunity 5(2):173–179

Kwong YL, Pang AW, Leung AY, Chim CS, Tse E (2014) Quantification of circulating Epstein-Barr virus DNA in NK/T-cell lymphoma treated with the SMILE protocol: diagnostic and prognostic significance. Leukemia 28(4):865–870. doi:10.1038/leu.2013.212

Lei KI, Chan LY, Chan WY, Johnson PJ, Lo YM (2001) Circulating cell-free Epstein-Barr virus DNA levels in patients with EBV-associated lymphoid malignancies. Ann NY Acad Sci 945:80–83

Lei KI, Chan LY, Chan WY, Johnson PJ, Lo YM (2002) Diagnostic and prognostic implications of circulating cell-free Epstein-Barr virus DNA in natural killer/T-cell lymphoma. Clinical Cancer Res Official J Am Assoc Cancer Res 8(1):29–34

Leung SF, Chan AT, Zee B, Ma B, Chan LY, Johnson PJ, Lo YM (2003) Pretherapy quantitative measurement of circulating Epstein-Barr virus DNA is predictive of posttherapy distant failure in patients with early-stage nasopharyngeal carcinoma of undifferentiated type. Cancer 98(2):288–291. doi:10.1002/cncr.11496

Leung SF, Chan KC, Ma BB, Hui EP, Mo F, Chow KC, Leung L, Chu KW, Zee B, Lo YM, Chan AT (2014) Plasma Epstein-Barr viral DNA load at midpoint of radiotherapy course predicts outcome in advanced-stage nasopharyngeal carcinoma. Annals Oncol Official J Eur Soc Med Oncol ESMO 25(6):1204–1208. doi:10.1093/annonc/mdu117

Li SW, Wang H, Xiang YQ, Zhang HB, Lv X, Xia WX, Zeng MS, Mai HQ, Hong MH, Guo X (2013) Prospective study of prognostic value of Raf kinase inhibitory protein and pretreatment plasma Epstein-Barr virus DNA for distant metastasis in locoregionally advanced nasopharyngeal carcinoma. Head Neck 35(4):579–591. doi:10.1002/hed.23009

Lin JC, Wang WY, Chen KY, Wei YH, Liang WM, Jan JS, Jiang RS (2004) Quantification of plasma Epstein-Barr virus DNA in patients with advanced nasopharyngeal carcinoma. The New Engl J Med 350(24):2461–2470. doi:10.1056/NEJMoa032260

Lin JC, Wang WY, Liang WM, Chou HY, Jan JS, Jiang RS, Wang JY, Twu CW, Liang KL, Chao J, Shen WC (2007) Long-term prognostic effects of plasma epstein-barr virus DNA by minor groove binder-probe real-time quantitative PCR on nasopharyngeal carcinoma patients receiving concurrent chemoradiotherapy. Int J Radiat Oncol Biol Phys 68(5):1342–1348. doi:10.1016/j.ijrobp.2007.02.012

Ling PD, Vilchez RA, Keitel WA, Poston DG, Peng RS, White ZS, Visnegarwala F, Lewis DE, Butel JS (2003) Epstein-Barr virus DNA loads in adult human immunodeficiency virus type

1-infected patients receiving highly active antiretroviral therapy. Clin Infect Dis Official Publ Infect Dis Soc Am 37(9):1244–1249. doi:10.1086/378808

Liu Y, Fang Z, Liu L, Yang S, Zhang L (2011) Detection of Epstein-Barr virus DNA in serum or plasma for nasopharyngeal cancer: a meta-analysis. Genetic Test Mol Biomark 15(7–8):495–502. doi:10.1089/gtmb.2011.0012

Lo JO, Cori DF, Norton ME, Caughey AB (2014) Noninvasive prenatal testing. Obstet Gynecol Surv 69(2):89–99. doi:10.1097/OGX.0000000000000029

Loginov R, Aalto S, Piiparinen H, Halme L, Arola J, Hedman K, Hockerstedt K, Lautenschlager I (2006) Monitoring of EBV-DNAemia by quantitative real-time PCR after adult liver transplantation. J Clin Virol Official Publ Pan Am Soc Clin Virol 37(2):104–108. doi:10.1016/j.jcv.2006.06.012

Ma AT, Ma BB, Teo PM, Chan AT (2008) A novel application of plasma and cerebrospinal fluid level of epstein barr virus DNA in the diagnosis of leptomeningeal metastasis from nasopharyngeal carcinoma. A case report. Oncology 74(1–2):119–122. doi:10.1159/000139140

Ma BB, King A, Lo YM, Yau YY, Zee B, Hui EP, Leung SF, Mo F, Kam MK, Ahuja A, Kwan WH, Chan AT (2006) Relationship between pretreatment level of plasma Epstein-Barr virus DNA, tumor burden, and metabolic activity in advanced nasopharyngeal carcinoma. Int J Radiat Oncol Biol Phys 66(3):714–720. doi:10.1016/j.ijrobp.2006.05.064

Machado AS, Da Silva Robaina MC, Magalhaes De Rezende LM, Apa AG, Amoedo ND, Bacchi CE, Klumb CE (2010) Circulating cell-free and Epstein-Barr virus DNA in pediatric B-non-Hodgkin lymphomas. Leuk Lymphoma 51(6):1020–1027. doi:10.3109/10428191003746331

Martelius T, Lappalainen M, Aalto SM, Nihtinen A, Hedman K, Anttila VJ (2010) Clinical characteristics, outcome and the role of viral load in nontransplant patients with Epstein-Barr viraemia. Clin Microbiol Infect Official Publ Eur Soc Clin Microbiol Infect Dis 16(6):657–662. doi:10.1111/j.1469-0691.2009.02922.x

Meerbach A, Wutzler P, Hafer R, Zintl F, Gruhn B (2008) Monitoring of Epstein-Barr virus load after hematopoietic stem cell transplantation for early intervention in post-transplant lymphoproliferative disease. J Med Virol 80(3):441–454. doi:10.1002/jmv.21096

Morishima S, Nakamura S, Yamamoto K, Miyauchi H, Kagami Y, Kinoshita T, Onoda H, Yatabe Y, Ito M, Miyamura K, Nagai H, Moritani S, Sugiura I, Tsushita K, Mihara H, Ohbayashi K, Iba S, Emi N, Okamoto M, Iwata S, Kimura H, Kuzushima K, Morishima Y (2014) Increased T-cell responses to Epstein-Barr virus with high viral load in patients with Epstein-Barr virus-positive diffuse large B-cell lymphoma. Leuk Lymphoma, 1–7. doi:10.3109/10428194.2014.938326

Musacchio JG, Carvalho Mda G, Morais JC, Silva NH, Scheliga A, Romano S, Spector N (2006) Detection of free circulating Epstein-Barr virus DNA in plasma of patients with Hodgkin's disease. Sao Paulo Med J Rev Paul de Med 124(3):154–157

O TM, Yu G, Hu K, Li JC (2007) Plasma Epstein-Barr virus immunoglobulin A and DNA for nasopharyngeal carcinoma screening in the United States. Otolaryngol Head Neck Surg Official J Am Acad Otolaryngol Head Neck Surg 136 (6):992–997. doi:10.1016/j.otohns.2006.11.053

Ogata M, Satou T, Kawano R, Yoshikawa T, Ikewaki J, Kohno K, Ando T, Miyazaki Y, Ohtsuka E, Saburi Y, Kikuchi H, Saikawa T, Kadota J (2011) High incidence of cytomegalovirus, human herpesvirus-6, and Epstein-Barr virus reactivation in patients receiving cytotoxic chemotherapy for adult T cell leukemia. J Med Virol 83(4):702–709. doi:10.1002/jmv.22013

Ono Y, Ito Y, Kaneko K, Shibata-Watanabe Y, Tainaka T, Sumida W, Nakamura T, Kamei H, Kiuchi T, Ando H, Kimura H (2008) Simultaneous monitoring by real-time polymerase chain reaction of epstein-barr virus, human cytomegalovirus, and human herpesvirus-6 in juvenile and adult liver transplant recipients. Transpl Proc 40(10):3578–3582. doi:10.1016/j.transproceed.2008.05.082

Ouedraogo DE, Makinson A, Vendrell JP, Casanova ML, Nagot N, Cezar R, Bollore K, Al Taaba Y, Foulongne V, Badiou S, Viljoen J, Reynes J, Van de Perre P, Tuaillon E (2013) Pivotal role of HIV and EBV replication in the long-term persistence of monoclonal

gammopathy in patients on antiretroviral therapy. Blood 122(17):3030–3033. doi:10.1182/blood-2012-12-470393
Pajand O, Pourakbari B, Mahjob F, Aghamohammadi A, Mamishi N, Mamishi S (2011) Detection of Epstein-Barr virus DNA in plasma and lymph node biopsy samples of pediatric and adult patients with Hodgkin lymphoma. Pediatr Hematol Oncol 28(1):10–15. doi:10.3109/08880018.2010.507691
Petrara MR, Cattelan AM, Zanchetta M, Sasset L, Freguja R, Gianesin K, Cecchetto MG, Carmona F, De Rossi A (2012) Epstein-Barr virus load and immune activation in human immunodeficiency virus type 1-infected patients. J Clin Virol Official Publ Pan Am Soc Clin Virol 53(3):195–200. doi:10.1016/j.jcv.2011.12.013
Pitetti RD, Laus S, Wadowsky RM (2003) Clinical evaluation of a quantitative real time polymerase chain reaction assay for diagnosis of primary Epstein-Barr virus infection in children. Pediatr Infect Dis J 22(8):736–739. doi:10.1097/01.inf.0000078157.90639.96
Righetti E, Ballon G, Ometto L, Cattelan AM, Menin C, Zanchetta M, Chieco-Bianchi L, De Rossi A (2002) Dynamics of Epstein-Barr virus in HIV-1-infected subjects on highly active antiretroviral therapy. AIDS 16(1):63–73
Ruf S, Behnke-Hall K, Gruhn B, Bauer J, Horn M, Beck J, Reiter A, Wagner HJ (2012) Comparison of six different specimen types for Epstein-Barr viral load quantification in peripheral blood of pediatric patients after heart transplantation or after allogeneic hematopoietic stem cell transplantation. J Clin Virol Official Publ Pan Am Soc Clin Virol 53(3):186–194. doi:10.1016/j.jcv.2011.11.010
Ryan JL, Fan H, Swinnen LJ, Schichman SA, Raab-Traub N, Covington M, Elmore S, Gulley ML (2004) Epstein-Barr Virus (EBV) DNA in plasma is not encapsidated in patients with EBV-related malignancies. Diagn Mol Pathol Am J Surg Pathol Part B 13(2):61–68
Sausen M, Parpart S, Diaz LA Jr (2014) Circulating tumor DNA moves further into the spotlight. Genome medicine 6(5):35. doi:10.1186/gm552
Savoldo B, Goss JA, Hammer MM, Zhang L, Lopez T, Gee AP, Lin YF, Quiros-Tejeira RE, Reinke P, Schubert S, Gottschalk S, Finegold MJ, Brenner MK, Rooney CM, Heslop HE (2006) Treatment of solid organ transplant recipients with autologous Epstein Barr virus-specific cytotoxic T lymphocytes (CTLs). Blood 108(9):2942–2949. doi:10.1182/blood-2006-05-021782
Shao JY, Zhang Y, Li YH, Gao HY, Feng HX, Wu QL, Cui NJ, Cheng G, Hu B, Hu LF, Ernberg I, Zeng YX (2004) Comparison of Epstein-Barr virus DNA level in plasma, peripheral blood cell and tumor tissue in nasopharyngeal carcinoma. Anticancer Res 24(6):4059–4066
Sinha M, Rao CR, Shafiulla M, Appaji L, Bs AK, Sumati BG, Avinash T, Jayshree RS (2013) Cell-free epstein-barr viral loads in childhood hodgkin lymphoma: a study from South India. Pediatr Hematol Oncol 30(6):537–543. doi:10.3109/08880018.2013.796026
Spacek M, Hubacek P, Markova J, Zajac M, Vernerova Z, Kamaradova K, Stuchly J, Kozak T (2011) Plasma EBV-DNA monitoring in Epstein-Barr virus-positive Hodgkin lymphoma patients. APMIS : Acta Pathologica, Microbiologica, et Immunologica Scandinavica 119(1):10–16. doi:10.1111/j.1600-0463.2010.02685.x
Stevens SJ, Blank BS, Smits PH, Meenhorst PL, Middeldorp JM (2002) High Epstein-Barr virus (EBV) DNA loads in HIV-infected patients: correlation with antiretroviral therapy and quantitative EBV serology. AIDS 16(7):993–1001
Stevens SJ, Smits PH, Verkuijlen SA, Rockx DA, van Gorp EC, Mulder JW, Middeldorp JM (2007) Aberrant Epstein-Barr virus persistence in HIV carriers is characterized by anti-Epstein-Barr virus IgA and high cellular viral loads with restricted transcription. AIDS 21(16):2141–2149. doi:10.1097/QAD.0b013e3282eeeba0
Stowe RP, Kozlova EV, Yetman DL, Walling DM, Goodwin JS, Glaser R (2007) Chronic herpesvirus reactivation occurs in aging. Exp Gerontol 42(6):563–570. doi:10.1016/j.exger.2007.01.005
Sun P, Chen C, Cheng YK, Zeng ZJ, Chen XL, Liu LZ, Gu MF (2014) Serologic biomarkers of Epstein-Barr virus correlate with TNM classification according to the seventh

edition of the UICC/AJCC staging system for nasopharyngeal carcinoma. Eur Arch otorhinolaryngol Official J Eur Fed OtoRhinoLaryngol Soc 271(9):2545–2554. doi:10.1007/s00405-013-2805-5

Suwiwat S, Pradutkanchana J, Ishida T, Mitarnun W (2007) Quantitative analysis of cell-free Epstein-Barr virus DNA in the plasma of patients with peripheral T-cell and NK-cell lymphomas and peripheral T-cell proliferative diseases. J Clin Virol Official Publ Pan Am Soc Clin Virol 40(4):277–283. doi:10.1016/j.jcv.2007.08.013

Suzuki R, Yamaguchi M, Izutsu K, Yamamoto G, Takada K, Harabuchi Y, Isobe Y, Gomyo H, Koike T, Okamoto M, Hyo R, Suzumiya J, Nakamura S, Kawa K, Oshimi K (2011) Prospective measurement of Epstein-Barr virus-DNA in plasma and peripheral blood mononuclear cells of extranodal NK/T-cell lymphoma, nasal type. Blood 118(23):6018–6022. doi:10.1182/blood-2011-05-354142

Tan EL, Looi LM, Sam CK (2006) Evaluation of plasma Epstein-Barr virus DNA load as a prognostic marker for nasopharyngeal carcinoma. Singapore Med J 47(9):803–807

Tang LQ, Chen QY, Fan W, Liu H, Zhang L, Guo L, Luo DH, Huang PY, Zhang X, Lin XP, Mo YX, Liu LZ, Mo HY, Li J, Zou RH, Cao Y, Xiang YQ, Qiu F, Sun R, Chen MY, Hua YJ, Lv X, Wang L, Zhao C, Guo X, Cao KJ, Qian CN, Zeng MS, Mai HQ (2013) Prospective study of tailoring whole-body dual-modality [18F]fluorodeoxyglucose positron emission tomography/computed tomography with plasma Epstein-Barr virus DNA for detecting distant metastasis in endemic nasopharyngeal carcinoma at initial staging. J Clin Oncol Official J Am Soc Clin Oncol 31(23):2861–2869. doi:10.1200/JCO.2012.46.0816

Teramura T, Tabata Y, Yagi T, Morimoto A, Hibi S, Imashuku S (2002) Quantitative analysis of cell-free Epstein-Barr virus genome copy number in patients with EBV-associated hemophagocytic lymphohistiocytosis. Leuk Lymphoma 43(1):173–179. doi:10.1080/10428190210176

Thorley-Lawson DA, Hawkins JB, Tracy SI, Shapiro M (2013) The pathogenesis of Epstein-Barr virus persistent infection. Curr Opin Virol 3(3):227–232. doi:10.1016/j.coviro.2013.04.005

Torre D, Tambini R (1999) Acyclovir for treatment of infectious mononucleosis: a meta-analysis. Scand J Infect Dis 31(6):543–547

Tsai DE, Douglas L, Andreadis C, Vogl DT, Arnoldi S, Kotloff R, Svoboda J, Bloom RD, Olthoff KM, Brozena SC, Schuster SJ, Stadtmauer EA, Robertson ES, Wasik MA, Ahya VN (2008) EBV PCR in the diagnosis and monitoring of posttransplant lymphoproliferative disorder: results of a two-arm prospective trial. Am J Trans Official J Am Soc Transpl Am Soc Transpl Surg 8(5):1016–1024. doi:10.1111/j.1600-6143.2008.02183.x

Twu CW, Wang WY, Liang WM, Jan JS, Jiang RS, Chao J, Jin YT, Lin JC (2007) Comparison of the prognostic impact of serum anti-EBV antibody and plasma EBV DNA assays in nasopharyngeal carcinoma. Int J Radiat Oncol Biol Phys 67(1):130–137. doi:10.1016/j.ijrobp.2006.07.012

van Esser JW, Niesters HG, van der Holt B, Meijer E, Osterhaus AD, Gratama JW, Verdonck LF, Lowenberg B, Cornelissen JJ (2002) Prevention of Epstein-Barr virus-lymphoproliferative disease by molecular monitoring and preemptive rituximab in high-risk patients after allogeneic stem cell transplantation. Blood 99(12):4364–4369

van Esser JW, van der Holt B, Meijer E, Niesters HG, Trenschel R, Thijsen SF, van Loon AM, Frassoni F, Bacigalupo A, Schaefer UW, Osterhaus AD, Gratama JW, Lowenberg B, Verdonck LF, Cornelissen JJ (2001) Epstein-Barr virus (EBV) reactivation is a frequent event after allogeneic stem cell transplantation (SCT) and quantitatively predicts EBV-lymphoproliferative disease following T-cell–depleted SCT. Blood 98(4):972–978

Vezina HE, Balfour HH Jr, Weller DR, Anderson BJ, Brundage RC (2010) Valacyclovir pharmacokinetics and exploratory pharmacodynamics in young adults with Epstein-Barr virus infectious mononucleosis. J Clin Pharmacol 50(7):734–742. doi:10.1177/0091270009351884

Wada K, Kubota N, Ito Y, Yagasaki H, Kato K, Yoshikawa T, Ono Y, Ando H, Fujimoto Y, Kiuchi T, Kojima S, Nishiyama Y, Kimura H (2007) Simultaneous quantification of Epstein-Barr virus, cytomegalovirus, and human herpesvirus 6 DNA in samples from transplant recipients by multiplex real-time PCR assay. J Clin Microbiol 45(5):1426–1432. doi:10.1128/JCM.01515-06

Wagner HJ, Fischer L, Jabs WJ, Holbe M, Pethig K, Bucsky P (2002) Longitudinal analysis of Epstein-Barr viral load in plasma and peripheral blood mononuclear cells of transplanted patients by real-time polymerase chain reaction. Transplantation 74(5):656–664

Wagner HJ, Wessel M, Jabs W, Smets F, Fischer L, Offner G, Bucsky P (2001) Patients at risk for development of posttransplant lymphoproliferative disorder: plasma versus peripheral blood mononuclear cells as material for quantification of Epstein-Barr viral load by using real-time quantitative polymerase chain reaction. Transplantation 72(6):1012–1019

Walton AH, Muenzer JT, Rasche D, Boomer JS, Sato B, Brownstein BH, Pachot A, Brooks TL, Deych E, Shannon WD, Green JM, Storch GA, Hotchkiss RS (2014) Reactivation of multiple viruses in patients with sepsis. PloS one 9(2):e98819. doi:10.1371/journal.pone.0098819

Wang WY, Twu CW, Chen HH, Jan JS, Jiang RS, Chao JY, Liang KL, Chen KW, Wu CT, Lin JC (2010) Plasma EBV DNA clearance rate as a novel prognostic marker for metastatic/recurrent nasopharyngeal carcinoma. Clin Cancer Res Official J Am Assoc Cancer Res 16(3):1016–1024. doi:10.1158/1078-0432.CCR-09-2796

Wang WY, Twu CW, Chen HH, Jiang RS, Wu CT, Liang KL, Shih YT, Chen CC, Lin PJ, Liu YC, Lin JC (2013) Long-term survival analysis of nasopharyngeal carcinoma by plasma Epstein-Barr virus DNA levels. Cancer 119(5):963–970. doi:10.1002/cncr.27853

Wang WY, Twu CW, Lin WY, Jiang RS, Liang KL, Chen KW, Wu CT, Shih YT, Lin JC (2011) Plasma Epstein-Barr virus DNA screening followed by (1)(8)F-fluoro-2-deoxy-D-glucose positron emission tomography in detecting posttreatment failures of nasopharyngeal carcinoma. Cancer 117(19):4452–4459. doi:10.1002/cncr.26069

Wang Z, Li L, Su X, Gao Z, Srivastava G, Murray PG, Ambinder R, Tao Q (2012a) Epigenetic silencing of the 3p22 tumor suppressor DLEC1 by promoter CpG methylation in non-Hodgkin and Hodgkin lymphomas. J Transl Med 10:209. doi:10.1186/1479-5876-10-209

Wang ZY, Liu QF, Wang H, Jin J, Wang WH, Wang SL, Song YW, Liu YP, Fang H, Ren H, Wu RY, Chen B, Zhang XM, Lu NN, Zhou LQ, Li YX (2012b) Clinical implications of plasma Epstein-Barr virus DNA in early-stage extranodal nasal-type NK/T-cell lymphoma patients receiving primary radiotherapy. Blood 120(10):2003–2010. doi:10.1182/blood-2012-06-435024

Wei W, Huang Z, Li S, Chen H, Zhang G, Li S, Hu W, Xu T (2014) Pretreatment Epstein-Barr virus DNA load and cumulative cisplatin dose intensity affect long-term outcome of nasopharyngeal carcinoma treated with concurrent chemotherapy: experience of an institute in an endemic area. Oncol Res Treat 37(3):88–95. doi:10.1159/000360178

Yamamoto M, Kimura H, Hironaka T, Hirai K, Hasegawa S, Kuzushima K, Shibata M, Morishima T (1995) Detection and quantification of virus DNA in plasma of patients with Epstein-Barr virus-associated diseases. J Clin Microbiol 33(7):1765–1768

Yang J, Tao Q, Flinn IW, Murray PG, Post LE, Ma H, Piantadosi S, Caligiuri MA, Ambinder RF (2000) Characterization of Epstein-Barr virus-infected B cells in patients with posttransplantation lymphoproliferative disease: disappearance after rituximab therapy does not predict clinical response. Blood 96(13):4055–4063

Yao QY, Rowe M, Martin B, Young LS, Rickinson AB (1991) The Epstein-Barr virus carrier state: dominance of a single growth-transforming isolate in the blood and in the oropharynx of healthy virus carriers. J Gen Virol 72(Pt 7):1579–1590

Yone CL, Kube D, Kremsner PG, Luty AJ (2006) Persistent Epstein-Barr viral reactivation in young African children with a history of severe Plasmodium falciparum malaria. Trans Roy Soc Trop Med Hyg 100(7):669–676. doi:10.1016/j.trstmh.2005.08.009

Zhang H, Meng Q, Yin W, Xu L, Lie L (2013) Adult aggressive natural killer cell leukemia. Am J Med Sci 346(1):56–63. doi:10.1097/MAJ.0b013e3182764b59

# Index




C. Münz (ed.), *Epstein Barr Virus Volume 2*, Current Topics in Microbiology and Immunology 391, DOI 10.1007/978-3-319-22834-1

Printed by Printforce, the Netherlands